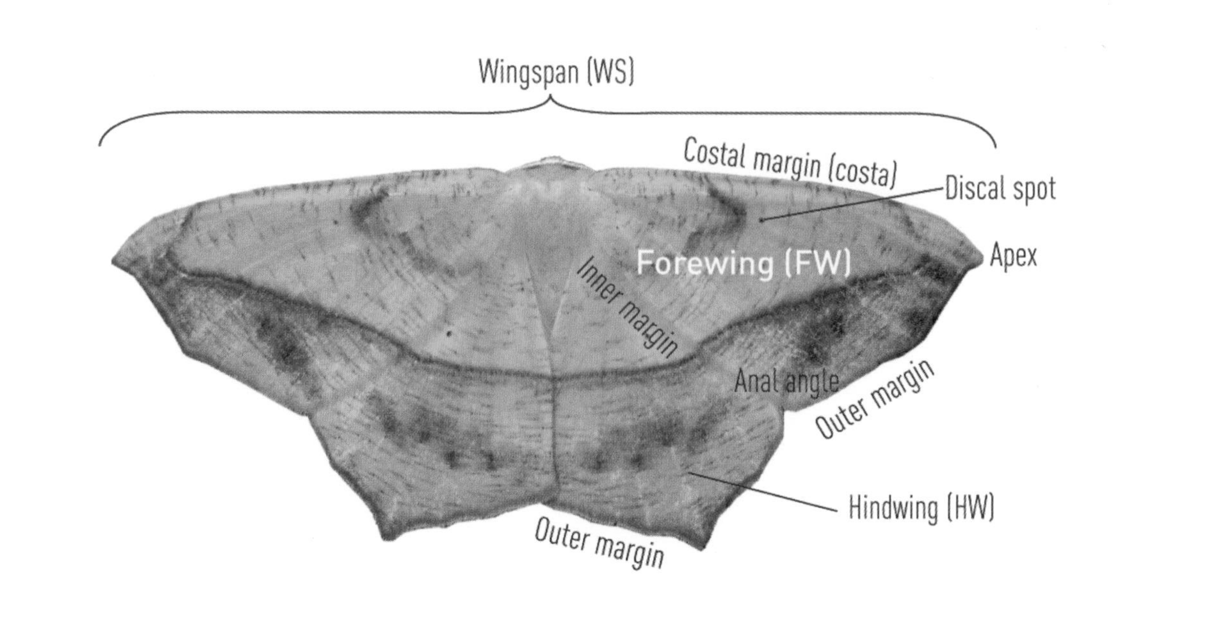
Wingspan (WS)
Costal margin (costa)
Discal spot
Apex
Forewing (FW)
Inner margin
Anal angle
Outer margin
Hindwing (HW)
Outer margin

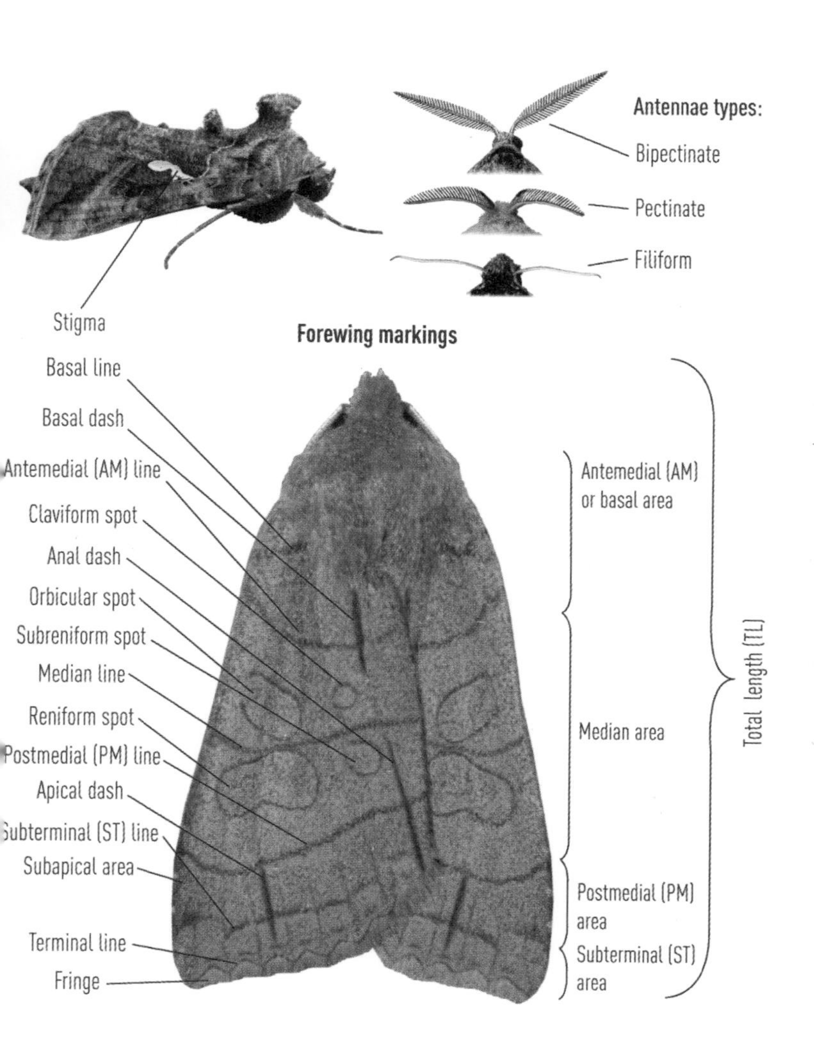
Antennae types:
Bipectinate
Pectinate
Filiform
Stigma
Forewing markings
Basal line
Basal dash
Antemedial (AM) line
Claviform spot
Anal dash
Orbicular spot
Subreniform spot
Median line
Reniform spot
Postmedial (PM) line
Apical dash
Subterminal (ST) line
Subapical area
Terminal line
Fringe
Antemedial (AM) or basal area
Median area
Postmedial (PM) area
Subterminal (ST) area
Total length (TL)

PETERSON FIELD GUIDE TO MOTHS of Southeastern North America

PETERSON FIELD GUIDE TO

MOTHS

of Southeastern North America

SEABROOKE LECKIE AND DAVID BEADLE

HOUGHTON MIFFLIN HARCOURT
BOSTON • NEW YORK • 2018

Sponsored by the
Roger Tory Peterson Institute
and the National Wildlife Federation

hmhco.com

Library of Congress Cataloging-in-Publication Data is available.

ISBN 978-0-544-25211-0

Book design by Eugenie S. Delaney

Printed in China

SCP 10 9 8 7 6 5 4 3 2 1

Preceding images:
Painted Flower Moth, *Schinia volupia*
Black-dotted Spragueia, *Spragueia onagrus*

To Dan,
for not cursing when I said I was
going to do another field guide
—SL

I would like to dedicate this book
to the memory of my sister Lynn Edwards
—DB

Slosson's Metalmark

ROGER TORY PETERSON INSTITUTE OF NATURAL HISTORY • JAMESTOWN, NEW YORK

ROGER TORY PETERSON INSTITUTE
OF NATURAL HISTORY

Continuing the work of Roger Tory Peterson through Art, Education, and Conservation

In 1984, the Roger Tory Peterson Institute of Natural History (RTPI) was founded in Peterson's hometown of Jamestown, New York, as an educational institution charged by Peterson with preserving his lifetime body of work and making it available to the world for educational purposes.

RTPI is the only official institutional steward of Roger Tory Peterson's body of work and his enduring legacy. It is our mission to foster understanding, appreciation, and protection of the natural world. By providing people with opportunities to engage in nature-focused art, education, and conservation projects, we promote the study of natural history and its connections to human health and economic prosperity.

Art—Using Art to Inspire Appreciation of Nature

The RTPI Archives contains the largest collection of Peterson's art in the world—iconic images that continue to inspire an awareness of and appreciation for nature.

Education—Explaining the Importance of Studying Natural History

We need to study, firsthand, the workings of the natural world and its importance to human life. Local surroundings can provide an engaging context for the study of natural history and its relationship to other disciplines such as math, science, and language. Environmental literacy is everybody's responsibility—not just experts and special interests.

Conservation—Sustaining and Restoring the Natural World

RTPI works to inspire people to choose action over inaction, and engages in meaningful conservation research and actions that transcend political and other boundaries. Our goal is to increase awareness and understanding of the natural connections between species, habitats, and people—connections that are critical to effective conservation.

For more information, and to support RTPI, please visit rtpi.org.

The Hebrew

Long-snouted Penthesilea

CONTENTS

PETERSON FIELD GUIDE TO MOTHS of Southeastern North America

Psychedelic Jones Moth

Coffee-loving Pyrausta

For years, moths have been plagued by bad press. The stereotype is of a drab brown creature that, at best, flutters aimlessly at your lights, and at worst, chews holes in your clothes. While there are species that do both of these things, they represent a tiny minority of the incredible diversity found in the world of moths.

More than 11,000 species of moths are currently recognized in North America. There are large moths and small ones, plain moths and bright ones, nocturnal moths and day-fliers, and moths inhabiting virtually every habitat niche you can think of. Our goal with this guide is to introduce you to some of the remarkable species that may inhabit your backyard and neighborhood. Once you've seen a spectacular Regal Moth or one of the incredible wasp-mimic clearwing borers, you won't be able to look at moths the same way again.

Eleven thousand species is a far greater number than even the most ambitious field guide might be able to cover in one volume. In order to offer the most useful field guide to moths that we could, we limited our scope to just southeastern North America. Even this much smaller region still contains thousands of species. Many of these are small, however, or very rare, or very localized in occurrence. We selected some 1,800 of the most common or most eye-catching of the moths in this area to include in this guide.

The majority of the species you are likely to encounter on any given night are present within these pages, and for those species that aren't, we have provided additional resources at the end of the book. For readers residing in the northern part of this book's range, particularly within the Appalachian Mountain ranges, we highly recommend also obtaining a copy of the *Peterson Field Guide to Moths of Northeastern*

North America. Many species are common in the Northeast but only just range into our southeastern area, and we chose to omit some of these in order to include a fuller coverage of southeastern residents and specialties.

The purpose of this guide is not to provide an exhaustive life history for each species, but rather to be an introduction to them. We have selected photographs that depict representative individuals for each species, and we have included basic information that will help in reaching an identification. If you wish to know more about any of these species, additional information can be found in the resources listed at the end of the guide.

For both of us, moths are our hobby, not our profession. While we have made every effort to ensure that the content of this guide is as accurate as possible, there may still be errors present. Moths have not had the benefit of decades of hobbyists collecting data the same way some other taxonomic groups, such as birds and butterflies, have, and so there are still many gaps in our knowledge. We hope that by providing a comprehensive, user-friendly guide we may be able to fill in some of these gaps through introducing new enthusiasts to this amazing group of insects. We would be interested in receiving feedback on errors and omissions so that future editions of this guide might be corrected and improved.

Texas Wasp Moth

How to See Moths

Moths are everywhere. You don't need to stray far from home to see them; even the tiniest urban lot has moths present, provided there are plants around for the caterpillars to feed on. Observing moths can be as easy as turning on the porch light on a warm summer evening and stepping out once in a while to see what might have come in.

Just as one's birdwatching experience is improved with a pair of binoculars, so too can a little bit of equipment greatly increase the enjoyment you take in looking for moths. You don't need to spend very much, either—a simple outlay of $15 can dramatically increase the number of moths you might see. Of course, as with any hobby, if you discover a passion for it, it's definitely possible to spend quite a bit more.

The Basics

Lightbulbs. While a simple incandescent light might draw in a few moths, the most effective bulbs are ones that project some light in the ultraviolet (UV) spectrum. There is some debate over why moths are attracted to light, and why UV light should make a difference. One theory suggests the insects navigate by the light of the moon and that artificial light disorients them; another holds that the wavelengths of UV light stimulate the receptors on moth antennae the same way pheromones from female moths do, thus luring them in. Whatever the reason, artificial lights will draw in the greatest number of species.

A black light is an inexpensive option for attracting moths. Black lights are often sold at home renovation centers or other stores that provide a wide array of bulbs as party accessories. Similar in nature are grow bulbs designed for plants or aquariums, and bug zappers (make sure you disable the zapper if you purchase one of these!). Regular household compact fluorescent light (CFL) bulbs can be surprisingly effective too, because of the mercury content, and are easy to find. A more expensive, but much more effective, option is a mercury or sodium vapor bulb. These powerful, high-wattage bulbs broadcast a very bright light in a broad spectrum of wavelengths, attracting more individuals and drawing moths in from farther away. They are the sort of bulbs found in typical outdoor security lights and usually need to be ballasted in an appropriate fixture (although some are sold as self-ballasted). Be careful using these—they get very hot and can present a fire risk if knocked over or if your sheet falls across them.

White cotton sheet. A lightbulb may be set up in front of a wall or other smooth surface that reflects the light and also provides a place for moths to settle. Pale surfaces work best, as they make it easier to detect the moths. A cotton sheet has the additional advantage of reflecting UV rays (synthetic fibers such as polyester do not), creating a much broader surface area for attracting moths. This is especially true when using a black light, which is much more effective when used in combination with a white sheet. Figure 1 shows two typical sheet and light setups.

Light trap. This is an essential piece of equipment for serious moth enthusiasts. A light trap holds the moths drawn to the light until the moth-er can return to check them. This has the great benefit of allowing you to get some sleep overnight! Light traps can be purchased online and can also be easily made from household components. Figure 2 shows the traps belonging to, and constructed by, the authors. At its most basic, a trap consists of a container in which to hold the moths, egg cartons or some other textured structure for the moths to hide in while inside, a lid to keep them from escaping, a funnel in the lid to let moths in, and a lightbulb to sit in the funnel. The exact pieces you use to accomplish these goals are determined by what's available to you and how much you want to spend. An image search

Fig 1. Two approaches to a sheet setup. Sheets can be freestanding (hung from a line or threaded over a pole, at left) or tacked to the side of a structure; the latter is easy to put up but restricts the range from which moths can see the sheet. The bulb can be clamped to a tripod or hung from a line in front of the sheet, or can be part of a light trap that's set in front. In all setups, be sure the sheet and bulb are firmly secured against wind.

on the Internet for "moth trap" will provide many different examples of this same general setup. If you choose to run a trap, be sure to check it first thing in the morning, or the birds will thank you for the breakfast buffet!

Sugar bait. Some species of moths are not very attracted to artificial light but are nectar-feeders and will come to sugar bait, as shown in Figure 3. The combination of ingredients you use is flexible and can be determined by what you have on hand in the kitchen. A particularly effective mixture is to blend one soft banana, a scoop of brown sugar, a dollop of molasses, and a glug or two of beer (flat or cheap is fine). Allow the mixture to sit at room temperature for a day for

Fig 2. Three styles of light traps. *From top to bottom:* commercially available Robinson trap with clear lid and circular funnel; David's homemade Skinner trap with clear lid and linear funnel-like gap; and Seabrooke's homemade Robinson-type trap with solid lid and circular funnel. The clear lids are thought to help moths settle more quickly within the trap, possibly avoiding damage to delicate wing scales, but many moth enthusiasts use solid-topped traps with few or no adverse effects. Note the egg cartons inside, which provide a resting substrate.

Fig 3. A selection of moths visiting sugar bait in September. Sugar bait is especially effective in cool weather but can be used year-round. Allowing the bait to sit for a few hours before darkness falls often helps attract more moths. Many other invertebrates are often attracted to sugar bait as well.

improved results. Paint this sticky concoction onto tree trunks with a brush. It may stain wood, so avoid using it on your deck or other structures. Another option is to soak a thick rope (at least 1.25 cm in diameter, so the moths can land on it) in the mixture and then string it between two supports. You will need to check the trees regularly, just as you would do with your cotton sheet, in order to see what the bait has attracted. This tool is particularly effective on cool nights.

Jars or containers. Many of the moth photographs in this guide were taken during daylight hours, which allows for more evenly lit photos without any of the scale glare caused by using a flash. Moths can be cooled in the refrigerator (*not* the freezer) for one night if necessary;

the low temperature puts them into a state of torpor, which keeps them calm and lowers their metabolism, resulting in more cooperative subjects. Pill bottles such as those provided by a pharmacist are ideal containers as they are small and don't take up much room in the fridge, and clear so that you can see what you have inside. You can order them online, or ask at your local pharmacy.

In "the Wild"

Not all moths are nocturnal. Some nectar-feeders can be seen supping at flowers during the daytime as well. A few species are almost exclusively diurnal and are rarely encountered at night. You can encourage day-fliers to visit your gardens by planting nectar-rich flowers, which are often sold at garden centers as appealing to butterflies or hummingbirds. Phlox and bee balm are particularly popular with sphinx moths; smaller moths with shorter tongues might benefit from butterfly weed and other smaller flowers.

Don't forget to keep your eyes peeled while out hiking. The moths that visit your garden plants can also be found on wildflowers. Many species will rest in the grass or the leaf litter on the forest floor, so watch for small pale shapes that rise up ahead of you. All of those moths that come to your sheet at night have to find somewhere to spend the day; examine tree trunks, bark crevices, rock ledges, and other protected places. Don't overlook your house and garage—moths can turn up in the strangest places!

Caterpillars and Cocoons

While this guide helps only with the identification of adults, moths actually spend a larger percentage of their lifetime in the larval stage. Caterpillars can be very cryptic, blending in with the plants on which they feed and often going unnoticed. Others have behaviors or patterns that make them instantly recognizable, such as the springtime tents made by the Eastern Tent Caterpillar, or the fuzzy black-and-brown Woolly Bears, which are caterpillars of the Isabella Tiger Moth. We do not provide any information on caterpillars or cocoons, but there are other publications that are very helpful. If you are interested in learning to identify the caterpillars of moths, check out the resources at the end of the book.

How to Identify Moths

With such a vast array of possibilities, it's easy to feel a bit overwhelmed when first learning to identify moths. Many of them seem to look the same, and even learning to tell one group from another can seem like a daunting challenge. It is only through constant practice and regular exposure that you will start to be able to discern differences, and practice takes time. Don't give up! It takes most moth enthusiasts a few years before they start feeling comfortable with their local species.

When you're beginning, it is easiest to select a subset of the moths at your lights and set about learning them first. Which subset you decide to start with is entirely up to you, but we recommend concentrating on learning the common species first, the ones that you see many of on a given night. Once you feel comfortable with the common species, it will be easier to spot the less common species. Another approach is to begin with the flashy or brightly colored species, as these are easily remembered and tend to stand out from the rest. In subsequent seasons you can build on the skills you developed the year before, and each year you'll be able to identify a few more species without having to refer to your guidebooks. We strongly recommend ignoring most of the smaller moths to begin with, and focusing just on the larger ones, which are usually easier to identify. When you feel comfortable with the larger moths, you can turn to the smaller ones without feeling overwhelmed.

Another good way to familiarize yourself with the various species is simply to spend some time flipping through your field guide. If you've often looked at an illustration in your book, when the species finally turns up at your light you'll already know what it is, or at least have a fair idea of where to find it in the guide.

Appearance

When trying to identify a particular individual, first make sure that the lepidopteran you're looking at is in fact a moth. Some diurnal moths resemble butterflies in color and habit, and some drab butterflies might bring to mind a moth. The easiest way to tell them apart is by the antennae: if they are clubbed at the tip, it's a butterfly; if they are threadlike or feathery, it's a moth.

There are many questions you should ask yourself, even before you begin looking at the patterns of lines and spots on the moth's wings. How large is it? How does it pose while at rest? What shape is it? What color is it? Answering these questions can help you narrow down which group a moth is in. A medium-sized moth that rests with its wings spread out to its sides is most likely in the family Geometridae, while one that sits with its wings folded tentlike over its back is probably in the Noctuidae. A noctuid that is long and thin is most likely in a different group than one that is more triangular. As with anything, there are exceptions, but it is a helpful place to start. The back endpapers of this guide show many different moth silhouettes.

Size is a very important key in identifying moths. Not only can it help you determine which group of moths an individual belongs to, it can also often help in separating two very similar species. All of the species on a photo plate are sized relative to each other, so larger species will be shown larger. We have placed a life-sized gray silhouette of one of the moths on each plate to give you a sense of actual size. The scale might be different from one plate to the next, even within the same family, so make sure to refer to the silhouette on each page.

Once you have narrowed down the possibilities to a few families, you can look more closely at the patterns. In this guide, we often refer to the various spots and lines (the latter are often called "bands" when they are wider) that can be seen on the wing. Each has its own unique name, and knowing the terminology for the structure and patterns of the wing will be very helpful when reading the species accounts. Figure 4 shows the different markings and parts of the wing that we refer to when describing species; it also shows different antennae types. This diagram is repeated on the front endpapers for convenience.

To conserve space, we frequently use abbreviations when referring to commonly referenced parts or structures. Forewing and hindwing are shortened to FW and HW, respectively. Likewise, the antemedial line, postmedial line, and subterminal line are the AM, PM, and ST lines. These abbreviations are also used when referring to the areas of the wing with the same name. Measurements are given as either wingspan (WS) or total length (TL), with the former being from wingtip to wingtip and the latter from head to tail.

Some additional moth-specific terminology is used throughout

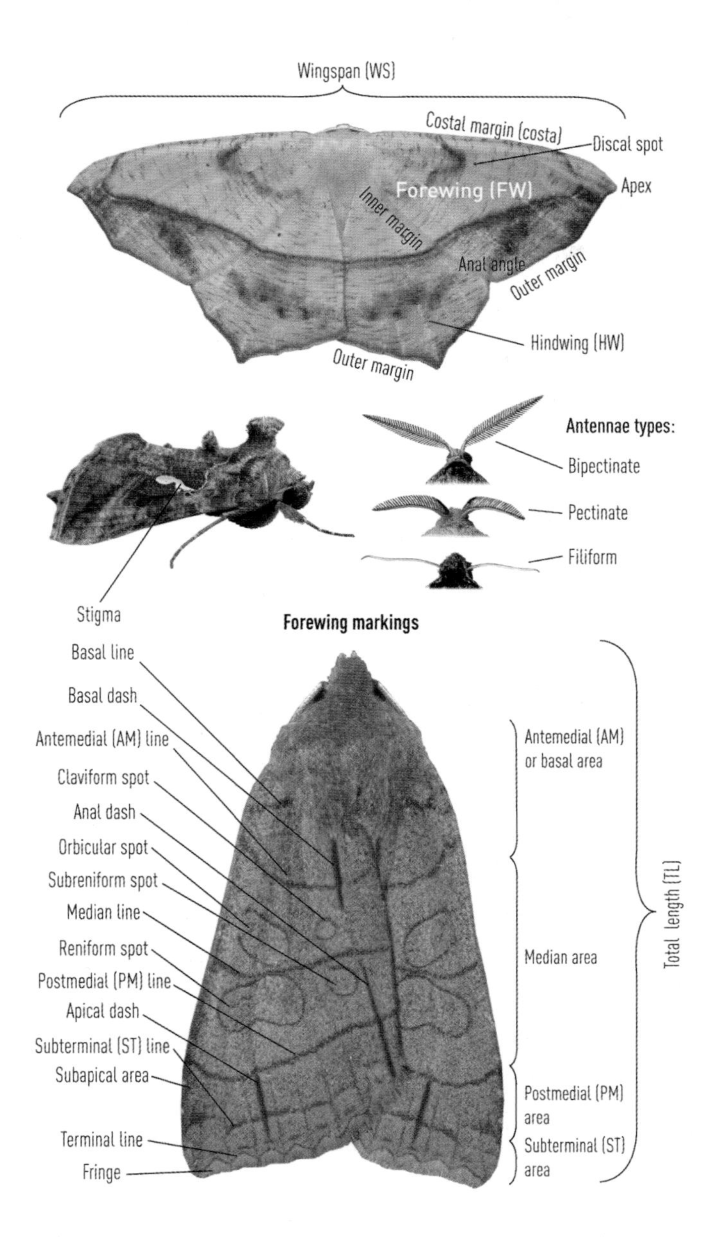

Fig 4. Wing parts, antennae types, and forewing markings referred to in the species accounts.

the accounts. These words and phrases are defined in the glossary at the back of the book.

Some species of moths, such as Harris's Three-Spot, are so distinctive that it's impossible to mistake them for any other species, while others, such as the datanas, are separable only by relatively subtle field marks. The markings that help identify a given species from similar ones are indicated on the plates using arrows, the unique and innovative system of identification pioneered by Roger Tory Peterson in his first field guide to birds. Clarification is also provided in the text. It is helpful to refer to both when trying to identify a moth.

Just as in humans, even within the same species of moth there is often quite a bit of individual variation. Some individuals might be lighter in color than others, or darker, or a different color altogether; some might be well marked while others are very faint. In some species there are multiple color patterns; in others, females look very different from males. And finally, some species of moths may live for a few weeks, and over time some scales may be rubbed off from the wings, affecting the appearance of their pattern. Some very worn individuals may not be identifiable at all.

Flight Periods

The species composition of the moths that come to your light will change over the course of a season. The majority of the moths you see in May will be different from those you see in September. Knowing when the adults of a certain species are on the wing can help narrow down an identification. If you are deciding between a species that flies in spring and one that flies in summer, and it's August, it's clear which of the two your moth is.

The length of time a species might be present varies among species. Some may be encountered nearly all year, while others might only fly during a three- or four-week window. Several species raise two broods, so you might encounter adults in two different non-overlapping periods; others emerge from their cocoons in the fall and then overwinter as adults, waking from hibernation to fly again in the spring.

Flight periods are illustrated with a colored graphic beside the species account, as shown in Figure 5. The four colored boxes rep-

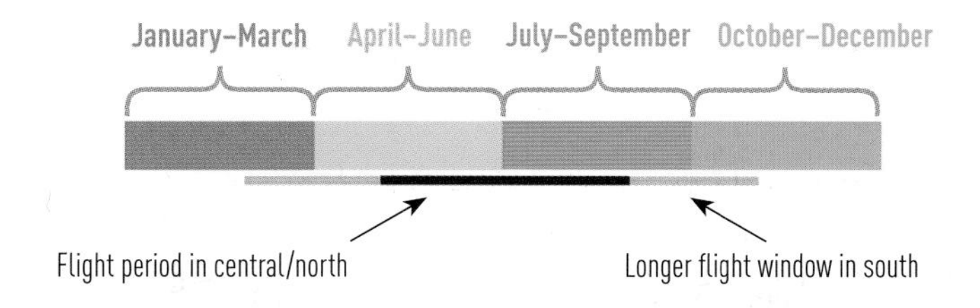

Fig 5. Flight periods are illustrated using four-color bars beside the species name. Each color represents a three-month period. The solid line underneath indicates when adults of the species are typically encountered. The flight periods given in this book reflect the average for the range of this book; in the northernmost part of this range the window may be shorter than shown, while in the far southern part it may be longer.

resent the seasons, each consisting of three months: January–March, April–June, July–September, and October–December. The black line below the boxes indicates when the adult moth is generally present. In the example shown here, the moth would be expected to fly from May through August. Some species are present for more of the year in the southern part of this book's range than they are in the central or northern parts; in such cases, we use a gray line to indicate this longer flight window. In a few cases, a moth occurs only in the very south, and so is represented by only a gray line. The flight periods shown represent the average for the region covered by this guide; on occasion you may encounter a species slightly outside the period shown. Flight periods are assumed to refer to a single brood unless otherwise indicated.

Range Maps

Knowing where a moth occurs is every bit as important as knowing when it occurs. Previous field guides to moths have typically provided only a written description of the range for each species. Such descriptions are vague at best, as much is left to the interpretation of the user. In this guide we have provided range maps for as many of the species as possible. The maps are easy to read and interpret and can quickly tell you whether to expect a species in your area.

The paucity of data on moth species compared with other, more familiar groups, such as birds and trees, presents a challenge when preparing range maps. While the range of many common species is well known, for most moths it is difficult to draw a map with any level of precision. To get around this difficulty, we relied heavily on the use of ecoregions in mapping ranges. Unlike vertebrates, moths are strongly tied to the food plants that their caterpillars eat—if the host plants aren't present, the moths won't be either. The host plants, in turn, are often restricted to certain environments, whether by temperature, soil type, drainage, and so on. Ecoregions are large areas that share similar environmental conditions and plant communities.

The ecoregion map we used to create the range maps in this guide is shown in Figure 6. This map was adapted from the North American Environmental Atlas, a project completed by the Commission for Environmental Cooperation in joint partnership with the governments of the United States, Canada, and Mexico.

While some plant species may be found in one part of an ecoregion but not another, in general this system allowed us to extrapolate a range from known data points. For instance, if a moth had been recorded in central Mississippi, there is a good chance it also occurs in central South Carolina, and we have depicted it as such, even if

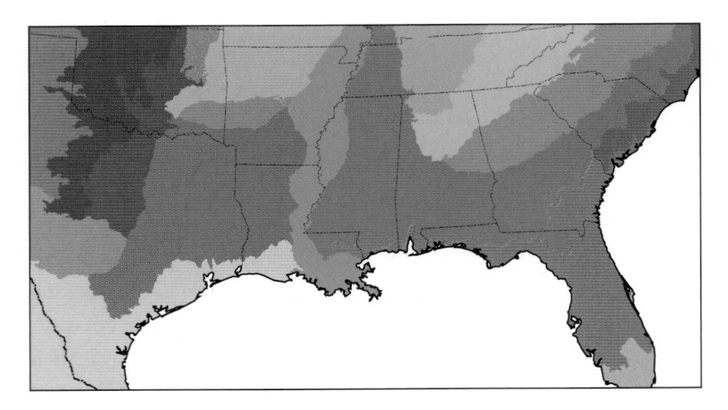

Fig 6. The map of ecoregions used to create the range maps in this guide. Species ranges were inferred by matching known data points to ecoregions and extrapolating to those ecoregions' boundaries. The map is adapted from the North American Environmental Atlas by the Commission for Environmental Cooperation and the governments of the United States, Canada, and Mexico. For a detailed explanation of the ecoregions, including region labels, visit www.cec.org.

Range Restrictions: The Variable Narrow-Wing (*Magusa divaricata*, shown here) and the nearly identical, smaller Orbed Narrow-Wing (*Magusa orbifera*) are examples of species best identified by range. The Variable Narrow-Wing will wander widely; the Orbed Narrow-Wing never does.

we had no data from the latter state. The one exception to this is for the Mississippi River basin ecoregion, which has limited occurrence data available. Since the status of most species in this ecoregion is unknown, we have typically included it when a species is known to occur in ecoregions on both sides, and excluded it when a species is known only from ecoregions on one side or the other.

This approach is of course not without its own problems, and undoubtedly there will be instances when the maps are inaccurate and either over- or underrepresent the actual distribution. Also, the transition between habitat types is rarely a sharp line, and depending on the requirements of a given species, it may be found slightly beyond the boundaries of its main ecoregion. Please bear this in mind when interpreting the range maps. Additionally, moths are winged creatures; many are capable of flying long distances, and they can also become caught up in traveling weather systems. A range map is never a hard-and-fast rule, as birdwatchers can attest—species may show up out of range from time to time.

Maps are depicted in one of two colors. The darker green shade indicates typical range, while the lighter green shade represents migratory or vagrant range. In some instances, when the lighter shade borders a darker one, it is an indication of an expanding range; this is usually noted in the text.

Habitat and Host Plants

A moth species is rarely distributed evenly across its range. With few exceptions, most species can be found only in or near habitats that contain their larval food plants. Even though a species may be shown to occur throughout an entire state, there will be locations within that state where the moth is abundant, and other spots where it cannot be found at all.

Habitat is another important tool in moth identification. Knowing that one species is found only in bogs and marshes while another frequents upland forests can make identification easier. Likewise, if a species feeds on bald cypress but there are no bald cypress trees anywhere near you, you are unlikely to have the species turn up (but see the caveat above regarding winged organisms!). For space reasons, we have not included habitat in the species accounts, but the larval food plant is given for all species when known, and this information can be used to extrapolate in what habitats the species will likely occur.

Some moth species are very specific in their requirements. Their larvae may feed exclusively on only one or two species of plants, and the distribution of such moths is often nearly identical to that of their host plants. Other species are generalists, whose caterpillars are not fussy about what they eat. These moths may be encountered across a broad range of habitats and ecoregions. Many (though not all) of the moths found throughout our coverage area are generalists, whereas those restricted to small ranges are often picky eaters.

Abundance

As you develop an interest in moths, one of the first things you are likely to notice is that some species are more abundant than others. This is true across all taxons—some species are just more common. We have tried to indicate relative abundance for each species, to give you a sense of the likelihood of occurrence. However, the relative abundance of the different species attracted to a light may change from one location to another; what is plentiful in one locale may not

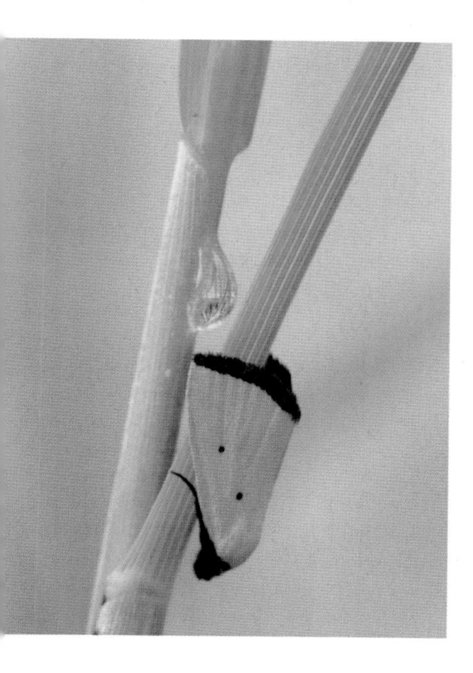

Host Plant Occurrence: Moth species that feed on widely occurring host plants, or are generalists on many different plant species, tend to be widespread and abundant. The caterpillars of this Black-bordered Lemon (*Marimatha nigrofimbria*) feed on grass, and the species is typically common for this reason.

be plentiful in another. This is partly determined by the food plants the larvae favor, with generalists being on average more abundant than specialists, and with specialists being more abundant when their host plant is more abundant. Abundance is also affected by flight periods. A species will be more abundant in the middle of its flight period than at either end. Dozens may be seen in a night at the peak of the flight period, whereas a few weeks later you might find only one in an entire evening.

A large part of species identification in any taxon is knowing what to expect. By combining information on appearance, flight period, range, habitat, and abundance, it is usually possible to reach a probable or definitive identification.

Moth Taxonomy

There are two basic approaches to organizing a field guide: by grouping species of similar appearance together, or by organizing species taxonomically. Although a beginner often finds it easier to use a book in which similar species are presented side by side, this approach is frustrating once you have enough experience to be able to recognize taxonomic groups. If you know the moth you are looking at is a dagger because of its size and shape, but you don't know which one, you want to be able to flip to the section on daggers and compare them all. Books organized by similar appearance will have you flipping back and forth and back again to compare the different daggers in order to determine which yours is. Becoming familiar with the different taxonomic groups may take some time, but in the long run it will greatly aid your ability to identify different species.

The organization of species accounts in this guide follows the order in the *Annotated taxonomic checklist of the Lepidoptera of North America, North of Mexico* by Pohl, Patterson, and Pelham (2016), which includes all of the most recent taxonomic updates as of this writing. Taxonomy is always changing as new research reveals more about the relationships among species and species are moved from one group to another, or are lumped or split to form new species. Major changes don't happen often, however, so the taxonomic organization used here should provide a firm base for learning.

Hodges and P3 Numbers

In addition to its scientific and common names, each species of moth in North America is represented by two unique identification numbers. The first series of numbers used originated from a 1983 publication by lepidopterist Ronald Hodges in which he assigned a number to every known species north of Mexico, and as such they are called Hodges numbers. Many taxonomic changes and updates have taken place since 1983, however. Some Hodges numbers no longer exist because the species they were assigned to were lumped with others; new species have been described, requiring new numbers that have resulted in decimals being used; and some species have been moved to new groups, resulting in the numbers being out of order taxonomically. Despite this, the Hodges numbers are often useful to know, as once assigned a number remains with a species permanently, even through taxonomic revisions, renamings, and reorderings. Most websites and publications still organize their lists by Hodges number.

More recently, a new numbering system was created by Bob Patterson of the Moth Photographers Group website. This system assigns a two-digit prefix to each superfamily, and to each species a four-digit suffix, unique within the superfamily. This allows superfamilies to be numbered independently, and quickly helps identify to which group a species belongs. First adopted by Don Lafontaine and Christian Schmidt in their 2010 revision of the Noctuoidea superfamily, this new numbering system has now been applied to all groups in the 2016 annotated checklist by Pohl, Patterson, and Pelham (P3). In the first edition of the *Peterson Field Guide to Moths of Northeastern North America*, we referred to this numbering system as "MPG numbers," but the accepted terminology is now "P3 numbers," after the new checklist's authors.

In this guide, we present both the P3 and Hodges numbers (with the latter in parentheses) after a species' common and scientific names. Although the Hodges numbers remain in widespread use, we recommend adopting the P3 numbers as your primary reference and labeling tool as these provide proper taxonomic order and will eventually replace the Hodges numbers on websites and in publications.

Common Names

A few taxonomic groups, such as birds and butterflies, have publications providing standardized common names, as decided by a committee of scientists with expertise in the field. No such publication yet exists for moths. While some species, such as the Luna Moth, are so familiar that they are known everywhere by a single name, many species have no commonly accepted name and can be found in different places with different names. To complicate matters, the common name of many species of moths includes the genus the moth belongs to; however, taxonomic changes that move the species into another genus require that the common name be changed to reflect this.

The common names we use in this guide come largely from existing publications, primarily *A Field Guide to Moths of Eastern North America* by Charles V. Covell, Jr. When multiple common names exist for a species, we have selected the one we feel is most representative or easiest to remember. In a few instances, we could find no evidence of a previously given common name and have coined our own.

Micros versus Macros

If you happen to attend a moth night with experienced moth enthusiasts, you might hear them refer to "macromoths" and "micromoths," or simply "macros" and "micros." For the most part, micromoths are relatively small moths and macromoths relatively large, but each group has a few exceptions that are closer in size to those of the other group.

The distinction between the two groups has more to do with taxonomy than it does with size, but the terms are handy labels. Taxonomy usually orders species according to their evolutionary age, with the oldest groups presented first, and the youngest—those that have diverged most recently—presented last. Strictly speaking, micromoths are those moths that appear on the first half of taxonomic lists; in this book, that means from the Goldcap Moss-Eater (P3 #01-0001) through the Glittering Magdalena (P3 #80-1560). The moths from Scalloped Sack-Bearer (P3 #83-0001) through to the end of the list are macromoths. It's just coincidence that these two broad taxonomic groups also conveniently split by size.

Moths and Conservation

Although we don't often see them, moths are an important part of the environment. Adult moths are a valuable food source for bats, many species of which feed almost exclusively while in flight, and caterpillars are a vital part of the diet that adult songbirds feed their nestlings. Some species of moths have periodic "outbreak" years, and certain types of birds are so tied to these outbreaks for their breeding success that populations are seminomadic. As moth outbreaks have been suppressed for the benefit of the forestry industry, so too have we seen their dependent bird populations decline. There is evidence that insect populations have been declining for several decades, especially those of flying insects, including moths. Most species of birds that feed exclusively on the wing, such as swallows and nighthawks, are also in sharp decline. Although we typically think of bees as the pollinators of our fruits and vegetables, many species of moths are also valuable pollinators, and their decline could have repercussions for us too.

A large part of the problem is that we simply don't know very much about our moths, or insects in general. Unlike for birds, which have millions of enthusiasts across the continent and for which there are several national and international monitoring schemes, there are few programs to track insect populations. Most of those that exist in North America are either local or limited to a few species. Great Britain has developed the National Moth Recording Scheme (NMRS), a citizen science project that invites anyone anywhere in Great Britain to submit their observations to a national database used to help track moth populations and learn more about their distribution. There are many excellent guides to the moths of Great Britain, and mothing is a popular activity as a result. Since the NMRS's inception in 2007, more than 11 million records have been collected for it; many of these records have been organized into maps and published as an atlas.

Currently in North America we lack even a solid grasp of the distribution of many moth species. While state and county lists exist for birds and butterflies and even dragonflies, there are fewer such inventories for moths, and many are incomplete. Because so little is known about our moths and their distribution, every time you put your lights out at night you have the potential to contribute something valuable—that dart could be new for the county, or your July date for that sallow could be the earliest on record.

The more we know, the better our understanding, and the better able we are to design and implement conservation efforts. Our hope is that this guide will not only open up the world of moths to budding naturalists and enthusiasts, but will also help ensure the future of these beautiful insects for generations to come.

Southern Purple Mint Moth

SPECIES ACCOUNTS

Moss-Eaters and Leaf-Miners
Families Micropterigidae and Eriocraniidae

Very small, metallic-looking moths with broad rounded wings. Among North America's smallest moths, the adult Goldcap Moss-Eater rarely comes to light and is best sought at liverwort beds during daytime. The larger and more golden Chinquapin Leaf-Miner readily visits lights in oak-dominated woodlands.

GOLDCAP MOSS-EATER

Epimartyria auricrinella 01-0001 (0001) **Local**

TL 3–4 mm. Metallic purple FW is peppered with golden scales but appears blackish in poor light. Fuzzy head is dull golden orange. **HOSTS:** Liverwort. **RANGE:** E. TN and w. NC.

CHINQUAPIN LEAF-MINER

Dyseriocrania griseocapitella 07-0001 (0003) **Uncommon**

TL 5–7 mm. Metallic golden FW is evenly peppered with metallic indigo scales. Fringe is gold. Hairy-looking head and thorax are brownish. **HOSTS:** Primarily black oak; also chinquapin and other oak species. **RANGE:** Widespread as far west as ne. TX.

Ghost Moths Family Hepialidae

Medium to large moths. Wings are covered with tiny spinelike scales that give some species a slightly hairy aspect. These moths rest with their wings held tight to the body in a compressed tent. Adults lack functional mouthparts and do not feed. Many species are crepuscular and can sometimes be found in mating swarms at dusk. Adults are rarely attracted to lights.

GRACEFUL GHOST MOTH

Korscheltellus gracilis 11-0011 (0031) **Uncommon**

TL 18–22 mm. Peppery brown FW is variably patterned with silvery gray patches and lines. Usually has an irregular silvery band slanting from inner margin to apex. **HOSTS:** Roots of balsam fir and red spruce; also aspen, birch, ferns, and white spruce. **RANGE:** E. TN and w. NC.

GOLD-SPOTTED GHOST MOTH

Sthenopis pretiosus 11-0019 (0022) **Uncommon**

TL 34–38 mm. Golden brown FW has a metallic brassy luster. Chunky gold spots mark AM, median, and ST areas. Costal streak and PM area are overlaid with pink lines. **HOSTS:** Ferns. **RANGE:** E. TN and w. NC.

MOSS-EATERS AND LEAF-MINERS

GOLDCAP MOSS-EATER

actual size

CHINQUAPIN LEAF-MINER

GHOST MOTHS

GRACEFUL GHOST MOTH

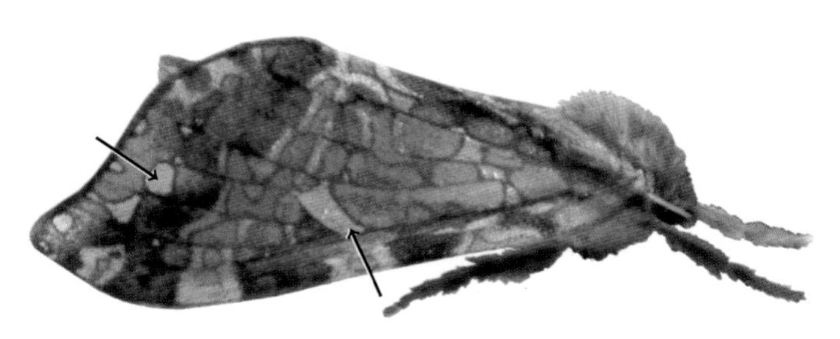

GOLD-SPOTTED GHOST MOTH

actual size

Fringed Miner Moths Families Nepticulidae, Opostegidae, Tischeriidae, Incurvariidae, and Prodoxidae

Tiny or small moths with broadly fringed lanceolate wings. The larvae of certain *Ectoedemia* species create the green spots on fallen poplar leaves in the fall. Those of the diminuitive Gooseberry Barkminer mine in stems and leaves of currants. The all-white Yucca Moth is often found in gardens with ornamental yucca plants. Adults are nocturnal and will come to lights in small numbers.

BROKEN-BANDED ECTOEDEMIA

Ectoedemia similella 16-0088 (0045) **Uncommon**

TL 3–4 mm. Metallic ink blue FW has small whitish spots at midpoints of inner margin and costa. Rounded apex is broadly fringed whitish. Note white eye-caps and orange tufted head. **HOSTS:** Pin oak. **RANGE:** Locally in GA and FL. **NOTE:** The genus *Ectoedemia* contains several look-alike species.

GOOSEBERRY BARKMINER

Pseudopostega quadristrigella 16-0107 (0122) **Uncommon**

TL 4–6 mm. White FW has a small brown smudge at midpoint of inner margin and a tiny dot at apex. **HOSTS:** Currants, including gooseberry. **RANGE:** W. NC west to OK, south to FL and ne. TX.

YUCCA MOTH *Tegeticula yuccasella* 21-0029 (0198) **Local**

TL 9–15 mm. Part of a complex of several look-alike species, each of which has a species-specific biology. Pure white FW has a rounded apex. Gray HW has a white fringe. White antennae have black tips. **HOSTS:** Yucca. **RANGE:** NC west to OK, south to FL and LA.

BOGUS YUCCA MOTH

Prodoxus decipiens 21-0044 (0200.1) **Local**

TL 8–12 mm. Resembles Yucca Moth, but white legs have rusty brown tips. **HOSTS:** Unknown. **RANGE:** SC west to OK, south to FL and cen. TX.

MAPLE LEAFCUTTER

Paraclemensia acerifoliella 21-0069 (0181) **Local**

TL 5–7 mm. Metallic ink blue FW looks blackish in poor light. Fuzzy head is bright orange. **HOSTS:** Maple, sometimes birch. **RANGE:** TN southwest to MS and e. TX.

OAK BLOTCH MINER

Tischeria quercitella 23-0001 (0144) **Local**

TL 4–5 mm. Bright tawny FW is lightly flecked with brown scales, especially toward apex. Note brown blotch at anal angle of FW. Frequently holds antennae outward when at rest. **HOSTS:** Oaks, mostly black oak group. **RANGE:** NC south to FL and LA.

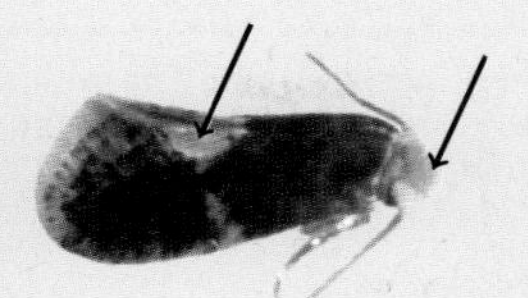

BROKEN-BANDED ECTOEDEMIA

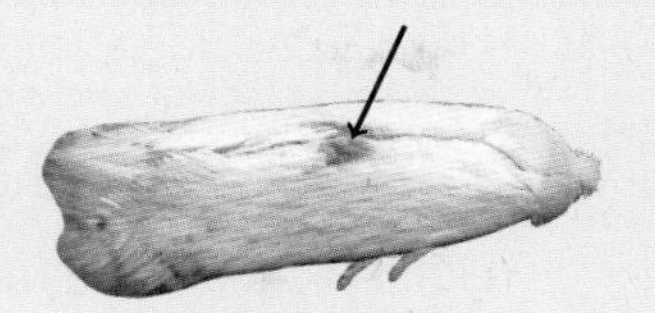

GOOSEBERRY BARKMINER

YUCCA MOTH

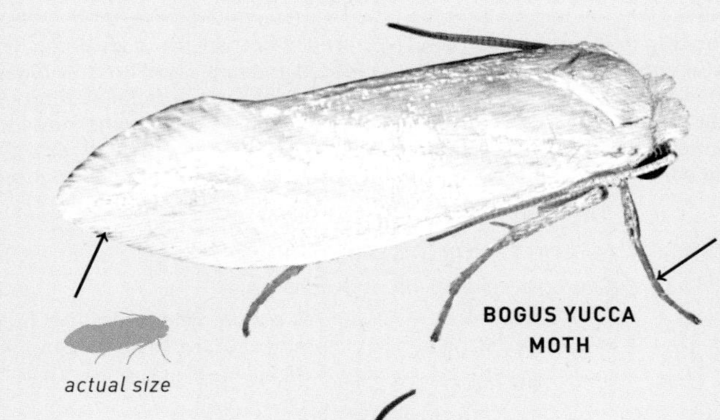

BOGUS YUCCA MOTH

actual size

MAPLE LEAFCUTTER

OAK BLOTCH MINER

Fairy (Longhorn) Moths Family Adelidae

Delicate moths with broad, often boldly patterned or metallic FW. The wings are held in a tent at rest. Most species have characteristically long antennae. They are active by day and can sometimes be found sitting on flower heads. Also nocturnal, and occasionally visit lights in small numbers.

SOUTHERN LONGHORN MOTH

Adela caeruleella 21-0117 (0227) **Common**

TL 6–8 mm. Uniform blackish FW has a strong purplish sheen in good light. Head is dull orange. Very long antennae of male are mostly whitish; shorter antennae of female have a thickened black basal half and white tips. **HOSTS:** Unknown, but adults have a preference for sitting on flowers of black snakeroot. **RANGE:** Widespread west to ne. TX.

RIDINGS' FAIRY MOTH

Adela ridingsella 21-0118 (0228) **Uncommon**

TL 4–6 mm. Golden brown FW has a straight silver median band. Inner ST area is boldly patterned with silver-edged black spots. Rounded apex is accented with silver dashes. **HOSTS:** Unknown. **RANGE:** W. NC and e. TN, southwest to MS.

Bagworm Moths Family Psychidae

Mostly small to medium-sized blackish or brown moths with broad wings (often translucent because of wear), a long tapered abdomen, and broadly bipectinate antennae. Females are wingless (or nearly so) and are confined to silken larval cases that are covered with debris; they attract males with pheromones. These moths occur in a variety of habitats. Males are sometimes attracted to lights, but some species are more commonly detected by the female's distinctive case.

KEARFOTT'S WHITE-BANDED MOTH

Kearfottia albifasciella 30-0006 (0319) **Local**

TL 5–7 mm. Velvety black FW has a wide white median band that widens toward costa. Legs are often striped white. **HOSTS:** Unknown. **RANGE:** Widespread west to OK, south to FL and e. TX.

NIGRITA BAGWORM MOTH

Cryptothelea nigrita 30-0011 (0441) **Common**

TL (male) 8–10 mm. Wings and body of male are sooty black without obvious markings. **HOSTS:** Unrecorded, but probably a variety of trees, shrubs, and low plants. **RANGE:** GA and FL west to cen. TX.

FAIRY (LONGHORN) MOTHS

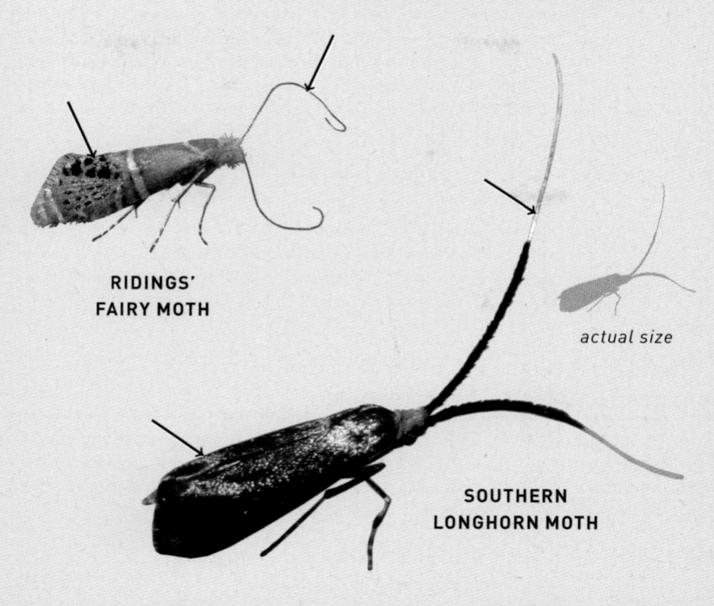

BAGWORM MOTHS

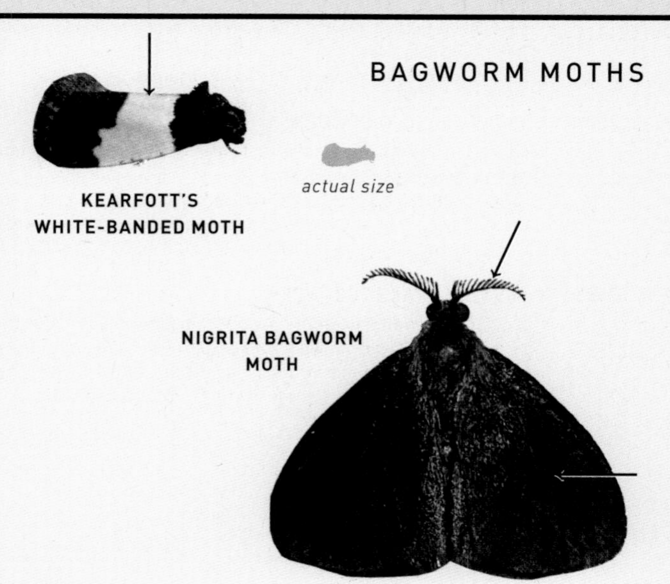

ABBOT'S BAGWORM MOTH

Oiketicus abbotii 30-0025 (0454) **Common**

TL (male) 16–18 mm. Male's FW is light tawny brown, shaded darker in distal half with a white-edged black patch in central median area. Tapering abdomen has dorsal row of black and white spots. **HOSTS:** A wide variety of trees, shrubs, and low plants. **RANGE:** Coastal NC west to MS, south to FL and e. TX.

EVERGREEN BAGWORM MOTH

Thyridopteryx ephemeraeformis 30-0028 (0457) **Common**

TL (male) 10–18 mm. Male has largely translucent wings. Chunky body and abdomen are sooty black. **HOSTS:** Various trees, shrubs, and low plants. **RANGE:** Widespread west to OK, south to FL and e. TX.

Fungus and Tube Moths

Families Tineidae and Acrolophidae

Small, streamlined moths with variably patterned FW markings. Many species have FW fringe slightly flared with hairlike scales. Larvae of many species are detritus feeders. Adults are mostly nocturnal and will come to lights in small numbers.

BROWN-BLOTCHED AMYDRIA

Amydria effrentella 30-0046 (0334) **Common**

TL 10–11 mm. Shiny light brown FW is blotched with dark brown spots in costal half of wing. Curved terminal line is checkered when fresh. Hornlike labial palps are densely hairy at base. **HOSTS:** Unknown. **RANGE:** Widespread west to AR, south to SC and MS.

TEXAS GRASS-TUBEWORM

Acrolophus texanella 30-0052 (0383) **Common**

TL 11–13 mm. Brindled grayish brown FW has darker shading distal to oblique PM line. Often has a black or brown patch in inner median area. Thorax has a backswept crest of hairlike scales. **HOSTS:** Grass. **RANGE:** Widespread west to OK, south to FL and e. TX.

EASTERN GRASS-TUBEWORM

Acrolophus plumifrontella 30-0058 (0372) **Common**

TL 14–16 mm. Rusty brown FW has faint black blotches in median area. Bristly looking thorax has a backswept bifurcate crest of hairlike scales. **HOSTS:** Unknown. **RANGE:** Widespread west to OK, south to FL and e. TX.

BAGWORM MOTHS

FUNGUS AND TUBE MOTHS

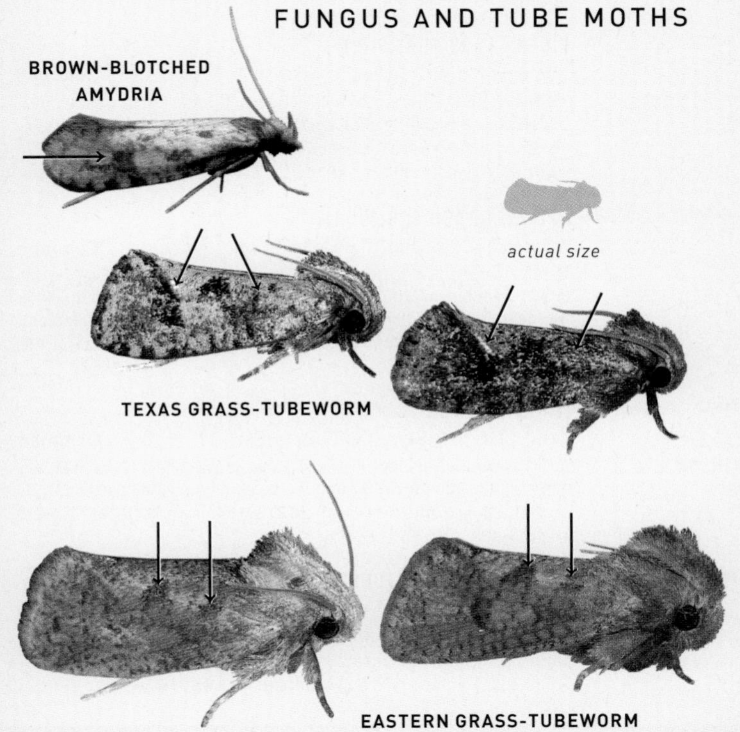

LONG-HORNED GRASS-TUBEWORM

Acrolophus mortipennella 30-0059 (0366) **Common**

TL 12–16 mm. Pale straw-colored FW is lightly speckled darker with irregular dark brown blotches in median and ST areas. Terminal line and fringe are weakly checkered when fresh. Long labial palps curve backward over head. **HOSTS:** Unknown. **RANGE:** TN west to OK, locally south to FL and TX.

CLEMENS' GRASS-TUBEWORM

Acrolophus popeanella 30-0060 (0373) **Common**

TL 13–17 mm. FW is variably straw-colored to dark gray, brindled with blackish lines. Often has black wedges in inner AM and median areas bordering paler streak along inner margin. **HOSTS:** Roots of red clover. **RANGE:** Widespread west to OK, south to FL and TX.

CRESSON'S GRASS-TUBEWORM

Acrolophus cressoni 30-0069 (0347) **Common**

TL 8–12 mm. Peppery brown FW has a marbled pattern of black-edged dark brown bands. Often has a black spot at midpoint of ST area. Inner margin is accented with tufts of raised scales. **HOSTS:** Grass. **RANGE:** Widespread west to OK, south to FL and TX. Local in northeast.

PIGER GRASS-TUBEWORM

Acrolophus piger 30-0071 (0371) **Common**

TL 10–12 mm. Silvery gray FW is lightly brindled with dusky lines. Usually has a brown median band that narrows to a point toward inner margin. Thorax has a tall crest of hairlike scales. **HOSTS:** Birch. **RANGE:** NC west to MS, south to FL and e. TX.

ARCANE GRASS-TUBEWORM

Acrolophus arcanella 30-0080 (0340) **Common**

TL 13–16 mm. FW is variably straw-colored to pale silvery gray, lightly brindled with dusky lines. Darker individuals have a blackish wedge in inner median area bordering irregular paler streak along inner margin. **HOSTS:** Unknown. **RANGE:** Widespread west to AR, south to FL and TX.

PANAMA GRASS-TUBEWORM

Acrolophus panamae 30-0083 (0368) **Common**

TL 8–10 mm. Silvery gray FW is variably patterned with warm brown AM, median, and partial ST bands, creating a marbled pattern. Thorax has a dusky bifurcate crest. **HOSTS:** Unknown. **RANGE:** NC west to MS, south to FL and LA.

HEPPNER'S GRASS-TUBEWORM

Acrolophus heppneri 30-0098 (0355.1) **Common**

TL 7–12 mm. Peppery brown FW is uniform apart from obscure darker blotches in central PM and inner median areas. **HOSTS:** Small-leaf climbing fern and sugarcane. **RANGE:** FL west to e. TX.

FUNGUS AND TUBE MOTHS

LONG-HORNED GRASS-TUBEWORM

actual size

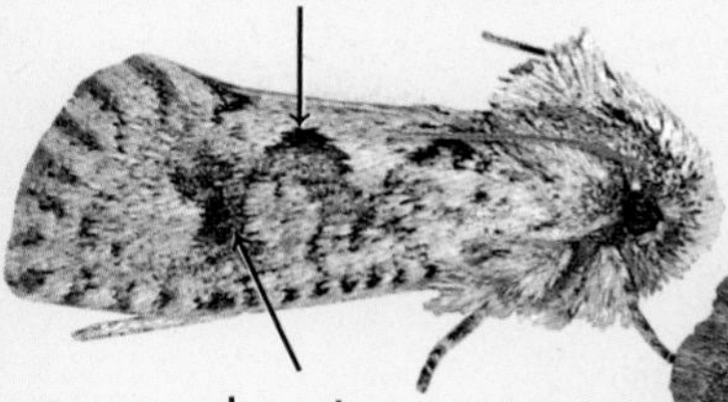

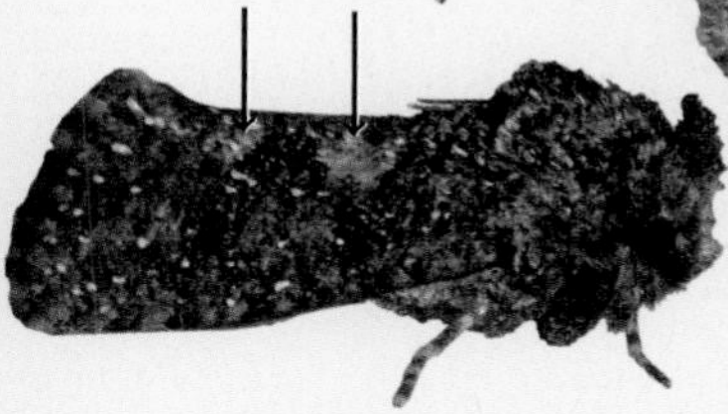

CLEMENS' GRASS-TUBEWORM

CRESSON'S GRASS-TUBEWORM

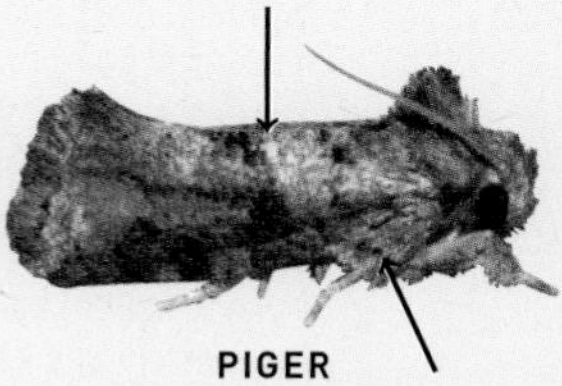

PIGER GRASS-TUBEWORM

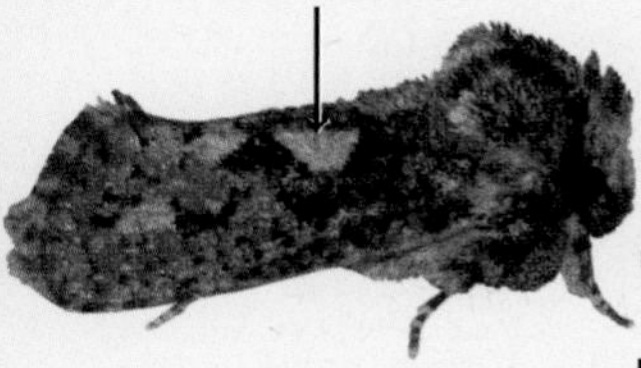

ARCANE GRASS-TUBEWORM

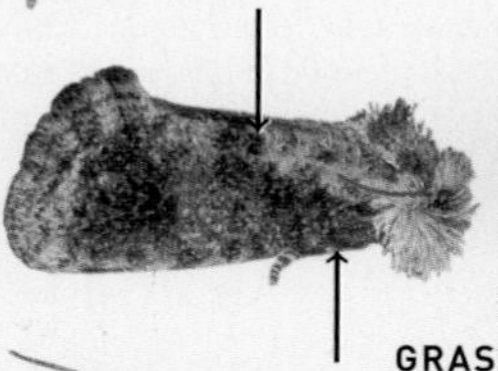

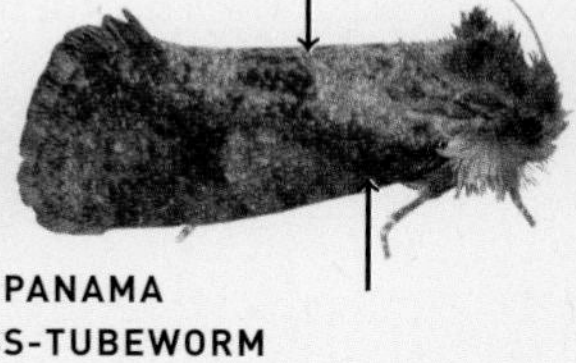

PANAMA GRASS-TUBEWORM

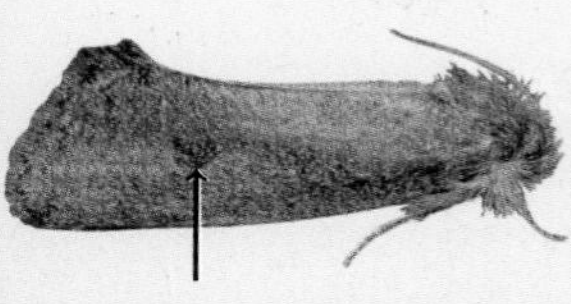

HEPPNER'S GRASS-TUBEWORM

FRILLY GRASS-TUBEWORM

Acrolophus mycetophagus 30-0100 (0367.1) **Common**

TL 7–13 mm. Whitish FW is variably shaded blackish in median area (sometimes forming a band) and ST area. Central pair of legs are adorned with large tufts of white hairlike scales. **HOSTS:** Bracket fungus. **RANGE:** Widespread west to MS, south to FL and e. TX.

BLACK-PATCHED NEMAPOGON

Nemapogon angulifasciella 30-0106 (0262) **Common**

TL 5–6 mm. Whitish FW is variably patterned with black lines and streaks. Always has a bold black triangle or sharply angled bar in central median area. Fuzzy head is whitish. **HOSTS:** Unknown. **RANGE:** NC west to TN, south to w. FL and LA.

MANY-LINED NEMAPOGON

Nemapogon multistriatella 30-0115 (0269) **Uncommon**

TL 4–6 mm. Whitish FW is finely peppered with brown. Note distinctive pattern of thin black streaks along veins. Whitish fringe is checkered with black. **HOSTS:** Unknown. **RANGE:** Locally from SC west to OK.

RILEY'S NEMAPOGON

Nemapogon rileyi 30-0118 (0272) **Common**

TL 7–8 mm. Black FW has a distinctive pattern of finely etched white lines passing from base to apex. Inner section of terminal line and fringe are white. Fuzzy head is black. **HOSTS:** Unknown. **RANGE:** SC west to MS, south to FL and e. TX.

OLD GOLD ISOCORYPHA

Isocorypha mediostriatella 30-0140 (0299) **Uncommon**

TL 4–6 mm. Shiny black FW has a wide golden streak through central basal and median areas and two isolated golden spots in inner and outer PM areas. Fuzzy head is yellow. Golden antennae are held angled upward. **HOSTS:** Unknown. **RANGE:** Widespread south to FL and west to OK and LA.

HOUSEHOLD CASEBEARER

Phereoeca uterella 30-0141 (0390) **Common**

TL 4–7 mm. Brownish FW is variably peppered with pale yellowish scales. Note bold blackish bars in AM and PM areas and two blackish spots in median area. Head is pale orange. Females are larger and more boldly patterned than males. **HOSTS:** Larvae reportedly feed on old spider webs and woolen items. **RANGE:** GA and AL south to FL and e. TX.

FUNGUS AND TUBE MOTHS

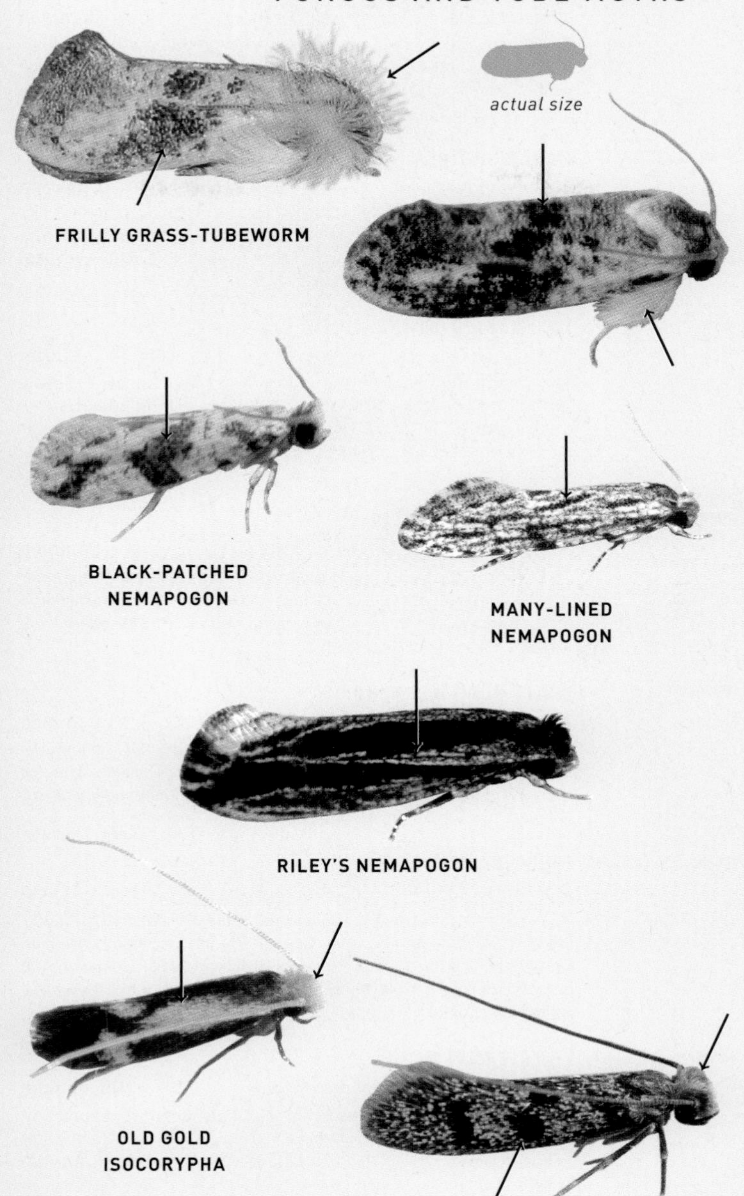

DARK-COLLARED TINEA

Tinea apicimaculella 30-0144 (0392) **Common**

TL 7–8 mm. Shiny light brown FW has a narrow black streak extending through central wing and a black dot in inner median area. Black collar behind yellowish head. **HOSTS:** Unknown. **RANGE:** Widespread west to OK, south to FL and e. TX.

MANDARIN TINEA

Tinea mandarinella 30-0152 (0400) **Common**

TL 4–5 mm. Peppery dark brown FW has an irregular pale yellowish stripe along inner margin that joins wide pale fringe. Usually has a large pale spot in outer ST area. Fuzzy head is pale yellow. **HOSTS:** Unknown. **RANGE:** NC west to AR, south to FL and e. TX.

CASEMAKING CLOTHES MOTH

Tinea pellionella 30-0157 (0405) **Local**

TL 5–8 mm. Shiny brown FW is uniform apart from two dusky dots in central median area and a larger spot in central PM area. Head is orange. **HOSTS:** Feathers, wool, leather, fur, and other animal products. **RANGE:** Locally in TN and MS, south to FL and e. TX. **NOTE:** Introduced from Eurasia.

SKUNKBACK MONOPIS

Monopis dorsistrigella 30-0171 (0416) **Common**

TL 7–8 mm. Blackish FW has an irregular white streak along inner margin. Midpoint of costa is marked with a white blotch. Dorsal surface of thorax is white. Head is pale yellow. **HOSTS:** Unknown. **RANGE:** TN south to GA and MS.

WHITE-BLOTCHED MONOPIS

Monopis marginistrigella 30-0172 (0417) **Common**

TL 7–8 mm. Blackish FW is lightly marbled with gray. Midpoints of costa and inner ST areas are blotched with white. Dorsal surface of thorax is black. Head is pale orange. **HOSTS:** Unknown. **RANGE:** W. NC west to AR, south to AL and LA.

PAVLOVSKI'S MONOPIS

Monopis pavlovski 30-0174 (0418.1) **Local**

TL 7–10 mm. Blue gray FW has a large white U-shaped patch covering outer median and ST area. Central PM area is tinged chestnut. Dorsal surface of thorax and head is white. **HOSTS:** Unknown, but probably detritus. **RANGE:** NC and TN, locally south to FL and MS. **NOTE:** Introduced from Asia.

YELLOW WAVE MOTH

Hybroma servulella 30-0182 (0300) **Uncommon**

TL 4–6 mm. Yellow FW has an irregular peppery brown patch at midpoint of inner margin. Costa is marked with brown triangles at midpoint and ST area. Head is yellow. **HOSTS:** Unknown. **RANGE:** Widespread south to FL and west to OK and LA.

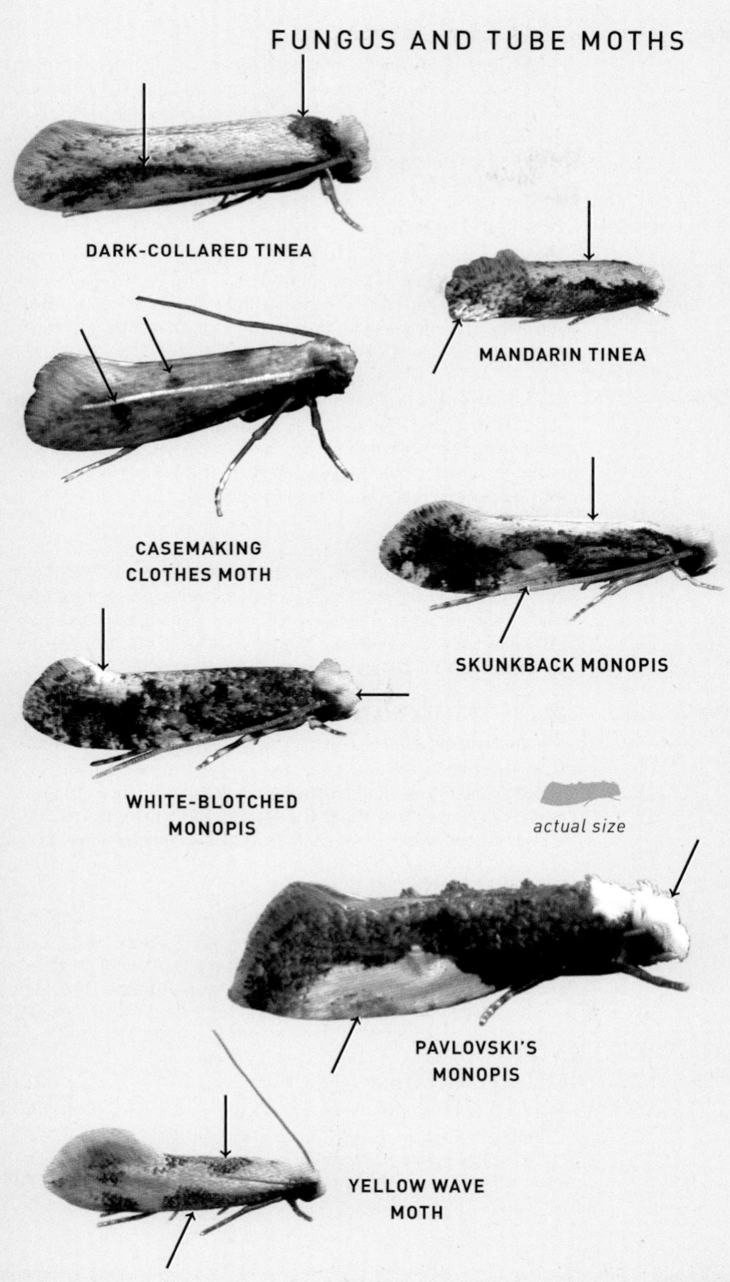
DARK-COLLARED TINEA
MANDARIN TINEA
CASEMAKING CLOTHES MOTH
SKUNKBACK MONOPIS
WHITE-BLOTCHED MONOPIS
actual size
PAVLOVSKI'S MONOPIS
YELLOW WAVE MOTH

CURVE-LINED HOMOSTINEA

Homostinea curviliniella 30-0184 (0301) **Uncommon**

TL 9–11 mm. Shiny light brown FW has a dark brown patch in inner basal area. Central median and ST areas are peppered with brown scales, sometimes forming a curved line or streak. Head is yellowish. **HOSTS:** Unknown. **RANGE:** Widespread south to FL and west to OK and e. TX.

TWO-SPOTTED MEA

Mea bipunctella 30-0186 (0305) **Common**

TL 5–6 mm. Whitish FW (sometimes tinged pale orange toward apex) has irregular black streaks along basal half of inner margin and costa. Top of head is white, but face is black. **HOSTS:** Unknown. **RANGE:** SC west to MS, south to FL and LA.

PIED SCARDIA *Scardia anatomella* 30-0203 (0311) **Common**

TL 12–14 mm. Black FW has an irregular whitish streak along inner margin and four white patches along costa. ST has a white patch at midpoint. Fringe is boldly checkered when fresh. **HOSTS:** Unknown. **RANGE:** SC west to AR, south to w. FL and e. TX.

BANDED SCARDIELLA

Scardiella approximatella 30-0205 (0308) **Common**

TL 8–9 mm. Peppery whitish FW has irregular blackish AM and median bands that meet in central part of wing. Sometimes has a partial ST band. Fringe is boldly checkered when fresh. **HOSTS:** Unknown. **RANGE:** W. NC and TN west to MS, south to FL.

BROWN-SPOTTED DIACHORISIA

Diachorisia velatella 30-0213 (0279) **Common**

TL 6–7 mm. White FW is variably peppered with brown scales. Large brown spots in basal and median areas are edged black. Fringe is checkered brown and white. Fuzzy head is whitish. **HOSTS:** Unknown. **RANGE:** Widespread south to FL and west to OK and e. TX.

CLEMENS' PHILONOME

Philonome clemensella 30-0220 (0462) **Common**

TL 7–10 mm. Rusty brown FW has a long, narrow white basal dash that connects white thorax to white median line. PM line is broken. Sometimes has black scales in terminal area. **HOSTS:** Unknown; possibly hickory and basswood. **RANGE:** NC west to OK, south to FL and LA.

SPECKLED XYLESTHIA

Xylesthia pruniramiella 30-0223 (0317) **Common**

TL 6–8 mm. Peppery brown FW is marked with pale bands. Tufts of raised scales in inner median area give the moth a lumpy appearance. **HOSTS:** Plum. **RANGE:** Widespread west to OK, south to FL and e. TX.

FUNGUS AND TUBE MOTHS

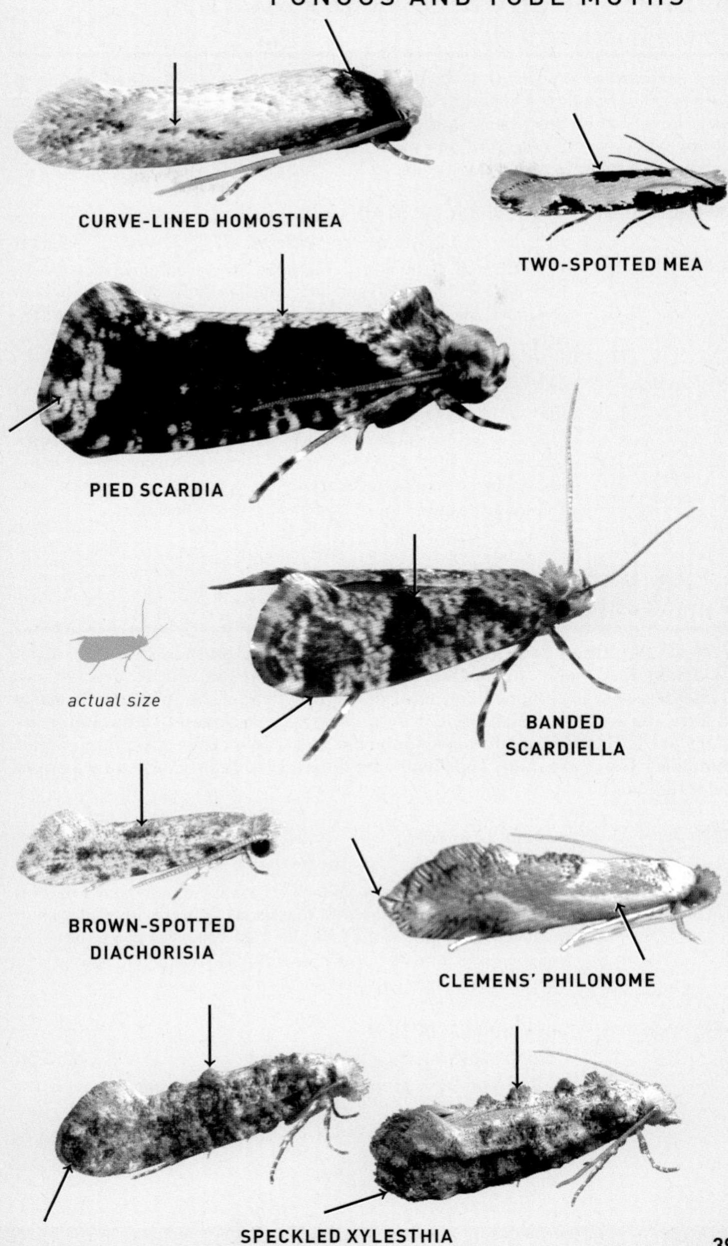

Ribbed Cocoon-Making Moths

Family Bucculatricidae

Tiny, streamlined moths that have flared, feathery tips to FW. Head is topped with a tuft of hairlike scales. Larvae are either leafminers or leaf skeletonizers on a variety of woody plants and deciduous trees. Some species will come to lights in very small numbers, but most adults are best observed by collecting and raising them from larval stages.

NARROW BUCCULATRIX

Bucculatrix angustata 33-0040 (0522) **Local**

TL 4–4.5 mm. Golden brown FW has an elongated whitish streak from base to median area. Whitish median and PM lines are broken chevrons, edged black. White apical patch has a black dash. **HOSTS:** Aster, fleabane, and goldenrod. **RANGE:** Locally in TN and s. AL.

CROWNED BUCCULATRIX

Bucculatrix coronatella 33-0079 (0559) **Local**

TL 5–7 mm. Pale orange FW has an obvious tuft of raised black scales in inner median area. Costa is marked with three oblique white lines. White apical patch contains a black dot. **HOSTS:** Unknown. **RANGE:** Locally in TN and south to GA and AL.

Leaf Blotch Miner Moths

Family Gracillariidae

Very small, streamlined moths that rest propped up on their forelegs, often twiddling their long, threadlike antennae until settled. *Phyllonorycter* and *Cameraria* rest in a more horizontal or slightly head-down posture. In early instars the larvae are leafminers on a variety of deciduous trees. Later instars are leafrollers. Adults are nocturnal and freely come to lights in small numbers. There are many species, some difficult to identify; we show a small selection here.

AZALEA LEAFMINER

Caloptilia azaleella 33-0113 (0592) **Common**

TL 5–7 mm. Dark brown FW has an elongated and poorly defined pale yellow patch along costa that widens toward inner margin in median area. Costa is often speckled with blackish scales. Front of head is white. **HOSTS:** Azalea. **RANGE:** SC west to OK, south to FL and LA. **NOTE:** Accidentally introduced from Japan.

DOGWOOD CALOPTILIA

Caloptilia belfragella 33-0115 (0594) **Common**

TL 5–7 mm. Dark brown FW has an elongated pale yellow patch along costa that widens toward inner margin in median area. Front of head is white. **HOSTS:** Dogwood. **RANGE:** Coastal NC west to OK, south to w. FL and LA.

RIBBED COCOON-MAKING MOTHS

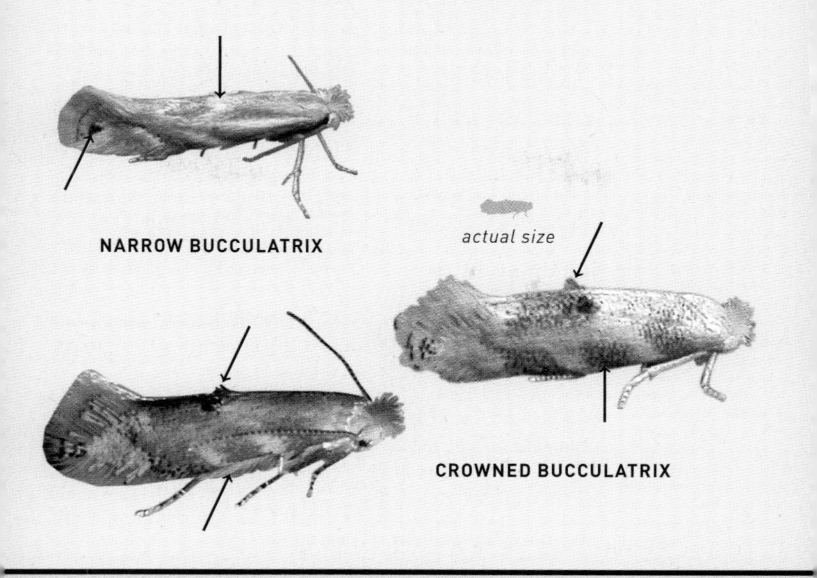

LEAF BLOTCH MINER MOTHS

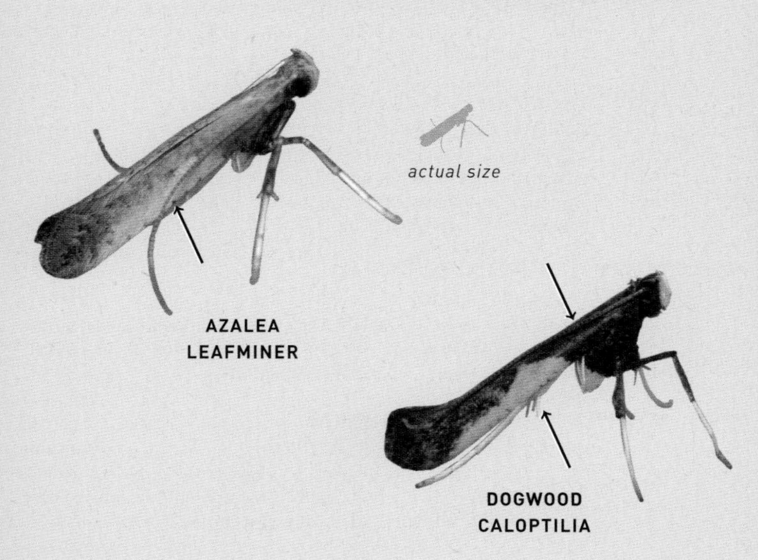

MAPLE CALOPTILIA

Caloptilia bimaculatella 33-0117 (0595) **Common**

TL 5–7 mm. Metallic purplish brown FW has two golden yellow triangles (largest in median area) jutting inward from costa. **HOSTS:** Maple. **RANGE:** TN, locally south to FL and MS. Absent from east.

WALNUT CALOPTILIA

Caloptilia blandella 33-0118 (0596) **Common**

TL 5–7 mm. Dark brown FW has a wide pale yellow costal streak interrupted by a slanting brown AM line. **HOSTS:** Black walnut and shagbark hickory. **RANGE:** NC west to OK, south to MS and LA.

SUMAC CALOPTILIA

Caloptilia rhoifoliella 33-0152 (0630) **Common**

TL 5–7 mm. Brown FW has a poorly defined pale costal streak (sometimes whitish) that is peppered with darker scales. **HOSTS:** Sumac. **RANGE:** NC west to AR, south to FL and e. TX.

SASSAFRAS CALOPTILIA

Caloptilia sassafrasella 33-0155 (0633) **Uncommon**

TL 5–7 mm. Warm brown FW is slightly paler along costa and lightly brindled darker. One or more distinct blackish spots sometimes mark outer median and ST areas. **HOSTS:** Sassafras. **RANGE:** W. NC locally west to OK, south to FL and MS.

POPLAR CALOPTILIA

Caloptilia stigmatella 33-0161 (0639) **Common**

TL 5–7 mm. Dark to rusty brown FW has a hooked pale yellow triangle at midpoint of costa. **HOSTS:** Poplar and willow. **RANGE:** SC west to OK, south to FL and e. TX.

WITCH-HAZEL CALOPTILIA

Caloptilia superbifrontella 33-0164 (0641) **Common**

TL 5–7 mm. Golden brown FW has a large rounded golden yellow patch covering much of outer median area. Inner basal area and thorax are yellow. **HOSTS:** Witch hazel. **RANGE:** SC west to AR, south to n. FL and MS.

TICK-TREFOIL CALOPTILIA

Caloptilia violacella 33-0168 (0644) **Common**

TL 5–7 mm. Bronzy brown FW has a broad pale yellow costal streak and a distinct black dot in central median area. **HOSTS:** Tick trefoil. **RANGE:** NC west to OK, south to FL and LA.

LOCUST DIGITATE LEAFMINER

Parectopa robiniella 33-0181 (0657) **Common**

TL 5–6 mm. Golden brown FW has a distinctive pattern of incomplete slanting white bands along costa and inner margin. Fuzzy head is white. **HOSTS:** Black locust. **RANGE:** TN, locally south to FL and LA.

LEAF BLOTCH MINER MOTHS

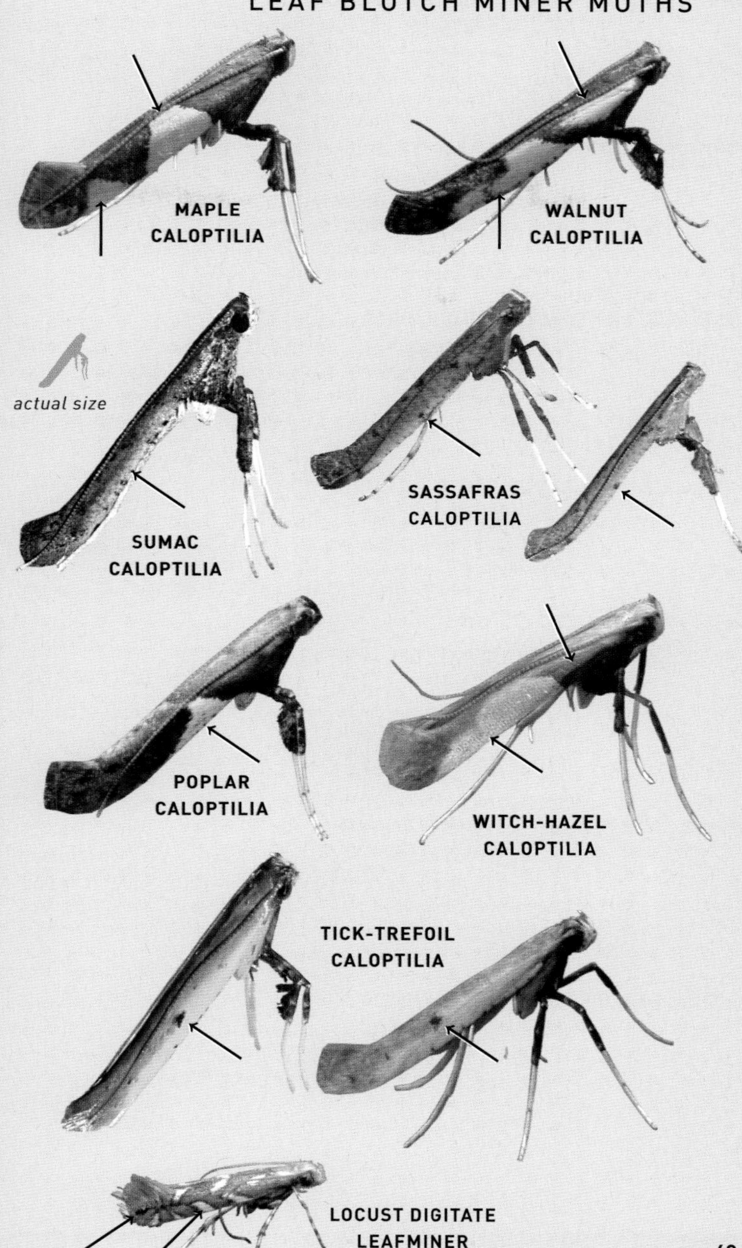

FINITE-CHANNELED LEAFMINER

Neurobathra strigifinitella 33-0187 (0663) **Common**

TL 5–6 mm. Grayish brown FW has a crisp pattern of angled, black-edged white lines. Apex is marked with a black dot. **HOSTS:** Oak. **RANGE:** NC west to OK, south to FL and LA.

WHITE OAK BLOTCH MINER

Phyllonorycter fitchella 33-0287 (0752) **Uncommon**

TL 4–5 mm. Golden brown FW has two large white patches along inner margin. Narrow, slanting white dashes mark costa. Note black dot at apex. Thorax and fuzzy head are white. **HOSTS:** White oak. **RANGE:** NC west to OK, south to FL and e. TX.

CHERRY BLOTCH MINER

Phyllonorycter propinquinella 33-0320 (0784) **Local**

TL 4–5 mm. Golden brown FW is boldly patterned with black-edged white lines and wedges. White basal dash and black apical dash are noticeable. **HOSTS:** Black cherry. **RANGE:** TN, locally south to n. AL and MS.

BLACK LOCUST LEAFMINER

Macrosaccus robiniella 33-0335 (0790) **Local**

TL 4–5 mm. Golden FW is boldly patterned with incomplete angled silver lines. Note short black bar in inner median area and black apical spot. **HOSTS:** Black locust. **RANGE:** W. NC west to TN, locally south to w. FL and e. TX.

CONGLOMERATE OAK LEAFMINER

Cameraria conglomeratella 33-0353 (0816) **Local**

TL 4–5 mm. Golden brown FW has a white stripe along inner margin ending as a short angled line in ST area. Three slanting dashes mark costa. Note peppered ST area. **HOSTS:** Oak. **RANGE:** NC west to OK, south to FL and e. TX.

SOLITARY OAK LEAFMINER

Cameraria hamadryadella 33-0361 (0823) **Uncommon**

TL 4–5 mm. Peppery whitish FW is boldly patterned with angled, black-edged brown median, PM, and ST bands. ST area is speckled with blackish scales. Rests in a shallow raised position. **HOSTS:** Oak. **RANGE:** NC west to OK, south to n. FL and LA.

RUSTY OAK LEAFMINER

Cameraria quercivorella 33-0378 (0835) **Local**

TL 4–5 mm. Golden brown FW has a white stripe along inner margin ending as a short angled line in ST area. Three slanting dashes mark costa. Note peppered ST area. Rests in a shallow raised position. **HOSTS:** Oak. **RANGE:** E. TN west to OK, locally south to FL.

LEAF BLOTCH MINER MOTHS

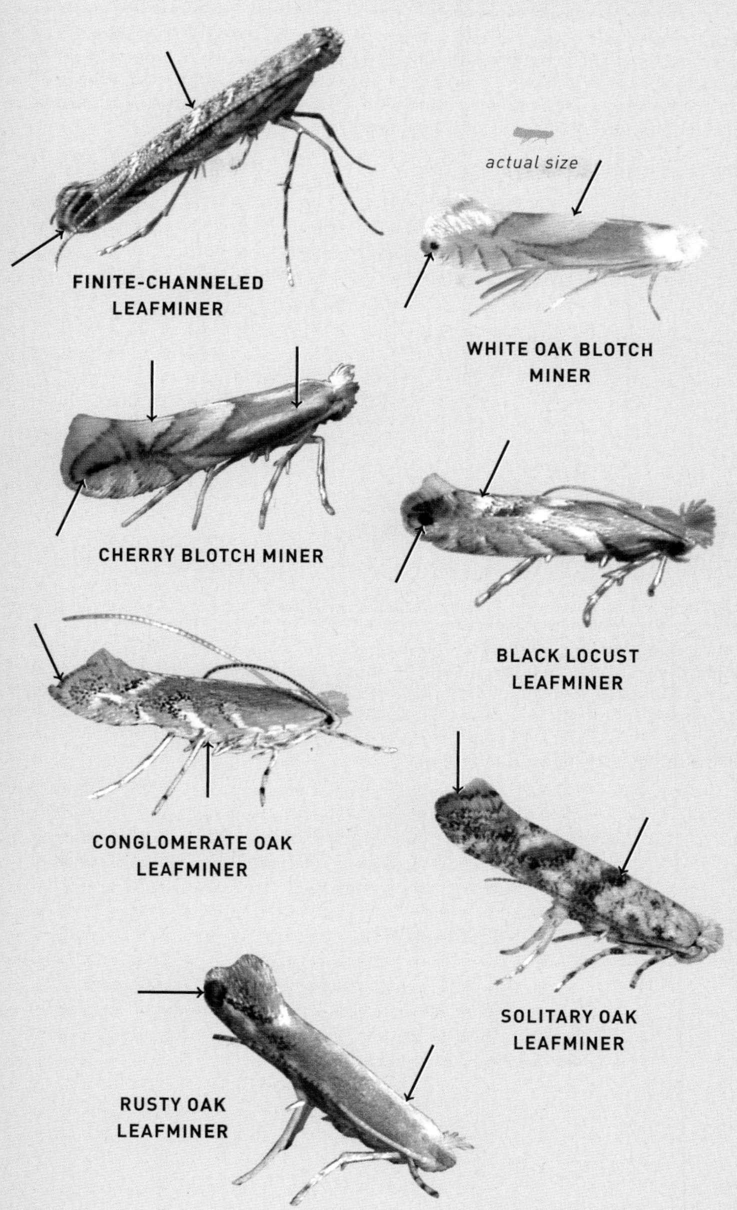

Ermine Moths Family Yponomeutidae

Small moths with long, narrow wings that often flare slightly upward at the outer margin. The true ermines of genus *Yponomeuta* are white with many small black spots that vary in number and placement. Members of other genera are variably gray or brown, often with a pale thorax. Adults are nocturnal and frequently visit lights in small numbers.

AMERICAN ERMINE

Yponomeuta multipunctella 36-0017 (2420) **Common**

TL 11–13 mm. White FW is evenly covered with three or four rows of tiny black dots. HW is pearly gray with a white fringe. **HOSTS:** Highbush cranberry and nannyberry. **RANGE:** NC west to MS, locally south to FL.

BRINDLED ZELLERIA

Zelleria retiniella 36-0026 (2431) **Common**

TL 7–8 mm. Shiny pale orange FW is variably brindled satin white in basal and median areas. Head and dorsal surface of thorax are white. Rests in a slightly head-down position. **HOSTS:** Unknown. **RANGE:** NC west to AR, south to FL and LA.

Sedge, Diamondback, and False Diamondback Moths Families Plutellidae and Glyphipterigidae

Small, narrow-winged moths. The *Plutella* species rest with their antennae held forward from the body; the flared wings may appear to curl up at the tips. Yellow Nutsedge Moth and Carrionflower Moth have a two-lobed outer margin. Adults are nocturnal and come to lights in small numbers.

DIAMONDBACK MOTH

Plutella xylostella 36-0083 (2366) **Common**

TL 7–8 mm. Sexually dimorphic. FW of male is brown with a jagged ochre stripe along inner margin. Paler female shows less contrast. **HOSTS:** Brassicaceae. **RANGE:** Widespread west to OK, south to FL and e. TX. **NOTE:** Introduced from Europe.

DAME'S ROCKET MOTH

Plutella porrectella 36-0084 (2363) **Local**

TL 7–8 mm. Straw-colored FW is streaked warm brown. Terminal line is blackish. Forward-facing antennae have three dusky bands near tips. **HOSTS:** Dame's rocket. **RANGE:** GA south to FL and LA. **NOTE:** Introduced from Europe.

ERMINE MOTHS

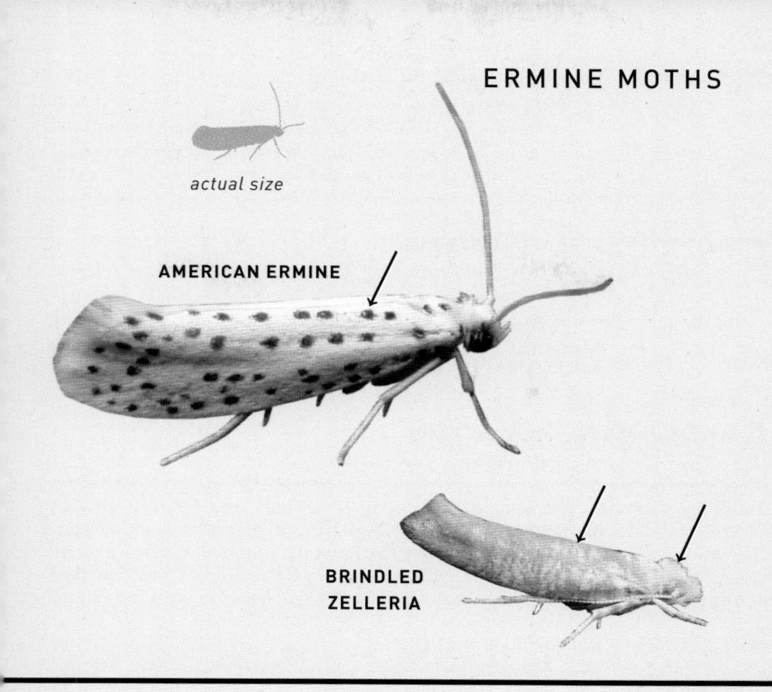

SEDGE, DIAMONDBACK, AND FALSE DIAMONDBACK MOTHS

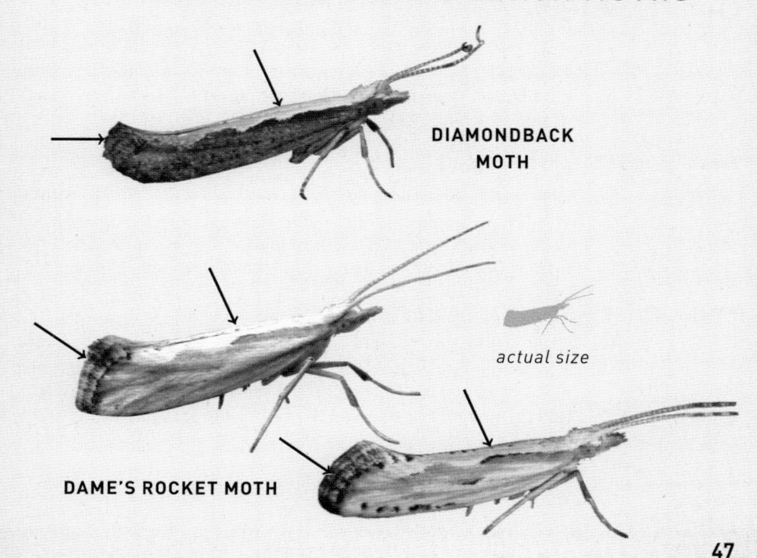

YELLOW NUTSEDGE MOTH

Diploschizia impigritella 36-0127 (2346) **Local**

TL 4–5 mm. Bronzy brown FW has a backward-pointing white crescent at midpoint of inner margin. Five short silver bars mark subapical area. **HOSTS:** Yellow nutsedge. **RANGE:** SC west to OK, south to FL and e. TX.

CARRIONFLOWER MOTH

Acrolepiopsis incertella 36-0134 (2490) **Local**

TL 5–6 mm. Bronzy brown FW has a contrasting white wedge at midpoint of inner margin. Midpoint of concave fringe is marked with a white spot. **HOSTS:** Bristly greenbrier. **RANGE:** E. TN, locally south to n. FL and MS.

Needleminer Moths

Families Argyresthiidae and Lyonetiidae

Small to tiny moths with long, narrow wings. Ailanthus Webworm resembles Ornate Moth (family Erebidae, tribe Arctiini). The *Argyresthia* are tiny species whose white and gold coloration is the most distinguishing feature when viewed with the naked eye; they typically rest in a head-down position. The adults of all species are nocturnal and frequently visit lights in small numbers.

HONEY-COMB MICRO

Argyresthia alternatella 36-0141 (2435) **Local**

TL 5–6 mm. Fawn-colored FW is intricately marked with a netlike pattern of brown lines. Dorsal surface of head is white. **HOSTS:** Berries of eastern red cedar. **RANGE:** N. MS west to OK, south to TX.

BRONZE ALDER MOTH

Argyresthia goedartella 36-0163 (2457) **Local**

TL 6–7 mm. Satin white FW is boldly patterned with metallic bronzy olive bands. Median band is Y-shaped. **HOSTS:** Catkins and shoots of alder and birch. **RANGE:** W. NC and e. TN.

CHERRY SHOOT BORER

Argyresthia oreasella 36-0174 (2467) **Common**

TL 6–7 mm. Similar to Bronze Alder Moth but lacks the bronzy basal band. Median band is often incomplete or narrows toward costa. **HOSTS:** Oak and *Prunus* species. **RANGE:** E. TN south to GA and MS.

SPECKLED ARGYRESTHIA

Argyresthia subreticulata 36-0187 (2479) **Local**

TL 4–6 mm. Satin white FW is patterned with dark gray marbling along costa and in ST area. Costa is sometimes tawny in basal half. **HOSTS:** Red maple. **RANGE:** TN west to OK, locally south to w. FL and e. TX.

SEDGE, DIAMONDBACK, AND FALSE DIAMONDBACK MOTHS

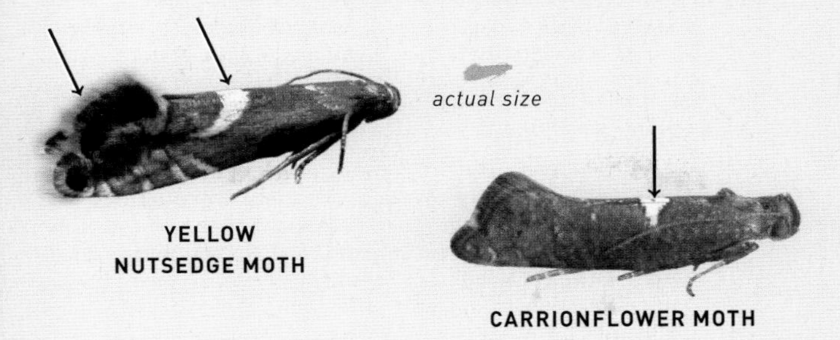

YELLOW NUTSEDGE MOTH

CARRIONFLOWER MOTH

NEEDLEMINER MOTHS

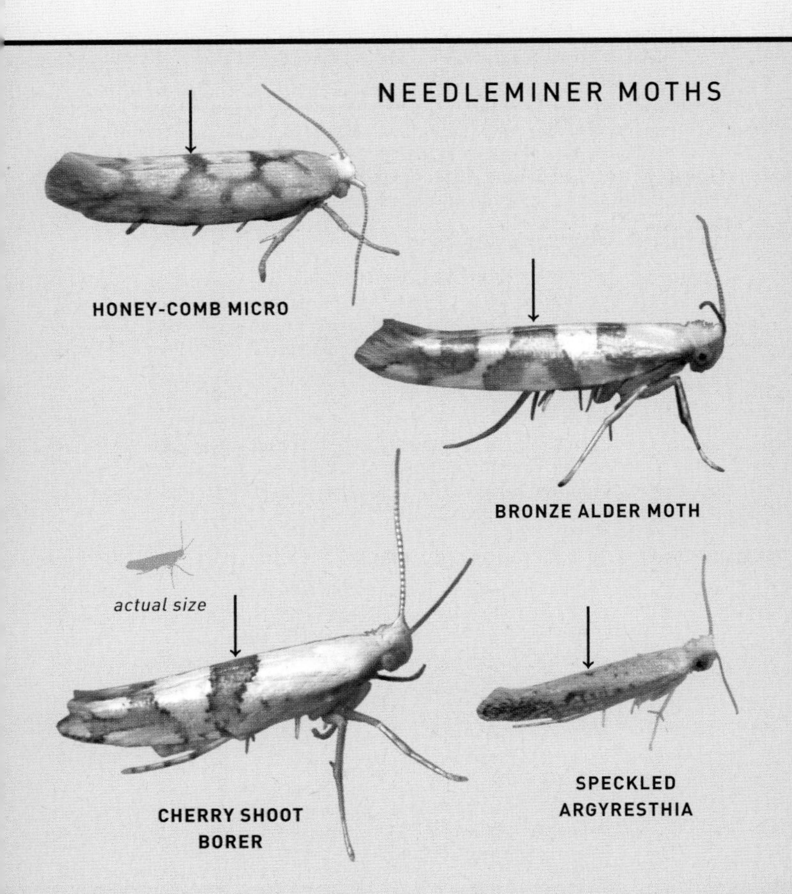

HONEY-COMB MICRO

BRONZE ALDER MOTH

CHERRY SHOOT BORER

SPECKLED ARGYRESTHIA

AILANTHUS WEBWORM

Atteva aurea 36-0211 (2401) **Common**

TL 10–16 mm. Reddish orange FW is boldly patterned with wide black-edged bands of pale yellow spots. HW is blackish. **HOSTS:** *Ailanthus*; also other deciduous trees and shrubs. **RANGE:** Widespread west to OK, south to FL and TX.

BRILLIANT SUNSPUR

Aetole tripunctella 36-0242 (2505) **Local**

TL 4–6 mm. Shiny orange FW has a wide, metallic gray terminal line that continues halfway up inner margin and raised, silvery spots. Spiny gray hindlegs are held upward at rest. **HOSTS:** Heartleaf four o'clock. **RANGE:** OK south to TX. **NOTE:** Beautiful Sunspur (*A. bella*, not shown) is similar, but terminal line pinches at midpoint of inner margin, and FW has silvery basal dash.

Concealer and Scavenger Moths

Families Autostichidae, Lecithoceridae, and Oecophoridae

Small moths with long narrow wings; those of a few species are flared. Many species are brightly or distinctively marked. Larvae of most species feed on dead leaves, detritus, and fungi. With the exception of the day-flying Newman's Mathildana, adults are nocturnal and will come to lights in small numbers.

STREAKED SPINITIBIA

Spinitibia hodgesi 42-0007 (1134.1) **Local**

TL 7–8 mm. Peppery brown FW is crisply streaked with pale orange and black lines along veins. Tiny black dots mark central median and PM areas. Curved ST line is dotted. **HOSTS:** Unknown. **RANGE:** SC west to OK, south to w. FL and LA.

GERDANA MOTH *Gerdana caritella* 42-0008 (1144) **Common**

TL 7–8 mm. Peppery yellow FW has a dark brown basal patch and curved, often fragmented, PM and ST lines. Central median area is marked with one or two dusky dots. **HOSTS:** Unknown. **RANGE:** TN west to OK, south to n. FL and e. TX.

TRIANGLE-MARKED TWIRLER

Taygete attributella 42-0009 (1842) **Common**

TL 5–6 mm. Speckled whitish FW has a small dusky basal patch and a bold blackish triangle at midpoint of costa. Apex is checkered when fresh. Appears flat when resting. **HOSTS:** Unknown. **RANGE:** W. NC west to OK, south to n. FL and e. TX.

SOUTHERN TAYGETE

Taygete gallaegenitella 42-0012 (1845) **Common**

TL 5–6 mm. Whitish FW is clouded with brown toward apex. Note black patches along costa in basal, median, and PM areas. A strongly angled white ST line is sometimes obvious. Appears flat when resting. **HOSTS:** Oak galls. **RANGE:** GA south to FL and e. TX.

NEEDLEMINER MOTHS

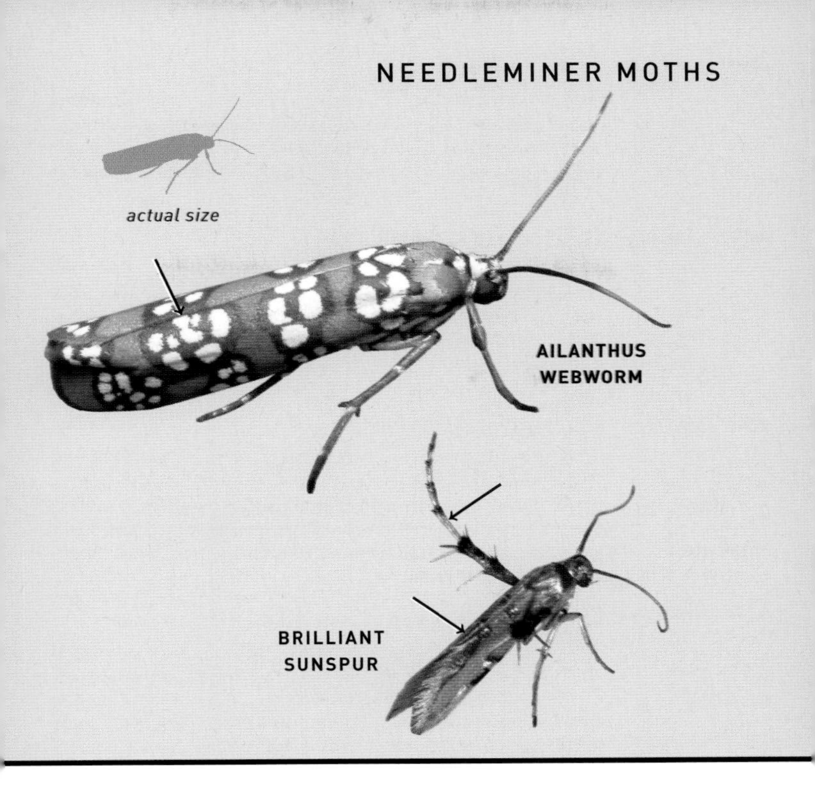

CONCEALER AND SCAVENGER MOTHS

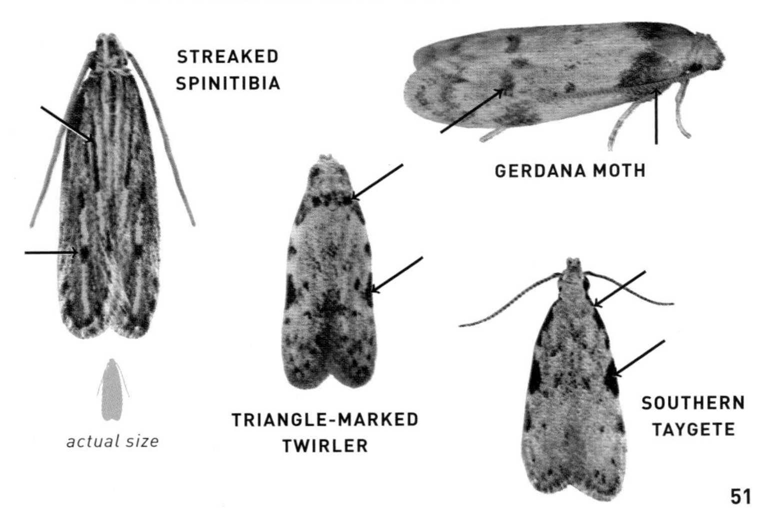

JUNIPER TIP MOTH

Glyphidocera juniperella 42-0020 (1136.1) **Common**

TL 7–8 mm. Light brown FW is densely peppered with dusky scales. Two dark brown spots mark central AM and PM areas. Rounded fringe is whitish. **HOSTS:** Juniper. **RANGE:** SC west to OK, south to FL and LA.

FIVE-SPOTTED GLYPHIDOCERA

Glyphidocera lactiflosella 42-0021 (1139) **Common**

TL 6–8 mm. Whitish to light gray FW is lightly peppered with black scales. Curved black AM and PM lines are obvious only in central area of wing. Curved black terminal line is obvious when fresh. **HOSTS:** Unknown. **RANGE:** Coastal NC west to AR, south to FL and e. TX.

SQUARE-SPOT MARTYRINGA

Martyringa latipennis 42-0027 (1065) **Common**

TL 9–12 mm. Broad, rounded FW is peppery grayish brown. Note two blackish square spots in central median area that are edged pale yellow. Curved PM line is pale yellow. Head and long labial palps are pale. **HOSTS:** Unknown. **RANGE:** NC west to AR, locally south to n. GA and w. FL.

HIMALAYAN GRAIN MOTH

Martyringa xeraula 42-0028 (1066) **Local**

TL 11–13 mm. Similar to Square-spot Martyringa but has narrower FW with a less distinct pattern. Blackish spots in central median area are not square. **HOSTS:** Detritus. **RANGE:** SC west to AR, south to n. FL and LA. **NOTE:** Introduced.

BLACK-MARKED INGA

Inga sparsiciliella 42-0029 (1034) **Common**

TL 8–10 mm. Broad, rounded FW is white with bold black patches in inner basal area and at midpoint of costa. Two small black dots mark midline of AM area. A curved ST line is sometimes obvious. **HOSTS:** Unknown. **RANGE:** NC west to n. MS, south to FL and e. TX.

CHALKY INGA *Inga cretacea* 42-0030 (1035) **Common**

TL 8–10 mm. Off-white FW is often peppered darker in distal half. Two or more tiny blackish dots mark central median area. Dotted ST line is sometimes obvious. **HOSTS:** Unknown. **RANGE:** TN west to OK, south to FL and TX.

RETICULATED DECANTHA

Decantha boreasella 42-0037 (1042) **Common**

TL 5–6 mm. Pale orange FW has a netlike pattern of white lines and patches of peppery brown shading. **HOSTS:** Unknown; possibly fungus on dead wood. **RANGE:** NC west to OK, south to FL and LA.

CONCEALER AND SCAVENGER MOTHS

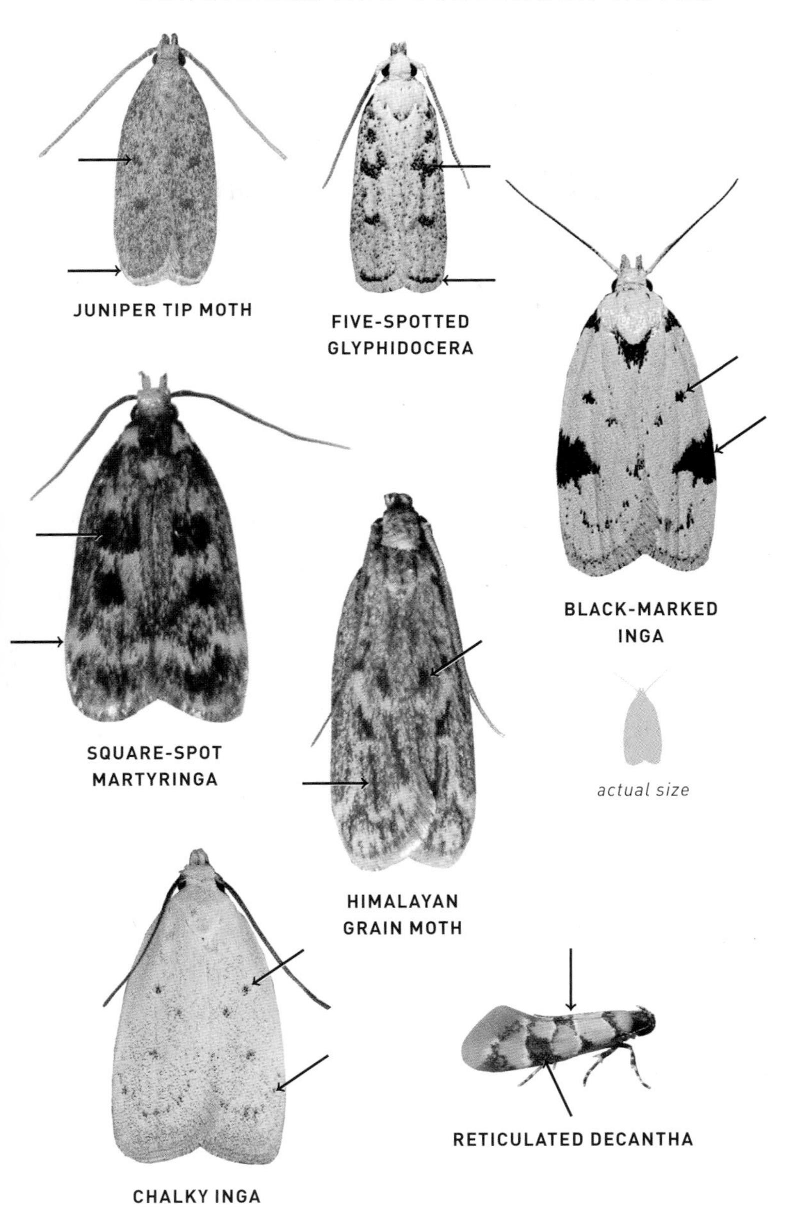

SMALL DECANTHA *Decantha stecia* 42-0038 (1043) **Common**

TL 4–5 mm. Similar to Reticulated Decantha but is slightly smaller and has basal area completely shaded brown. **HOSTS:** Unknown. **RANGE:** TN south to FL and LA.

ORANGE-HEADED EPICALLIMA

Epicallima argenticinctella 42-0041 (1046) **Common**

TL 6–7 mm. Yellowish FW has wide brown AM band and shading in inner ST area. Silver basal and AM lines are slightly bent. Fragmented PM line is sharply kinked at midpoint and widens toward costa. **HOSTS:** Elm. **RANGE:** NC west to OK, south to n. FL and e. TX.

NEWMAN'S MATHILDANA

Mathildana newmanella 42-0055 (1059) **Common**

TL 8–10 mm. This attractive day-flying moth has a black FW with a purplish sheen when the light catches it. Note two orange streaks in central basal and median areas that almost touch. Labial palps are orange. Antennae are tipped white. **HOSTS:** Larvae feed from webbing under bark of dead trees. **RANGE:** NC west to AR, south to n. GA and MS.

THREE-SPOTTED CONCEALER MOTH

Eido trimaculella 42-0062 (1068) **Common**

TL 6–7 mm. Grayish brown FW is lightly peppered with whitish scales. Two white costal patches and checkered fringe are noticeable when fresh. Rests in a head-down position with broad wings held flat. **HOSTS:** Reported on bracket fungus and elm. **RANGE:** W. NC west to AR, south to n. FL and MS.

JANE'S YMELDIA *Ymeldia janae* 42-0065 (2216) **Common**

TL 3–4 mm. Golden brown FW has a whitish streak from base almost to apex, narrowly edged with a black line that is broken in median area. Dorsal surface of thorax is marked with black and white lines. **HOSTS:** Unknown. **RANGE:** GA south to FL and s. LA.

Grass Miner Moths Family Depressariidae

A distinctive group of small, flattish moths with long, upward-curving labial palps. Several species overwinter as adults and can be encountered in early spring. The more boldly patterned *Ethmia* and *Antaeotricha* hold their narrow wings tight to the body. All species are nocturnal and visit lights in small numbers.

FEATHERDUSTER AGONOPTERIX

Agonopterix pulvipennella 42-0079 (0867) **Local**

TL 10–12 mm. Brown FW has paler shoulders and costa. Median and ST areas are streaked and mottled blackish. Dusky patch in central median area is accented with a white dot. **HOSTS:** Goldenrod and nettle. **RANGE:** W. NC west to OK, south to n. GA and MS.

CONCEALER AND SCAVENGER MOTHS

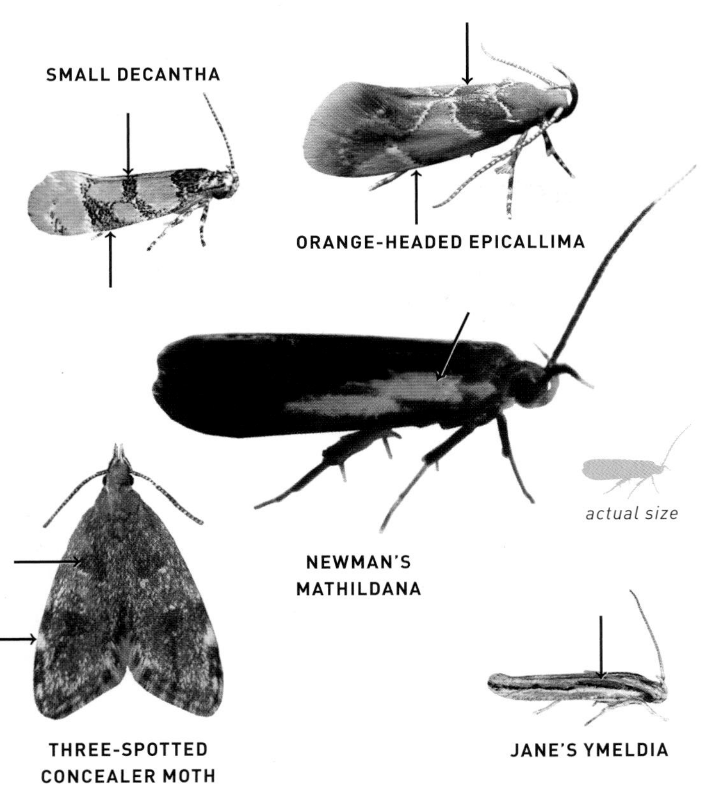

GRASS MINER MOTHS

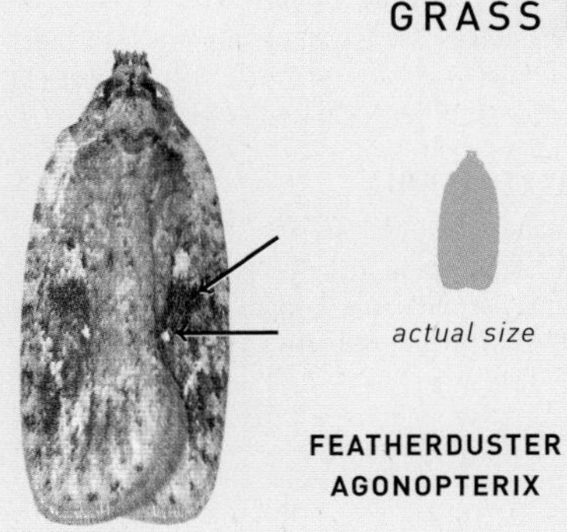

FOUR-DOTTED AGONOPTERIX

Agonopterix robiniella 42-0095 (0882) **Common**

TL 8–12 mm. Pale orange FW is peppered with rusty scales. Darker zigzag bar extends through inner median area. A black dot marks central AM area. **HOSTS:** Black locust. **RANGE:** W. NC west to AR, south to GA and MS.

CLAY-COLORED AGONOPTERIX

Agonopterix argillacea 42-0102 (0889) **Common**

TL 8–12 mm. Grayish brown FW has pale shoulders and basal half of costa. A prominent dusky gray blotch in central median area is highlighted with one or (usually) two white specks along its inner edge. **HOSTS:** Common hoptree, desert false indigo, and willow. **RANGE:** GA and n. MS south to FL and e. TX.

PACKARD'S SEMIOSCOPIS

Semioscopis packardella 42-0126 (0912) **Local**

TL 12–15 mm. Silvery gray FW has a sinuous black line extending from base to central median area. Sometimes has broad ochre shading along upper costa. **HOSTS:** Hawthorn, mountain ash, and *Prunus* species. **RANGE:** W. NC and TN south to GA.

MERRICK'S SEMIOSCOPIS

Semioscopis merriccella 42-0127 (0913) **Local**

TL 12–15 mm. Resembles Packard's Semioscopis, but sinuous black line is thicker and does not reach base of FW. Often has brownish shading in basal and PM areas. **HOSTS:** Unknown. **RANGE:** W. NC and TN south to GA and MS.

MOURNING ETHMIA *Ethmia semilugens* 42-0183 (0976) **Local**

TL 9–13 mm. Black FW has an irregular white line along inner margin and white patch at apex. Curved terminal line is a row of black dots. White thorax and head are patterned with black spots. **HOSTS:** Phacelias. **RANGE:** Cen. TX.

ZELLER'S ETHMIA

Ethmia zelleriella 42-0199 (0992) **Common**

TL 11–14 mm. White FW is boldly marked with black spots and streaks. Abdomen and legs are pale yellow. **HOSTS:** Phacelias. **RANGE:** W. NC west to OK, locally south to GA and e. TX.

LADDER-BACKED ETHMIA

Ethmia delliella 42-0200 (0993) **Local**

TL 10–13 mm. White FW is strikingly marked with dark metallic blue lines and spots. Terminal line is gold. Abdomen has black and gold bands. **HOSTS:** Sandpaper tree. **RANGE:** Se. TX.

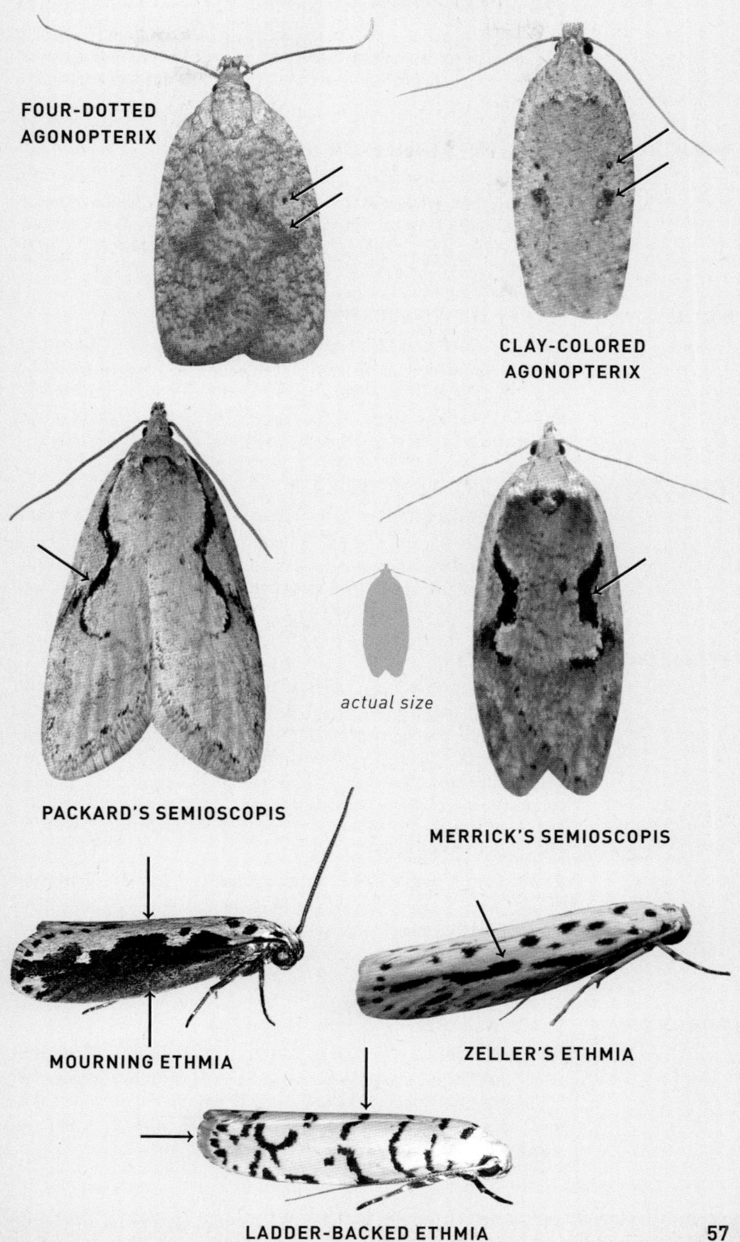
FOUR-DOTTED AGONOPTERIX
CLAY-COLORED AGONOPTERIX
actual size
PACKARD'S SEMIOSCOPIS
MERRICK'S SEMIOSCOPIS
MOURNING ETHMIA
ZELLER'S ETHMIA
LADDER-BACKED ETHMIA

BOLD-STREAKED ETHMIA

Ethmia trifurcella 42-0215 (1003) **Common**

TL 8–10 mm. White FW has an irregular blackish streak passing through central area from base to apex. Costa is smudged with brown, especially near apex. **HOSTS:** Unknown. **RANGE:** NC west to OK, south to w. FL and MS.

SCHLAEGER'S FRUITWORM MOTH

Antaeotricha schlaegeri 42-0224 (1011) **Common**

TL 10–15 mm. White FW has incomplete light gray bands in basal, median, and ST areas. Thorax has an erect tuft of metallic blue-black scales. **HOSTS:** Oak and possibly beech. **RANGE:** NC west to OK, south to FL and e. TX.

PALE GRAY BIRD-DROPPING MOTH

Antaeotricha leucillana 42-0227 (1014) **Common**

TL 8–12 mm. Similar to larger Schlaeger's Fruitworm Moth but tends to have a smaller dark thoracic patch tinted bronze brown. **HOSTS:** Deciduous trees, including ash, basswood, birch, elm, and oak. **RANGE:** NC west to OK, south to FL and e. TX.

DOTTED BIRD-DROPPING MOTH

Antaeotricha humilis 42-0232 (1019) **Common**

TL 8–10 mm. Silvery gray FW is lightly patterned with fragmented black AM and PM lines and dots. Note inverted U-shape formed by inner AM line. **HOSTS:** Oak. **RANGE:** NC west to OK, south to FL and e. TX.

VESTAL MOTH

Antaeotricha albulella 42-0237 (1024) **Common**

TL 8–10 mm. White to pale grayish brown FW can be uniform or marked with one or two blackish dots in central PM area. Note rounded costal shape. Legs and antennae are tinted pinkish brown. **HOSTS:** Unknown. **RANGE:** E. NC west to n. MS, south to FL and LA. **NOTE:** Previously *A. vestalis.*

YELLOW-VESTED MOTH

Rectiostoma xanthobasis 42-0245 (1026) **Common**

TL 6–7 mm. Shiny blackish FW is glossed with indigo and violet sheens. Bright yellow basal area has black patches along costa and inner margin. Yellow head contrasts with black thorax. **HOSTS:** Unknown. **RANGE:** NC west to OK, south to FL and MS.

BLACK MENESTA

Menesta melanella 42-0249 (1031) **Common**

TL 6–7 mm. Shiny black FW has a contrasting chestnut-edged whitish triangle at midpoint of costa. Note tiny white dot in central median area. Fringe is white. Legs are white. **HOSTS:** Unknown. **RANGE:** E. NC west to MS, south to FL.

GRASS MINER MOTHS

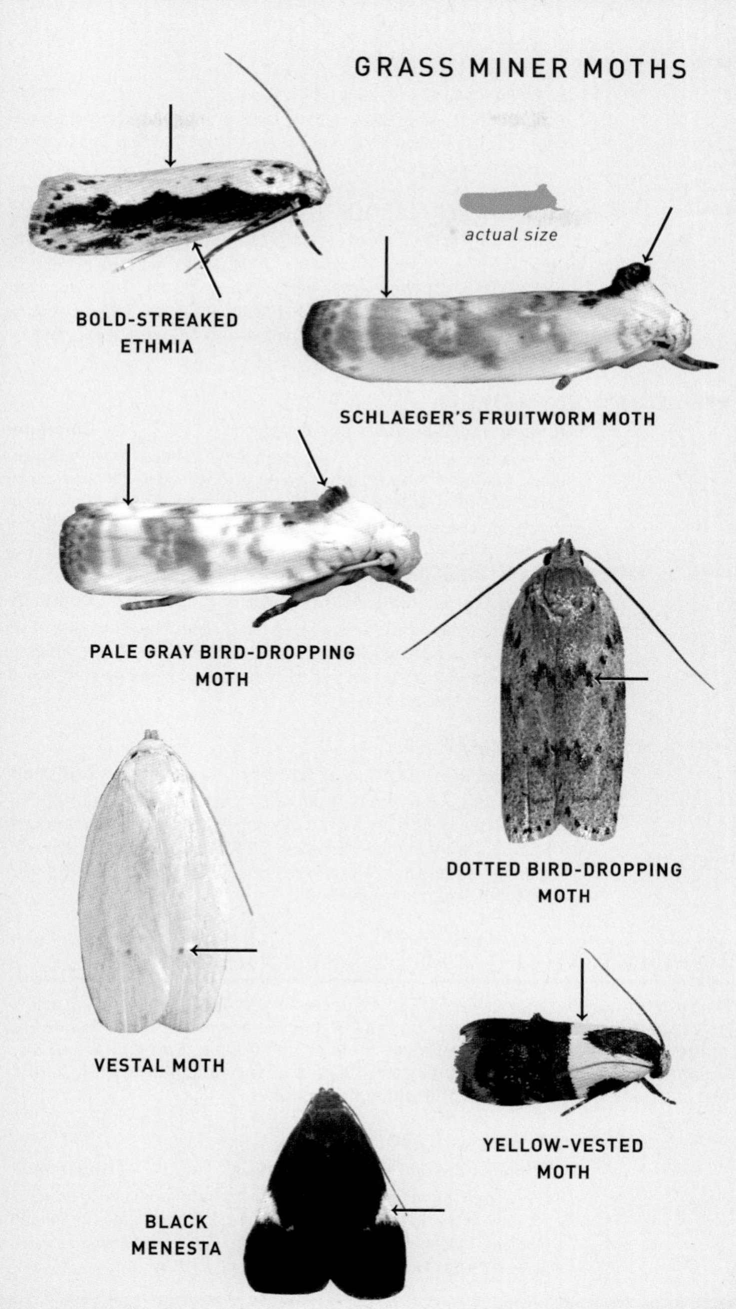

STREAKED EUPRAGIA

Eupragia hospita 42-0255 (0953) **Common**

TL 9–11 mm. Brown FW is boldly streaked whitish. Note white-edged black crescent in central PM area. **HOSTS:** Unknown. **RANGE:** SC west to n. MS, south to FL and e. TX.

GOLD-STRIPED LEAFTIER

Machimia tentoriferella 42-0257 (0951) **Common**

TL 12–14 mm. Light brown FW is marked with black dots in central AM area and with fragmented median band. Zigzag ST line appears dotted. Apex is pointed. **HOSTS:** Deciduous trees, including ash, birch, elm, oak, maple, and poplar. **RANGE:** NC west to OK, locally south to MS and e. TX.

OAK LEAFTIER

Psilocorsis quercicella 42-0259 (0955) **Common**

TL 7–9 mm. Light brown FW is coarsely brindled with thin brown lines. Central AM and median areas are accented with tiny black dots. Inner PM area is often smudged dusky. **HOSTS:** Oak; also beech and chestnut. **RANGE:** NC west to OK, south to FL and e. TX.

BLACK-FRINGED LEAFTIER

Psilocorsis cryptolechiella 42-0260 (0956) **Common**

TL 7–9 mm. Similar to Oak Leaftier but lacks the black dots on FW and has an obvious black-dotted terminal line when fresh. **HOSTS:** Oak; also chestnut, locust, and northern bayberry. **RANGE:** NC west to OK, south to n. FL and ne. TX.

DOTTED LEAFTIER

Psilocorsis reflexella 42-0261 (0957) **Common**

TL 7–12 mm. Similar to Oak and Black-fringed leaftiers and can be hard to identify with certainty. Paler individuals are lightly brindled with a black dot in central median area. **HOSTS:** Deciduous trees, including beech, birch, hickory, oak, maple, and poplar. **RANGE:** NC west to OK, south to FL and LA.

Cosmet Moths Family Cosmopterigidae

Small moths with narrow wings. Some species are boldly patterned with colorful patches and metallic bands on the FW. Larvae are predominantly miners of leaves or stems, or feed on flower buds or seedheads. Caterpillars of Shy Cosmet are responsible for fluffy cattail heads observed in midwinter. Adults are predominantly nocturnal and will come to lights.

SWEETCLOVER ROOT BORER

Walshia miscecolorella 42-0321 (1615) **Uncommon**

TL 6–10 mm. Light brown FW has a dusky basal area and shading at apex. Two tufts of raised black scales mark inner margin in AM and PM areas. **HOSTS:** Lupine, sweet clover, and other legumes; also thistle. **RANGE:** TN west to OK, south to FL and LA.

GRASS MINER MOTHS

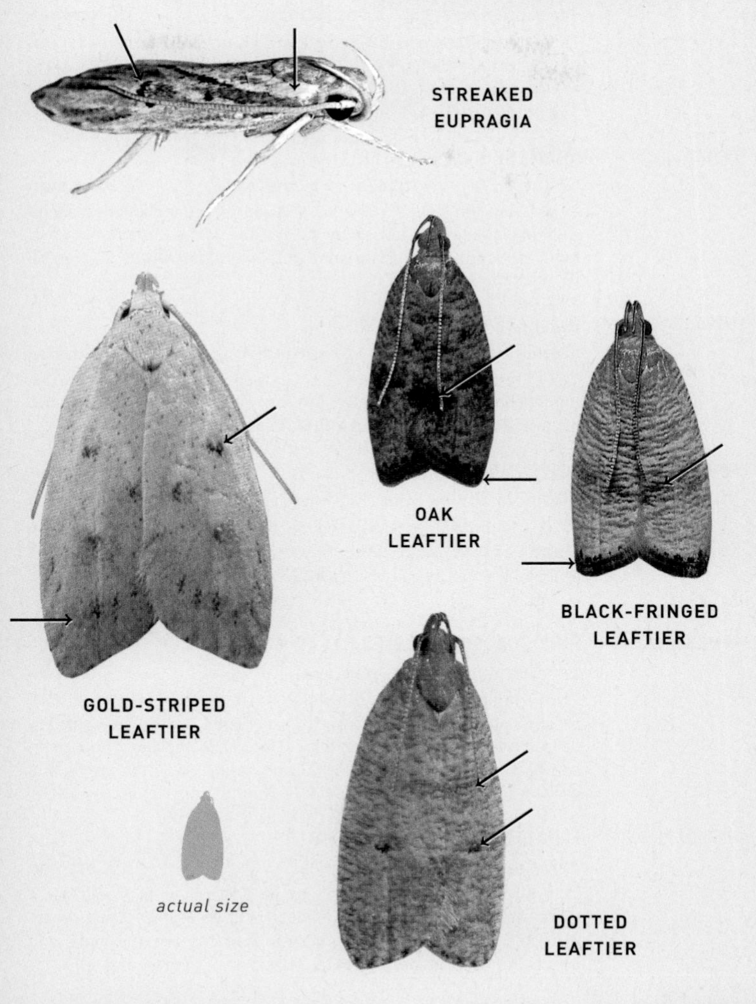

COSMET MOTHS

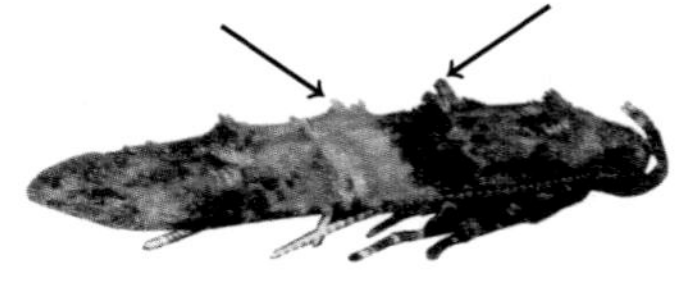
SWEETCLOVER ROOT BORER

actual size

GRAY PERIMEDE

Perimede erransella 42-0329 (1623) Uncommon

TL 5–7 mm. Dark gray FW is flecked with tiny white dots in central median area and at apex. Note tiny black dot in point of apex. Usually rests in a head-down position. **HOSTS:** Elm, hickory, oak, and tulip tree. **RANGE:** TN and n. MS south to FL and e. TX.

CHAMBERS' COSMOPTERIX

Cosmopterix pulchrimella 42-0355 (1472) Common

TL 4–5 mm. Dark brown FW has a pale orange median band edged with straight silver bands. Four silvery lines pass through AM area. Eyes are red. **HOSTS:** Clearweed and pellitory. **RANGE:** SC west to OK, south to FL and e. TX.

DELICATE COSMOPTERIX

Cosmopterix delicatella 42-0362 (1480) Local

TL 4–5 mm. Similar to Chambers' Cosmopterix, but the pale orange median band is edged with two offset metallic spots basally and two distally. **HOSTS:** Unknown. **RANGE:** NC west to n. MS, south to FL.

LINED MELANOCINCLIS

Melanocinclis lineigera 42-0387 (1503) Common

TL 3–4 mm. Whitish FW is boldly patterned with crisp blackish brown lines in central basal, median, and ST areas. **HOSTS:** Cones of loblolly and slash pines. **RANGE:** NC west to n. MS, south to FL and e. TX.

PINK SCAVENGER CATERPILLAR MOTH

Pyroderces rileyi 42-0398 (1512) Common

TL 5–7 mm. Light pinkish brown FW has a marbled pattern of dark brown streaks broken by wavy, black-edged white lines. **HOSTS:** Larvae are scavengers on plants, including castor bean, bushmint, cotton, and pineapple. **RANGE:** SC west to cen. MS, south to FL and e. TX.

FLORIDA PINK SCAVENGER MOTH

Pyroderces badia 42-0399 (1513) Common

TL 4–6 mm. Similar to Pink Scavenger Caterpillar Moth but has a more muted FW pattern with less distinct whitish cross lines. **HOSTS:** Seedpods of coffee senna and fruit of grapefruit, lime, and peach; also pinecones. **RANGE:** SC and GA south to FL and e. TX.

SHY COSMET

Limnaecia phragmitella 42-0401 (1515) Common

TL 7–11 mm. Shiny light brown FW has a darker central streak interrupted by two oval white-ringed spots in AM and PM areas. Apex is pointed. **HOSTS:** Flowers and developing seeds of cattail. **RANGE:** TN west to OK, locally south to FL and LA.

COSMET MOTHS

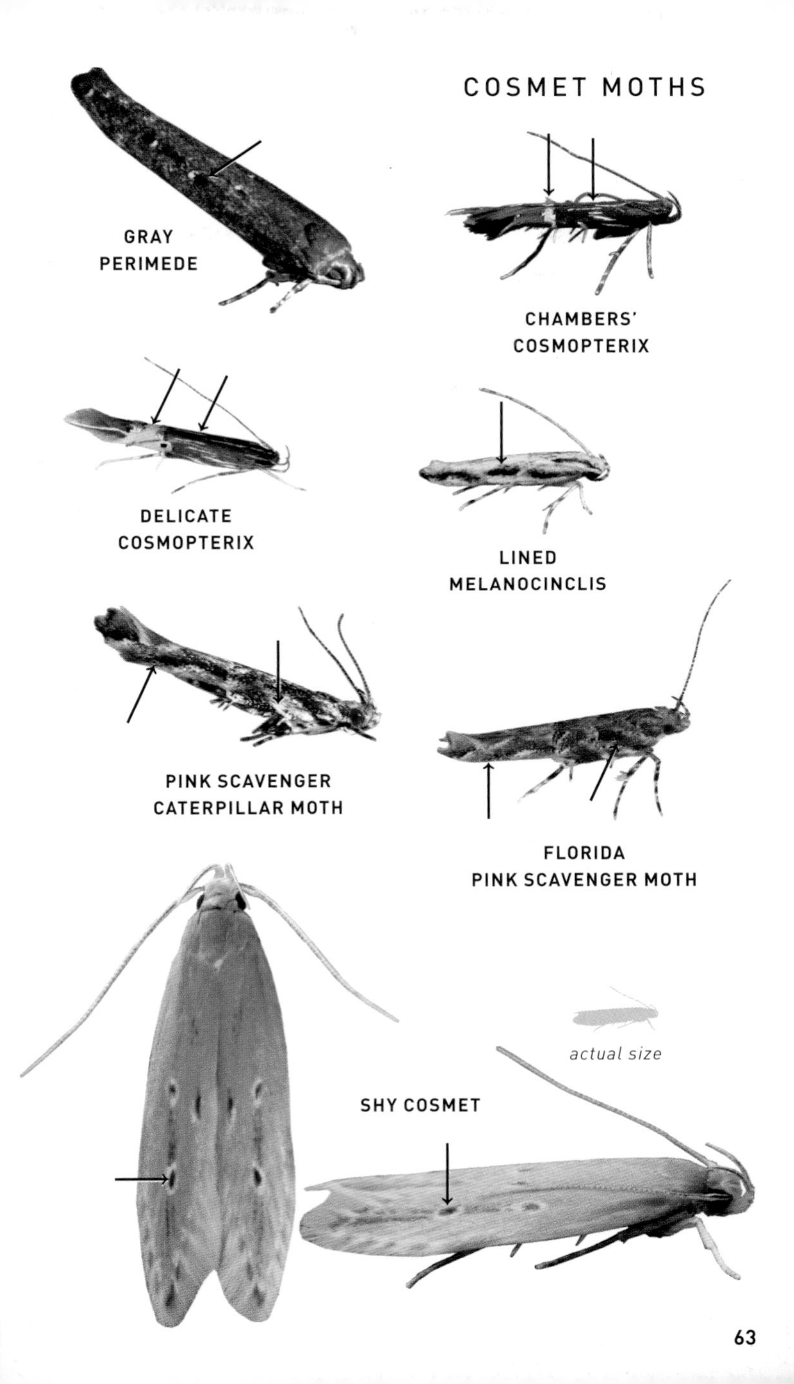

BUSH-CLOVER TRICLONELLA

Triclonella pergandeella 42-0410 (1524) **Common**

TL 5–7 mm. Creamy yellow FW has a peppery brown terminal line beyond oblique white PM line. Note tiny black dot in central AM area. **HOSTS:** Bush clover and tick trefoil. **RANGE:** NC west to OK, south to n. FL and ne. TX.

ORANGE-BANDED TRICLONELLA

Triclonella determinatella 42-0413 (1527) **Common**

TL 4–6 mm. Dark grayish brown FW is boldly patterned with a pale orange AM band. A pale orange spot in central PM area is joined to costa by a white streak (separated in far south). **HOSTS:** Unknown. **RANGE:** GA west to OK, south to FL and cen. TX.

KERMES SCALE MOTH

Euclemensia bassettella 42-0448 (1467) **Common**

TL 5–7 mm. Black FW has a broad orange (sometimes yellow) stripe along costa and basal area. Raised tufts of silvery scales mark AM and PM areas. Black antennae are tipped white. **HOSTS:** Larvae parasitize oak-inhabiting scale insects in the genus *Kermes*. **RANGE:** NC west to OK, south to FL and e. TX.

TWIRLER MOTHS Family Gelechiidae

A huge assemblage of small and very small moths, with upward-pointing labial palps that curve over the head like tiny horns. A varied group; some are remarkably colorful or metallic, while others are relatively plain and difficult to identify. They are found in a wide variety of habitats, including gardens in large cities. Adults of most species come to lights in small to moderate numbers. Some, such as Pink-washed Aristotelia, are also commonly observed during daytime.

PEACH TWIG BORER *Anarsia lineatella* 42-0453 (2257) **Local**

TL 6–7 mm. Peppery gray FW is marked with short blackish streaks between veins. Rests with stubby labial palps slightly raised. **HOSTS:** Fruit trees, including peach and plum. **RANGE:** GA and FL. **NOTE:** Introduced from Europe.

BLACK-FACED TWIRLER

Battaristis nigratomella 42-0468 (2227) **Common**

TL 5–6 mm. Light brown FW has a strongly angled white PM line. Black costa has short oblique white streaks at midpoint and ST area. Small black spots mark inner median and apical areas. Underside of head is black. **HOSTS:** Unknown. **RANGE:** W. NC west to OK, south to n. FL and LA.

COSMET MOTHS

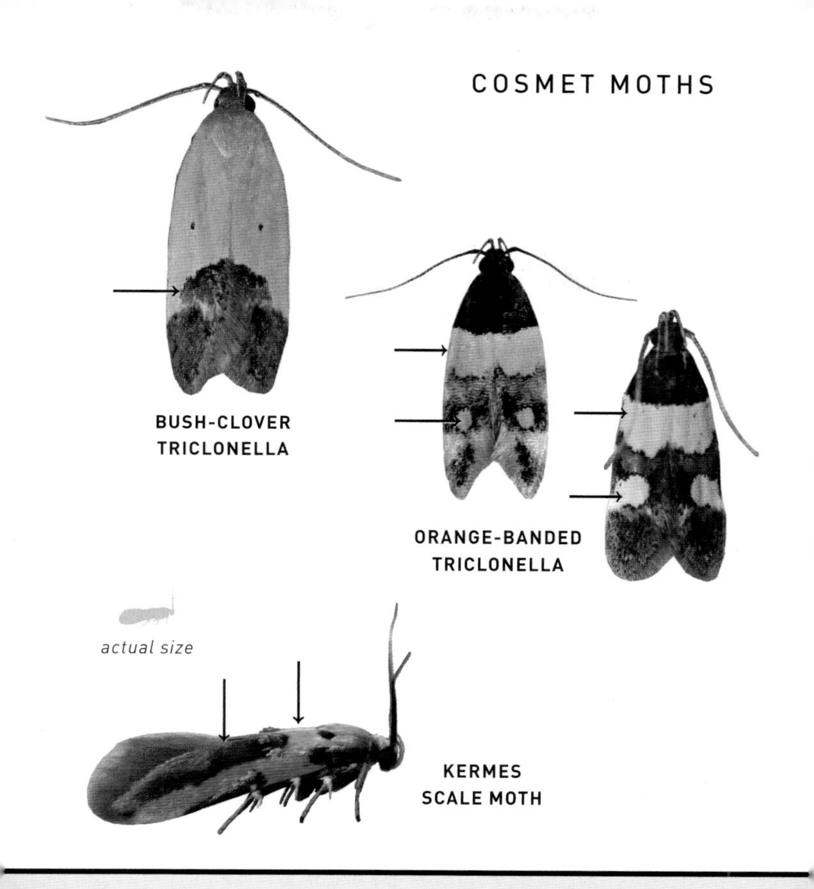

TWIRLER MOTHS

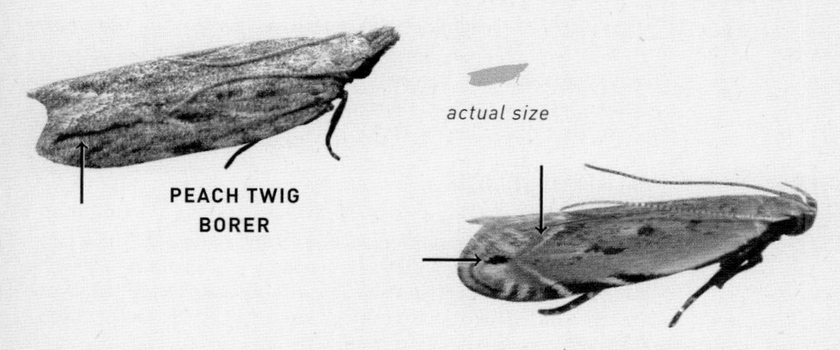

ORANGE STRIPE-BACKED MOTH

Battaristis vittella 42-0470 (2229) **Common**

TL 5–6 mm. Rusty orange FW is boldly patterned with three silvery gray bands. A tiny black dot marks inner apical area. Eyes are red. **HOSTS:** Unknown. **RANGE:** NC west to OK, south to FL and e. TX.

AGRIMONY ANACAMPSIS

Anacampsis agrimoniella 42-0471 (2230) **Uncommon**

TL 6–7 mm. Purplish gray FW grades to blackish beyond midpoint. A bold, straight white PM band widens toward apex at costa. Eyes are red. **HOSTS:** Possibly agrimony. **RANGE:** TN west to AR, locally south to n. and w. FL.

COVERDALE'S ANACAMPSIS

Anacampsis coverdalella 42-0476 (2234) **Common**

TL 5–6 mm. Strongly bicolored FW is mostly pale orange, especially toward inner margin, with peppery blackish base and ST area. Head and thorax are blackish. **HOSTS:** Unknown. **RANGE:** NC west to AR, south to FL and e. TX.

SILVER-DASHED ANACAMPSIS

Anacampsis levipedella 42-0483 (2241) **Uncommon**

TL 12 mm. Bicolored FW is grayish brown basally, blending to dark brown beyond midpoint. Distinct silvery white dashes mark inner median and ST areas and outer PM and ST lines at costa. Eyes are red. **HOSTS:** Unknown. **RANGE:** NC west to OK, locally south to n. FL.

IRIDESCENT STROBISIA

Strobisia iridipennella 42-0496 (2253) **Common**

TL 5–6 mm. Narrow chocolate brown FW has iridescent bronze highlights. Three oblique silvery blue lines fall short of inner margin. Basal and ST areas are marked with metallic blue dashes. Fringe is silvery blue. **HOSTS:** Unknown. **RANGE:** TN west to OK, south to n. FL and e. TX.

CHAMBERS' TWIRLER

Helcystogramma chambersella 42-0508 (2265) **Common**

TL 4–6 mm. Peppery cream-colored to brown FW is variably streaked dark brown between the veins. Tiny black discal dots in median area are partly outlined orange. **HOSTS:** Ragweed. **RANGE:** SC west to OK, south to FL and e. TX.

PALMERWORM MOTH

Dichomeris ligulella 42-0510 (2281) **Common**

TL 8–11 mm. Sexually dimorphic. Male's FW is blackish with a jagged cream-colored costal streak; female's is peppery brown with black dots in median area. Wing shape is narrow and pointed. Eyes are red. **HOSTS:** A general feeder with a preference for oak. **RANGE:** NC west to OK, south to FL and e. TX.

TWIRLER MOTHS

ORANGE STRIPE-BACKED MOTH

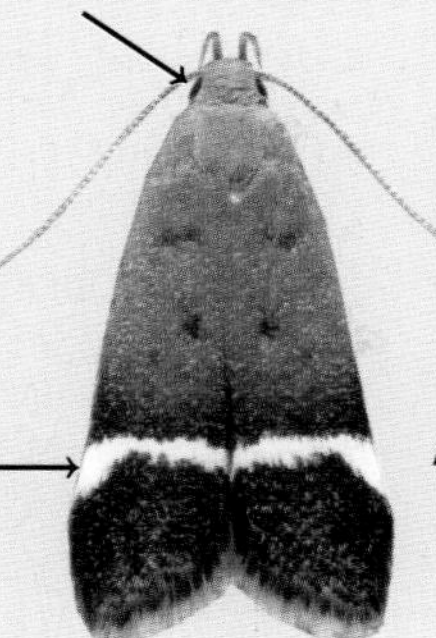

AGRIMONY ANACAMPSIS

COVERDALE'S ANACAMPSIS

actual size

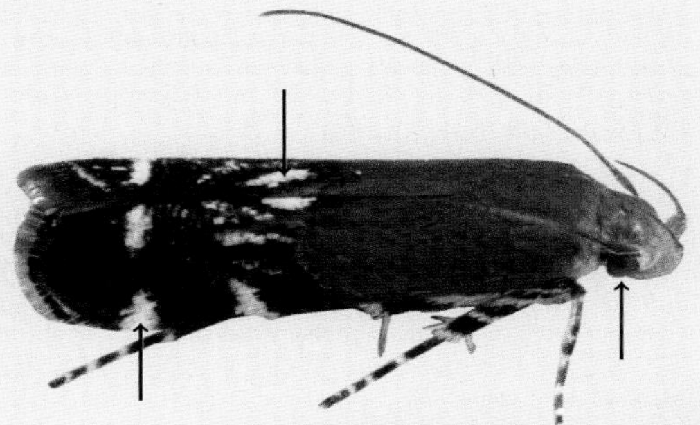

SILVER-DASHED ANACAMPSIS

IRIDESCENT STROBISIA

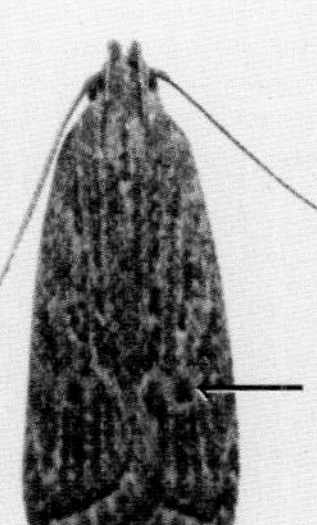

CHAMBERS' TWIRLER

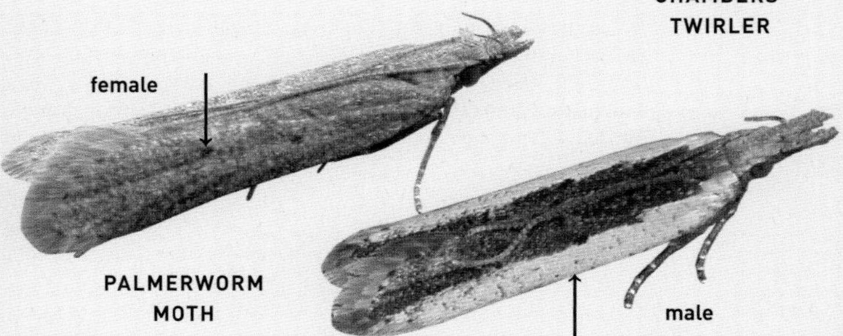

PALMERWORM MOTH

SPOTTED DICHOMERIS

Dichomeris punctidiscella 42-0521 (2283) **Common**

TL 7–9 mm. Pale yellowish FW is often peppered darker beyond curved PM line. Central median area is variably patterned with brown spots. **HOSTS:** Unknown. **RANGE:** NC west to AR, south to FL and LA.

CREAM-EDGED DICHOMERIS

Dichomeris flavocostella 42-0529 (2295) **Common**

TL 8–9 mm. Black FW has a broad creamy costal stripe that has a spur extending into PM area. Curved labial palps are pale orange. **HOSTS:** Goldenrod and aster. **RANGE:** NC west to AR, south to FL and LA.

INVERSED/KIMBALL'S DICHOMERIS *Dichomeris inversella/kimballi* 42-0531/42-0532 (2310/2310.1) **Common**

TL 6–8 mm. These two species are often inseparable in the field. Peppery silvery gray FW has faint darker bands and a curved whitish ST line. Long labial palps are usually brushy. Kimball's Dichomeris averages more uniformly darker with a less well defined ST line. **HOSTS:** Possibly hickory. **RANGE:** NC west to OK, south to FL and LA.

GEORGIA DICHOMERIS

Dichomeris georgiella 42-0534 (2277) **Common**

TL 9–12 mm. Peppery pale yellowish FW is variably marked with white-accented dusky spots along inner margin. Dotted terminal line is present when fresh. **HOSTS:** Possibly oak. **RANGE:** TN west to OK, south to FL and e. TX.

HAIRY DICHOMERIS

Dichomeris setosella 42-0537 (2302) **Common**

TL 6–8 mm. Inner FW is black at base, gray in median area, and darker gray in ST area. Cream-colored costal streak is wide and jagged-edged basally but fades before reaching apex. An irregular whitish ST line is usually present. **HOSTS:** Arborvitae, white pine, ironweed, white snakeroot, and others. **RANGE:** W. NC west to OK, south to FL and LA.

SHINING DICHOMERIS

Dichomeris ochripalpella 42-0554 (2289) **Local**

TL 7–8 mm. Blackish FW has an irregular, metallic, silvery blue costal streak, inner PM spot, and terminal line. PM line widens into a small pale orange triangle at costa. Base of labial palps is orange. **HOSTS:** Aster and goldenrod. **RANGE:** W. NC west to n. MS, south to w. FL and LA.

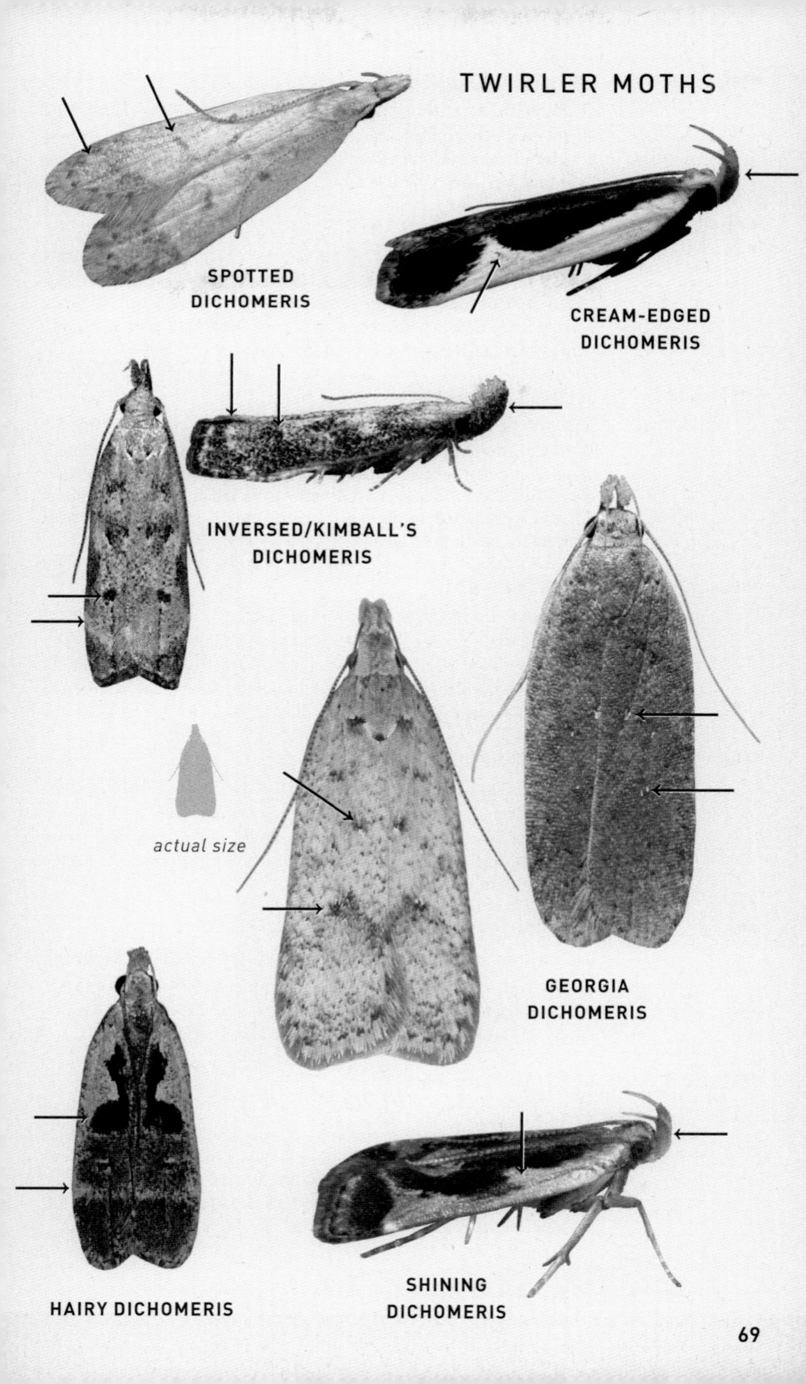
SPOTTED DICHOMERIS
CREAM-EDGED DICHOMERIS
INVERSED/KIMBALL'S DICHOMERIS
actual size
GEORGIA DICHOMERIS
HAIRY DICHOMERIS
SHINING DICHOMERIS

INDENTED DICHOMERIS

Dichomeris inserrata 42-0556 (2297) Common

TL 7–9 mm. Resembles Cream-edged Dichomeris, but wavy-edged creamy costal streak lacks PM spur. **HOSTS:** Goldenrod. **RANGE:** SC west to n. MS, south to FL and LA.

ORANGE-DOTTED DICHOMERIS

Dichomeris juncidella 42-0576 (2298) Common

TL 6–8 mm. Slate gray FW has a variable pattern of three black-edged, dull orange spots in central median area. Black ST line is usually faint. Labial palps are dull orange. **HOSTS:** Aster and goldenrod. **RANGE:** TN west to OK, locally south to n. FL and LA.

GLENN'S DICHOMERIS

Dichomeris glenni 42-0577 (2278) Uncommon

TL 9–13 mm. Broad, light gray FW has two almost-touching black spots in median area. Inconspicuous PM line has two elongated black spots at midpoint. Remarkably long labial palps are brushy basally, with diverging curved tips. **HOSTS:** Unknown. **RANGE:** GA west to OK, south to FL and ne. TX.

POLYHYMNO MOTH

Polyhymno luteostrigella 42-0594 (2211) Common

TL 5–6 mm. Straw-colored FW has two brown streaks that converge in orange-tinted ST area. Terminal line is accented with black dots near inner margin. Note the upward-pointing spike at apex. **HOSTS:** Unknown. **RANGE:** Coastal NC west to OK, south to FL and e. TX.

BURDOCK SEEDHEAD MOTH

Metzneria lappella 42-0600 (1685) Uncommon

TL 8–10 mm. Pale golden brown FW is streaked rusty brown with a tiny black discal spot. Large labial palps curve over the head. **HOSTS:** Burdock seeds. **RANGE:** GA west to MS, south to FL. **NOTE:** Introduced from Eurasia.

FIVE-SPOTTED TWIRLER

Monochroa quinquepunctella 42-0631 (1716) Common

TL 4–6 mm. Pale grayish brown FW is patterned with widely spaced, black elongated dots. Fringe around apex is checkered. **HOSTS:** Unknown. **RANGE:** NC west to AR, south to s. MS and LA.

CONSTRICTED TWIRLER

Theisoa constrictella 42-0635 (1722) Common

TL 4–6 mm. Pale tan FW has an earth brown basal patch sharply edged with black and white lines. Blackish triangle at midpoint of costa is edged whitish. Tiny black discal dots mark median area. Rests in a slightly head-down position. **HOSTS:** Elm. **RANGE:** SC west to OK, locally south to FL and TX.

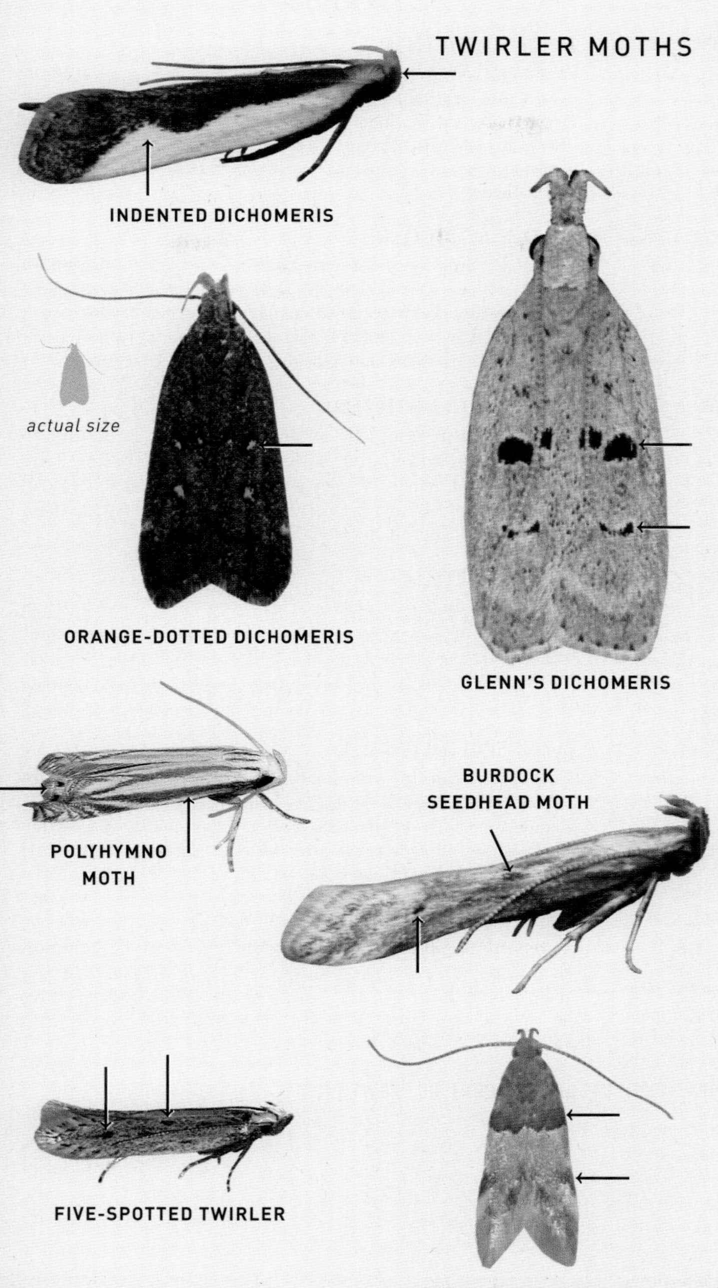
INDENTED DICHOMERIS
actual size
ORANGE-DOTTED DICHOMERIS
GLENN'S DICHOMERIS
POLYHYMNO MOTH
BURDOCK SEEDHEAD MOTH
FIVE-SPOTTED TWIRLER
CONSTRICTED TWIRLER

ACACIA ARISTOTELIA

Aristotelia corallina 42-0647 (1733.1) **Common**

TL 4–5 mm. Blackish brown FW has a sharply contrasting cream-colored streak along inner margin. Costa near apex is patterned with pale dashes. Dorsal surface of thorax and head is pale. **HOSTS:** Sweet acacia and sensitive partridge pea. **RANGE:** TN south to FL and e. TX.

ELEGANT ARISTOTELIA

Aristotelia elegantella 42-0650 (1736) **Common**

TL 6–7 mm. Reddish brown FW is boldly patterned with black-edged white patches. Note row of five raised, silver-edged black spots in central median area. **HOSTS:** Has been reared on willow herb. **RANGE:** MS west to OK, south to TX.

PINK-WASHED ARISTOTELIA

Aristotelia roseosuffusella 42-0670 (1761) **Common**

TL 7–8 mm. Brown FW has a bold pattern of oblique black and white bars. Light brown inner margin is marked with pink spots. **HOSTS:** Clover. **RANGE:** NC west to OK, south to FL and e. TX. **NOTE:** There are several similar-looking moths in this genus.

RUBY ARISTOTELIA

Aristotelia rubidella 42-0671 (1762) **Common**

TL 5–7 mm. Peppery brown FW has a bold pattern of oblique, fragmented blackish bars and pink spots without any obvious white edging. **HOSTS:** Unknown. **RANGE:** NC west to OK, south to FL and e. TX.

BLACK-SPOTTED TWIRLER

Deltophora sella 42-0681 (1928) **Common**

TL 5–7 mm. Silvery gray FW has a distinctive tear-shaped black spot in inner median area. Also note black crescent in inner basal area and black dot in central PM area. **HOSTS:** Unknown. **RANGE:** TN west to OK, south to FL and e. TX.

SKUNK TWIRLER

Agnippe prunifoliella 42-0698 (1771) **Common**

TL 5–6 mm. Black FW has a bold, wavy-edged white streak along inner margin. Costa usually shows a small white subapical patch. Top of head and labial palps are white. **HOSTS:** *Prunus* species. **RANGE:** NC west to OK, south to FL and se. TX.

COLEOTECHNITES FLOWER MOTH

Coleotechnites florae 42-0727 (1809) **Common**

TL 5–6 mm. Silvery gray FW is boldly patterned with black bars along costa that are connected with dusky shading. Raised tufts of scales near inner margin are tipped orange. **HOSTS:** Unknown. **RANGE:** W. NC west to OK, south to s. AL and e. TX. **NOTE:** There are several similar-looking moths in this genus.

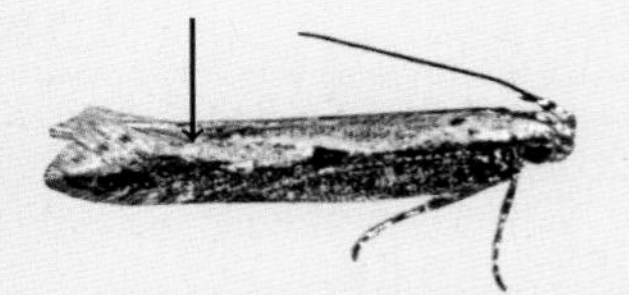

ACACIA ARISTOTELIA

ELEGANT ARISTOTELIA

actual size

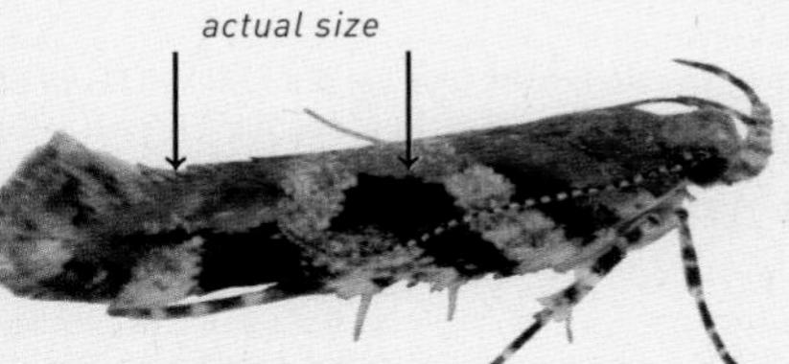

PINK-WASHED ARISTOTELIA

RUBY ARISTOTELIA

BLACK-SPOTTED TWIRLER

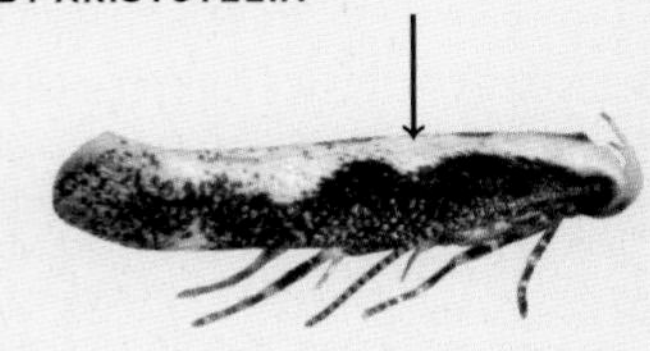

SKUNK TWIRLER

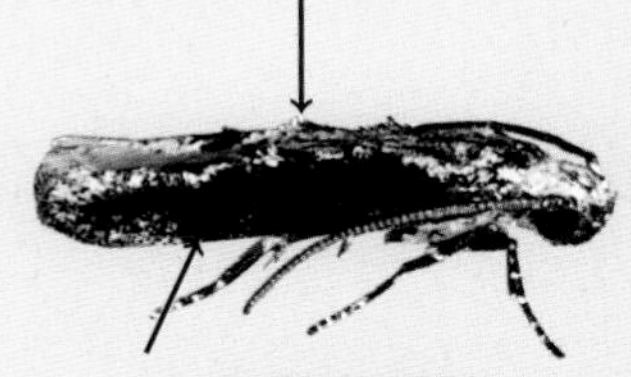

COLEOTECHNITES FLOWER MOTH

PINE NEEDLEMINER

Exoteleia pinifoliella 42-0762 (1840) Common

TL 4–5 mm. Orange brown FW is boldly marked with broad silvery bands. Silvery inner margin is accented with tiny raised black tufts. **HOSTS:** Pine. **RANGE:** NC west to OK, south to FL and e. TX.

WHITE STRIPE-BACKED MOTH

Arogalea cristifasciella 42-0765 (1851) Common

TL 5–6 mm. Whitish FW has an oblique black AM band that ends as a raised tuft near inner margin. Note black blotch in outer PM area. Terminal line is dotted. **HOSTS:** Unknown. **RANGE:** NC west to OK, south to FL and e. TX.

Y-BACKED TELPHUSA

Telphusa longifasciella 42-0769 (1858) Common

TL 8–9 mm. Black FW has sharply contrasting white streak along inner margin joining an oblique white AM band. Top of head is white. **HOSTS:** Unknown. **RANGE:** TN west to n. LA, south to FL and s. MS.

WHITE-BANDED TELPHUSA

Pubitelphusa latifasciella 42-0773 (1857) Common

TL 5–7 mm. Blackish gray FW typically has a black basal area and wide white median band. Sometimes FW is mostly dusky gray with a ghost of typical pattern. **HOSTS:** Unknown. **RANGE:** W. NC west to OK, south to n. FL and e. TX.

WALSINGHAM'S MOTH

Pseudochelaria walsinghami 42-0805 (1864) Common

TL 7–8 mm. Gray FW is boldly patterned with black patches in basal half of inner margin and outer PM area. A black streak passes through central ST area. **HOSTS:** Unknown. **RANGE:** NC west to AR, locally south to n. FL and MS.

WHITE-SPOTTED CHIONODES

Chionodes fuscomaculella 42-0889 (2079) Common

TL 5–8 mm. Light gray FW is variably mottled blackish in median area. Costa is marked with whitish blotches in AM and PM areas. Note obvious white spot in inner basal area. **HOSTS:** Oak. **RANGE:** NC west to OK, south to FL and e. TX. **NOTE:** Two broods.

TWO-SPOTTED CHIONODES

Chionodes bicostomaculella 42-0890 (2064) Common

TL 6–8 mm. Peppery gray FW is variably mottled with blackish patches. Costa has blurry pale patches in AM and PM areas. **HOSTS:** Oak. **RANGE:** W. NC west to OK, south to FL and e. TX. **NOTE:** There are several similar-looking moths in this genus.

TWIRLER MOTHS

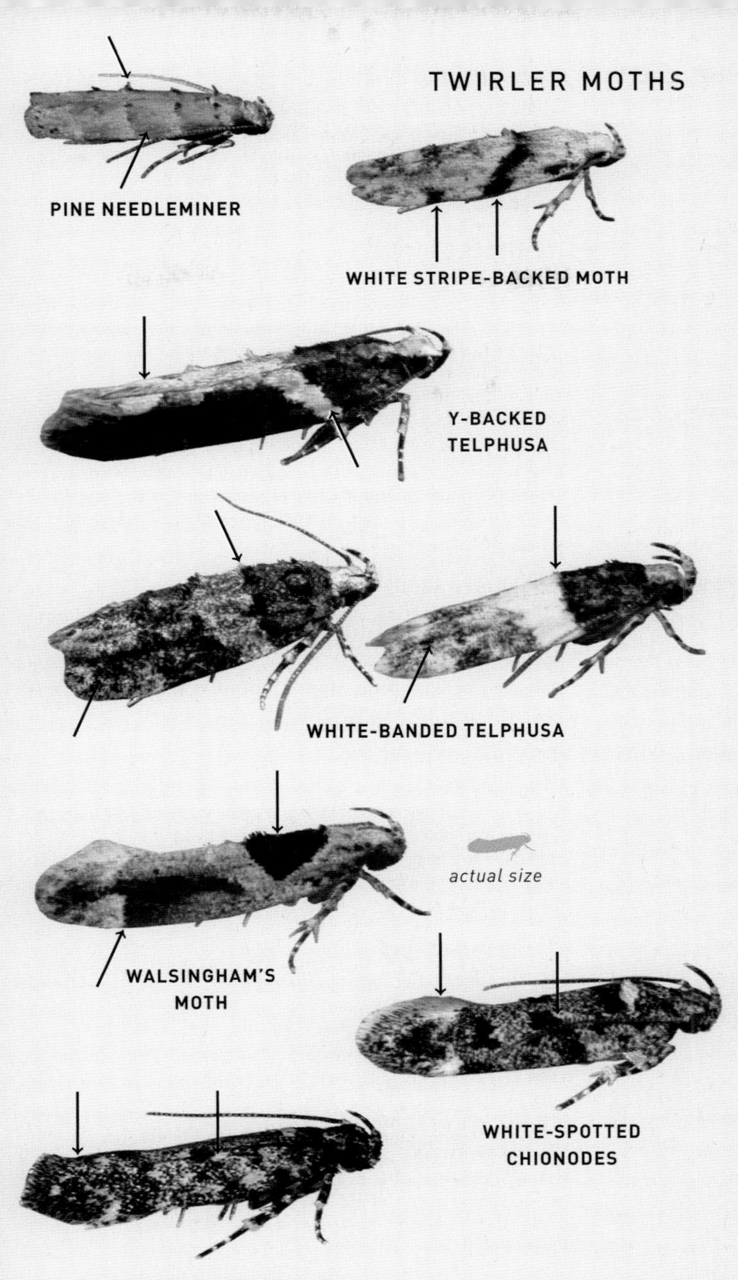

BLACK-SMUDGED CHIONODES

Chionodes mediofuscella 42-0971 (2093) **Common**

TL 5–8 mm. Light brown FW has pale basal area sharply contrasting with extensive blackish shading in outer median area. ST area and angled terminal line are dotted black. **HOSTS:** Giant ragweed. **RANGE:** NC west to OK, south to FL and LA.

EYE-RINGED CHIONODES

Chionodes discoocellella 42-1006 (2072) **Common**

TL 5–8 mm. Dark chocolate brown FW has blackish streaks along veins in ST area. Tiny black discal spot is edged with tufts of yellowish scales. **HOSTS:** *Polygonum* species. **RANGE:** NC west to OK, south to FL and e. TX.

SMALL-SPOTTED FILATIMA

Filatima biminimaculella 42-1086 (2136) **Common**

TL 6–7 mm. Blackish FW has a white patch in inner median area. Note small white smudge in inner AM area and slightly curved white ST line. **HOSTS:** Unknown. **RANGE:** TN west to OK, south to s. AL and LA.

SIX-SPOTTED AROGA

Aroga compositella 42-1140 (2187) **Common**

TL 7–9 mm. Black FW is boldly patterned with white spots of varying size. White spot in AM area is barlike. Black head contrasts with white base of labial palps. **HOSTS:** Unknown. **RANGE:** Coastal NC west to OK, south to FL and e. TX.

RED-STRIPED FIREWORM

Aroga trialbamaculella 42-1151 (2198) **Common**

TL 6–9 mm. Peppery black FW has fragmented white ST line, obvious only at costa and inner margin. Black head contrasts with orange base to labial palps. **HOSTS:** Various woody plants, including blueberry, locust, myrtle, oak, and sweet fern. **RANGE:** GA west to OK, south to FL and LA.

REDBUD LEAFFOLDER

Fascista cercerisella 42-1157 (2204) **Common**

TL 7–9 mm. Shiny blackish FW is boldly patterned with white patches along costa (largest near base) and anal angle. Dorsal surface of head and labial palps are white. **HOSTS:** Eastern redbud. **RANGE:** NC west to OK, south to FL and e. TX.

RED-NECKED PEANUTWORM MOTH

Stegasta bosqueella 42-1168 (2209) **Common**

TL 6–8 mm. Black FW is boldly marked with an irregular orange or cream-colored patch along inner margin. Large white subapical spot is obvious. **HOSTS:** Unknown. **RANGE:** NC west to OK, south to FL and e. TX.

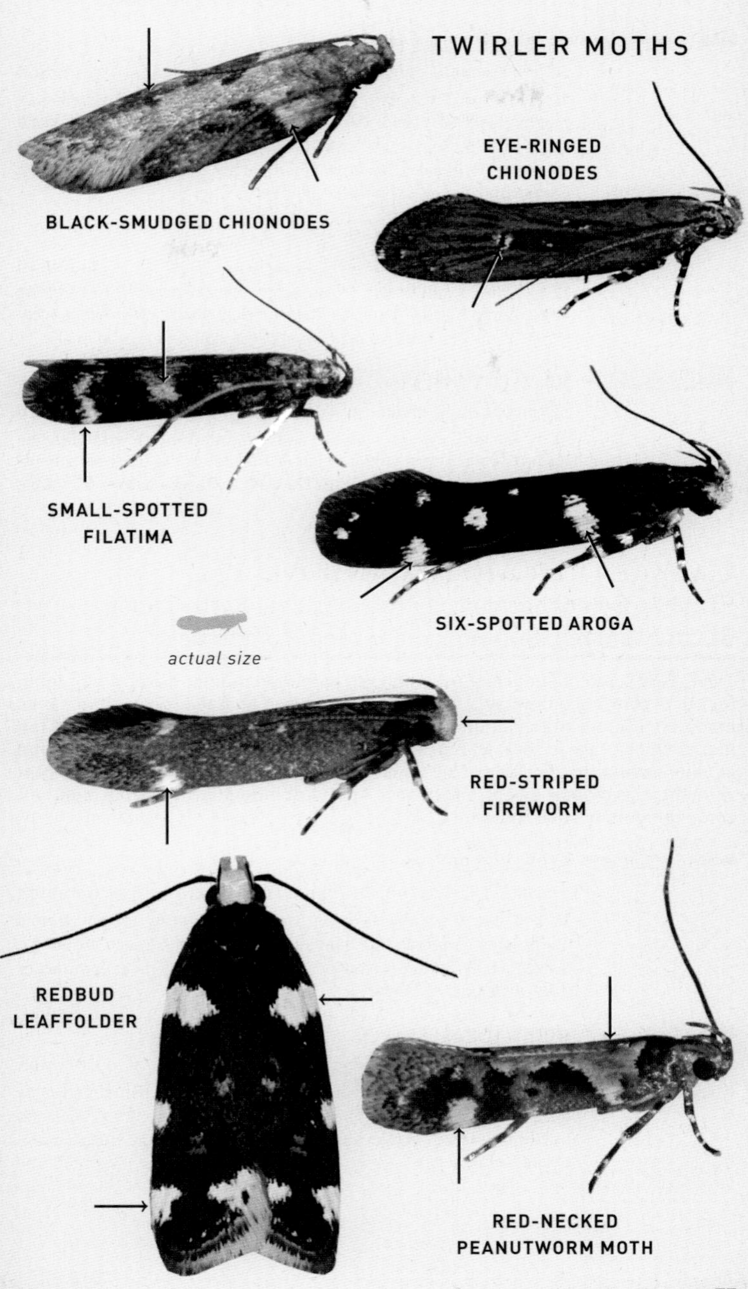
TWIRLER MOTHS
BLACK-SMUDGED CHIONODES
EYE-RINGED CHIONODES
SMALL-SPOTTED FILATIMA
SIX-SPOTTED AROGA
actual size
RED-STRIPED FIREWORM
REDBUD LEAFFOLDER
RED-NECKED PEANUTWORM MOTH

GOLDENROD GALL MOTH

Gnorimoschema gallaesolidaginis 42-1204 (1986) **Common**

TL 12–14 mm. Frosty gray FW has chocolate brown patches in basal, median, and ST areas. Large whitish labial palps curve over the head. **HOSTS:** Goldenrod. **RANGE:** NC west to OK, south to FL and e. TX. **NOTE:** There are several similar-looking moths in this genus. Larvae make elliptical galls on goldenrod stems.

STREAKED TWIRLER

Symmetrischema striatella 42-1330 (2039) **Common**

TL 6–7 mm. Light brown FW is boldly streaked rusty brown along veins. A thin black line passes through median area to pointed apex. **HOSTS:** Unknown. **RANGE:** SC west to OK, south to FL and e. TX.

MOTTLED TWIRLER

Frumenta nundinella 42-1352 (2052) **Common**

TL 10–12 mm. Powdery whitish FW is mottled with brown blotches toward costa. Fringe appears checkered when fresh. Head is whitish. **HOSTS:** Unknown. **RANGE:** NC west to OK, south to n. FL and e. TX.

Casebearer Moths and Allies

Families Coleophoridae, Batrachedridae, Scythrididae, Blastobasidae, Stathmopodidae, and Momphidae

The *Coleophora* casebearers are small, streamlined moths, usually with long, forward-pointing antennae. Larvae mine in leaves and seeds, living in cases made from plant material and frass. The *Homaledra* are narrow-winged tan moths whose larvae feed on palms. Momphas are often boldly patterned and usually have tufts of raised scales along the inner margin of the FW; they rest with their antennae swept backward. Adults are mostly nocturnal and will come to lights in small numbers.

STREAKED COLEOPHORA

Coleophora cratipennella 42-1622 (1365) **Common**

TL 6–8 mm. Pale FW is boldly striated with rusty brown streaks along veins. Antennae are banded at base. **HOSTS:** Seeds of rushes. **RANGE:** TN west to OK, south to s. MS. **NOTE:** The most common of a few look-alike species.

METALLIC CASEBEARER

Coleophora mayrella 42-1646 (1387) **Common**

TL 6–8 mm. Metallic bronzy green FW has a reddish sheen near apex. White-tipped antennae are thickened at base. **HOSTS:** Seeds of clover. **RANGE:** W. NC and TN, south to GA and LA. **NOTE:** Introduced from Europe.

TWIRLER MOTHS

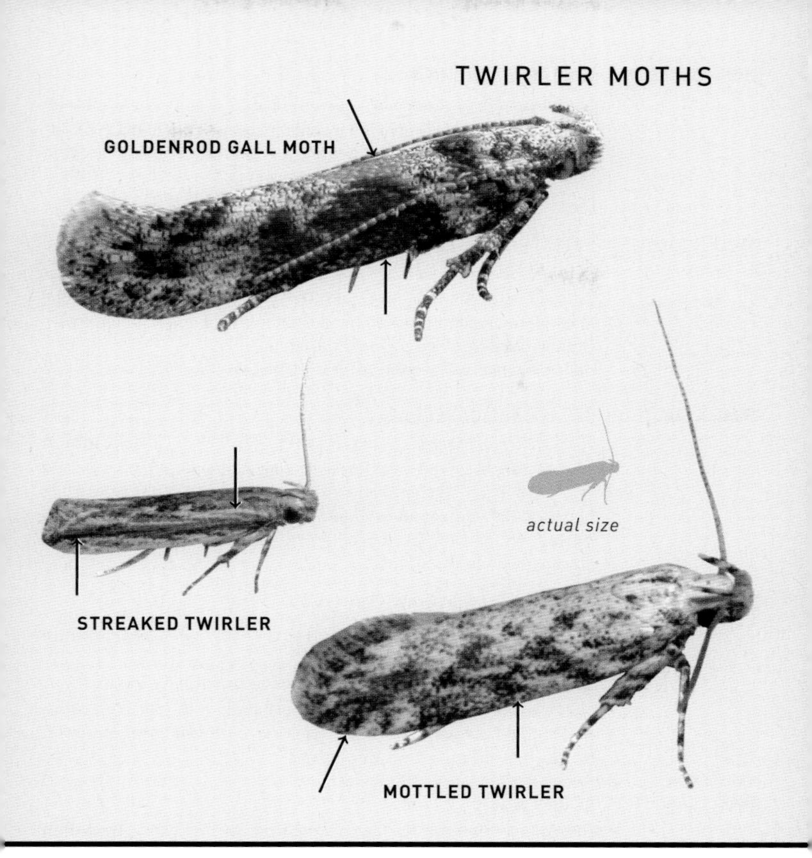

CASEBEARER MOTHS AND ALLIES

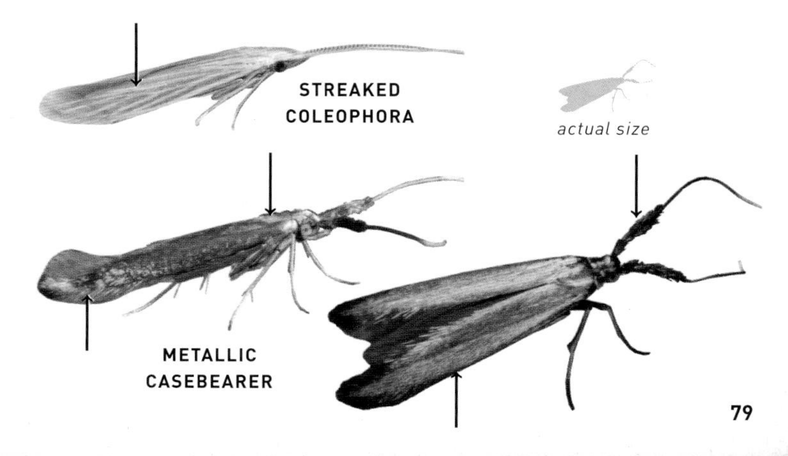

EXCLAMATION MOTH

Homaledra heptathalama 42-1686 (1421) Local

TL 10–13 mm. Light brown FW is outlined rusty orange. Bold "exclamation mark" pattern is formed by silvery white streak and dot in central median area. Bluntly pointed wings are held flat. **HOSTS:** Leaves of cabbage palmetto. **RANGE:** SC and GA south to FL.

PALM LEAF SKELETONIZER

Homaledra sabalella 42-1687 (1422) Local

TL 9–11 mm. Shiny light brown FW is marked with two elliptical black dots along inner margin. Terminal line is dotted when fresh. Bluntly pointed wings are held flat. **HOSTS:** Leaves of cabbage palmetto. **RANGE:** SC and GA south to FL and se. TX.

OCTAGONAL CASEBEARER

Homaledra octagonella 42-1688 (1398) Uncommon

TL 5–7 mm. Whitish FW is sparsely speckled with blackish scales. Two crisp black dashes mark AM and PM areas of inner margin. Long whitish antennae are held swept back. **HOSTS:** Unknown. **RANGE:** NC and TN, locally south to FL and e. TX. **NOTE:** Previously placed in genus *Coleophora*.

BANDED SCYTHRIS

Scythris trivinctella 42-1713 (1678) Uncommon

TL 6–7 mm. Shiny, dark bronzy brown FW has slightly irregular whitish basal, AM, and PM bands. PM band divides, forming a Y-shape at inner margin. Tip of FW is flared, with hairlike scales. **HOSTS:** Slim amaranth. **RANGE:** GA west to OK, south to FL and TX. **NOTE:** This species is mostly diurnal.

ACORN MOTH

Blastobasis glandulella 42-1766 (1162) Common

TL 8–13 mm. Gray FW has an angled whitish AM band bordered black distally. Two black dots in central PM area are usually noticeable. Pointed terminal line is dotted. **HOSTS:** Larvae feed inside acorns and chestnuts. **RANGE:** NC west to OK, south to FL and e. TX.

GOLDEN STATHMOPODA

Stathmopoda elyella 42-1807 (1070) Local

TL 4–6 mm. Wings, body, and legs are shiny metallic gold. Spiky-looking central pair of legs are distinctively held outward and often slightly elevated. **HOSTS:** Ferns. **RANGE:** GA south to cen. and w. FL.

BOTTIMER'S MOMPHA

Mompha bottimeri 42-1818 (1429) Local

TL 5–6 mm. White FW has chocolate brown patch covering distal half of wing, edged with angled black lines. Note obvious raised tufts of black and white scales along inner margin. Head is white. **HOSTS:** *Helianthemun* species. **RANGE:** SC and GA south to FL and s. MS.

CASEBEARER MOTHS AND ALLIES

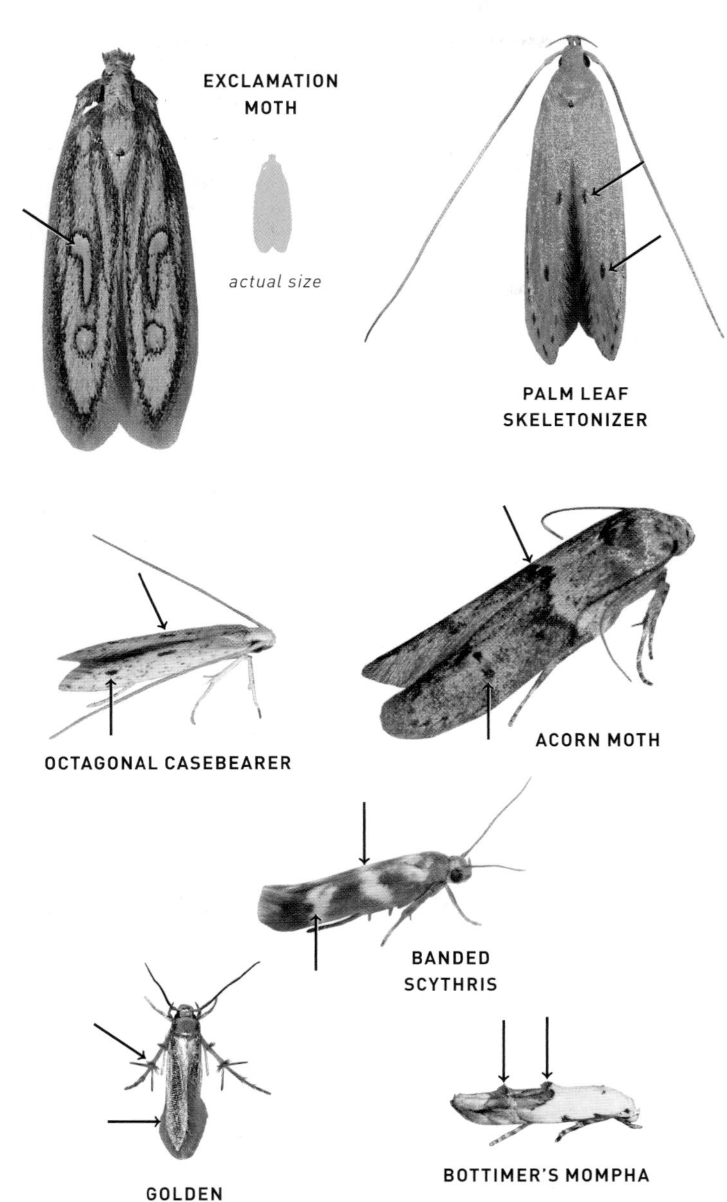

CIRCUMSCRIPT MOMPHA

Mompha circumscriptella 42-1823 (1434) **Common**

TL 6–7 mm. White FW has a peppery rusty brown distal half and large triangle in outer AM area. Short raised tufts of black scales mark inner margin in median and PM areas. Head is white. **HOSTS:** Evening primrose. **RANGE:** E. NC west to OK, south to FL and e. TX.

RED-STREAKED MOMPHA

Mompha eloisella 42-1833 (1443) **Common**

TL 6–8 mm. White FW has reddish V-shaped lines and tufts of gray scales toward pointed apex. Black spots mark median area, inner margin, and thorax. **HOSTS:** Stems of evening primrose. **RANGE:** W. NC west to OK, south to FL and TX.

WHITE-BASED MOMPHA

Mompha murtfeldtella 42-1841 (1448) **Common**

TL 5–7 mm. Peppery brown FW has an angled white basal patch. Large tufts of white scales mark inner margin in median and PM areas. Small white subapical patch is usually obvious. **HOSTS:** Evening primrose. **RANGE:** NC west to OK, south to FL and e. TX.

Many-Plumed Moths Family Alucitidae

These unusual small moths have wings made up of multiple feathery plumes, which are usually held spread while at rest. Only one species occurs in our region and is sometimes found on walls or in woodpiles during daytime. It is nocturnal and will come to light in small numbers.

SIX-PLUME MOTH *Alucita montana* 44-0001 (2313) **Local**

WS 12–14 mm. All wings are divided into six featherlike branches. Darker brown bands on wings create a zigzag pattern. Plume shafts on paler HW are banded black and white. **HOSTS:** Honeysuckle and snowberry. **RANGE:** GA and nw. FL.

Plume Moths Family Pterophoridae

These spindly-legged moths have a characteristic "airplane" posture at rest. A notch at the tip of the FW divides the wing into two lobes. FW patterns are often very similar, and species can sometimes be difficult to tell apart. Most are nocturnal and will often visit lights in small numbers. Sometimes they can be found resting outdoors on walls or among plants during the daytime.

ARTICHOKE PLUME MOTH

Platyptilia carduidactylus 46-0005 (6109) **Common**

WS 18–27 mm. Grayish brown FW has a triangular brown patch along costa in PM area. Terminal line on both lobes is blackish with a whitish fringe. Abdomen is whitish at base with two dark, inverted V-shaped bands on inner segments. **HOSTS:** Artichoke and thistle. **RANGE:** W. NC west to OK, locally south to FL and e. TX.

CASEBEARER MOTHS AND ALLIES

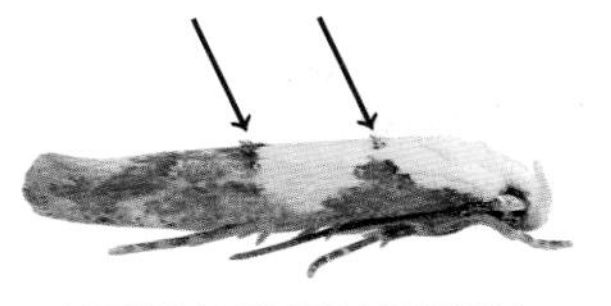

CIRCUMSCRIPT MOMPHA

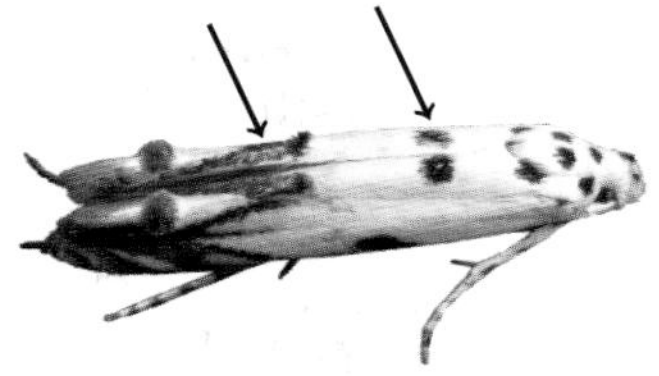

RED-STREAKED MOMPHA

WHITE-BASED MOMPHA

actual size

MANY-PLUMED MOTHS

SIX-PLUME MOTH

actual size

PLUME MOTHS

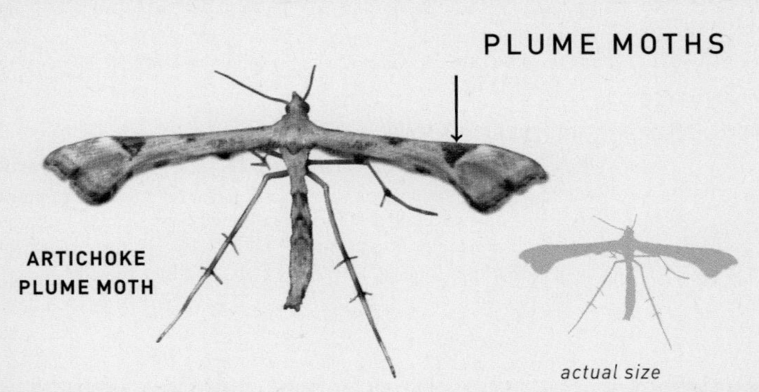

ARTICHOKE PLUME MOTH

actual size

LANTANA PLUME MOTH

Lantanophaga pusillidactylus 46-0015 (6119) **Local**

WS 10–14 mm. Light brown FW has a white-edged brown ST band on both lobes and a brown triangle along costa in PM area. Light brown abdomen has brown bands at base and midpoint. **HOSTS:** Presumably lantana. **RANGE:** Coastal SC south to FL and se. TX.

NARROW PLUME MOTH COMPLEX *Stenoptilodes brevipennis/taprobanes* 46-0016/46-0018 (6122/6121.1) **Common**

WS 16–20 mm. These two virtually identical species cannot be reliably separated in the field, though *S. taprobanes* may tend toward darker, more defined markings. Light grayish brown FW has a white ST line edged with an inward-pointing blackish wedge on upper lobe. Brown patch in PM area curves around FW notch. AM area has dark spots at costa and near inner margin. Abdomen has dorsal line of tiny black dots. **HOSTS:** Unknown. **RANGE:** SC and TN south to FL and s. MS.

MOUSY PLUME MOTH

Lioptilodes albistriolatus 46-0054 (6120) **Common**

WS 10–18 mm. Light grayish brown FW has a weak dusky smudge at PM area of costa. A tiny blackish fleck marks outer AM area. Abdomen has slightly paler basal segments and thin blackish dorsal line. **HOSTS:** Unknown. **RANGE:** TN west to OK, locally south to FL and e. TX.

DWARF PLUME MOTH

Exelastis pumilio 46-0060 (6099.1) **Common**

WS 12–14 mm. Uniform, light golden brown FW is slightly darker near mouth of notch and has dotting along outer and inner margins. Pale thorax blends into darker abdomen, which is often weakly marked with a row of dark dots. **HOSTS:** Legumes, especially tick trefoil. **RANGE:** SC west to MS, south to FL and LA.

GRAPE PLUME MOTH

Geina periscelidactylus 46-0062 (6091) **Common**

WS 16–18 mm. Peppery golden brown FW has a darker PM band bordered by thin white lines. Brown abdomen has a darker band about midway along its length, and white lateral lines on most segments. **HOSTS:** Grape and Virginia creepers. **RANGE:** W. NC west to AR, south to GA.

SUNDEW PLUME MOTH

Buckleria parvulus 46-0072 (6098) **Local**

WS 10–12 mm. Light golden brown FW has a bold pattern of black and white dashes along costa near tip. Abdomen is striated with thin black and white lines down center. **HOSTS:** Sticky hairs and leaves of sundew. **RANGE:** Coastal NC west to n. AL, south to FL.

PLUME MOTHS

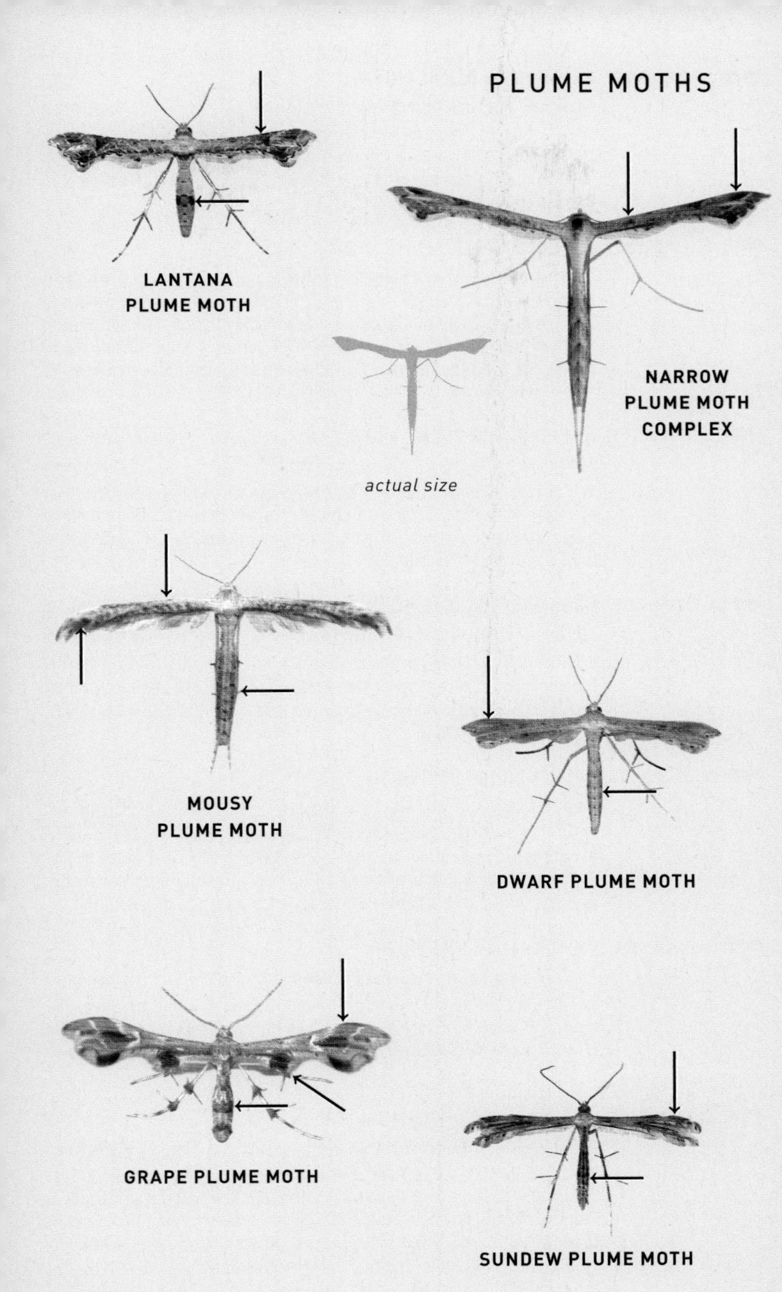
LANTANA PLUME MOTH
NARROW PLUME MOTH COMPLEX
actual size
MOUSY PLUME MOTH
DWARF PLUME MOTH
GRAPE PLUME MOTH
SUNDEW PLUME MOTH

BELFRAGE'S PLUME MOTH

Pselnophorus belfragei 46-0075 (6154) **Common**

WS 14–18 mm. Whitish FW has blackish dashes along costa in AM, PM, and ST areas and in mouth of notch. Abdomen is finely striated with thin white lines. **HOSTS:** Carolina ponysfoot. **RANGE:** E. NC west to OK, south to FL and e. TX.

BLACK-MARKED PLUME MOTH

Hellinsia inquinatus 46-0087 (6186) **Common**

WS 16–18 mm. Peppery light gray FW has short black and white streaks along costa near apex and at mouth of notch. Median area is marked with a blackish dot. Grayish abdomen has a dorsal row of paired, pale-edged, blackish streaks. **HOSTS:** Unknown. **RANGE:** E. NC west to OK, south to FL and e. TX.

ONE-SPOTTED PLUME MOTH

Hellinsia paleaceus 46-0107 (6207) **Common**

WS 20 mm. Cream-colored FW is marked with a single blackish dot at mouth of notch. Central median area is sometimes tinged warm brown. Apex is pointed. **HOSTS:** Unknown. **RANGE:** TN west to OK, locally south to n. FL and MS.

GROUNDSEL PLUME MOTH

Hellinsia balanotes 46-0110 (6210) **Common**

WS 32–45 mm. Note large size. Light brown to whitish FW has a tiny blackish dot at mouth of notch. Apex is pointed. **HOSTS:** Groundsel tree and other *Baccharis* species. **RANGE:** Coastal NC west to AR, south to FL and e. TX.

GOLDENROD PLUME MOTH

Hellinsia kellicottii 46-0112 (6212) **Common**

WS 22–25 mm. Cream-colored FW has small blackish dot at mouth of notch with a smaller dark fleck on edge of notch near apex. Abdomen has brownish lateral stripes and a very thin, faint dorsal stripe. **HOSTS:** Goldenrod. **RANGE:** SC west to AR, south to FL and e. TX.

UNMARKED PLUME MOTH

Hellinsia unicolor 46-0125 (6226) **Common**

WS 20 mm. Whitish wings are tinted light brown without any markings. Abdomen has narrow pale brown dorsal stripe and thin lateral stripes. **HOSTS:** Unknown. **RANGE:** SC west to MS, south to FL and s. LA.

MORNING-GLORY PLUME MOTH

Emmelina monodactyla 46-0150 (6234) **Common**

WS 21–25 mm. Light gray to pinkish brown FW has darker dashes at ST area of costa and at mouth of notch, and a black dot in AM area. Pale thorax appears strongly triangular. Abdomen is patterned with parallel lines of short black dashes. **HOSTS:** Convolvulaceae, including morning glory and bindweed. **RANGE:** NC west to OK, south to FL and TX.

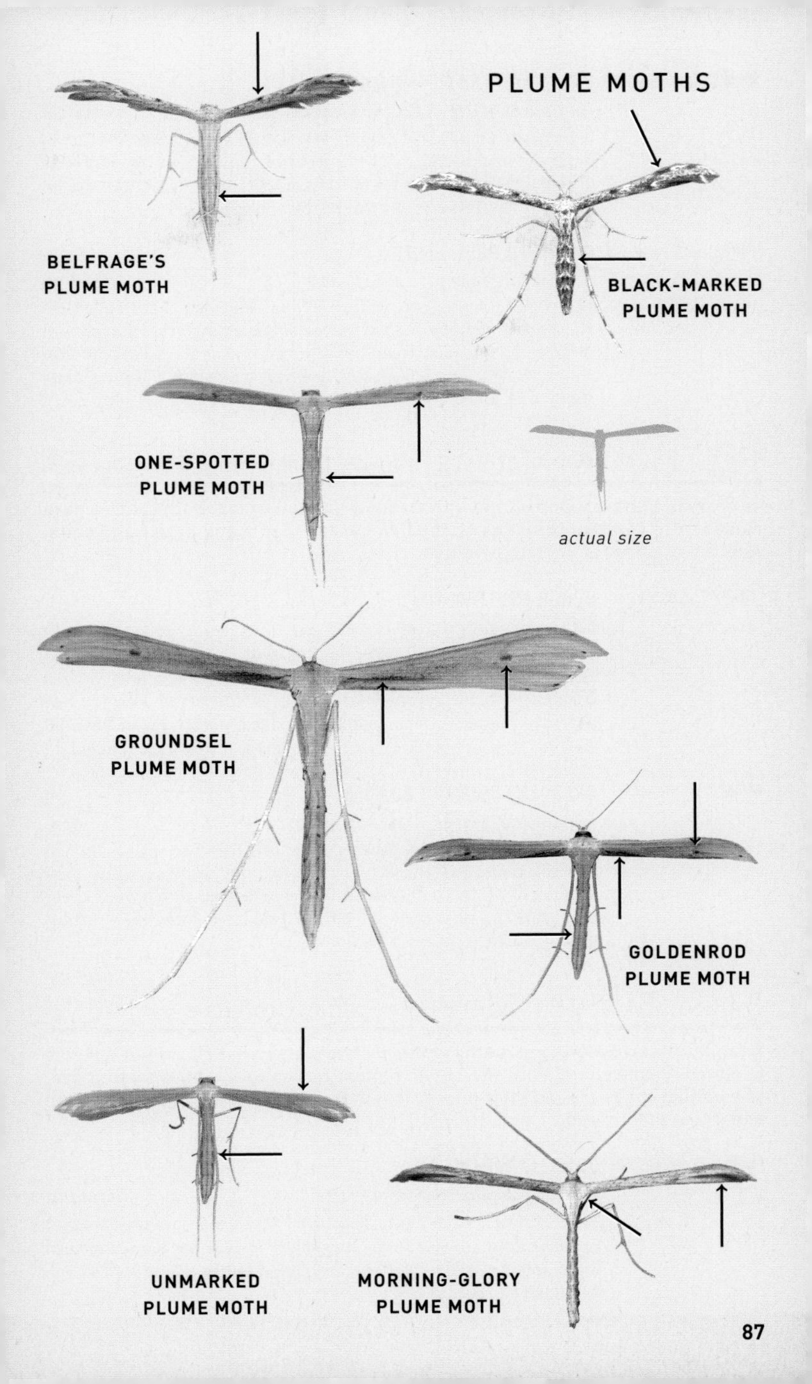
PLUME MOTHS
BELFRAGE'S PLUME MOTH
BLACK-MARKED PLUME MOTH
ONE-SPOTTED PLUME MOTH
actual size
GROUNDSEL PLUME MOTH
GOLDENROD PLUME MOTH
UNMARKED PLUME MOTH
MORNING-GLORY PLUME MOTH

THREE-DOTTED PLUME MOTH

Adaina simplicius 46-0153 (6155.1) **Local**

WS 10 mm. Light straw-colored FW has a blackish spot in mouth of notch. Additional dots mark PM area of costa and near apex on edge of notch. Pale abdomen is weakly marked with thin brown dorsal line. **HOSTS:** Unknown. **RANGE:** S. FL and w. MS.

RAGWEED PLUME MOTH

Adaina ambrosiae 46-0157 (6160) **Common**

WS 15–16 mm. Peppery whitish FW has blackish dashes and dots along costa and at mouth of notch. Note single black dot in central AM area. Whitish abdomen has a dorsal row of black dots. **HOSTS:** *Ambrosia* species, aster, and ragweed. **RANGE:** SC west to MS, south to FL and e. TX.

FRUITWORM MOTHS Family Carposinidae

Very small delta-shaped moths with strongly pointed wings. Both species have raised tufts of metallic scales on the FW. Larvae feed within fruits and galls. Adults are rarely attracted to lights.

PEACH FRUIT MOTH

Carposina sasakii 48-0006 (2314) **Uncommon**

TL 7–9 mm. Light gray FW is mottled with white along inconspicuous AM and PM lines. A black dash in central median area is often clouded with dusky shading. Fringe is checkered when fresh. **HOSTS:** Fruit trees, including apple, peach, and plum. **RANGE:** SC west to OK, south to nw. FL and e. TX. **NOTE:** Introduced from Asia.

CRESCENT-MARKED BONDIA

Bondia crescentella 48-0011 (2319) **Local**

TL 7–9 mm. Silvery gray FW has black tufts of raised scales outlined in silver. Hourglass-shaped orbicular and reniform spots have dark shading at center; outer half of reniform is filled with a white crescent. **HOSTS:** Unknown; presumably fruits of trees and shrubs. **RANGE:** W. NC southwest to cen. MS.

FALSE BURNET MOTHS Family Urodidae

A family of narrow-winged moths with only two, somewhat dissimilar species in North America, and only one in our region. Both species make distinctive, tiny netlike cages around their pupae; that of Bumelia Webworm dangles from a leaf by a thin thread.

BUMELIA WEBWORM

Urodus parvula 54-0001 (2415) **Common**

TL 11–15 mm. Smoky grayish brown FW has a greasy sheen. Head and legs have a bluish tinge. **HOSTS:** Various woody plants, including bumelia, hibiscus, oak, and redbay. **RANGE:** NC west to w. TN, south to FL and e. TX.

PLUME MOTHS

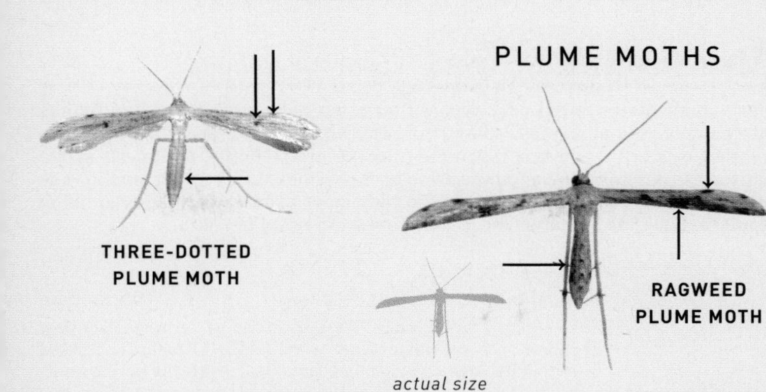

FRUITWORM MOTHS

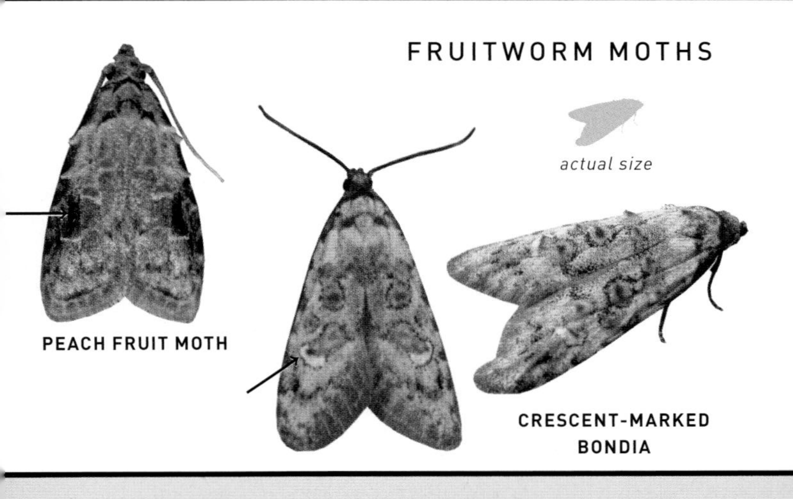

FALSE BURNET MOTHS

actual size

BUMELIA WEBWORM

METALMARK MOTHS Family Choreutidae

Very small moths with broad wings that often bear metallic scales. Wings are usually held at an angle above the body while at rest. Some appear more tentlike or tubular and may resemble colorful tortrix moths. Adults are active during the day and can be found sitting on flower heads or resting among vegetation near larval food plants, where they move around like fidgity jumping spiders. Skullcap Skeletonizer is sometimes attracted to lights.

PEACOCK BRENTHIA

Brenthia pavonacella 58-0002 (2627) **Uncommon**

TL 4–5 mm. Grayish brown FW is sprinkled white in median area. Reniform spot is partly outlined white. Black ST band is accented with metallic blue and violet wedges. Rests with HW splayed outward, revealing a white-spotted pattern. **HOSTS:** Tick trefoil. **RANGE:** W. NC west to OK, south to n. FL and MS.

SKULLCAP SKELETONIZER

Prochoreutis inflatella 58-0006 (2629) **Uncommon**

TL 5–6 mm. Orange FW has a largely dark brown median area sprinkled with silvery scales. Costa is marked with a white wedge between silver PM and ST lines. Fringe is silvery white. **HOSTS:** Skullcap. **RANGE:** E. NC west to n. MS, south to n. FL and LA.

EVERLASTING TEBENNA

Tebenna gnaphaliella 58-0024 (2647) **Uncommon**

TL 4–5 mm. Peppery silvery gray FW has an orange basal area. Three black-outlined silver spots form a triangle in inner PM area. **HOSTS:** Low plants, including cudweed, pussytoes, and strawflower. **RANGE:** NC and TN south to FL and e. TX.

SLOSSON'S METALMARK

Tortyra slossonia 58-0031 (2653) **Local**

TL 6–7 mm. Metallic gold and purple FW is boldly patterned with wide black-edged silver- and black-speckled AM and PM bands. Rests with wings tightly closed. **HOSTS:** Fig. **RANGE:** S. FL.

DIVA HEMEROPHILA

Hemerophila diva 58-0034 (2655) **Local**

TL 6–7 mm. Unmistakable. Metallic indigo FW is strikingly patterned with a straight yellow AM band and curving yellow PM line that fades before reaching inner margin. Terminal line is vermillion. HW is orange. **HOSTS:** Unknown. **RANGE:** S. FL.

METALMARK MOTHS

PEACOCK BRENTHIA

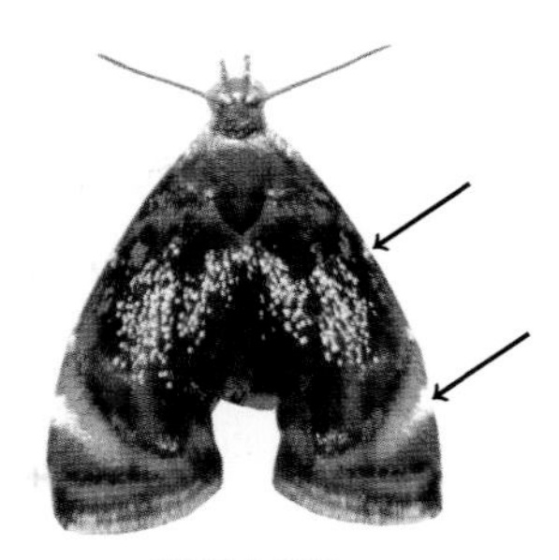

SKULLCAP SKELETONIZER

EVERLASTING TEBENNA

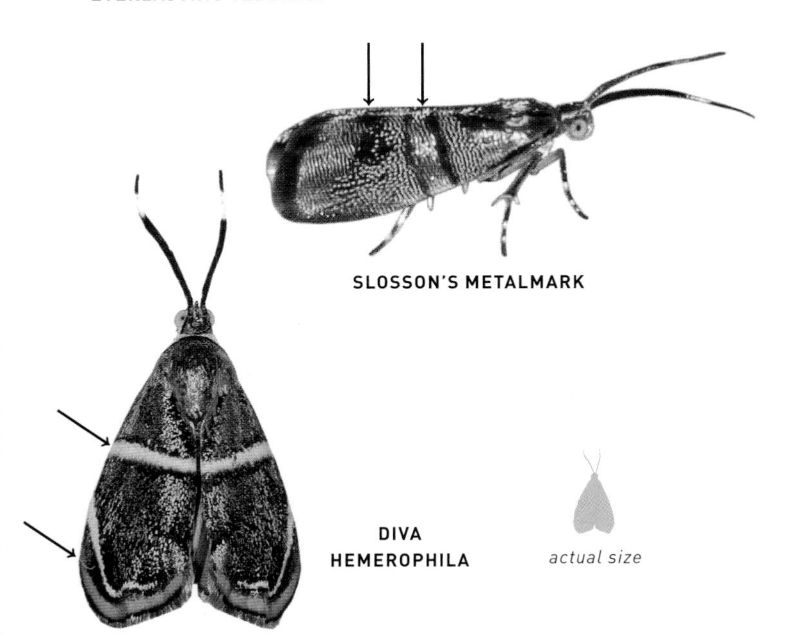

SLOSSON'S METALMARK

DIVA HEMEROPHILA

actual size

Mimosa Webworm Family Galacticidae

An introduced species native to east Asia, first discovered in 1943 in Washington, D.C., and now expanding its range through much of the e. U.S. Caterpillars live in communal webs in the fall. Adults are nocturnal and come to light.

MIMOSA WEBWORM

Homadaula anisocentra 60-0001 (2353) **Local**

TL 7–9 mm. Shiny gray FW is speckled with black dots that increase in size, but decrease in number, toward inner margin. **HOSTS:** Mimosa and honey locust. **RANGE:** Widespread south to n. FL, west to MS.

Tortrix Leafrollers Family Tortricidae, Subfamily Tortricinae, Tribes Tortricini and Cnephasiini

Small, flat moths with straight or sightly rounded wings that usually form a shallow point at the apex. Many species show a dark triangle on the costal edge of the FW. A few, such as Multiform Leafroller, are incredibly variable with many phenotypes. Larvae are typically leafrollers. Some species hibernate as adults and fly early the following spring. Adults are nocturnal and will visit lights in small numbers.

OAK LEAFSHREDDER

Acleris semipurpurana 62-0003 (3503) **Common**

TL 7–8 mm. Pale yellow FW typically has chocolate brown shading covering inner median area. Metallic AM and PM bands fade before reaching costa. Some individuals are paler with obscure light orange lines. **HOSTS:** Oak. **RANGE:** W. NC west to OK, south to FL and e. TX.

BLUEBERRY LEAFTIER

Acleris curvalana 62-0004 (3504) **Common**

TL 7–8 mm. Pale yellow FW typically has orange brown shading covering inner median area. Metallic gray AM and PM bands (if present) fade before reaching costa. Some individuals are paler with faint light orange lines or shading. **HOSTS:** Blueberry, huckleberry, oak, and rose. **RANGE:** W. NC west to AR, south to FL.

SNOWY-SHOULDERED ACLERIS

Acleris nivisellana 62-0010 (3510) **Local**

TL 7–8 mm. Marbled FW has conspicuous snow-white basal costal streak that forms a collar behind blackish head. Reddish tufts mark inner AM and ST areas. Slightly indented costa has a dark brown triangle at midpoint. **HOSTS:** Pin cherry. **RANGE:** W. NC and TN south to GA.

MIMOSA WEBWORM

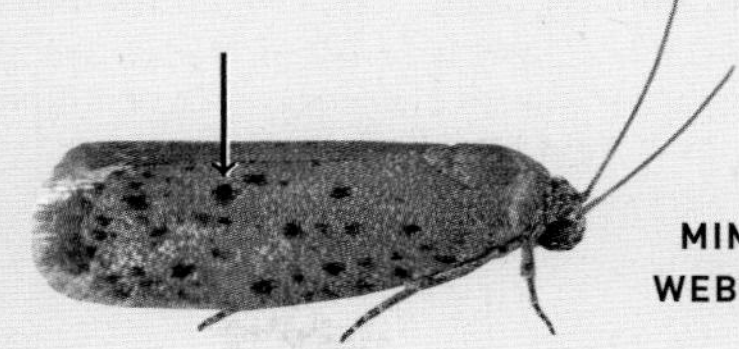

TORTRIX LEAFROLLERS

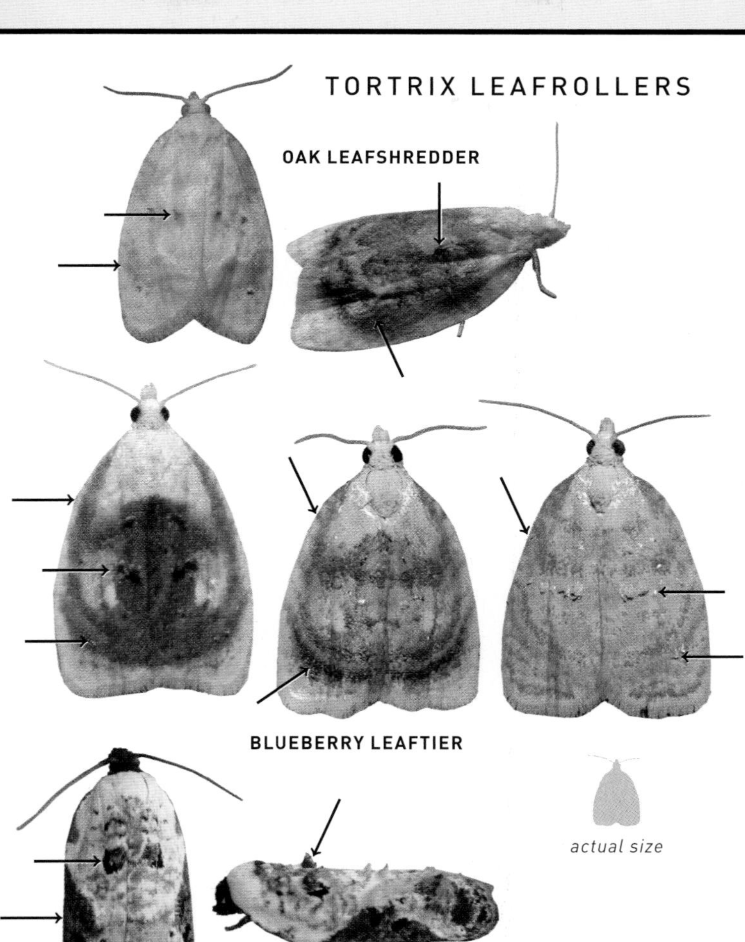

COMMON ACLERIS

Acleris subnivana 62-0016 (3517) **Uncommon**

TL 7–8 mm. White FW is variably mottled brown in median and ST areas. Note the grayish brown triangle (sometimes fragmented) at midpoint of indented costa. **HOSTS:** Red oak. **RANGE:** NC west to AR, south to s. MS.

SCHALLER'S ACLERIS

Acleris schalleriana 62-0027 (3527) **Local**

TL 8–10 mm. Light grayish brown FW usually has a brown triangle at median area. Some individuals are lightly speckled or streaked blackish. Obscure lines are accented with raised metallic dots. **HOSTS:** Viburnum. **RANGE:** W. NC and TN, south to SC.

LESSER MAPLE LEAFROLLER

Acleris chalybeana 62-0039 (3539) **Common**

TL 9–11 mm. Silvery gray FW is variably marked with indistinct blackish lines. Sometimes has a fragmented triangle at midpoint of costa. Some individuals have a blackish bar in inner basal area. **HOSTS:** Deciduous trees, including apple, beech, birch, maple, and oak. **RANGE:** W. NC west to AR, south to n. FL.

BLACK-HEADED BIRCH LEAFROLLER

Acleris logiana 62-0040 (3540) **Common**

TL 9–11 mm. Silvery gray FW is dotted with black-tipped tufts of raised dusky scales, especially along AM line. Blackish costal triangle (sometimes absent) is often reduced to a bar in central median area. **HOSTS:** Alder, birch, and viburnum. **RANGE:** W. NC west to MS, south to nw. FL.

MULTIFORM LEAFROLLER

Acleris flavivittana 62-0043 (3542) **Local**

TL 10–11 mm. Polymorphic. Gray or brown FW can have fine or bold white streaks along inner margin. Sometimes has a black basal dash. Some individuals are mostly black with mottled white AM band and basal area. **HOSTS:** Alder, birch, and viburnum. **RANGE:** W. NC and TN, south to n. GA.

STAINED-BACK LEAFROLLER

Acleris maculidorsana 62-0044 (3543) **Local**

TL 8–10 mm. Silvery gray FW is variably shaded warm brown in outer half as far as dotted AM line. Usually has a brown dash (sometimes absent) in inner basal area. **HOSTS:** Unknown. **RANGE:** E. NC west to AR, south to n. FL and s. MS.

GRAY-MARKED TORTRICID

Decodes basiplagana 62-0072 (3573) **Common**

TL 8–10 mm. Light gray FW is lightly brindled with black lines. Darker gray bands in basal, median, and ST areas are often fragmented. FW apex is rounded. **HOSTS:** Oak. **RANGE:** W. NC west to TN, south to coastal GA and s. MS.

TORTRIX LEAFROLLERS

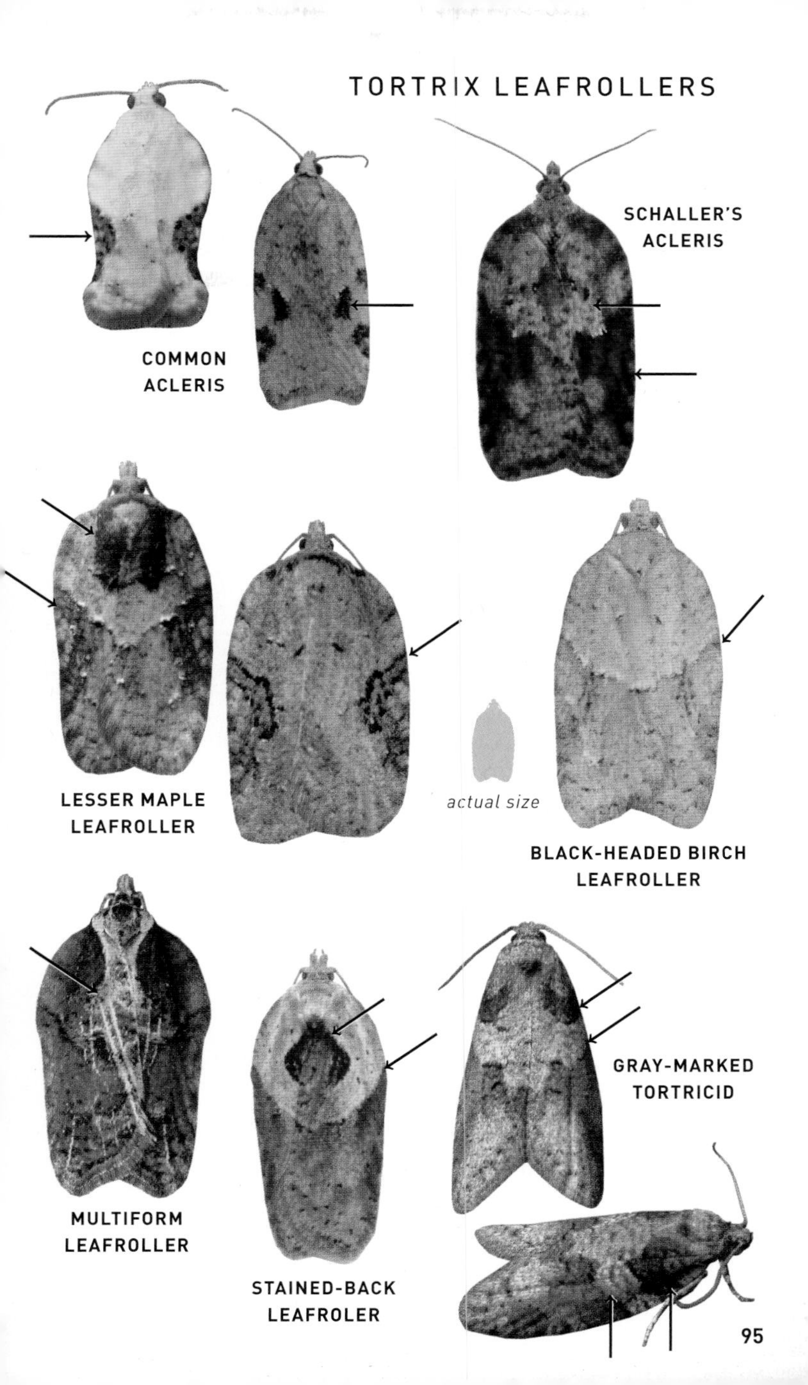

Cochylid Moths

Family Tortricidae, Subfamily Tortricinae, Tribe Cochylini

Small moths with flared wings that are folded against the body at rest. The head, with its fuzzy labial palps, is usually tucked downward, creating a snouty, hunchbacked appearance. Larvae are seed, flower, and stem borers. Adults are mostly nocturnal and will come to light in small numbers.

RAZOWSKI'S AETHES

Aethes razowskii 62-0112 (3759.3) **Local**

TL 6–9 mm. Cream-colored FW is boldly patterned with fragmented oblique brown bands in inner basal and PM areas. Bands are partly edged black and white. **HOSTS:** Unknown. **RANGE:** TN south to GA and s. AL.

SERIATED AETHES

Aethes seriatana 62-0114 (3760.1) **Common**

TL 6–7 mm. Cream-colored FW is boldly patterned with fragmented, oblique, warm brown bands in inner basal and PM areas. Bands are edged with silvery scales. **HOSTS:** Unknown. **RANGE:** NC west to OK, locally south to FL and e. TX.

BANDED SUNFLOWER MOTH

Cochylis hospes 62-0135 (3777) **Local**

TL 6–7 mm. Pale yellow FW has silvery blue studded, brown median band that widens toward inner margin. Incomplete ST line is dark brown. **HOSTS:** Developing seeds of sunflower. **RANGE:** NC west to OK, locally south to s. AL and e. TX.

TWO-SPOTTED COCHYLID

Eugnosta bimaculana 62-0144 (3763) **Common**

TL 7–8 mm. Light brown FW is boldly marked with incomplete chestnut brown bands in inner median and ST areas, forming two distinct spots. **HOSTS:** Unknown. **RANGE:** NC and TN south to FL and e. TX.

DECEPTIVE COCHYLID

Eugnosta deceptana 62-0149 (3789) **Local**

TL 7–8 mm. Peppery light gray FW is clouded with darker grayish brown shading beyond incomplete white PM line. A large rounded black patch covers central median area. **HOSTS:** Unknown. **RANGE:** S. TX.

FLEABANE COCHYLID

Eugnosta erigeronana 62-0150 (3790) **Common**

TL 6–7 mm. Strongly bicolored FW has a sharply demarcated whitish basal area. A broad black median band contrasts slightly with chestnut-spangled ST area. Some individuals have colors inversed, pale in distal half. **HOSTS:** Associated with fly galls in fleabane. **RANGE:** Coastal NC west to OK, south to FL and e. TX.

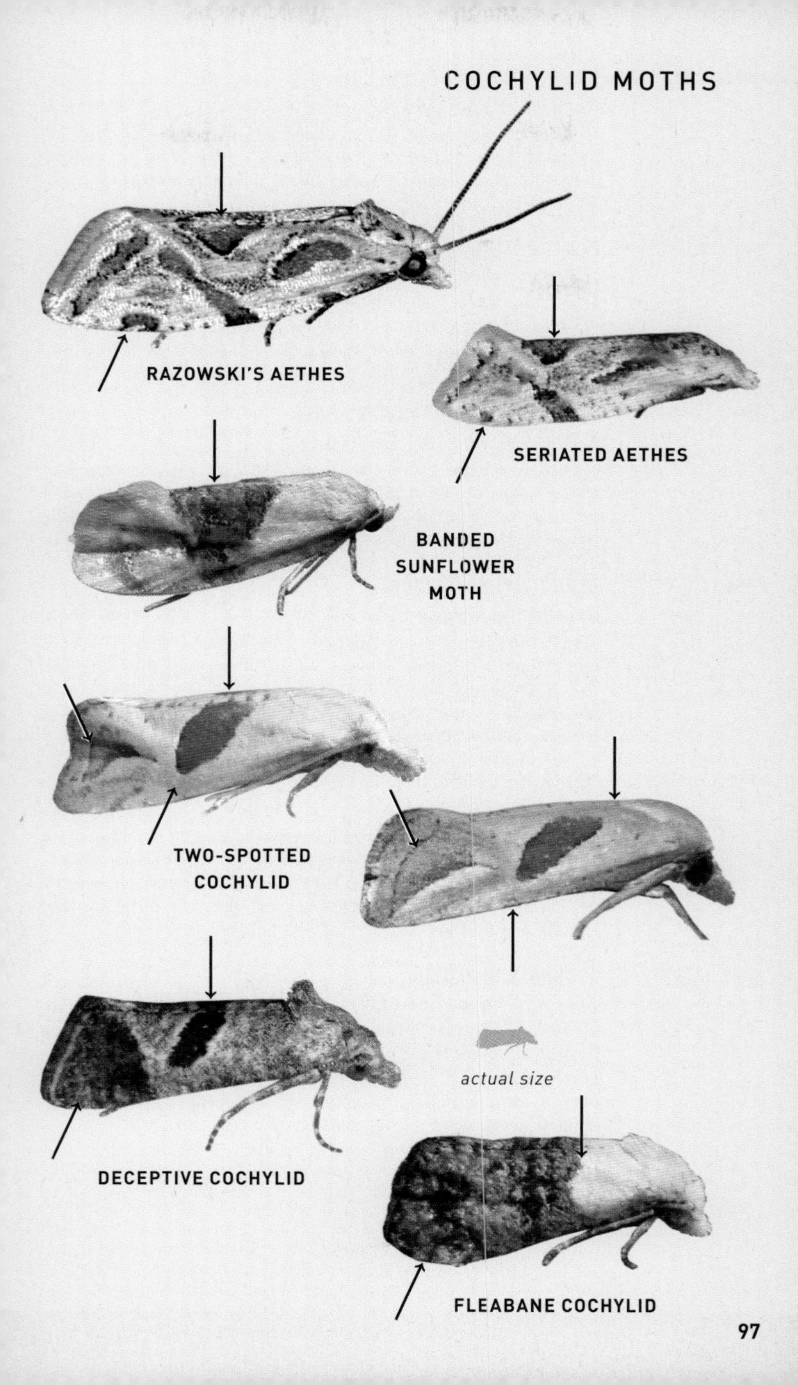
COCHYLID MOTHS
RAZOWSKI'S AETHES
SERIATED AETHES
BANDED SUNFLOWER MOTH
TWO-SPOTTED COCHYLID
actual size
DECEPTIVE COCHYLID
FLEABANE COCHYLID

BROWN-PATCHED COCHYLID

Carolella sartana 62-0152 (3764) **Common**

TL 6–8 mm. Similar to Two-spotted Cochylid, but median band is more complete, sometimes reaching costa. Usually has a smaller, chestnut brown subapical dash. **HOSTS:** Unknown. **RANGE:** NC and TN south to FL and e. TX.

CONTRASTING HENRICUS

Henricus contrastana 62-0158 (3796) **Local**

TL 6–7 mm. White FW has a silvery-banded black patch covering inner median and basal area. ST area is lightly patterned with gray bands. **HOSTS:** Associated with fly galls in fleabane. **RANGE:** W. NC and TN, south to n. FL.

CHRYSANTHEMUM FLOWER BORER

Lorita scarificata 62-0164 (3803) **Local**

TL 4–6 mm. Light brown FW is marbled with fragmented brown bands in AM, median, and ST areas. Note small black patches at inner margin, central median, and ST areas. **HOSTS:** Associated with fly galls in fleabane. **RANGE:** FL and s. AL.

BLACK-TIPPED RUDENIA

Rudenia leguminana 62-0200 (3839) **Uncommon**

TL 6–9 mm. Gray FW has a broad white AM band, sometimes extending into basal area. Median and ST areas are lightly brindled with blackish lines. Note tuft of brown scales at base of inner margin. **HOSTS:** Legumes, including acacia, locust, and mesquite. **RANGE:** Coastal SC west to OK, south to FL and TX.

PRIMROSE COCHYLID

[Atroposia] oenotherana 62-0213 (3848) **Uncommon**

TL 5–6 mm. Rose pink FW has a large yellow patch covering inner basal area. ST area is sometimes marked with a yellow spot of variable size. Fringe is yellowish. **HOSTS:** Unknown. **RANGE:** NC west to OK, south to FL and e. TX. **NOTE:** Previously placed in *Atroposia*, this species now belongs in its own, as yet unnamed, new genus.

FERRUGINOUS EULIA

Eulia ministrana 62-0227 (3565) **Local**

TL 9–11 mm. Rusty FW has a pale yellow costal streak that sweeps through ST area. Midpoint of inner margin is marked with a lilac patch. Rests with wings tented. **HOSTS:** Deciduous trees and shrubs, including birch, rose, and willow. **RANGE:** W. NC and TN, south to n. GA. **NOTE:** Introduced from Europe.

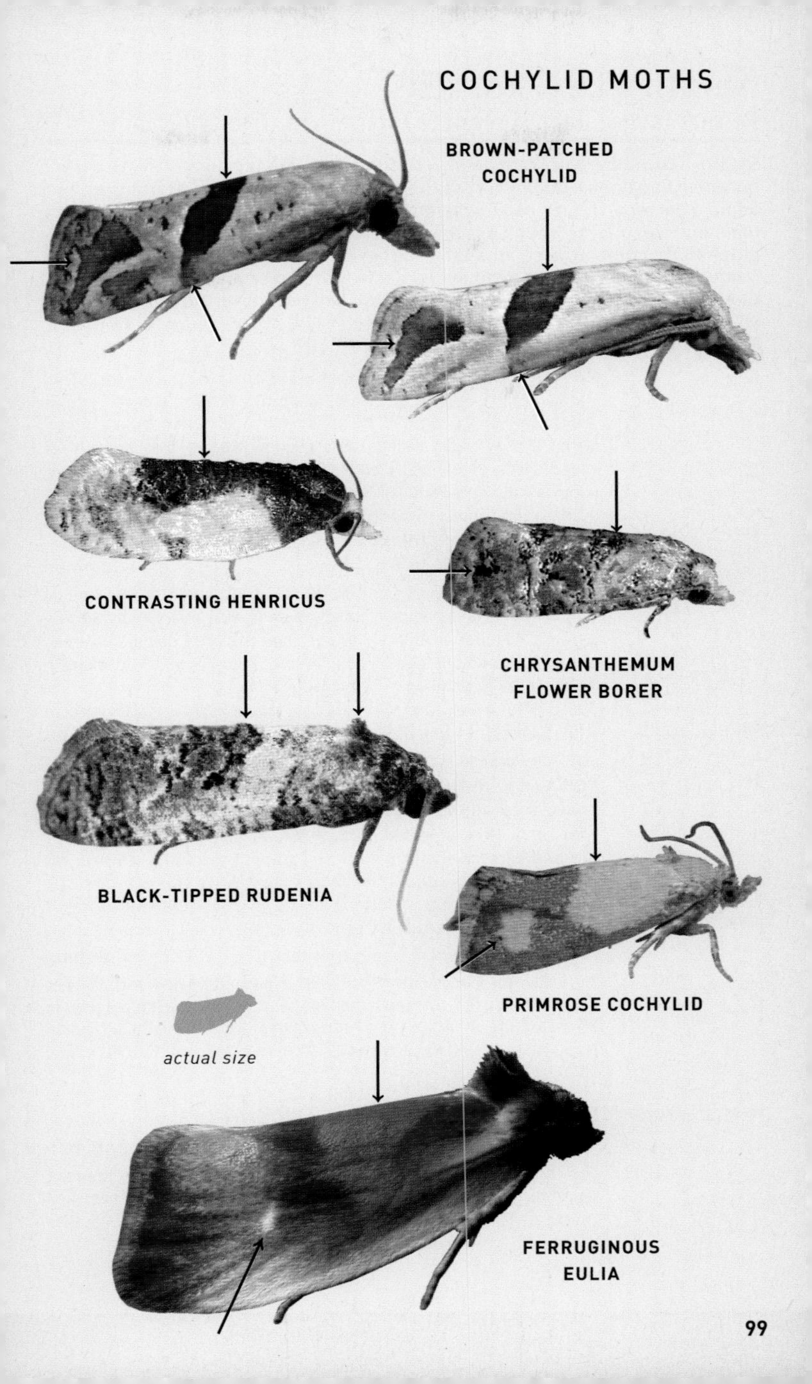
COCHYLID MOTHS
BROWN-PATCHED COCHYLID
CONTRASTING HENRICUS
CHRYSANTHEMUM FLOWER BORER
BLACK-TIPPED RUDENIA
PRIMROSE COCHYLID
actual size
FERRUGINOUS EULIA

Archips Leafrollers

Family Tortricidae, Subfamily Tortricinae, Tribe Archipini

Small, flat moths with rounded or curvy wings. Many species show darker bands or fine lines running across the wing, though some are mottled with no defined pattern. In many species, females are larger than males. Larvae are typically leafrollers, though a few, such as those of Southern Ugly-nest Caterpillar Moth, build large communal webs resembling those of tent caterpillars. Some species are serious crop pests, particularly in orchards or pine plantations. Adults are nocturnal and will come to lights in small numbers.

WOODGRAIN LEAFROLLER MOTH

Pandemis lamprosana 62-0248 (3593) **Common**

TL 10–12 mm. Slightly brindled, light brown FW has a darker brown basal area, oblique median band, and poorly defined subapical patch. Darker areas are edged pale yellow. **HOSTS:** Deciduous trees and shrubs, including apple, beech, chokecherry, hawthorn, oak, and sycamore. **RANGE:** NC west to OK, south to SC and cen. MS.

THREE-LINED LEAFROLLER

Pandemis limitata 62-0249 (3594) **Common**

TL 10–12 mm. Peppery light brown FW has a darker base, oblique median band, and subapical patch. Slightly wavy edges to darker areas are edged with thin yellowish lines. **HOSTS:** Deciduous trees and shrubs, including alder, apple, birch, maple, and oak. **RANGE:** NC west to OK, south to n. FL and LA.

RED-BANDED LEAFROLLER

Argyrotaenia velutinana 62-0255 (3597) **Common**

TL 7–9 mm. Whitish FW is mottled gray and brown in basal and ST areas, often with a blackish patch in inner basal area. Chestnut median band sometimes has a black bar at midpoint. **HOSTS:** Trees and shrubs, including apple, cherry, grape, and spruce. **RANGE:** NC west to OK, south to n. FL and LA.

HODGES' LEAFROLLER MOTH

Argyrotaenia hodgesi 62-0257 (3598.1) **Common**

TL 6–8 mm. Pinkish brown FW is boldly patterned with darker brown basal area, irregular median band, and two large patches in ST area. Note short spur projecting inward from midpoint of median band. **HOSTS:** Unknown. **RANGE:** SC west to MS, south to FL and e. TX.

FLORIDA LEAFROLLER

Argyrotaenia floridana 62-0258 (3599) **Common**

TL 7–9 mm. Similar to Red-banded Leafroller, but shiny FW appears more variegated with alternating wavy bands of light brown and chestnut. Paler individuals have weaker bands. **HOSTS:** Unknown. **RANGE:** NC west to AR, south to FL and e. TX.

ARCHIPS LEAFROLLERS

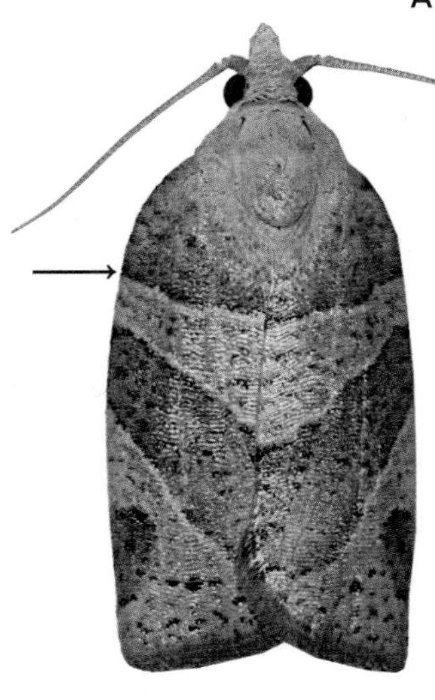

WOODGRAIN LEAFROLLER MOTH

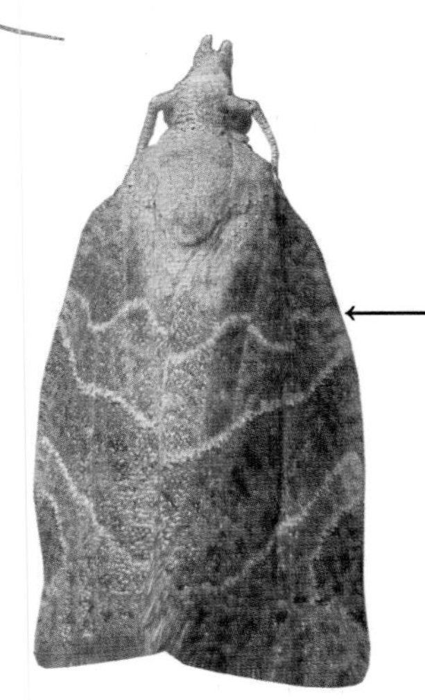

THREE-LINED LEAFROLLER

actual size

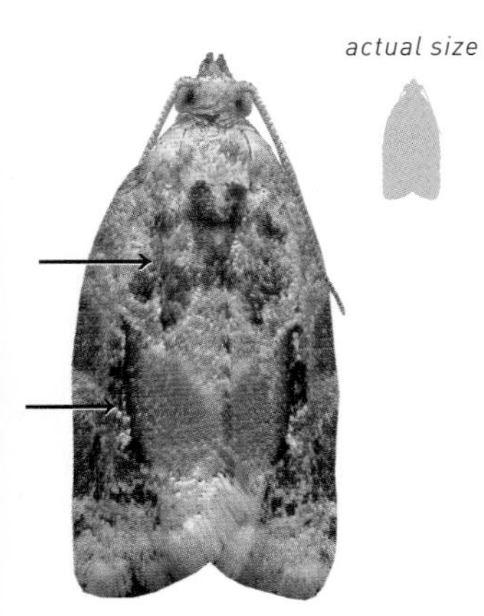

RED-BANDED LEAFROLLER

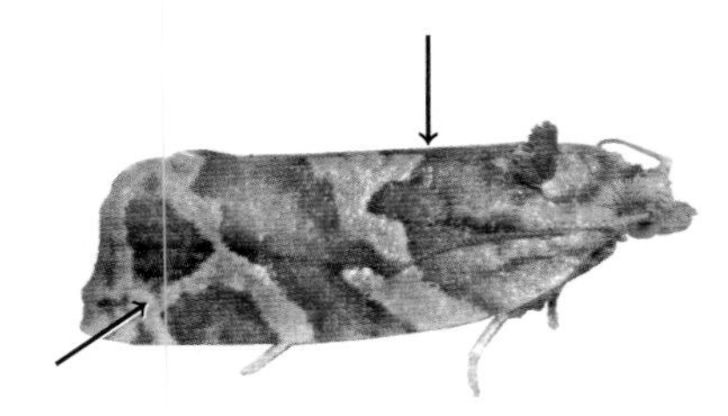

HODGES' LEAFROLLER MOTH

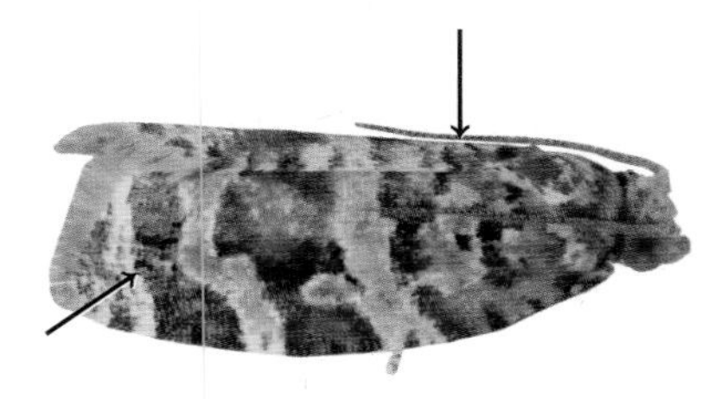

FLORIDA LEAFROLLER

KIMBALL'S LEAFROLLER

Argyrotaenia kimballi 62-0259 (3600) **Common**

TL 8–10 mm. Whitish FW is mottled black and chestnut in basal and ST areas. Chestnut median band becomes pale orange near inner margin. Fringe is pale orange. **HOSTS:** Unknown. **RANGE:** NC and TN south to FL and e. TX.

JACK PINE TUBE MOTH

Argyrotaenia tabulana 62-0262 (3603) **Common**

TL 8–10 mm. Pale orange FW has irregular white-edged outlines to rusty basal area, PM band, and fragmented ST band. Some individuals are more grayish overall. **HOSTS:** Pine. **RANGE:** NC west to e. OK, south to FL and LA.

HICKORY LEAFROLLER

Argyrotaenia juglandana 62-0281 (3622) **Common**

TL 9–12 mm. Peppery rusty brown FW has slightly oblique dark brown median and PM lines that widen toward costa. A small whitish spot at base of inner margin forms a distinct patch on closed wings. **HOSTS:** Hickory. **RANGE:** NC west to AR, south to FL and e. TX.

LINED OAK LEAFROLLER

Argyrotaenia quercifoliana 62-0282 (3623) **Common**

TL 9–13 mm. Cream-colored FW is heavily speckled light brown, creating a netlike pattern. Oblique brown AM and PM lines are usually obvious. PM and ST lines are joined at midpoint. **HOSTS:** Oak and witch hazel. **RANGE:** NC west to OK, south to FL and e. TX.

WHITE-SPOTTED LEAFROLLER

Argyrotaenia alisellana 62-0283 (3624) **Common**

TL 10–12 mm. Chocolate brown FW is boldly marked with large creamy patches in basal area, along costa, and in ST area. **HOSTS:** Oak. **RANGE:** NC west to OK, south to nw. FL and LA.

OBSOLETE LEAFROLLER

Choristoneura obsoletana 62-0296 (3631) **Common**

TL 11–13 mm. Light yellowish tan FW is faintly brindled with rusty brown lines. Brown median and ST bands are more obvious toward costa. Some individuals lack any obvious markings. **HOSTS:** Unknown. **RANGE:** NC west to OK, south to FL and LA.

BROKEN-BANDED LEAFROLLER

Choristoneura fractivittana 62-0297 (3632) **Common**

TL 11–14 mm. Yellowish tan FW is overlaid with a faint network of brown scales. Brown median band is faint or broken at midpoint. Males are usually smaller and darker than females. **HOSTS:** Apple, birch, elm, oak, and raspberry. **RANGE:** NC west to e. OK, south to n. FL and LA.

ARCHIPS LEAFROLLERS

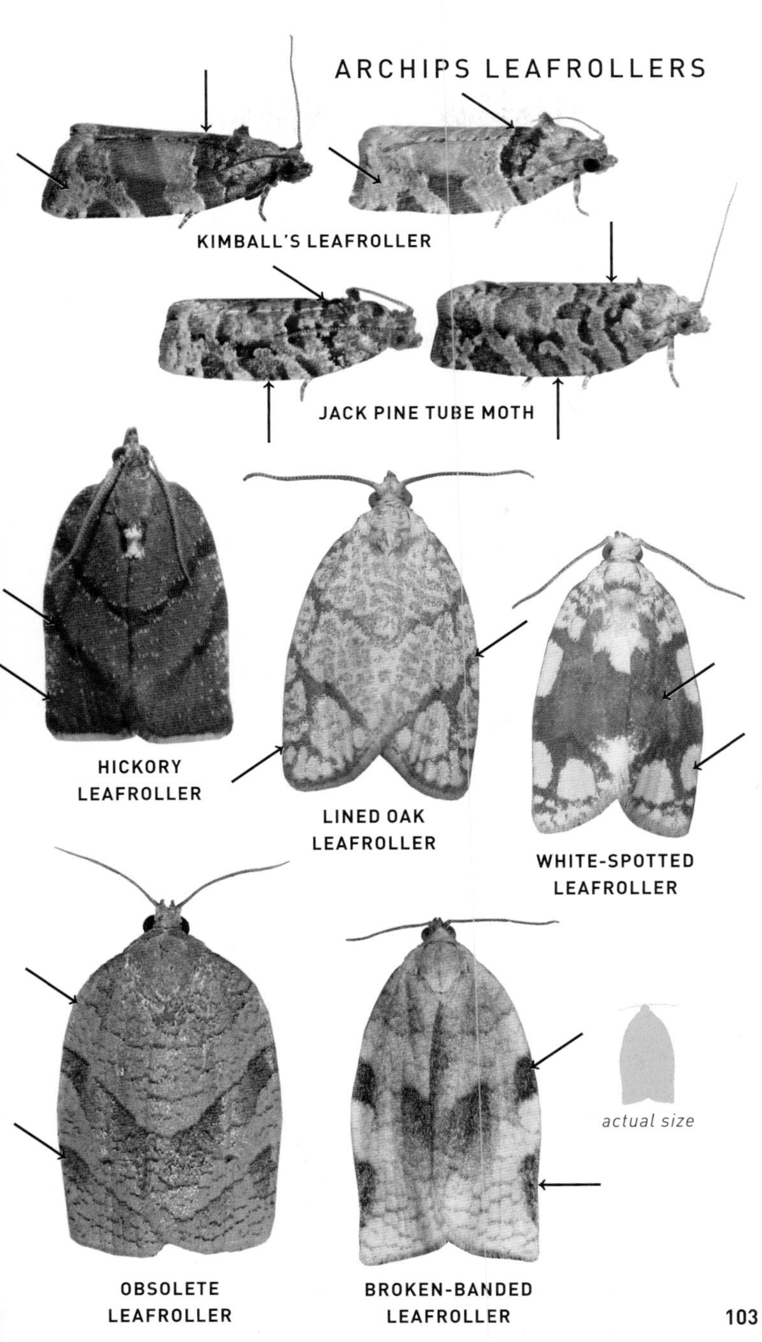

PARALLEL-BANDED LEAFROLLER

Choristoneura parallela 62-0298 (3633) **Common**

TL 11–14 mm. Bell-shaped, yellowish tan FW is faintly brindled with fine brown lines. Oblique AM, median, and PM lines are roughly parallel. Purplish brown median band is usually complete. **HOSTS:** Unknown. **RANGE:** NC and TN south to n. FL and MS.

OBLIQUE-BANDED LEAFROLLER

Choristoneura rosaceana 62-0300 (3635) **Common**

TL 8–14 mm. Resembles Parallel-banded Leafroller, but AM line is straight. Male has a costal fold near base of wing and is usually smaller than female. **HOSTS:** Woody plants, including apple, blueberry, oak, and pine. **RANGE:** NC west to OK, south to FL and e. TX.

JACK PINE BUDWORM

Choristoneura pinus 62-0308 (3643) **Common**

TL 8–15 mm. Pale orange FW is brindled with rusty lines. Irregular silvery gray median and ST bands (sometimes obvious only along costa) create a marbled pattern. **HOSTS:** Jack pine and Scotch pine. **RANGE:** NC and TN, locally south to n. FL and e. TX.

JUNIPER BUDWORM MOTH

Choristoneura houstonana 62-0313 (3647) **Common**

TL 10–12 mm. Light orange FW is marked with a netlike pattern of rusty brown lines, creating a marbled pattern. Obscure AM, PM, and ST bands are sometimes tinged silvery gray. **HOSTS:** Juniper. **RANGE:** TN west to OK, sw. to TX.

FRUIT-TREE LEAFROLLER

Archips argyrospila 62-0323 (3648) **Common**

TL 8–13 mm. Reddish brown FW is variably patterned with irregular bluish gray bands. Two cream-colored spots mark costa in AM and PM areas. **HOSTS:** Fruit-bearing trees and plants, including apple, blueberry, peach, and pear. **RANGE:** NC west to OK, south to FL and TX.

WHITE-SPOTTED OAK LEAFROLLER

Archips semiferanus 62-0329 (3653) **Common**

TL 9–12 mm. Pinkish brown FW is patterned with irregular whitish bands that are most obvious toward costa. Darker individuals have a more contrasting pattern. **HOSTS:** Oak; also apple and witch hazel. **RANGE:** Coastal NC west to OK, south to FL and e. TX.

GEORGIA LEAFROLLER

Archips georgiana 62-0331 (3656) **Common**

TL 9–12 mm. Light orange FW has a marbled pattern of brown lines and irregular pale bands. Fragmented brown median band is sometimes boldest marking. **HOSTS:** Unknown. **RANGE:** SC west to AR, south to FL and e. TX.

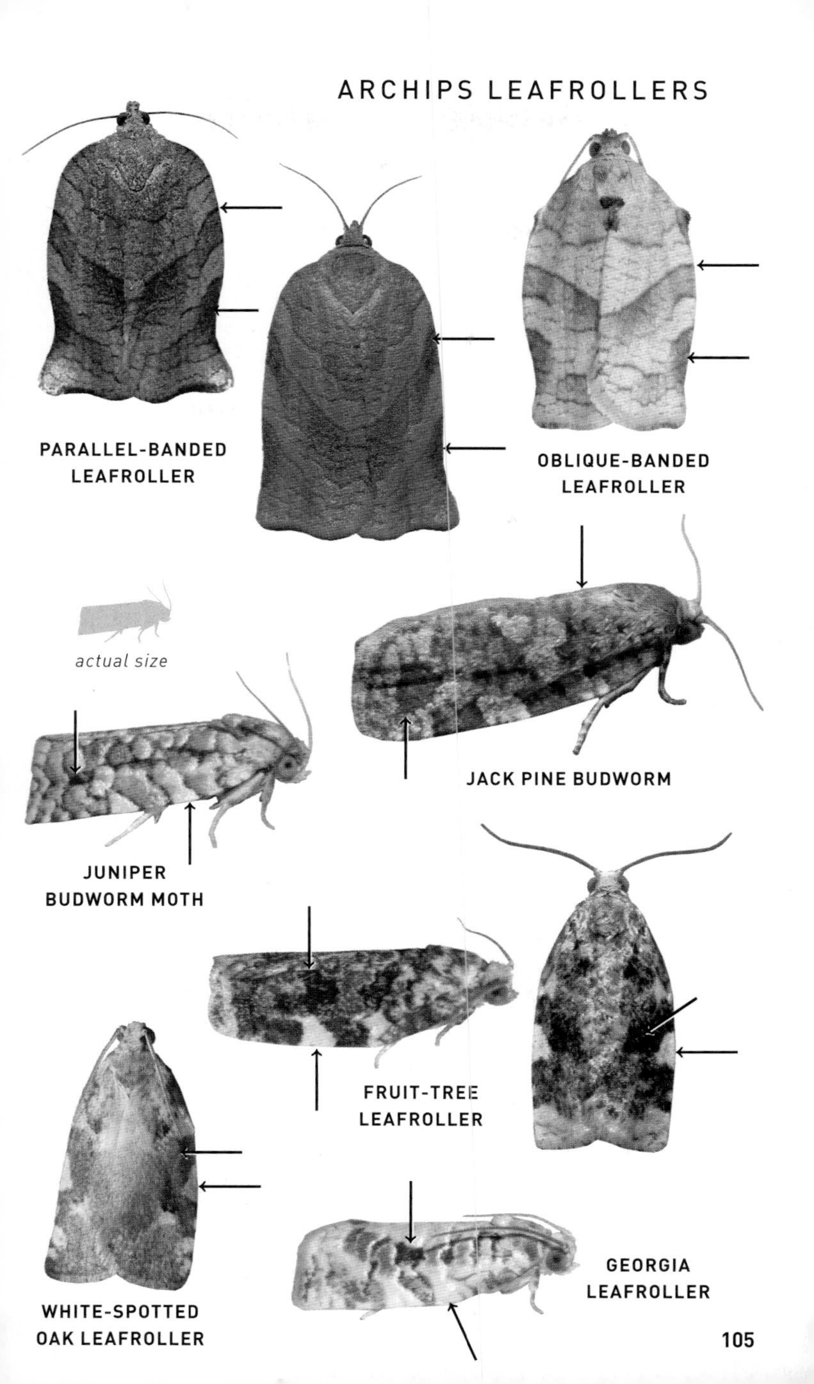

PARALLEL-BANDED LEAFROLLER

OBLIQUE-BANDED LEAFROLLER

JACK PINE BUDWORM

JUNIPER BUDWORM MOTH

FRUIT-TREE LEAFROLLER

GEORGIA LEAFROLLER

WHITE-SPOTTED OAK LEAFROLLER

GRAY ARCHIPS *Archips grisea* 62-0333 (3660) Common

TL 9–12 mm. Light gray FW is boldly patterned with black patches in inner basal, central median, and outer PM areas. Basal section of costa, inner median area, and ST band are tinged warm brown. **HOSTS:** Hickory. **RANGE:** NC west to OK, south to FL and e. TX.

SOUTHERN UGLY-NEST CATERPILLAR MOTH

Archips rileyana 62-0337 (3662) Common

TL 9–12 mm. Pinkish orange FW is lightly marked with chestnut spots along inconspicuous AM, median, and ST lines. Some individuals are mostly orange with spots evident only along costa. **HOSTS:** Woody plants, including hickory and buckeye. **RANGE:** W. NC and TN, south to n. FL and MS.

GARDEN TORTRIX *Clepsis peritana* 62-0364 (3688) Common

TL 6–8 mm. Light pinkish brown FW is lightly speckled darker. Note oblique brown median band and rounded blackish subapical patch. **HOSTS:** Low plants, including strawberry. **RANGE:** NC west to OK, south to FL and e. TX.

GREENISH APPLE MOTH

Clepsis virescana 62-0366 (3689) Common

TL 6–9 mm. Pale straw-colored FW is lightly brindled darker. Oblique brown median band is often poorly defined and fragmented. Rounded dusky subapical patch is usually most obvious marking. **HOSTS:** Fresh and decaying leaves of *Prunus* and *Rosa*. **RANGE:** W. NC west to OK, locally south to MS and TX.

SPARGANOTHID LEAFROLLERS Family Tortricidae, Subfamily Tortricinae, Tribe Sparganothini

Small moths, similar to other Tortricinae, but with longer labial palps that give them a snouty appearance. Many species are brightly colored, often yellowish with chestnut bands, streaks, or netlike patterns. The larvae are typically leafrollers. Adults are nocturnal and readily come to light.

WHITE-LINE LEAFROLLER

Amorbia humerosana 62-0374 (3748) Common

TL 12–17 mm. Gray FW is evenly speckled with blackish scales. Inner margin is often shaded chocolate brown. Sometimes has a dusky triangle along costa. **HOSTS:** Various trees and shrubs, including apple, conifers, huckleberry, poison ivy, and sumac. **RANGE:** NC west to AR, south to FL and e. TX.

THE BATMAN MOTH

Coelostathma discopunctana 62-0379 (3747) Common

TL 7–8 mm. Pale tan FW has gently curved, rusty AM and PM lines edged with warm brown shading. Note the tiny dot in central median area. Apex is pointed. **HOSTS:** Clover, impatiens, and strawberry. **RANGE:** NC west to OK, south to FL and e. TX.

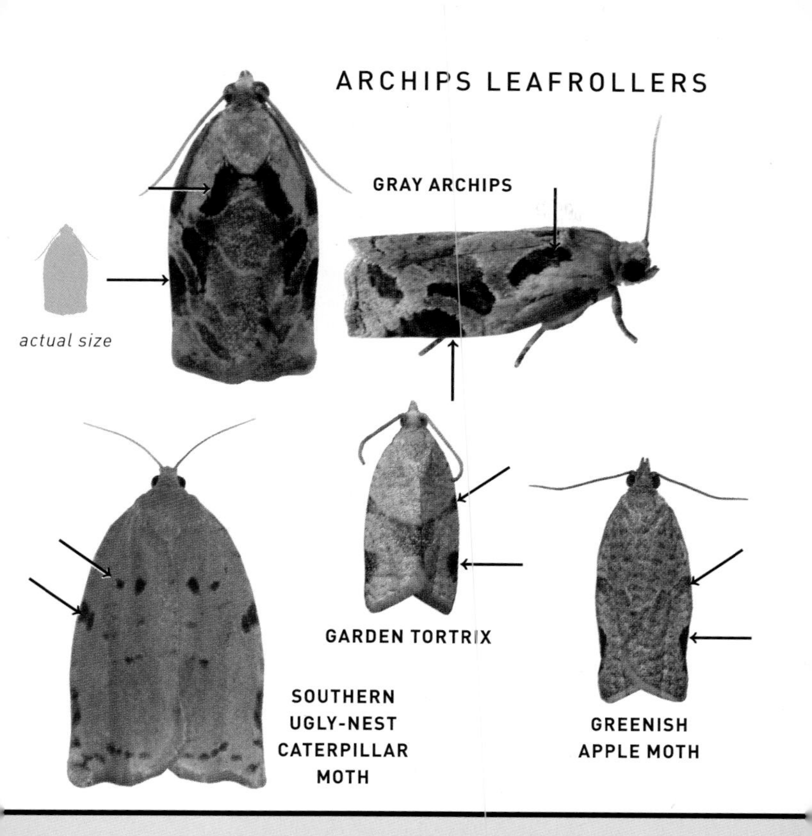
ARCHIPS LEAFROLLERS
GRAY ARCHIPS
actual size
GARDEN TORTRIX
SOUTHERN
UGLY-NEST
CATERPILLAR
MOTH
GREENISH
APPLE MOTH

SPARGANOTHID LEAFROLLERS
actual size
THE BATMAN
MOTH
WHITE-LINE
LEAFROLLER

LENTIGINOS MOTH

Sparganothoides lentiginosana 62-0384 (3731) **Common**

TL 7–8 mm. Brown FW is lightly sprinkled with darker brown scales. A blackish smudge in inner PM area is usually obvious but is reduced or lacking in paler individuals. **HOSTS:** Unknown. **RANGE:** NC west to OK, south to FL and e. TX.

NETTED SPARGANOTHIS

Sparganothis caryae 62-0386 (3700) **Common**

TL 8–11 mm. Pale yellow FW is boldly marked with chestnut lines, forming a distinctive netlike pattern. Markings are usually densest in central median area. HW is white. **HOSTS:** Unknown; possibly hickory. **RANGE:** SC west to OK, south to FL and e. TX.

SPARGANOTHIS FRUITWORM

Sparganothis sulfureana 62-0390 (3695) **Common**

TL 7–12 mm. Yellow FW is variably overlaid with a network of orange lines. If present, reddish AM and PM lines merge at inner margin, forming an X-shape on folded wings. **HOSTS:** Various trees and plants, including apple, clover, corn, cranberry, pine, and willow. **RANGE:** NC west to OK, south to FL and e. TX.

TWO-STRIPED SPARGANOTHIS

Sparganothis bistriata 62-0393 (3698) **Common**

TL 9–11 mm. Yellow FW has striking reddish streaks along costa and through central median area, which are studded with metallic silvery gray scales. All streaks end in PM area. **HOSTS:** Unknown; possibly conifers. **RANGE:** E. NC west to MS, south to FL and LA.

THREE-STREAKED SPARGANOTHIS

Sparganothis tristriata 62-0394 (3699) **Local**

TL 9–11 mm. Resembles Two-striped Sparganothis but has a third streak along inner margin. All streaks reach at least to ST area, usually to outer margin. **HOSTS:** Unknown. **RANGE:** NC and TN south to FL and MS.

DISTINCT SPARGANOTHIS

Sparganothis distincta 62-0412 (3704) **Common**

TL 9–12 mm. Pale orange to rusty FW has a faint netlike pattern of slightly raised paler scales. HW is orange brown. **HOSTS:** Goldenrod. **RANGE:** NC west to OK, south to FL and e. TX.

BEAUTIFUL SPARGANOTHIS

Sparganothis pulcherrimana 62-0413 (3701) **Common**

TL 12–14 mm. Chocolate brown FW has a large triangular pale yellow patch in outer median area with a smaller spot at base of inner margin. ST band is pale yellow. **HOSTS:** Hackberry. **RANGE:** SC west to OK, south to n. FL and e. TX.

SPARGANOTHID LEAFROLLERS

LENTIGINOS MOTH

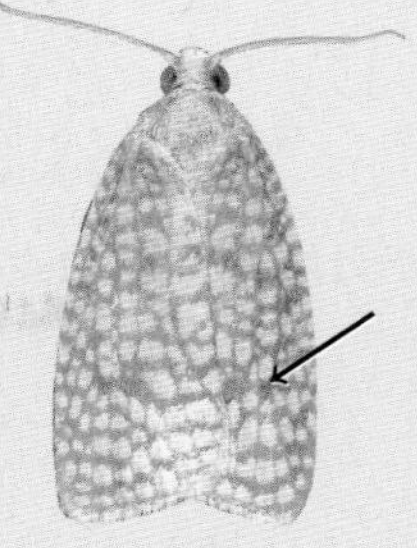

NETTED SPARGANOTHIS

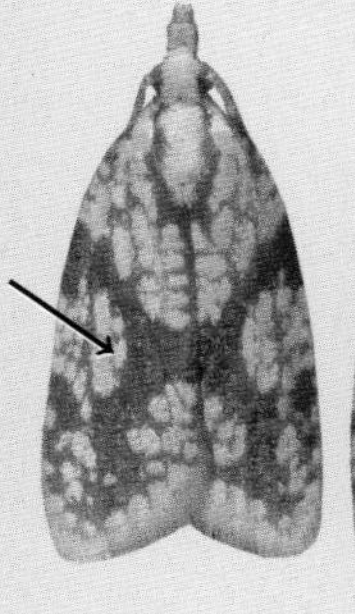

SPARGANOTHIS FRUITWORM

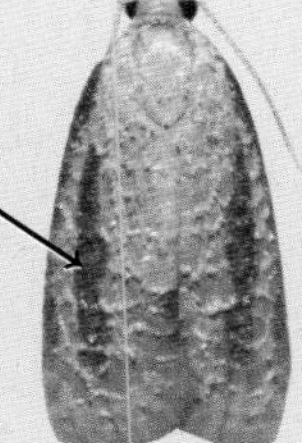

TWO-STRIPED SPARGANOTHIS

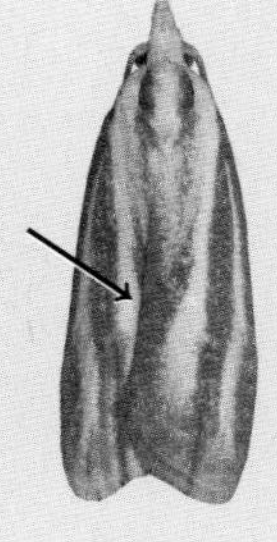

THREE-STREAKED SPARGANOTHIS

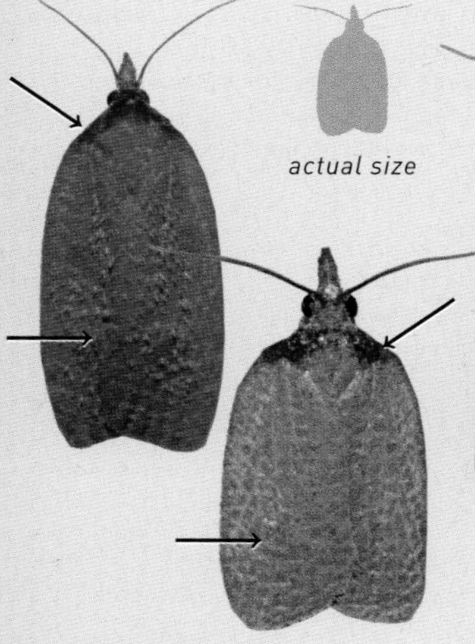

actual size

DISTINCT SPARGANOTHIS

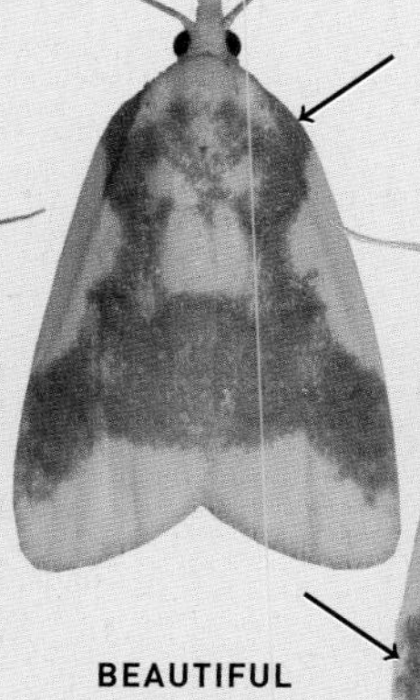

BEAUTIFUL SPARGANOTHIS

MAPLE-BASSWOOD LEAFROLLER

Cenopis pettitana 62-0417 (3725) **Common**

TL 10–15 mm. Shiny FW is variably whitish to pale yellow. Fragmented brown AM and PM lines usually widen toward costa but are sometimes absent. **HOSTS:** Deciduous trees, primarily basswood and maple. **RANGE:** Coastal NC west to OK, south to n. FL and e. TX.

APRONED CENOPIS

Cenopis niveana 62-0418 (3727) **Common**

TL 9–11 mm. Chocolate brown FW is typically patterned with pale yellow patches at base, midpoint of costa, and band along outer margin. Some individuals have inner median area mostly pale yellow. **HOSTS:** Eastern hornbeam, ironwood, and maple. **RANGE:** Coastal NC west to OK, south to cen. FL and e. TX.

RETICULATED FRUITWORM

Cenopis reticulatana 62-0419 (3720) **Common**

TL 9–12 mm. Pale yellow FW is overlaid with a fine orange netlike pattern. An oblique purplish brown median band is fused to thinner Y-shaped ST line. **HOSTS:** Various trees, shrubs, and low plants, including alder, apple, aster, blueberry, maple, and oak. **RANGE:** NC west to OK, south to n. FL and e. TX.

OAK CENOPIS *Cenopis diluticostana* 62-0423 (3716) **Common**

TL 7–8 mm. Shiny FW is variably light brown to chestnut. Darker purplish brown AM and PM bands cut through yellowish costal margin. **HOSTS:** Various deciduous trees, including beech, apple, ash, cherry, and oak. **RANGE:** NC west to MS, south to FL and e. TX.

CHOKECHERRY LEAFROLLER

Cenopis directana 62-0425 (3722) **Common**

TL 9–11 mm. Creamy yellow to light brown FW is variably overlaid with a faint orange netlike pattern. Note rusty brown smudges in AM and PM areas along costa (absent on some individuals). **HOSTS:** Chokecherry; also hickory in FL. **RANGE:** NC west to OK, south to FL and e. TX.

TUFTED APPLE BUD MOTH

Platynota idaeusalis 62-0433 (3740) **Common**

TL 10–13 mm. Silvery gray FW is variably shaded chocolate brown in basal area and along distal half of inner margin. Three lines of raised scales cross median area. **HOSTS:** Various trees and low plants, including apple, black walnut, box elder, clover, and pine. **RANGE:** NC west to OK, south to FL and e. TX.

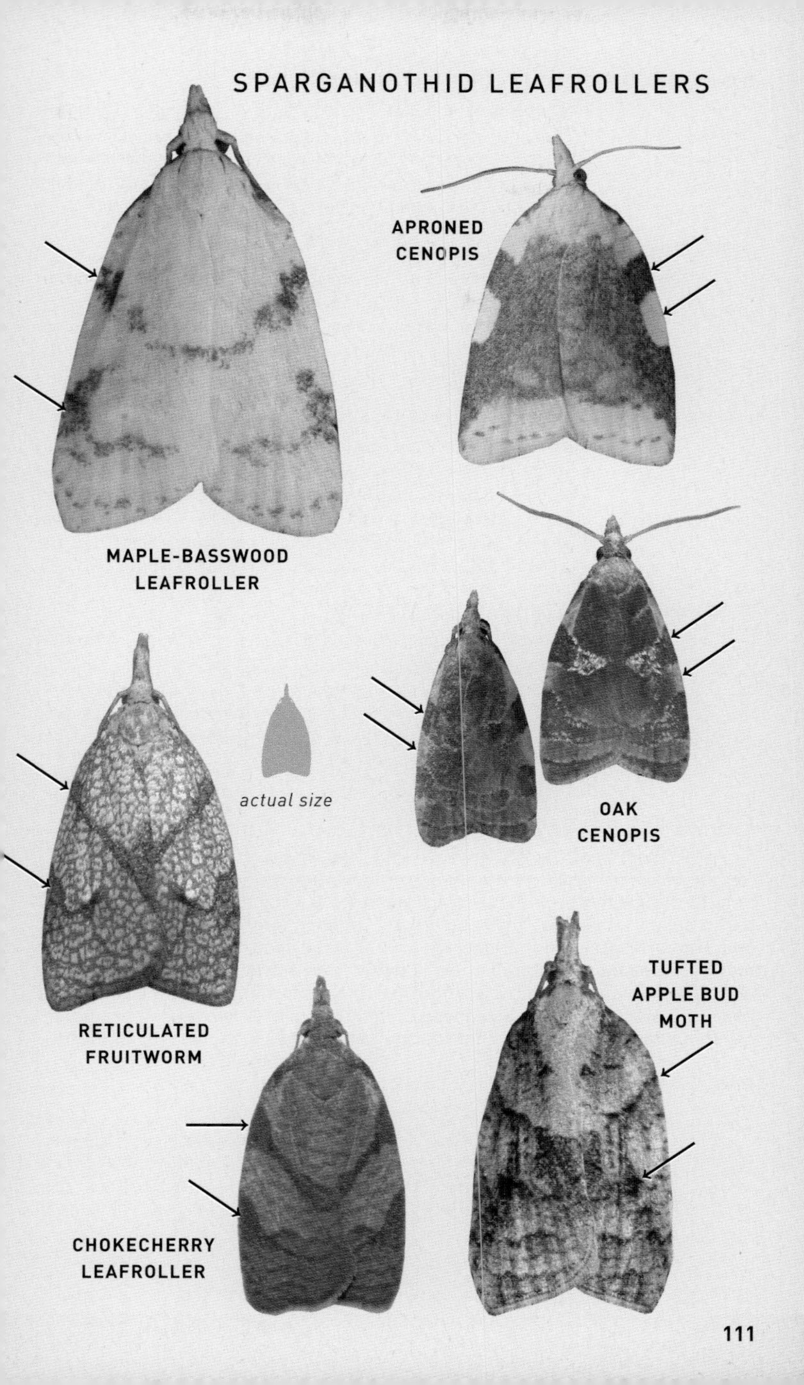
SPARGANOTHID LEAFROLLERS
APRONED
CENOPIS
MAPLE-BASSWOOD
LEAFROLLER
actual size
OAK
CENOPIS
RETICULATED
FRUITWORM
TUFTED
APPLE BUD
MOTH
CHOKECHERRY
LEAFROLLER

EXASPERATING PLATYNOTA

Platynota exasperatana 62-0434 (3743) **Common**

TL 7–9 mm. Light gray FW is brindled with black lines and variably shaded dark gray along costa and in median area. Often has whitish inner basal area and terminal line. **HOSTS:** Unknown; possibly grass. **RANGE:** NC west to OK, south to FL and e. TX.

SINGED PLATYNOTA

Platynota semiustana 62-0435 (3741) **Common**

TL 6–9 mm. Sooty FW has a contrastingly paler terminal line. Raised tufts of scales in median area are sometimes tipped whitish. Some individuals have a pale basal patch. **HOSTS:** Unknown. **RANGE:** NC and TN south to FL and e. TX.

OMNIVOROUS PLATYNOTA

Platynota rostrana 62-0436 (3745) **Common**

TL 7–10 mm. Light brown to reddish brown FW has oblique AM and PM lines edged with bands of darker shading that widen toward costa. Sometimes has a paler ST band. **HOSTS:** Avocado, bag-pod sesbania, grapefruit, orange, and others. **RANGE:** Coastal NC and TN, south to FL and e. TX.

BLACK-SHADED PLATYNOTA

Platynota flavedana 62-0443 (3732) **Common**

TL 7–10 mm. Light to reddish brown FW is variably shaded blackish in median area with a pale terminal line. Paler individuals have an oblique blackish AM band and patch in outer PM area. **HOSTS:** Low plants, including aster, blueberry, clover, and strawberry. **RANGE:** NC west to OK, south to FL and e. TX.

Olethreutine Moths

Family Tortricidae, Subfamily Olethreutinae

Small tortricid moths with a FW that narrows toward the base, creating a tapered appearance. They often hold their wings in a tent or rolled tight to the body, imparting a tubular appearance. The raised head posture is distinct, and many species have fuzzy labial palps that create a "snout." Larvae are either leafrollers or leaftiers, but a few bore into roots and stems of their woody plant hosts. Some, such as the introduced Codling Moth, are crop pests of commercial importance. Adults will come to light, sometimes in moderate numbers.

WHITE-DOTTED CRYPTASPASMA

Cryptaspasma bipenicilla 62-0459 (2704.1) **Local**

TL 10–12 mm. Dark purplish brown FW has a blackish wedge in inner median area and shading along ST line. An orange-ringed white discal dot is most obvious marking. **HOSTS:** Possibly *Persea* species, such as redbay and swamp bay. **RANGE:** Coastal NC south to FL and s. MS.

SPARGANOTHID LEAFROLLERS

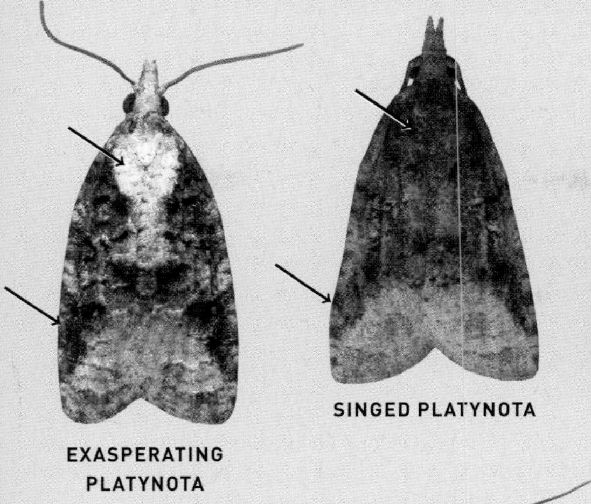

EXASPERATING PLATYNOTA

SINGED PLATYNOTA

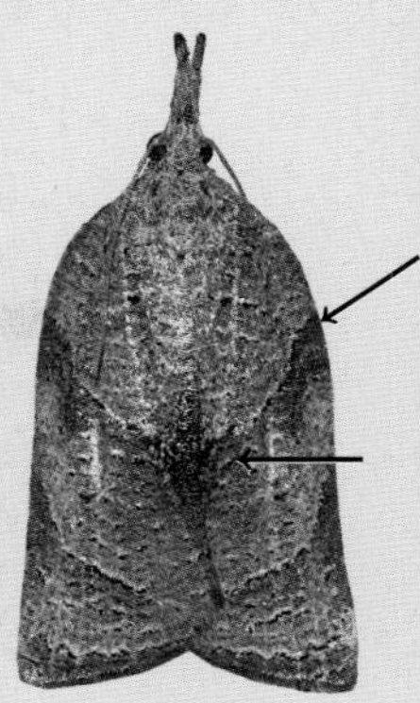

OMNIVOROUS PLATYNOTA

actual size

BLACK-SHADED PLATYNOTA

OLETHREUTINE MOTHS

actual size

WHITE-DOTTED CRYPTASPASMA

VERBENA BUD MOTH

Endothenia hebesana 62-0466 (2738) **Common**

TL 6–9 mm. Light brown FW has a marbled pattern of gray and brown. Note the two-pronged dusky median band. Raised thoracic tuft is bright chestnut. **HOSTS:** Seeds of low plants, including iris, speedwell, and vervain. **RANGE:** NC west to OK, south to FL and e. TX.

IMPUDENT HULDA

Hulda impudens 62-0475 (2747) **Common**

TL 6–8 mm. Pale pinkish FW has a contrasting dark gray basal patch that extends to midpoint of costa. Inner median area is marked with a two-lobed warm brown patch. **HOSTS:** Unknown. **RANGE:** W. NC and TN, southwest to MS.

RUSH BACTRA *Bactra furfurana* 62-0477 (2706) **Common**

TL 6–8 mm. Light brown FW has a marbled pattern of pale-edged dusky lines. Darker brown median band curves outward toward inner ST area. **HOSTS:** Rush and bulrush. **RANGE:** TN west to OK, south to n. FL and e. TX.

JAVELIN MOTH *Bactra verutana* 62-0478 (2707) **Common**

TL 6–10 mm. Pale tan FW is often tinted reddish brown along inner margin. Central median area is marked with two obvious blackish blotches. **HOSTS:** Flatsedge. **RANGE:** NC west to OK, south to FL and e. TX.

MAPLE TIP BORER *Episimus tyrius* 62-0483 (2703) **Common**

TL 6–7 mm. Light orange FW has outer basal and median areas shaded chestnut. Basal half of inner margin is mottled white, forming a distinct streak. **HOSTS:** Maple. **RANGE:** NC and TN south to FL and LA.

SUMAC LEAFTIER

Episimus argutana 62-0485 (2701) **Common**

TL 7–8 mm. Light brown FW is tinted reddish in distal half and finely patterned with black and silvery gray brindling. **HOSTS:** Spurge, poison ivy, sumac, and witch hazel. **RANGE:** W. NC west to OK, south to FL and LA.

TULIP-TREE LEAFTIER

Paralobesia liriodendrana 62-0493 (2711) **Common**

TL 6–8 mm. Gray FW has an orange basal patch. Shiny bronze median band ends in a point at inner margin, and blunted projection almost touches a bronze patch in central ST area. **HOSTS:** Tulip tree; also southern magnolia and sweet bay. **RANGE:** W. NC and TN, south to FL and LA.

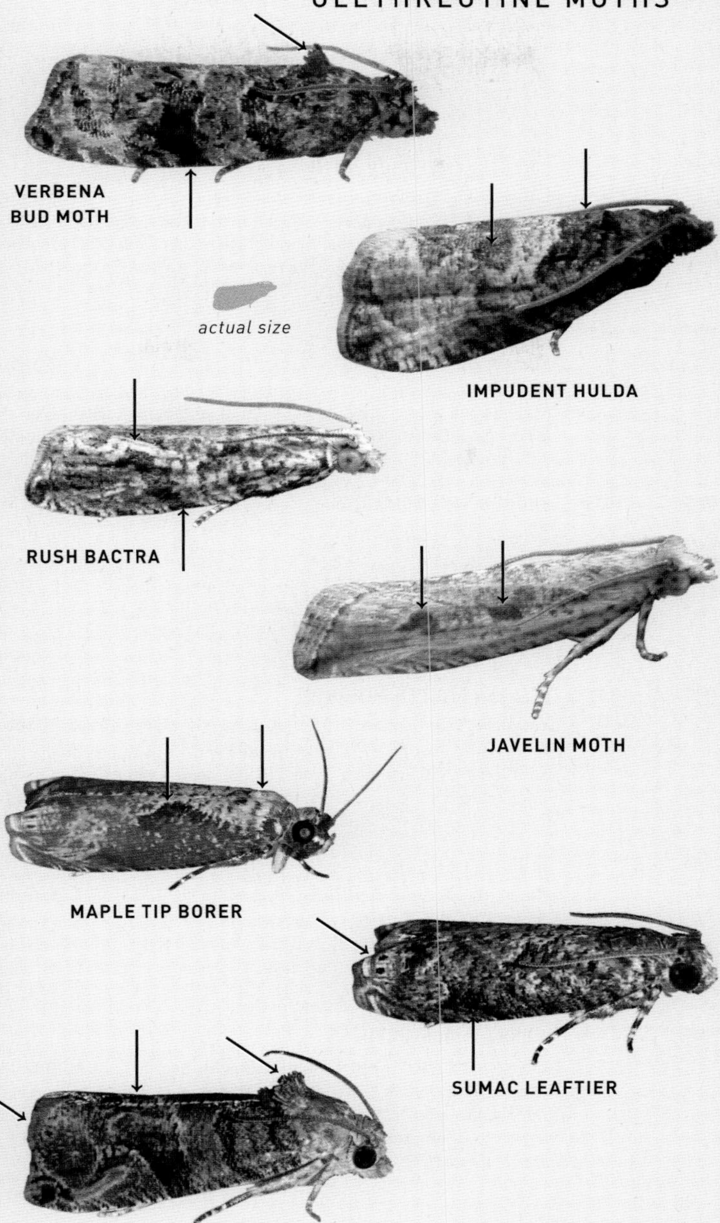
VERBENA BUD MOTH
actual size
IMPUDENT HULDA
RUSH BACTRA
JAVELIN MOTH
MAPLE TIP BORER
SUMAC LEAFTIER
TULIP-TREE LEAFTIER

GRAPE BERRY MOTH

Paralobesia viteana 62-0494 (2712) **Common**

TL 4–6 mm. Similar to larger Tulip-tree Leaftier, but projection of bronze median band is usually pointed, and patch in ST area is often irregular or fragmented. **HOSTS:** Grape. **RANGE:** TN south to FL and LA.

SWEETBAY SEEDPOD MOTH

Paralobesia cyclopiana 62-0495 (2727) **Common**

TL 7–9 mm. Brindled gray FW has a brown basal patch and straw-colored distal half. Note distinctive black and brown eyespot in central ST area. **HOSTS:** Seedpods of sweet bay. **RANGE:** SC south to FL and MS.

SCULPTURED MOTH

Eumarozia malachitana 62-0517 (2749) **Common**

TL 7–8 mm. Lilac gray FW is tinged rosy red distally. Large, rounded, white-edged mossy green median patch (sometimes fawn-colored) is conspicuous. **HOSTS:** Persimmon. **RANGE:** NC west to OK, south to FL and e. TX.

BROKEN-LINE ZOMARIA

Zomaria interruptolineana 62-0518 (2750) **Common**

TL 6–8 mm. Straw-colored FW is tinged orange along inner margin. Oblique chestnut AM band ends at midpoint. A dark gray patch marks midpoint of costa. Note thin black lines in central median and ST areas. **HOSTS:** Blueberry and huckleberry. **RANGE:** NC and TN south to FL and s. AL.

YELLOW-STRIPED ZOMARIA

Zomaria rosaochreana 62-0519 (2751) **Local**

TL 6–7 mm. Pinkish red FW is streaked with chestnut through central median and subapical areas. Bright golden yellow stripe along inner margin ends abruptly in PM area. **HOSTS:** Unknown. **RANGE:** Coastal NC south to GA and FL.

LABYRINTH MOTH

Phaecasiophora niveiguttana 62-0541 (2772) **Common**

TL 7–9 mm. Cream-colored FW has a finely etched pattern of irregular black and orange lines. Fragmented AM, median, and ST bands are filled with brown shading. **HOSTS:** Sassafras. **RANGE:** NC and TN south to FL and MS.

INORNATE OLETHREUTES

Olethreutes inornatana 62-0557 (2788) **Common**

TL 9–11 mm. Finely brindled pale FW has dusky shading in inner basal area, along costa (especially at midpoint), and in ST area. Note raised chestnut thoracic tuft. **HOSTS:** Black cherry and chokecherry. **RANGE:** W. NC and TN, south to cen. AL and MS.

OLETHREUTINE MOTHS

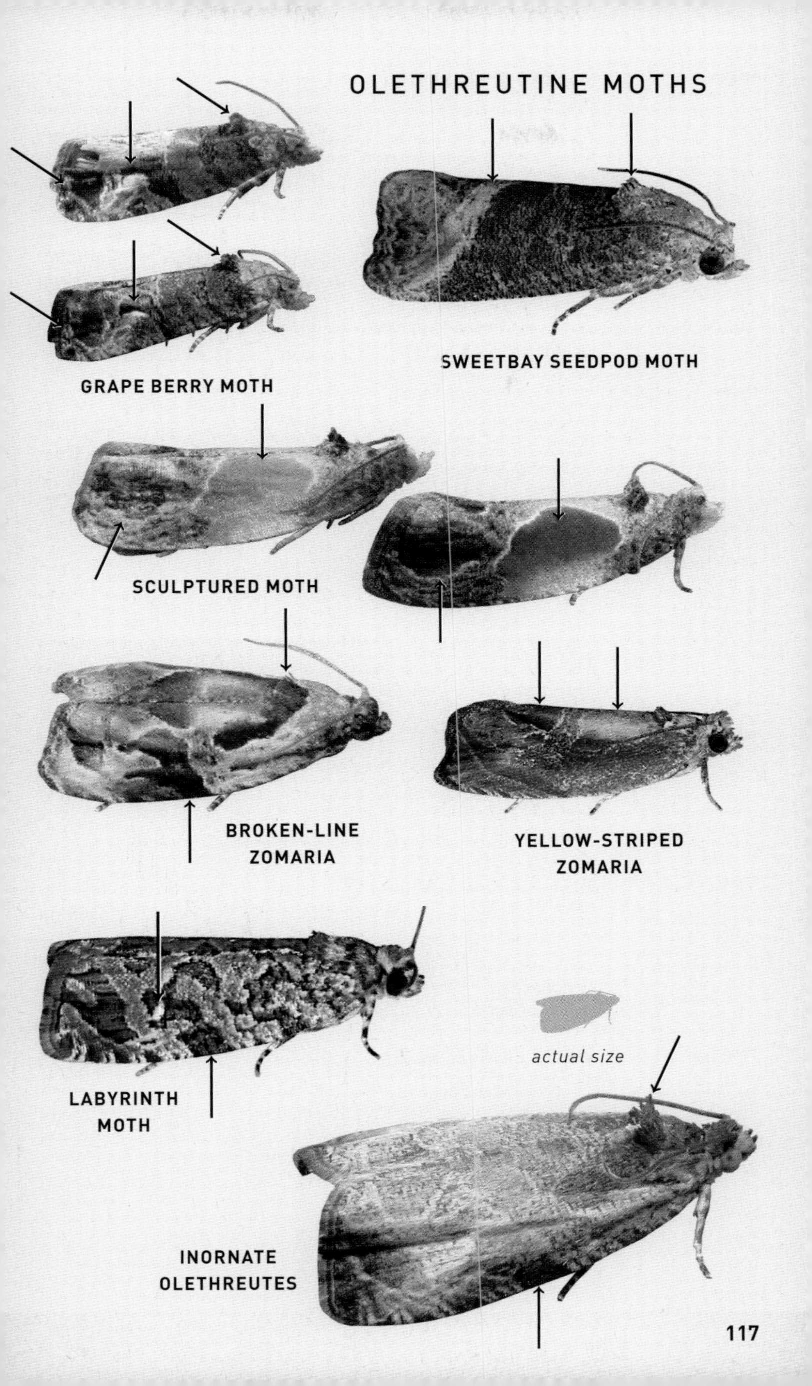

BANDED OLETHREUTES

Olethreutes fasciatana 62-0591 (2823) **Common**

TL 6–9 mm. Dark brown FW has bronzy olive highlights along silvery lines. Note broad white AM band and mottling in ST area. **HOSTS:** Poplar and willow. **RANGE:** NC and TN south to n. FL and AL.

HYDRANGEA LEAFTIER

Olethreutes ferriferana 62-0595 (2827) **Common**

TL 8–10 mm. Silvery, violet gray FW has contrasting chestnut basal area and trapezoidal patch in outer median area that slants toward outer margin. **HOSTS:** Wild hydrangea. **RANGE:** W. NC and TN, south to cen. AL and MS.

PUTTY-PATCHED MOTH

Olethreutes griseoalbana 62-0596 (2828) **Common**

TL 7–8 mm. Whitish FW has contrasting black patches in inner basal area (including raised thoracic tuft), midpoint of costa, and inner ST area. Note contrasting olive patch in inner median area. **HOSTS:** Unknown. **RANGE:** NC and TN south to n. FL and AL.

FERN OLETHREUTES

Olethreutes osmundana 62-0597 (2829) **Local**

TL 6–8 mm. Variably light to dark grayish brown FW has a large, irregularly shaped orange patch in inner half of median area. Some individuals are entirely mottled orange outside of patch. **HOSTS:** Ferns. **RANGE:** TN and SC, south to cen. FL and AL.

PIED OLETHREUTES

Olethreutes devotana 62-0622 (2857) **Local**

TL 8–10 mm. Blackish FW has a contrasting white patch in inner median area, forming a saddle when wings are folded. ST area is mostly whitish. **HOSTS:** Unknown. **RANGE:** Cen. GA south to FL.

PINK-WASHED LEAFROLLER

Hedya separatana 62-0634 (2860) **Common**

TL 6–8 mm. Pinkish FW has blue gray shading in between irregular black AM and median bands. Isolated black spot in central PM area is noticeable. Head is blackish. **HOSTS:** Black cherry, blackberry, and rose. **RANGE:** W. NC and TN, south to nw. FL and AL.

WHITE-SPOTTED HEDYA

Hedya chionosema 62-0637 (2863) **Local**

TL 7–8 mm. Silvery gray FW has a bold white semicircle at midpoint of costa. Obscure brown-tinged AM, median, and ST bands are partly edged black. Can sometimes be found settled on food plants during daytime. **HOSTS:** Apple, crab apple, and hawthorn. **RANGE:** W. NC and TN, south to n. FL.

OLETHREUTINE MOTHS

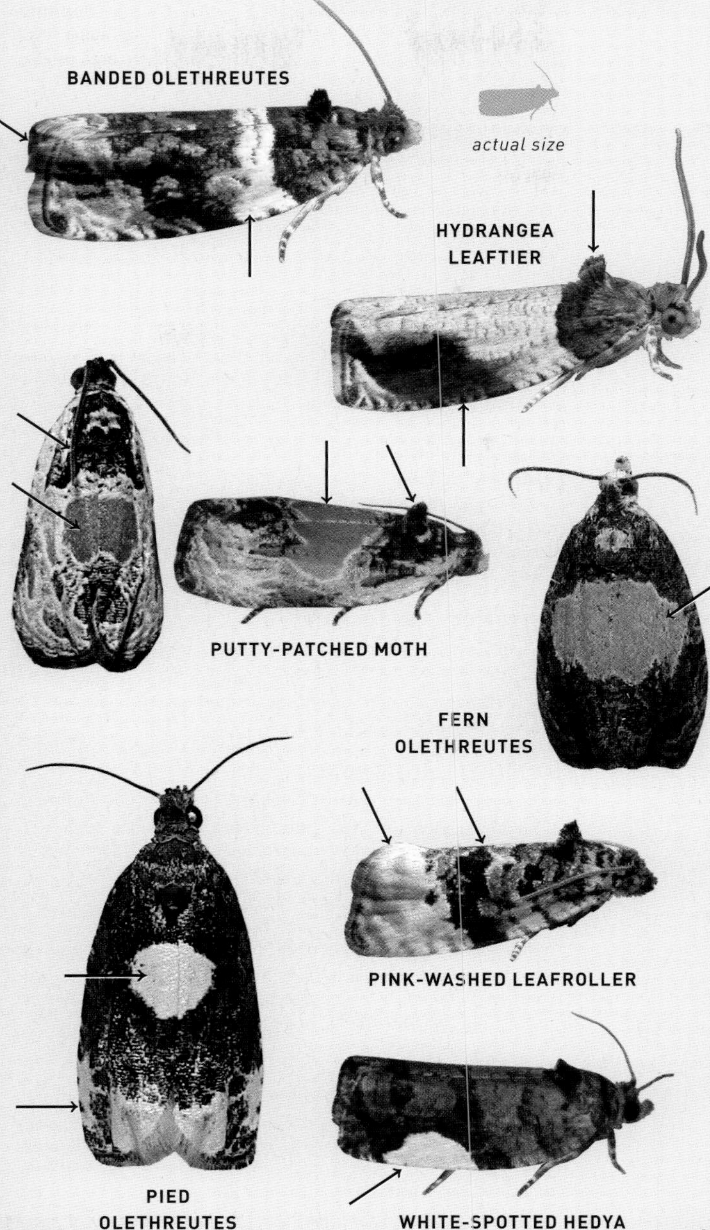

SYCAMORE LEAFFOLDER

Ancylis platanana 62-0658 (3370) **Common**

TL 6–8 mm. Fawn-colored FW is lightly striated with rusty lines. Midpoint of whitish costa is marked with oblique rusty bar. **HOSTS:** Sycamore. **RANGE:** NC west to OK, south to FL and LA.

STRAWBERRY LEAFFOLDER

Ancylis comptana 62-0662 (3374) **Common**

TL 6–8 mm. Fawn-colored FW has contrasting rusty brown inner basal area edged with oblique white line. Inner ST area is streaked silver. **HOSTS:** *Rubus* species and strawberry. **RANGE:** NC west to OK, south to FL and e. TX.

TWO-TONED ANCYLIS

Ancylis divisana 62-0663 (3375) **Common**

TL 6–8 mm. Pale orange FW has silvery gray basal half of costa. White-edged inner basal area is rusty brown. **HOSTS:** Oak and sycamore. **RANGE:** NC and TN south to FL and e. TX.

PITCH PINE TIP MOTH

Rhyacionia rigidana 62-0695 (2868) **Common**

TL 7–10 mm. Brindled orange FW has pale median area and is crossed by faint silvery gray double lines (often raised when fresh). Head is whitish. Note slightly flared wing shape. **HOSTS:** Pine. **RANGE:** TN and SC south to FL and MS.

SUBTROPICAL PINE TIP MOTH

Rhyacionia subtropica 62-0696 (2869) **Local**

TL 6–10 mm. Resembles Pitch Pine Tip Moth but is usually whiter with reduced orange brindling and fainter silvery lines. **HOSTS:** Pine. **RANGE:** SC south to FL and s. AL.

NANTUCKET PINE TIP MOTH

Rhyacionia frustrana 62-0710 (2882) **Common**

TL 5–8 mm. Resembles Pitch Pine Tip Moth, but median area is orange, and silvery lines are stronger; lines may be fragmented or solid. **HOSTS:** Pine. **RANGE:** NC and TN south to FL and AL.

GRAY RETINIA

Retinia gemistrigulana 62-0727 (2898) **Uncommon**

TL 9–11 mm. Ash gray FW is finely brindled with crisp black lines. Slightly darker gray AM and PM bands are sometimes evident. **HOSTS:** Unknown. **RANGE:** NC and TN south to FL and s. MS.

BAYBERRY LEAFTIER

Strepsicrates smithiana 62-0737 (2907) **Common**

TL 6–8 mm. Brown FW has a striate appearance. Silvery gray streak along inner margin has a jagged edge bordered with thin black lines. Slanting black apical dash extends into central PM area. **HOSTS:** Bayberry. **RANGE:** E. NC south to FL and LA.

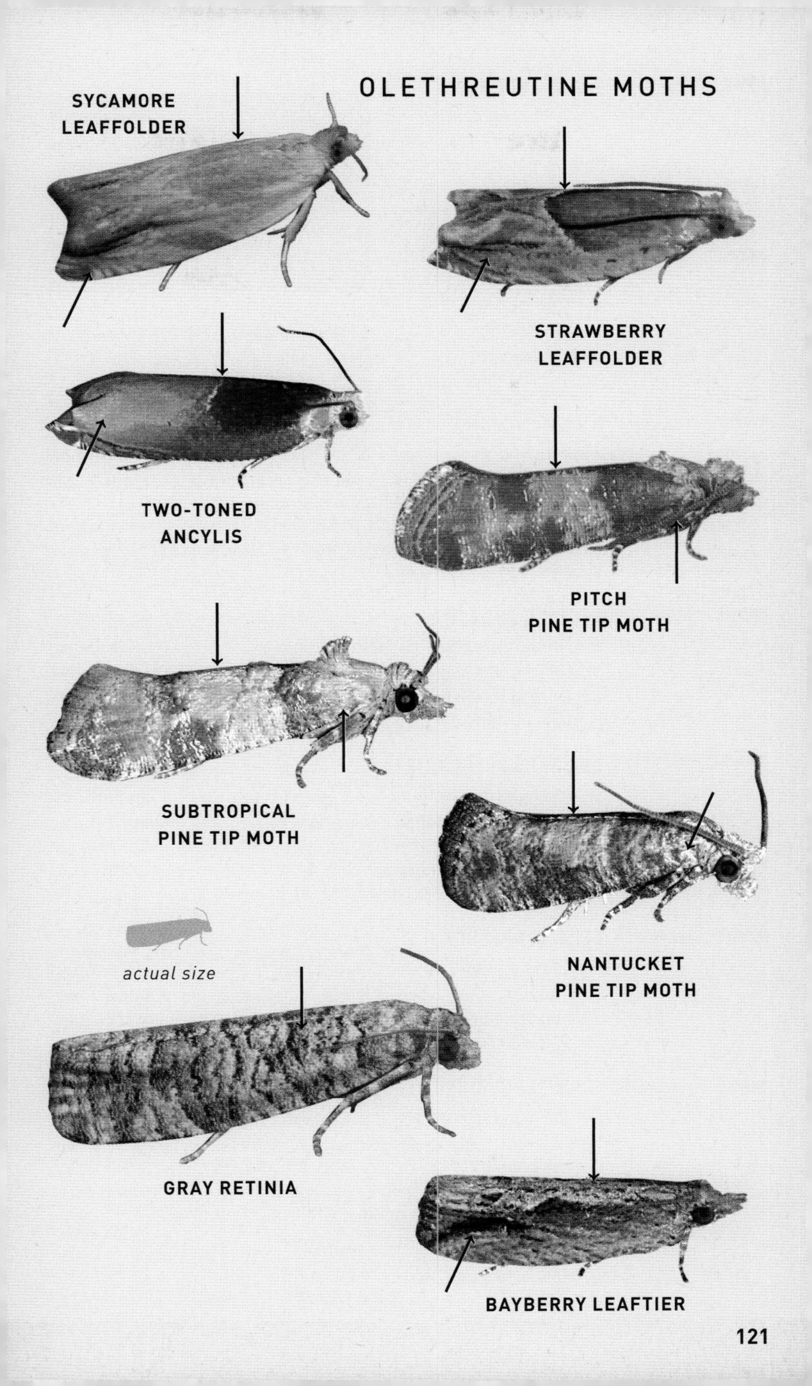
OLETHREUTINE MOTHS
SYCAMORE LEAFFOLDER
STRAWBERRY LEAFFOLDER
TWO-TONED ANCYLIS
PITCH PINE TIP MOTH
SUBTROPICAL PINE TIP MOTH
NANTUCKET PINE TIP MOTH
actual size
GRAY RETINIA
BAYBERRY LEAFTIER

GIANT EUCOSMA

Eucosma giganteana 62-0745 (3098) **Uncommon**

TL 10–18 mm. White FW has brown patch in inner PM and ST areas that is banded with silvery lines. Inner basal area is sometimes lightly patterned gray. **HOSTS:** Cup plant and prairie rosinweed. **RANGE:** NC west to AR, south to FL and e. TX.

COASTAL EUCOSMA

Eucosma argutipunctana 62-0795 (3005.1) **Local**

TL 7–9 mm. White FW is conspicuously barred with fragmented black lines. ST area is sometimes tinged warm brown. Head is white. **HOSTS:** Unknown. **RANGE:** Coastal NC, locally south to FL and e. TX.

ASTER EUCOSMA

Eucosma parmatana 62-0832 (2937) **Common**

TL 5–7 mm. Grayish brown FW has an angled black AM line and slanting lines along costa. Note contrasting white patches in inner median and inner ST areas. Form "crispina" is dark with thin to faint white lines in place of patches. **HOSTS:** Aster. **RANGE:** NC and TN, locally south to FL and e. TX.

REDDISH EUCOSMA

Eucosma raracana 62-0837 (2928) **Common**

TL 6–7 mm. Brick red FW has contrasting whitish patch in inner ST area. Head is white. **HOSTS:** Goldenrod. **RANGE:** TN west to AR, south to FL and e. TX.

SOLIDAGO PELOCHRISTA

Pelochrista cataclystiana 62-0909 (3142) **Uncommon**

TL 8–10 mm. Pale orange FW is finely striated with thin brown lines. Median area is sometimes shaded with rusty brown. FW is creased in inner ST area, forming a slightly falcate wing shape. **HOSTS:** Flat-top goldenrod. **RANGE:** NC west to OK, south to FL and AL.

DERELICT PELOCHRISTA

Pelochrista derelicta 62-0926 (3120) **Common**

TL 8–10 mm. Pinkish brown FW is lightly brindled with brown lines. Note slanting brown-edged white PM line. **HOSTS:** Roots of goldenrod. **RANGE:** TN, locally south to SC and FL.

TRIANGLE-BACKED PELOCHRISTA

Pelochrista dorsisignatana 62-0930 (3116) **Common**

TL 8–12 mm. Light grayish brown FW is finely brindled with brown lines. A brown triangle marks inner AM area. Rounded brown patch in inner PM area fades before reaching costa. **HOSTS:** Roots of goldenrod. **RANGE:** NC west to OK, south to nw. FL and e. TX.

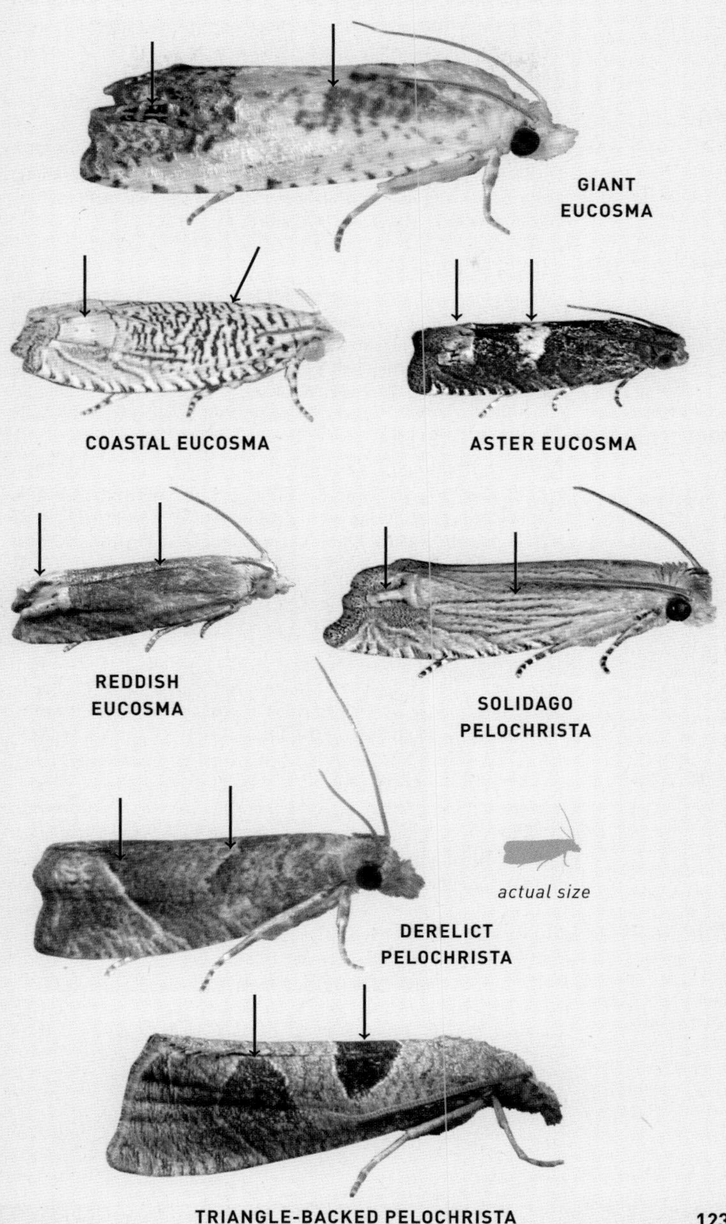
GIANT
EUCOSMA
COASTAL EUCOSMA
ASTER EUCOSMA
REDDISH
EUCOSMA
SOLIDAGO
PELOCHRISTA
actual size
DERELICT
PELOCHRISTA
TRIANGLE-BACKED PELOCHRISTA

FIVE-SPOTTED PELOCHRISTA

Pelochrista quinquemaculana 62-1015 (3008) **Common**

TL 8–10 mm. Golden brown FW is boldly patterned with five large, black-edged, satin white spots. Head is brown. **HOSTS:** Unknown. **RANGE:** SC and GA, south to FL.

ROBINSON'S PELOCHRISTA

Pelochrista robinsonana 62-1021 (3009) **Common**

TL 6–8 mm. Brown FW is boldly patterned with narrow, black-edged white bands and spots. White head is tinged brown. **HOSTS:** Unknown. **RANGE:** NC west to OK, south to FL and MS.

SPANGLED PELOCHRISTA

Pelochrista scintillana 62-1028 (3151) **Uncommon**

TL 7–15 mm. Silvery gray FW is finely peppered darker. Pale orange bands in median, PM, and ST areas are incomplete. Silvery inner ST area is boldly patterned with black dots. **HOSTS:** Sunflower heads. **RANGE:** TN west to OK, south to cen. AL and e. TX.

RAGWEED EPIBLEMA

Epiblema strenuana 62-1065 (3172) **Common**

TL 6–9 mm. Grayish brown FW is finely peppered white. Costa is striated with slanting brown and white dashes. Whitish inner ST area is creased. **HOSTS:** Stems of annual ragweed. **RANGE:** NC west to OK, south to FL and e. TX.

ABRUPT EPIBLEMA

Epiblema abruptana 62-1066 (3173) **Common**

TL 6–8 mm. Light brown FW has a mottled pattern of white-edged dusky lines. Dark brown patch in inner AM area is sometimes obvious. Whitish inner ST area is creased. **HOSTS:** Unknown. **RANGE:** NC west to OK, south to FL and e. TX.

THREE-PARTED EPIBLEMA

Epiblema tripartitana 62-1078 (3184) **Common**

TL 5–10 mm. Mottled grayish brown FW has a contrasting white median band that narrows slightly toward costa. **HOSTS:** Flower heads of great coneflower. **RANGE:** NC west to OK, south to FL and e. TX.

SCUDDER'S EPIBLEMA

Epiblema scudderiana 62-1082 (3186) **Common**

TL 7–11 mm. White FW has a dark bluish gray basal patch. Grayish inner PM area is marked with a black dot. Apex is accented with chestnut dashes. **HOSTS:** Stems of goldenrod, causing elongate galls. **RANGE:** NC west to OK, south to FL and e. TX.

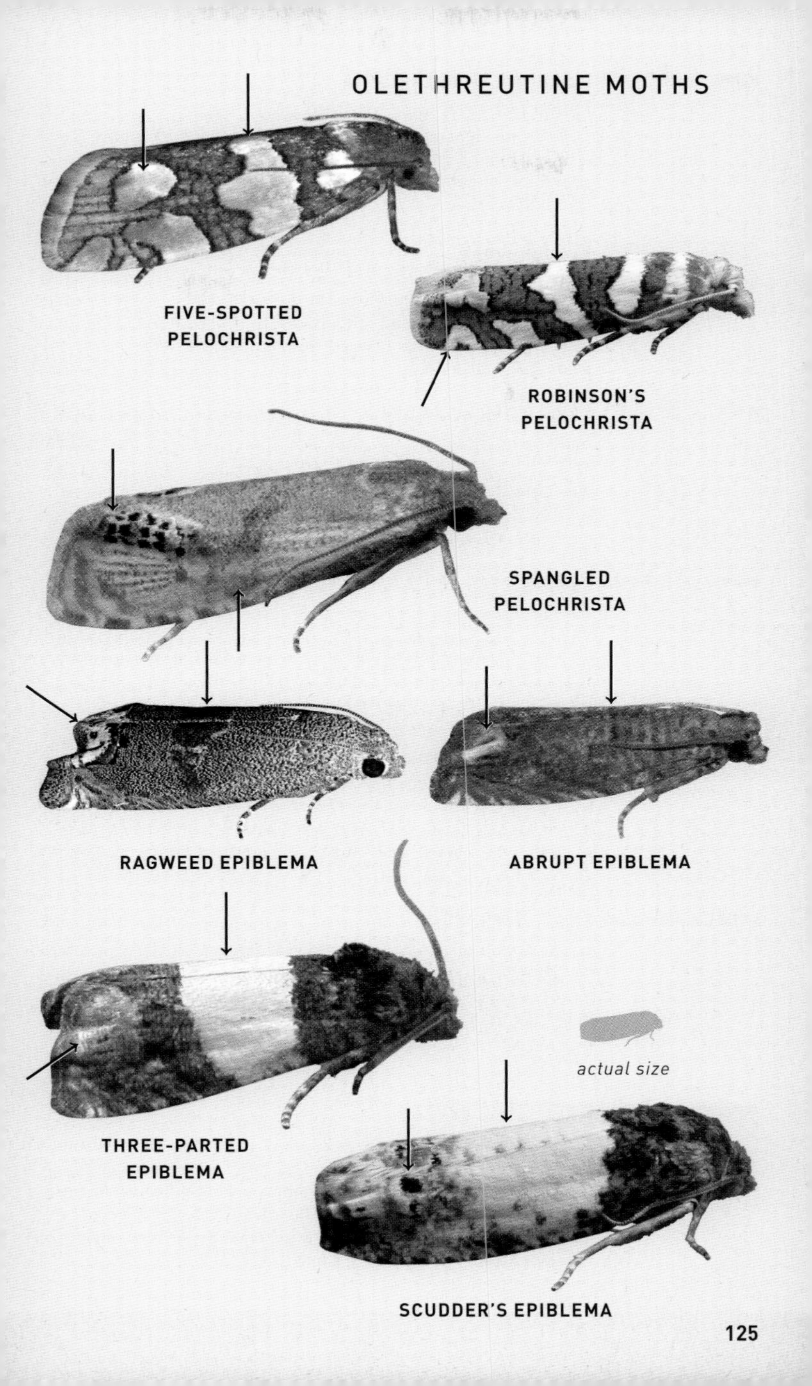

FIVE-SPOTTED PELOCHRISTA

ROBINSON'S PELOCHRISTA

SPANGLED PELOCHRISTA

RAGWEED EPIBLEMA

ABRUPT EPIBLEMA

actual size

THREE-PARTED EPIBLEMA

SCUDDER'S EPIBLEMA

DISCRIMINATING EPIBLEMA

Epiblema discretivana 62-1083 (3188) **Local**

TL 7–9 mm. Light pinkish FW is variably brindled darker gray. Basal area and inner ST areas are contrastingly darker gray. Thorax has a raised tuft of dark gray scales. **HOSTS:** Unknown. **RANGE:** SC and GA south to FL.

BIDENS BORER *Epiblema otiosana* 62-1098 (3202) **Common**

TL 6–10 mm. Dark grayish brown FW has a large U-shaped white patch at midpoint of inner margin. Creased inner ST area is pale gray. Thorax has a raised tuft of blackish scales. **HOSTS:** Stems of devil's beggar tick and annual ragweed. **RANGE:** NC west to OK, south to FL and e. TX.

SUNFLOWER BUD MOTH

Suleima helianthana 62-1110 (3212) **Local**

TL 5–7 mm. Brindled light gray FW has dark gray patch in inner median area, forming a crescent when wings are folded. ST area is whitish, bordered by dark gray shading. **HOSTS:** Stems and buds of sunflower. **RANGE:** Cen. GA, locally south to FL and e. TX.

CONSTRICTED SONIA

Sonia constrictana 62-1116 (3218) **Common**

TL 5–7 mm. Light gray FW has darker brown basal area sharply demarcated with straight, white-edged, black AM line. PM area is accented with large brown patches at midpoint and at inner margin. **HOSTS:** Goldenrod. **RANGE:** NC and TN south to FL and e. TX.

HEBREW SONIA *Sonia paraplesiana* 62-1117 (3218.1) **Common**

TL 7–9 mm. Grayish brown FW has rusty inner basal area bordered with a curved white AM line. Rusty triangle in inner PM area is edged whitish. Costa is striated with slanting rusty dashes. Apex is marked with a black spot. **HOSTS:** Unknown. **RANGE:** NC and TN south to FL.

SMALL GYPSONOMA

Gypsonoma salicicolana 62-1129 (3228) **Local**

TL 5–6 mm. Light brown to grayish FW has darker basal area bordered with a strongly curved, thin whitish AM line. Curved PM band fades before reaching darker triangle at inner margin. Apex is marked with a black spot. **HOSTS:** Willow. **RANGE:** TN, locally south to FL and s. LA.

MAPLE TWIG BORER

Proteoteras aesculana 62-1133 (3230) **Common**

TL 7–10 mm. Gray FW is mottled darker with yellowish green shading along AM and PM lines; often washed mint green when fresh. Blackish line from midpoint of costa arcs toward apex. Raised tufts of scales give FW a lumpy appearance. **HOSTS:** Terminal shoots of maple. **RANGE:** W. NC and TN, south to n. FL.

OLETHREUTINE MOTHS

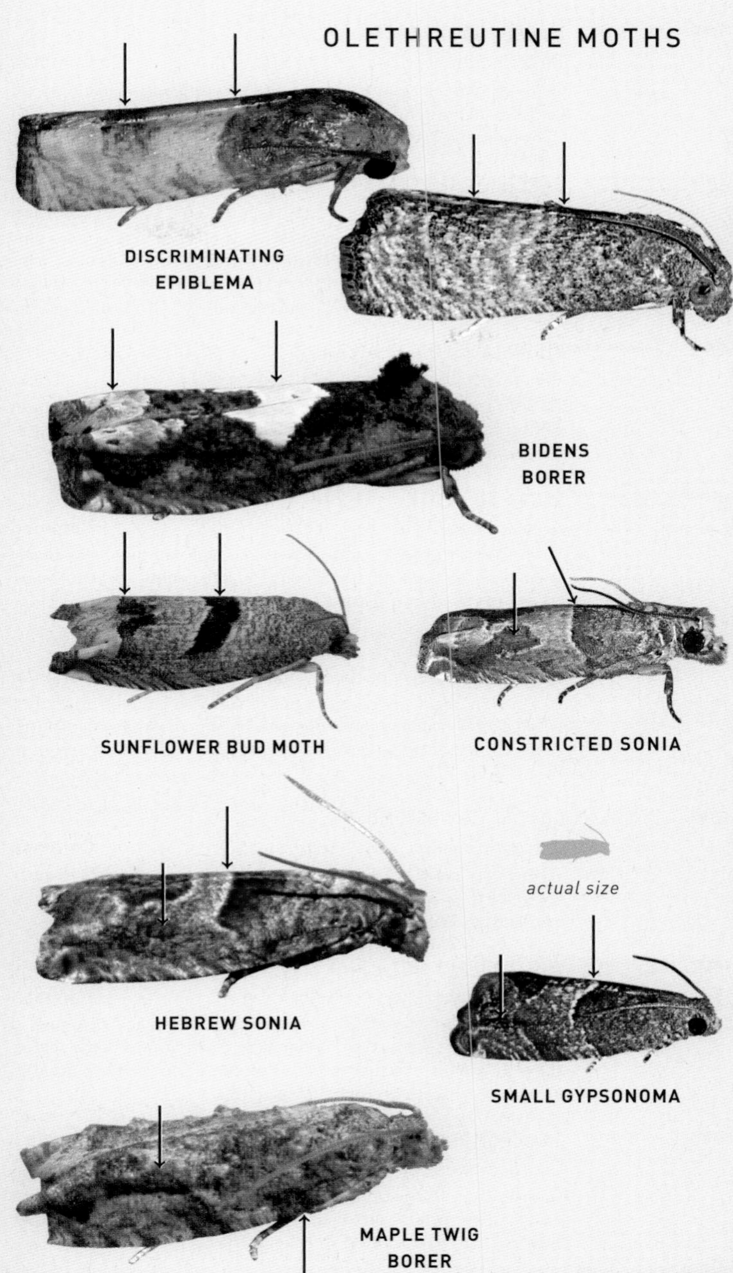

MAPLE SHOOT BORER

Proteoteras moffatiana 62-1138 (3235) Common

TL 8–10 mm. Resembles Maple Twig Borer, but blackish apical crescent and fragmented AM band are usually well defined. **HOSTS:** Buds of maple. **RANGE:** NC and TN south to nw. FL.

BARE-PATCHED LEAFROLLER

Pseudexentera spoliana 62-1157 (3251) Common

TL 7–10 mm. Light silvery gray FW is variably brindled darker. Incomplete blackish AM and median bands are obvious only near inner margin. Costa is marked with thin black and white dashes toward apex. **HOSTS:** Oak. **RANGE:** NC and TN south to cen. FL and s. AL.

VIRGINIA PSEUDEXENTERA

Pseudexentera virginiana 62-1166 (3258) Local

TL 8–10 mm. Silvery gray FW has a large rounded brown patch in central median area that extends along costa toward apex. **HOSTS:** Unknown. **RANGE:** NC and TN, locally south to FL and LA.

PECAN BUD MOTH

Gretchena bolliana 62-1171 (3263) Common

TL 8–10 mm. Grayish brown FW is mottled with irregular dusky lines. A zigzag black streak extends through central median and ST areas. **HOSTS:** Pecan. **RANGE:** SC west to OK, south to FL and e. TX.

FILIGREED MOTH

Chimoptesis pennsylvaniana 62-1181 (3273) Common

TL 7–9 mm. Blackish FW has a boldly serrated, green-tinted white stripe extending along inner margin. Costa is marked with slanting white dashes. **HOSTS:** Unknown. **RANGE:** NC west to OK, south to FL and e. TX.

COTTON TIPWORM MOTH

Crocidosema plebejana 62-1182 (3274) Common

TL 6–8 mm. Light brown FW has mottled blackish patches in inner AM and PM areas. ST area is obviously creased. **HOSTS:** Tree-mallow. **RANGE:** TN and SC south to FL.

WINTERBERRY MOTH

Rhopobota dietziana 62-1189 (3277) Common

TL 6–8 mm. White FW has a mottled gray and black AM band. A curved black line arcs from midpoint of costa toward black spot at apex. ST area is obviously creased. **HOSTS:** Common winterberry. **RANGE:** NC west to OK, south to FL and LA.

OAK TRUMPET SKELETONIZER

Catastega timidella 62-1267 (3333) Local

TL 8–10 mm. Peppery light gray FW has shading through central median area. Whitish streak along inner margin has a zigzag lower edge. **HOSTS:** Oak. **RANGE:** W. NC south to GA and FL.

OLETHREUTINE MOTHS

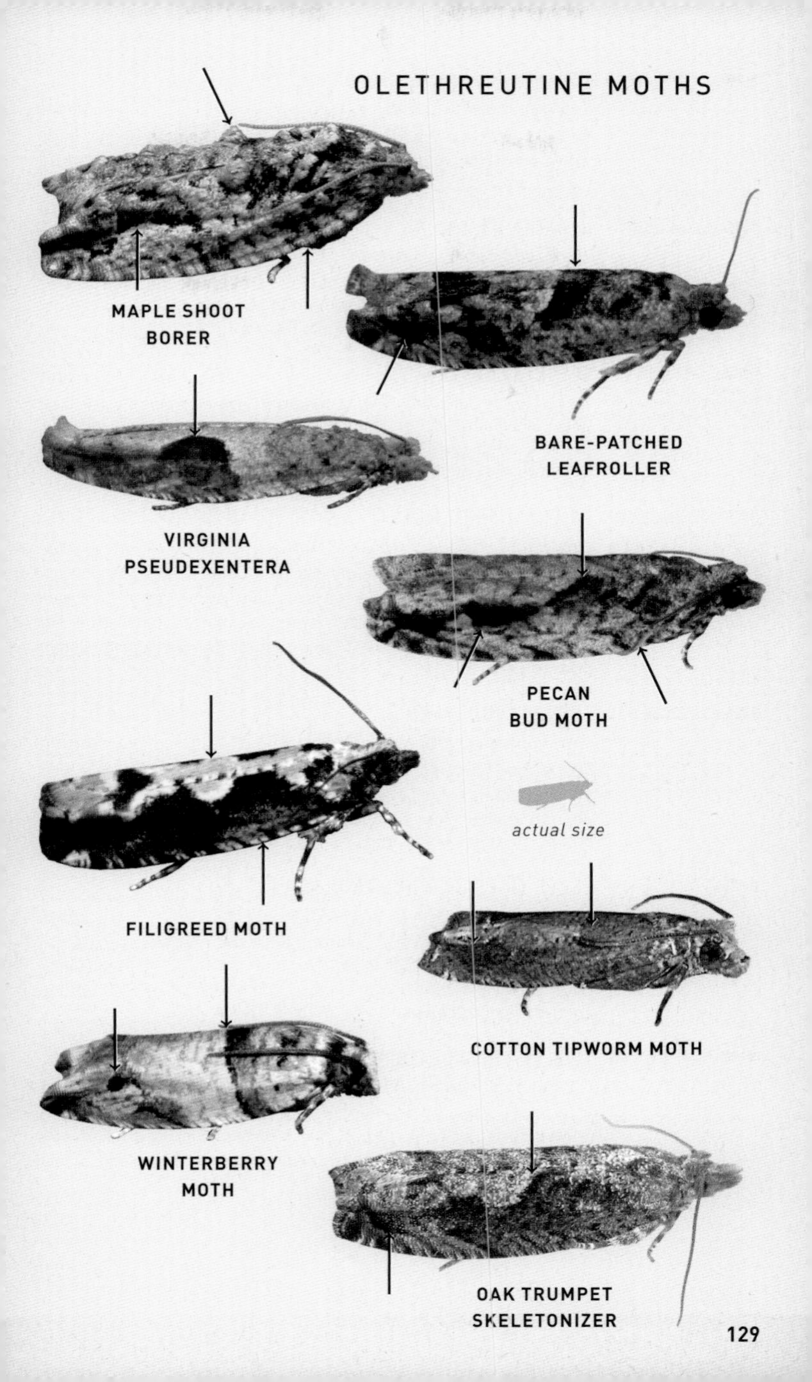

SOUTHERN PINE CATKINWORM MOTH

Satronia tantilla 62-1292 (3415) Local

TL 7–9 mm. Grayish brown FW is tinted bronze distally and crossed with silvery gray lines. **HOSTS:** Pine. **RANGE:** TN and SC south to cen. FL and AL.

DARK-BASED LARISA

Larisa subsolana 62-1302 (3423) Common

TL 5–7 mm. Light gray FW has darker basal area edged with a gently curved AM line. Black PM line is often edged with dusky shading. **HOSTS:** Pecan. **RANGE:** TN south to FL and e. TX.

SPECKLED SEREDA *Sereda tautana* 62-1304 (3425) Local

TL 5–7 mm. Dark gray FW is patterned with angled silvery blue lines. Blackish inner median area is edged silvery white distally. Black-spotted terminal line contrasts with silver fringe. **HOSTS:** Northern red oak. **RANGE:** NC and TN south to GA and nw. FL.

ORIENTAL FRUIT MOTH

Grapholita molesta 62-1305 (3426) Local

TL 6–7 mm. Gray FW has indistinct pattern of wavy black lines. A whitish dot in central PM area is usually evident. **HOSTS:** Fruit-bearing trees, including apple, cherry, pear, and peach. **RANGE:** TN and SC south to cen. FL. **NOTE:** Accidentally introduced from Japan prior to 1915.

CHERRY FRUITWORM

Grapholita packardi 62-1307 (3428) Common

TL 4–6 mm. Gray FW has wavy black lines accented with silvery blue bands. Usually has an obscure darker median band. **HOSTS:** Fruit-bearing trees, including apple, cherry, pear, and peach. **RANGE:** TN and SC south to cen. FL and LA.

THREE-LINED GRAPHOLITA

Grapholita tristrigana 62-1324 (3443) Local

TL 5–8 mm. Gray FW has lustrous blue and purple sheen. White patch at midpoint of inner margin contains three slanting black lines. Costa is accented with slanting white dashes. **HOSTS:** Lupine and wild indigo. **RANGE:** GA south to FL.

TWELVE-LINED OFATULENA

Ofatulena duodecemstriata 62-1327 (3444) Local

TL 6–9 mm. Gray FW has an intricate pattern of silvery bands and finely etched black lines. A brownish patch in inner PM area forms a distinct dark triangle. **HOSTS:** Seedpods of mesquite and screw bean. **RANGE:** Cen. TX.

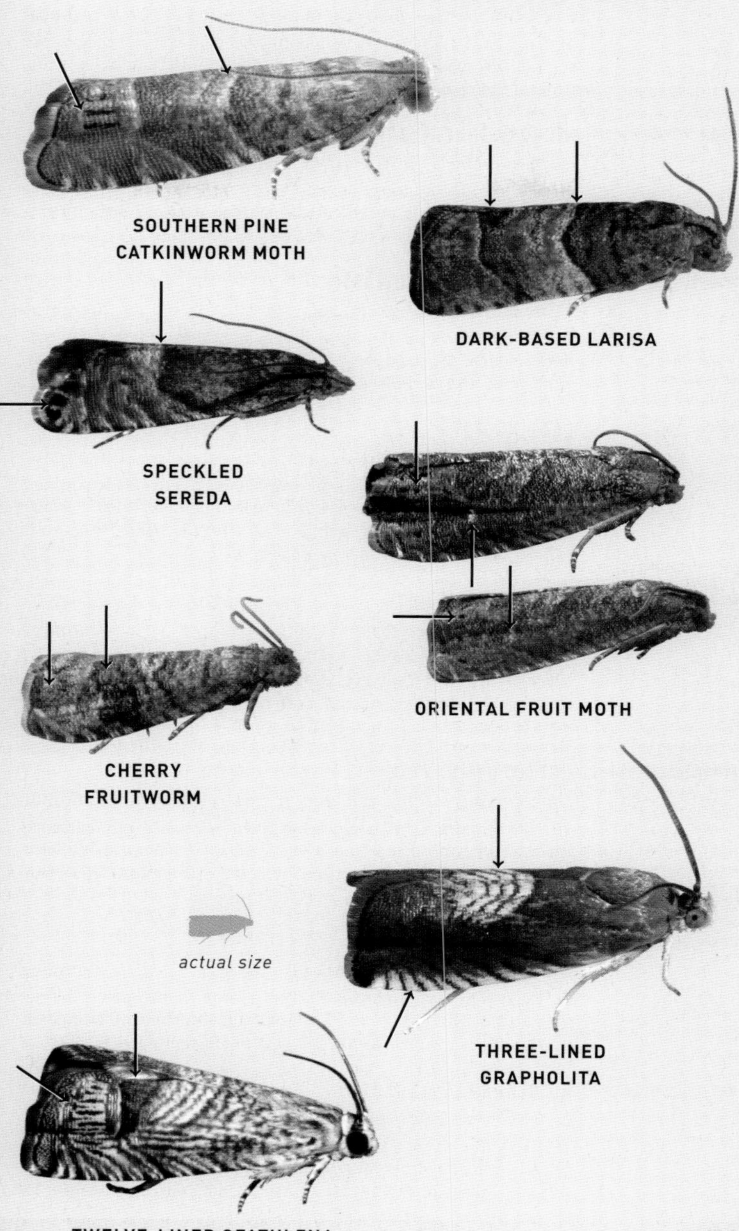
SOUTHERN PINE
CATKINWORM MOTH
DARK-BASED LARISA
SPECKLED
SEREDA
ORIENTAL FRUIT MOTH
CHERRY
FRUITWORM
actual size
THREE-LINED
GRAPHOLITA
TWELVE-LINED OFATULENA

BRONZE-BANDED CORTICIVORA

Corticivora parva 62-1330 (3446.2) **Local**

TL 5–6 mm. Pearly white FW is patterned with shiny bronzy brown AM and PM bands. Terminal line is dotted. Fuzzy head is white. **HOSTS:** Unknown. **RANGE:** W. NC west to OK, south to FL and s. MS.

HICKORY SHUCKWORM MOTH

Cydia caryana 62-1357 (3471) **Common**

TL 6–8 mm. Dark bronzy brown FW is crossed with metallic blue gray bands. Costa is accented with slanted white dashes. **HOSTS:** Hickory and pecan. **RANGE:** W. NC west to OK, south to FL and e. TX.

EASTERN PINE SEEDWORM

Cydia toreuta 62-1372 (3486) **Common**

TL 6–8 mm. Dark purplish gray FW is finely vermiculated with gold flecks. Thick, black-edged, silvery gray median and PM bands are noticeable. **HOSTS:** Pine. **RANGE:** NC and TN south to cen. FL.

CODLING MOTH

Cydia pomonella 62-1380 (3492) **Uncommon**

TL 8–12 mm. Ash gray FW is densely brindled with dark gray lines. Darker patch in inner ST area contains fragmented metallic bronze bars. **HOSTS:** Apple, pear, and plum. **RANGE:** NC and TN south to cen. FL. **NOTE:** Introduced from Europe.

FILBERTWORM MOTH

Cydia latiferreana 62-1383 (3494) **Common**

TL 7–11 mm. Peppery tan to reddish brown FW has broad, metallic silver median and PM bands. Costa is patterned with slanting metallic dashes. **HOSTS:** Beech, filbert, hazelnut, and oak. **RANGE:** NC west to OK, south to FL and e. TX.

DOTTED GYMNAST

Gymnandrosoma punctidiscanum 62-1385 (3495) **Common**

TL 9–12 mm. Dark, mottled grayish brown FW has an angled blackish median band containing a white dot at midpoint. ST area is sometimes whitish. **HOSTS:** Locust and red oak. **RANGE:** NC west to OK, south to FL and s. AL.

LOCUST TWIG BORER

Ecdytolopha insiticiana 62-1387 (3497) **Common**

TL 10–14 mm. Grayish brown FW has a large whitish patch in inner median and ST areas. Inner ST line is fragmented into black dashes. **HOSTS:** Black locust. **RANGE:** NC west to AR, south to n. FL.

BICOLORED ECDYTOLOPHA

Ecdytolopha mana 62-1388 (3498) **Common**

TL 10–14 mm. Bicolored FW is brown and whitish with fragmented, raised silvery bands, creating a lumpy appearance. Apex is rounded. **HOSTS:** Unknown. **RANGE:** TN west to OK, south to cen. FL and e. TX.

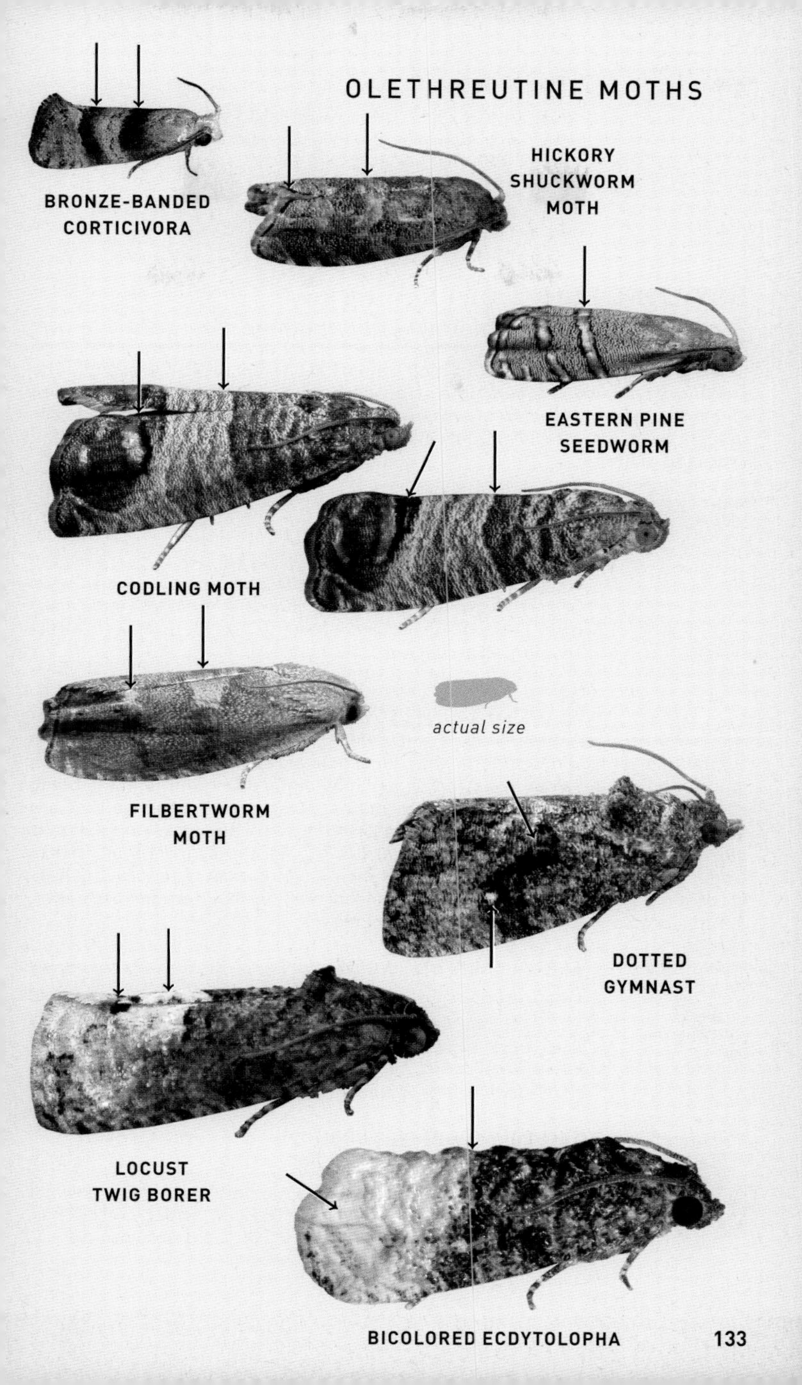
OLETHREUTINE MOTHS
BRONZE-BANDED CORTICIVORA
HICKORY SHUCKWORM MOTH
EASTERN PINE SEEDWORM
CODLING MOTH
FILBERTWORM MOTH
actual size
DOTTED GYMNAST
LOCUST TWIG BORER
BICOLORED ECDYTOLOPHA

INIMICAL BORER

Pseudogalleria inimicella 62-1391 (3500) **Uncommon**

TL 9–12 mm. Light blue gray FW has a pale orange basal area and is streaked with rust along costa. Crescent-shaped chestnut ST line cuts through midpoint of pale yellow fringe. **HOSTS:** Stems and roots of carrion flower. **RANGE:** Coastal NC west to AR, locally south to FL and e. TX.

PSYCHEDELIC LEAFROLLERS
Family Tortricidae, Subfamily Chlidanotinae

Small totricid moths noted for their remarkably colorful FW patterns. The striking Psychedelic Jones Moth is an uncommon inhabitant of old-growth pine forest with a well-developed understory. Members of this small clade are thought to be mostly diurnal; however, this species is also active at night and attracted to lights.

PSYCHEDELIC JONES MOTH

Thaumatographa jonesi 62-1395 (3751) **Local**

TL 7–8 mm. Unmistakable. Metallic inky blue FW is mostly reddish orange in distal half. Orange basal dash ends in median area. Oblique silvery blue lines sweep backward from inner margin and costa. Legs are banded black and white. **HOSTS:** Unknown. **RANGE:** TN and n. MS, south to cen. FL and s. LA.

CARPENTERWORM MOTHS Family Cossidae

Exceptionally large micromoths that superficially resemble prominent moths in size and shape. Most are strongly sexually dimorphic, with females being considerably larger and bulkier than males. Most have thin wings that are semi-translucent and fray easily. Males (and in some species females) have bipectinate antennae. Larvae are borers in trunks and branches of various trees, taking several years to complete their development. The larval galleries of some species may weaken the host tree, making it vulnerable to drought. Adults occasionally visit lights in small numbers.

ARBELA CARPENTERWORM

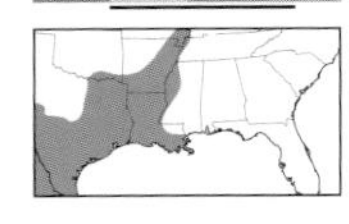

Givira arbeloides 64-0009 (2661) **Local**

TL 15–17 mm. Grayish brown FW is lightly patterned with darker lines. Outer median area is often shaded with a dusky patch. Pale costal streak is blotched with black. Rusty fringes on all wings are checkered when fresh. **HOSTS:** Unknown.

THEODORE CARPENTERWORM

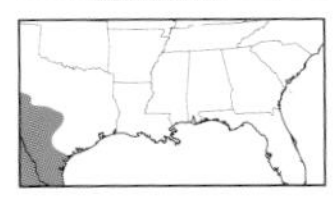

Givira theodori 64-0010 (2662) **Local**

TL 12–18 mm. White FW has a broad blackish PM band that does not reach inner margin. ST area is mottled gray between white veins. Tufted thorax is blackish distally. Black-tipped abdomen extends beyond wings at rest. **HOSTS:** Unknown.

OLETHREUTINE MOTHS

INIMICAL BORER

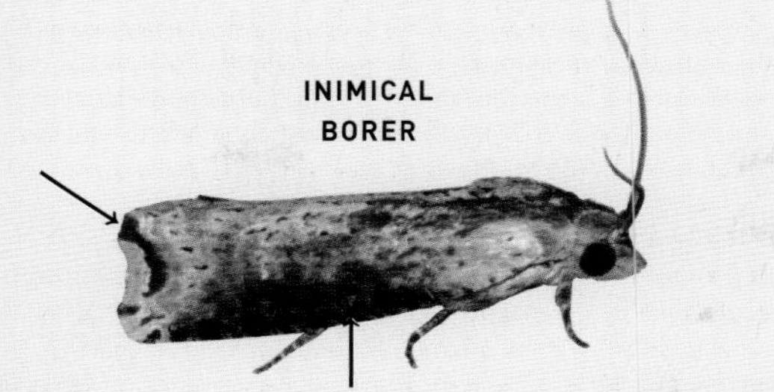

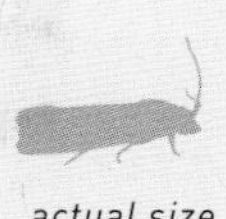

actual size

PSYCHEDELIC LEAFROLLERS

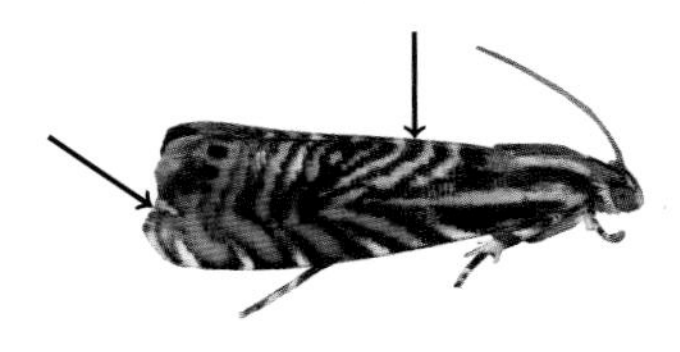

PSYCHEDELIC JONES MOTH

actual size

CARPENTERWORM MOTHS

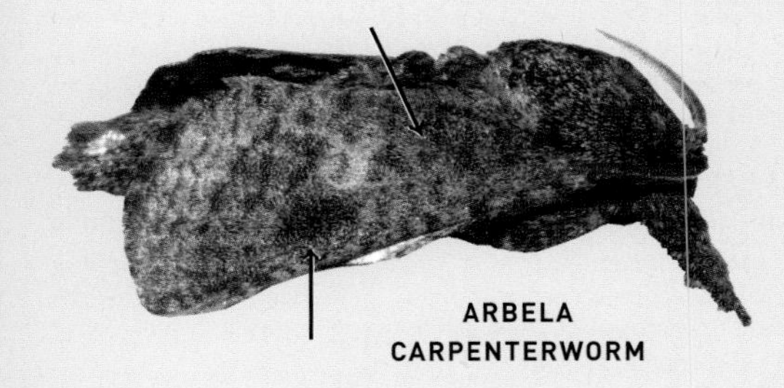

ARBELA CARPENTERWORM

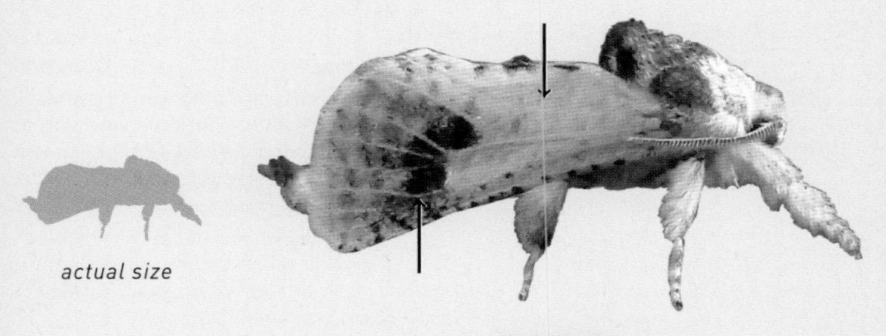

actual size

THEODORE CARPENTERWORM

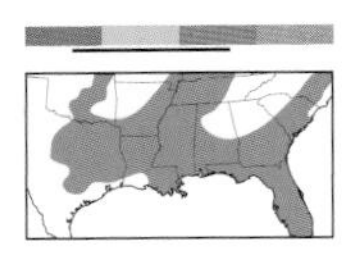

ANNA CARPENTERWORM

Givira anna 64-0016 (2668) **Common**

TL 14–20 mm. Peppery light gray FW is variably shaded brown along inner margin and PM band. A short, dusky-edged white crescent in central PM area is most obvious marking. Sometimes has white veins passing through ST area. Two-pronged abdomen extends beyond wings at rest. **HOSTS:** Pine.

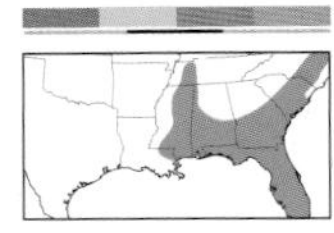

FRANCESCA CARPENTERWORM

Givira francesca 64-0019 (2671) **Common**

TL 14–20 mm. Peppery light gray FW is densely patterned with wavy, dusky gray lines and a faint dark median band. A short white crescent in outer PM area is inconspicuous. Fringe is checkered. Tufted abdomen extends slightly beyond wings at rest. **HOSTS:** Unknown.

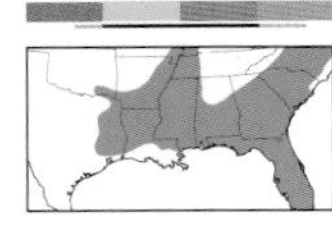

BLACK-LINED CARPENTERWORM

Inguromorpha basalis 64-0022 (2659) **Local**

TL 15–20 mm. Light gray FW is dusted white in median area and tinged warm brown in basal and ST areas. Much of wing is covered with a cobweb of fine black lines, with a bold straight AM line. A short black crescent edges a dusky apical spot. Rests with tip of abdomen raised above wings. **HOSTS:** Unknown.

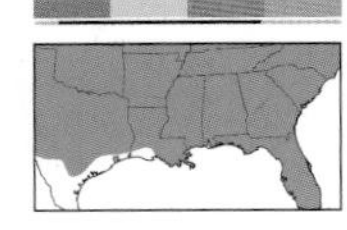

ROBIN'S CARPENTERWORM

Prionoxystus robiniae 64-0029 (2693) **Uncommon**

TL 27–45 mm. Semi-translucent, light gray FW is patterned with a dense network of black veins and lines. Variable blackish blotches mark basal and median areas. HW of smaller male has a broad golden yellow ST band. **HOSTS:** Ash, chestnut, locust, oak, poplar, and willow.

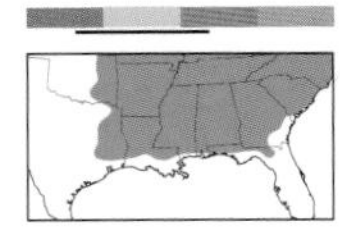

LITTLE CARPENTERWORM

Prionoxystus macmurtrei 64-0030 (2694) **Uncommon**

TL 23–40 mm. Resembles Robin's Carpenterworm but is smaller and less densely patterned with black lines. Semi-translucent HW is usually grayish in both sexes. Thorax has a whitish half-collar. **HOSTS:** Ash, maple, and oak.

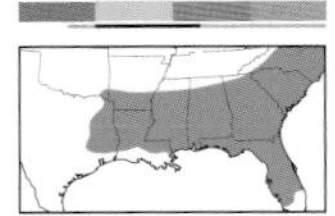

PECAN CARPENTERWORM

Cossula magnifica 64-0047 (2674) **Common**

TL 18–24 mm. Silvery gray FW is finely brindled with crisp black lines with a thicker, fragmented AM line. Black ST line curves around streaky fawn-colored terminal patch. This blunt-faced moth often rests in a head-down position like some *Schizura* prominents. **HOSTS:** Hickory, oak, pecan, and persimmon.

CARPENTERWORM MOTHS

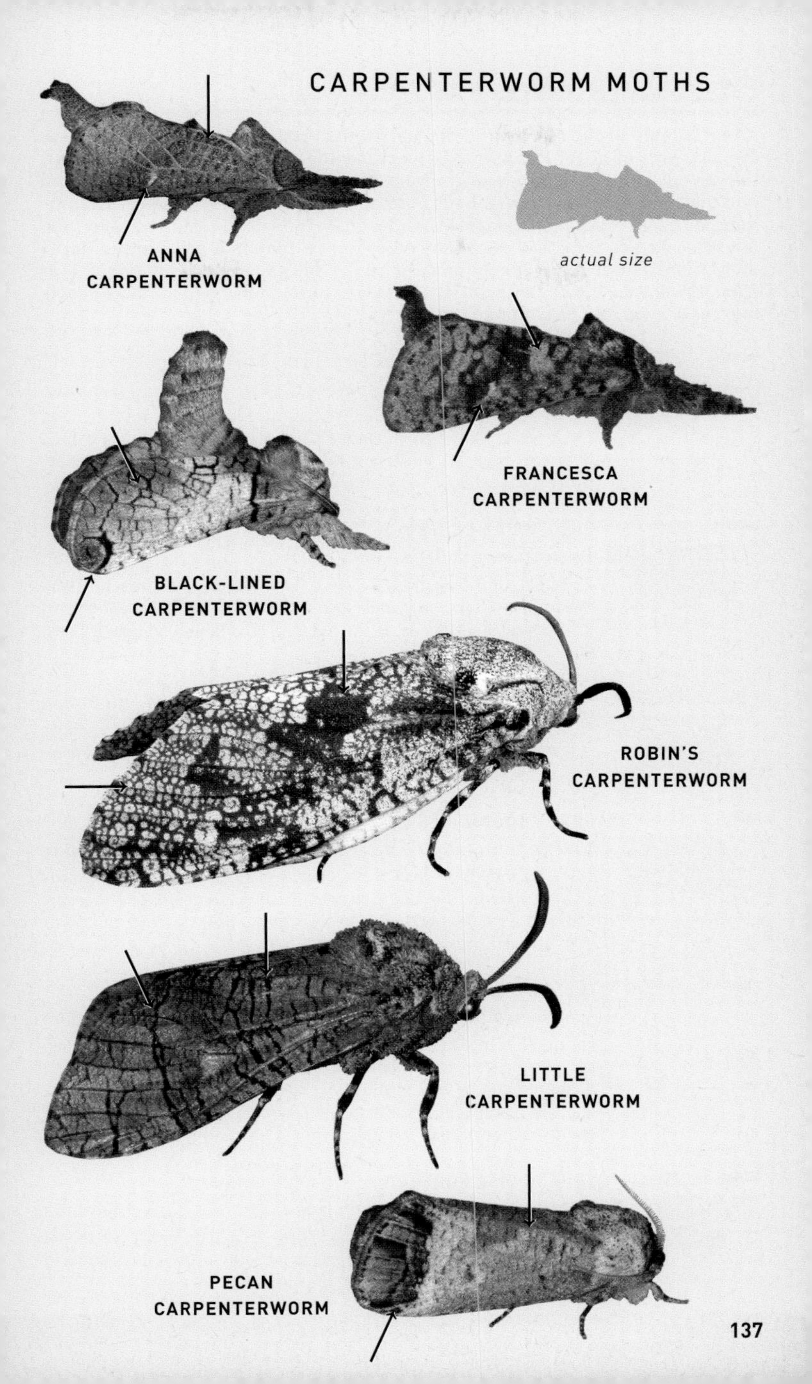

Clearwing Borers Family Sesiidae

A remarkable group of colorful, diurnal moths that closely resemble hornets and wasps, but differ in antennal structure and lack of a constricted waist. Larvae bore into the roots, branches, and trunks of a wide variety of trees and into the stems of many herbaceous plants. Several species are pests of commercial fruit trees, such as apple and peach. Inconspicuous and hard to find; few species are regularly encountered at lights, though Maple Callus Borer and Arkansas Borer are exceptions. Some species can be found taking nectar from flowers during daytime, but the best way to see most is to purchase commercially available pheromone lures.

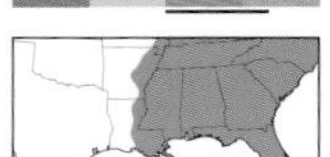

RASPBERRY CROWN BORER

Pennisetia marginatum 64-0057 (2513) **Common**

TL 13–16 mm. Transparent FW has a reddish brown border and black veins. Abdomen is banded black and yellow with raised band of black scales on third segment. Male has bipectinate antennae. **HOSTS:** Blackberry, raspberry, and boysenberry.

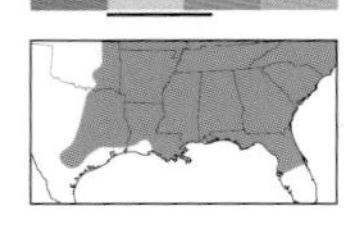

GOLDEN OAK BORER

Paranthrene simulans 64-0059 (2527) **Common**

TL 16–24 mm. Transparent FW has blackish costa and veins. Black thorax has yellow collar and lateral stripes. Golden yellow abdomen has narrow black bands. Black antennae are tipped orange. **HOSTS:** Oak.

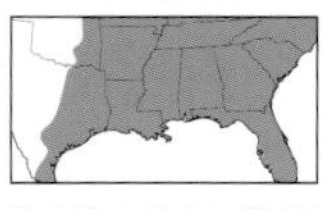

POPLAR BORER *Paranthrene dollii* 64-0062 (2523) **Common**

TL 12–16 mm. Blackish FW has a few reddish scales at base. Thorax is reddish. Abdomen is red with thin yellow bands. Black antennae are yellow at base and tips. **HOSTS:** Poplar and willow.

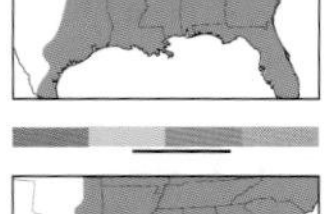

GRAPE ROOT BORER

Vitacea polistiformis 64-0065 (2530) **Common**

TL 16–24 mm. Blackish FW has rusty base and veins. Thorax is reddish brown with thin black stripes. Rusty abdomen has thin yellow and black bands. Legs are mostly orange. **HOSTS:** Wild and cultivated grape.

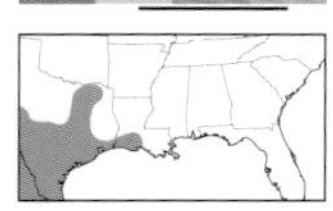

LESSER GRAPE BORER

Vitacea admiranda 64-0067 (2528) **Uncommon**

TL 14–20 mm. Blackish FW sometimes has rusty veins. Reddish brown thorax is edged yellow and often has black lateral stripes. Abdomen can be black with rusty rings or rusty with black rings, and two broad yellow bands. Legs are mostly orange. **HOSTS:** Wild and cultivated grape.

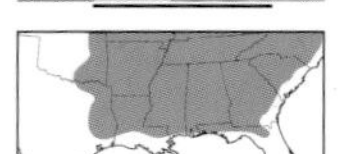

SQUASH VINE BORER

Melittia cucurbitae 64-0081 (2536) **Common**

TL 17–19 mm. Black FW. Thorax is gray. Red abdomen has a row of black dorsal dots. Hindlegs are adorned with large tufts of reddish scales. **HOSTS:** Squash, gourds, and pumpkin.

CLEARWING BORERS
male
female
actual size
RASPBERRY CROWN BORER
GOLDEN OAK BORER
POPLAR BORER
GRAPE ROOT BORER
LESSER GRAPE BORER
SQUASH VINE BORER

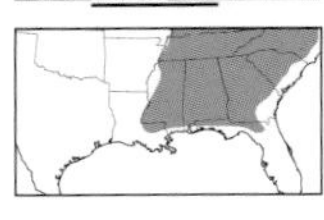

RED MAPLE BORER

Synanthedon acerrubri 64-0087 (2546) **Uncommon**

TL 8–12 mm. Transparent FW has a black border, veins, and PM band. Blue-black abdomen has an orange-tipped, fanned tuft at tip. **HOSTS:** Red and sugar maples.

PECAN BARK BORER

Synanthedon geliformis 64-0088 (2547) **Local**

TL 8–12 mm. Blackish FW is shaded chestnut near tip. Thorax is blue-black. Bright red abdomen has a black-edged tuft at tip. **HOSTS:** Deciduous trees, including elm, hickory, oak, and pecan.

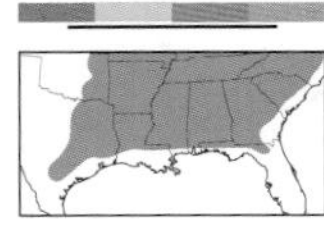

DOGWOOD BORER

Synanthedon scitula 64-0090 (2549) **Common**

TL 8–12 mm. Transparent FW has black border and PM band with yellow patch near tip and thin yellow streaks inside costa. Thorax has yellow lateral stripes. Banded black and yellow abdomen has a fanned tuft at tip. **HOSTS:** Deciduous trees, including apple, birch, dogwood, oak, and willow.

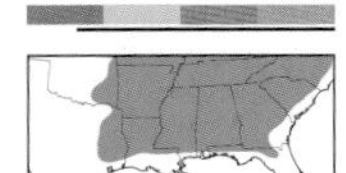

LESSER PEACHTREE BORER

Synanthedon pictipes 64-0091 (2550) **Common**

TL 10–14 mm. Transparent FW has thin black border, veins, and PM band. Thorax has white sides to thin black collar. Head has small white dots in front of eyes. Blue-black abdomen has a tapered tuft at tip. **HOSTS:** Fruit-bearing trees, including cherry and plum.

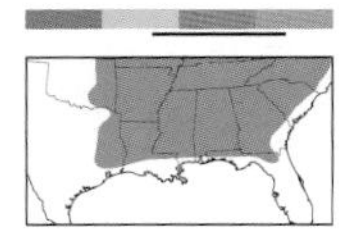

RILEY'S CLEARWING

Synanthedon rileyana 64-0093 (2552) **Common**

TL 10–15 mm. Transparent FW has a narrow reddish brown border and a red PM band. Thorax has yellow patches at base of wings and a narrow yellow collar. Black abdomen has yellow bands on each segment. **HOSTS:** Carolina horse nettle.

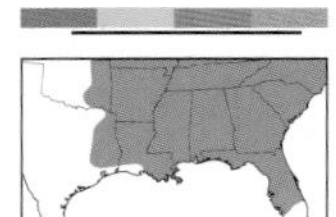

MAPLE CALLUS BORER

Synanthedon acerni 64-0095 (2554) **Common**

TL 10–15 mm. Transparent FW has a black border and PM band and pale yellow streaks at tip. Head is orange. Thorax is variably blackish, orange, or red. Slender abdomen has a fanned orange tuft at tip. **HOSTS:** Maple. **NOTE:** Regularly encountered at lights.

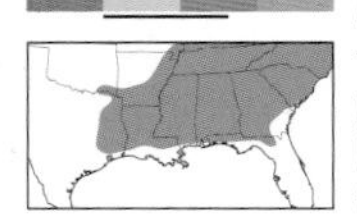

HOLLY BORER *Synanthedon kathyae* 64-0099 (2579) **Local**

TL 10–14 mm. Transparent FW has a thin black border and PM band, and some yellow scales at base and along costa. Black thorax has yellow collar and lateral stripes. Black abdomen has two or three yellow bands near tip with a pointed black tuft at tip. **HOSTS:** Holly.

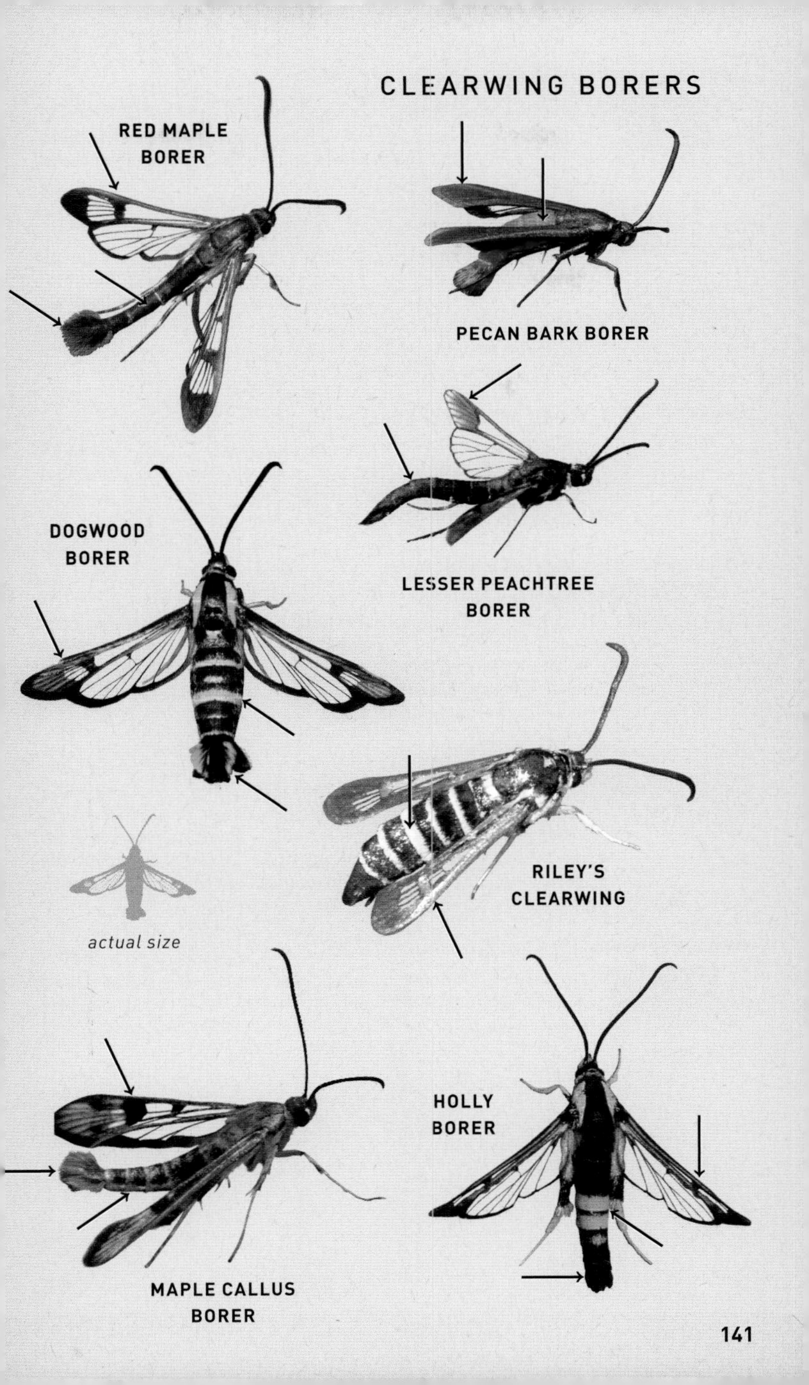
CLEARWING BORERS
RED MAPLE BORER
PECAN BARK BORER
LESSER PEACHTREE BORER
DOGWOOD BORER
RILEY'S CLEARWING
actual size
HOLLY BORER
MAPLE CALLUS BORER

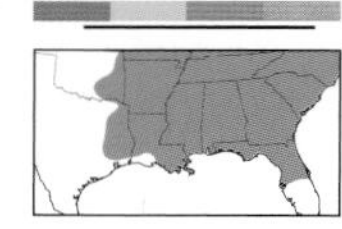

TUPELO CLEARWING

Synanthedon rubrofascia 64-0109 (2567) **Common**

TL 10–14 mm. Transparent FW (sometimes black) has thin black border, veins, and PM band. Head and thorax are black. Black abdomen has a red band at midpoint and a pointed black tuft at tip. **HOSTS:** Tupelo, including black gum.

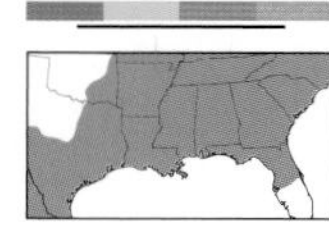

OAKGALL BORER

Synanthedon decipiens 64-0114 (2571) **Common**

TL 12–14 mm. Transparent FW has black and chestnut border and red PM band. Black thorax has a yellow collar. Black abdomen is banded yellow with a fanned, yellow-edged black tuft at tip. **HOSTS:** Galls of cynipid wasps and bark of oak.

FLORIDA OAKGALL BORER

Synanthedon sapygaeformis 64-0115 (2573) **Local**

TL 10–14 mm. Mostly black FW is streaked red through central area. Black thorax has a red collar. Black abdomen is banded red with a fanned black tuft at tip. **HOSTS:** Galls of cynipid wasps.

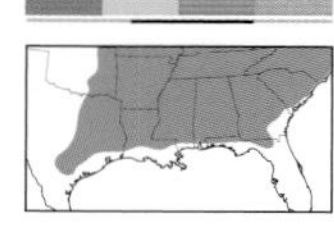

ARKANSAS BORER

Synanthedon arkansasensis 64-0117 (2575) **Common**

TL 10–14 mm. Transparent FW has a black border and PM band, and reddish streaks along costa and in ST area. Black thorax has yellow collar and lateral stripes. Black abdomen is banded yellow with a fanned, yellow-edged black tuft at tip. **HOSTS:** Unknown. **NOTE:** Regularly encountered at lights.

PEACHTREE BORER

Synanthedon exitiosa 64-0124 (2583) **Common**

TL 15–20 mm. Sexually dimorphic. Male resembles Lesser Peachtree Borer but is larger, with amber-tinted transparent wings. Head is dusted with white scales dorsally. Thorax has whitish collar. Female has black FW and a wide orange band on abdomen. **HOSTS:** Fruit-bearing trees, including cherry, peach, and plum.

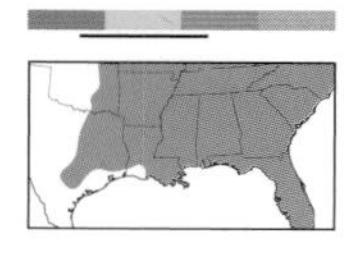

LILAC BORER *Podosesia syringae* 64-0130 (2589) **Common**

TL 16–18 mm. Purplish blue FW has some red scales at base. Black thorax has a red collar. Slightly constricted abdomen is black. Long legs are tipped yellow. Antennae are reddish. **HOSTS:** Ash, fringe tree, lilac, and related species (family Oleaceae).

CLEARWING BORERS

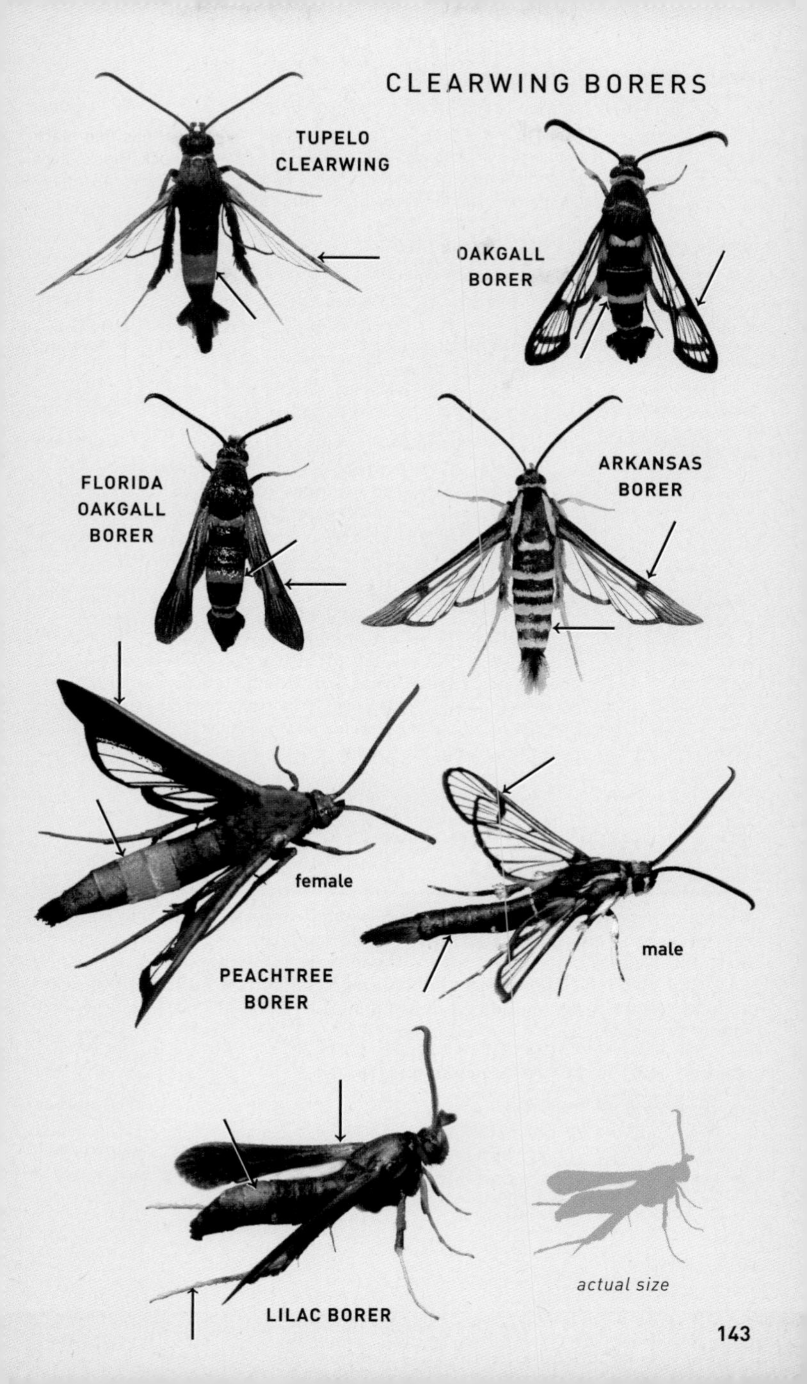

PERSIMMON BORER

Sannina uroceriformis 64-0132 (2590) **Common**

TL 16–18 mm. Black FW is tinged brown toward tip. Shiny black abdomen has an orange band (sometimes lacking) at midpoint. Often has orange epaulets and/or collar. Male has long, narrow tufts at tip of abdomen. Antennae are black. **HOSTS:** Persimmon.

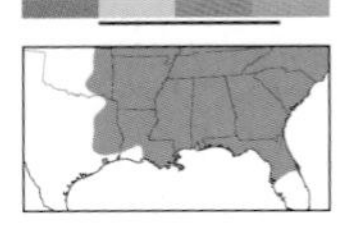

BLAZING STAR BORER

Carmenta anthracipennis 64-0134 (2592) **Common**

TL 10–14 mm. Blackish FW has a yellow-dusted transparent patch in central PM area. Black thorax has narrow yellow collar and lateral stripes. Blackish abdomen has thin yellow bands at base and midpoint. **HOSTS:** Blazing star.

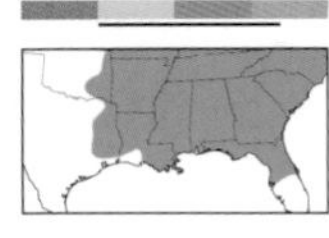

BONESET BORER

Carmenta pyralidiformis 64-0155 (2608) **Common**

TL 10–14 mm. Blackish FW has some yellow scales along veins in ST area. Black thorax has narrow yellow collar and lateral stripes. Blackish abdomen has a bold yellow band at midpoint with a narrow line at tip. Pointed brushy tip has a dusting of yellow scales. **HOSTS:** Boneset.

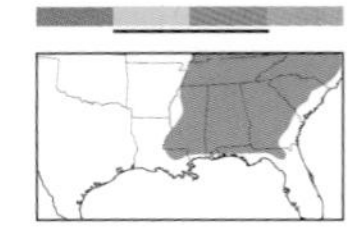

CLEMATIS CLEARWING

Alcathoe caudata 64-0172 (2623) **Uncommon**

TL 10–18 mm. Sexually dimorphic. Blackish FW is variably dusted reddish toward apex. Thorax and abdomen are reddish brown. Antennae are mostly orange. Hindlegs are adorned with brushy orange tufts. Male has a long yellow appendage trailing beyond tip of abdomen. Larger female is mostly blackish. **HOSTS:** Clematis and *Ribes* species.

Planthopper Parasite Moth

Family Epipyropidae

A small, sooty black moth that often rests with wings held outward. Eggs are laid on plants frequented by planthoppers (Homoptera, superfamily Fulgoroidea). First-instar caterpillars attach themselves to planthoppers, extracting body fluids from the abdomen beneath the wings. Adults are similar in appearance to some bagworm moths, but are generally smaller and antennae have longer branches.

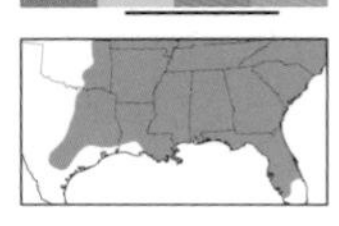

PLANTHOPPER PARASITE MOTH

Fulgoraecia exigua 66-0001 (4701) **Common**

TL 5–7 mm. Rounded wings and body are uniform sooty black. Sometimes has a very faint paler dot in central PM area of FW. Both sexes have broadly bipectinate antennae. **HOSTS:** Body fluids of planthoppers.

CLEARWING BORERS

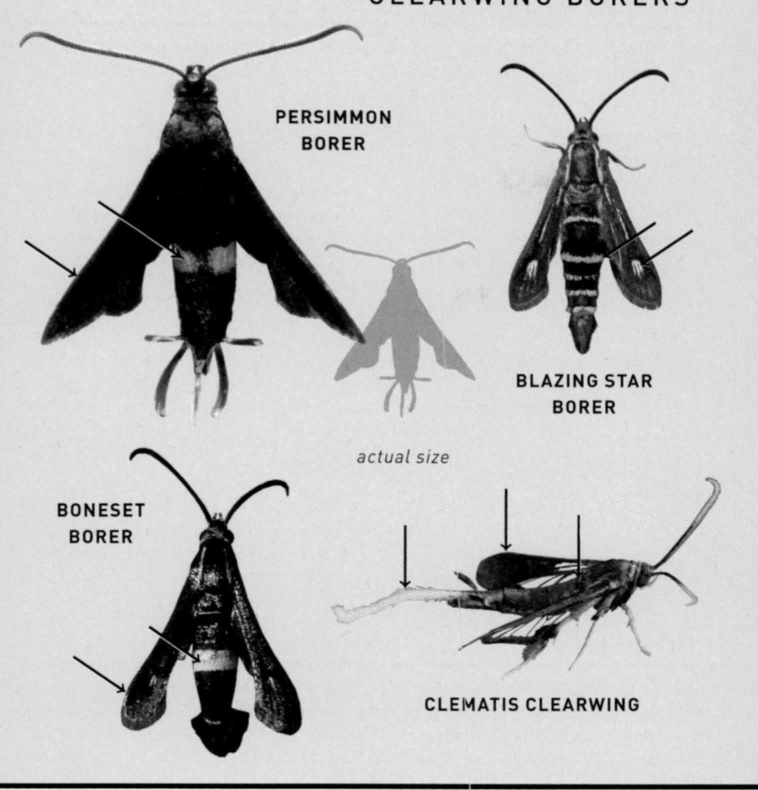

PLANTHOPPER PARASITE MOTH

Tropical Burnet Moths Family Lacturidae

Small white moths with black dots on the FW and pinkish orange thighs, superficially resembling ermine moths. Some species also have a pinkish orange fringe. Adults visit lights in small numbers.

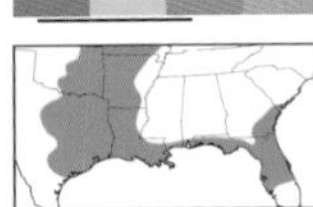

RED-SPOTTED LACTURA

Lactura basistriga 66-0003 (2404) **Common**

TL 11–13 mm. White FW is finely speckled with reddish dashes, especially toward apex. AM and PM lines are slanting rows of dark red dots. Patch at base of wing and fringe is pinkish red. Abdomen, HW, legs, and antennae are tinged orange. **HOSTS:** Unknown.

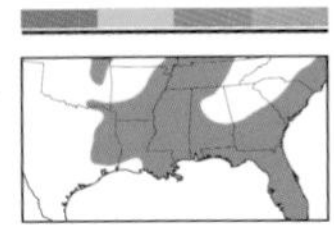

STREAKED LACTURA

Lactura pupula 66-0004 (2405) **Common**

TL 9–14 mm. White FW is strikingly patterned with thin black streaks along costa and in ST area, and black spots in inner AM and PM areas. Fringe, HW, head, and inner leg segments are orange. **HOSTS:** Saffron plum.

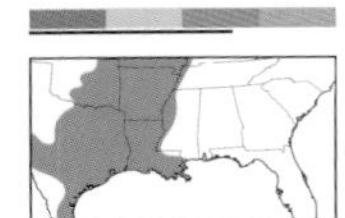

SPECKLED LACTURA

Lactura subfervens 66-0006 (2407) **Common**

TL 11–13 mm. Whitish FW is evenly speckled with reddish scales, sometimes with bolder spots along AM and PM lines. HW is orange. Legs and antennae are tinged pink. **HOSTS:** Unknown.

Slug Moths Family Limacodidae

Small, chunky moths that hold their rounded wings in a tentlike position when at rest. Some species curl their abdomen above the level of the wings. Larvae of many species are bizarre in form and color and often have stinging hairs. Found mostly in deciduous woodlands. The adults are strictly nocturnal and visit lights in small numbers.

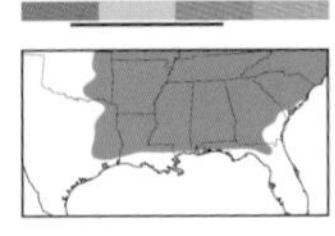

EARLY BUTTON SLUG MOTH

Tortricidia testacea 66-0010 (4652) **Common**

TL 8–12 mm. Pale orange FW has darker veins in ST area. Slanting rusty median band ends at apex. **HOSTS:** Deciduous trees, including beech, birch, black cherry, chestnut, and oak.

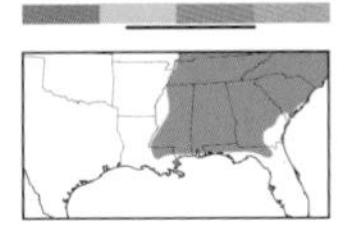

RED-CROSSED BUTTON SLUG MOTH

Tortricidia pallida 66-0011 (4653) **Local**

TL 8–12 mm. Many individuals are almost identical to Abbreviated Button Slug Moth. Paler individuals that lack obvious lines on the FW may possibly be identified with caution as this species. **HOSTS:** Deciduous trees, including beech, cherry, oak, and willow.

TROPICAL BURNET MOTHS

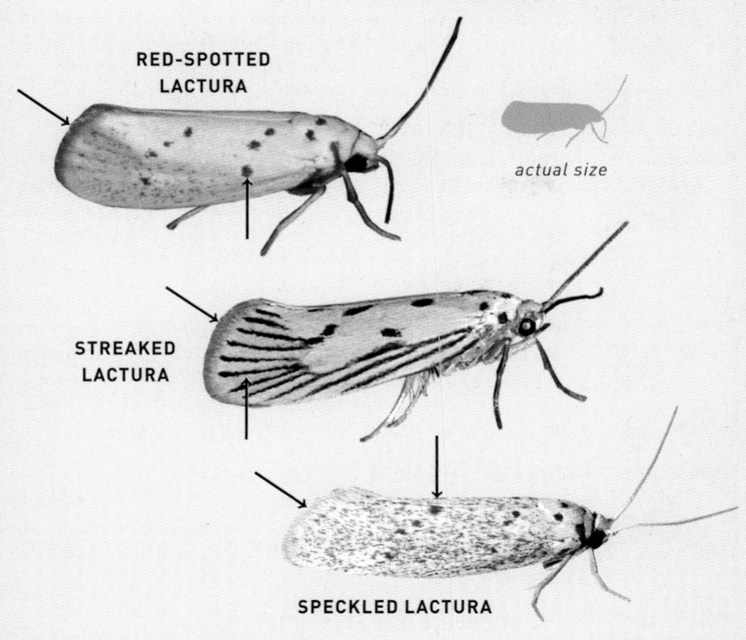

SLUG MOTHS

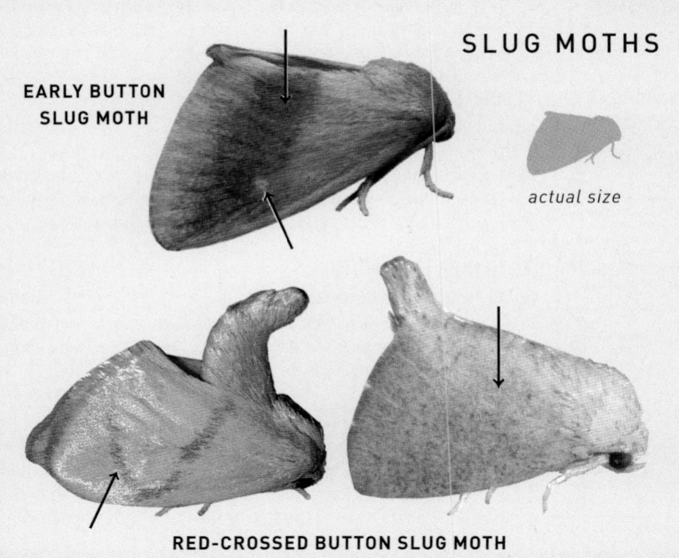

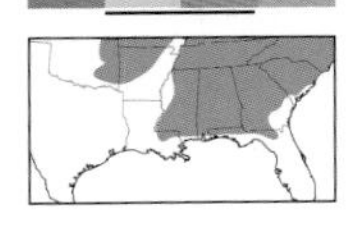

ABBREVIATED BUTTON SLUG MOTH

Tortricidia flexuosa 66-0012 (4654) **Common**

TL 8–12 mm. Yellowish FW has indistinct PM and ST lines fused at costa, forming a U-shape. Median area is sometimes shaded brown. **HOSTS:** Deciduous trees, including apple, black cherry, chestnut, oak, and plum.

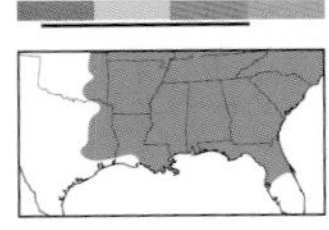

RED-EYED BUTTON SLUG MOTH

Heterogenea shurtleffi 66-0015 (4657) **Common**

TL 8–10 mm. Yellowish to rusty brown FW has fragmented brown PM and ST lines fused at costa, forming a U-shape. Often has some dusky mottling in central PM area. **HOSTS:** Unknown, but probably various woody plants.

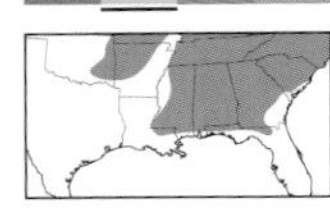

JEWELED TAILED SLUG MOTH

Packardia geminata 66-0017 (4659) **Common**

TL 8–12 mm. Peppery light brown FW has darker brown median area. Inner ST line is marked with a string of three white dots. **HOSTS:** Trees and shrubs, including birch, hickory, oak, and spruce.

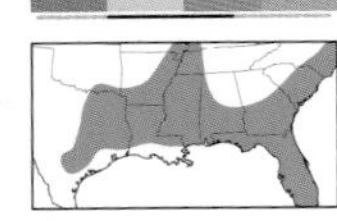

GRACEFUL SLUG MOTH

Lithacodes gracea 66-0022 (4664) **Common**

TL 12–14 mm. Orange FW has a crisp white median line that usually has a slight kink at midpoint and brown shading distally. **HOSTS:** Unknown.

YELLOW-SHOULDERED SLUG MOTH

Lithacodes fasciola 66-0023 (4665) **Common**

TL 10–13 mm. Orange to rusty FW has a kinked white median line edged gray and black. ST line angles across apex. **HOSTS:** Deciduous trees and shrubs, including apple, beech, elm, oak, and willow.

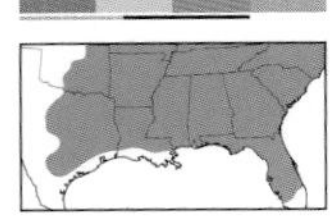

YELLOW-COLLARED SLUG MOTH

Apoda y-inversum 66-0025 (4667) **Common**

TL 11–15 mm. Pale orange FW has brown median and ST lines that converge toward costa. Two indistinct lines form an X-shape near anal angle. Darker individuals have brown shading within lines, forming a Y-shape. **HOSTS:** Beech, hickory, ironwood, and oak.

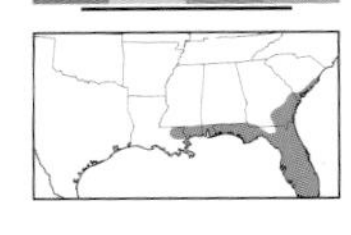

RECTILINEA SLUG MOTH

Apoda rectilinea 66-0026 (4668) **Local**

TL 10–12 mm. Light brown FW has slightly darker basal half. Black-edged, white median line fades before reaching costa. ST line across apex is faint on some individuals. **HOSTS:** Unknown.

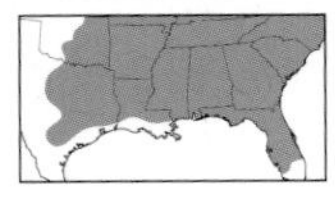

SHAGREENED SLUG MOTH

Apoda biguttata 66-0027 (4669) **Common**

TL 10–15 mm. Grayish brown FW is marked with white-edged chestnut patches at apex and anal angle. Basal area is often darker within oblique white AM line. **HOSTS:** Ironwood, hickory, and oak.

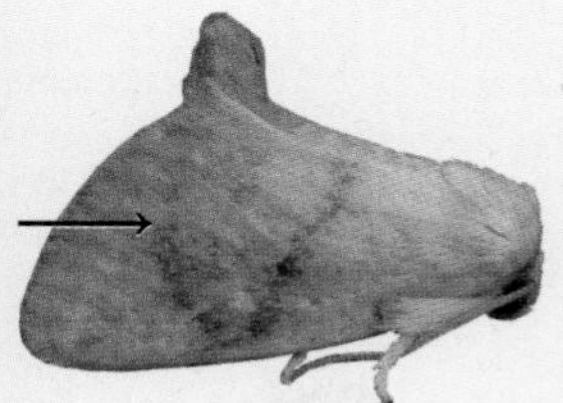

ABBREVIATED BUTTON SLUG MOTH

RED-EYED BUTTON SLUG MOTH

actual size

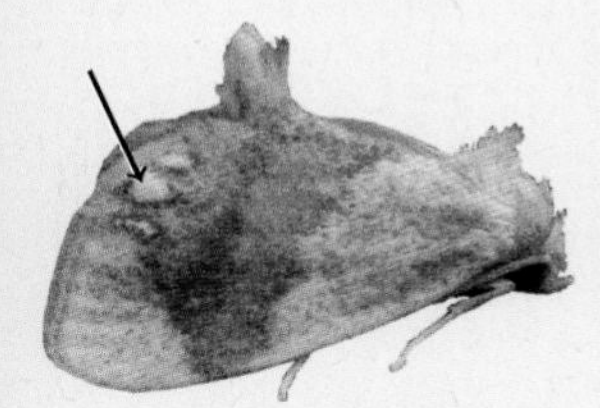

JEWELED TAILED SLUG MOTH

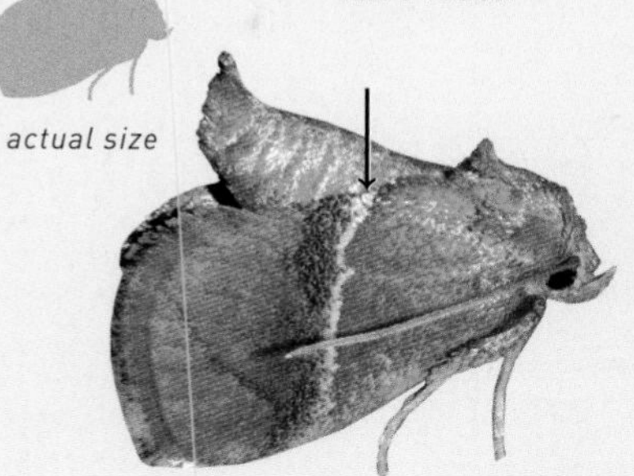

GRACEFUL SLUG MOTH

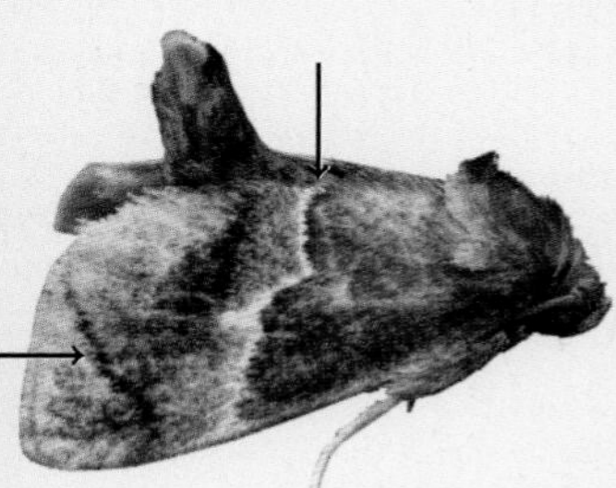

YELLOW-SHOULDERED SLUG MOTH

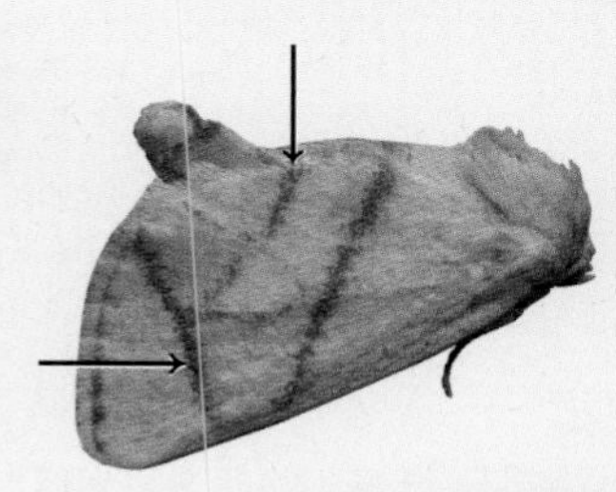

YELLOW-COLLARED SLUG MOTH

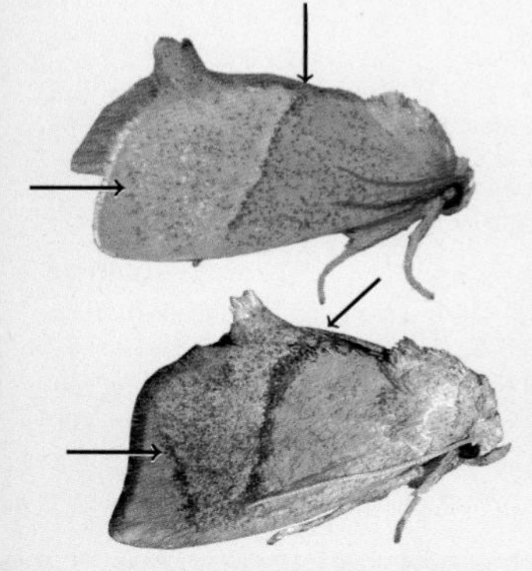

RECTILINEA SLUG MOTH

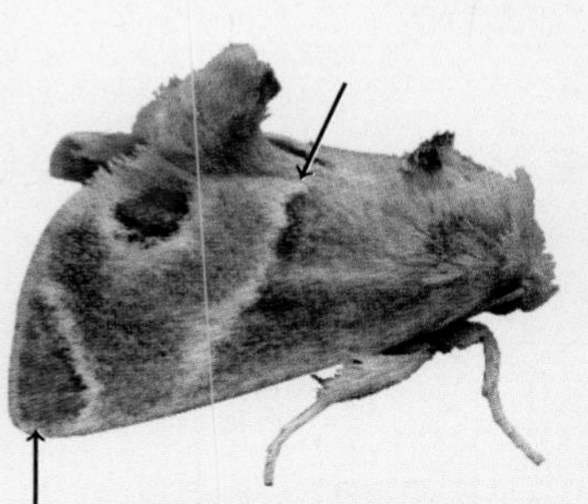

SHAGREENED SLUG MOTH

SKIFF MOTH

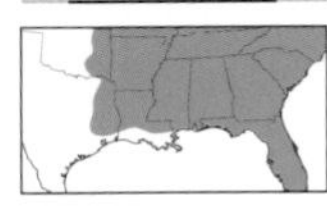

Prolimacodes badia 66-0029 (4671) **Common**

TL 12–17 mm. Milky coffee-colored FW has a rounded, white-edged brown median patch that continues along costa to base of wing. **HOSTS:** Trees and woody plants, including birch, blueberry, oak, poplar, and willow.

PACKARD'S WHITE SLUG MOTH

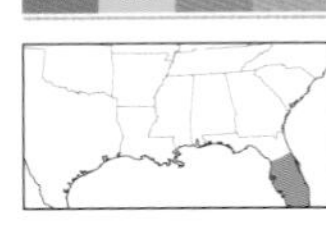

Alarodia slossoniae 66-0031 (4673) **Local**

TL 10–12 mm. Uniform satin white FW is marked only by a tiny brown dot in inner PM area. Often rests with hairy white forelegs splayed out. **HOSTS:** Unknown. **NOTE:** Commonly called Packard's White Flannel Moth, but unrelated to moths of family Megalopygidae.

SPUN GLASS SLUG MOTH

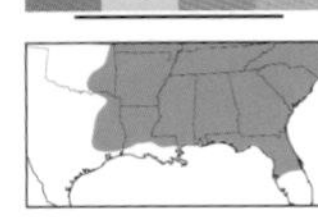

Isochaetes beutenmuelleri 66-0033 (4675) **Common**

TL 10–12 mm. Orange brown FW has a marbled pattern of indistinct brown lines and silvery gray patches. Rests with heavily tufted legs splayed outward. **HOSTS:** Swamp oak.

HAG MOTH

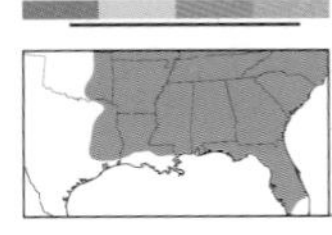

Phobetron pithecium 66-0035 (4677) **Common**

TL 10–15 mm. Sexually dimorphic. FW of male is largely transparent with black and gold veins, discal spot, and fringe. Female has purplish gray FW with irregular black and yellow lines. Both sexes have central pair of legs adorned with large pale tufts. **HOSTS:** Woody plants, including apple, ash, dogwood, oak, and willow. **NOTE:** Also known as Monkey Slug Moth.

NASON'S SLUG MOTH

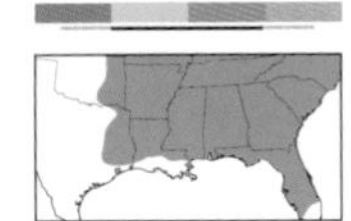

Natada nasoni 66-0037 (4679) **Common**

TL 10–15 mm. Light brown FW has an oblique white-edged AM line that almost touches ST line near apex. **HOSTS:** Beech, hickory, and eastern hornbeam.

CROWNED SLUG MOTH

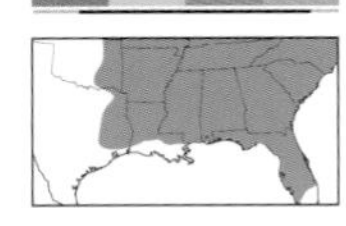

Isa textula 66-0039 (4681) **Common**

TL 9–16 mm. Peppery light brown FW has a diffuse pale gray median band and apical patch. **HOSTS:** Trees and shrubs, including elm, hickory, maple, and oak.

PURPLE-CRESTED SLUG MOTH

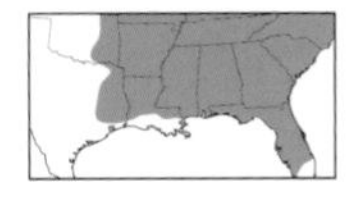

Adoneta spinuloides 66-0043 (4685) **Common**

TL 8–12 mm. Rusty brown FW has a wide blackish streak through median area and black-dotted ST line. White basal dash almost joins incomplete peppery white median line near inner margin. **HOSTS:** Trees and shrubs, including beech, birch, chestnut, basswood, and willow.

SLUG MOTHS

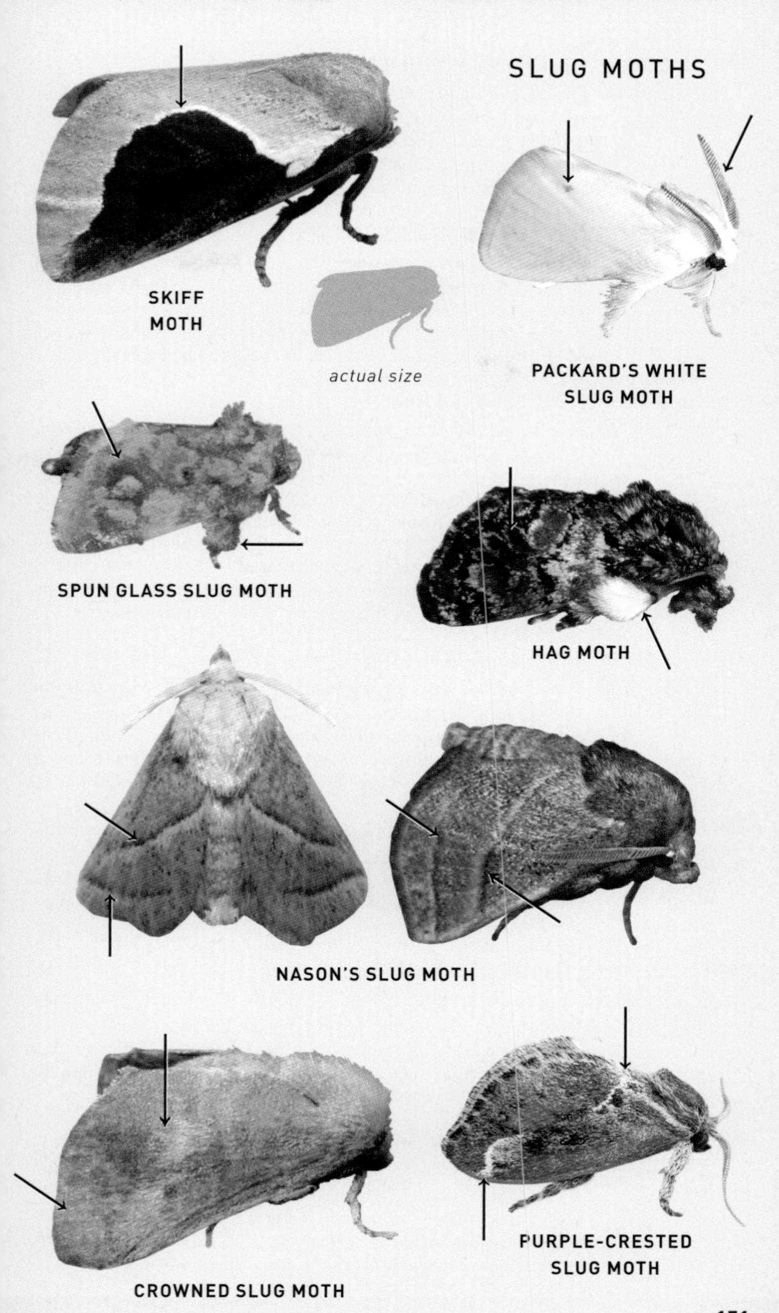
SKIFF MOTH
actual size
PACKARD'S WHITE SLUG MOTH
SPUN GLASS SLUG MOTH
HAG MOTH
NASON'S SLUG MOTH
CROWNED SLUG MOTH
PURPLE-CRESTED SLUG MOTH

PIN-STRIPED VERMILION SLUG MOTH

Euclea semifascia 66-0047 (4691) **Common**

TL 8–12 mm. Reddish brown FW has a wavy black-edged white AM line that starts at inner margin and ends in central part of wing. Rests with slightly tufted legs splayed outward. **HOSTS:** Cherry, oak, pecan, and persimmon.

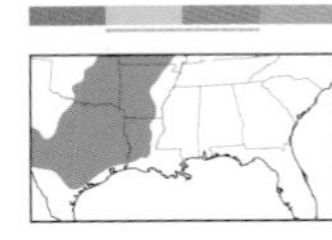

GREEN OAK-SLUG MOTH

Euclea incisa 66-0050 (4696) **Local**

TL 10–15 mm. Apple green FW has a chocolate brown ST band that wraps around anal angle, usually forming a point into median area. Brown edging along costa is very narrow. Thorax and legs are brown. **HOSTS:** Unknown, but presumably deciduous trees.

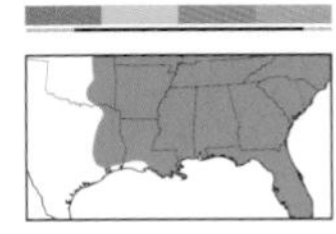

SPINY OAK-SLUG MOTH

Euclea delphinii 66-0051 (4697) **Common**

TL 10–15 mm. Chocolate brown FW is marked with variably sized apple green patches in inner median and outer PM areas that can sometimes be fused or even mostly absent. Chestnut shading borders green patch in inner median and outer ST areas. Black reniform spot is sometimes large. Some individuals resemble Green Oak-Slug Moth, except projection into median area is rounded and costa is more broadly edged brown, especially basally. **HOSTS:** Trees and woody plants, including apple, beech, chestnut, maple, and oak.

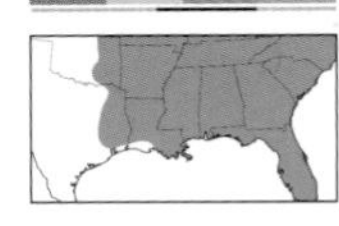

NANINA OAK-SLUG MOTH

Euclea nanina 66-0052 (4697.1) **Uncommon**

TL 9–10 mm. Resembles larger Spiny Oak-Slug Moth, and green patches are similarly variable in size, but median patch is generally larger than ST patch and is broader basally, creating a more square shape. **HOSTS:** Unknown, but presumably deciduous trees.

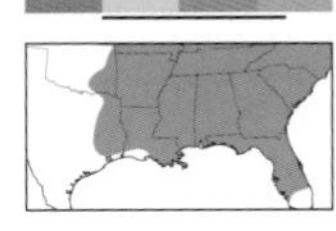

SMALLER PARASA *Parasa chloris* 66-0054 (4698) **Common**

TL 10–14 mm. Resembles green form of the oak-slug moths, but thorax is green and basal area is brown. Broad, brown ST band does not wrap around anal angle. **HOSTS:** Deciduous trees, including apple, dogwood, elm, and oak.

STINGING ROSE CATERPILLAR MOTH

Parasa indetermina 66-0055 (4699) **Common**

TL 12–15 mm. Resembles Smaller Parasa but has a broader green median area. Midpoint of narrow brown terminal line is often marked with diffuse blackish shading. **HOSTS:** Trees and shrubs, including apple, dogwood, hickory, maple, and rose.

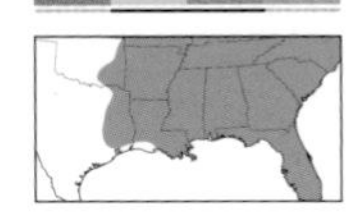

SADDLEBACK CATERPILLAR MOTH

Acharia stimulea 66-0055 (4700) **Common**

TL 14–22 mm. Chocolate brown FW is patterned with peppery ash gray streaks along costa, inner margin, and central median area. Has small white dots in basal area and along outer PM area. **HOSTS:** Various, including apple, blueberry, elm, maple, and oak.

SLUG MOTHS

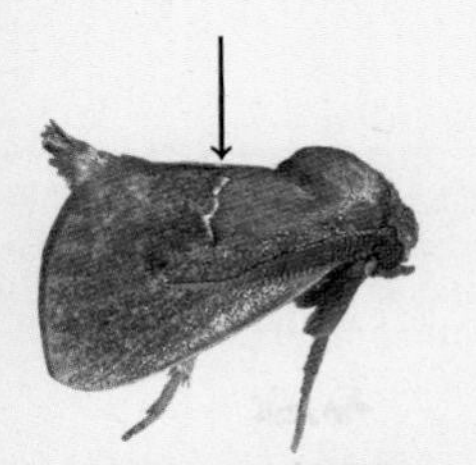

PIN-STRIPED VERMILION SLUG MOTH

actual size

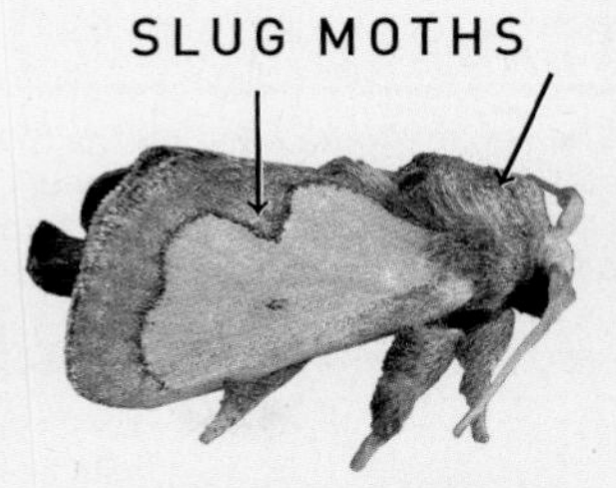

GREEN OAK-SLUG MOTH

SPINY OAK-SLUG MOTH

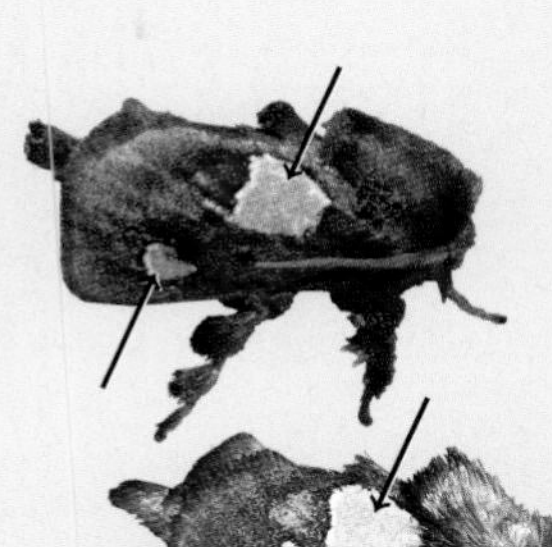

NANINA OAK-SLUG MOTH

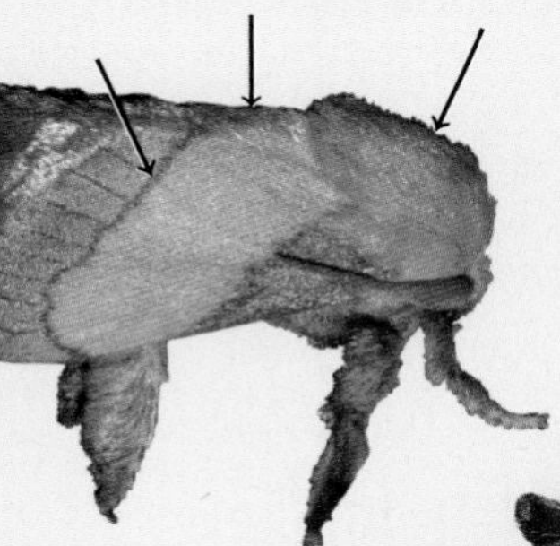

SMALLER PARASA

STINGING ROSE CATERPILLAR MOTH

SADDLEBACK CATERPILLAR MOTH

FLANNEL MOTHS Family Megalopygidae

Chunky, medium-sized moths that have a thickly plush thorax and hold their soft, velvety wings in a tent when resting. Males have broad, comblike antennae. Caterpillars bear stinging setae; those of *Megalopyge* are densely hairy. Adults are nocturnal and regularly visit lights in small numbers.

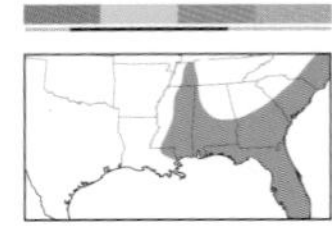

YELLOW FLANNEL MOTH

Megalopyge pyxidifera 66-0059 (4642) **Common**

TL 14–17 mm. Pale orange FW has a rippled texture, creating parallel lines. Orange thorax is densely hairy. Stout abdomen is tipped whitish. **HOSTS:** Blueberry, oak, and plum.

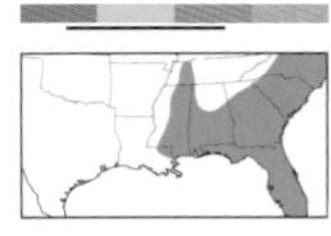

FLORIDA FLANNEL MOTH

Megalopyge lacyi 66-0060 (4643) **Local**

TL 13–15 mm. Cream-colored FW is often rippled. Note blackish dot in outer median area. Sometimes has a broad, fragmented, faint brown PM line. **HOSTS:** Unknown.

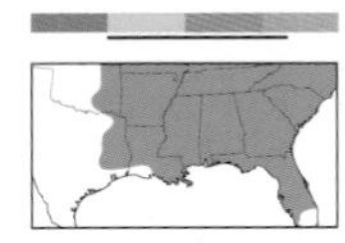

BLACK-WAVED FLANNEL MOTH

Megalopyge crispata 66-0061 (4644) **Common**

TL 13–20 mm. Sexually dimorphic. Male has cream-colored FW patterned with distinctive rippled black and chestnut lines through median area. Paler, faintly patterned female has filiform antennae. **HOSTS:** Many woody plants, including apple, birch, oak, poplar, and willow.

SOUTHERN FLANNEL MOTH

Megalopyge opercularis 66-0063 (4647) **Common**

TL 13–19 mm. Cream-colored FW is shaded chesnut in basal area and black along upper costa, with white veins in median area; it is sometimes rippled in texture, diffusing white markings. Orange thorax has a yellow semi-collar. Legs are boldly tipped black. **HOSTS:** Woody plants, including almond, apple, hackberry, orange, and pecan.

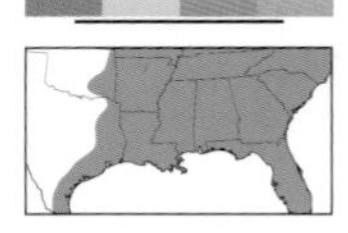

WHITE FLANNEL MOTH

Norape ovina 66-0068 (4650) **Common**

TL 12–18 mm. Body and wings are uniform satin white. Comblike antennae of male are contrastingly orange. **HOSTS:** Black locust, hackberry, and redbud.

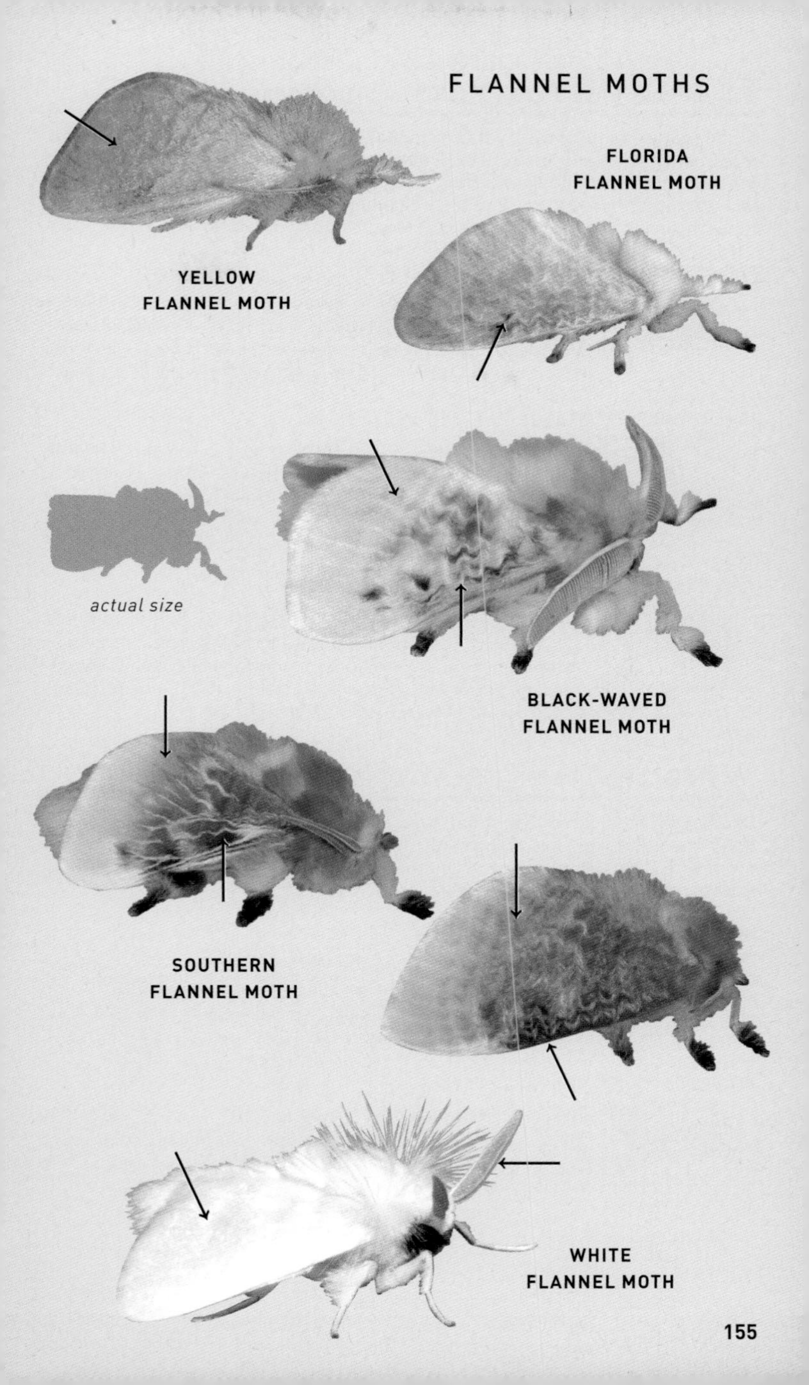
FLANNEL MOTHS
FLORIDA
FLANNEL MOTH
YELLOW
FLANNEL MOTH
actual size
BLACK-WAVED
FLANNEL MOTH
SOUTHERN
FLANNEL MOTH
WHITE
FLANNEL MOTH

Leaf Skeletonizers Family Zygaenidae

Small moths, some of which may be mistaken for lycid beetles in the genus *Calopteron*, or smaller members of the tiger moths (family Erebidae, tribe Arctiini). Moths in this family produce hydrogen cyanide, and their bright coloration is a warning to predators. Found in deciduous woodlands, most are active at flowers during the day, but will also come to lights at night.

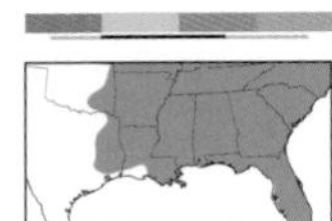

CLEMENS' FALSE SKELETONIZER

Acoloithus falsarius 66-0069 (4629) Common

TL 7–9 mm. Blackish wings and body are relieved only by an incomplete orange collar. Pectinate antennae are black. Rests with broad, rounded wings held flatly together. **HOSTS:** Grape and peppervine.

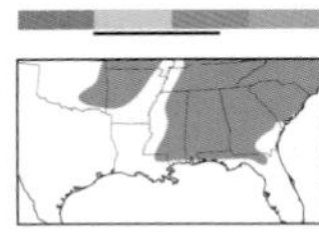

ORANGE-PATCHED SMOKY MOTH

Pyromorpha dimidiata 66-0083 (4639) Common

TL 8–12 mm. Semi-translucent smoky black FW has a contrasting pale orange patch covering outer basal area. Pectinate antennae are black. Rests with ample, rounded wings held flatly together. **HOSTS:** Dead oak leaves.

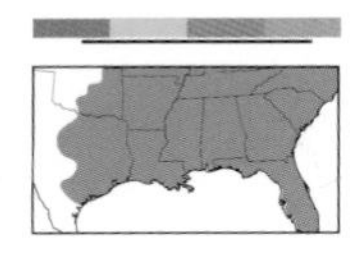

GRAPELEAF SKELETONIZER

Harrisina americana 66-0092 (4624) Common

TL 10–15 mm. Semi-translucent wings and tufted abdomen are smoky black. Thorax has a bright orange collar. Pectinate antennae are black. Rests with narrow wings held outward. **HOSTS:** Grape, redbud, and Virginia creeper.

Window-Winged Moths Family Thyrididae

Chunky day-flying moths that habitually spread their wings when alighting on flower heads or wet sand along forest tracks. The wings have spotted patterns, often with translucent panels or spots forming windows. The larvae are leafrollers or borers. Adults take nectar from flowers.

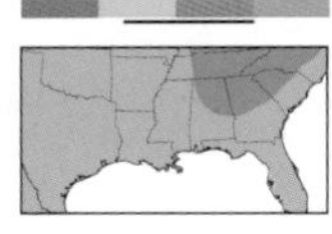

SPOTTED THYRIS

Thyris maculata 70-0003 (6076) Common

WS 12–15 mm. Blackish wings and body are speckled orange. Wings have small semi-translucent white spots in median area. Fringes are partly white. **HOSTS:** Clematis and *Houstonia* species.

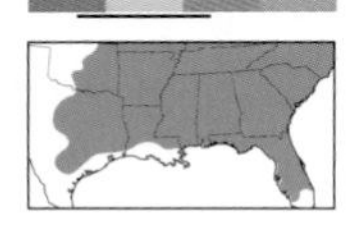

MOURNFUL THYRIS

Pseudothyris sepulchralis 70-0004 (6077) Common

WS 15–23 mm. Black wings and body are spotted with white. Patches in median areas of wings are sometimes fused. Fringes have isolated white spots. **HOSTS:** Clematis and grape.

LEAF SKELETONIZERS

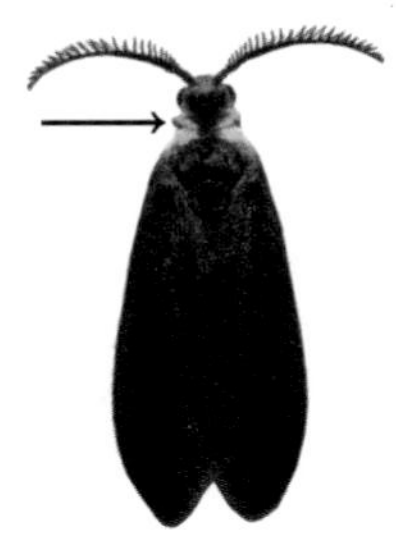

CLEMENS' FALSE SKELETONIZER

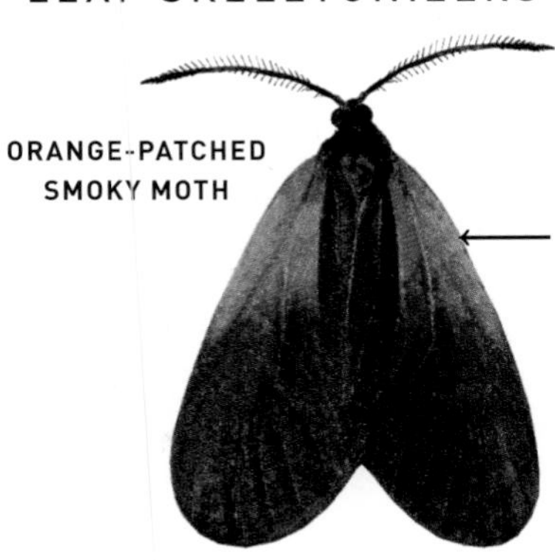

ORANGE-PATCHED SMOKY MOTH

actual size

GRAPELEAF SKELETONIZER

WINDOW-WINGED MOTHS

SPOTTED THYRIS

MOURNFUL THYRIS

actual size

EYED DYSODIA

Dysodia oculatana 70-0005 (6078) **Local**

WS 18–21 mm. Brown wings and body are spangled with orange and gold. FW has a small translucent reniform spot. A large tooth-shaped translucent spot marks central median area of HW. **HOSTS:** Bean and *Eupatorium* species.

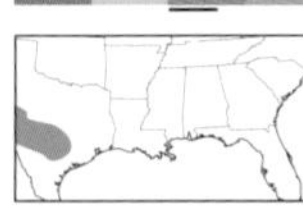

NETTED DYSODIA

Dysodia granulata 70-0006 (6079) **Local**

WS 20–26 mm. Orange to brown wings are covered with a fine net-like pattern of darker lines and veins. FW has a small translucent reniform spot. A large double-toothed translucent spot marks central median area of HW. **HOSTS:** Possibly bean.

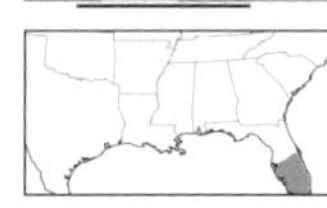

SEAGRAPE BORER

Hexeris enhydris 70-0009 (6082) **Local**

WS 34–38 mm. Golden wings are crisply patterned with a network of fine rusty lines and veins. Costa has slightly concave contour. Labial palps are long. **HOSTS:** Sea grape and pigeon-plum.

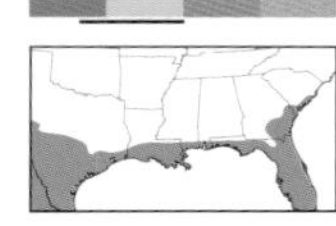

WINDOW-WINGED MOTH

Meskea dyspteraria 70-0010 (6085) **Local**

WS 38 mm. Unique. The tightly rolled FW exposes the pointed HW and raised abdomen. Streaky tan FW is speckled darker. HW has black and white median bands. **HOSTS:** Mallow.

ASSORTED PYRALIDS Family Pyralidae, Subfamilies Galleriinae, Pyralinae, and Epipaschiinae

Small, broad-winged, delta-shaped pyralid moths. A few species rest with their wings spread open and abdomen slightly raised. The larvae of some species live mostly indoors and feed on stored grains, dried vegetable matter, or even dead animals. Many are nocturnal and come freely to lights. Some can be found indoors as adults and may be encountered year-round.

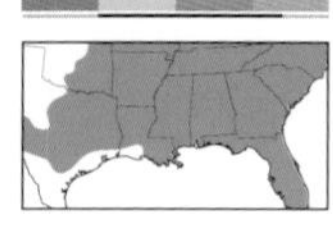

GREATER WAX MOTH

Galleria mellonella 80-0001 (5622) **Common**

TL 13–18 mm. Grayish brown FW has a streaky pattern of black-and-white dashes, especially along curved PM line. Outer margin of FW is concave. Labial palps are short and pointed. Females are larger than males. **HOSTS:** Beeswax, dried fruit, pollen, and dead insects.

LESSER WAX MOTH *Achroia grisella* 80-0002 (5623) **Local**

TL 8–12 mm. Rounded FW is uniformly mousy gray. Head is pale yellow. Females are larger than males. **HOSTS:** Old wax and debris of vacated beehives. **NOTE:** Introduced.

WINDOW-WINGED MOTHS

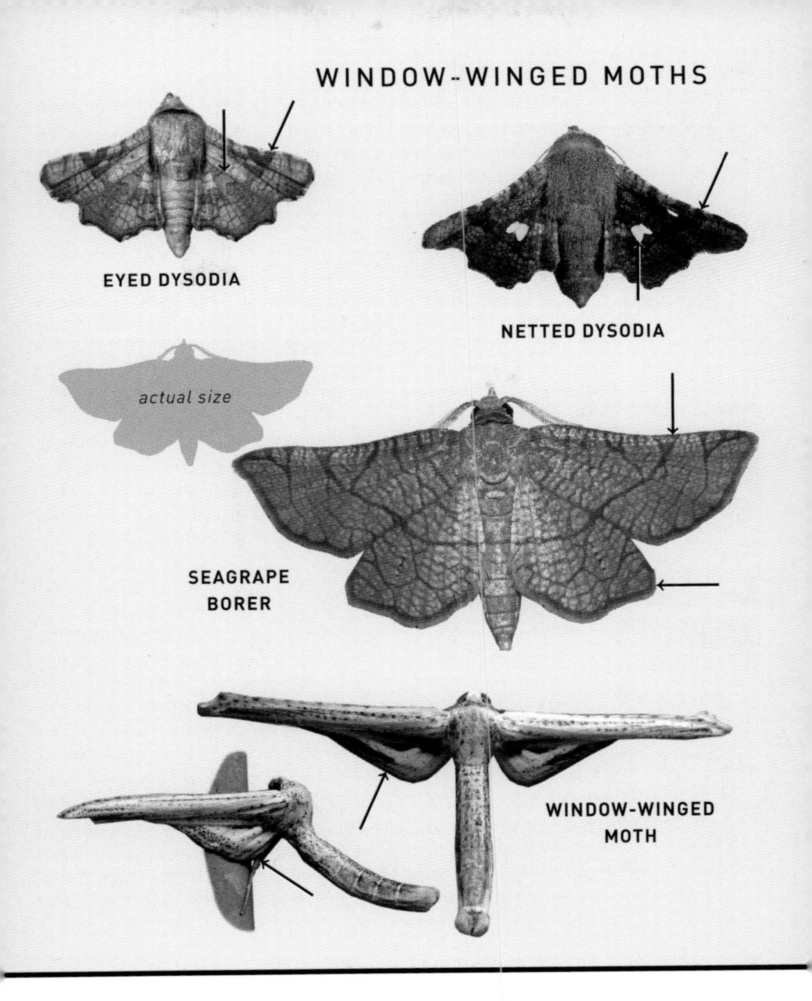

ASSORTED PYRALIDS

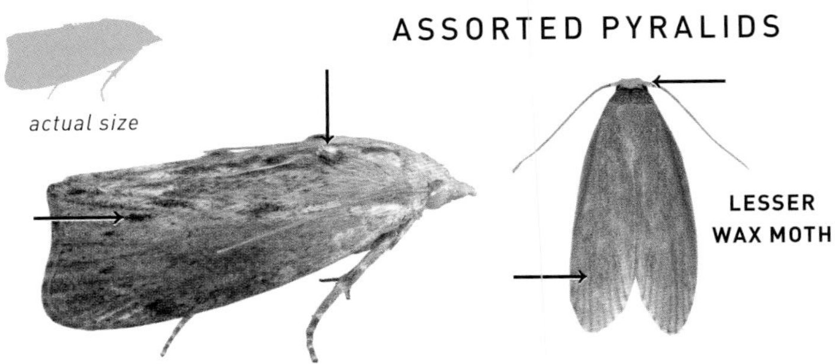

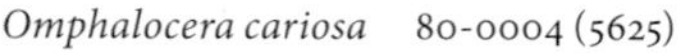

MOONSEED WEBWORM

Omphalocera cariosa 80-0004 (5625) **Common**

TL 15–20 mm. Grayish brown FW has outer median area shaded dark brown. Jagged AM and PM lines are inconspicuous except toward costa. ST area is finely streaked black along veins. **HOSTS:** Moonseed.

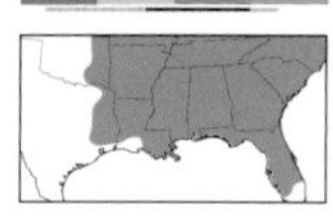

ASIMINA WEBWORM

Omphalocera munroei 80-0006 (5627) **Common**

TL 13–18 mm. Chestnut brown FW has a contrasting light brown basal area. Jagged AM and PM lines are inconspicuous. Central median area is marked with two whitish dots. Some individuals are light brown with darker shading in median area. **HOSTS:** Pawpaw.

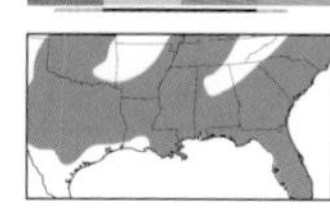

OCHRE PARACHMA

Parachma ochracealis 80-0033 (5538) **Common**

TL 9–10 mm. Yellowish to reddish brown FW has almost straight, pale yellow AM and PM lines. Fringe is pale yellow beyond thin, dark brown terminal line. Typically rests with body and flat wings held almost vertically, propped up by long tufted forelegs. **HOSTS:** Unknown.

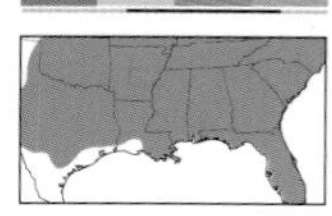

BOXWOOD LEAFTIER

Galasa nigrinodes 80-0048 (5552) **Common**

TL 9–11 mm. Brick red FW has scarlet basal area and fragmented pale lines that converge at costa, forming a pale patch. Costal margin is concave at midpoint. Rests with flat wings held at a slight angle, supported by long tufted legs. **HOSTS:** Boxwood.

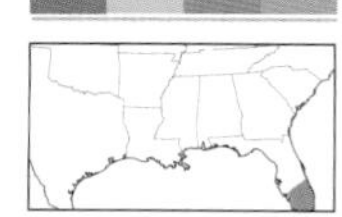

AMAZON QUEEN MOTH

Penthesilea difficilis 80-0050 (5554) **Local**

TL 10–12 mm. Dark pinkish red FW is shaded darker in outer basal and central median areas. White AM line curves inward toward costa. PM line is straight. Rests with slightly creased wings at a steep angle, supported by long forelegs. **HOSTS:** Golden dewdrop and light-blue snakeweed.

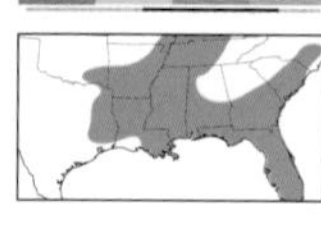

LONG-SNOUTED PENTHESILEA

Penthesilea sacculalis 80-0051 (5555) **Uncommon**

TL 8–9 mm. Reddish brown FW has an almost straight, white AM line and a weakly defined, curved PM line. Note fiery orange spots in median and subapical areas. Long downward-pointing labial palps create a snoutlike appearance. Rests at a slight angle with abdomen raised. **HOSTS:** Unknown.

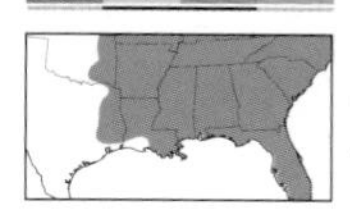

DIMORPHIC TOSALE

Tosale oviplagalis 80-0052 (5556) **Common**

TL 8–9 mm. Sexually dimorphic. Male has pinkish brown FW with a wide dark brown AM band and ST shading, edged with white lines. Female has similar pattern but is paler and grayer. Rests at a slight angle with abdomen raised. **HOSTS:** Unknown.

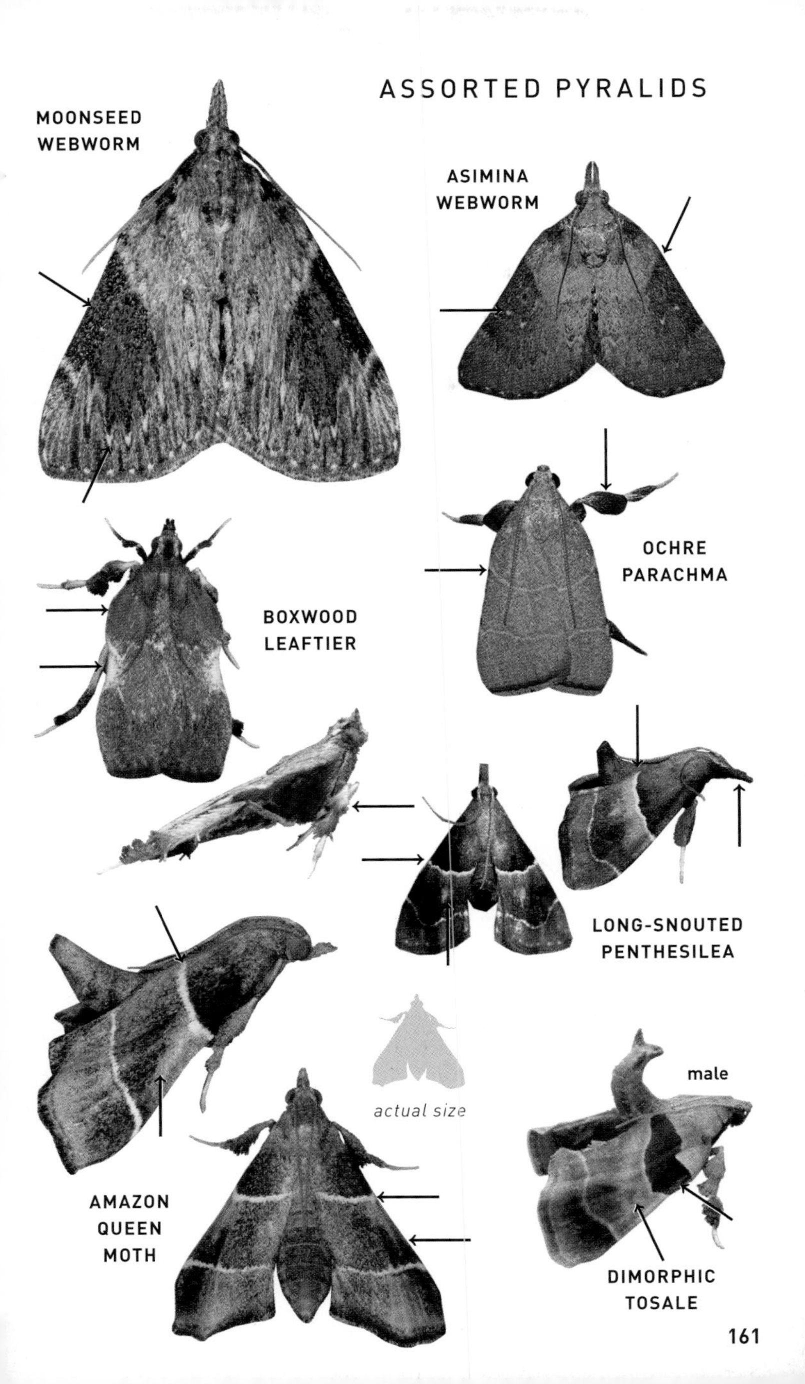
ASSORTED PYRALIDS
MOONSEED WEBWORM
ASIMINA WEBWORM
OCHRE PARACHMA
BOXWOOD LEAFTIER
LONG-SNOUTED PENTHESILEA
actual size
male
AMAZON QUEEN MOTH
DIMORPHIC TOSALE

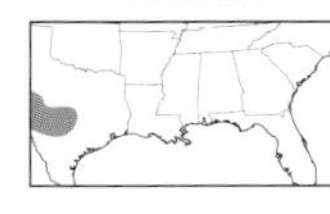

SINCERE SALOBRENA

Salobrena sincera 80-0055 (5559) **Local**

TL 6–7 mm. Ample brown FW has a crinkled-looking indentation at midpoint of costa. Paler AM and PM lines are poorly defined. Middle pair of legs (perhaps just in male) are adorned with brushy tufts. **HOSTS:** Unknown.

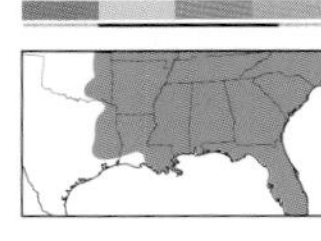

TRUMPET VINE MOTH

Clydonopteron sacculana 80-0061 (5563) **Common**

TL 8–13 mm. Chestnut brown FW is silvery gray beyond double PM line. Fiery orange AM band and spot in median area glow brightly. Note wavy indentation at midpoint of costa. Long black labial palps diverge at tip. **HOSTS:** Seedpods of trumpet creeper.

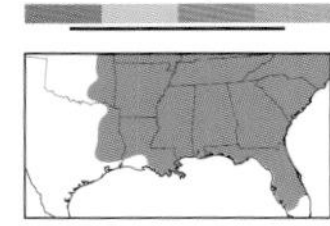

POSTURING ARTA *Arta statalis* 80-0062 (5566) **Common**

TL 8–9 mm. Rusty brown or tawny FW has slightly wavy, parallel pale yellow AM and PM lines. ST line is dotted. Fringe is reddish beyond black terminal line. Rests with flat wings held almost vertically, supported by long forelegs. **HOSTS:** Unknown.

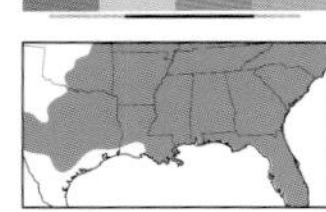

OLIVE ARTA

Arta olivalis 80-0064 (5568) **Common**

TL 8–9 mm. Similar to Posturing Arta, but FW is olive green with faint AM and PM lines. Terminal line and fringe are chestnut brown. **HOSTS:** Unknown.

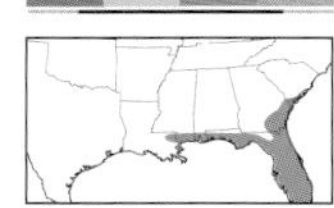

RUSTY HELIADES

Heliades mulleolella 80-0069 (5574) **Local**

TL 6–8 mm. Rusty brown FW is marked with slightly jagged, black-edged, pale yellow AM and PM lines. Dashed terminal line contrasts with silvery gray fringe. **HOSTS:** Unknown.

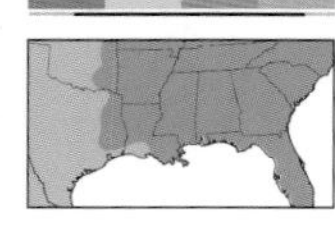

COMMON MEAL MOTH

Pyralis farinalis 80-0072 (5510) **Common**

TL 14–16 mm. Chestnut brown FW has a wide fawn brown median area bordered by wavy white AM and PM lines. **HOSTS:** Stored grain products.

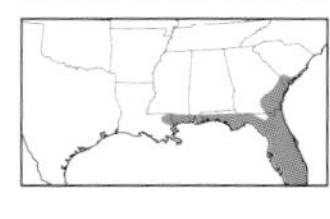

TROPICAL MEAL MOTH

Pyralis manihotalis 80-0073 (5515) **Local**

TL 8–10 mm. Peppery light brown FW has darker brown basal area bordered by wavy white AM line. Wavy white PM line widens at costa. Black discal spot is usually present. **HOSTS:** Unknown.

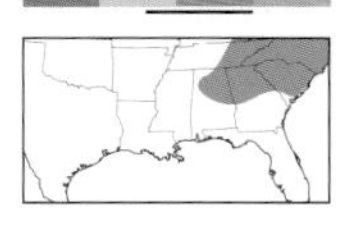

CALICO PYRALID

Aglossa costiferalis 80-0074 (5511) **Uncommon**

TL 10–11 mm. Shiny pinkish tan FW has blackish basal area bordered by angled white AM line. Black costal streak is accented with white dots. Black discal and subapical spots are usually obvious. **HOSTS:** Unknown.

ASSORTED PYRALIDS

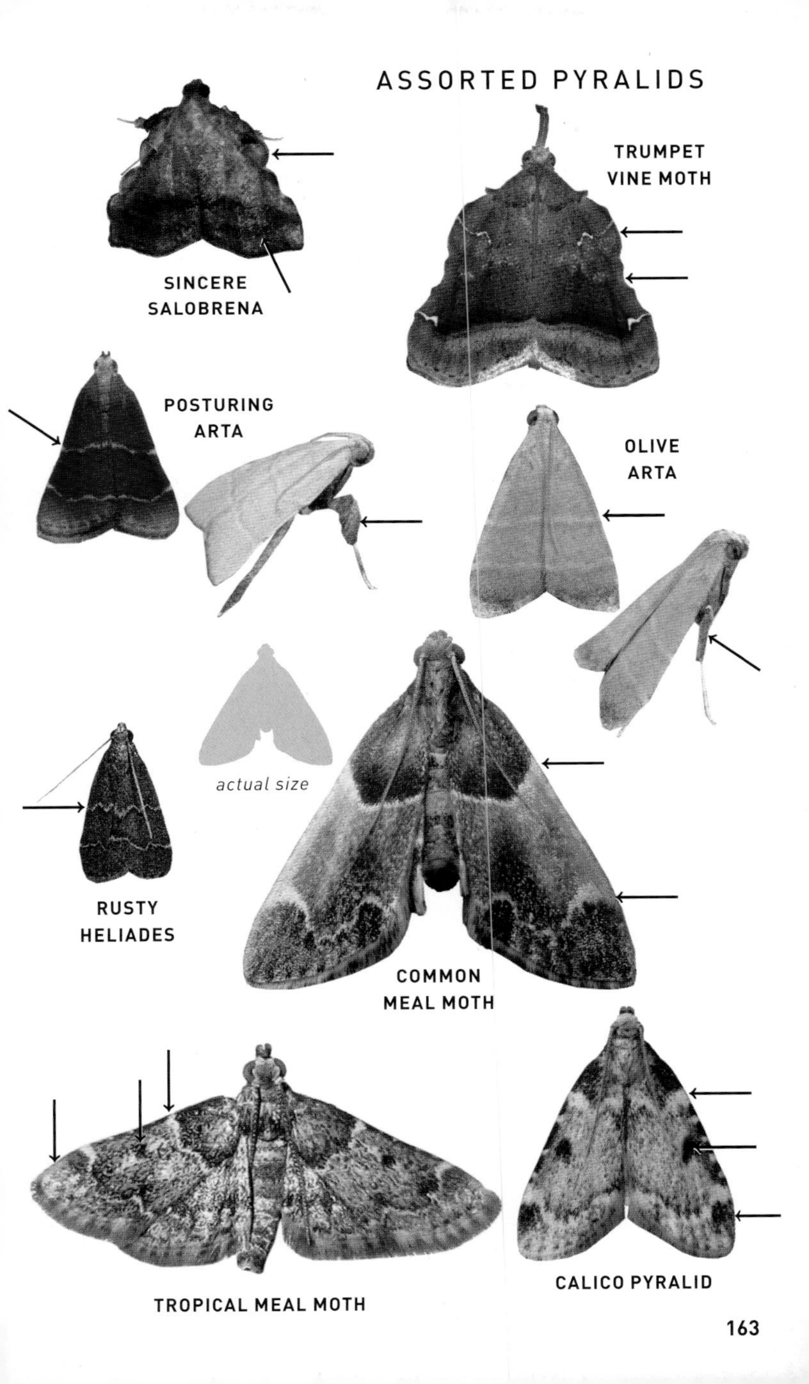

PINK-MASKED PYRALID

Aglossa disciferalis 80-0075 (5512) **Common**

TL 10–11 mm. Shiny tan to pinkish red FW is boldly patterned with jagged black AM and PM lines edged with pale yellow bands. Black discal spot is usually well defined. **HOSTS:** Unknown.

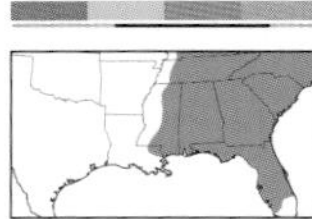

STORED GRAIN MOTH

Aglossa caprealis 80-0079 (5517) **Common**

TL 12–14 mm. Shiny brown FW is marked with jagged pale yellow AM and ST lines. Central median area is smudged with pale yellow spots, creating a mottled appearance. **HOSTS:** Stored foods, grain chaff, fungi, and dead animals.

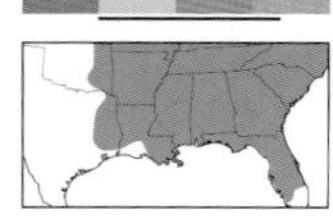

GREASE MOTH

Aglossa cuprina 80-0080 (5518) **Common**

TL 15–17 mm. Shiny pinkish brown FW is marked with pale-edged jagged black lines. Note large bulge in AM line and pale outline to black discal spot. **HOSTS:** Stored foods, grain chaff, fungi, and dead animals.

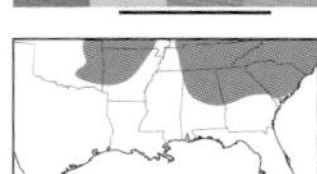

CLOVER HAYWORM

Hypsopygia costalis 80-0086 (5524) **Common**

TL 9–10 mm. Peppery brick red wings have broad yellow fringe when fresh. Outer margin is yellow. AM and PM lines widen into large yellow triangles at costa. Often rests with wings spread open. **HOSTS:** Mostly dried plant material and stored hay.

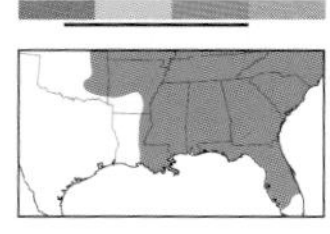

RED-SHAWLED MOTH

Hypsopygia intermedialis 80-0088 (5526) **Common**

TL 10–12 mm. Reddish brown FW has slightly jagged black AM and PM lines edged pale yellow; edging to lines widens at costa. **HOSTS:** Unknown.

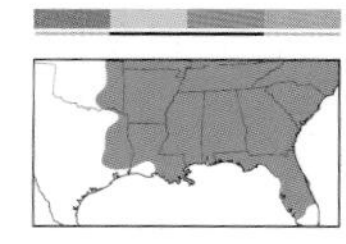

PINK-FRINGED HYPSOPYGIA

Hypsopygia binodulalis 80-0092 (5530) **Common**

TL 10–12 mm. Resembles Yellow-fringed Hypsopygia, but purplish olive FW has a bright pinkish red costal streak and fringe. Sometimes rests with wings spread open. **HOSTS:** Unknown.

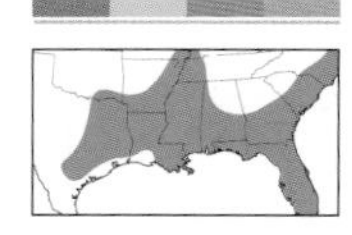

SOUTHERN HYPSOPYGIA

Hypsopygia nostralis 80-0093 (5531) **Local**

TL 10–12 mm. Resembles Pink-fringed Hypsopygia, but almost-straight AM and PM lines are weakly edged paler, and yellow costal spots are reduced or nearly absent. Costa and fringe are lightly tinged pink. **HOSTS:** Unknown.

ASSORTED PYRALIDS

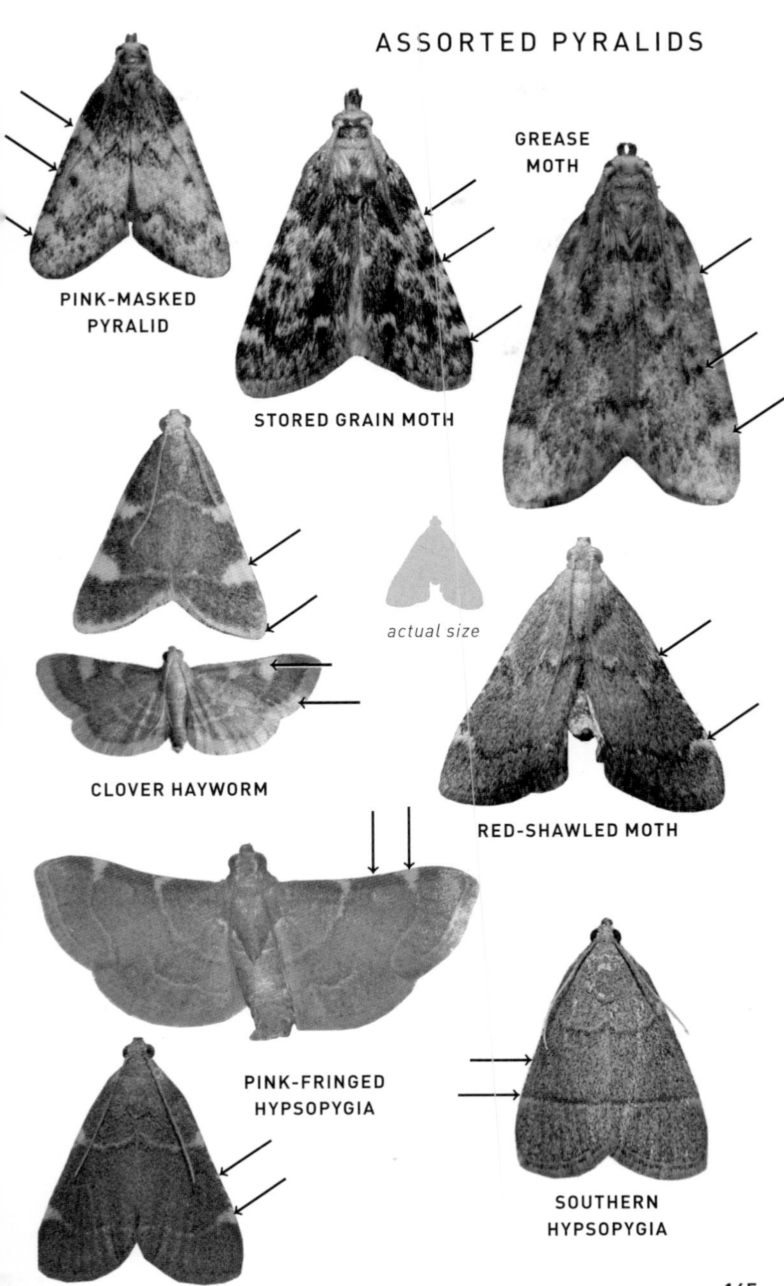

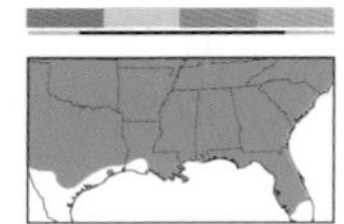

YELLOW-FRINGED HYPSOPYGIA

Hypsopygia olinalis 80-0094 (5533) **Common**

TL 10–12 mm. Resembles Clover Hayworm, but purplish red wings have narrow yellow fringe, and outer margin is dark purple. Triangles at costa are smaller. Sometimes rests with wings spread open and abdomen raised. **HOSTS:** Oak.

ZELLER'S MACALLA

Macalla zelleri 80-0100 (5579) **Common**

TL 12–14 mm. Light brown FW has a contrasting dark brown basal area and ST band. Median area is sprinkled white, especially along bold white AM line. Black terminal line is accented with white dots. **HOSTS:** Poison ivy.

DIMORPHIC EPIPASCHIA

Epipaschia superatalis 80-0102 (5577) **Common**

TL 10–12 mm. Straw-colored FW (often tinted green) has contrasting warm brown shading beyond kinked black PM line. Usually has a small black discal spot in median area. **HOSTS:** Unknown. **NOTE:** Despite widely used common name, not actually dimorphic.

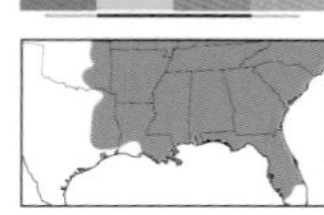

ORANGE-TUFTED ONEIDA

Oneida lunulalis 80-0112 (5588) **Common**

TL 12–14 mm. Silvery gray FW has AM line bordered with a ridge of raised orange scales. Median area is tinged green toward AM area. Curved section of orange-edged outer PM line borders streaky blackish apical patch. **HOSTS:** Oak.

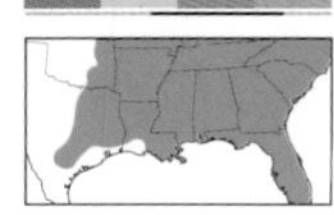

WATSON'S TALLULA

Tallula watsoni 80-0117 (5592) **Common**

TL 10–13 mm. Whitish FW is variably shaded light gray in median and ST areas. Wide, dark brown AM band is tinted warm brown at inner margin and usually constricted at midpoint. White PM line contrasts with blackish subapical patch. **HOSTS:** Unknown.

PINE WEBWORM

Pococera robustella 80-0122 (5595) **Common**

TL 12–14 mm. Gray FW is shaded warm brown in AM, inner median, and ST areas. Broad blackish AM band is usually most obvious marking. Median area is sprinkled with white scales. **HOSTS:** Needles of pine.

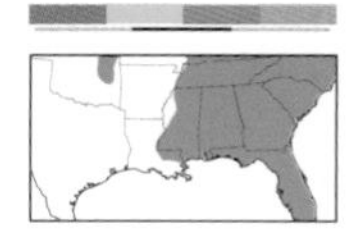

LESPEDEZA WEBWORM

Pococera scortealis 80-0123 (5596) **Common**

TL 12–14 mm. Similar to Pine Webworm, but black AM band is even wider and blends into grayish basal area. PM line is more evenly edged with brown shading. **HOSTS:** Unknown; probably bush clover.

ASSORTED PYRALIDS

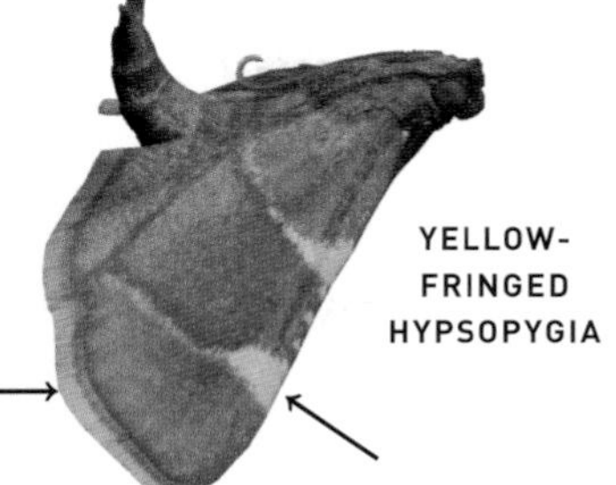

YELLOW-FRINGED HYPSOPYGIA

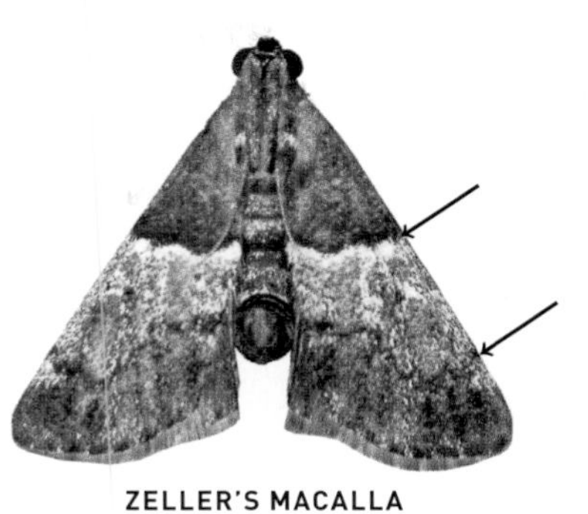

ZELLER'S MACALLA

actual size

DIMORPHIC EPIPASCHIA

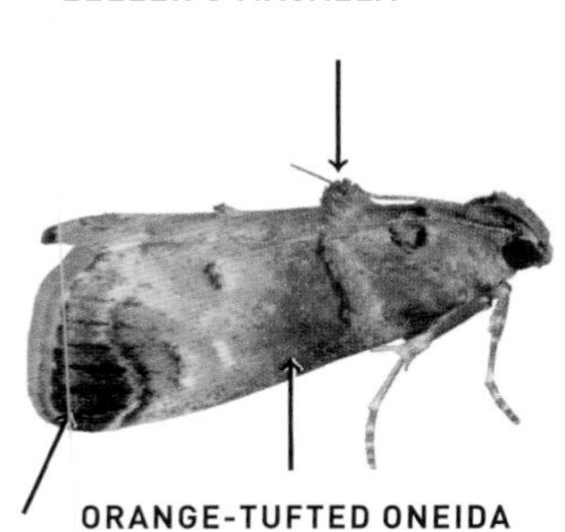

ORANGE-TUFTED ONEIDA

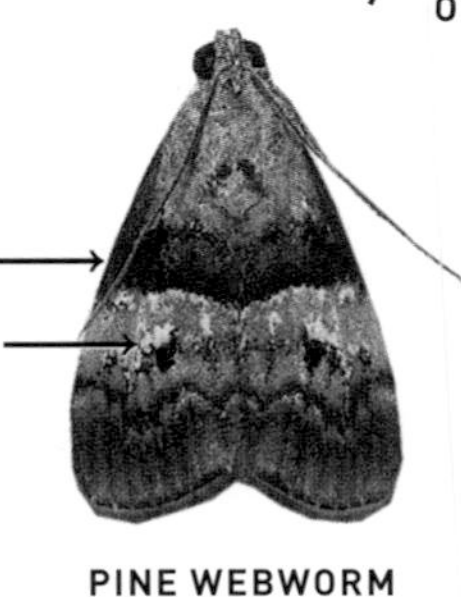

PINE WEBWORM

WATSON'S TALLULA

LESPEDEZA WEBWORM

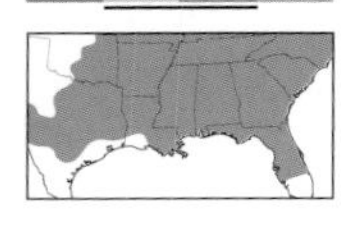

GRAY-BANDED POCOCERA

Pococera maritimalis 80-0130 (5603) **Local**

TL 14–15 mm. Brown FW has a grayish median band bounded by gray-edged black AM and PM lines. Black median line is visible only along inner half. Thorax is tan. **HOSTS:** Unknown.

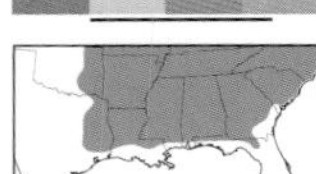

SYCAMORE WEBWORM

Pococera militella 80-0131 (5604) **Common**

TL 12–14 mm. Pale pinkish brown FW has a whitish median band and double black AM line interrupted by translucent panel. Central basal area has black tuft of raised scales. **HOSTS:** Sycamore.

MAPLE WEBWORM

Pococera asperatella 80-0133 (5606) **Common**

TL 12–14 mm. Peppery gray FW has a whitish median band that fades before reaching costa. Black AM and PM lines are double and filled light gray. **HOSTS:** Maple.

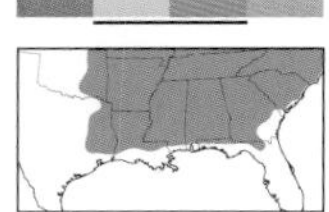

STRIPED OAK WEBWORM

Pococera expandens 80-0135 (5608) **Common**

TL 14–15 mm. Gray FW has contrasting straw-colored basal area tinged rust at inner margin. Double AM and PM lines are filled light gray. **HOSTS:** Unknown; probably oak.

Phycitinine Moths

Family Pyralidae, Subfamily Phycitinae

A large, homogenous group of narrow-winged pyralid moths that typically appear streamlined and small-headed at rest. Most species have short, upturned labial palps. Some species, most notably in the genera *Acrobasis* and *Dioryctria*, have rows of raised scales on the FW, usually adjacent to the AM line. Most are nocturnal and will come to lights. A few are prone to northward migrations during periods of drought or favorable winds, sometimes appearing far out of their normal ranges.

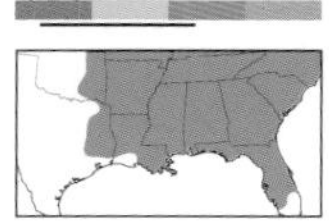

CRANBERRY FRUITWORM

Acrobasis vaccinii 80-0150 (5653) **Common**

TL 9–12 mm. Gray FW has white costal streak interrupted by an incomplete blackish median band. Basal area is blackish. **HOSTS:** Blueberry, cranberry, and huckleberry.

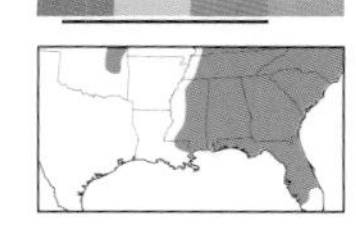

LEAF CRUMPLER MOTH

Acrobasis indigenella 80-0152 (5651) **Common**

TL 9–12 mm. Whitish FW has dark brown median and PM lines that converge toward inner margin, forming a dusky patch. Incomplete light brown median band is edged black basally. **HOSTS:** Apple, cherry, hawthorn, and plum.

ASSORTED PYRALIDS

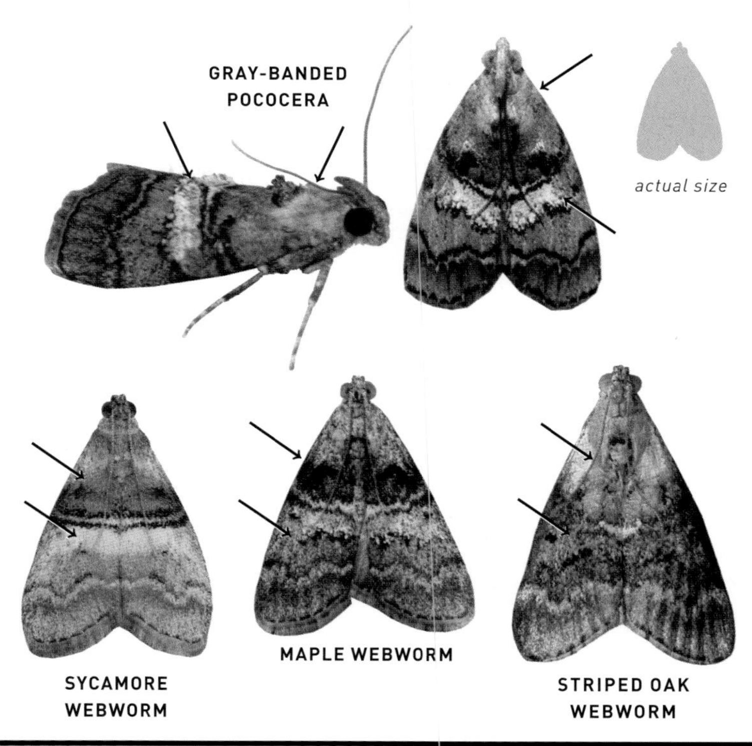

PHYCITININE MOTHS

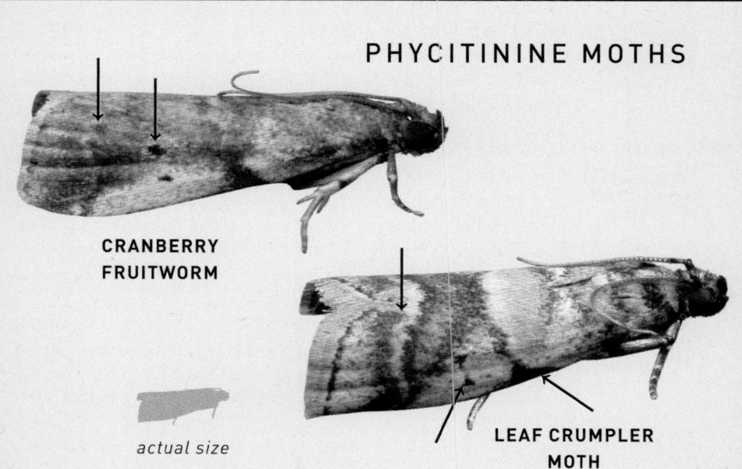

HICKORY SHOOT BORER

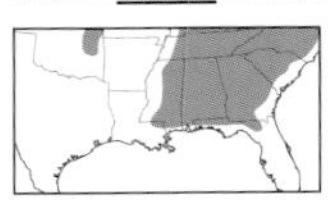

Acrobasis caryae 80-0156 (5664) **Common**

TL 10–12 mm. Uniform gray FW has incomplete narrow reddish bands on either side of black-and-white scale ridge. Jagged ST line is sometimes obvious. **HOSTS:** Hickory.

CORDOVAN ACROBASIS

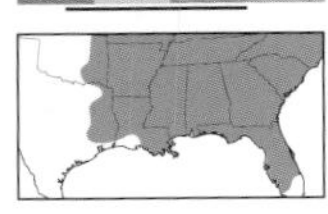

Acrobasis exsulella 80-0165 (5672) **Common**

TL 8–11 mm. Dark gray FW has contrasting reddish basal area and ST band. Inner median area is whitish, bordered with a thick black band basally. **HOSTS:** Hickory.

WALNUT SHOOT MOTH

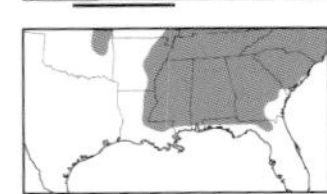

Acrobasis demotella 80-0166 (5674) **Uncommon**

TL 8–10 mm. Dark gray FW has contrasting light brown basal area. Reddish-edged scale ridge is bordered with a curved white line distally. **HOSTS:** Black walnut.

AMERICAN PLUM BORER

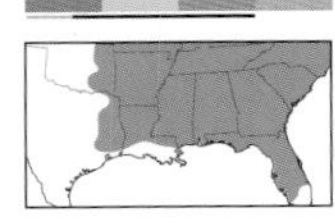

Euzophera semifuneralis 80-0215 (5995) **Common**

TL 9–13 mm. Gray FW has reddish inner basal and ST areas. Broad blackish median band is edged with whitish zigzag AM and PM lines. Terminal line is dotted. **HOSTS:** Deciduous trees, including apple, cherry, peach, pear, sweet gum, and walnut.

ROOT COLLAR BORER

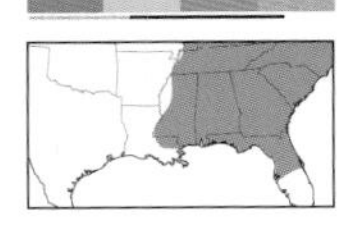

Euzophera ostricolorella 80-0220 (5997) **Common**

TL 13–20 mm. Resembles American Plum Borer but is considerably larger. Basal and ST areas are brighter red. White PM line is jagged. **HOSTS:** Tulip tree.

BROAD-BANDED EULOGIA

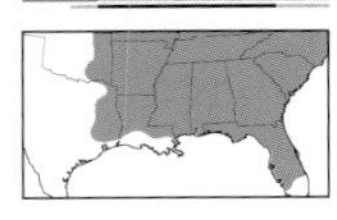

Eulogia ochrifrontella 80-0222 (5999) **Common**

TL 6–8 mm. Reddish brown FW is marked with a broad yellow-edged blackish median band. All lines are slightly wavy. **HOSTS:** Reported on apple, oak, and pecan.

REDDISH EPHESTIODES

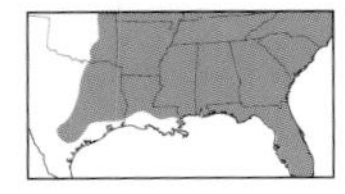

Ephestiodes infimella 80-0225 (6001) **Common**

TL 6–8 mm. Bicolored FW has a yellowish brown basal area and is reddish distal to white-edged AM line. Central median area is often streaked black and peppered white. **HOSTS:** Cherry and greenbrier.

DARKER MOODNA

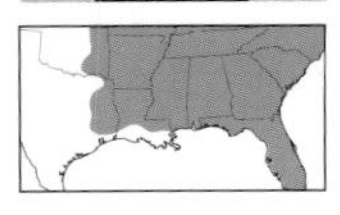

Moodna ostrinella 80-0232 (6005) **Common**

TL 6–9 mm. Brick red FW has a blackish median band broadly edged white. Reniform area is speckled white. ST band is reddish. **HOSTS:** Trees and low plants, including apple, birch, cotton, iris, oak, pine, and sumac.

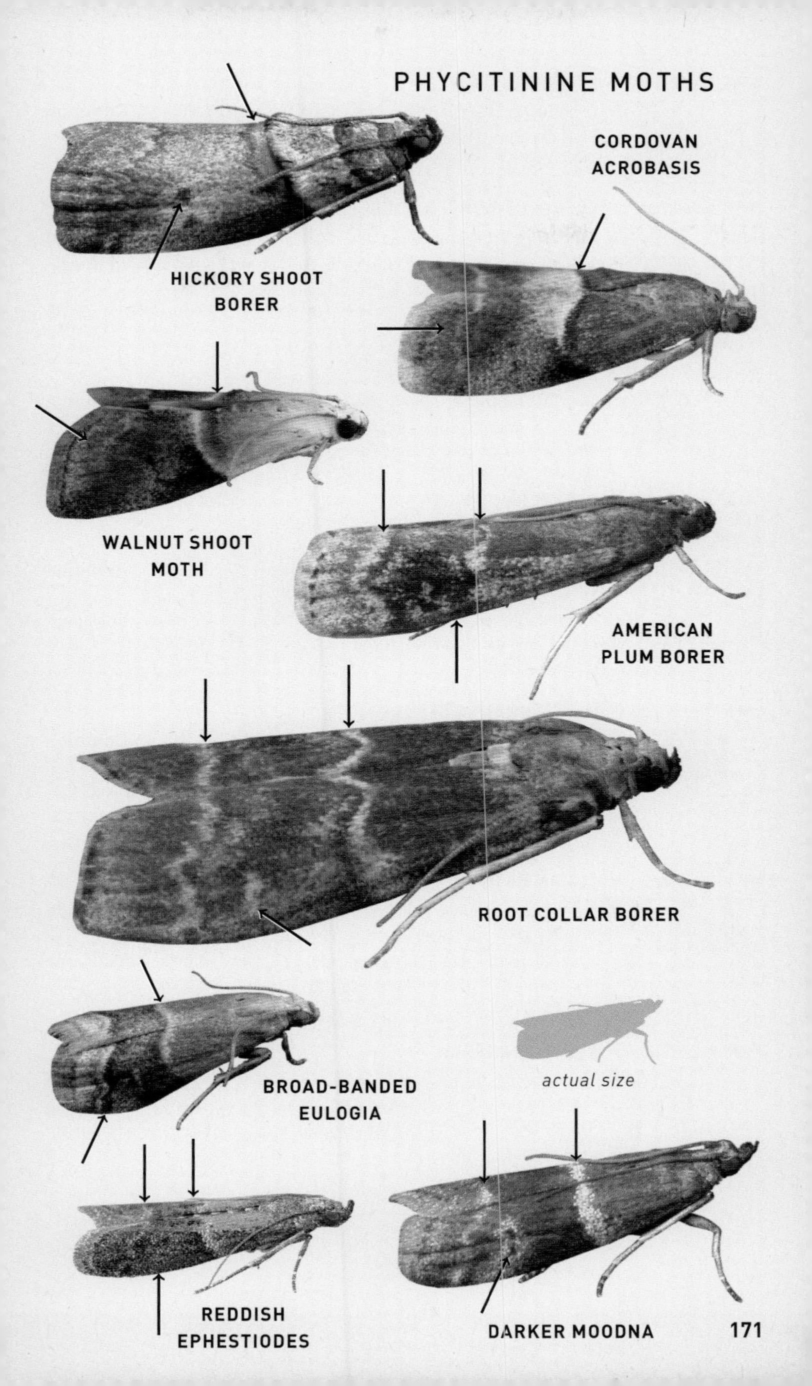
CORDOVAN ACROBASIS
HICKORY SHOOT BORER
WALNUT SHOOT MOTH
AMERICAN PLUM BORER
ROOT COLLAR BORER
BROAD-BANDED EULOGIA
actual size
REDDISH EPHESTIODES
DARKER MOODNA

PALER MOODNA

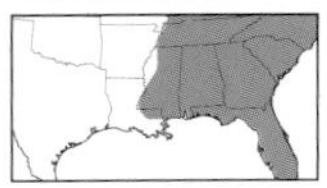

Moodna pallidostrinella 80-0233 (6005.1) **Common**

TL 6–9 mm. Similar to Darker Moodna, but basal area is more orange red. ST band is blackish. **HOSTS:** Reported on shortleaf pine.

CRESCENT-WINGED CAUDELLIA

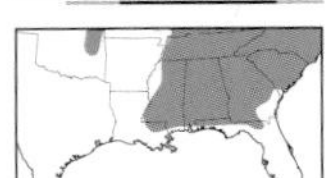

Caudellia apyrella 80-0235 (6012) **Uncommon**

TL 6–8 mm. Blackish FW has an oblique white AM band that narrows to a point at inner margin. Central median area is tinted reddish and peppered with white scales toward costa. **HOSTS:** Dried seeds of dodder.

DRIED-FRUIT MOTH

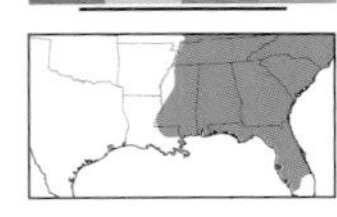

Vitula edmandsii 80-0239 (6007) **Common**

TL 9–12 mm. Peppery pale gray FW has fragmented jagged black lines and a crescent-shaped reniform spot. Inner margin is tinted reddish. **HOSTS:** Pollen, honey, and immature hymenopterans in colonies of bees.

INDIAN MEAL MOTH

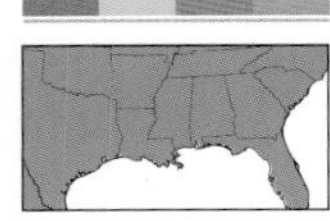

Plodia interpunctella 80-0258 (6019) **Common**

TL 6–9 mm. Shiny FW has a pale yellow basal area and orange red distal half of wing. AM, median, and PM bands are silvery gray, peppered darker. **HOSTS:** Stored food products, including flour, oatmeal, and seeds.

MEDITERRANEAN FLOUR MOTH

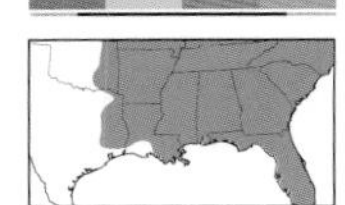

Ephestia kuehniella 80-0260 (6020) **Common**

TL 8–14 mm. Peppery pale gray FW has a bold white zigzag AM line and wide black median band. Reniform area is marked with two distinct black dots. **HOSTS:** Stored food products, including seeds of barley, corn, oats, and rice.

TAMPA MOTH

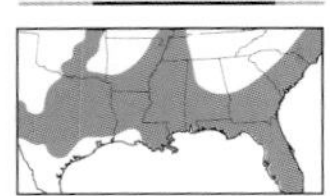

Tampa dimediatella 80-0269 (6028) **Common**

TL 8–10 mm. Bicolored FW has a strongly curved costa and rounded apex. Note thin black streak dividing peppery whitish costal streak from pinkish purple inner margin. Labial palps point upward. **HOSTS:** Unknown.

RUSTY VARNERIA

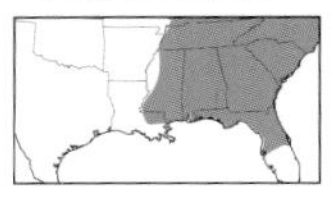

Varneria postremella 80-0270 (6029) **Common**

TL 4–6 mm. Uniform rusty red FW has a silvery gray basal patch and fringe when fresh. Short labial palps impart a blunt-faced appearance. **HOSTS:** Unknown.

PHYCITININE MOTHS

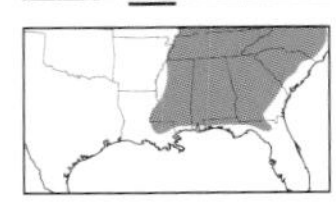

BLACK-BANDED IMMYRLA

Immyrla nigrovittella 80-0300 (5766) **Uncommon**

TL 11–12 mm. Light gray FW has slightly paler median area. Raised black scale ridge is edged with parallel black and white lines distally. **HOSTS:** Hickory and eastern hornbeam.

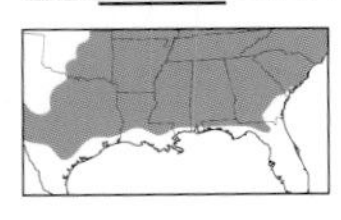

ENGEL'S SALEBRIARIA

Salebriaria engeli 80-0310 (5773) **Common**

TL 8–10 mm. Dark, dusky brown FW has a contrasting white patch at midpoint of inner margin. **HOSTS:** Oak.

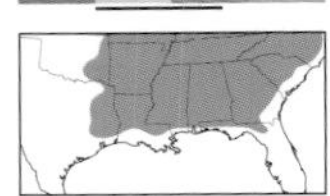

WHITE-BANDED SALEBRIARIA

Salebriaria rufimaculatella 80-0313 (5775.2) **Local**

TL 8–11 mm. Dark, dusky brown FW has a contrasting broad white AM band. Black dots in central PM area are partly edged white. **HOSTS:** Unknown.

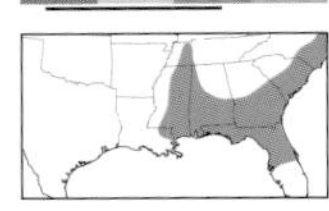

WHITE-PATCHED SALEBRIARIA

Salebriaria squamopalpiella 80-0321 (5775.3) **Local**

TL 8–9 mm. Grayish brown FW has a large whitish patch covering much of outer half of FW, liberally peppered with black scales. White extends toward inner margin in AM and PM areas. **HOSTS:** Unknown.

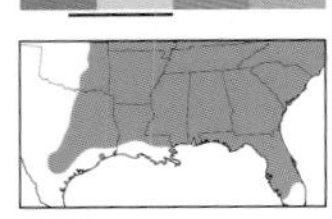

BLACK-PATCHED SALEBRIARIA

Salebriaria annulosella 80-0324 (5774) **Common**

TL 8–10 mm. Light gray FW has a contrasting white AM line that is edged with bold black patches distally along costa and basally at inner margin. Two well-defined black dots in central PM area are edged white. **HOSTS:** Unknown.

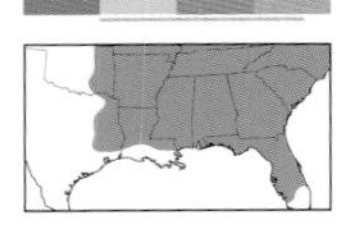

VIBURNUM GLYPTOCERA

Glyptocera consobrinella 80-0342 (5745) **Local**

TL 10–12 mm. Light gray FW is tinged reddish along inner margin and boldly patterned with double wavy median and ST lines. A broad blackish AM band contrasts with whitish basal area. **HOSTS:** Viburnum.

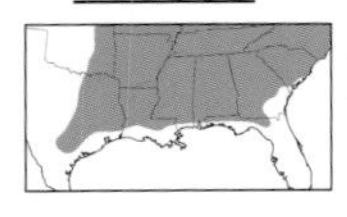

RED-SPLASHED LEAFROLLER

Sciota rubrisparsella 80-0349 (5804) **Local**

TL 7–9 mm. Reddish FW has a broad black AM band that extends along costa almost to apex. Jagged white median line fades before reaching costa. Curved white ST line is accented with tiny black wedges. **HOSTS:** Hackberry.

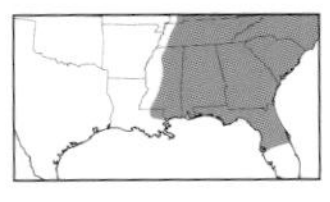

BELTED LEAFROLLER

Sciota vetustella 80-0365 (5794) **Common**

TL 12–13 mm. Light gray FW has contrasting reddish basal area. Thick black median band is edged with pale gray and black lines distally. **HOSTS:** Basswood.

PHYCITININE MOTHS

actual size

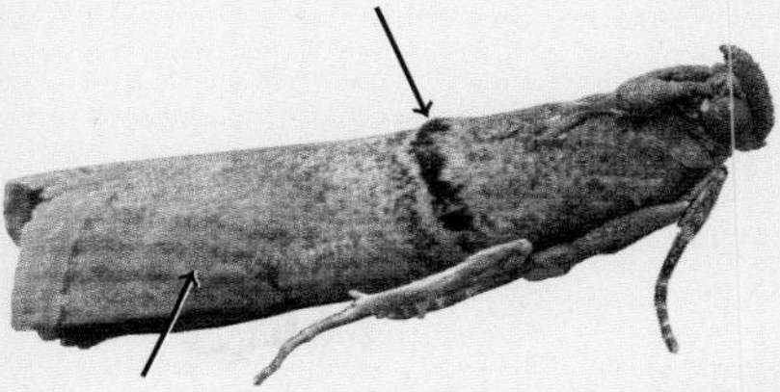

BLACK-BANDED IMMYRLA

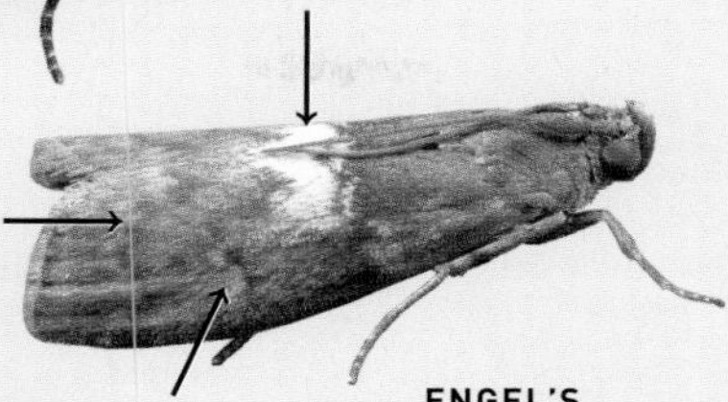

ENGEL'S SALEBRIARIA

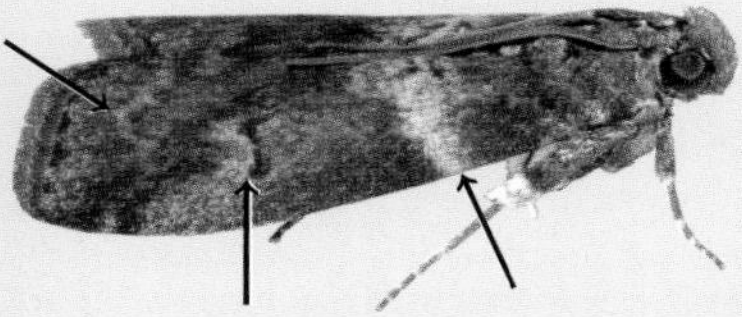

WHITE-BANDED SALEBRIARIA

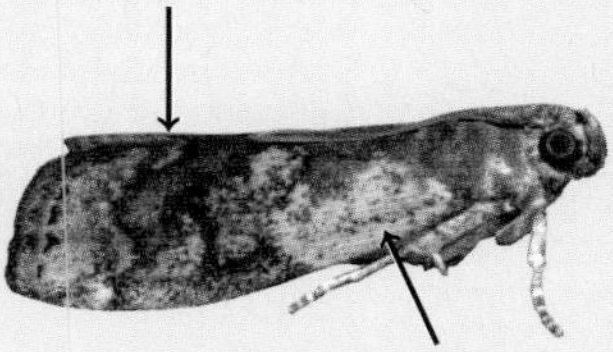

WHITE-PATCHED SALEBRIARIA

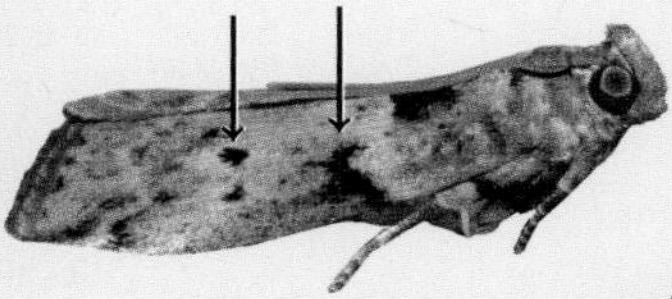

BLACK-PATCHED SALEBRIARIA

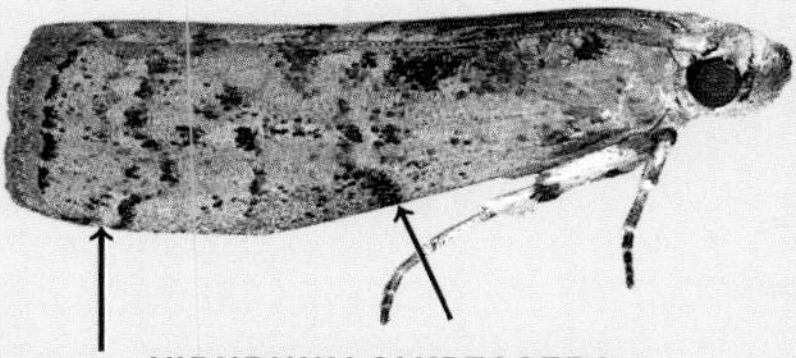

VIBURNUM GLYPTOCERA

RED-SPLASHED LEAFROLLER

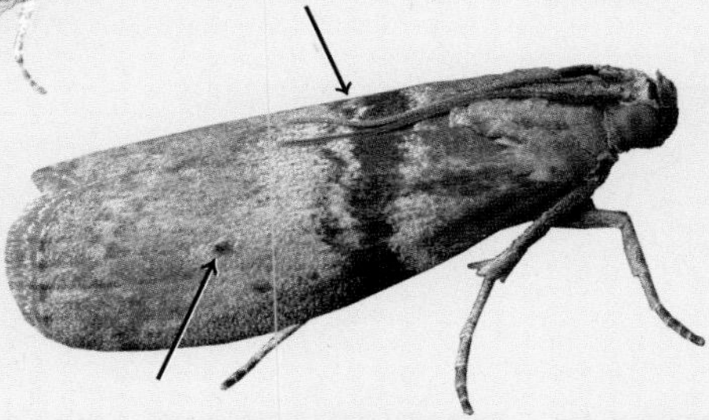

BELTED LEAFROLLER

BLACK-SPOTTED LEAFROLLER

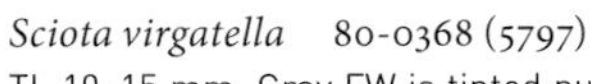

Sciota virgatella 80-0368 (5797) **Common**

TL 10–15 mm. Gray FW is tinted purplish brown in basal and inner median areas. Jagged black AM band is sometimes present. Fragmented lines create a spotted appearance. **HOSTS:** Black locust.

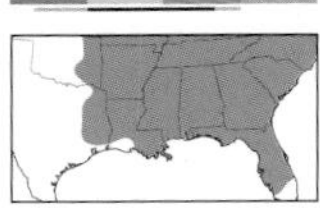

HACKBERRY LEAFROLLER

Sciota celtidella 80-0370 (5803) **Common**

TL 10–12 mm. Gray FW has light brown basal area and reddish patches along inner margin. Fragmented black and white lines create a speckled appearance. **HOSTS:** Hackberry.

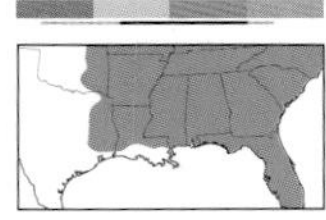

STRIPED SUMAC LEAFROLLER

Sciota subfuscella 80-0371 (5789) **Common**

TL 12–13 mm. Peppery gray FW has contrasting black AM band bisected by a whitish zigzag line near inner margin. Basal area is pale distally with a reddish patch at inner margin. **HOSTS:** Sumac.

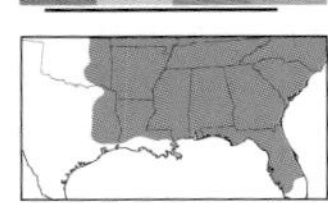

SWEETGUM LEAFROLLER

Sciota uvinella 80-0372 (5802) **Common**

TL 7–9 mm. Light gray FW has a thick purplish brown AM band separated from brown basal area by whitish band. Double PM lines are filled white. **HOSTS:** Sweet gum.

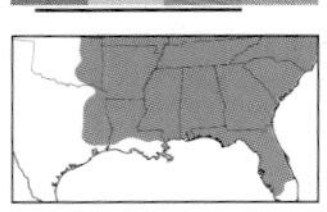

HONEY LOCUST LEAFROLLER

Tlascala reductella 80-0374 (5808) **Common**

TL 8–10 mm. Pale silvery gray FW is marked with three parallel black AM lines. Basal and ST areas are tinged reddish. **HOSTS:** Honey locust.

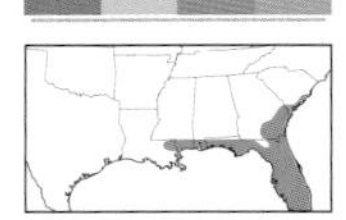

SOUTH COASTAL CONEWORM

Dioryctria ebeli 80-0410 (5863.4) **Local**

TL 9–13 mm. Whitish FW has a complex pattern of zigzag, white-edged lines. Chestnut bands mark median and ST areas. Note pale green bar in inner AM area. **HOSTS:** Slash, longleaf, and loblolly pines.

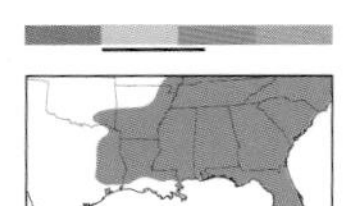

WEBBING CONEWORM

Dioryctria disclusa 80-0414 (5847) **Common**

TL 11–15 mm. Orange FW is marked with three jagged white lines. Median and ST areas are tinged rusty. **HOSTS:** Pine.

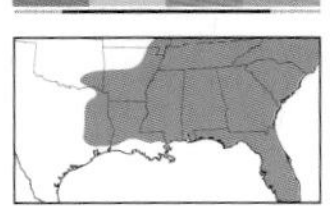

BLISTER CONEWORM

Dioryctria clarioralis 80-0428 (5863.1) **Common**

TL 11–14 mm. Light gray FW has wavy, black-edged white AM and PM lines. Note oblique black-edged chestnut AM band. Median and ST areas are tinged reddish. **HOSTS:** Pine.

PHYCITININE MOTHS

SOUTHERN PINE CONEWORM

Dioryctria amatella 80-0435 (5853) **Common**

TL 12–16 mm. Rusty brown FW is boldly patterned with black-edged, white zigzag lines. Basal and median areas are marked with white spots. **HOSTS:** Longleaf pine.

BALD CYPRESS CONEWORM

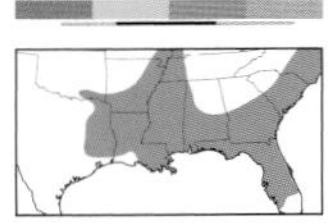

Dioryctria pygmaeella 80-0447 (5849) **Local**

TL 7–10 mm. Light gray FW has fragmented, rusty-edged black and white lines forming a marbled pattern. **HOSTS:** Bald cypress.

ELM LEAFTIER

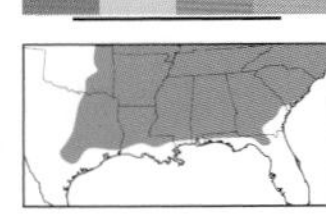

Canarsia ulmiarrosorella 80-0479 (5926) **Uncommon**

TL 8–10 mm. Grayish brown FW is marked with jagged white-edged lines. Sometimes has a whitish patch distal to inner section of median line. Note black crescent in reniform area. **HOSTS:** American elm.

RUSTY-BANDED ADELPHIA

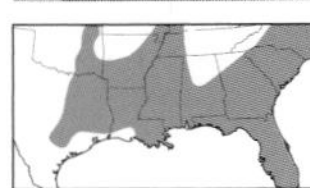

Adelphia petrella 80-0482 (5890) **Common**

TL 10–13 mm. Grayish brown FW is tinged purple and patterned with jagged white-edged lines. Median band and basal area are light orange and rust. **HOSTS:** Cassia.

RED-WASHED UFA

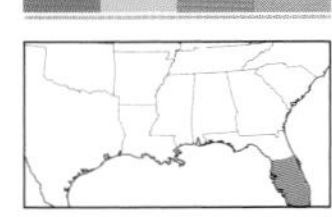

Ufa rubedinella 80-0483 (5895) **Local**

TL 9–10 mm. Straw-colored FW is washed with variable reddish shading (stronger in females). Often has contrasting pale costal streak. Slanting PM line and two blackish dots in median area are only markings. **HOSTS:** Cowpea and lima bean.

LESSER CORNSTALK BORER

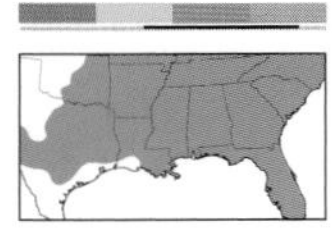

Elasmopalpus lignosellus 80-0486 (5896) **Common**

TL 7–11 mm. Shiny straw-colored FW (darker in females) has peppery brown shading along costa and inner margin. Short black labial palps are upturned. **HOSTS:** Grasses and legumes.

REDDISH MACRORRHINIA

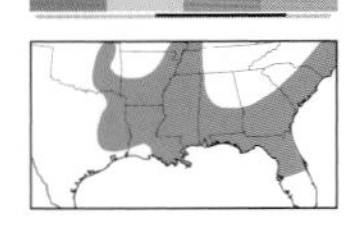

Macrorrhinia endonephele 80-0506 (5913) **Uncommon**

TL 9–12 mm. Straw-colored FW is variably washed reddish. Dusky lines and spots are inconspicuous. Long, brushy labial palps are slightly upturned. **HOSTS:** Unknown.

GOLD-BANDED ETIELLA

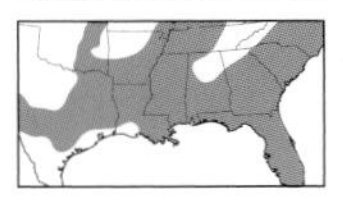

Etiella zinckenella 80-0523 (5744) **Local**

TL 9–13 mm. Peppery gray FW has a white costal streak. Narrow chestnut scale ridge is edged with a wide ochre median band. Upturned labial palps are brushy and pointed. **HOSTS:** Immature seeds of legumes.

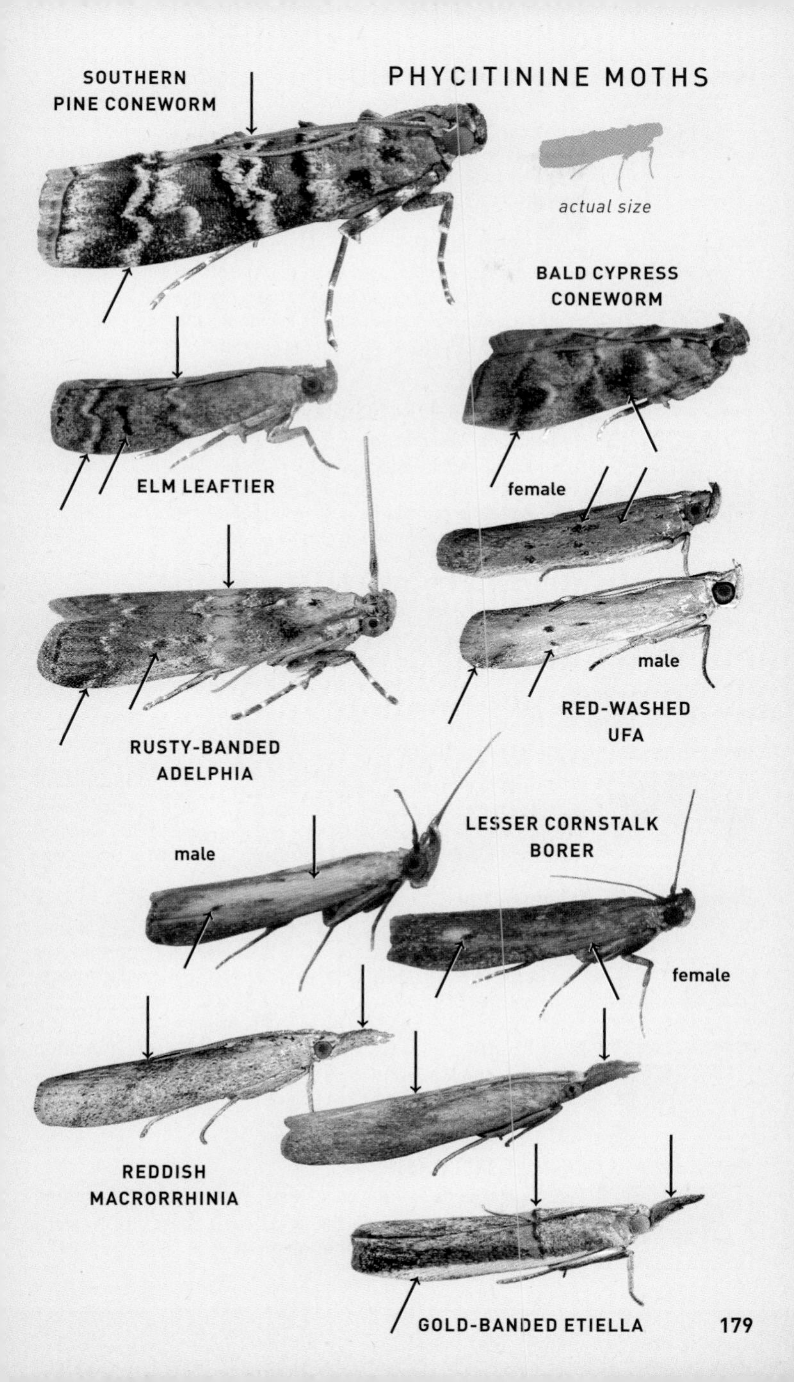
SOUTHERN PINE CONEWORM
actual size
BALD CYPRESS CONEWORM
ELM LEAFTIER
female
male
RED-WASHED UFA
RUSTY-BANDED ADELPHIA
LESSER CORNSTALK BORER
male
female
REDDISH MACRORRHINIA
GOLD-BANDED ETIELLA

SCALE-FEEDING SNOUT MOTH

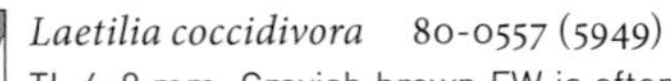

Laetilia coccidivora 80-0557 (5949) **Common**

TL 6–9 mm. Grayish brown FW is often profusely speckled reddish along inner margin. Broad, peppery white costal streak is divided by oblique reddish brown median line. Kinked white ST line is edged black. **HOSTS:** Larvae are predators of scale insects and mealybugs.

SUNFLOWER MOTH

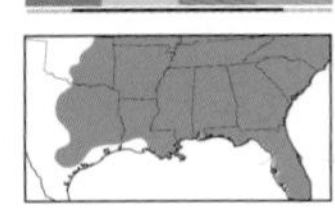

Homoeosoma electella 80-0626 (5935) **Common**

TL 8–13 mm. Speckled light gray FW has an ill-defined whitish costal streak. Note short black dashes in inner median and central PM areas. **HOSTS:** Mostly aster; sometimes a pest of commercial sunflowers.

BLACK-BANDED HOMOEOSOMA

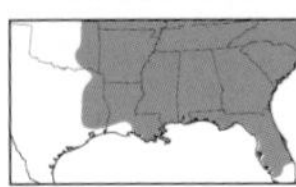

Homoeosoma deceptorium 80-0641 (5944) **Uncommon**

TL 11–14 mm. Whitish to light gray FW has a wide black AM band that fades before costa. Two black dots mark reniform area. White costal streak is obvious on darker individuals. **HOSTS:** Unknown. **NOTE:** Restricted to sandy and coastal areas.

WHITE-EDGED PHYCITODES

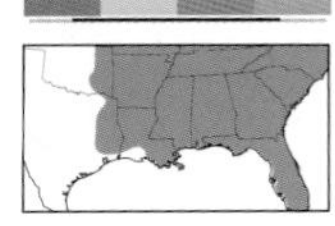

Phycitodes reliquellum 80-0651 (5946.2) **Common**

TL 7–9 mm. Peppery light gray FW has an ill-defined whitish costal streak. Midpoint of fragmented black AM line often forms an irregular blotch. Two black dots mark reniform area. **HOSTS:** Aster, lettuce, yellow thistle, and others.

LONG-PALPS PEORIA

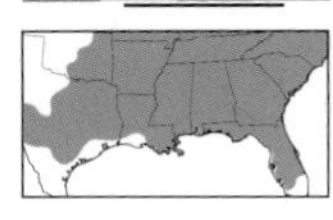

Peoria longipalpella 80-0658 (6042) **Uncommon**

TL 10–12 mm. Peppery grayish brown FW has a whitish costal streak. Oblique blackish AM and PM lines fade before reaching costa. Upturned labial palps are noticeably long. **HOSTS:** Unknown.

FLORIDA PEORIA *Peoria floridella* 80-0662 (6046) **Local**

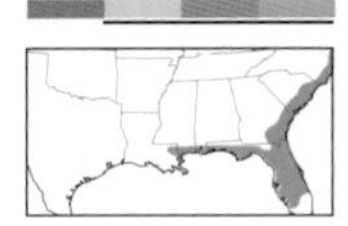

TL 13–15 mm. Tawny brown FW is boldly patterned with whitish veins, especially toward costa. Main white vein in central median area is edged black. Long labial palps are brushy-looking. **HOSTS:** Unknown. **NOTE:** Restricted to coastal environs.

ROSY PEORIA *Peoria roseotinctella* 80-0666 (6049) **Common**

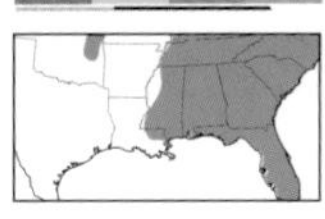

TL 8–9 mm. Rosy pink FW has a sharply demarcated peppery white costal streak. PM line is a row of black dashes. Kinked labial palps are brushy-looking. **HOSTS:** Unknown.

CARMINE SNOUT MOTH

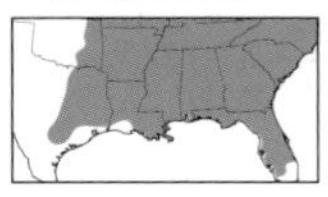

Peoria approximella 80-0670 (6053) **Common**

TL 8–9 mm. Bright carmine FW has cream-colored streaks along costa and inner margin. Slightly upturned labial palps are brushy-looking. **HOSTS:** Unknown.

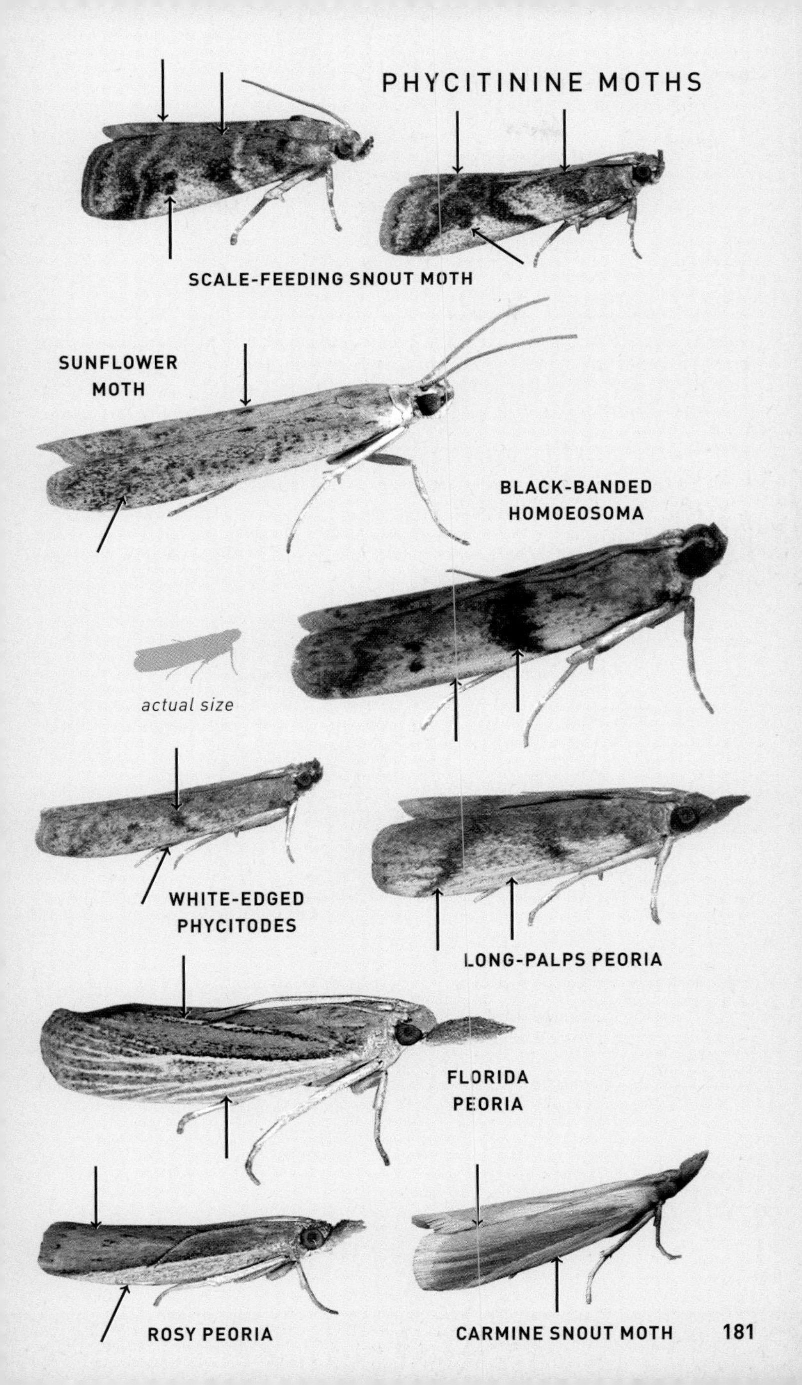
PHYCITININE MOTHS
SCALE-FEEDING SNOUT MOTH
SUNFLOWER MOTH
BLACK-BANDED HOMOEOSOMA
actual size
WHITE-EDGED PHYCITODES
LONG-PALPS PEORIA
FLORIDA PEORIA
ROSY PEORIA
CARMINE SNOUT MOTH

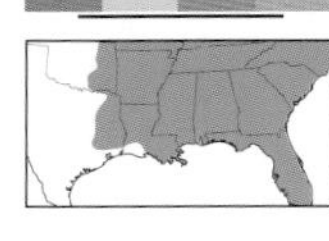

ROSY ATASCOSA

Atascosa glareosella 80-0685 (6067) **Common**

TL 8–9 mm. Brown FW has a pinkish tint when fresh. Whitish costal streak is finely peppered with rusty scales. Sometimes has contrasting blackish streak through center of wing and oblique ST line. Upturned labial palps are brushy. **HOSTS:** Unknown.

DONACAULAS AND ALLIES

Family Crambidae, Subfamily Schoenobiinae

Very similar to grass-veneers, the donacaulas typically have pointier wings and are generally a more uniform orange brown, though some species are completely white or patterned with contrasting spots and streaks. Adults regularly visit lights.

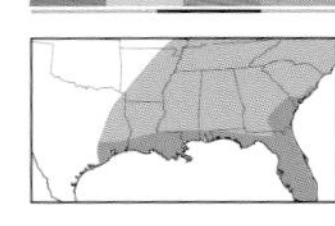

REED-BORING CRAMBID

Carectocultus perstrialis 80-0695 (5307) **Local**

TL 14–15 mm. Shiny chocolate brown FW has a bold white central streak that slants outward at apex. Some individuals lack white and simply have dark shading through middle of FW. HW is white. Rests with forelegs splayed out in front. **HOSTS:** Aquatic plants, including sedge.

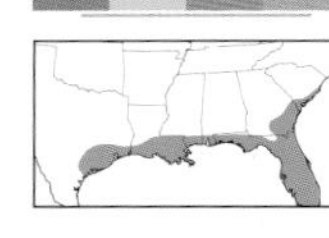

YELLOW-BANDED LEPTOSTEGES

Leptosteges flavicostella 80-0698 (5301) **Local**

TL 7–8 mm. White FW has wavy dark yellow AM and PM bands, and reniform spot, edged blackish. ST area has a single dark yellow dash at midpoint. Rare individuals may be peppered blackish. **HOSTS:** Unknown.

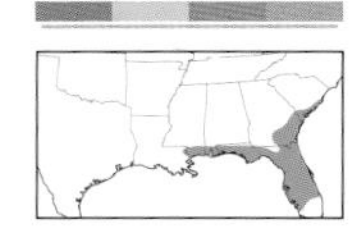

GOLDEN-BANDED LEPTOSTEGES

Leptosteges flavifascialis 80-0699 (5302) **Local**

TL 7–8 mm. Whitish FW is variably peppered with brown scales. Wide golden yellow AM and PM bands are almost connected by yellow spot in central median area. ST area is streaked yellow. Head is white. **HOSTS:** Unknown.

SATIN RUPELA *Rupela tinctella* 80-0705 (5311) **Uncommon**

TL 14–22 mm. Satin white wings are sometimes tinged brownish in males, which average smaller than females. Abdomen has a line of brown dorsal spots. Long spindly legs distinguish this moth as *Rupela* from other all-white moths. **HOSTS:** Unknown. **NOTE:** Other *Rupela* species are nearly identical but less common.

PHYCITININE MOTHS

DONACAULAS AND ALLIES

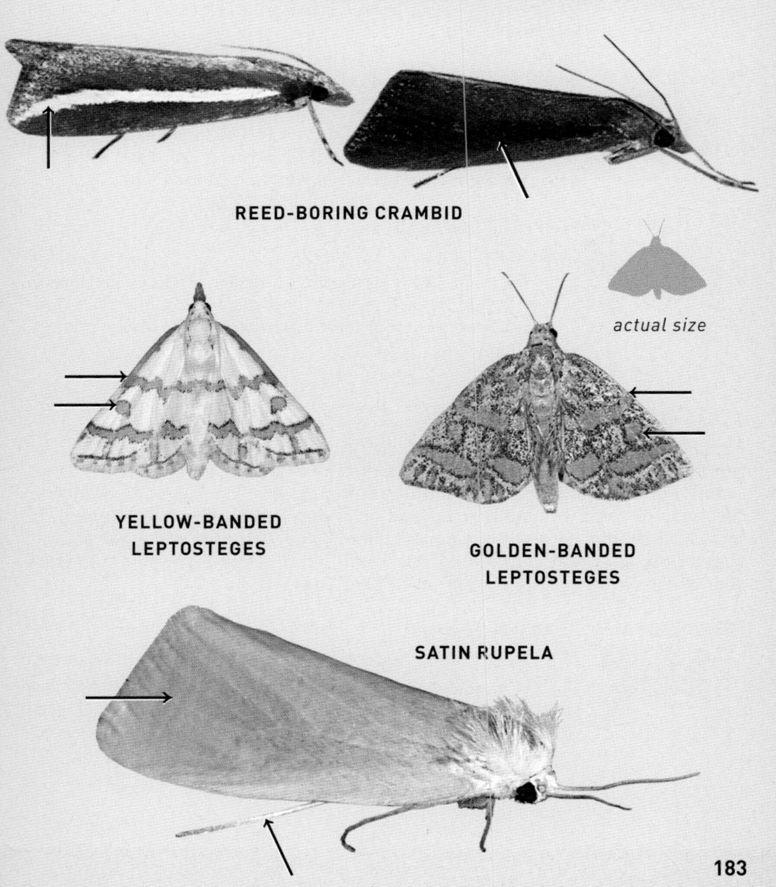

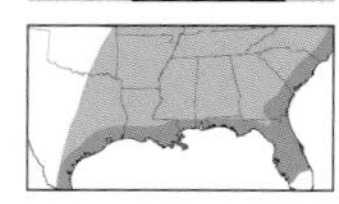

SORDID DONACAULA

Donacaula sordidella 80-0707 (5313) **Common**

TL 15–17 mm. Tawny brown FW has darker shading down center of FW. Usually has a slanting, dark brown apical dash. Sometimes has faint dotting in PM and median areas. **HOSTS:** Unknown.

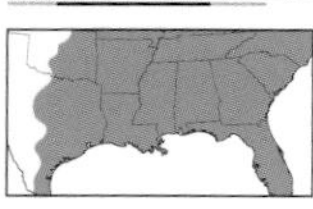

DELIGHTFUL DONACAULA

Donacaula melinellus 80-0710 (5316) **Common**

TL 18–20 mm. Pale orange to tan FW has a brown streak down center of FW. Slanting apical dash and reniform dot are usually noticeable. Sometimes has a paler costal streak. **HOSTS:** Unknown.

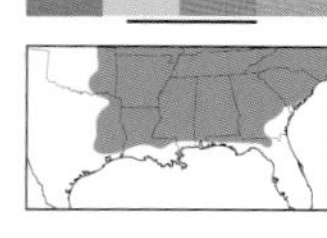

LONG-BEAKED DONACAULA

Donacaula longirostrallus 80-0715 (5319) **Uncommon**

TL 12–15 mm. Tawny brown FW has darker shading along inner costa and ST area. Slanting apical dash is usually well defined. Central median area is often mottled with brown streaks or spots. **HOSTS:** Unknown.

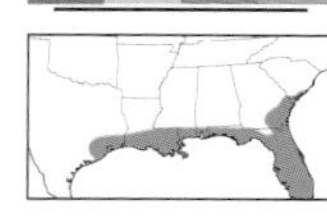

BROWN DONACAULA

Donacaula roscidellus 80-0716 (5321) **Uncommon**

TL 12–15 mm. Uniform tawny brown FW has a faint pattern of paler veins, especially in ST area. Blackish dot in reniform area is the only noticeable marking. HW is white. **HOSTS:** Unknown.

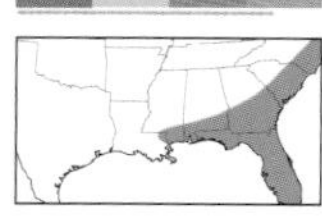

PEPPERED DONACAULA

Donacaula uxorialis 80-0718 (5323) **Local**

TL 10–12 mm. Tawny brown FW is lightly peppered with dusky scales. Curved PM and terminal lines are rows of tiny black dots. Often has a row of three or more black dots parallel to inner margin. **HOSTS:** Unknown.

Aquatic Crambids

Family Crambidae, Subfamily Acentropinae

Small moths that often have beautifully complex wing markings. Resting posture may be with wings either spread or folded (sometimes quite tightly) to form a delta shape. Larvae are aquatic, feeding on water lily and other aquatic vegetation. They will come to light in small numbers and can sometimes be seen fluttering above lily pads or other pond vegetation.

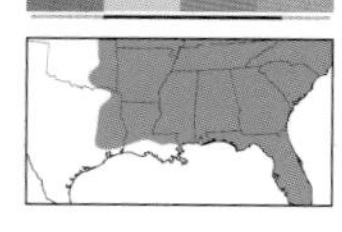

PONDSIDE CRAMBID

Elophila icciusalis 80-0724 (4748) **Common**

TL 7–13 mm. Fawn-colored or orange wings have an intricate pattern of brown-edged white lines and spots. Fringes are checkered when fresh. **HOSTS:** Aquatic plants, including buckbean, duckweed, eelgrass, and sedge.

DONACAULAS AND ALLIES

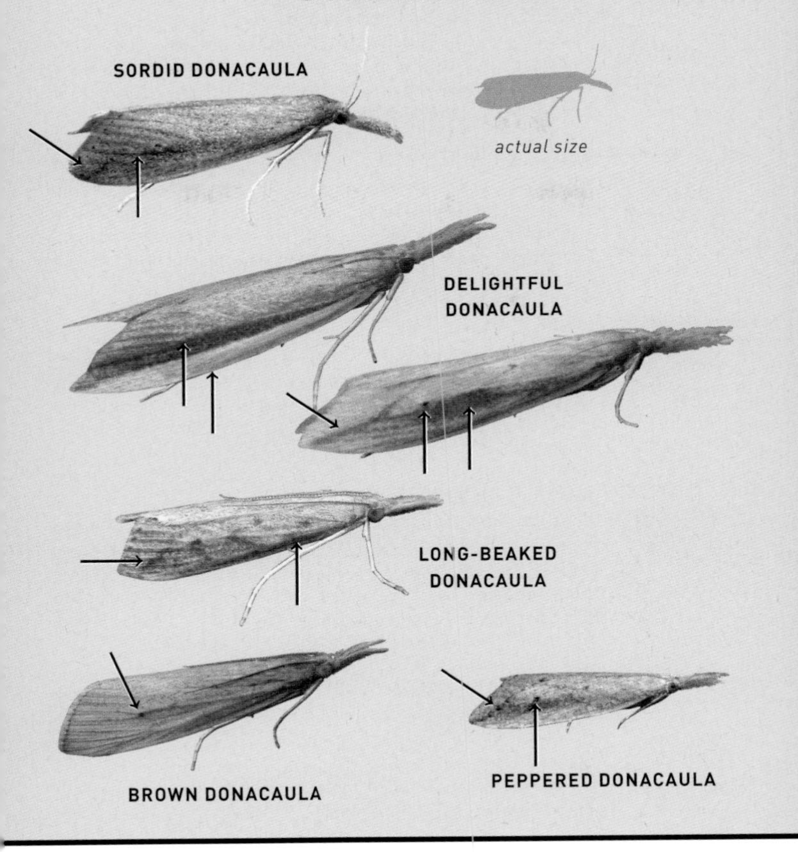

AQUATIC CRAMBIDS

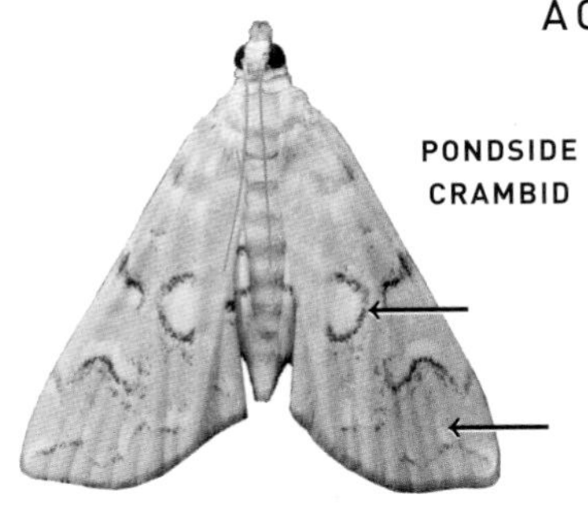

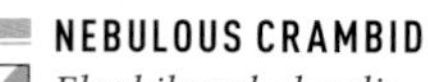

NEBULOUS CRAMBID

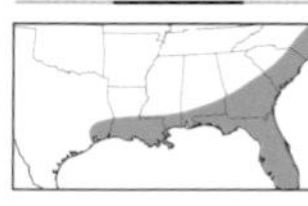

Elophila nebulosalis 80-0726 (4750) **Local**

TL 6–9 mm. Light orange FW is patterned with wavy black-edged white lines and outlines to spots. Central median and ST areas are variably shaded brown. Median band of HW has large white patches. **HOSTS:** Unknown, but probably aquatic plants.

WATERLILY BORER

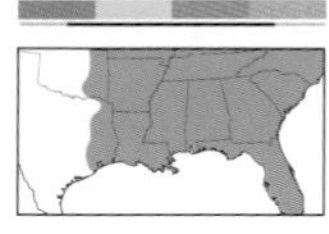

Elophila gyralis 80-0727 (4751) **Common**

TL 10–16 mm. Grayish brown (sometimes orange or dusky) FW has a complex pattern of white-edged lines and spots. Dark patch at midpoint of inner margin often contains a white dot. **HOSTS:** Water lily.

BLACK DUCKWEED MOTH

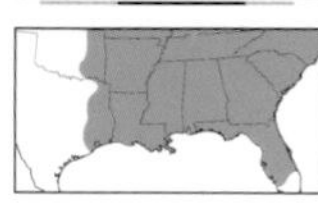

Elophila tinealis 80-0728 (4754) **Uncommon**

TL 4–6 mm. Blackish FW is lightly peppered with paler scales. Often has a diffuse whitish spot in median area. Stout blackish abdomen is ringed whitish at each segment. Antennae are short. **HOSTS:** Duckweed.

WATERLILY LEAFCUTTER

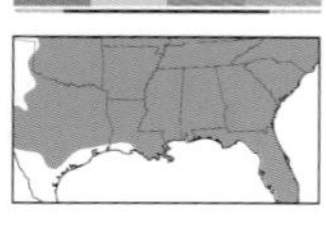

Elophila obliteralis 80-0729 (4755) **Common**

TL 7–11 mm. Mottled brown FW has jagged, white-edged median and ST lines. Blackish median band is obvious at inner margin. White reniform spot is connected to whitish shading at midpoint of costa. Females are larger and paler than males. **HOSTS:** Aquatic plants, including duckweed, pondweed, and water lily.

POLYMORPHIC PONDWEED MOTH

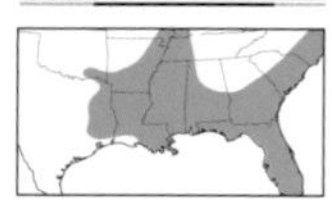

Parapoynx maculalis 80-0734 (4759) **Common**

TL 8–15 mm. Whitish FW is boldly marked with dusky blocklike spots in basal, median, and ST areas. Darker individuals are uniformly gray or brown, often with a pale spot in reniform area. HW is satin white. **HOSTS:** Aquatic plants.

OBSCURE PONDWEED MOTH

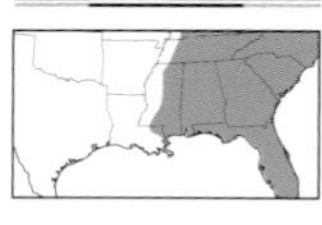

Parapoynx obscuralis 80-0735 (4760) **Common**

WS 16–26 mm. Peppery whitish FW has black blotches in median area and a white-edged dusky brown ST band. Banded HW has broad golden terminal line. Wings are usually held spread open. **HOSTS:** Aquatic plants, including eelgrass, pondweed, and yellow water lily.

CHESTNUT-MARKED PONDWEED MOTH

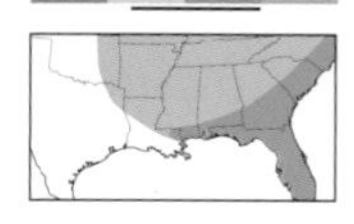

Parapoynx badiusalis 80-0736 (4761) **Local**

WS 16–24 mm. White wings are marked with bold blackish lines. Terminal line on all wings is golden yellow. Wings are usually held spread open. **HOSTS:** Aquatic plants, including pondweed.

AQUATIC CRAMBIDS

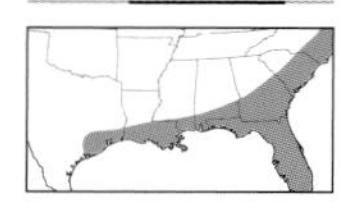

FLOATING-HEART WATERLILY MOTH

Parapoynx seminealis 80-0738 (4763) **Local**

TL 7–11 mm. Chocolate brown FW has a bold white ST band that angles inward along inner margin. Banded HW has golden terminal line accented with black dots near midpoint. Wings are usually held slightly open. Some individuals have gray instead of white bands on FW. **HOSTS:** Floating heart.

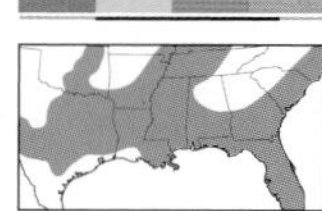

WATERMILFOIL LEAFCUTTER

Parapoynx allionealis 80-0739 (4764) **Common**

WS 14–26 mm. Whitish wings are variably streaked brown and marked with wavy black lines. Double terminal line is brown on all wings. Wings are usually held spread open. Some individuals are uniformly brown. **HOSTS:** Aquatic plants, including broadleaf water milfoil and spikerush.

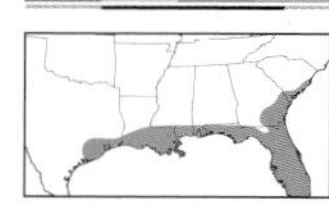

HYDRILLA LEAFCUTTER

Parapoynx diminutalis 80-0740 (4765) **Common**

WS 15–20 mm. Peppery white wings are boldly patterned with wavy golden yellow bands. Note smudgy blackish spots in median area of FW and inner ST band on HW. Wings are usually held spread open. **HOSTS:** *Hydrilla* species.

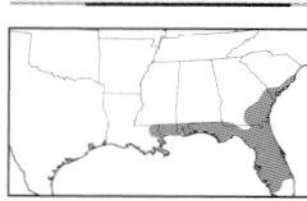

SCROLLWORK PYRALID

Neocataclysta magnificalis 80-0741 (4743) **Uncommon**

TL 10–12 mm. Creamy FW has an intricate pattern of brown lines and streaks. Oblique AM line is straight. PM line has a sharp point near inner margin. Note metallic spots in orange terminal line of HW. **HOSTS:** Unknown, but probably aquatic plants.

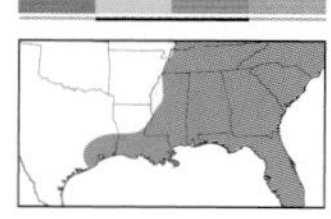

BOLD MEDICINE MOTH

Chrysendeton medicinalis 80-0742 (4744) **Common**

TL 7–8 mm. Golden brown FW is boldly marked with brown-edged satin white bands and spots. Complete white AM band widens toward inner margin. Orange terminal line of HW is studded with metallic spots. **HOSTS:** Unknown, but probably aquatic plants.

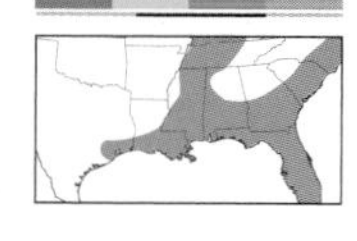

IMITABLE CRAMBID

Chrysendeton imitabilis 80-0744 (4746) **Common**

TL 6–7 mm. Bronzy brown FW has large white patches in inner basal and median areas. Incomplete white PM and ST lines converge but do not touch. Mostly white HW has a brown median band and black studded terminal line. **HOSTS:** Unknown, but probably aquatic plants.

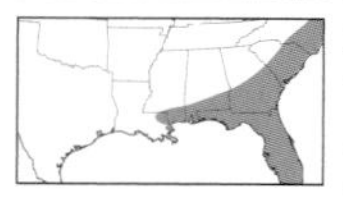

DIMORPHIC LEAFCUTTER

Neargyractis slossonalis 80-0747 (4769) **Local**

TL 7–8 mm. Shiny brown wings have a broad black-edged whitish AM band. White ST line on FW is edged orange on both sides. Terminal line on HW is accented with metallic black spots. Wings are usually held slightly open. **HOSTS:** Unknown.

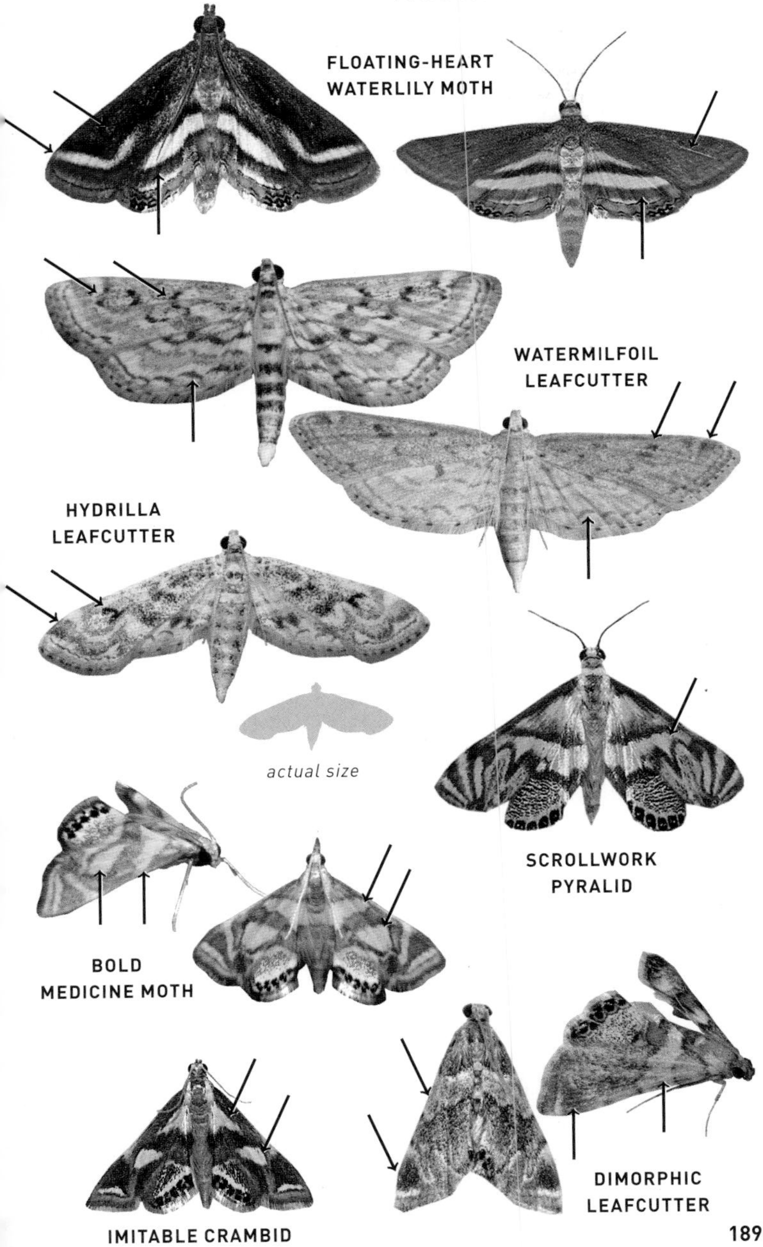

FLOATING-HEART WATERLILY MOTH

WATERMILFOIL LEAFCUTTER

HYDRILLA LEAFCUTTER

actual size

SCROLLWORK PYRALID

BOLD MEDICINE MOTH

DIMORPHIC LEAFCUTTER

IMITABLE CRAMBID

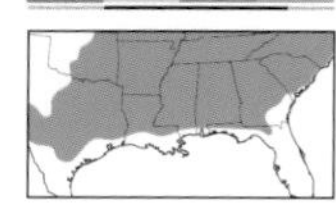

TWO-BANDED PETROPHILA

Petrophila bifascialis 80-0754 (4774) **Common**

TL 7–12 mm. Peppery white FW has double orange brown median band and three angled dashes at apex. Whitish HW has a row of black and silver spots along outer margin. **HOSTS:** Diatoms and algae scraped from rocks.

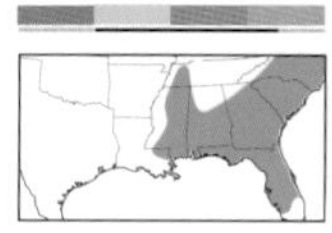

DEWY LEAFCUTTER

Eoparargyractis irroratalis 80-0766 (4785) **Common**

TL 6–8 mm. Peppery whitish FW has a brown AM band. Oblique median line angles inward toward costa. Wedge-shaped white ST line is edged orange. Terminal line on HW is accented with metallic black spots. Wings are usually held slightly open. **HOSTS:** Unknown.

Grass-Veneers

Family Crambidae, Subfamily Crambinae

Small, narrow moths that are predominantly golden brown, often with satin white streaks. They rest with their wings tight to the body, forming a tubular shape. Long, fuzzy palps give them a snouty look. Common in old-fields and grassy woodlands; adults can frequently be flushed from low vegetation during the daytime but are also regular visitors to lights at night.

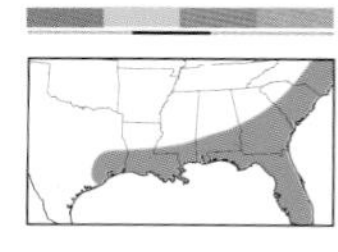

SERPENTINE GRASS-VENEER

Prionapteryx serpentella 80-0779 (5336) **Common**

TL 15–16 mm. Chocolate brown FW is streaked white along costa and through central part of wing. Angled AM, PM, and ST lines cut through the white sections. Central ST area is creased, forming a point. **HOSTS:** Unknown.

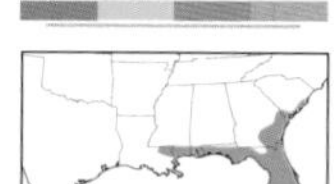

X-LINEAR GRASS-VENEER

Xubida linearella 80-0787 (5499) **Local**

TL 12–15 mm. Similar to Wainscot Grass-Veneer, but FW is more boldly streaked along veins and usually has some trace of wavy PM line, especially toward inner margin. **HOSTS:** Probably grass.

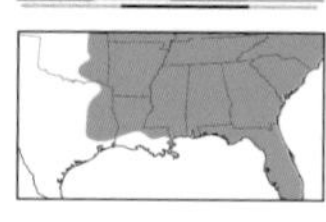

MANY-LINED GRASS-VENEER

Xubida panalope 80-0788 (5500) **Common**

TL 12–15 mm. Similar to X-linear Grass-Veneer, but more peppery FW lacks any trace of wavy PM line. **HOSTS:** Probably grass.

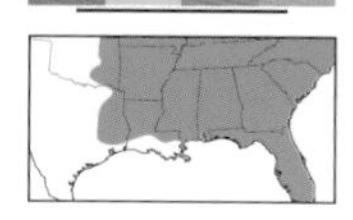

SCALED HAIMBACHIA

Haimbachia squamulella 80-0795 (5482) **Common**

TL 9–11 mm. Whitish FW is finely sprinkled with dusky scales. Thin, wavy golden PM line doubles as it curves inward toward costa. Double ST line is evenly curved and streaked golden brown. Double terminal line is black. **HOSTS:** Probably grass.

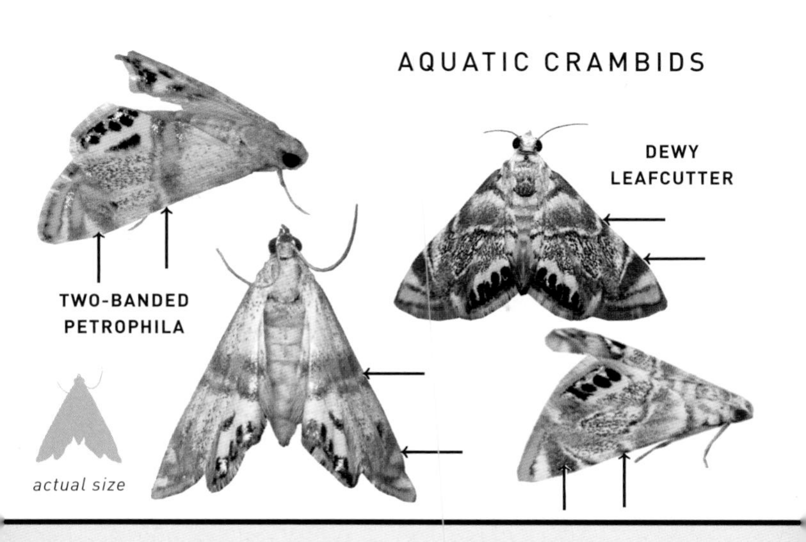
AQUATIC CRAMBIDS
DEWY
LEAFCUTTER
TWO-BANDED
PETROPHILA
actual size

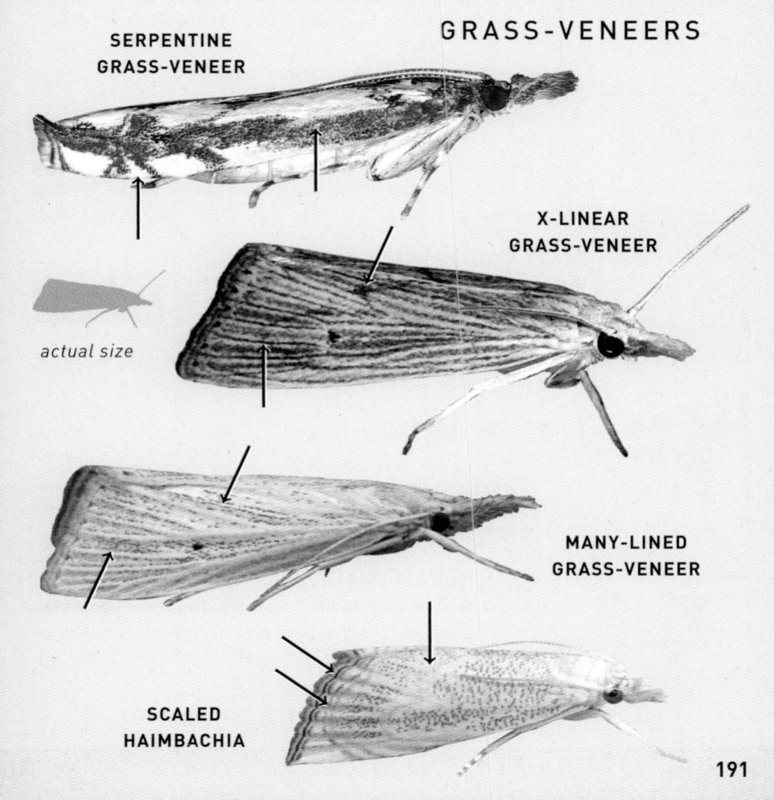
GRASS-VENEERS
SERPENTINE
GRASS-VENEER
X-LINEAR
GRASS-VENEER
actual size
MANY-LINED
GRASS-VENEER
SCALED
HAIMBACHIA

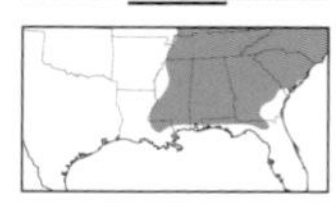

PEPPERED HAIMBACHIA

Haimbachia placidella 80-0802 (5489) **Uncommon**

TL 9–11 mm. Similar to Scaled Haimbachia, but PM line is thicker. Double ST line is evenly curved and shaded orange. Inner part of double terminal line is dotted. **HOSTS:** Probably grass.

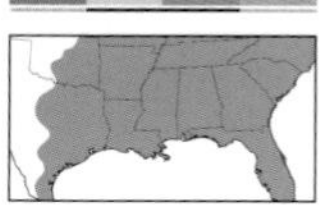

WAINSCOT GRASS-VENEER

Eoreuma densella 80-0805 (5492) **Common**

TL 10–12 mm. Light brown FW is finely streaked whitish along veins. A tiny black dot marks central median area. Dotted terminal line is often reduced. Rests with forelegs splayed out in front. **HOSTS:** Probably grass.

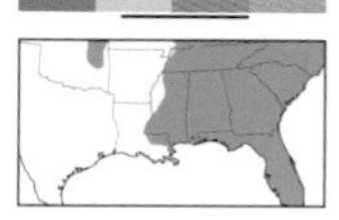

SATIN ARGYRIA

Argyria nummulalis 80-0817 (5460) **Uncommon**

TL 7–8 mm. Satin white FW has golden brown inner margin and fringe. Terminal line is dark brown. Head and palps are rusty. Some individuals have an orange spot at midpoint of inner margin or, rarely, traces of median and ST lines along costa. **HOSTS:** Grass. **NOTE:** Many individuals have often been erroneously labeled as the very similar but much rarer Mother-of-Pearl Moth (*A. rufisignella*), which has fine striations near apex of costa and lacks strong golden edge along inner margin.

MILKY ARGYRIA

Argyria lacteella 80-0815 (5463) **Common**

TL 8–10 mm. Satin white FW has traces of rusty median and PM lines, especially along costa. Midpoint of median line is often marked with a blackish dot. Terminal line is a row of black dashes. Fringe is orange. **HOSTS:** Probably grass.

BLACK-SPOTTED ARGYRIA

Argyria tripsacas 80-0816 (5463.1) **Common**

TL 8–10 mm. Satin white FW is boldly patterned with three black spots in inner median area. Black-dashed terminal line fades before apex. Fringe is orange. Rusty brown thorax and head have broad white central stripe. **HOSTS:** Probably grass.

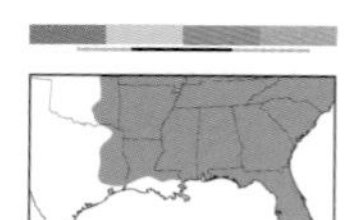

CURVE-LINED ARGYRIA

Argyria auratella 80-0819 (5465) **Common**

TL 8–10 mm. Satin white FW has golden orange median and terminal lines that connect along inner margin. Thorax has orange lateral stripes. **HOSTS:** Unknown.

STRAIGHT-LINED ARGYRIA

Argyria critica 80-0820 (5466) **Common**

TL 8–12 mm. Resembles Curve-lined Argyria, but median and terminal lines are not connected along inner margin. **HOSTS:** Unknown.

GRASS-VENEERS

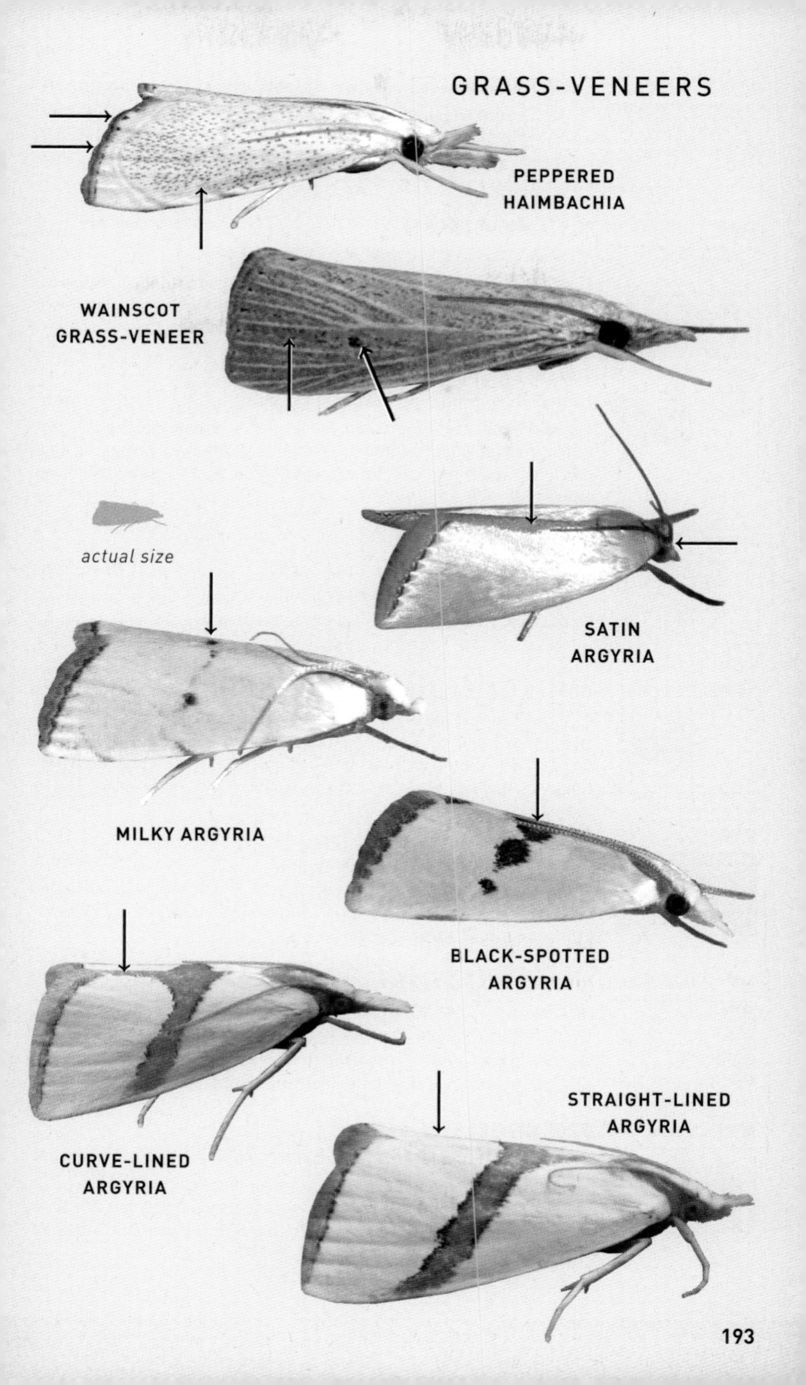

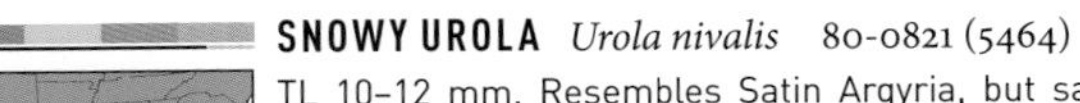

SNOWY UROLA *Urola nivalis* 80-0821 (5464) **Common**

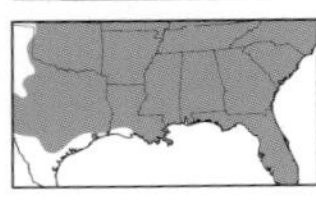

TL 10–12 mm. Resembles Satin Argyria, but satin white FW has white inner margin, often with tiny black spot near midpoint. Dark brown terminal line is dashed. Rusty brown head and palps have broad white central stripe. **HOSTS:** Grass.

RUSTY MICROCAUSTA

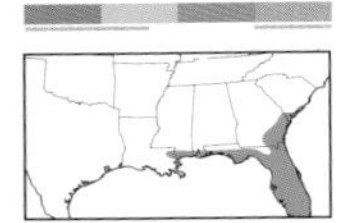

Microcausta flavipunctalis 80-0824 (5456) **Local**

TL 6–8 mm. Peppery brick red FW is marked with orange patches in basal, reniform, and subapical areas. Blackish AM line thickens toward inner margin. PM line is dotted. **HOSTS:** Unknown.

SOUTHERN CORNSTALK BORER

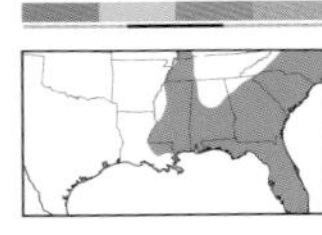

Diatraea crambidoides 80-0835 (5476) **Common**

TL 10–12 mm. Straw-colored FW is finely streaked brown along veins. Inconspicuous slanting ST line is a series of dashes. A tiny black discal dot marks central median area. **HOSTS:** Probably corn and grass.

BLACK-DOT DIATRAEA

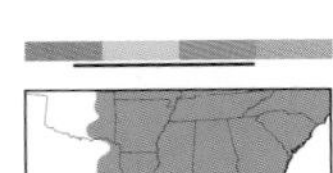

Diatraea evanescens 80-0838 (5478) **Common**

TL 10–12 mm. Straw-colored FW is finely streaked light brown along veins. A tiny blackish dot in central median area is only obvious marking. **HOSTS:** Probably grass.

DOTTED DIATRAEA *Diatraea lisetta* 80-0841 (5481) **Common**

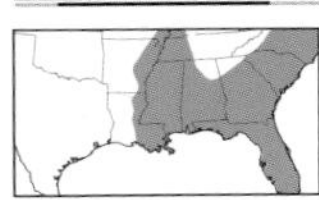

TL 10–12 mm. Pale straw-colored FW is weakly streaked brown. Oblique PM and ST lines appear as rows of brown dashes along veins. Black-dotted terminal line and dot in central median area are boldest markings. **HOSTS:** Probably grass.

BELTED GRASS-VENEER

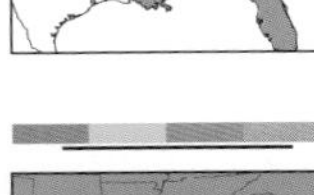

Euchromius ocellea 80-0842 (5454) **Uncommon**

TL 12–14 mm. Peppery straw-colored FW is crossed by two golden yellow bands in PM area. Angled ST line is accented with black dots along inner half. **HOSTS:** Grain products.

WOOLLY GRASS-VENEER

Thaumatopsis pexellus 80-0850 (5439) **Local**

TL 12–18 mm. Grayish FW has tawny streaks along costa and inner margin. Dark streaks in median area extend to inconspicuous PM line. Male has large bipectinate antennae. **HOSTS:** Grass.

PROFANE GRASS-VENEER

Fissicrambus profanellus 80-0866 (5431) **Common**

TL 10–12 mm. Golden brown FW is streaked gray along veins. A bold whitish streak from base fades at small blackish discal dot. Median and ST lines are jagged. Rests in a head-down position. **HOSTS:** Probably grass.

GRASS-VENEERS

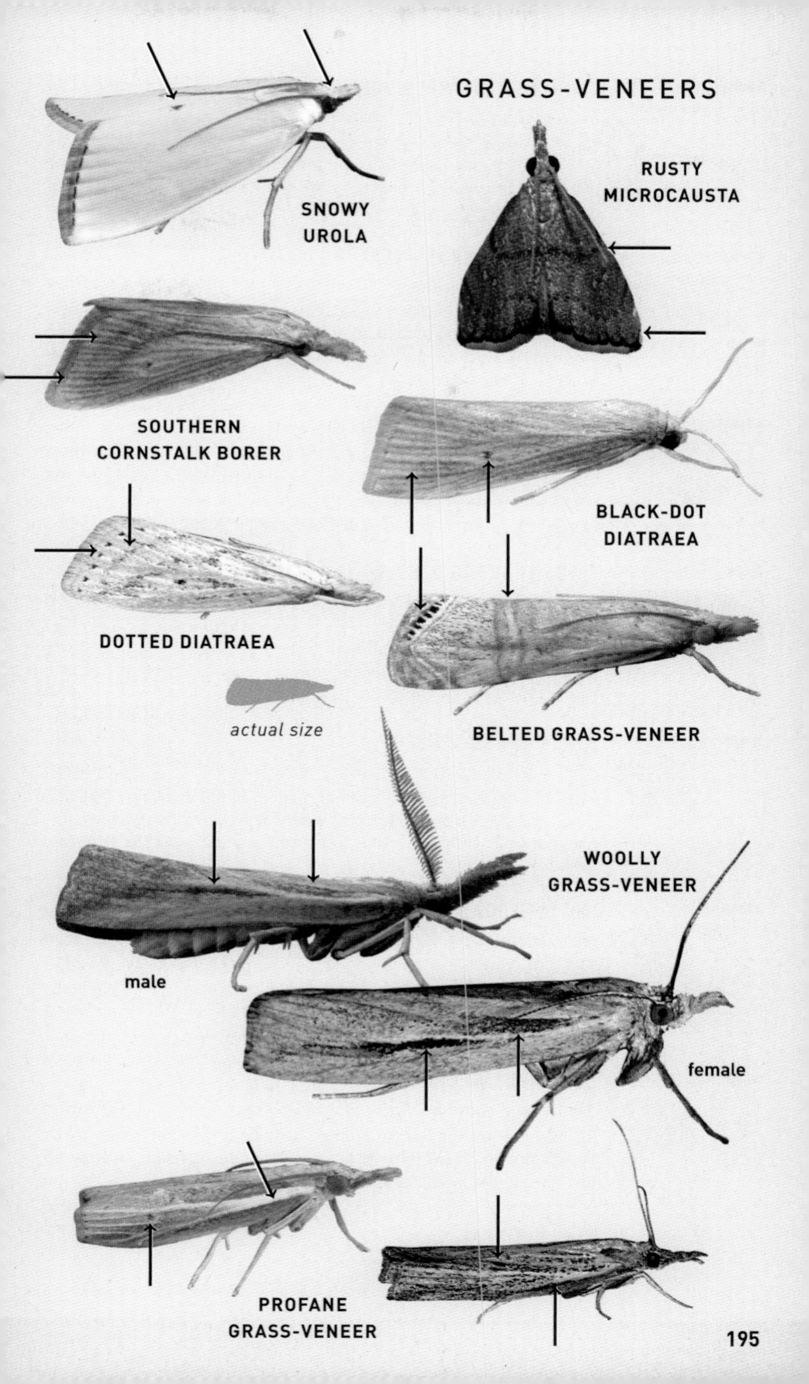

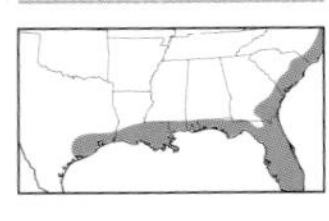

CARPET-GRASS WEBWORM

Fissicrambus haytiellis 80-0868 (5433) **Local**

TL 10–12 mm. Similar to Profane Grass-Veneer (some individuals may not be easily separable in the field), but FW is slightly darker with a bolder whitish central streak. Costa is shaded dark brown. **HOSTS:** Probably grass.

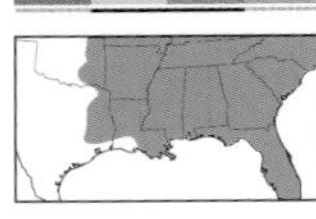

CHANGEABLE GRASS-VENEER

Fissicrambus mutabilis 80-0870 (5435) **Common**

TL 10–12 mm. Light gray FW has a tawny streak along inner margin and whitish shading in median area. Median and ST lines are jagged and usually thickened at midpoint. Rests in a head-down position. **HOSTS:** Grass.

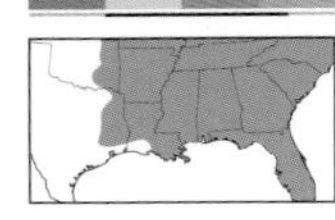

GOLD-STRIPE GRASS-VENEER

Microcrambus biguttellus 80-0874 (5419) **Common**

TL 8–10 mm. Satin white FW has angled brown median and PM lines, sometimes obvious only near costa. Median line is accented with black dots at midpoint and inner margin. **HOSTS:** Grass.

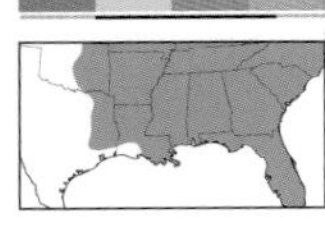

ELEGANT GRASS-VENEER

Microcrambus elegans 80-0875 (5420) **Common**

TL 8–10 mm. White FW is variably peppered brown along inner margin and beyond angled PM line. Inner median and PM lines form dusky patches. White-edged ST line is straight. Terminal line is a row of tiny black dots. **HOSTS:** Grass.

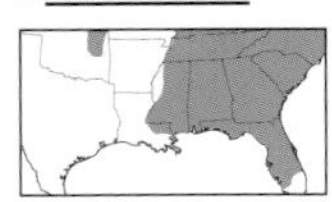

MINOR GRASS-VENEER

Microcrambus minor 80-0877 (5422) **Common**

TL 8–10 mm. White FW is sparingly peppered brown along costa and inner margin. Blackish blotch at inner median area is usually boldest marking. Brown ST line is slightly angled. **HOSTS:** Probably grass.

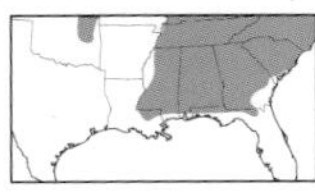

MOTTLED GRASS-VENEER

Neodactria luteolellus 80-0887 (5379) **Common**

TL 10–15 mm. Brownish FW is variably peppered silvery gray, often with dusky blotches in median area. Angled brown lines are usually fragmented. Often not possible to tell apart from other *Neodactria* species of the region. **HOSTS:** Probably grass.

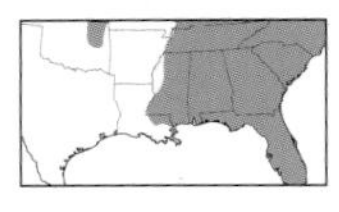

CORN ROOT WEBWORM

Neodactria caliginosellus 80-0889 (5381) **Common**

TL 10–15 mm. Similar to Mottled Grass-Veneer but tends to be darker brown on FW with inconspicuous lines. If visible, angled ST line has jagged teeth. **HOSTS:** Probably grass.

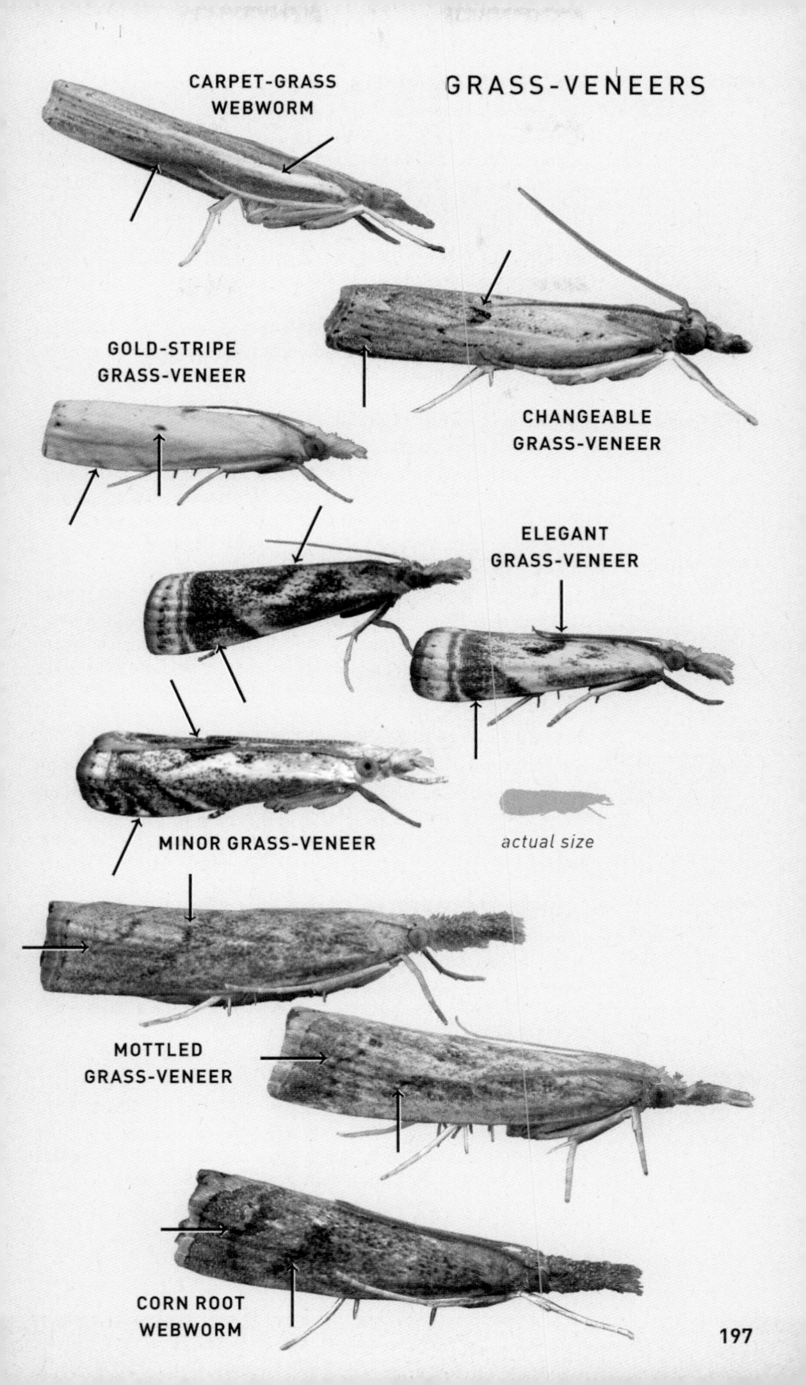
GRASS-VENEERS
CARPET-GRASS WEBWORM
CHANGEABLE GRASS-VENEER
GOLD-STRIPE GRASS-VENEER
ELEGANT GRASS-VENEER
MINOR GRASS-VENEER
actual size
MOTTLED GRASS-VENEER
CORN ROOT WEBWORM

GRACEFUL GRASS-VENEER

Parapediasia decorellus 80-0906 (5450) **Common**

TL 12–14 mm. Pearly gray FW is streaked white along veins. Golden orange median and PM lines are slightly bent at midpoint. Orange ST line is accented with black dots. Fringe is metallic gold. **HOSTS:** Unknown.

BLUEGRASS WEBWORM

Parapediasia teterrellus 80-0907 (5451) **Common**

TL 12–14 mm. Tawny brown FW is streaked silvery gray between veins as far as angled rusty ST line. Often shows blackish blotches inside median line. **HOSTS:** Grass, including bluegrass and tall fescue.

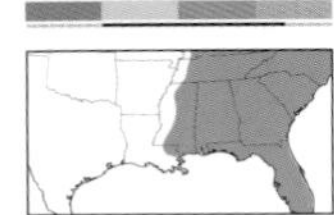

DIMINUTIVE GRASS-VENEER

Raphiptera argillaceellus 80-0913 (5393) **Common**

TL 8–9 mm. Grayish brown FW has a pointed white central stripe extending to broadly jagged PM line. Note white apical patch beyond angled ST line. Apex is sharply pointed. **HOSTS:** Probably grass.

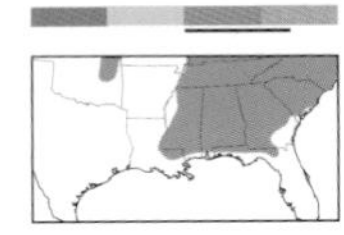

LESSER VAGABOND CRAMBUS

Agriphila ruricolellus 80-0918 (5399) **Common**

TL 10–12 mm. Peppery tan FW has faint brown streaks along veins. Angled smudgy brown median and ST lines are usually conspicuous. Terminal line is dotted. Fringe is metallic gold. **HOSTS:** Grass and common sheep sorrel.

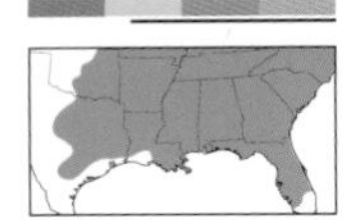

VAGABOND CRAMBUS

Agriphila vulgivagellus 80-0922 (5403) **Common**

TL 15–20 mm. Pale tan FW has bold streaky pattern without obvious cross lines. Terminal line is accented with black and gold dots. Fringe is metallic gold. **HOSTS:** Grass and grains, including wheat and rye.

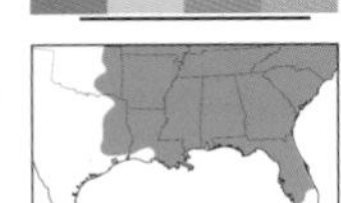

COMMON GRASS-VENEER

Crambus praefectellus 80-0943 (5355) **Common**

TL 15–16 mm. Shiny golden brown FW has pointed white streak extending through central part of wing, but it does not touch costa. Sometimes has a tiny spur projecting from inner edge at midpoint of streak. **HOSTS:** Grass and cereal grains.

LEACH'S GRASS-VENEER

Crambus leachellus 80-0945 (5357) **Uncommon**

TL 15–16 mm. Similar to Common Grass-Veneer but has a broader white streak that extends to basal section of costa. Often has a more obvious spur projecting from white streak. **HOSTS:** Grass.

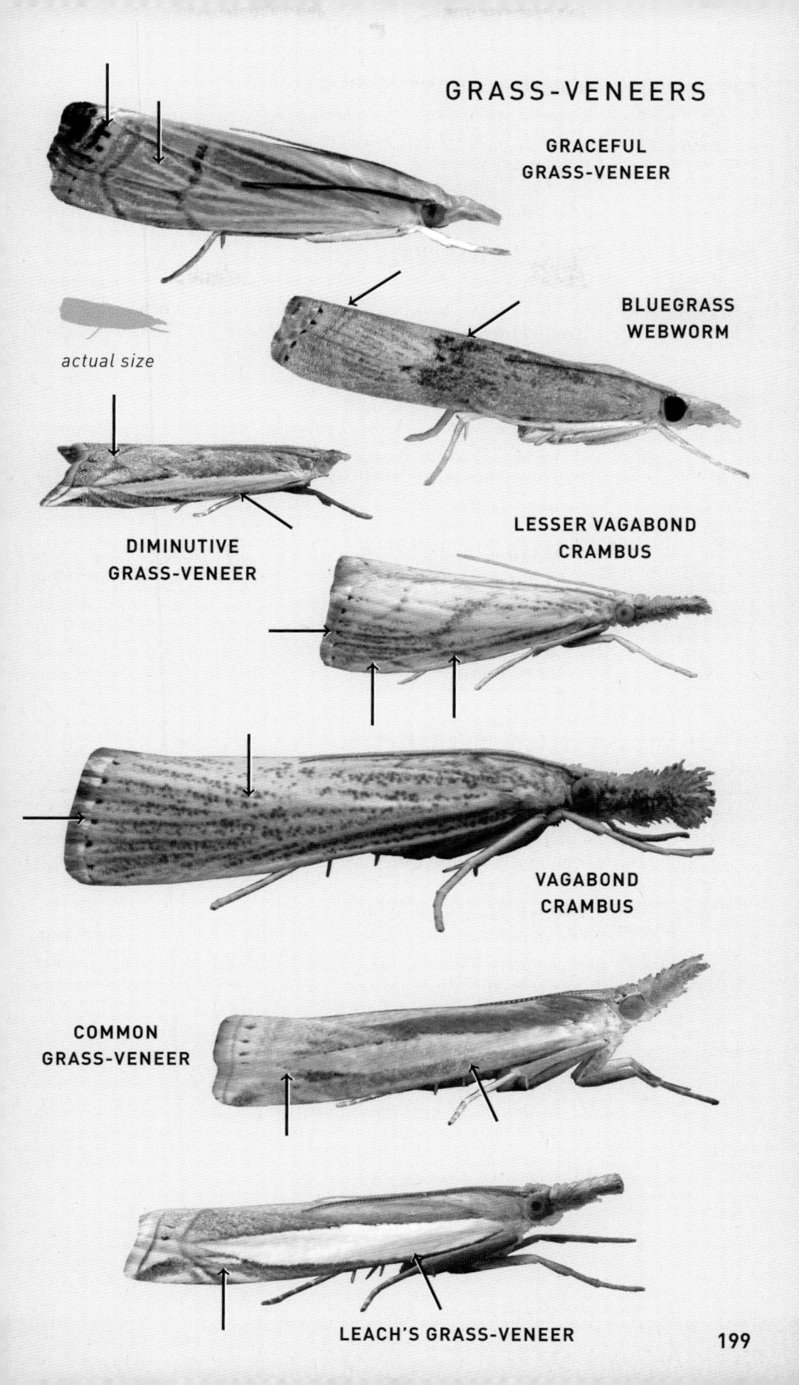
GRASS-VENEERS
GRACEFUL GRASS-VENEER
BLUEGRASS WEBWORM
actual size
DIMINUTIVE GRASS-VENEER
LESSER VAGABOND CRAMBUS
VAGABOND CRAMBUS
COMMON GRASS-VENEER
LEACH'S GRASS-VENEER

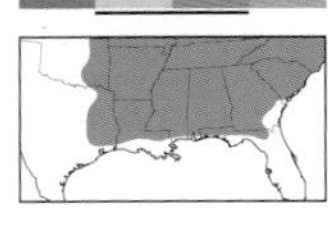

DOUBLE-BANDED GRASS-VENEER

Crambus agitatellus 80-0950 (5362) **Common**

TL 13–14 mm. Golden brown FW has a pointed white patch (sometimes faintly bisected) covering outer basal half of wing. ST area is streaked black and silver along veins with white patches at midpoint and costa. **HOSTS:** Grass and low plants.

LARGE-STRIPED GRASS-VENEER

Crambus quinquareatus 80-0957 (5369) **Common**

TL 13–15 mm. Shiny tan FW has broad white central streak from base to apex, interrupted by angled ST line. Thin black lines pass through grayish ST area. **HOSTS:** Grass.

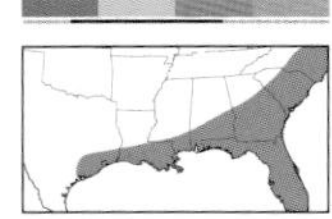

BEAUTIFUL GRASS-VENEER

Crambus satrapellus 80-0960 (5372) **Common**

TL 14–18 mm. White central streak on FW has a long narrow spur projecting from midpoint of inner edge. Note elliptical spot below tip of streak. Apex is sharply pointed. **HOSTS:** Probably grass.

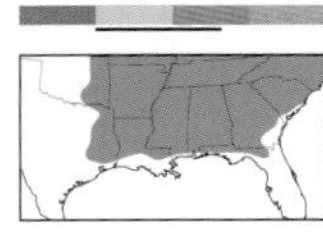

EASTERN GRASS-VENEER

Crambus laqueatellus 80-0966 (5378) **Common**

TL 12–14 mm. Shiny golden brown FW has narrow white streaks passing through central area and along costa. Rest of wing has thin black lines along veins. Silvery gray ST area has a white stripe at midpoint. **HOSTS:** Grass.

Moss-Eating Crambids

Family Crambidae, Subfamily Scopariinae

Small, narrowly delta-shaped moths with monochromatic, pointed wings. Indistinct dark tufts of scales are present on the FW; they are absent in other crambids. Adults frequently visit lights.

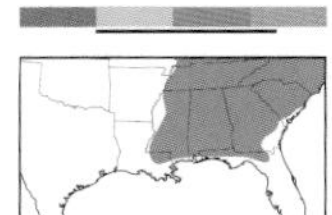

DOUBLE-STRIPED SCOPARIA

Scoparia biplagialis 80-0982 (4716) **Common**

TL 7–9 mm. Peppery gray FW has black-edged brown AM and PM bands. Note white AM and angled ST lines. Appears rather narrow when at rest. **HOSTS:** Unknown; possibly mosses.

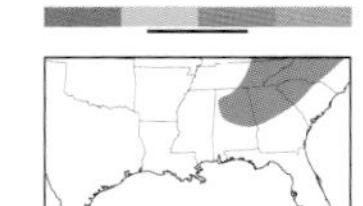

SOOTY SCOPARIA

Scoparia cinereomedia 80-0984 (4718) **Uncommon**

TL 7–9 mm. Light gray FW has a darker gray basal area and shading along ST line. Note short black dashes projecting from AM and PM lines. Terminal line is dotted. **HOSTS:** Unknown; possibly mosses.

GRASS-VENEERS

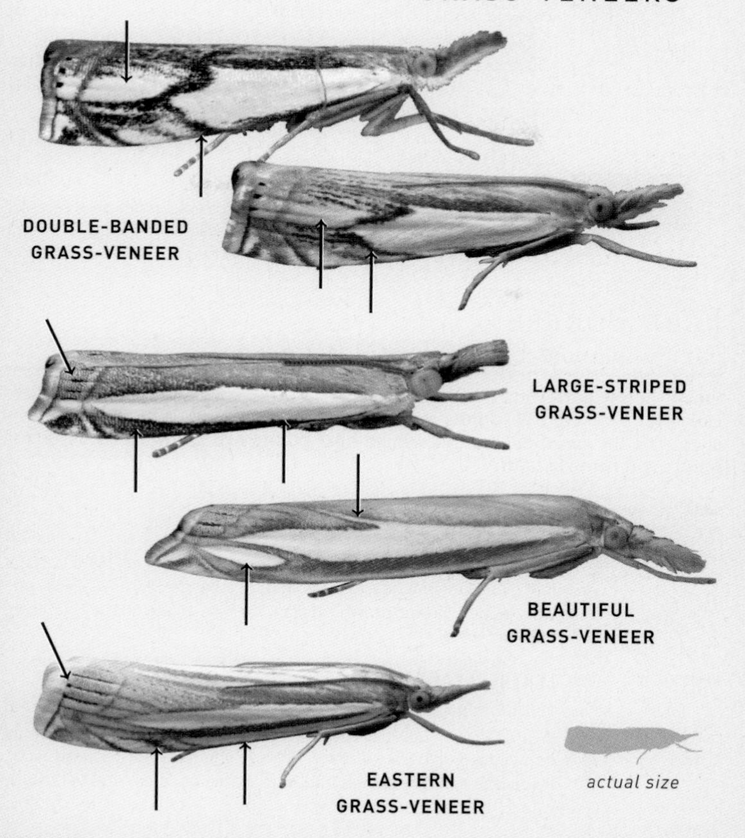

MOSS-EATING CRAMBIDS

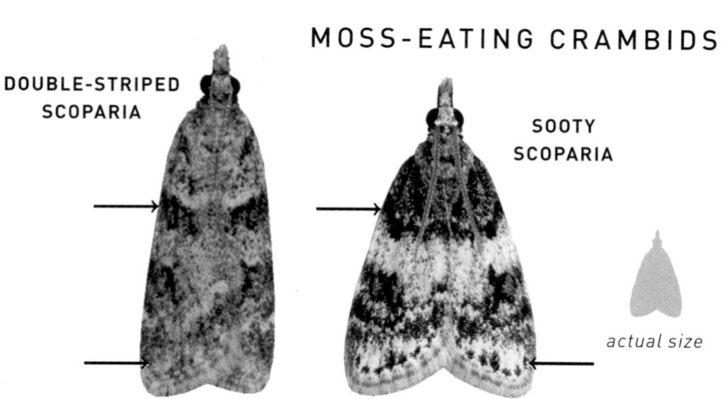

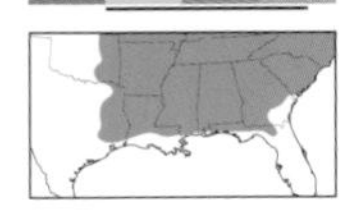

MANY-SPOTTED SCOPARIA

Scoparia basalis 80-0985 (4719) **Common**

TL 7–9 mm. Peppery light grayish brown FW is patterned with inconspicuous white-edged AM and PM lines. Black dashes in median area are often boldest markings. **HOSTS:** Unknown; possibly mosses.

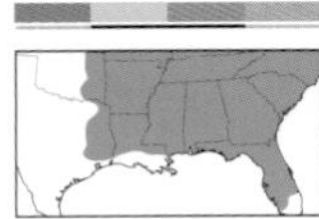

STRIPED EUDONIA

Eudonia strigalis 80-1004 (4738) **Common**

TL 6–8 mm. Peppery whitish FW has thin black longitudinal streaks extending through median and ST areas. Whitish AM and PM lines are indistinct. **HOSTS:** Unknown; possibly mosses.

Fern Borers

Family Crambidae, Subfamily Musotiminae

Small, boldly patterned moths that rest with wings spread. *Undulambia* species often hold the FW slightly apart from the HW, creating a notched appearance. The larvae of all species whose host plants are known feed on ferns. Adults will come to light.

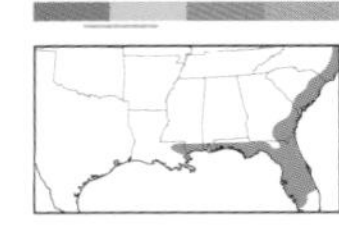

BEAUTIFUL UNDULAMBIA

Undulambia striatalis 80-1006 (4740) **Local**

WS 18–20 mm. Light brown wings are boldly streaked white in distal half. Note broad white AM line and incomplete PM line of FW, and wavy-edged darker brown median band on HW. **HOSTS:** Unknown; possibly ferns.

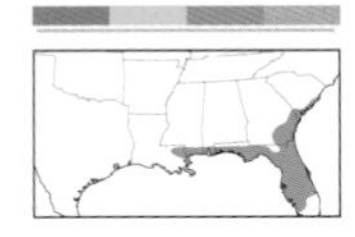

LEATHERLEAF FERN BORER

Undulambia polystichalis 80-1007 (4741) **Local**

WS 18–20 mm. Golden brown wings are crossed by thick, wavy white lines edged with dark brown. FW has oblong white reniform spot. Median line of HW widens into white patch at costa. **HOSTS:** Leatherleaf fern.

GOLD-LINED UNDULAMBIA

Undulambia rarissima 80-1008 (4742) **Local**

WS 13–15 mm. Satin white wings have a complex pattern of golden orange lines and streaks. All wings have falcate tips. White abdomen has an orange dorsal stripe. **HOSTS:** Unknown; possibly ferns.

LYGODIUM DEFOLIATOR

Neomusotima conspurcatalis 80-1009 (4896.5) **Local**

WS 13–15 mm. Brown to grayish wings have diffuse dark brown AM and PM lines. FW has broad white patches along ST line and narrow white reniform spot. HW has a row of black dots along scalloped outer margin. **HOSTS:** *Lygodium microphyllum*. **NOTE:** Introduced to FL in 2008 for biological control of Asian climbing fern (*L. microphyllum*).

MOSS-EATING CRAMBIDS

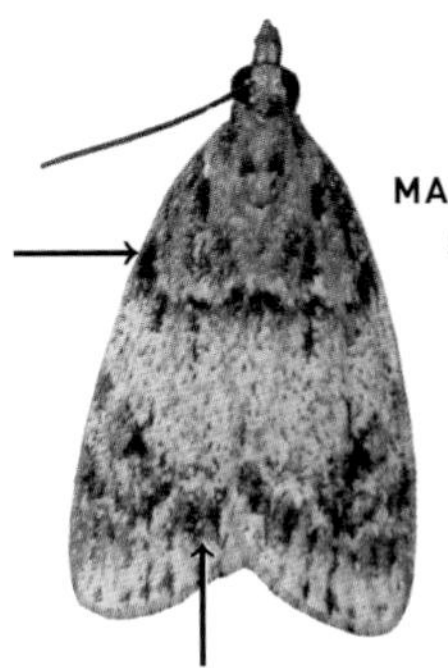

MANY-SPOTTED SCOPARIA

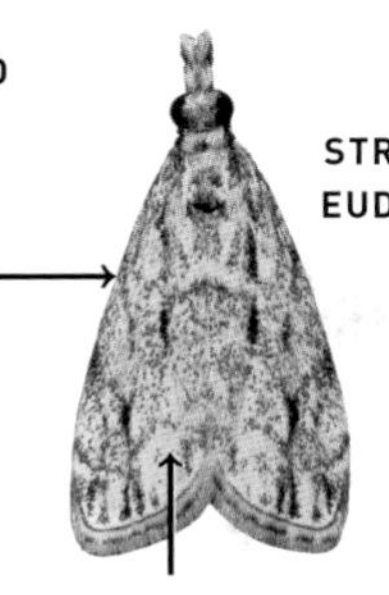

STRIPED EUDONIA

actual size

FERN BORERS

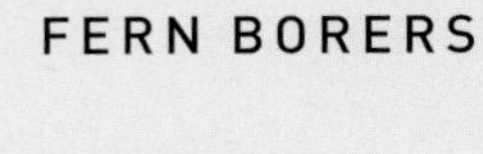

BEAUTIFUL UNDULAMBIA

actual size

LEATHERLEAF FERN BORER

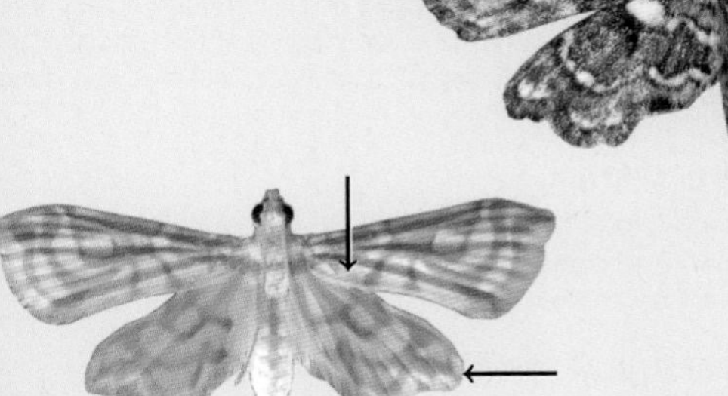

GOLD-LINED UNDULAMBIA

LYGODIUM DEFOLIATOR

Assorted Crambids

Family Crambidae, Subfamilies Glaphyriinae and Odontiinae

Small moths that superficially resemble members of the subfamily Pyraustinae. Broad, triangluar wings are usually held flat but in a few species may be curved over the body. Larvae of a few species, such as Cabbage Webworm, feed on Brassicaceae and may be crop pests. Adults will visit lights in small numbers.

CAPER-LEAF WEBWORM

Dichogama redtenbacheri 80-1014 (4790) **Local**

TL 14–16 mm. Straw-colored FW is distinctively patterned with a swirling AM line and strongly curved PM line ending as black patches at inner margin. Large hollow reniform spot has a black crescent in center. **HOSTS:** Unknown.

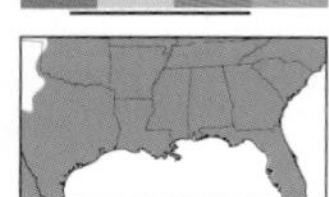

SPOTTED PEPPERGRASS MOTH

Eustixia pupula 80-1017 (4794) **Common**

TL 8–9 mm. White FW is patterned with evenly spaced black dots. Face and labial palps are black. **HOSTS:** Peppergrass, field pennycress, and cabbage.

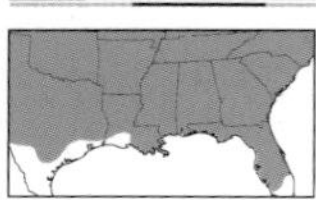

CABBAGE WEBWORM

Hellula rogatalis 80-1018 (4846) **Common**

TL 11–15 mm. Peppery light brown FW has darker median area between wavy white AM and PM lines. Rounded reniform spot is blackish. ST line is a row of thin black dashes. **HOSTS:** Brassicaceae, including cabbage, kale, rape, and horseradish.

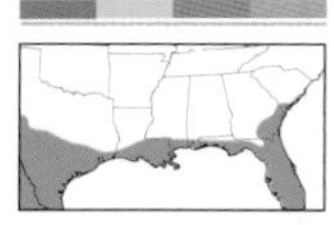

CABBAGE BUDWORM

Hellula phidilealis 80-1019 (4847) **Local**

TL 8–12 mm. Similar to Cabbage Webworm but is mottled white in basal area, has an irregularly shaped blackish reniform spot, and lacks ST line. Incomplete terminal line consists of three black dots. **HOSTS:** Brassicaceae, including cabbage, kale, mustard, and rape; also beet.

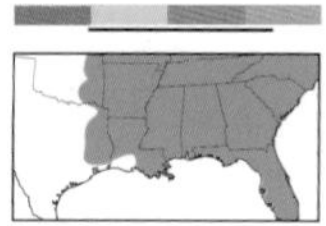

COMMON GLAPHYRIA

Glaphyria glaphyralis 80-1023 (4869) **Common**

TL 7–9 mm. Pale orange FW is marked with wavy white lines that curve inward toward costa. Fringe is dull orange. **HOSTS:** Unknown.

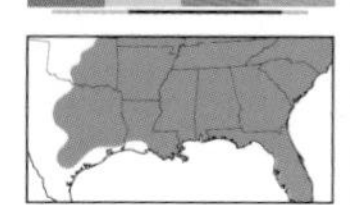

WHITE-ROPED GLAPHYRIA

Glaphyria sequistrialis 80-1024 (4870) **Common**

TL 8–9 mm. Peppery orange FW has wavy double AM and PM lines filled white. Fringe is orange brown beyond black-dotted terminal line. **HOSTS:** Live oak.

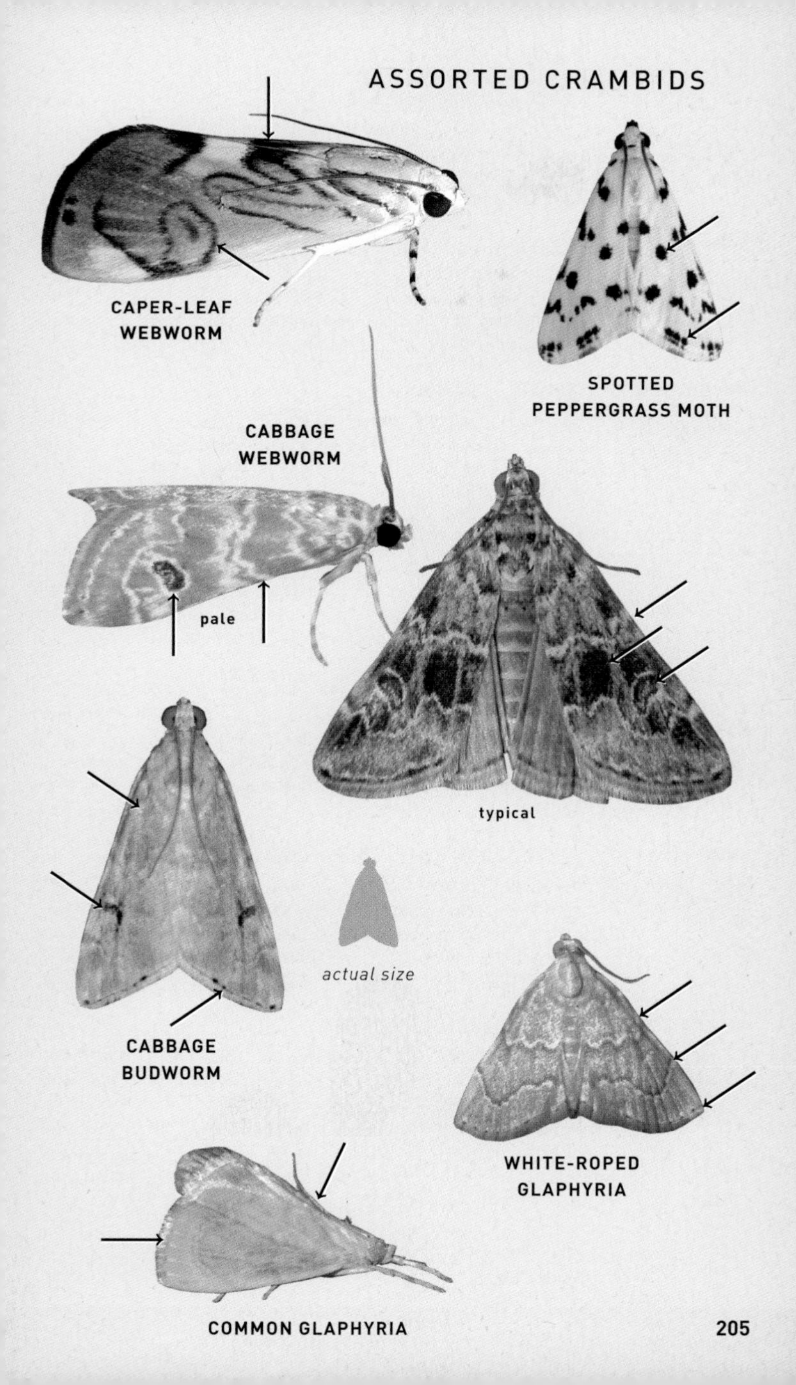
ASSORTED CRAMBIDS
CAPER-LEAF
WEBWORM
SPOTTED
PEPPERGRASS MOTH
CABBAGE
WEBWORM
pale
typical
actual size
CABBAGE
BUDWORM
WHITE-ROPED
GLAPHYRIA
COMMON GLAPHYRIA

BASAL-DASH GLAPHYRIA

Glaphyria basiflavalis 80-1025 (4871) **Local**

TL 8–10 mm. Light brown FW has a marbled pattern of white lines and chestnut patches in basal AM and outer ST areas. Note short yellow dash in lower AM area. Often rests with abdomen raised. **HOSTS:** Unknown.

BLACK-PATCHED GLAPHYRIA

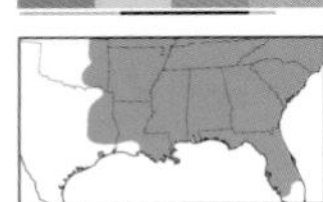

Glaphyria fulminalis 80-1027 (4873) **Common**

TL 6–7 mm. Yellowish FW has faint white zigzag median and PM lines passing through dark brown median area. Fringe is pale gold. Rests with forelegs held out in front of head. **HOSTS:** Unknown.

INVISIBLE CRAMBID

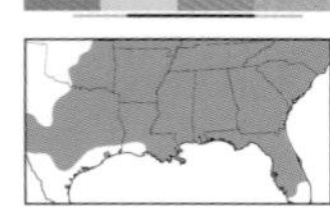

Aethiophysa invisalis 80-1032 (4877) **Common**

TL 6–7 mm. Orange FW has faint, pale-edged brown AM and PM lines. Terminal line is brown. Fringe is whitish when fresh. Rests with forelegs held out in front of head. **HOSTS:** Unknown.

XANTHOPHYSA MOTH

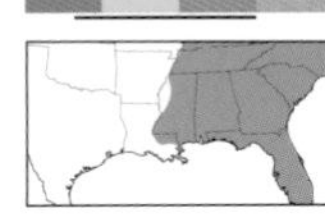

Xanthophysa psychialis 80-1035 (4879) **Common**

TL 7–9 mm. Pale orange FW is patterned with zigzag silvery gray AM and PM lines. Dusky fringe is accented with tiny black and silvery dots. Fringe is silvery gray when fresh. **HOSTS:** Unknown.

ORANGE-HEADED STEGEA

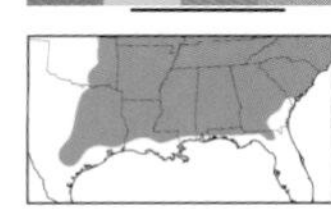

Stegea eripalis 80-1044 (4864) **Common**

TL 8–10 mm. Peppery brown FW has paler orange shading in basal area. Wavy white AM and PM lines curve inward toward costa. White fringe is slightly checkered. Head and abdomen are pale orange. **HOSTS:** Unknown.

WHITE-SPOTTED NEPHROGRAMMA

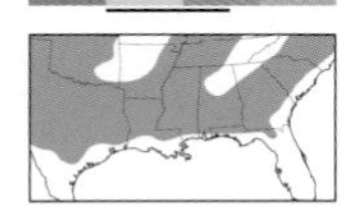

Nephrogramma reniculalis 80-1047 (4857) **Local**

TL 7–9 mm. Light brown FW is densely peppered with dusky scales. Kidney-shaped reniform and smaller subreniform spots are white. Curvy black-edged white AM and PM lines are inconspicuous. **HOSTS:** Unknown.

BROWN-BANDED LIPOCOSMA

Lipocosma sicalis 80-1055 (4881) **Common**

TL 6–8 mm. Whitish FW is shaded warm brown in median area. Wavy blackish AM and PM lines are often inconspicuous. HW has a similar pattern but with a blackish patch in inner median area. **HOSTS:** Unknown.

ASSORTED CRAMBIDS

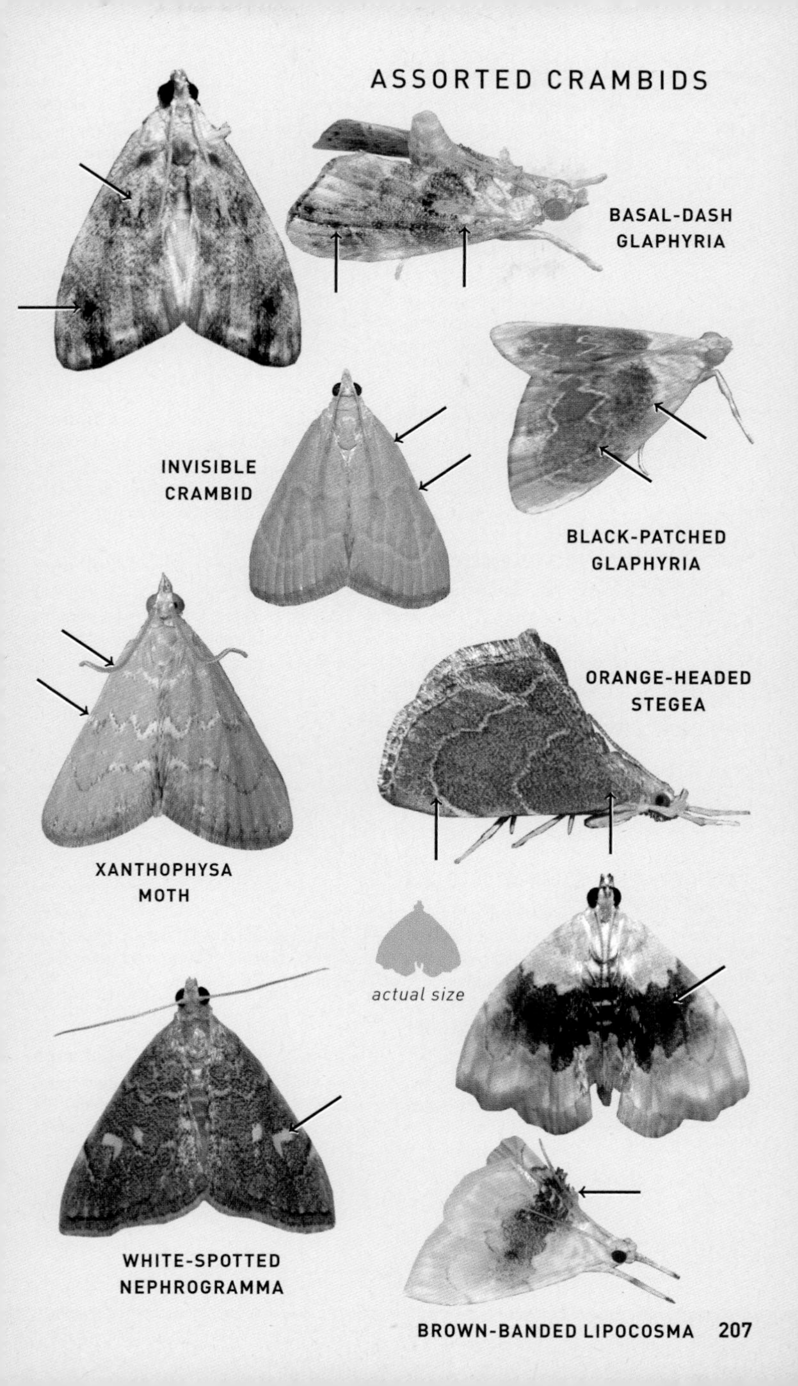

EXPOSED LIPOCOSMA

Lipocosma septa 80-1059 (4885) **Local**

TL 6–7 mm. White FW is variably shaded with warm brown bands in AM and ST areas. Wavy blackish AM and PM lines are often faint. Small black discal dot is often present in median area. **HOSTS:** Unknown.

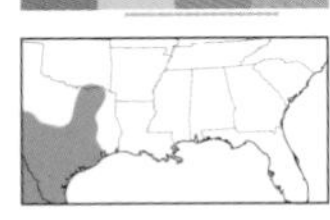

MANTLED LIPOCOSMA

Lipocosma polingi 80-1061 (4887) **Local**

TL 6–8 mm. Chocolate brown FW has a whitish basal area and peppery gray ST band. Wavy whitish AM, median, and PM lines are often faint. Rests with forelegs held out in front of head. **HOSTS:** Unknown; possibly oak.

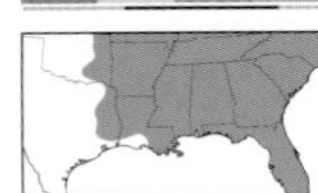

SOOTY LIPOCOSMODES

Lipocosmodes fuliginosalis 80-1062 (4888) **Common**

TL 6–8 mm. Dusky brown FW has a whitish basal patch and shading in ST area. Wavy blackish lines are inconspicuous. Often has a diffuse whitish spot in inner median area. **HOSTS:** Unknown.

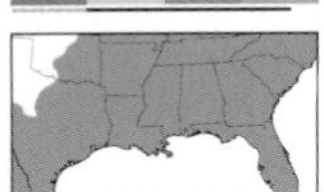

JULIA'S DICYMOLOMIA

Dicymolomia julianalis 80-1063 (4889) **Common**

TL 8–11 mm. Pale orange FW has a peppery gray median area edged with chestnut AM and ST bands. Crescent-shaped reniform spot is white. **HOSTS:** Cattail heads, dead cotton bolls, prickly pear, and eggs of bagworm moths.

GRAY DICYMOLOMIA

Dicymolomia grisea 80-1067 (4893) **Local**

TL 8–11 mm. Whitish FW is peppered with silvery gray scales in AM and ST areas. Black AM band is conspicuous. Basal AM area is streaked pale orange. **HOSTS:** Unknown.

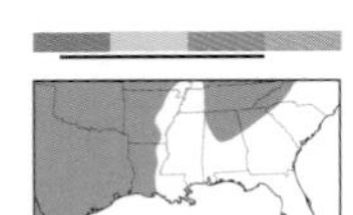

SOOTY-WINGED CHALCOELA

Chalcoela iphitalis 80-1069 (4895) **Local**

TL 8–10 mm. Pale orange FW has large silvery gray patch covering most of median and ST area. HW has a row of black and silver dots along terminal edge. **HOSTS:** Larvae of paper wasps.

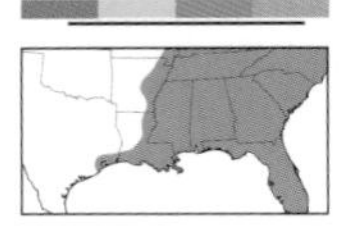

PEGASUS CHALCOELA

Chalcoela pegasalis 80-1070 (4896) **Common**

TL 8–10 mm. Peppery silvery gray FW has chocolate brown basal area and patches at either end of ST band. Curved white AM and PM lines are conspicuous. HW has a row of metallic black and silver dots along inner terminal line. **HOSTS:** Larvae of paper wasps.

ASSORTED CRAMBIDS

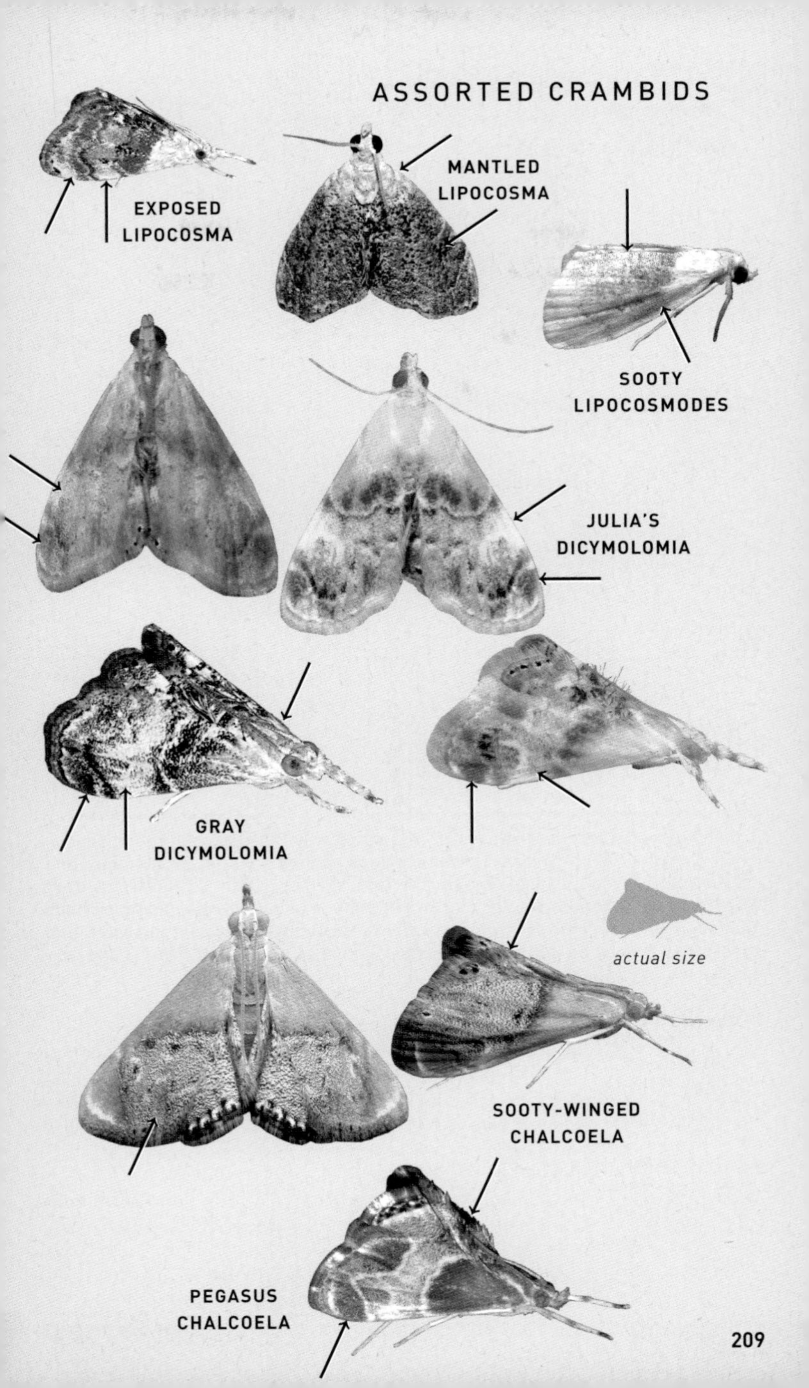

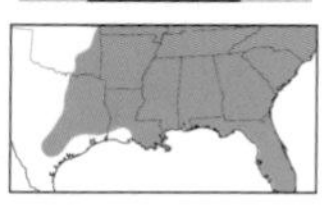

CROSS-STRIPED CABBAGEWORM

Evergestis rimosalis 80-1072 (4898) **Common**

TL 13–15 mm. Straw-colored FW has dusky brown shading along oblique blackish apical dash. Wavy AM and PM lines are inconspicuous. Inner ST area is often warm brown. **HOSTS:** Brassicaceae, including cabbage and Brussels sprouts.

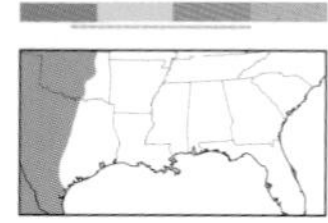

WHIP-MARKED SNOUT MOTH

Microtheoris vibicalis 80-1131 (4795) **Local**

TL 6–7 mm. Pale yellowish FW is boldly marked with oblique pinkish red AM and PM bands that are usually connected along inner margin. Fringe is pinkish red. **HOSTS:** Unknown.

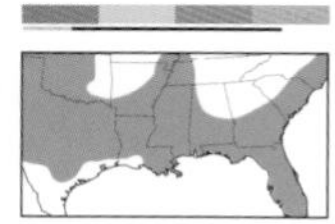

YELLOW-VEINED MOTH

Microtheoris ophionalis 80-1132 (4796) **Common**

TL 6–7 mm. Brick red FW has a distinct yellow ST line. Yellow veins extend from basal area to dusky PM line. Sometimes has a blackish discal dot. **HOSTS:** Unknown.

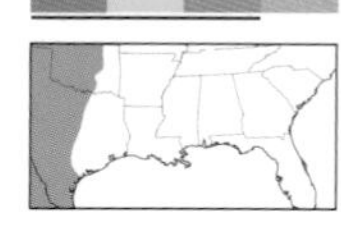

RUFOUS-BANDED CRAMBID

Mimoschinia rufofascialis 80-1155 (4826) **Local**

TL 7–12 mm. Pale cream-colored FW is boldly marked with broad, slightly wavy reddish AM and PM bands. Note reddish patches at midpoint of costa and at apex. Some individuals have more extensive reddish shading in median and terminal areas. **HOSTS:** Seeds of mallow.

Spilomeline Moths

Family Crambidae, Subfamily Spilomelinae

Previously considered a tribe within the subfamily Pyraustinae, these broad-winged moths typically rest with their wings spread wide or partly open, and are usually noticably long-legged. Certain species, such as Hawaiian Beet Webworm, are prone to wander far north of their usual ranges during periods of drought or favorable migration weather. They commonly come to light, and some species may be abundant, but a few can also be flushed from grassy areas during daylight hours.

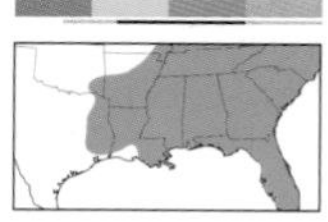

WHITE-SPOTTED ORANGE MOTH

Diastictis argyralis 80-1164 (5253) **Uncommon**

TL 12–13 mm. Bright orange FW is marked with fragmented white chunks along obscure lines. Fringe is silvery gray when fresh. Rests with wings folded. **HOSTS:** Unknown.

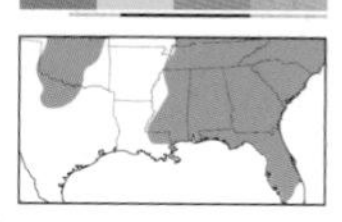

WHITE-SPOTTED BROWN MOTH

Diastictis ventralis 80-1166 (5255) **Uncommon**

TL 12–15 mm. Chocolate brown FW is variably patterned with fragmented white chunks along obscure PM and ST lines. Fringe is silvery gray when fresh. Rests with wings folded. **HOSTS:** Unknown.

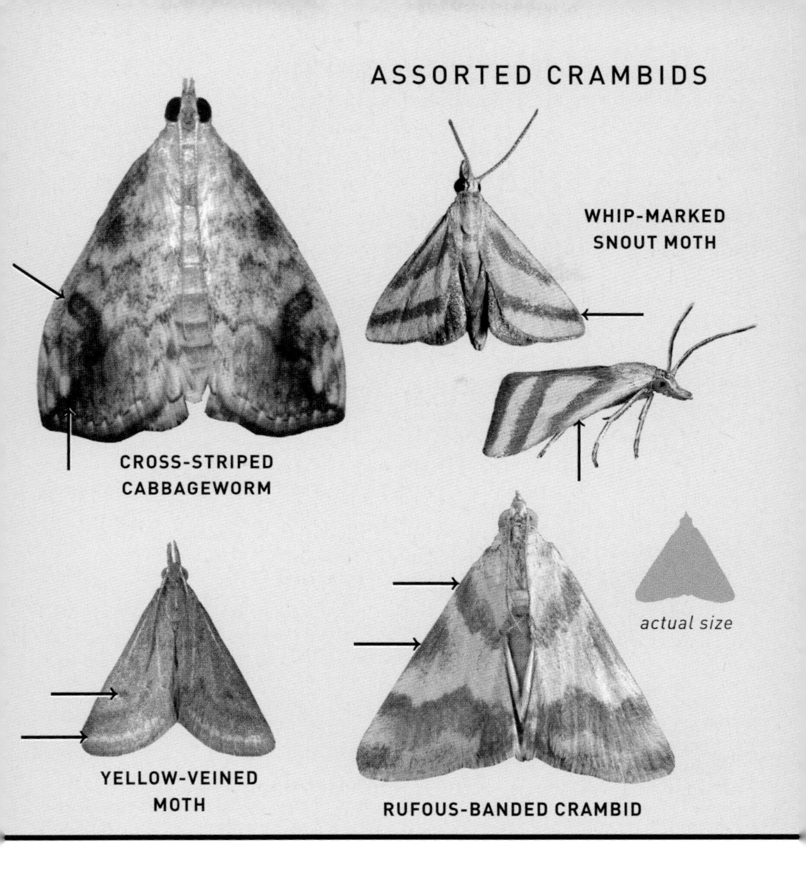

SPILOMELINE MOTHS

actual size

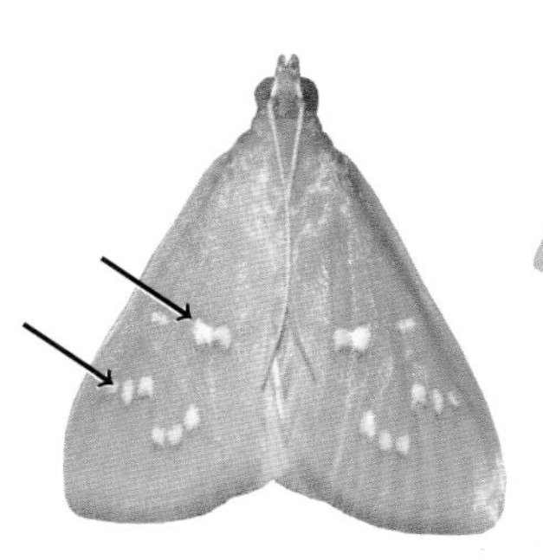

WHITE-SPOTTED ORANGE MOTH

WHITE-SPOTTED BROWN MOTH

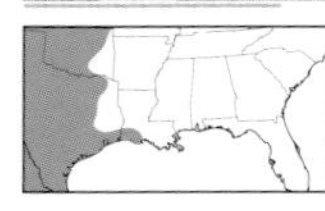

FRACTURED WESTERN SNOUT MOTH

Diastictis fracturalis 80-1167 (5256) **Local**

TL 11–16 mm. Golden brown FW is variably patterned with irregular satin white bands that do not reach costa. Fringe is silvery gray when fresh. Rests with wings folded. **HOSTS:** Unknown.

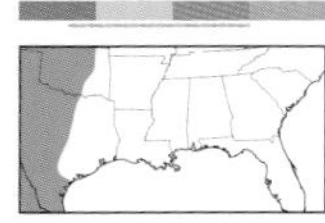

EIGHT-BARRED LYGROPIA

Lygropia octonalis 80-1178 (5251) **Local**

TL 8–10 mm. Snow-white FW is boldly patterned with fragmented lines that form large golden brown patches along costa. Outer half of fringe is shaded dark brown. Rests with wings folded. **HOSTS:** *Heliotrophium curassivicum.*

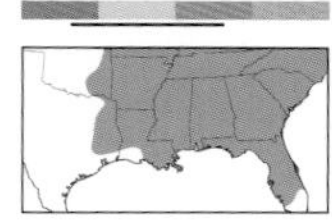

BASSWOOD LEAFROLLER

Pantographa limata 80-1184 (5241) **Common**

WS 36–38 mm. Shiny straw-colored FW is marked with a crisp pattern of brown lines and outlines to spots. Inner margin and central ST area are shaded metallic purplish brown. **HOSTS:** Basswood, oak, and rock elm.

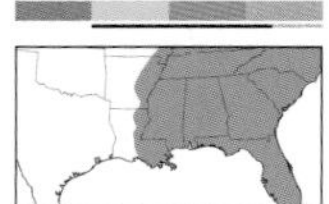

HERBIVOROUS PLEUROPTYA

Pleuroptya silicalis 80-1188 (5243) **Common**

WS 24–27 mm. Semi-translucent, pale yellow FW has a violet sheen and a pattern of slightly scalloped brown lines. PM line is wildly sinuous. Tip of pointed abdomen projects beyond wings at rest. **HOSTS:** Convolvulaceae and rougeplant.

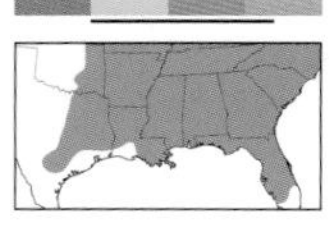

SERPENTINE WEBWORM

Herpetogramma aeglealis 80-1191 (5280) **Common**

WS 25–27 mm. Resembles Herbivorous Pleuroptya, but darker brown FW has a simple pattern of pale-edged dusky lines. Less obviously jagged ST line is poorly defined or absent. Darker individuals often have a pale rectangle in central median area. **HOSTS:** Ferns and pokeweed.

TWO-SPOTTED HERPETOGRAMMA

Herpetogramma bipunctalis 80-1193 (5272) **Common**

WS 23–25 mm. Shiny light brown FW is marked with brown lines and blackish orbicular and reniform spots. Inner costa is tinged brown. Abdomen has a pair of tiny black dots on second segment. **HOSTS:** Generalist on low plants.

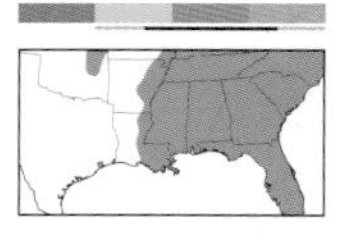

GREATER SWEETPOTATO WEBWORM

Herpetogramma fluctuosalis 80-1195 (5244) **Common**

WS 22–25 mm. Pale tan FW is shaded brown along inner costa and in ST area. Slightly scalloped AM and PM lines are wavy. Orbicular and larger reniform spots are black. **HOSTS:** Smallspike false nettle and sweet potato.

SPILOMELINE MOTHS

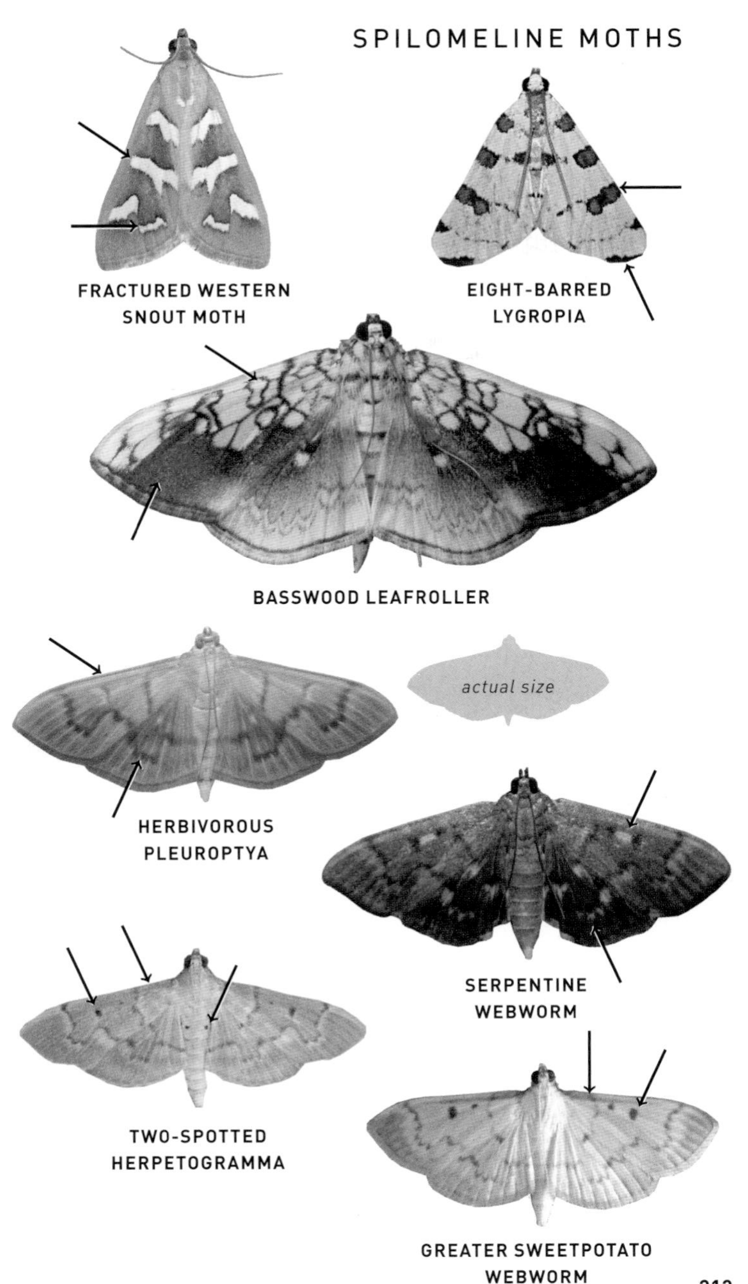

DUSKY HERPETOGRAMMA

Herpetogramma phaeopteralis 80-1196 (5274) **Common**

WS 18–24 mm. Dull, grayish brown wings are weakly patterned with dusky lines. Tiny black orbicular spot and larger crescent-shaped reniform spot can be obvious. **HOSTS:** Grass.

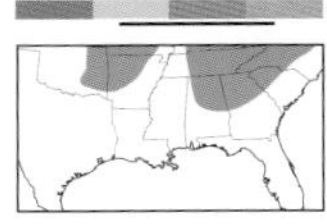

BOLD-FEATHERED GRASS MOTH

Herpetogramma pertextalis 80-1197 (5275) **Common**

WS 22–28 mm. Semi-translucent straw-colored wings have a violet sheen when fresh. FW has a pattern of scalloped brown lines. Costa is tinged brown. Note pale square in outer median area and deeply zigzagged ST line. **HOSTS:** Low plants, including violets.

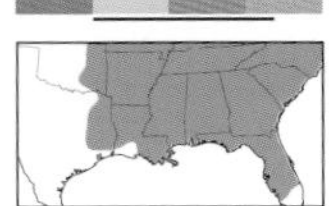

ZIGZAG HERPETOGRAMMA

Herpetogramma thestealis 80-1199 (5277) **Common**

WS 29–35 mm. Resembles Bold-feathered Grass Moth, but darker FW has bolder jagged lines with more obvious whitish ST band and rectangular spots in median area. **HOSTS:** Reported on hazelnut, basswood, and strawberry bush.

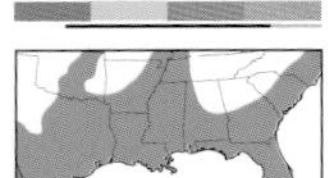

MAGICIAN MOTH

Hileithia magualis 80-1202 (5187) **Uncommon**

WS 14–15 mm. Golden brown FW has a bold pattern of crisp brown lines and outlines to white spots. White AM and PM bands widen toward inner margin. Costa is tinged brown basally. Terminal line is solidly brown. **HOSTS:** Unknown.

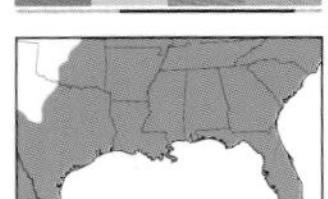

SCRAPED PILOCROCIS

Pilocrocis ramentalis 80-1207 (5281) **Common**

WS 24–29 mm. Brown wings have a bronzy sheen when fresh. Black-edged white PM line on FW is strongly curved. Crescent-shaped reniform spot is edged black. **HOSTS:** False nettle and cardinal's guard.

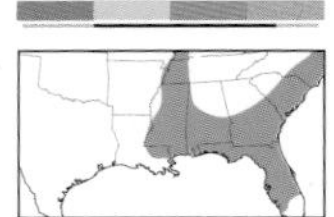

LESSER CANNA LEAFROLLER

Geshna cannalis 80-1215 (5126) **Local**

WS 20–25 mm. Light brown FW is weakly marked with zigzag AM and PM lines. Whitish discal spot is partly edged blackish. **HOSTS:** *Canna* species.

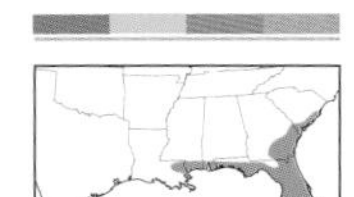

ORNATE HYDRIRIS

Hydriris ornatalis 80-1216 (5127) **Local**

WS 16–20 mm. Light golden brown FW has silvery gray bands along curved AM and PM lines. HW has a similar pattern with a tiny black discal dot in central median area. Note long abdomen. **HOSTS:** Morning glory.

SPILOMELINE MOTHS

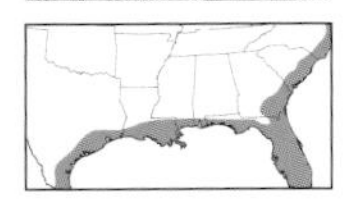

EASTERN LINEODES

Lineodes fontella 80-1217 (5106) **Common**

TL 9–10 mm. Distinctive stilted appearance, with abdomen curled high above head. Brown FW has a narrow, wildly sinuous black PM line edged in white. Black-edged brown median band has a long downward-reaching spur near costa. **HOSTS:** Ground cherry.

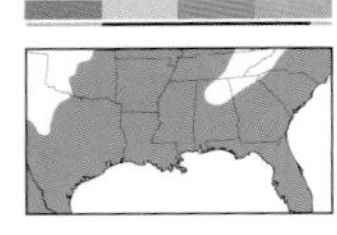

EGGPLANT LINEODES

Lineodes integra 80-1218 (5107) **Common**

TL 9–10 mm. Closely resembles Eastern Lineodes, but brown median band adjoins PM line in inner half and together they form a sinuous curve through middle of FW. **HOSTS:** Eggplant, ground cherry, pepper, and others in nightshade family.

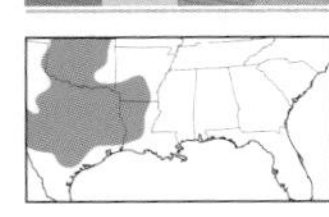

INTERRUPTED LINEODES

Lineodes interrupta 80-1219 (5108) **Common**

TL 9–10 mm. Closely resembles Eastern Lineodes, but PM line adjoins median line in inner half, and line and median band are broken near costa. **HOSTS:** Unknown.

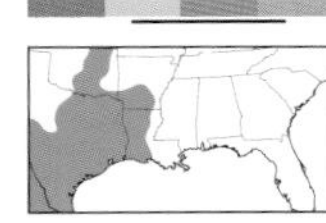

VICTORIAN LAMPLIGHTER

Lamprosema victoriae 80-1225 (5104) **Local**

TL 9–11 mm. Earth brown FW has black-edged white AM and PM lines. PM line bulges at midpoint and widens to form a white triangle at costa. Note prominent white triangle in central median area. **HOSTS:** Unknown.

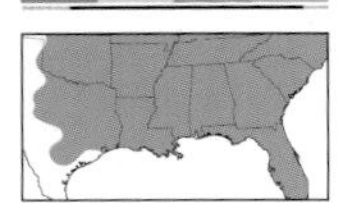

CELERY LEAFTIER *Udea rubigalis* 80-1230 (5079) **Common**

TL 7–10 mm. FW is variably pale tan to reddish brown with crisp dusky lines and outlines to spots. Reniform spot is figure eight-shaped. **HOSTS:** Low plants and crops, including bean, beet, celery, and spinach.

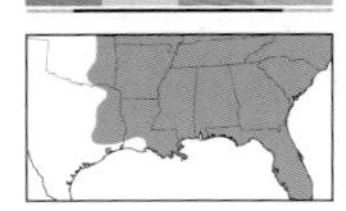

YELLOW-SPOTTED WEBWORM

Anageshna primordialis 80-1254 (5176) **Common**

WS 12–15 mm. Dusty brown FW is marked with black-edged yellow lines and spots. HW has a relatively wide, pale yellow median band. **HOSTS:** Unknown; reported on broadleaf lady palm.

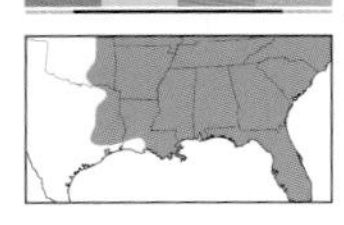

CHECKERED APOGESHNA

Apogeshna stenialis 80-1255 (5177) **Common**

WS 14–17 mm. Peppery chocolate brown FW is boldly patterned with yellowish edges to AM, median, and PM lines that flare toward costa. HW has pale yellow bands. **HOSTS:** Unknown.

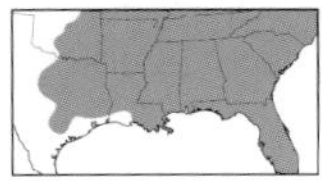

HOLLOW-SPOTTED BLEPHAROMASTIX

Blepharomastix ranalis 80-1256 (5182) **Common**

WS 16–20 mm. Shiny yellowish FW has a bold pattern of crisp brown lines and outlines to spots. PM line is sharply kinked. Costa is tinged brown basally. Terminal line is a row of dashes. **HOSTS:** Goosefoot.

SPILOMELINE MOTHS

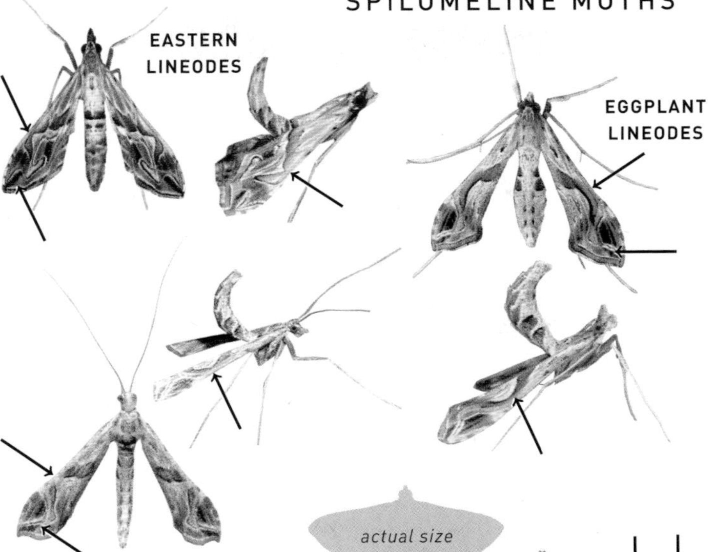

EASTERN LINEODES

EGGPLANT LINEODES

INTERRUPTED LINEODES

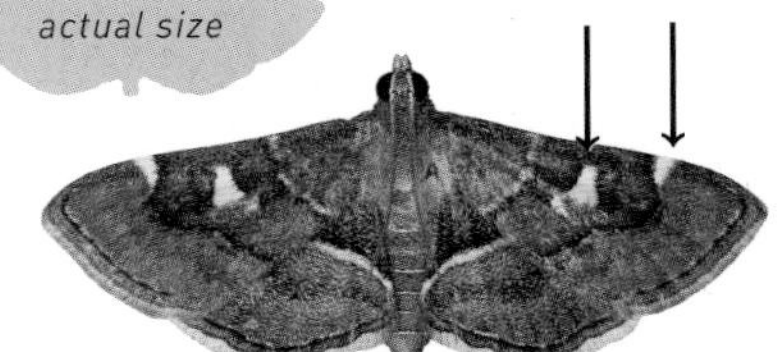

VICTORIAN LAMPLIGHTER

CELERY LEAFTIER

YELLOW-SPOTTED WEBWORM

CHECKERED APOGESHNA

HOLLOW-SPOTTED BLEPHAROMASTIX

CALICO BLEPHAROMASTIX

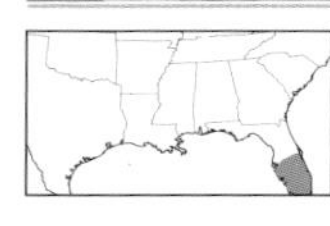

Blepharomastix schistisemalis 80-1261 (5191) **Local**

TL 8–10 mm. Pale straw-colored FW has a complex pattern of crisp brown lines and outlines to spots. Often has bold blackish patches forming a fragmented ST band. Pale abdomen is banded brown. **HOSTS:** Unknown.

GRAPE LEAFFOLDER

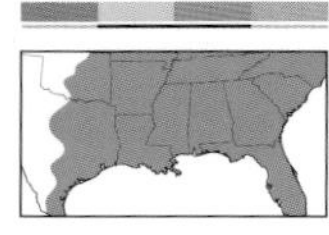

Desmia funeralis 80-1262 (5159) **Common**

WS 25–30 mm. Sexually dimorphic. Black FW is marked with two large white spots in median area. HW has a broad white median band, which is usually pinched into two in females. Males have a distinct "joint" at midpoint of antenna. Underside of abdomen has a solid white patch. **HOSTS:** Evening primrose, grape, and redbud.

WHITE-HEADED GRAPE LEAFFOLDER

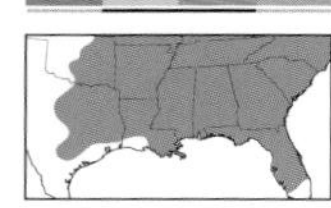

Desmia maculalis 80-1263 (5160) **Common**

WS 20–25 mm. Virtually identical to larger Grape Leaffolder, but underside of abdomen has white and black bands. Large females may approach small male Grape Leaffolders in size, so sex is important to identification by size. **HOSTS:** Evening primrose, grape, and redbud.

DEPLORING DESMIA

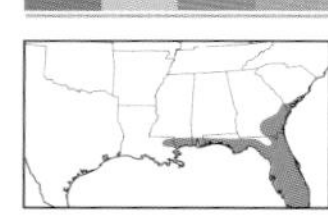

Desmia deploralis 80-1268 (5166) **Local**

WS 20–25 mm. Similar to larger Grape Leaffolder, but white spots on FW are smaller and more rectangular, with inner spot often divided. HW has white median band divided by a black line. **HOSTS:** Unknown.

PIED SHAWL

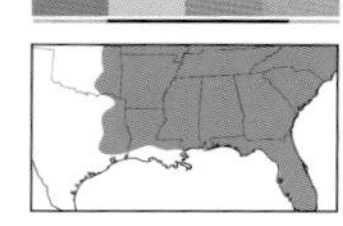

Diasemiodes janassialis 80-1272 (5172) **Common**

TL 10–15 mm. Black wings are patterned with sinuous and slightly fragmented thin white lines. Terminal line is checkered. White fringe on FW is shaded black at apex and midpoint. **HOSTS:** Unknown.

BLACK SHAWL

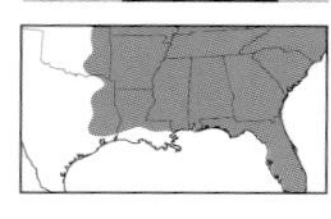

Diasemiodes nigralis 80-1273 (5173) **Common**

TL 8–10 mm. Black FW is liberally patterned with white spots in AM area and outer section of PM line. Grayish fringe has white sections near apex and anal angle. Black HW (when visible) has a broken white median band. **HOSTS:** Unknown.

BRINDLED SHAWL

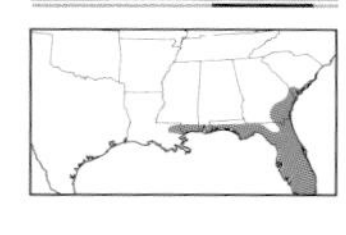

Diasemiopsis leodocusalis 80-1275 (5171) **Local**

TL 10–15 mm. Dusky wings are strongly brindled orange and white with an indistinct pattern of whitish bands (bolder on HW). Terminal line is checkered. White fringe on FW is weakly shaded dusky at apex and midpoint. **HOSTS:** Unknown.

SPILOMELINE MOTHS

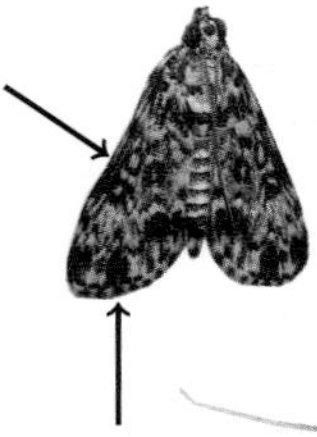

CALICO BLEPHAROMASTIX

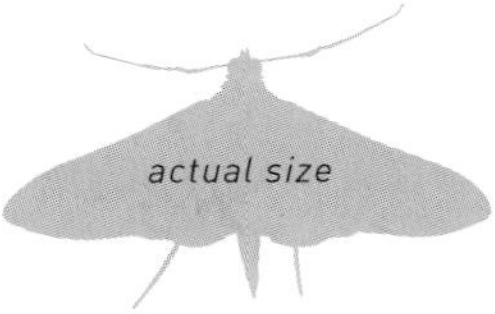

actual size

dorsal

GRAPE LEAFFOLDER

ventral

dorsal

WHITE-HEADED GRAPE LEAFFOLDER

ventral

DEPLORING DESMIA

PIED SHAWL

BLACK SHAWL

BRINDLED SHAWL

RECONDITE WEBWORM

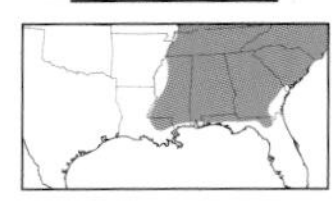

Diathrausta reconditalis 80-1276 (5174) **Local**

WS 13–18 mm. Very similar to Harlequin Webworm, and some individuals may not be reliably separated, but orange AM and PM lines are usually very thin and may fade away entirely through midsection of FW. Broken median band on HW may be connected on some individuals. **HOSTS:** Unknown.

HARLEQUIN WEBWORM

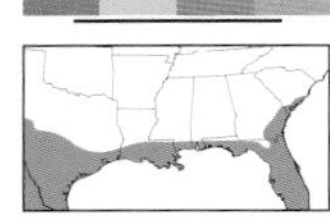

Diathrausta harlequinalis 80-1277 (5175) **Local**

WS 13–18 mm. Blackish FW has three rectangular white spots in median area. AM and PM lines are orange but may be pale when worn. Basal and ST areas are sometimes shaded orange. Median band of HW consists of a large white spot separated from a narrower white line. **HOSTS:** Unknown.

SPOTTED BEET WEBWORM

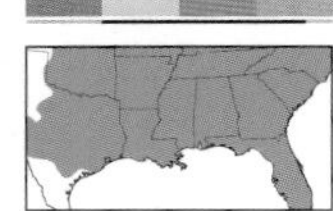

Hymenia perspectalis 80-1279 (5169) **Common**

TL 10–12 mm. Brown FW has a broken white median band and a long white patch at costal end of PM line. Median band on HW is narrow and toothed. **HOSTS:** Low plants, including beet, chard, and potato.

HAWAIIAN BEET WEBWORM

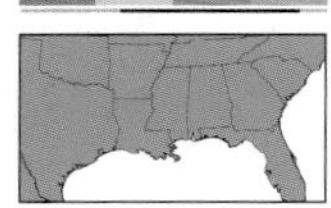

Spoladea recurvalis 80-1280 (5170) **Common**

TL 9–11 mm. Resembles Spotted Beet Webworm, but median band is wider and more even on both wings. **HOSTS:** Low plants, including beet, chard, and spinach.

MERRICK'S CRAMBID

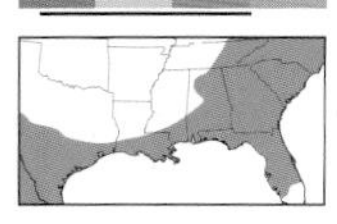

Loxostegopsis merrickalis 80-1283 (5117) **Uncommon**

TL 9–12 mm. Peppery dark gray FW has a diffuse pale yellow discal spot and subapical patch. Head and brushy labial palps are pale yellow. **HOSTS:** Unknown.

BLACK PENESTOLA *Penestola bufalis* 80-1286 (5179) **Local**

TL 9–11 mm. Peppery dusky brown FW has a mottled pattern of smudgy dark lines with paler edges and spots. Apex is rounded. Dusky HW has pale yellow ST line. **HOSTS:** Unknown.

DISTINGUISHED COLOMYCHUS

Colomychus talis 80-1292 (5200) **Common**

WS 13–20 mm. Shiny fuchsia FW is boldly marked with brassy patches in outer PM and inner median areas. HW has a bulging brassy median band. Fringes are pale gold. **HOSTS:** Unknown.

ALAMO MOTH *Condylorrhiza vestigialis* 80-1293 (5215) **Local**

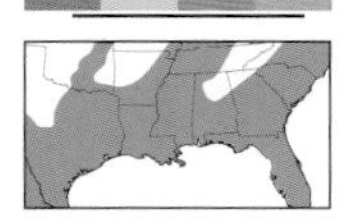

WS 28–32 mm. Satin white to bright yellow wings are lightly patterned with wavy brown lines. Fringe is white. Pointed abdomen projects beyond wings when at rest. **HOSTS:** Unknown.

SPILOMELINE MOTHS

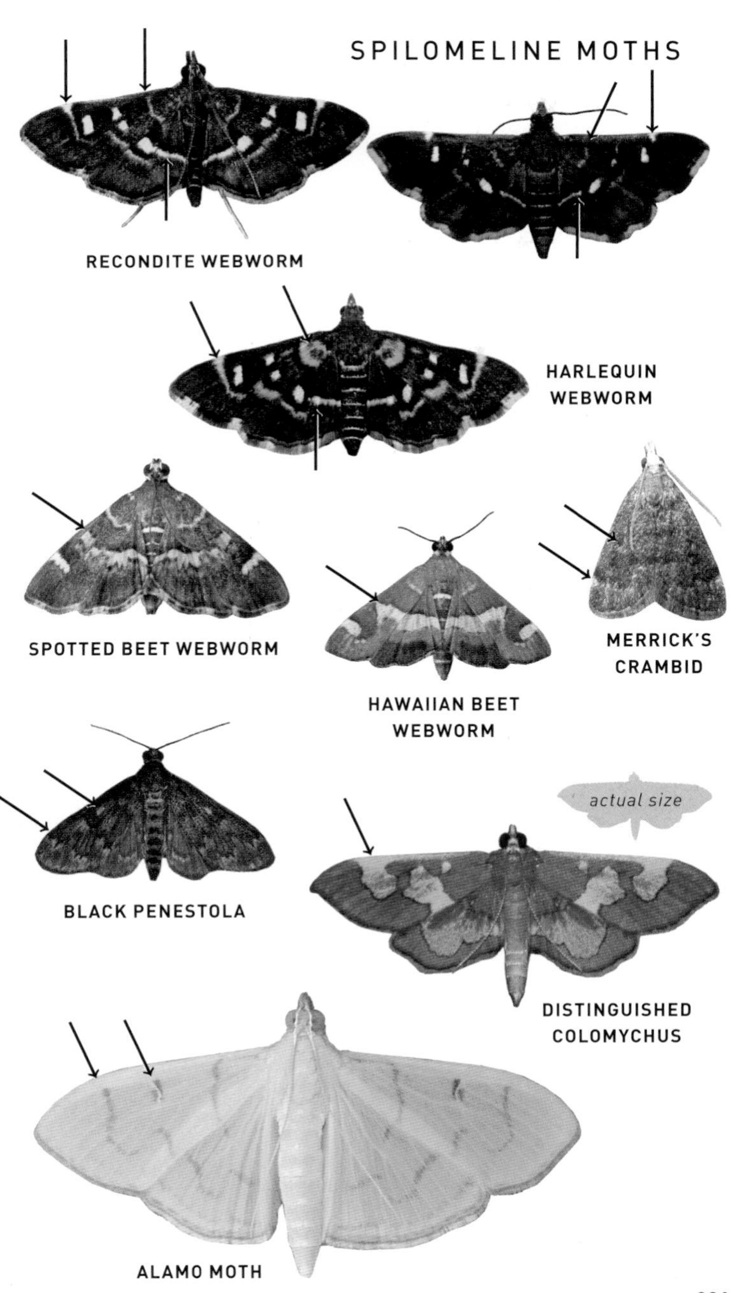

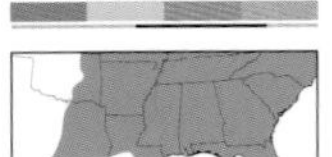

PICKLEWORM MOTH

Diaphania nitidalis 80-1295 (5202) **Common**

WS 25–32 mm. Shiny purplish brown FW has an irregular pale yellow patch in inner PM area. Largely pale yellow HW has a broad purplish brown terminal line. Tip of brown abdomen has a brushy yellow tuft. **HOSTS:** Goosefoot.

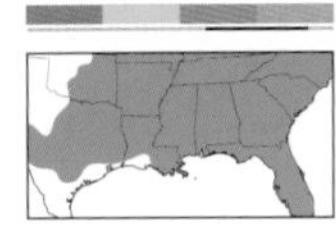

MELONWORM MOTH

Diaphania hyalinata 80-1297 (5204) **Common**

WS 27–30 mm. Semi-translucent satin white FW is narrowly bordered chocolate brown along costa and outer margin. HW has a brown outer margin. Tip of white abdomen ends with a brushy orange tuft. **HOSTS:** Cucumber, melon, and squash.

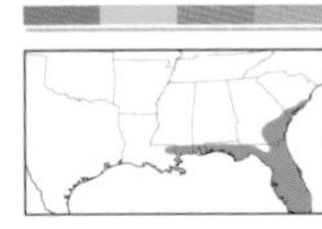

CUCUMBERWORM MOTH

Diaphania modialis 80-1298 (5205) **Local**

WS 20–22 mm. Similar to Melonworm Moth, but bronzy FW has a narrow satin white median bar that tapers to a sharp point toward apex. Abdomen is bronzy brown. **HOSTS:** Possibly creeping cucumber.

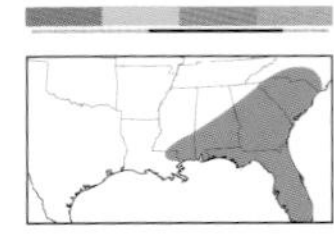

MULBERRY LEAFTIER

Glyphodes sibillalis 80-1304 (5198) **Local**

WS 22–24 mm. Rusty brown FW is boldly patterned with wide, black-edged, satin white (sometimes yellowish) median and incomplete PM bands. Yellowish HW is warm brown beyond black PM line. **HOSTS:** Unknown; presumably mulberry.

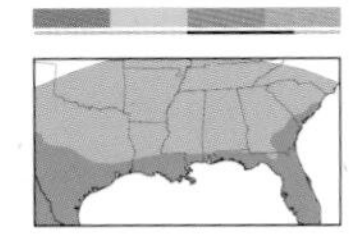

BEAN-LEAF WEBWORM

Omiodes indicata 80-1311 (5212) **Local**

WS 20–22 mm. Golden brown FW is crisply marked with dark brown lines and crescent-shaped discal spot. Sinuous PM line blends with median line on HW. Fringe is whitish beyond double terminal line. **HOSTS:** Unknown.

SATIN WHITE PALPITA

Palpita flegia 80-1315 (5217) **Local**

WS 34–38 mm. Semi-translucent wings have white veins and white clouding on much of HW. Costa of FW is darker gray. Pointed abdomen projects beyond wings when at rest. **HOSTS:** Yellow oleander.

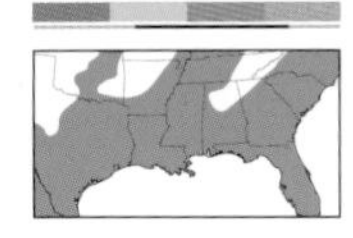

FOUR-SPOTTED PALPITA

Palpita quadristigmalis 80-1316 (5218) **Common**

WS 30–34 mm. Satin white wings are marked with tiny black discal dots. Costa of FW is broadly edged bronzy brown. Pointed abdomen projects beyond wings when at rest. **HOSTS:** Unknown.

SPILOMELINE MOTHS

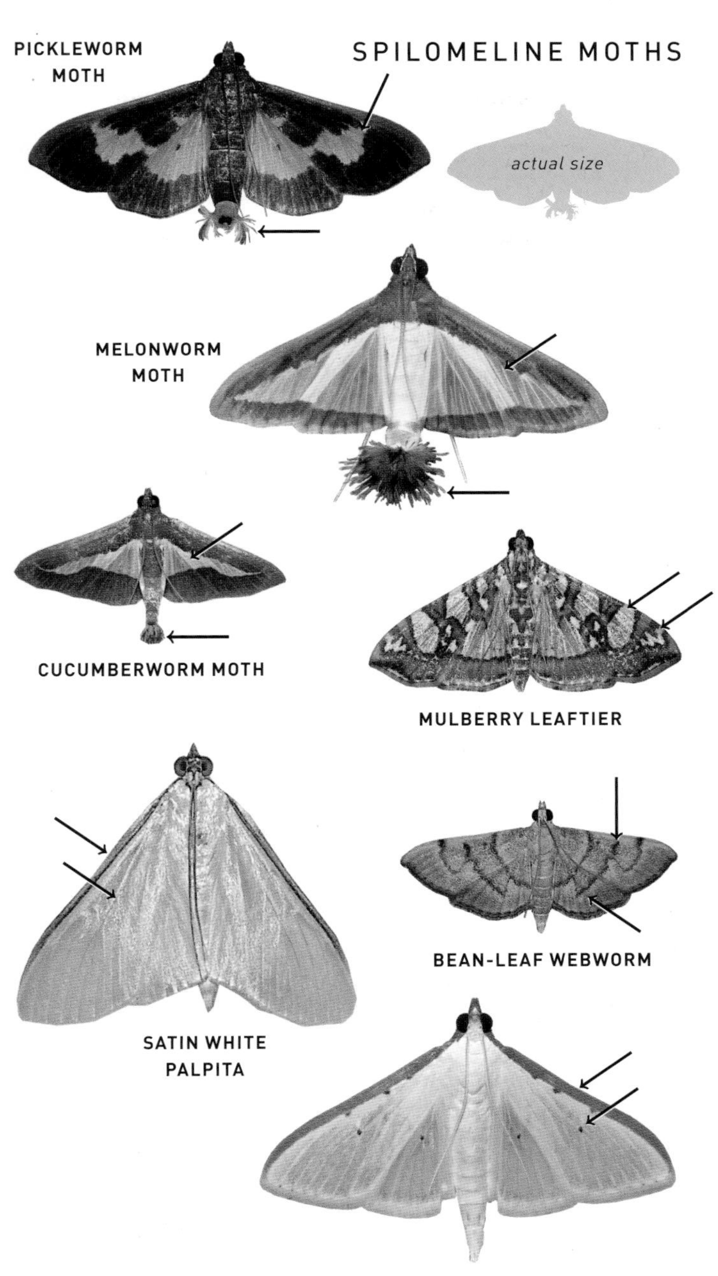

GRACILE PALPITA

Palpita atrisquamalis 80-1319 (5220) **Local**

WS 22–26 mm. Satin white, semi-translucent FW has a contrasting golden brown costal streak. Note diffuse, peppery black median and ST bands, and black crescent in outer PM area. Translucent white HW has a dusky ST line. **HOSTS:** Can be a pest on ornamental privet.

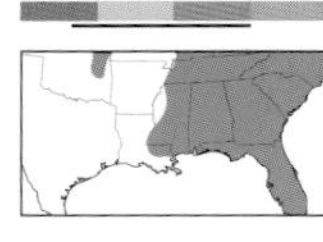

INKBLOT PALPITA

Palpita illibalis 80-1322 (5223) **Common**

WS 23–27 mm. Light gray to whitish FW has a dusky costal streak and is variably speckled dusky brown. Blackish orbicular dot and barlike reniform spot are usually obvious. Translucent HW is lightly speckled in ST area. **HOSTS:** Unknown.

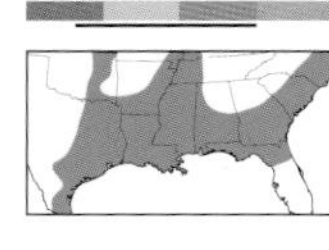

FREEMAN'S PALPITA

Palpita freemanalis 80-1324 (5225) **Common**

WS 23–27 mm. Dimorphic. Early spring individuals have pale wings with dark brownish gray speckling, sometimes dense enough to isolate white patches in outer median area. Summer individuals are slightly smaller with rusty brown to orange speckling. **HOSTS:** Unknown.

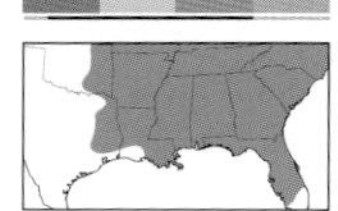

SPLENDID PALPITA

Palpita magniferalis 80-1325 (5226) **Common**

WS 23–27 mm. Whitish FW is variably speckled and blotched dusky brown and blackish. Slightly fragmented median band is usually obvious. Uniform HW is lightly speckled in ST area. **HOSTS:** Ash.

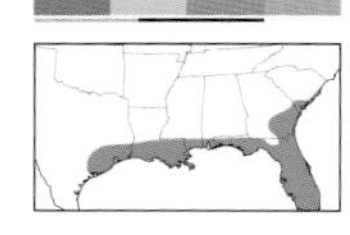

SKY-POINTING MOTH

Agathodes designalis 80-1328 (5240) **Local**

TL 14–20 mm. Pale olive FW is shaded orange in inner ST area. Broad, oblique, purplish median band swirls toward apex. Costal streak is white. Rests with wings folded and pointed abdomen curved upward. **HOSTS:** *Erythrina*, *Inga*, and *Citharexylum* species.

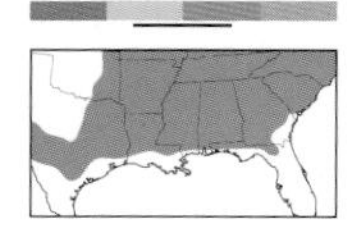

ORNATE COMPACTA

Compacta capitalis 80-1330 (5233) **Local**

WS 26–28 mm. Satin white FW is shaded warm brown along inner margin. All wings are crisply patterned with thin black lines and outlines to spots. Often has blackish basal area and apical patch. **HOSTS:** Unknown.

IRONWEED ROOT MOTH

Polygrammodes flavidalis 80-1334 (5228) **Common**

WS 24–35 mm. Shiny, pale yellow FW has wavy warm brown lines and orbicular spot. Pattern rapidly fades with wear. **HOSTS:** Ironweed.

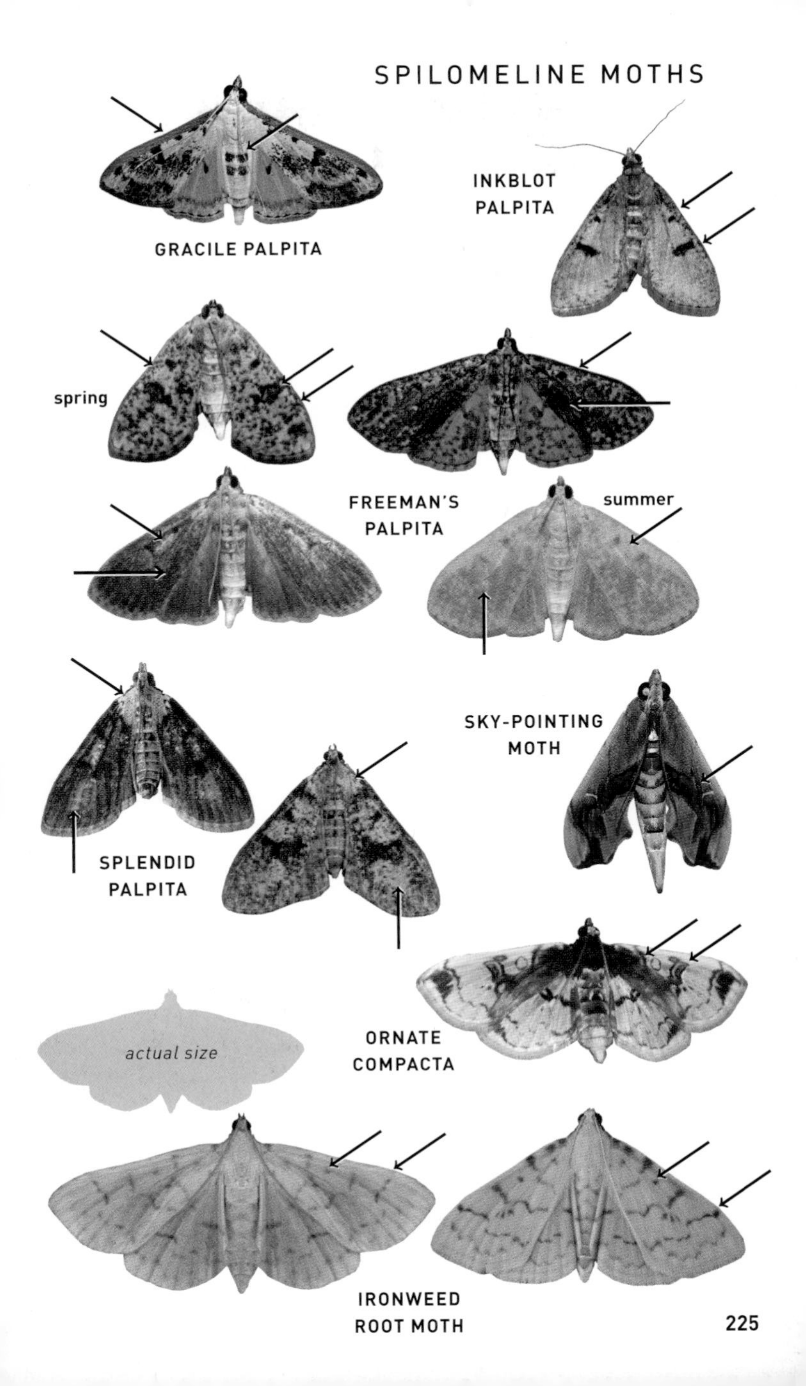
GRACILE PALPITA
INKBLOT PALPITA
spring
FREEMAN'S PALPITA
summer
SPLENDID PALPITA
SKY-POINTING MOTH
ORNATE COMPACTA
actual size
IRONWEED ROOT MOTH

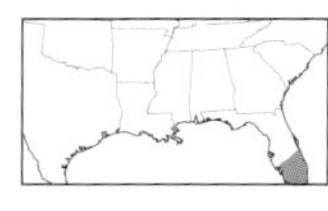

RED-SPOTTED SWEETPOTATO MOTH

Polygrammodes elevata 80-1337 (5230) **Local**

WS 22–24 mm. Golden yellow wings are boldly patterned with purplish red spots. Terminal line is silver. Golden yellow abdomen is marked with rows of purplish red spots on each segment. **HOSTS:** Unknown; possibly sweet potato.

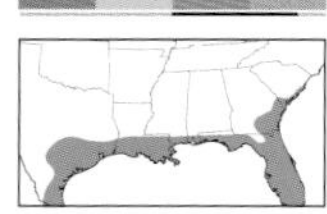

ERYTHRINA BORER

Terastia meticulosalis 80-1339 (5239) **Local**

TL 17–24 mm. Light brown FW is shaded darker brown in basal and ST areas. Note thick, straight, blackish median band. Rests with wings folded and lumpy-looking abdomen slightly raised. **HOSTS:** *Erythrina* species.

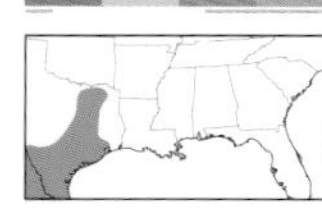

ORANGE SILVER-SPOT

Ommatospila narcaeusalis 80-1345 (5294) **Local**

WS 13–15 mm. Orange wings have an unmistakable pattern of silvery gray bands and spots outlined with black and white. Thin black terminal line is edged white on both sides. **HOSTS:** Unknown.

STAINED ATEGUMIA

Ategumia ebulealis 80-1348 (5158) **Local**

TL 18–20 mm. Resembles Paler Diacme, but bright golden yellow wings have smaller, rounded orbicular and reniform spots, and brown terminal line is narrower. **HOSTS:** Melastomaceae species, including *Heterotrichum umbellatum*.

PALER DIACME *Diacme elealis* 80-1349 (5142) **Common**

WS 17–23 mm. Pale yellow FW has square purplish brown orbicular and reniform spots abutting brown costa. All wings have narrow, broken AM and PM bands and a broad purplish brown terminal line. **HOSTS:** Unknown.

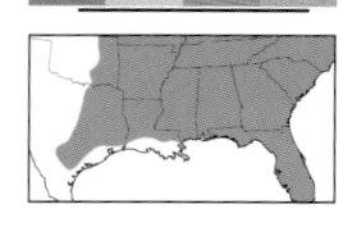

DARK DIACME

Diacme adipaloides 80-1350 (5143) **Common**

WS 18–20 mm. Resembles Paler Diacme but has peppery shading along veins and inner margin, and thicker AM and PM bands, creating overall darker look. **HOSTS:** Unknown.

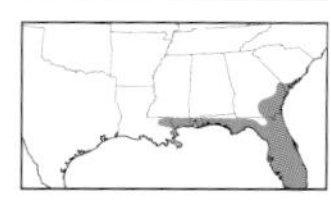

RUSTY DIACME

Diacme phyllisalis 80-1351 (5144) **Local**

WS 18–20 mm. Resembles Paler Diacme, but FW is darker tan with smaller, rounded orbicular and reniform spots, an obvious claviform spot, and solid AM and PM lines. **HOSTS:** Unknown.

YELLOW DIACME *Diacme mopsalis* 80-1352 (5145) **Local**

WS 18–20 mm. Resembles Paler Diacme, but upper spot is broken into separate orbicular and claviform spots. Terminal line on HW is noticeably narrow. **HOSTS:** Unknown.

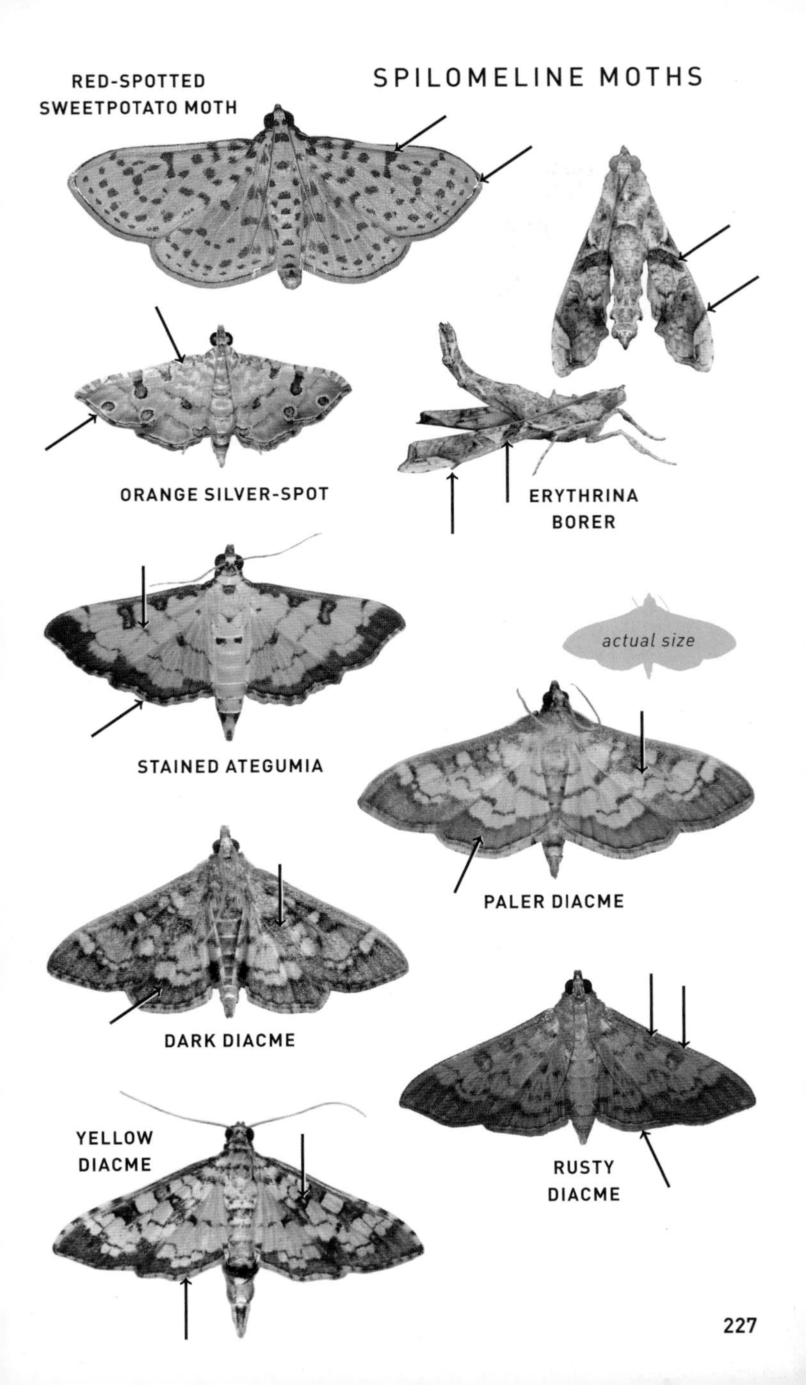
RED-SPOTTED SWEETPOTATO MOTH
ORANGE SILVER-SPOT
ERYTHRINA BORER
actual size
STAINED ATEGUMIA
PALER DIACME
DARK DIACME
RUSTY DIACME
YELLOW DIACME

FORSYTH'S EPIPAGIS

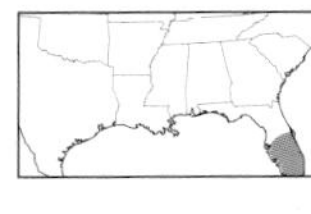

Epipagis forsythae 80-1353 (5146) **Local**

WS 18–20 mm. Dark brown FW is boldly patterned with chunky white patches. Sinuous PM line widens toward costa. HW is mostly white basally. All wings have a checkered fringe when fresh. **HOSTS:** Unknown.

ORANGE EPIPAGIS

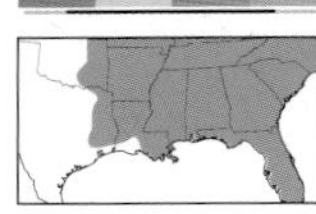

Epipagis fenestralis 80-1354 (5147) **Common**

WS 18–20 mm. Resembles Forsyth's Epipagis, but wing markings are dull orange. White HW is crossed with narrow brown bands. **HOSTS:** Unknown.

LUCERNE MOTH

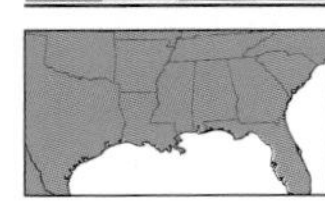

Nomophila nearctica 80-1365 (5156) **Common**

TL 14–18 mm. Very long and narrow brown FW has large dark brown orbicular, claviform, and reniform spots that appear to form two bars across wings. Costa has four dark squares toward apex. **HOSTS:** Low plants, including alfalfa, celery, clover, and smartweed. **NOTE:** Frequently flushed from grass and low vegetation during daytime.

ASSEMBLY MOTH

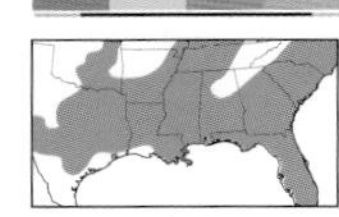

Samea ecclesialis 80-1366 (5150) **Common**

WS 16–20 mm. Resembles *Diacme* species, but FW is white, and dark veins and bold lines create a strongly spotted appearance. Terminal line and fringe are unevenly checkered. Males are smaller and tend to have larger pale spots on FW. **HOSTS:** Unknown.

SALVINIA STEM-BORER

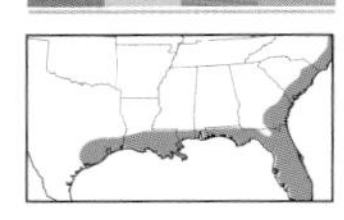

Samea multiplicalis 80-1367 (5151) **Common**

WS 16–20 mm. Resembles Assembly Moth, but dark bands and spotting are usually more extensive. Terminal line is evenly checkered and fringe lacks any dark shading. **HOSTS:** Water lettuce, water fern, and duckweed.

MEDIA MOTH

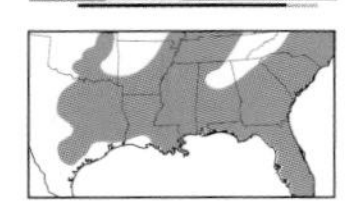

Samea baccatalis 80-1368 (5152) **Common**

WS 22–24 mm. Warm brown FW has three rows of large, dark-edged white spots. Whitish HW has a golden band beyond sinuous PM line. **HOSTS:** Unknown.

WATERHYACINTH MOTH

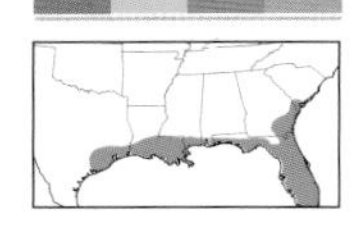

Niphograpta albiguttalis 80-1370 (5149) **Local**

WS 18–21 mm. Golden brown FW is shaded darker brown along costa and in ST area. Sinuous PM line and central median area are accented with chunky white spots. **HOSTS:** Water hyacinth. **NOTE:** An Amazonian species introduced in 1976 as a biological control agent.

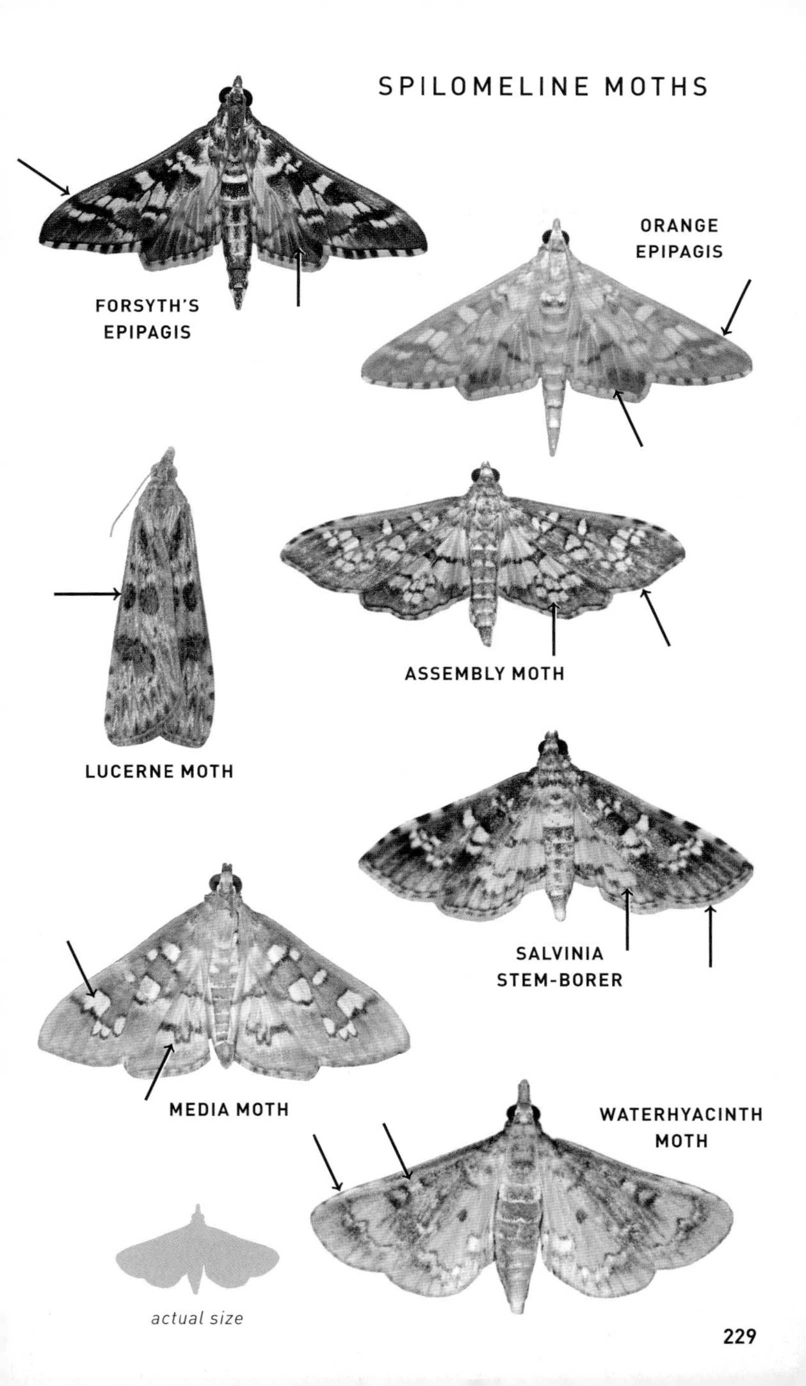
FORSYTH'S EPIPAGIS
ORANGE EPIPAGIS
LUCERNE MOTH
ASSEMBLY MOTH
SALVINIA STEM-BORER
MEDIA MOTH
WATERHYACINTH MOTH
actual size

OBSCURE PSARA

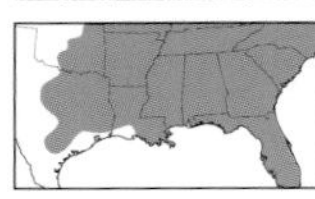

Psara obscuralis 80-1373 (5268) **Common**

WS 19–23 mm. Shiny pale tan to brown FW has brown lines and outlines to spots. Sinuous PM line has a scalloped bulge at midpoint. Darker individuals can show a pale square in central median area. HW has a similar pattern. **HOSTS:** *Petiveria alliacea*.

BLUSH CONCHYLODES

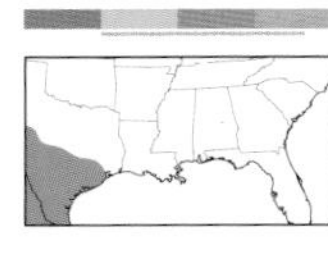

Conchylodes salamisalis 80-1377 (5291) **Local**

WS 26–27 mm. Satin white, semi-translucent wings have a pale fawn ST band and violet sheen when fresh. FW is unmistakably patterned with crisp black lines and outline to large hollow spots in median area. Black-and-white banded abdomen has tan shading on terminal segments. **HOSTS:** Unknown.

ZEBRA CONCHYLODES

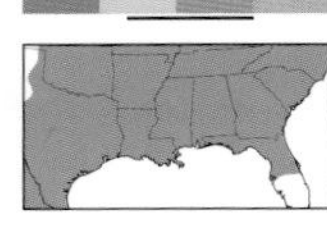

Conchylodes ovulalis 80-1378 (5292) **Common**

WS 23–30 mm. White wings have a violet sheen when fresh. FW is unmistakably patterned with crisp black lines and outline to large hollow reniform spot. Black-banded white abdomen has yellow on terminal segments. **HOSTS:** Aster.

TRAPEZE MOTH

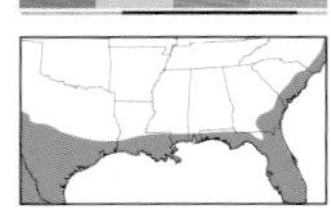

Cnaphalocrocis trapezalis 80-1380 (5288) **Local**

WS 13–15 mm. Pale FW is shaded brown toward costa and in ST area. Bold AM and PM lines are blackish; PM line is broken, with inner half level with reniform spot. Note double black terminal line. Tapered white-tipped abdomen projects beyond wings at rest. **HOSTS:** Grains, including corn and rice.

ACROBAT MOTH

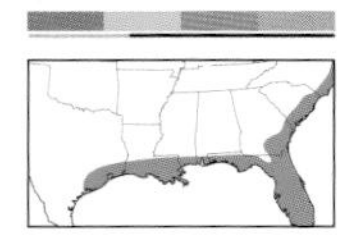

Cnaphalocrocis cochrusalis 80-1381 (5289) **Local**

WS 13–15 mm. Similar to Trapeze Moth, but inner and outer sections of PM line are connected and closer together. Light brown wings show less contrast, and lines are browner. Lines on HW almost touch as they converge toward anal angle. **HOSTS:** Unknown.

LANTANA LEAFTIER

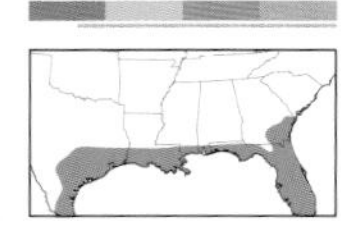

Salbia haemorrhoidalis 80-1384 (5287) **Local**

WS 14–18 mm. Yellowish wings are shaded warm brown in ST area. FW is patterned with crisp dark brown lines and barlike reniform spot. Inner costa is tinged brown. Fringe is silvery gray when fresh. **HOSTS:** Lantana.

ORANGE-SPOTTED FLOWER MOTH

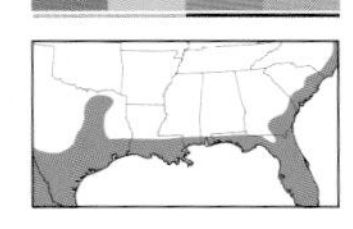

Syngamia florella 80-1386 (5284) **Local**

WS 15–19 mm. Unmistakable! Purplish brown wings are boldly patterned with large yellow spots. Yellow thorax has a black dorsal stripe. Abdomen is mostly scarlet-orange. **HOSTS:** False nettle.

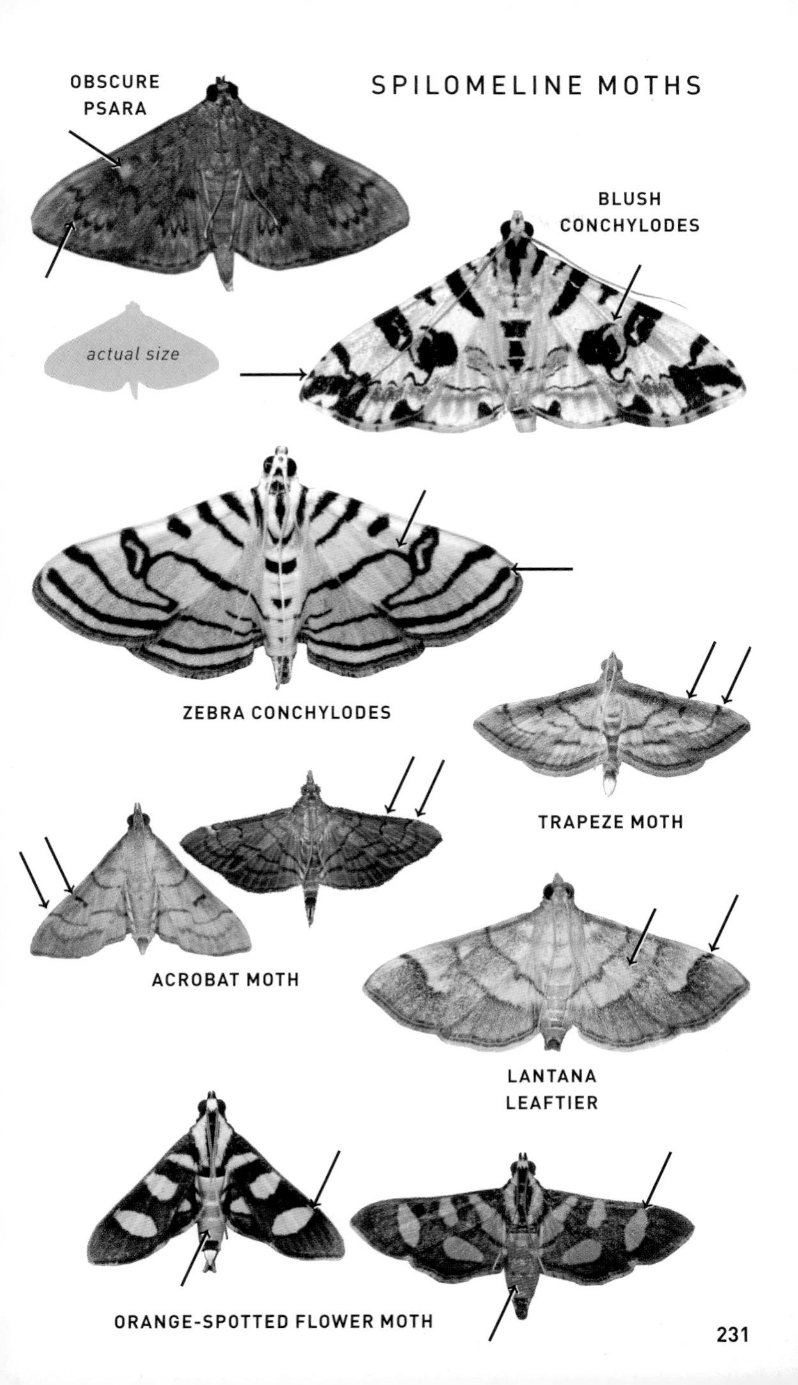

SPILOMELINE MOTHS
OBSCURE PSARA
actual size
BLUSH CONCHYLODES
ZEBRA CONCHYLODES
TRAPEZE MOTH
ACROBAT MOTH
LANTANA LEAFTIER
ORANGE-SPOTTED FLOWER MOTH

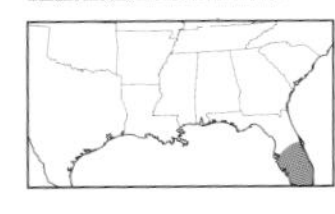

ONE-WEDGED DEUTEROPHYSA

Deuterophysa fernaldi 80-1391 (5123) **Local**

TL 7–8 mm. Resembles Clover Hayworm and Yellow-fringed Hypsopygia (both Pyralinae), but purplish FW has only one yellow triangle. Palps are long, creating a snouted appearance, and FW apex is pointed. **HOSTS:** Wild coffee.

SPANGLED EMBERWING

Eurrhyparodes lygdamis 80-1393 (5122) **Local**

WS 20–24 mm. Dark brown wings are accented with metallic blue scales. Pale orange PM line is fragmented. HW has variable pale orange basal patch. **HOSTS:** Unknown.

Pyraustine Moths

Family Crambidae, Subfamily Pyraustinae

Small moths that generally rest with their wings folded in a delta shape. Most are relatively plain, but those in the group's namesake genus, *Pyrausta*, can be quite colorful. They are primarily nocturnal, but some species can also be encountered in low vegetation during the daytime.

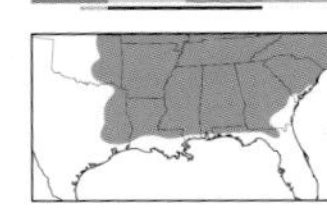

DOGBANE SAUCROBOTYS

Saucrobotys futilalis 80-1407 (4936) **Common**

TL 13–14 mm. Peppery light brown to orange FW is faintly marked with pale-edged scalloped brown AM and PM lines. Terminal line is accented with pale dots. **HOSTS:** Dogbane and milkweed.

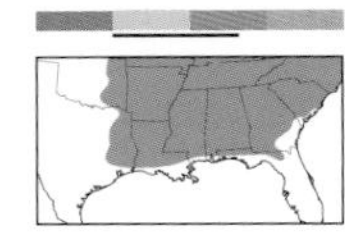

STREAKED ORANGE MOTH

Nascia acutellus 80-1409 (4937) **Uncommon**

TL 10–12 mm. Pale orange FW is distinctively streaked with brown along veins. Fringe is whitish beyond brown terminal line. **HOSTS:** Possibly grass and sedge.

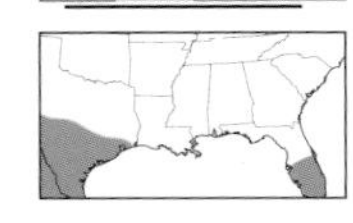

BEAUTIFUL PSEUDOPYRAUSTA

Pseudopyrausta santatalis 80-1413 (4939) **Local**

TL 8–9 mm. Shiny satin white wings are boldly marked with swirling brown lines edged with bands of pale orange. Note brown-ringed pale orange spot in outer AM area and brown collar behind head. **HOSTS:** Unknown.

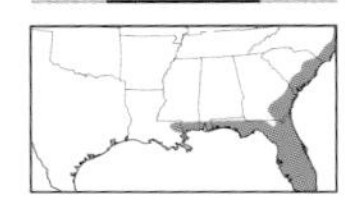

WINE-TINTED OENOBOTYS

Oenobotys vinotinctalis 80-1414 (4940) **Local**

TL 8–10 mm. Warm brown to maroon FW is patterned with slightly fragmented scalloped lines. Orbicular spot is a tiny black dot; larger reniform spot is usually conspicuous. **HOSTS:** Unknown.

SPILOMELINE MOTHS

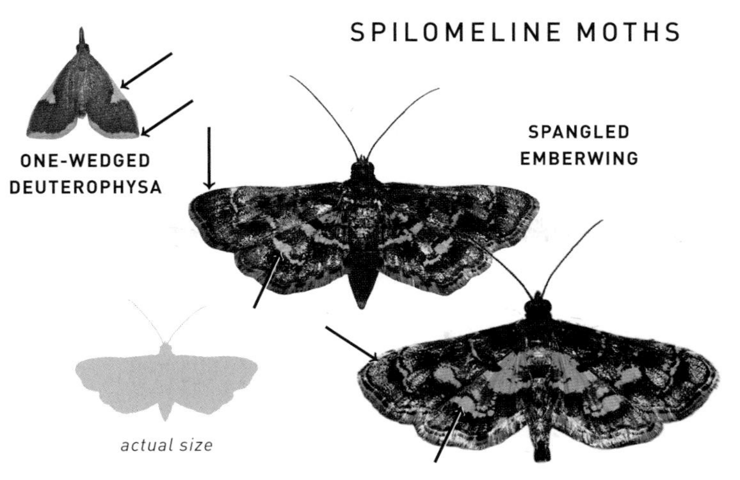

PYRAUSTINE MOTHS

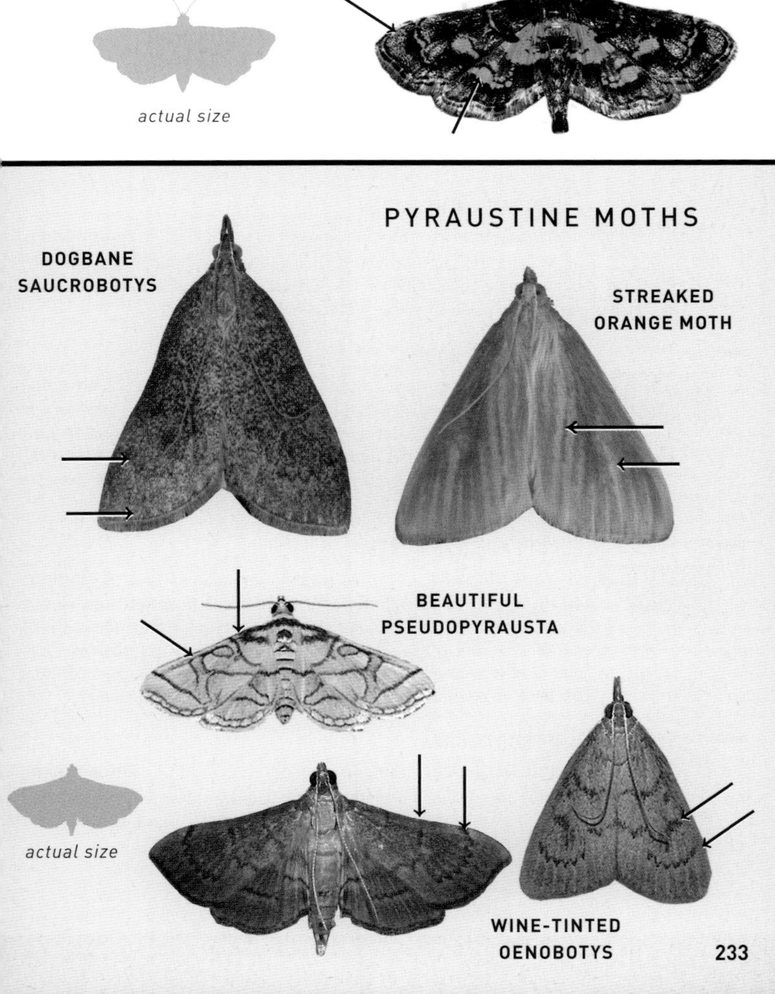

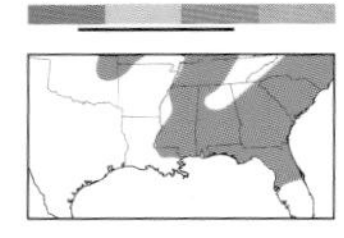

SCALLOPED CROCIDOPHORA

Crocidophora pustuliferalis 80-1417 (4943) **Uncommon**

TL 8–10 mm. Mottled, pale tan FW is marked with deeply scalloped PM and faint ST lines. Male has sharply elliptical fovea in central median area. Slightly kinked costa is accented with a darker brown streak. **HOSTS:** Cane.

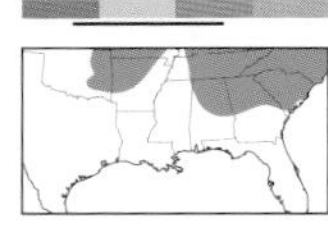

ANGELIC CROCIDOPHORA

Crocidophora serratissimalis 80-1418 (4944) **Common**

TL 10–13 mm. Shiny, pale tan wings are crisply marked with deeply scalloped ST lines. Costa of FW is unmarked. Note pronounced outward bulge at midpoint of PM line of FW. **HOSTS:** Has been reared on rice cutgrass.

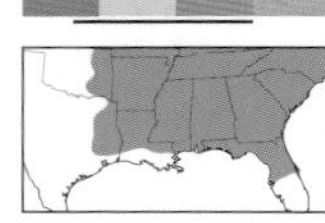

PALE-WINGED CROCIDOPHORA

Crocidophora tuberculalis 80-1419 (4945) **Common**

TL 8–10 mm. Shiny, pale tan FW has broad, evenly curved brown ST line and narrow, stepped AM and PM lines. Male has an unusual creased appearance around semi-transparent fovea in central median area. **HOSTS:** Unknown.

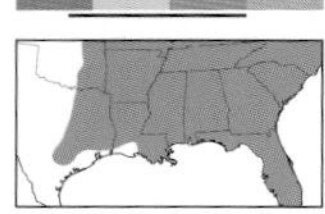

AMERICAN LOTUS BORER

Ostrinia penitalis 80-1420 (4946) **Common**

TL 14–16 mm. Light brown to orange FW of male is patterned with deeply zigzagged brown lines. Often has a large blotch of darker shading in reniform area. **HOSTS:** Water lily.

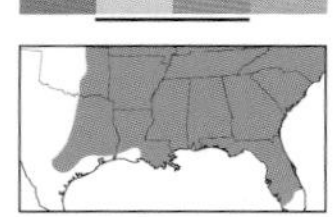

SMARTWEED BORER

Ostrinia obumbratalis 80-1421 (4947) **Common**

TL 11–12 mm. Straw-colored FW is intricately patterned with irregular brown lines. AM line is strongly zigzagged. Scalloped PM and ST lines are roughly parallel. Often has a dark blotch in central PM area and a small orbicular dot. **HOSTS:** Boneset, ragweed, smartweed, and others.

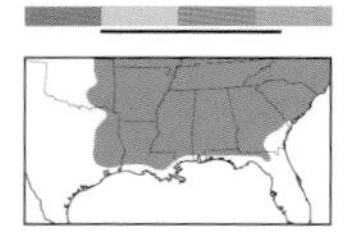

EUROPEAN CORN BORER

Ostrinia nubilalis 80-1423 (4949) **Common**

TL 14–16 mm. Sexually dimorphic. FW of male is mostly brown with a yellowish band beyond jagged PM line. Yellowish female is often darker in median area and has jagged brown lines. **HOSTS:** Low plants and crops, including aster, bean, corn, millet, and potato. **NOTE:** This abundant crop pest was introduced from Europe.

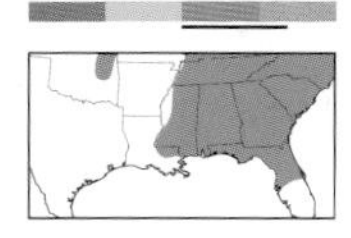

MINT ROOT BORER

Fumibotys fumalis 80-1424 (4950) **Uncommon**

TL 11–13 mm. Rusty orange FW has slightly scalloped brown lines and a brown blotch in reniform area. Typically rests in a slightly tented position with abdomen raised. **HOSTS:** Mint.

PYRAUSTINE MOTHS

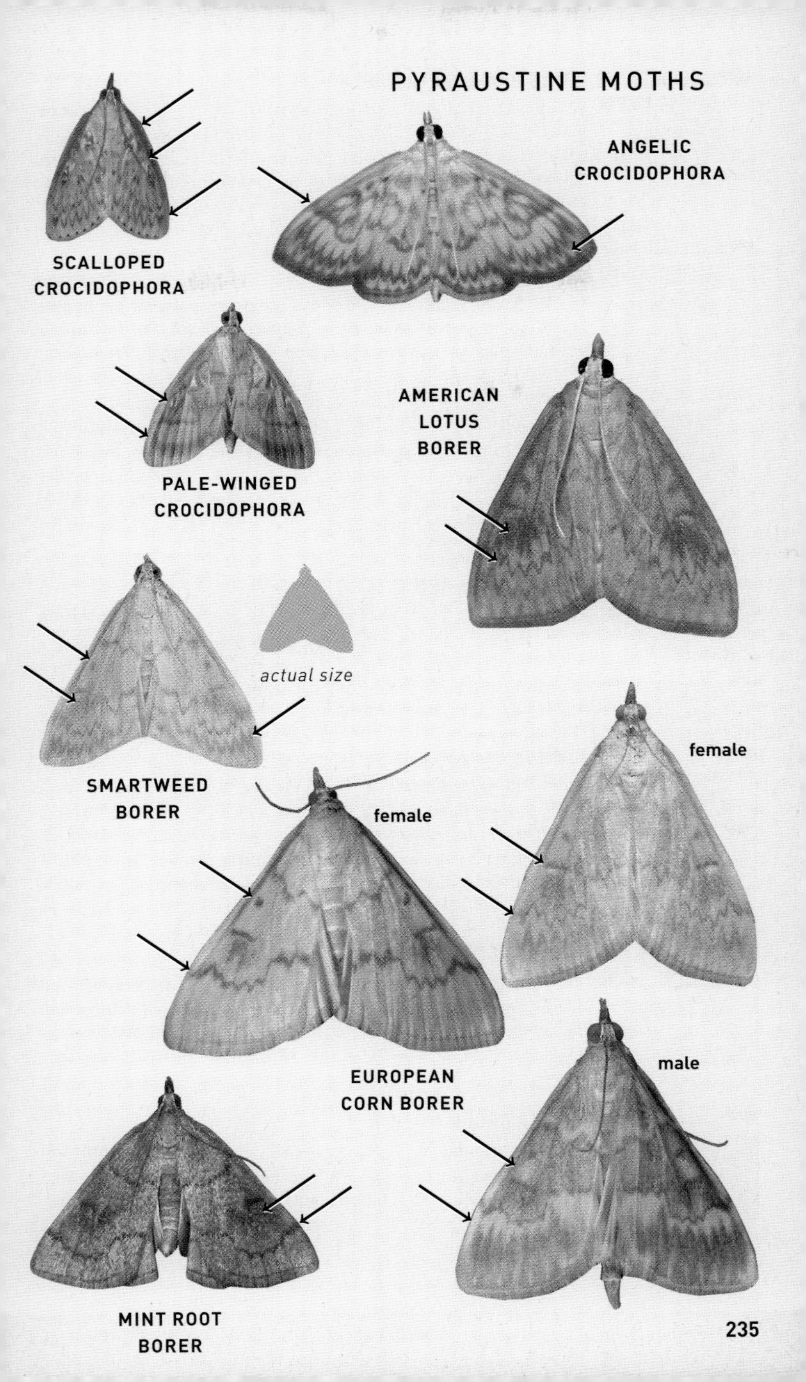

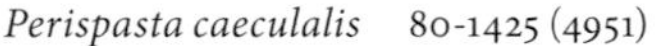

TITIAN PEALE'S PYRALID

Perispasta caeculalis 80-1425 (4951) **Common**

TL 7–10 mm. Broad, shiny violet brown FW has indistinct blackish lines. Male has a rectangular whitish bar in outer median area. Slightly concave outer margin near apex is accented with white fringe when fresh. **HOSTS:** Unknown.

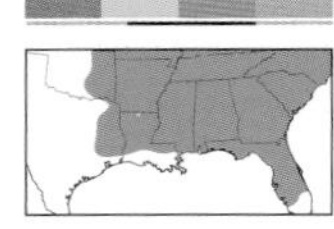

CROWNED ANANIA

Anania tertialis 80-1427 (4953) **Common**

TL 11–13 mm. Grayish brown wings are boldly patterned with pale yellow spots and accents along jagged lines. **HOSTS:** Deciduous trees and plants, including alder, elderberry, hickory, and viburnum.

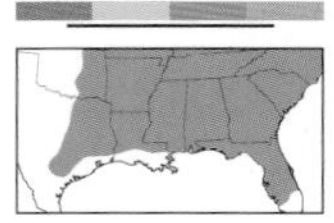

COMMON HAHNCAPPSIA

Hahncappsia mancalis 80-1444 (4967) **Common**

TL 12–14 mm. Shiny, pale orange FW has evenly curved, brown AM and PM lines. Inner section of PM line has a sharp kink before reaching inner margin. Reniform spot is crescent-shaped. Note that all *Hahncappsia* species are very similar. **HOSTS:** Various low plants, including amaranth, dock, morning glory, and tobacco.

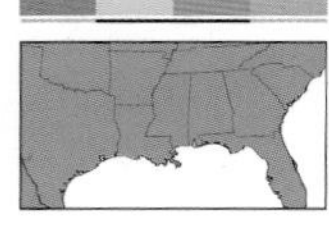

GARDEN WEBWORM

Achyra rantalis 80-1452 (4975) **Common**

TL 10–11 mm. FW is variably pale tan to rusty brown, often with a pale patch between dusky orbicular and reniform spots. Sometimes has paler band in ST area. **HOSTS:** Low plants and crops, including alfalfa, bean, corn, and strawberry.

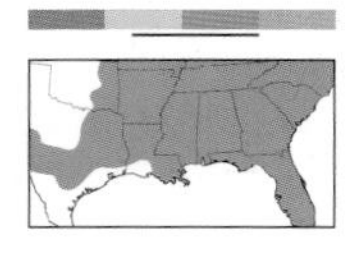

COMMON CARAMEL

Neohelvibotys neohelvialis 80-1454 (4977) **Uncommon**

TL 10–11 mm. Shiny straw-colored FW is marked with thick, evenly curved brown to rusty AM and PM lines. Reniform spot is crescent-shaped. Costal streak is often darker brown to orange. **HOSTS:** Unknown. **NOTE:** Moths in the genera *Neohelvibotys* and *Helvibotys* are frequently difficult to separate in the field.

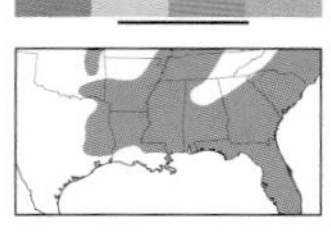

FALCATE SERICOPLAGA

Sericoplaga externalis 80-1469 (4991) **Uncommon**

TL 10–13 mm. Grayish brown FW is weakly patterned with jagged lines and spots. Outer margin is slightly concave near apex. Fringe is white beyond dark terminal line when fresh. **HOSTS:** Osage orange.

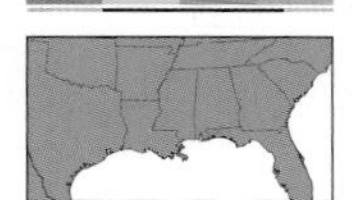

GENISTA BROOM MOTH

Uresiphita reversalis 80-1470 (4992) **Common**

TL 15–18 mm. Rusty brown FW has dotted brown lines and bold blackish orbicular and reniform spots. HW is distinctively bright orange when visible. **HOSTS:** Pea family shrubs, including acacia, genista, and Texas mountain laurel.

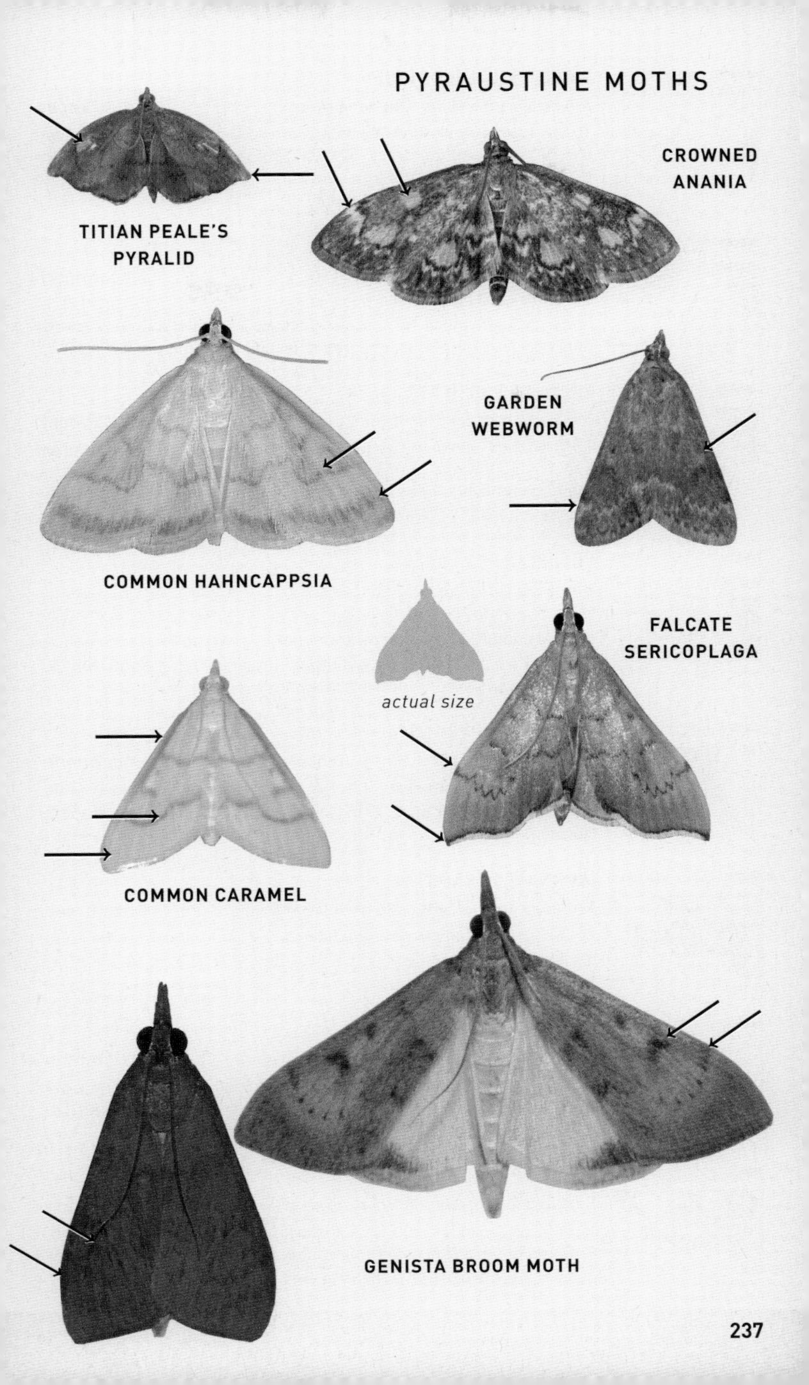
PYRAUSTINE MOTHS
TITIAN PEALE'S
PYRALID
CROWNED
ANANIA
GARDEN
WEBWORM
COMMON HAHNCAPPSIA
actual size
FALCATE
SERICOPLAGA
COMMON CARAMEL
GENISTA BROOM MOTH

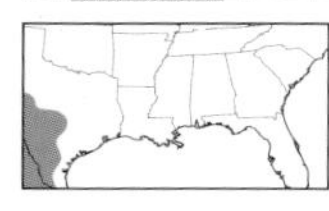

WOLFBERRY LOXOSTEGE

Loxostege allectalis 80-1479 (5000) **Local**

TL 10–13 mm. Pale tan FW has a distinct blue gray tinge to median and ST areas. Crisp black PM line bulges outward at midpoint. Dusky orbicular and reniform spots are separated by a pale rectangle. **HOSTS:** Berlandier's wolfberry.

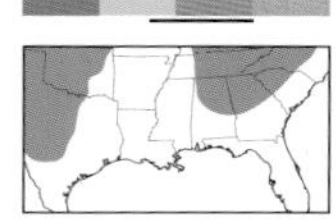

ALFALFA WEBWORM

Loxostege cereralis 80-1496 (5017) **Uncommon**

TL 10–13 mm. Pale tan FW has a streaky pattern of brown and black lines. Straight black ST line is edged with a pale yellow terminal line. **HOSTS:** Low plants and crops, including alfalfa.

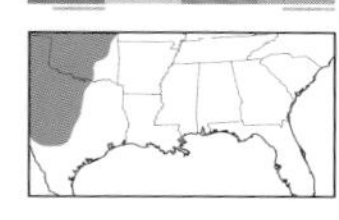

FULVOUS-EDGED PYRAUSTA

Pyrausta nexalis 80-1498 (5019) **Local**

TL 7–10 mm. Strikingly patterned FW has bright fulvous basal area and costal streak. Peppery brown inner median area is boldly marked with white lines along veins. ST band is speckled with whitish scales. **HOSTS:** Unknown.

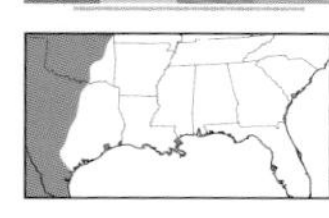

VOLUPIAL PYRAUSTA

Pyrausta volupialis 80-1508 (5029) **Local**

TL 9–11 mm. Rose pink to brownish FW has contrasting cream-colored AM and PM lines and reniform dot. Zigzag AM line fades before reaching costa. Fringe is yellowish brown when fresh. **HOSTS:** Rosemary and other members of mint family.

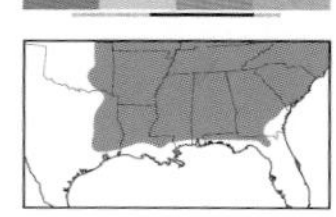

RASPBERRY PYRAUSTA

Pyrausta signatalis 80-1513 (5034) **Uncommon**

TL 8–11 mm. Bright rose pink FW has fragmented cream-colored lines and reniform spot. Fringe is yellowish when fresh. **HOSTS:** Horsemint.

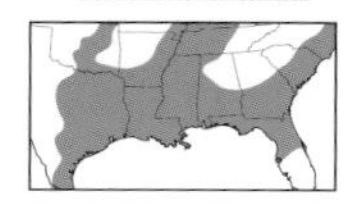

INORNATE PYRAUSTA

Pyrausta inornatalis 80-1516 (5037) **Uncommon**

TL 7–8 mm. Sharply pointed fuchsia FW is uniform apart from darker terminal line. Fringe is pale yellow when fresh. **HOSTS:** Unknown.

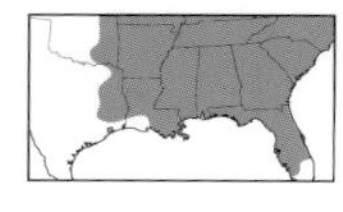

BICOLORED PYRAUSTA

Pyrausta bicoloralis 80-1519 (5040) **Common**

TL 9–11 mm. Golden yellow wings are shaded purple beyond wavy PM line. Fringe is checkered black and white when fresh. **HOSTS:** Unknown.

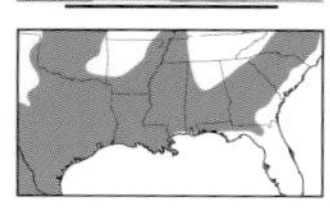

GOLDEN PYRAUSTA

Pyrausta onythesalis 80-1521 (5042) **Common**

TL 9–12 mm. Golden yellow wings have a bold purplish red ST band. Median area of FW is sometimes shaded reddish between curvy AM and PM lines. **HOSTS:** Has been reared on *Salvia* species.

PYRAUSTINE MOTHS

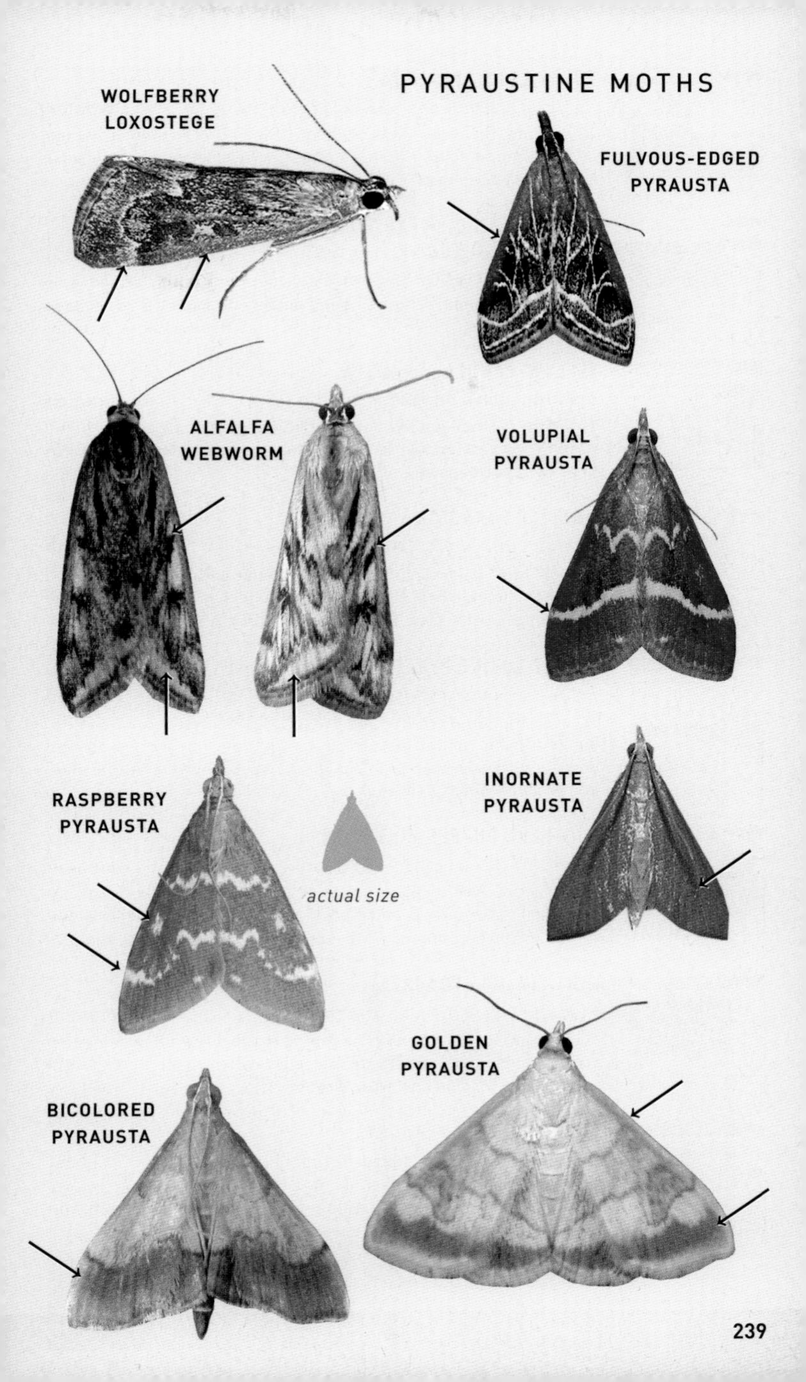

PHOENICEAN PYRAUSTA

Pyrausta phoenicealis 80-1529 (5049) **Common**

TL 8–9 mm. Brick red FW has golden yellow patches in median area and beyond sinuous PM line. Reddish HW has a single yellow ST band. **HOSTS:** Unknown.

VARIABLE REDDISH PYRAUSTA

Pyrausta rubricalis 80-1531 (5051) **Uncommon**

TL 7–8 mm. Rusty FW is weakly patterned with slightly jagged lines and blackish spots. Yellow ST band is present on all wings. **HOSTS:** Unknown.

YELLOW-BANDED PYRAUSTA

Pyrausta pseuderosnealis 80-1533 (5053) **Uncommon**

TL 7–9 mm. Brick red FW has a slightly wavy yellow PM band and a variably faint yellow AM band. HW has a yellow patch at inner PM line. **HOSTS:** Unknown.

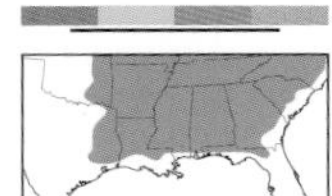

MOTTLED PYRAUSTA

Pyrausta insequalis 80-1540 (5060) **Common**

TL 7–10 mm. Yellow to orange FW is variably patterned with brownish lines and blotches. HW has a bold pattern of even, pale orange bands edged with black. **HOSTS:** Reported on thistle.

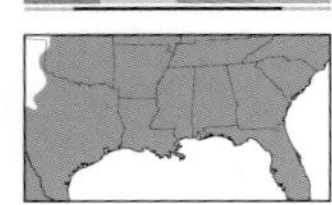

COFFEE-LOVING PYRAUSTA

Pyrausta tyralis 80-1550 (5069) **Common**

TL 7–8 mm. Wine red FW has irregular golden yellow AM and PM bands. Golden yellow discal spot can be large or almost absent. Some individuals have pink bands and spots. **HOSTS:** Wild coffee, and probably other plants.

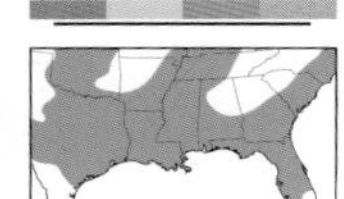

SOUTHERN PURPLE MINT MOTH

Pyrausta laticlavia 80-1551 (5070) **Common**

TL 8–12 mm. Bright golden yellow FW is boldly patterned with purplish pink costal streak, and median and ST bands of variable width. Yellow spot in outer median area is variable in size. **HOSTS:** Mint.

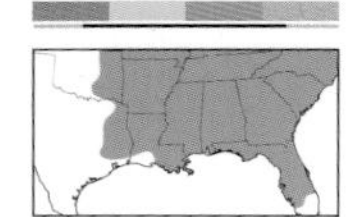

MINT-LOVING PYRAUSTA

Pyrausta acrionalis 80-1552 (5071) **Common**

TL 7–9 mm. Purplish red FW has variable yellow mottling in median area. Terminal line and fringe are yellow when fresh. Reddish HW has a yellow median line. **HOSTS:** Mint.

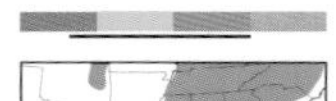

WHITE-FRINGED PYRAUSTA

Pyrausta niveicilialis 80-1554 (5073) **Local**

TL 11–13 mm. Sooty black wings have contrasting white fringes when fresh. Sometimes has faint trace of pale PM line near costa. **HOSTS:** Unknown.

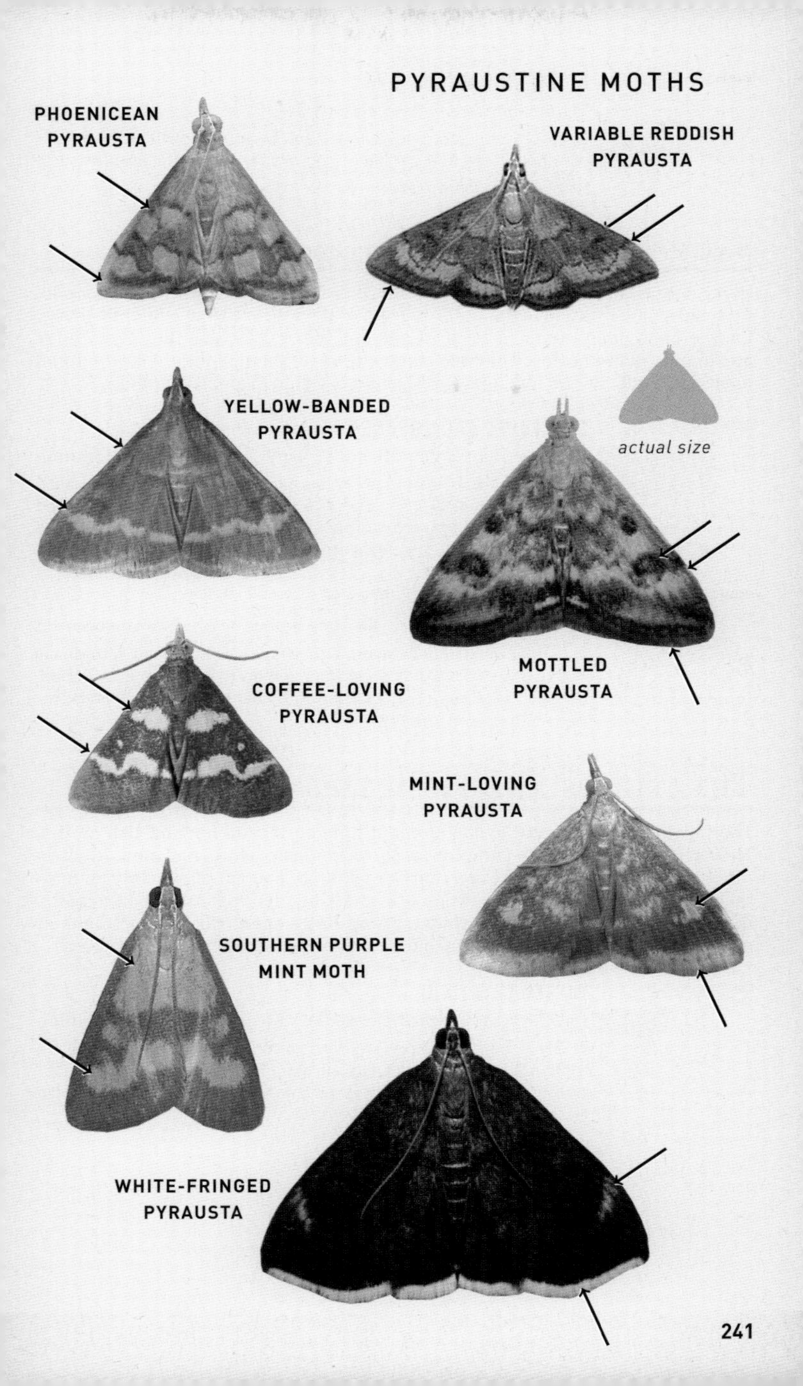

PYRAUSTINE MOTHS
PHOENICEAN PYRAUSTA
VARIABLE REDDISH PYRAUSTA
YELLOW-BANDED PYRAUSTA
actual size
MOTTLED PYRAUSTA
COFFEE-LOVING PYRAUSTA
MINT-LOVING PYRAUSTA
SOUTHERN PURPLE MINT MOTH
WHITE-FRINGED PYRAUSTA

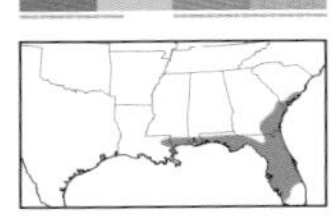

GLITTERING MAGDALENA

Daulia magdalena 80-1560 (5295) **Local**

TL 9–11 mm. Light orange FW is boldly patterned with metallic silvery gold lines and streaks. Long brushy palps project in front of head. **HOSTS:** Unknown.

SACK-BEARERS Family Mimallonidae

Hairy, stout-bodied moths that rest with their wings flat or slightly drooped. All have an indentation in the outer margin of the FW, creating a falcate shape. Larvae build open-ended "sacks" of silk and leaves in which they overwinter and pupate in the spring. They are generally uncommon in their oak woodland habitat. Strictly nocturnal; adults will come to lights in small numbers.

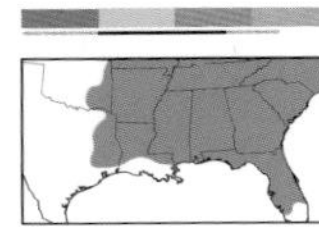

SCALLOPED SACK-BEARER

Lacosoma chiridota 83-0001 (7659) **Uncommon**

WS 20–32 mm. Pale orange FW has darker shading (brown in males, orange in females) in median area. Brown PM line is diffuse and sinuous. Elongated reniform spot is marked in black. Outer margins of wings are boldly scalloped. **HOSTS:** Oaks.

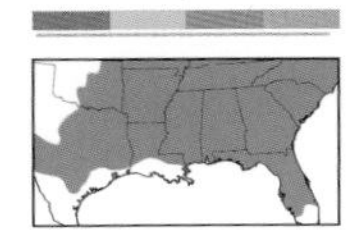

MELSHEIMER'S SACK-BEARER

Cicinnus melsheimeri 83-0004 (7662) **Uncommon**

TL 21–26 mm. Ghostly pearl gray wings are sprinkled with dusky scales and flushed pinkish orange in median area and along veins. Blackish PM line on FW curves inward near costa. Bar-shaped reniform spot is black. **HOSTS:** Oaks.

THYATIRIDS AND HOOKTIPS Family Drepanidae

The thyatirids are a small group of beautifully patterned noctuid-like moths. They are mostly gray with complex patterns in black, white, and pink. All have a raised thoracic crest, rest in a tentlike position, and could possibly be mistaken for prominents. The hooktips are small-headed, geometrid-like moths that typically rest with their wings spread flatly open. Both groups inhabit mostly woodlands; they are nocturnal and will visit lights in small numbers.

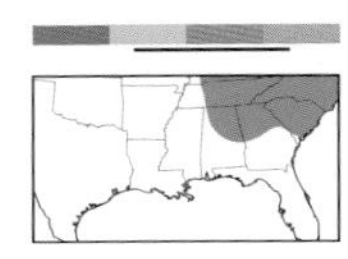

TUFTED THYATIRID

Pseudothyatira cymatophoroides 85-0005 (6237) **Common**

TL 23–25 mm. Gray to white FW has a contrasting black basal patch, triple AM line, and dark spots along inner ST line. Basal area is usually shaded pinkish. In form "expultrix," black markings are reduced or absent.

PYRAUSTINE MOTHS

SACK-BEARERS

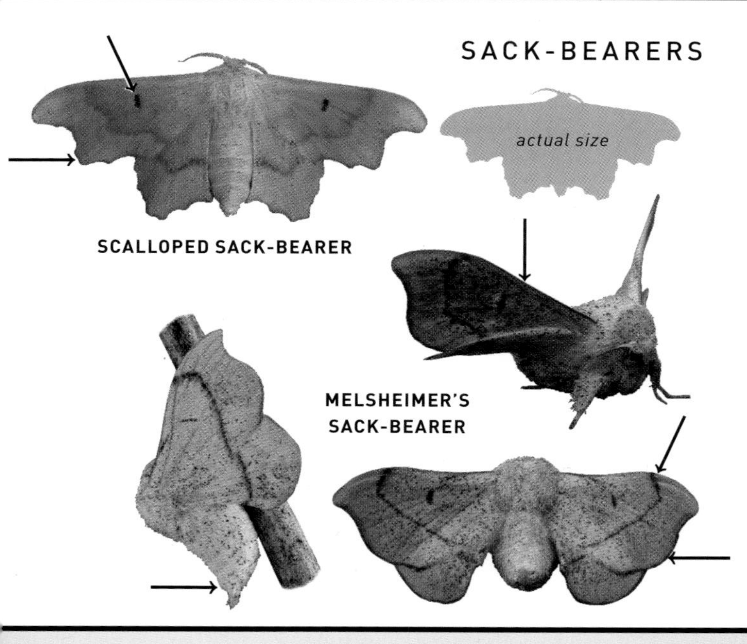

THYATIRIDS AND HOOKTIPS

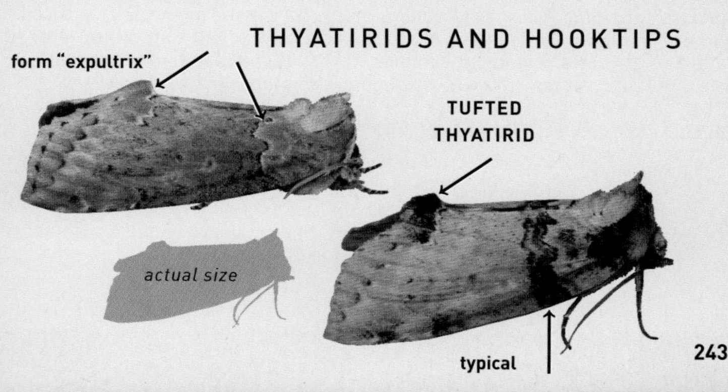

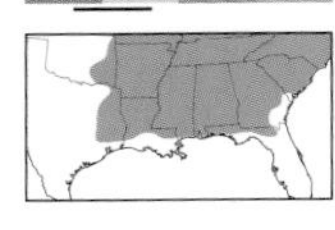

DOGWOOD THYATIRID

Euthyatira pudens 85-0008 (6240) **Uncommon**

TL 23–25 mm. Gray FW is distinctively patterned with large pink patches in basal, costal, and apical areas. Faint blackish lines are sprinkled with white scales. Individuals of the uncommon form "pennsylvanicus" are mostly gray with ghostlike markings of the typical form. **HOSTS:** Flowering dogwood.

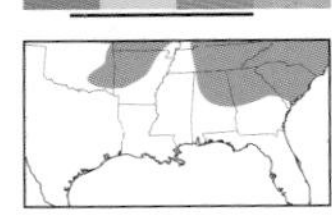

ARCHED HOOKTIP

Drepana arcuata 85-0019 (6251) **Common**

WS 27–40 mm. Pale orange FW has a bold rusty PM line that curves toward falcate apex. Reniform area is marked with two black, sometimes hollow, spots. **HOSTS:** Alder and birch.

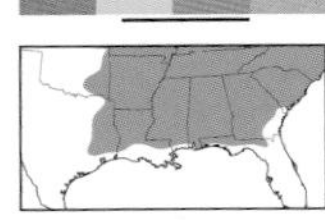

NORTHERN EUDEILINIA

Eudeilinia herminiata 85-0021 (6253) **Uncommon**

WS 25–30 mm. White wings have dotted brown AM and PM lines of varying intensity. Labial palps are black. Forelegs are black along inner length. Antennae are filiform. **HOSTS:** Dogwood.

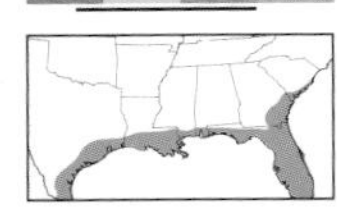

SOUTHERN EUDEILINIA

Eudeilinia luteifera 85-0022 (6254) **Local**

WS 25–30 mm. Similar to Northern Eudeilinia, but bolder AM and PM lines are orange brown. **HOSTS:** Flowering dogwood.

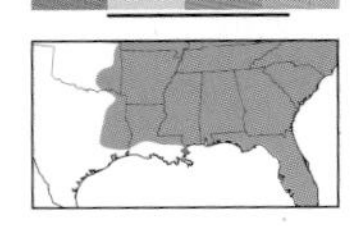

ROSE HOOKTIP *Oreta rosea* 85-0023 (6255) **Common**

WS 25–35 mm. Apex of FW is falcate. Wings vary in color but typically are rusty brown with a broad yellow band between PM and ST lines. Some individuals are uniformly pinkish or rusty brown, often with fragmented blackish lines. Two tiny white dots mark central median area. **HOSTS:** Birch and viburnum.

TENT CATERPILLAR AND LAPPET MOTHS
Family Lasiocampidae

Small to medium-sized, extremely hairy moths. Females are often considerably larger than males. They adopt a tentlike position while resting, with some species sprawling their hairy forelegs in front of them. The tolypes have a distinctive "mohawk" of blue and black scales. Eastern Tent Caterpillar Moth is well known for its prominent communal webs that house the developing larvae. Adults do not feed. All will visit lights, sometimes in large numbers.

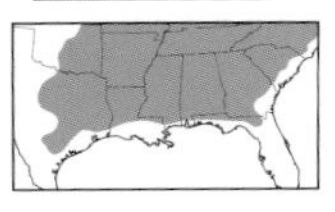

RILEY'S LAPPET MOTH

Heteropacha rileyana 87-0001 (7685) **Uncommon**

TL 15–19 mm. Peppery grayish brown FW is slightly translucent. Median area and dotted ST line are darker gray. Jagged whitish AM line borders paler basal area. Thorax has dark dorsal stripe. **HOSTS:** Honey locust. **NOTE:** Two broods.

THYATIRIDS AND HOOKTIPS

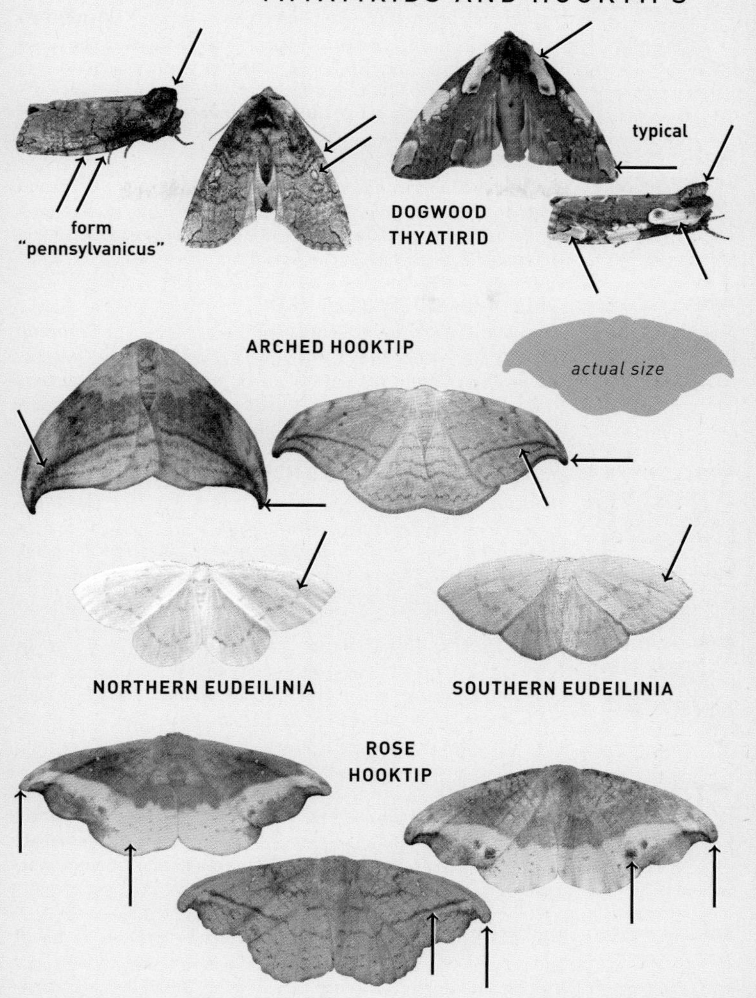

TENT CATERPILLAR AND LAPPET MOTHS

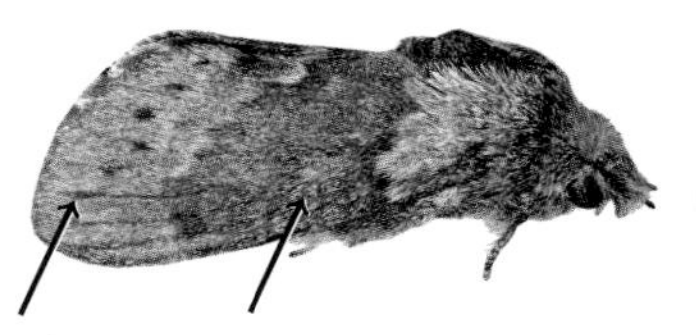

RILEY'S
LAPPET MOTH

SOUTHERN LAPPET MOTH

Phyllodesma occidentis 87-0002 (7686) **Uncommon**

TL 15–25 mm. Reddish brown FW has light bluish gray edging to PM line and terminal line. Scalloped lines are inconspicuous. Large white reniform spot is usually noticeable. At rest, scallop-edged HW protrudes below costa of FW. **HOSTS:** Unknown.

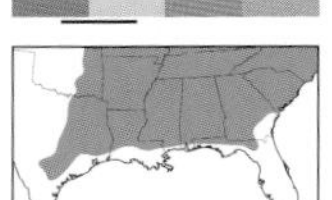

LAPPET MOTH

Phyllodesma americana 87-0003 (7687) **Common**

TL 15–25 mm. Similar to Southern Lappet Moth but always lacks white reniform spot. Outer margin of FW is scalloped when fresh. **HOSTS:** Alder, birch, oak, poplar, and rose.

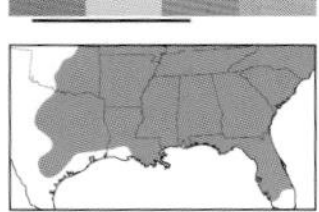

FOREST TENT CATERPILLAR MOTH

Malacosoma disstria 87-0014 (7698) **Common**

TL 17–21 mm. Yellowish to rusty brown FW appears furred. Median area between parallel AM and PM lines is sometimes darker brown. Fringe is unevenly checkered. **HOSTS:** Alder, aspen, basswood, birch, cherry, maple, oak, and other deciduous trees.

EASTERN TENT CATERPILLAR MOTH

Malacosoma americana 87-0017 (7701) **Common**

TL 15–24 mm. Furry-looking brown FW has broad cream-colored AM and PM lines. Some individuals have a paler median band. Fringe is unevenly checkered. **HOSTS:** Deciduous trees, especially apple, cherry, and crab apple.

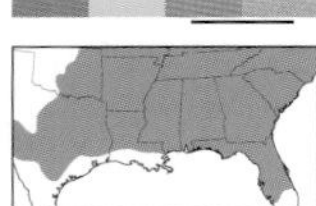

LARGE TOLYPE *Tolype velleda* 87-0021 (7670) **Common**

TL 17–28 mm. Gray FW is boldly patterned with contrasting white veins, gently wavy AM and double PM lines. White ST line has a shallow bulge at midpoint. **HOSTS:** Apple, ash, birch, elm, oak, plum, and other trees.

SMALL TOLYPE

Tolype notialis 87-0025 (7674) **Common**

TL 16–20 mm. Similar to larger and paler Large Tolype, but PM and ST lines have two noticeable bulges along their length. Some males can be almost blackish. **HOSTS:** Coniferous trees.

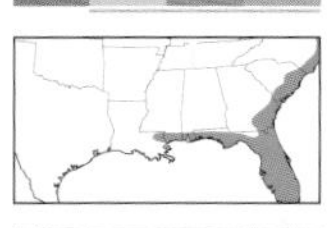

SOUTHERN TOLYPE *Tolype minta* 87-0026 (7675) **Local**

TL 14–16 mm. Similar to Small Tolype but has a whiter FW with contrasting and slightly wavy dark gray ST band. Other lines are inconspicuous except along darker inner margin. **HOSTS:** Unknown.

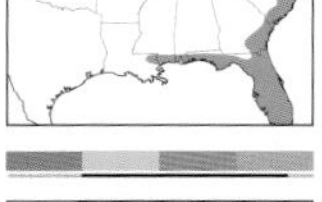

DOT-LINED WHITE MOTH

Artace cribrarius 87-0036 (7683) **Common**

TL 14–31 mm. Pearly gray FW is patterned with contrasting white veins and dotted black lines that often form costal spots. Terminal line is dashed black. Woolly thorax and abdomen are gleaming white. **HOSTS:** Oak; also cherry and rose.

TENT CATERPILLAR AND LAPPET MOTHS

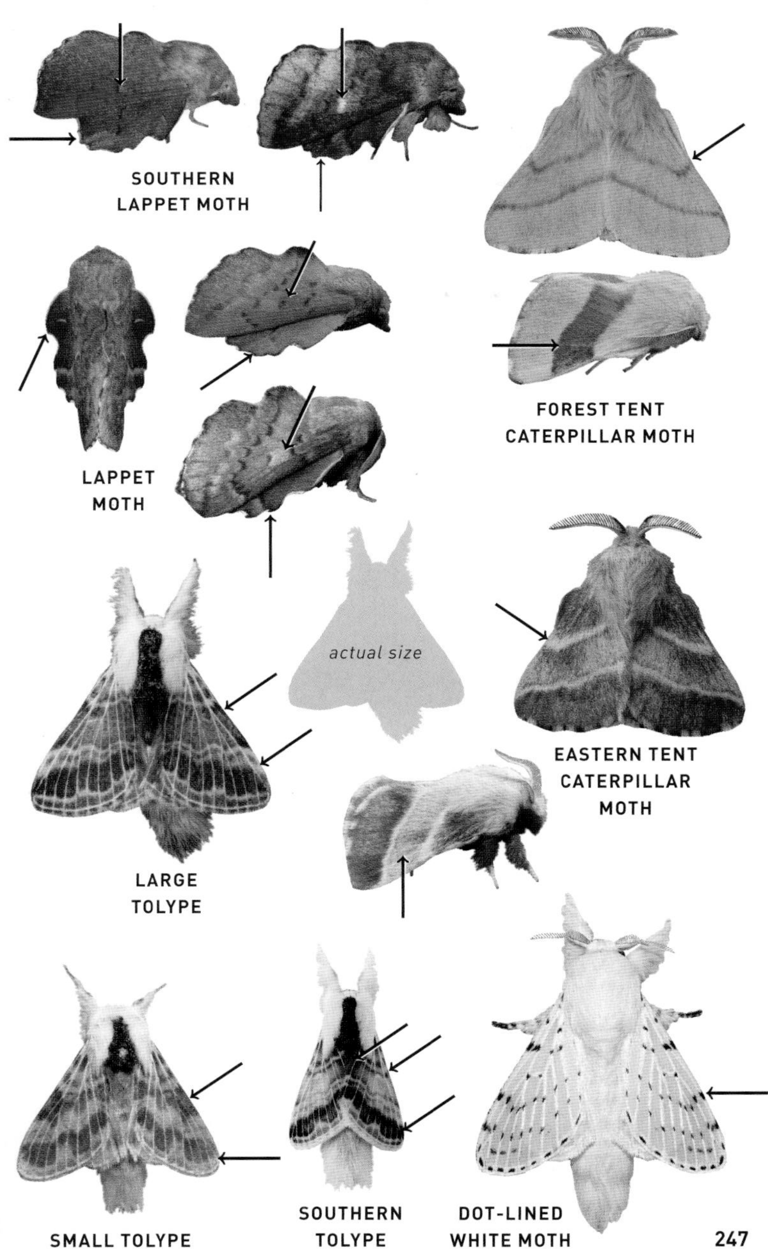

SOUTHERN LAPPET MOTH

FOREST TENT CATERPILLAR MOTH

LAPPET MOTH

EASTERN TENT CATERPILLAR MOTH

LARGE TOLYPE

SMALL TOLYPE

SOUTHERN TOLYPE

DOT-LINED WHITE MOTH

Apatelodid Moths Family Apatelodidae

Medium-sized woodland moths that rest in acrobatic headstand positions with the abdomen raised high above the thorax. The Angel and The Seraph hold their wings in a delta shape, while Spotted Apatelodes fans its wings apart. They are nocturnal and will visit lights in small numbers.

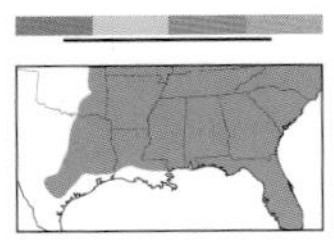

SPOTTED APATELODES

Apatelodes torrefacta 89-0001 (7663) **Common**

TL 18–22 mm. Gray FW is marked with faint blackish lines that widen near inner margin. Black basal patch continues across rear edge of thorax, forming a collar. Translucent dot near apex is more obvious from underside. **HOSTS:** Ash, cherry, maple, and oak. **NOTE:** Two broods.

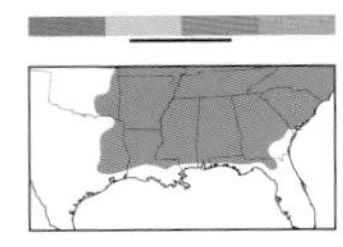

THE ANGEL *Olceclostera angelica* 89-0004 (7665) **Common**

TL 18–22 mm. Gray FW has brown bands along dotted lines. Midpoint of ST line is marked with two large translucent spots. Outer margin is slightly scalloped. Thorax has a sharp brown dorsal ridge. **HOSTS:** Ash and lilac.

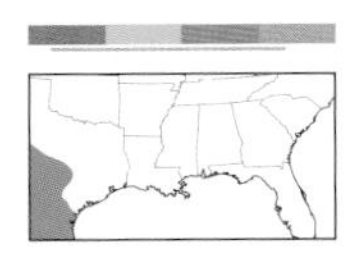

THE SERAPH *Olceclostera seraphica* 89-0006 (7667) **Local**

TL 20–25 mm. Similar to The Angel but tends to be purer gray on the wings with less brown shading on cross bands. The translucent spots at midpoint of ST line are smaller. Outer margin of FW is less scalloped when fresh. **HOSTS:** Ash and desert willow.

Royal Silkworm Moths
Family Saturniidae, Subfamily Ceratocampinae

Medium to large, brightly colored moths. These silkmoths are more narrow-winged than the other groups, and most rest with their wings folded back in a delta shape. Adults do not feed. Larvae of all species but Pine Devil feed on deciduous trees and shrubs. Adults are nocturnal and will come to lights in small numbers.

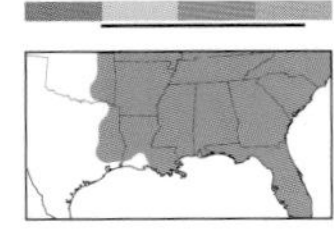

REGAL MOTH *Citheronia regalis* 89-0009 (7706) **Common**

WS 95–155 mm. Lead gray FW is marked with bright orange veins and bold cream-colored spots. HW is orange with pale yellow patches at base and inner margin. Orange thorax has pale yellow lateral stripes. **HOSTS:** Ash, butternut, hickory, sycamore, walnut, and other deciduous trees. **NOTE:** One or two broods.

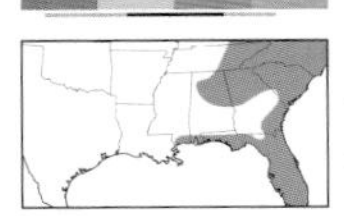

PINE DEVIL

Citheronia sepulcralis 89-0010 (7708) **Uncommon**

WS 70–135 mm. Dark violet gray FW is faintly marked with rosy veins, dusky reniform spot, and inconspicuous ST line. Violet gray HW has a rosy basal area and veins. **HOSTS:** Pine, including pitch pine and white pine. **NOTE:** Two or more broods.

APATELODID MOTHS

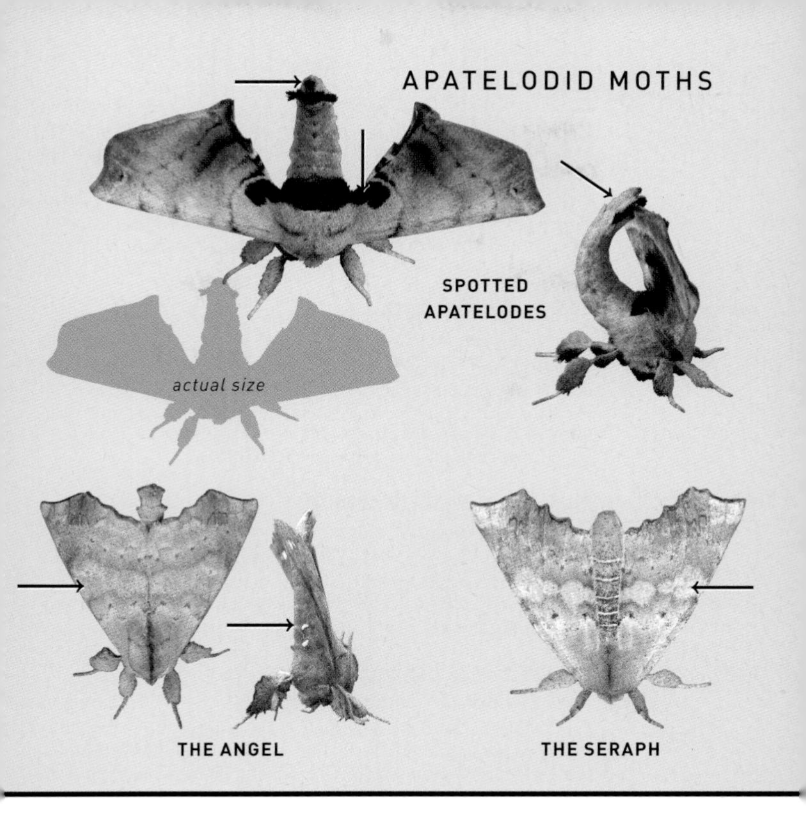

ROYAL SILKWORM MOTHS

IMPERIAL MOTH

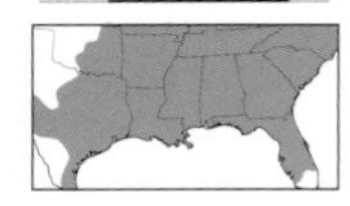

Eacles imperialis 89-0012 (7704) **Common**

WS 80–174 mm. Mustard yellow wings are shaded purplish brown in basal area and sometimes also ST area. Males are generally more strongly marked than females. Some individuals in Texas are completely shaded. Brown-ringed discal spots are usually noticeable. Purplish brown thorax has a yellow collar. **HOSTS:** Basswood, birch, alder, cedar, elm, hickory, maple, oak, walnut, and other deciduous trees. **NOTE:** Two broods.

SPINY OAKWORM MOTH

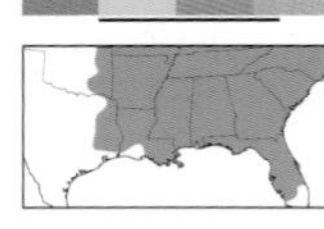

Anisota stigma 89-0014 (7716) **Common**

TL 22–38 mm. Yellow or russet brown FW is variably speckled darker. AM line curves inward before reaching costa. Straight PM line extends from inner margin to apex. FW has a round white reniform spot. Females are larger but otherwise nearly identical to males. **HOSTS:** Oak; also basswood and hazelnut.

PINK-STRIPED OAKWORM MOTH

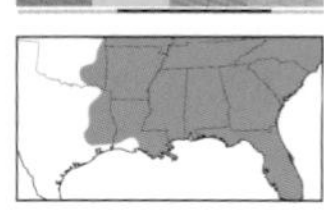

Anisota virginiensis 89-0017 (7723) **Common**

TL 20–35 mm. Sexually dimorphic. Russet FW of female has a pink terminal line beyond PM line. Smaller male has partly translucent reddish FW and straight outer margins to all wings. Both sexes have a round white reniform spot. **HOSTS:** Oak. **NOTE:** Up to three broods.

ORANGE-TIPPED OAKWORM MOTH

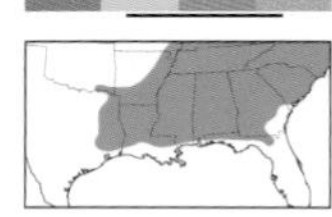

Anisota senatoria 89-0018 (7719) **Uncommon**

TL 17–28 mm. Sexually dimorphic. Pale orange FW of female is lightly speckled and has a slanting, dusky PM line. Smaller male has partly translucent reddish FW and straight outer margins to all wings. Both sexes have a round white reniform spot. **HOSTS:** Oak. **NOTE:** One or two broods.

ROSY MAPLE MOTH

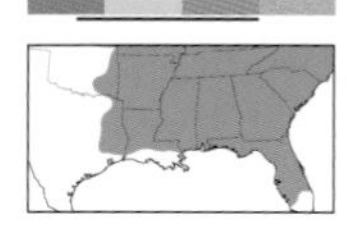

Dryocampa rubicunda 89-0022 (7715) **Common**

TL 26–28 mm. Bright pink FW has a broad pale yellow median band that reaches apex. Variable; FW can range from mostly pink to nearly whitish. Plush yellow thorax is densely hairy. **HOSTS:** Maple and oak. **NOTE:** Two broods.

BISECTED HONEY LOCUST MOTH

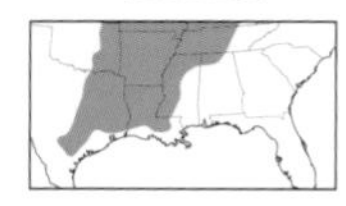

Sphingicampa bisecta 89-0023 (7712) **Local**

TL 26–35 mm. Pale orange FW is variably speckled with brown scales. Angled AM line can be faint. Oblique PM line sweeps across wing from midpoint of inner margin to apex. Orange HW has a bold pink patch through central area. **HOSTS:** Honey locust and Kentucky coffee tree. **NOTE:** Two broods.

ROYAL SILKWORM MOTHS

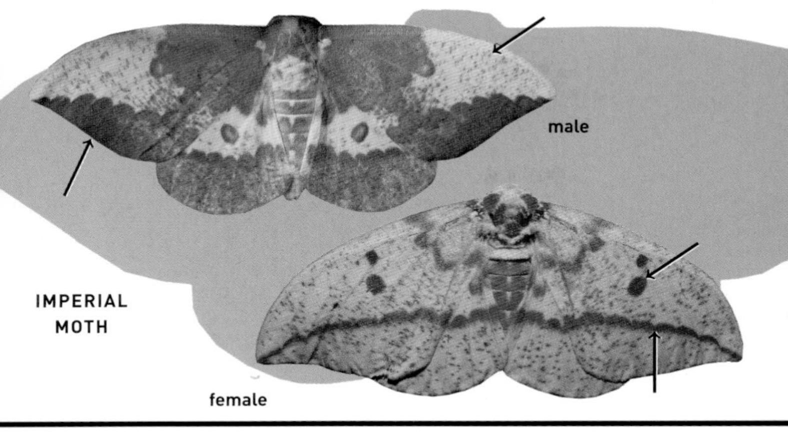

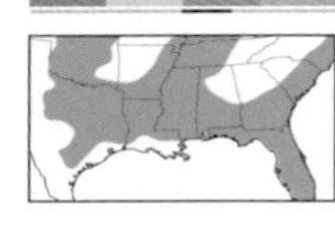

HONEY LOCUST MOTH

Sphingicampa bicolor 89-0024 (7709) **Uncommon**

TL 26–35 mm. FW is variably gray or orange, lightly speckled darker. Blurry AM and PM lines are inconspicuous. Two white dots in reniform area may be reduced or absent. Pink HW has a crimson flush along inner margin. **HOSTS:** Honey locust and Kentucky coffee tree. **NOTE:** Up to three broods.

HEILIGBRODT'S MESQUITE MOTH

Sphingicampa heiligbrodti 89-0025 (7710) **Local**

TL 28–38 mm. Peppered light gray FW has an angled AM line and scalloped PM line. Note two white dots within dark blotch in central median area. Gray HW has bold pink patch covering central area, accented with a large black spot. **HOSTS:** Honey mesquite.

Silkworm Moths

Family Saturniidae, Subfamily Hemileucinae

Medium large, boldly patterned moths. Eastern Buck Moth is diurnal and can sometimes be flushed or encountered flying rapidly in oak forests on warm late-fall days. At rest it folds its wings tented over its abdomen. Io Moth typically holds its wings closed and flat at rest; the large eyespots on the HW are likely flashed to frighten potential predators. Adults do not feed. Io Moths are commonly encountered at lights.

EASTERN BUCK MOTH

Hemileuca maia 89-0040 (7730) **Common**

WS 50–75 mm. Black wings have a white median band. Black reniform spot on FW is fused to black basal area. Densely hairy thorax is black with a white collar. Hairy abdomen is tipped orange in male. Forelegs have hairy orange tufts. **HOSTS:** Scrub oak and other oaks.

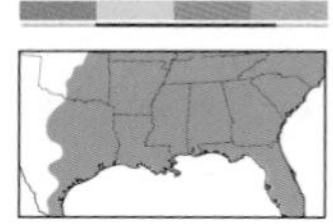

IO MOTH *Automeris io* 89-0055 (7746) **Common**

WS 50–80 mm. Sexually dimorphic. FW is largely yellow in male, bronzy gray to purplish pink in female. Dark reniform spot appears broken or speckled. Yellow HW has a large blue eyespot boldly outlined black. **HOSTS:** Birch, clover, corn, maple, oak, willow, and many other trees, shrubs, and plants.

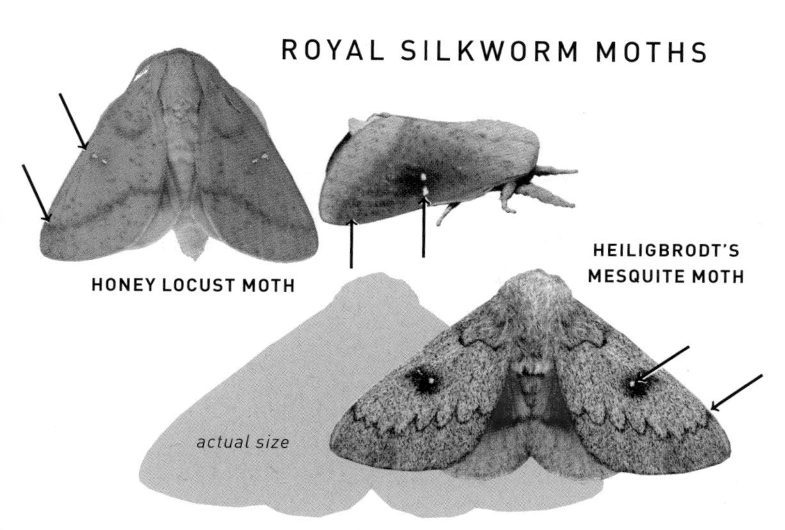
ROYAL SILKWORM MOTHS
HONEY LOCUST MOTH
HEILIGBRODT'S
MESQUITE MOTH
actual size

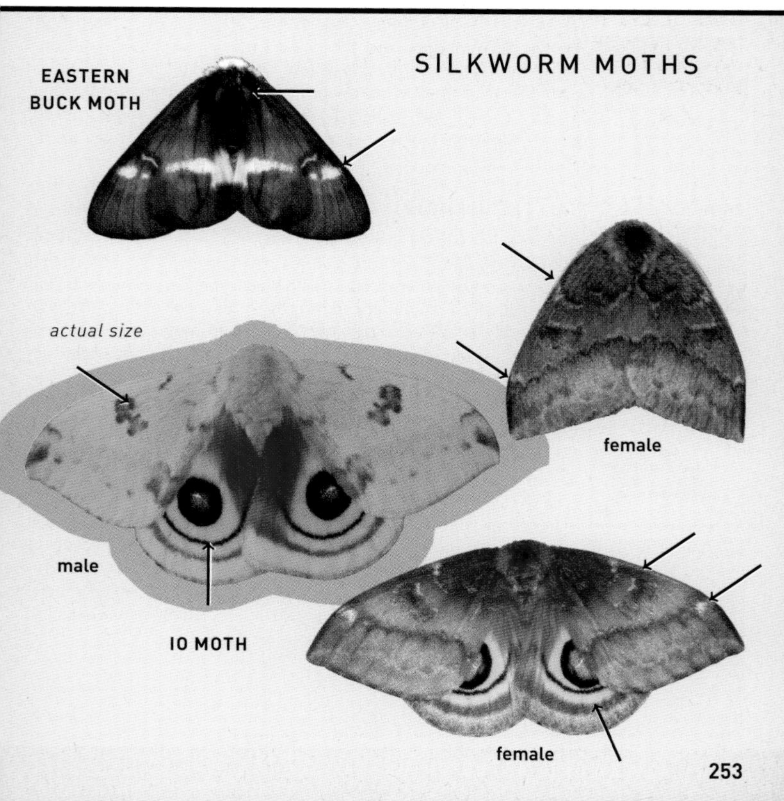
SILKWORM MOTHS
EASTERN
BUCK MOTH
actual size
female
male
IO MOTH
female

Silkworm Moths

Family Saturniidae, Subfamily Saturniinae

Large to very large, flat-winged, woodland-dwelling moths. They are characterized by their strikingly beautiful and colorful wing patterns, often with eyespots. Males have feathery, bipectinate antennae that are used to detect female pheromones. Female antennae shape varies with species. Some species rest with their wings closed tight over their abdomen and will drop to the ground like a fallen leaf if disturbed. Others, such as the Luna Moth, rest with flat wings. Adults do not feed. They are mostly nocturnal, often appearing a few hours after dusk, and are attracted to lights in small numbers. Male Calleta Silkmoths are active during midmorning.

POLYPHEMUS MOTH

Antheraea polyphemus 89-0070 (7757) **Common**

WS 100–150 mm. Rosy brown wings have pink-edged black AM and PM lines, grayish costa, and cinnamon ST band. Transparent eyespots are outlined with yellow, blue, and black. **HOSTS:** Ash, birch, grape, hickory, maple, oak, pine, and other woody plants. **NOTE:** Two broods.

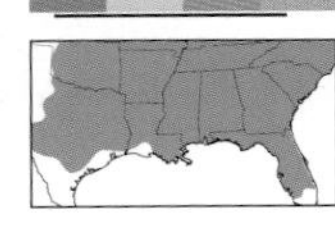

LUNA MOTH *Actias luna* 89-0072 (7758) **Common**

WS 75–105 mm. Apple green wings are marked with sleepy-looking elliptical eyespots outlined in yellow, white, and black. Costa is purplish brown. Thorax is yellow to white. HW has dramatically long, slightly twisted tail, especially on males. **HOSTS:** Alder, beech, cherry, hazelnut, hickory, willow, and other deciduous trees. **NOTE:** Three broods.

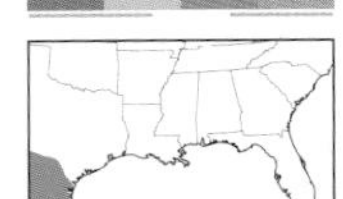

CALLETA SILKMOTH

Eupackardia calleta 89-0078 (7763) **Local**

WS 100–115 mm. Dark brown wings have creamy PM and terminal lines. ST area of FW is accented with black-edged blue eyespots, the largest of which is clouded red near apex. All wings are marked with pale wedge-shaped spots in central median area (sometimes reduced or absent, especially on females). **HOSTS:** Cenizo, ash, willow, and others. **NOTE:** Two broods.

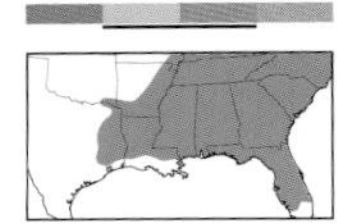

PROMETHEA MOTH

Callosamia promethea 89-0079 (7764) **Uncommon**

WS 75–95 mm. Sexually dimorphic. Blackish wings of male have a cream-colored band beyond ST line. Female has wings largely reddish brown with a wavy white PM line. Small, irregularly shaped pale reniform spots on all wings are sometimes faint. Pale border of HW is accented with a lacy pattern of blackish (male) or reddish (female) spots. **HOSTS:** Apple, ash, basswood, birch, cherry, maple, spicebush, sweet gum, tulip tree, and others. **NOTE:** Two broods.

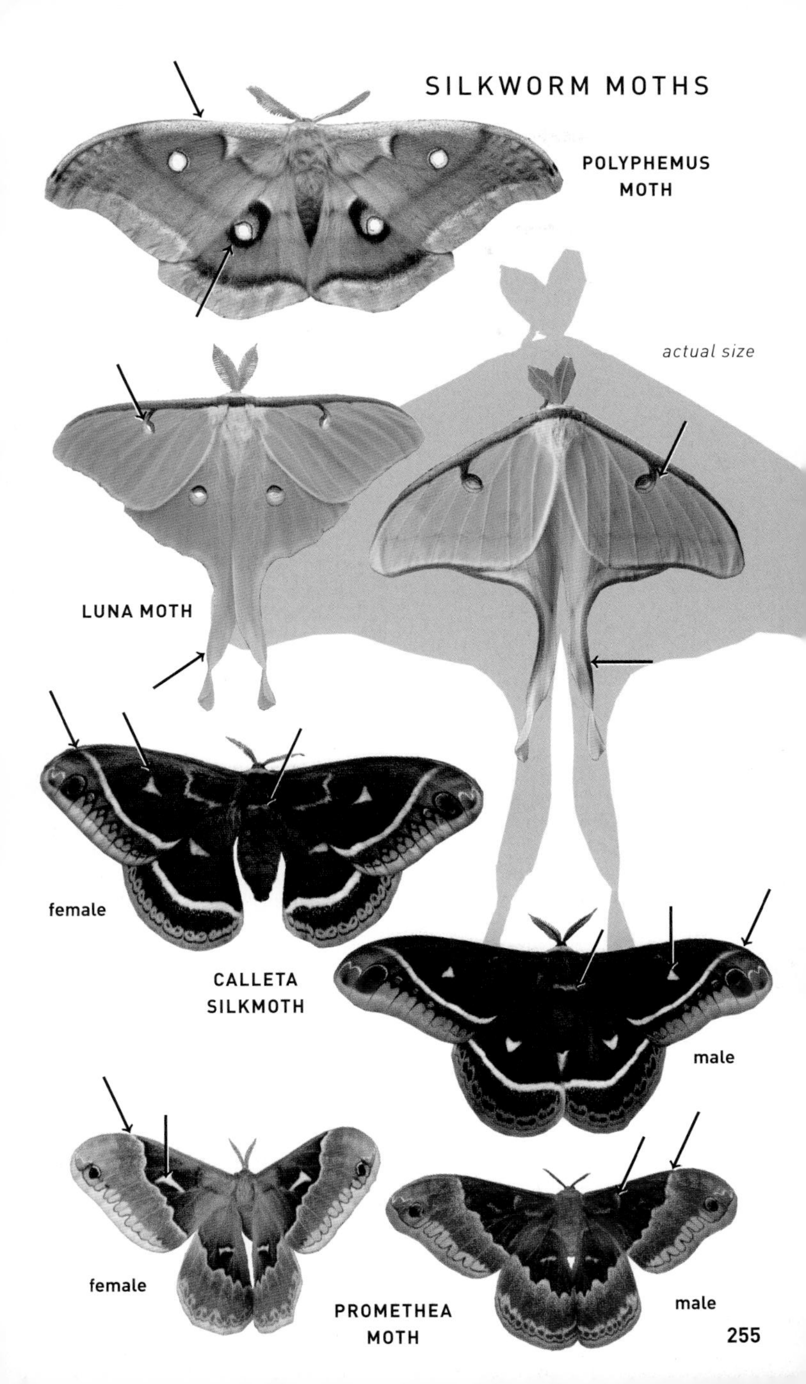
SILKWORM MOTHS
POLYPHEMUS MOTH
actual size
LUNA MOTH
female
CALLETA SILKMOTH
male
female
PROMETHEA MOTH
male

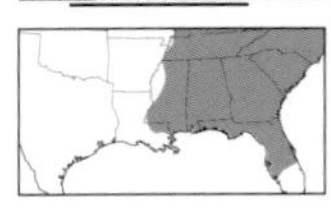

TULIP-TREE SILKMOTH

Callosamia angulifera 89-0080 (7765) **Common**

WS 80–110 mm. Both sexes resemble female Promethea Moth, but male is darker overall and female is paler orange. Large T-shaped reniform spots mark all wings of both sexes. Pale border of HW is accented with a lacy pattern of dark (male) or orange (female) spots. **HOSTS:** Tulip tree; also black cherry and sassafras. **NOTE:** Two broods.

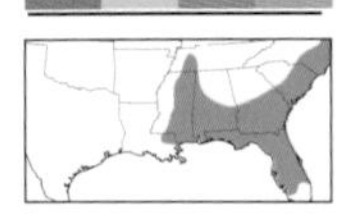

SWEETBAY SILKMOTH

Callosamia securifera 89-0081 (7766) **Common**

WS 80–110 mm. Both sexes resemble Tulip-tree Silkmoth, except the pale reniform spot on HW is reduced or absent. Females appear more washed out and yellowish, especially in spring brood. **HOSTS:** Sweet bay. **NOTE:** Two broods.

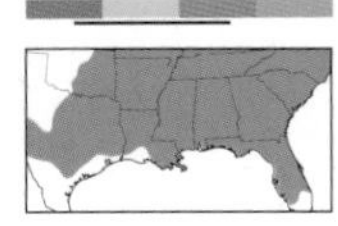

CECROPIA MOTH

Hyalophora cecropia 89-0082 (7767) **Common**

WS 110–150 mm. Grayish brown wings have tear-shaped reniform spots and broad red and white PM bands. Basal area is reddish inside curved whitish AM line. Red thorax has a white frontlet. **HOSTS:** Apple, ash, beech, birch, elm, maple, poplar, oak, willow, and other deciduous trees.

Large Sphinx Moths

Family Sphingidae, Subfamily Sphinginae

Robust, medium-sized to large moths. Most species have long pointed wings and a long tapering abdomen. They are powerful fliers. Adults feed on nectar taken from tubular flowers such as phlox and bergamot and can often be seen at dusk. Most species are nocturnal and are attracted to lights.

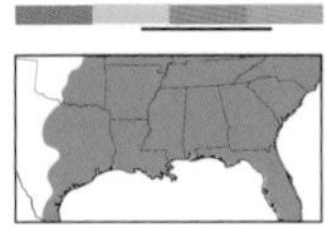

PINK-SPOTTED HAWKMOTH

Agrius cingulata 89-0086 (7771) **Common**

TL 55–65 mm. FW has variegated pattern of gray and brown. Darker median patch surrounds small white discal spot. HW is pale gray with black bands and pink basal patch. Abdomen has pairs of pink and black lateral spots. Base of thorax is usually bordered by thick black line set with thin orange spots and a hint of blue. **HOSTS:** Low plants and shrubs, including sweet potato and jimsonweed.

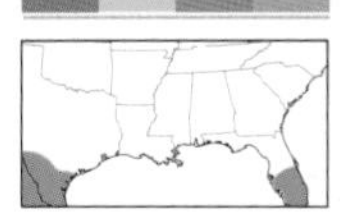

GIANT SPHINX *Cocytius antaeus* 89-0087 (7772) **Local**

TL 83 mm. Large. FW has mottled pattern of brown and buff marked with paired thick black lines. Two white discal spots on FW. HW has yellow patch at base and a transparent midsection crossed by black veins. Abdomen has pairs of yellow lateral spots. Thorax is plain or with a pair of tiny black-edged white spots. **HOSTS:** *Annona* species, including custard apple.

SILKWORM MOTHS

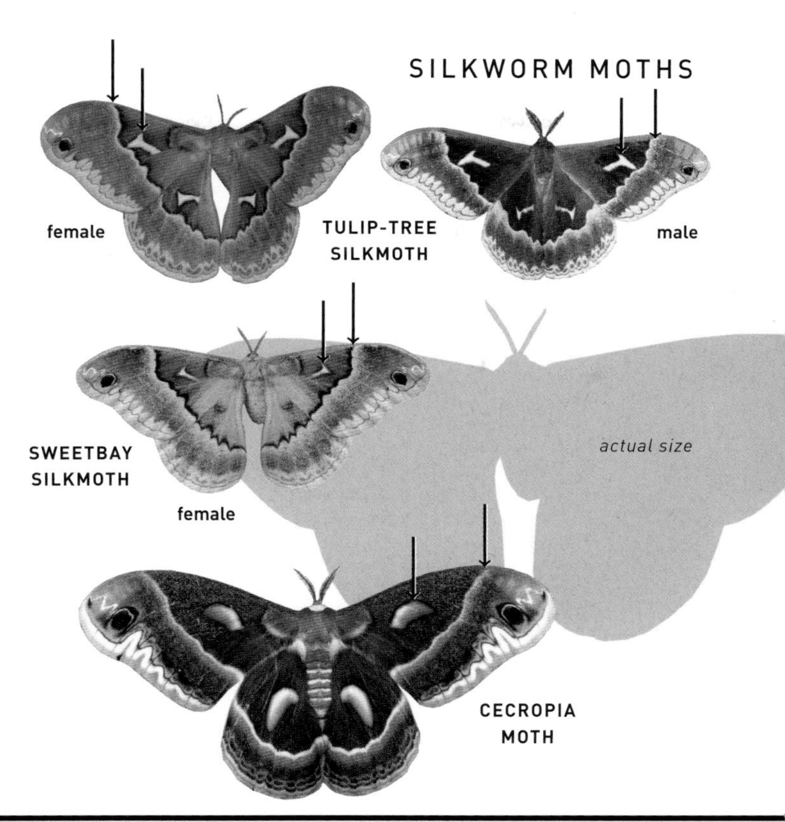

LARGE SPHINX MOTHS

CAROLINA SPHINX

Manduca sexta 89-0090 (7775) **Common**

TL 55–65 mm. Mottled gray FW has obscure blackish lines, brown shading, and small white discal spot. HW is gray with black-and-white bands. Fringe is checkered. Abdomen has six pairs of yellow spots along sides. Dark thorax often has two indistinct black lines and/or tiny white dots. **HOSTS:** Crops, including potato, tobacco, and tomato. **NOTE:** Caterpillar is known as Tobacco Hornworm.

FIVE-SPOTTED HAWKMOTH

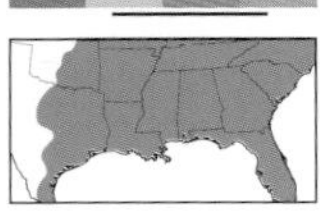

Manduca quinquemaculatus 89-0091 (7776) **Common**

TL 50–70 mm. Smooth gray FW has wavy black lines and brownish shading in a vaguely woodgrain pattern. HW is pale gray with widely separated, jagged black bands. Abdomen has five or six pairs of yellow spots laterally. Back of thorax has two black-outlined rings. **HOSTS:** Crops, including potato, tobacco, and tomato. **NOTE:** Caterpillar is known as Tomato Hornworm.

RUSTIC SPHINX *Manduca rustica* 89-0092 (7778) **Common**

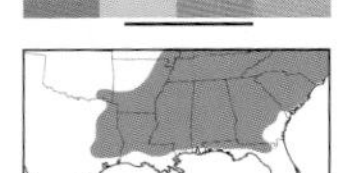

TL 50–80 mm. FW is white with bold, jagged black lines and variably gray or brown patches in median and basal areas. FW has small white discal spot. HW is gray with blackish bands and whitish spots. Abdomen has three pairs of yellow spots near base. Thorax has a white "face" pattern. **HOSTS:** Fringe tree, jasmine, and bignonia. **NOTE:** Two broods.

ASH SPHINX

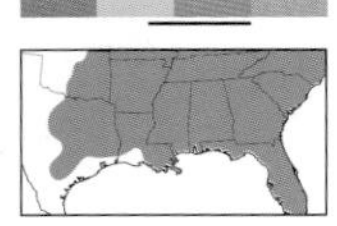

Manduca jasminearum 89-0096 (7783) **Common**

TL 48–58 mm. Gray FW is lightly peppered. Thick black line extends through median area from costa to midway along outer margin. Warm brown patch covers area behind small white discal spot. HW is black with pale gray shading along inner margin. Thorax is plain. **HOSTS:** Ash. **NOTE:** Two broods.

PAWPAW SPHINX *Dolba hyloeus* 89-0100 (7784) **Common**

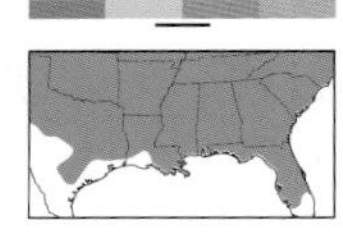

TL 30–35 mm. Similar to Rustic Sphinx, but wing pattern is less contrasting and dark median patch is often larger, often blending into rest of wing. Jagged black lines are more diffuse. HW is dusky with whitish bands. Thorax is unmarked or has two small white spots. **HOSTS:** Pawpaw, holly, sweet fern, possum haw, and inkberry.

ELM SPHINX *Ceratomia amyntor* 89-0102 (7786) **Common**

TL 50–60 mm. Brown FW has thin black streaks, often set in a dark band running from base to apex. Costa is pale. Parallel thin ST lines are most visible near costa. FW has small white discal spot. HW is brown with dusky bands. Brown thorax is bordered by dark bands. **HOSTS:** Elm, birch, basswood, and cherry.

LARGE SPHINX MOTHS

CAROLINA SPHINX

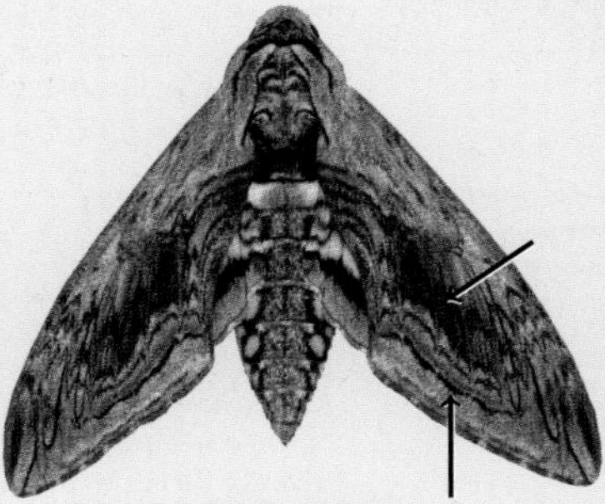

FIVE-SPOTTED HAWKMOTH

RUSTIC SPHINX

ASH SPHINX

actual size

PAWPAW SPHINX

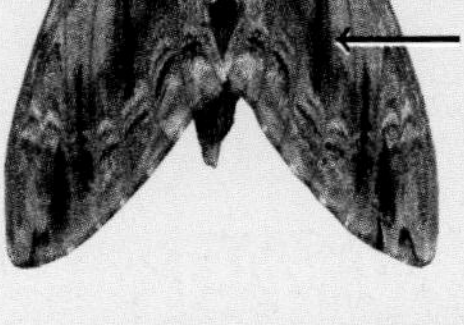

ELM SPHINX

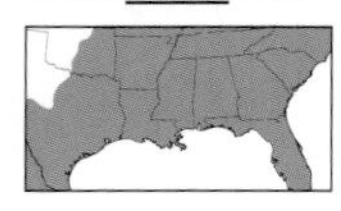

WAVED SPHINX

Ceratomia undulosa 89-0103 (7787) **Common**

TL 45–60 mm. Pale gray FW is crossed by narrow, jagged black lines. Thin black streaks run from base to apex, never set in a dark band. White discal spot is conspicuous. HW is pale gray with faint dusky bands. Gray thorax is circled by a thin black ring, edged bluish white at base. **HOSTS:** Ash, privet, oak, hawthorn, and fringe tree.

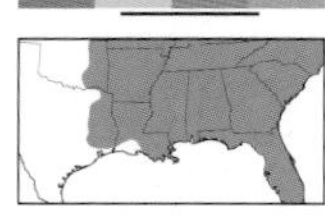

CATALPA SPHINX

Ceratomia catalpae 89-0105 (7789) **Common**

TL 42–47 mm. Plain, yellowish brown FW has a scalloped, double ST line and short black apical dash; other lines are indistinct. Inconspicuous discal spot is edged black. HW is yellowish brown with slightly darker bands and pale fringe. Thorax is bordered with black-edged pale bands and a thin bluish white line at base. **HOSTS:** Catalpa. **NOTE:** Two broods.

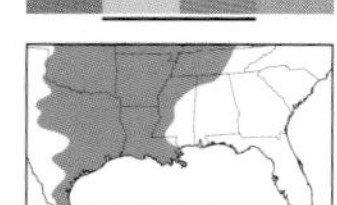

HAGEN'S SPHINX *Ceratomia hageni* 89-0106 (7790) **Local**

TL 42–48 mm. Light gray FW is tinged variably mossy green or brown along costa and in ST area. Dark zigzag AM and PM bands are connected by dusky patch in inner median area. Shows pale patches in outer median and apical areas. FW has two discal spots. Gray HW has a wide dusky ST band. Pale thorax is bordered by two dark bands. **HOSTS:** Osage orange. **NOTE:** Three or four broods.

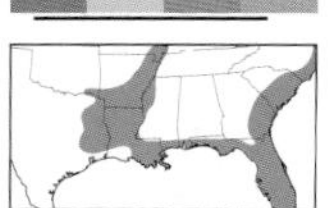

BALD CYPRESS SPHINX

Isoparce cupressi 89-0108 (7791) **Local**

TL 30–35 mm. Smooth gray FW has a series of black dashes from near anal angle to basal costa, and at apex. Dashes are sometimes edged brown in median area. HW is uniformly dark gray. Thorax has multiple narrow vertical stripes. **HOSTS:** Bald cypress. **NOTE:** Up to four broods.

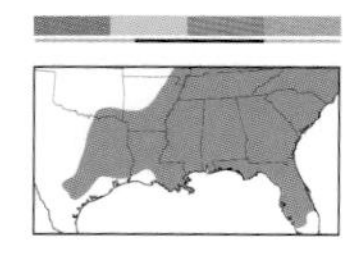

PLEBIAN SPHINX *Paratrea plebeja* 89-0110 (7793) **Common**

TL 33–38 mm. Gray FW has a series of short black dashes from basal area to apex, and white dashes in ST area. Double AM and PM lines are indistinct. White dash connects to discal spot on FW. HW is dark gray with indistinct bands. Thorax is bordered by white-edged black bands. **HOSTS:** Trumpet creeper, yellow trumpetbush, passionflower, and lilac. **NOTE:** Two broods.

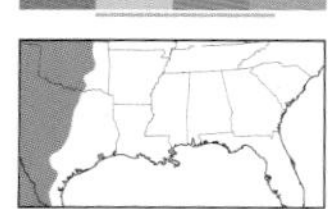

INCENSE CEDAR SPHINX

Sphinx libocedrus 89-0113 (7804) **Local**

TL 32–38 mm. Gray FW has narrow dashes from apex to basal area. Diffuse AM and PM lines are indistinct. White edge to ST line widens toward inner margin. Discal spot is faint or absent. Gray HW has two black bands. Abdomen is banded black and white along sides. Thorax has thin black lines. **HOSTS:** Ash.

LARGE SPHINX MOTHS

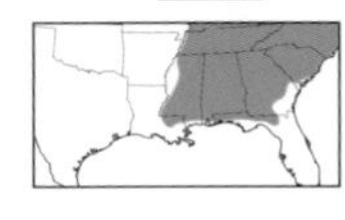

FRANCK'S SPHINX *Sphinx franckii* 89-0117 (7808) **Local**

TL 55–65 mm. Bicolored FW is ashy gray along costa and tawny along inner margin. Apical dash is indistinct. Dark ST lines create a broken-bark effect at outer margin. HW has two broad black bands. Gray thorax has narrow tawny-edged black stripes down center. **HOSTS:** Ash. **NOTE:** One or two broods.

LAUREL SPHINX *Sphinx kalmiae* 89-0118 (7809) **Common**

TL 40–55 mm. Fawn-colored FW has dark brown shading along inner margin and outer apex. White streaks blend together through ST area. Terminal line is white. Black discal spot is tiny. HW has thick blackish bands. Thorax has broad white-edged black stripes. Abdomen has pairs of black-and-white spots along sides. **HOSTS:** Laurel, lilac, ash, poplar, and others. **NOTE:** Two broods.

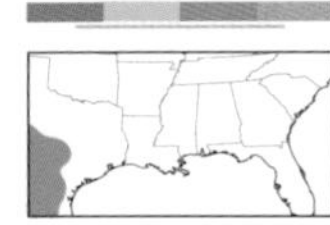

ISTAR SPHINX *Lintneria istar* 89-0132 (7799) **Local**

TL 55–60 mm. Light gray FW is usually tinged warm brown. An irregular dark band with thin black streaks extends from inner basal area to outer margin. Narrow apical dash fades in median area. Jagged AM and PM lines are double and sometimes indistinct. FW has two discal spots, the larger one often shaded with gray. Blackish HW has two wavy pale bands. Thorax is bordered by broad white and dark stripes. **HOSTS:** Mint.

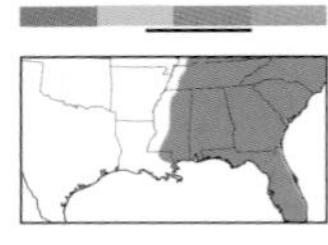

SOUTHERN PINE SPHINX

Lapara coniferarum 89-0135 (7816) **Common**

TL 30–35 mm. Grayish FW has warm brown shading along inner margin. Two black dashes in inner median area. Black-edged pale PM line is sharply toothed. Thorax has pale mushroom-shaped central stripe. **HOSTS:** Pine, including loblolly pine and longleaf pine.

Eyed Sphinx Moths

Family Sphingidae, Subfamily Smerinthinae

Medium-sized to large sphinx moths with scalloped wings that are held elevated and slightly away from the body. In most species the HW has a blue-filled eyespot. All are nocturnal and will regularly visit lights in small numbers.

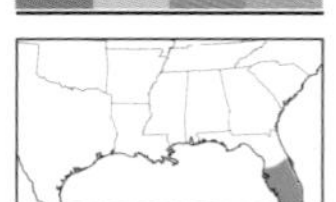

STREAKED SPHINX

Protambulyx strigilis 89-0138 (7818) **Local**

TL 62–70 mm. Long, narrow tawny FW has a faint netlike pattern and three or four short brown dashes along costa. Indented inner margin is marked with blackish blotches near anal angle. Fiery orange HW is marked with narrow brown lines. Thorax has black lateral stripes. **HOSTS:** Brazilian pepper.

LARGE SPHINX MOTHS

EYED SPHINX MOTHS

TWIN-SPOTTED SPHINX

Smerinthus jamaicensis 89-0140 (7821) **Common**

TL 38–45 mm. Pale lilac gray to brownish FW is crossed with noticeable lines. AM line angles up on inside of dark median bar, forming a Y. Black spot at apex is edged in white. Discal spot is white. Rosy pink HW has a black-edged blue eyespot divided by black line. Thorax has blackish dorsal patch. **HOSTS:** Deciduous trees, including apple, ash, elm, poplar, and birch. **NOTE:** Several broods.

BLINDED SPHINX

Paonias excaecata 89-0144 (7824) **Common**

TL 35–50 mm. Tan FW has darker brown and violet shading in median area and along scalloped outer margin. AM line angles down on inside of dark median bar. Discal spot is black. Rosy pink HW has a black-edged blue eyespot. Thorax has brown dorsal stripe. **HOSTS:** Deciduous trees, including basswood, willow, birch, and poplar. **NOTE:** Several broods.

SMALL-EYED SPHINX

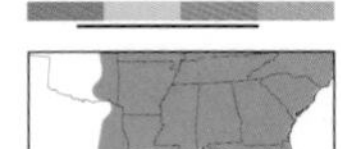

Paonias myops 89-0145 (7825) **Common**

TL 32–35 mm. Violet gray FW is washed with orange and has orange spots near apex and anal angle. AM and PM lines are edged violet. HW has yellow patch surrounding black-edged, blue eyespot. Thorax has a violet-edged orange dorsal stripe. Usually rests with HW protruding beyond costa. **HOSTS:** Deciduous trees, including black cherry, serviceberry, and basswood. **NOTE:** Several broods.

HUCKLEBERRY SPHINX

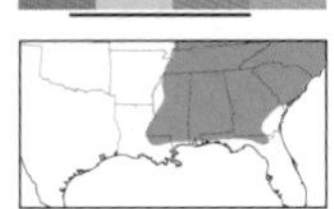

Paonias astylus 89-0146 (7826) **Uncommon**

TL 30–40 mm. Orange brown FW has lilac basal patch and ST line. Brown lines slant across basal area and inward from apex. HW has yellow patch surrounding black-edged, blue eyespot. Thorax has a dull orange dorsal stripe. Usually rests with HW protruding beyond costa. **HOSTS:** Blueberry, huckleberry, cherry, and willow.

WALNUT SPHINX

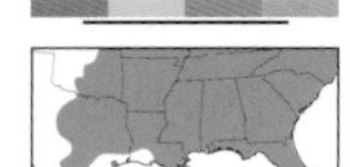

Amorpha juglandis 89-0147 (7827) **Common**

TL 30–40 mm. FW is variably pale pinkish gray to brown, often with darker shading in inner median area and beyond PM line. AM and PM lines are double. Discal spot is dark. Scalloped HW usually protrudes beyond costa while at rest. **HOSTS:** Deciduous trees, including walnut, butternut, hickories, alder, and beech. **NOTE:** Several broods.

MODEST SPHINX

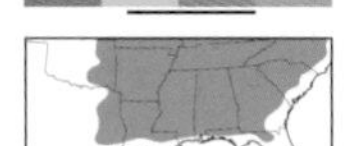

Pachysphinx modesta 89-0148 (7828) **Uncommon**

TL 45–65 mm. Large. Violet gray FW has white-edged, dark median band surrounding small white discal spot. Outer margin is scalloped. Rosy HW has a blue patch bordered by black at anal angle. **HOSTS:** Cottonwood, poplar, and willow. **NOTE:** Also known as Big Poplar Sphinx.

EYED SPHINX MOTHS

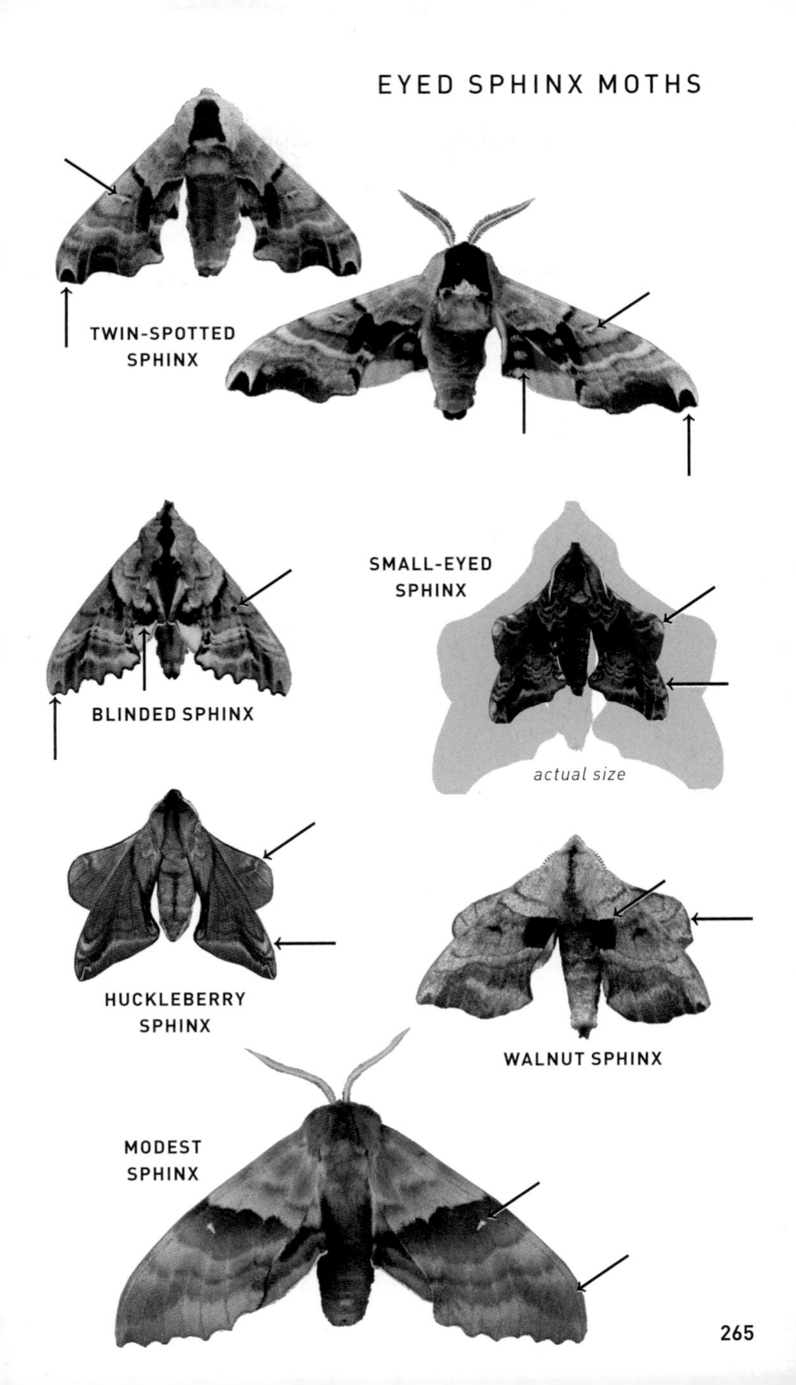

Small Sphinx Moths

Family Sphingidae, Subfamily Macroglossinae

Similar in shape and habit to the large sphinx moths, this group is generally more colorful and varied. Some are crepuscular and can be found feeding on flowers at dusk. A small number are diurnal and will visit patches of flowers in meadows and gardens on sunny afternoons. All but the clearwings and Nessus Sphinx may come to lights.

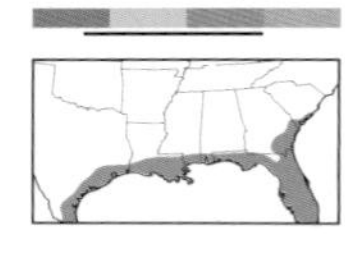

TETRIO SPHINX *Pseudosphinx tetrio* 89-0150 (7830) **Local**

TL 58–62 mm. Gray FW is sometimes shaded with brown in median area. Costa is marked with a black-outlined triangle and semicircle. Whitish ST line isolates darker apical area. Brown basal patch is edged black. Brown HW is pale gray at anal angle. Abdomen is banded black and white. Females are larger than males. **HOSTS:** Dogbane. **NOTE:** Several broods.

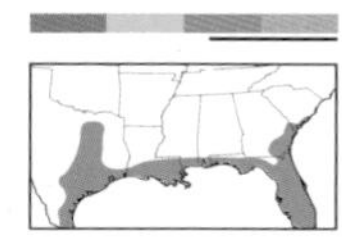

LASSAUX'S SPHINX *Erinnyis lassauxii* 89-0153 (7833) **Local**

TL 40–52 mm. Grayish brown FW has a streaky barklike pattern. Inner margin has whitish patches at base. Apex is tipped pale. Dull orange HW has a diffuse dusky ST band. Abdomen is banded black and white. Dark thorax has a pale stripe down center. **HOSTS:** Papaya, spurge, and milkweed.

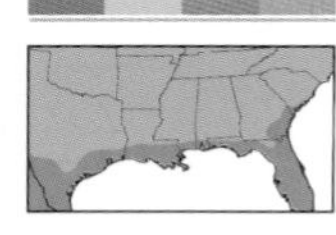

ELLO SPHINX *Erinnyis ello* 89-0154 (7834) **Common**

TL 53–59 mm. Dimorphic. Ashy gray FW may either be uniform or have a bold dark streak from base to apex. Inner half of FW may be paler. Burnt orange HW has a blackish ST band. Tapered abdomen is banded black and gray. Thorax is plain or banded, depending on wing pattern. **HOSTS:** Cassava, poinsettia, and a variety of woody plants, including guava and willow bustic.

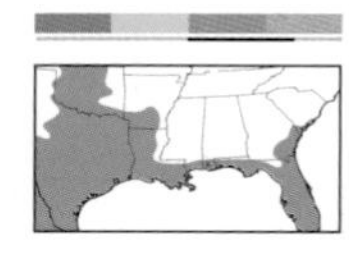

OBSCURE SPHINX *Erinnyis obscura* 89-0156 (7837) **Common**

TL 29–37 mm. Dimorphic. Resembles much larger Ello Sphinx but is usually more contrasting, with pale patch at anal angle. On band-winged form, thorax has two dark stripes. Orange HW has an incomplete dusky ST band. Lacks pale bands on abdomen. Male has a distinct blackish streak through FW from base to apex. **HOSTS:** Dogbane, papaya, and spurge.

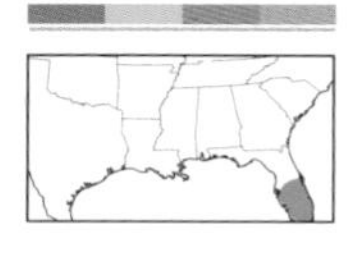

CAICUS SPHINX *Phryxus caicus* 89-0160 (7840) **Local**

TL 40–50 mm. Fawn-colored FW has a bold blackish central streak edged whitish and is paler along inner margin. Orange HW has an ST band of black spots along veins. Abdomen is boldly banded black and white. **HOSTS:** Mangrove rubber vine.

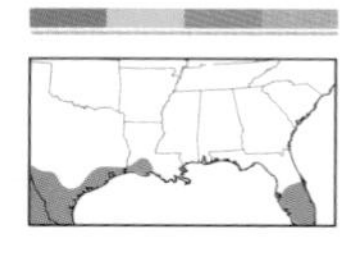

FIG SPHINX *Pachylia ficus* 89-0161 (7841) **Local**

TL 75–83 mm. Large. Dark grayish to brown FW is lightly marked with darker bands and discal dot. Apex is dusky with large white patch. Diffusely banded HW has tiny white spot at anal angle. **HOSTS:** Fig.

SMALL SPHINX MOTHS

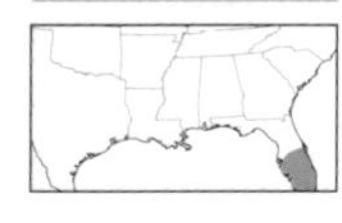

FALSE-WINDOWED SPHINX

Madoryx pseudothyreus 89-0163 (7843) **Local**

TL 35–42 mm. Grayish brown FW is slightly darker in basal and median areas and patterned with crisp black lines. Large white comma-shaped reniform spot is flanked by two smaller dots. Wing margins are scalloped. Brushy-tipped abdomen is curled upward at rest. **HOSTS:** Black mangrove.

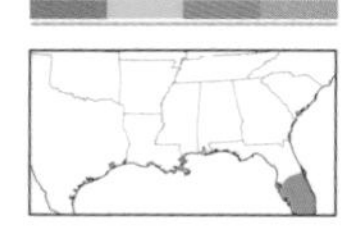

HALF-BLIND SPHINX *Perigonia lusca* 89-0166 (7846) **Local**

TL 30–35 mm. Brown FW has pale, straight AM and PM lines bordering a dark median band. Terminal area is pale below gently curving ST line. FW has a pair of tiny black discal spots. HW has a broad yellow basal patch and spot at anal angle. Wide abdomen has a fanned tip. **HOSTS:** Tawnyberry holly and rough velvetseed.

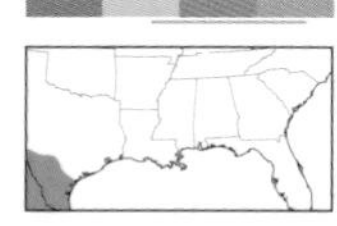

CLAVIPES SPHINX *Aellopos clavipes* 89-0169 (7848) **Local**

TL 30–35 mm. Dark olive brown FW has three white spots at midpoint of ST line. Paler median band has a small black discal spot. HW is largely blackish. Wide abdomen has a white band on fourth segment and tan patch in middle of brushy tip. **HOSTS:** Madder, indigoberry, and velvetseed. **NOTE:** Diurnal.

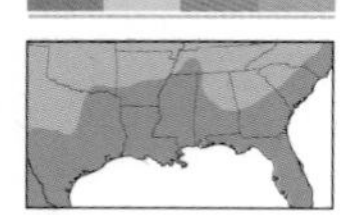

TITAN SPHINX *Aellopos titan* 89-0170 (7849) **Common**

TL 30–35 mm. Similar to Clavipes Sphinx, but pale median band is reduced, and usually has four white spots along ST line. HW has patch of whitish scales at anal angle (visible in flight). **HOSTS:** Madder, including seven-year apple; also pond apple. **NOTE:** Diurnal.

MOURNFUL SPHINX *Enyo lugubris* 89-0172 (7851) **Common**

TL 38–42 mm. Dark violet brown FW has a straight yellowish, black-edged AM. Thin yellowish ST line is sharply pointed at midpoint. Black discal spot on FW is outlined in yellow. Outer margin is scalloped. Rests with brushy-tipped (or flanged in male) abdomen curled upward. **HOSTS:** Possum vine, peppervine, and members of Vitaceae family. **NOTE:** Active both day and night.

GROTE'S SPHINX *Cautethia grotei* 89-0174 (7867) **Local**

TL 20–25 mm. Very small. Pale gray FW has indistinct black lines and a diffuse whitish ST band. Small discal spot is white. A black dash angles inward from anal angle. Orange HW has a broad black ST band. **HOSTS:** Milkberry.

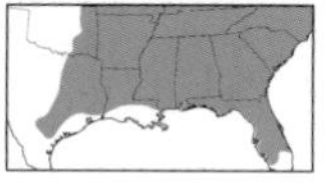

HUMMINGBIRD CLEARWING

Hemaris thysbe 89-0177 (7853) **Common**

TL 25–30 mm. Mostly transparent wings have a broad reddish border and veins. Thorax is greenish brown above, white below. Abdomen has yellow segments at base and a reddish band around middle; brushy tip has tan center. Legs pale. **HOSTS:** Honeysuckle, snowberry, hawthorn, cherry, and plum. **NOTE:** Two broods. Diurnal.

SMALL SPHINX MOTHS

SLENDER CLEARWING

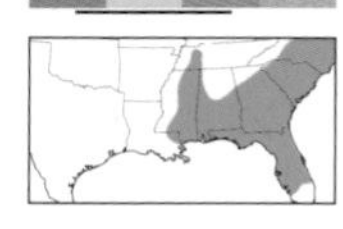

Hemaris gracilis 89-0178 (7854) **Local**

TL 25–30 mm. Resembles Hummingbird Clearwing, but reddish border of FW has white inner edge. Sides of thorax (under wings) have a red stripe. Abdomen has a red band edged white and with white spots down center; middle of brushy tip is rusty. Green thorax has thin white stripes laterally. **HOSTS:** Heaths, including blueberry and laurel. **NOTE:** Two broods. Diurnal.

SNOWBERRY CLEARWING

Hemaris diffinis 89-0179 (7855) **Common**

TL 22–30 mm. Resembles Hummingbird Clearwing, but olive thorax has cream-colored bands laterally. Black abdomen has a yellowish band near tip that is often split in two; brushy tip may be all black or have tan center. Legs dark. **HOSTS:** Snowberry, dogbane, and honeysuckle. **NOTE:** Two broods. Diurnal.

PANDORUS SPHINX

Eumorpha pandorus 89-0182 (7859) **Common**

TL 45–60 mm. Pale green FW has dark green patches along basal area of inner margin, in outer median area, and near apex and anal angle. Pink veins extend through inner ST area. Discal spot is broken into three. Pale HW has two large blackish patches. Thorax has dark green triangles laterally. **HOSTS:** Grape, peppervine, and Virginia creeper. **NOTE:** Two or more broods.

ACHEMON SPHINX

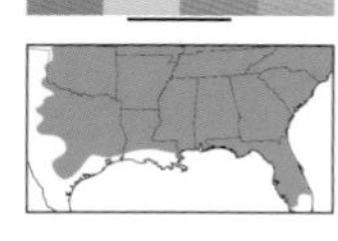

Eumorpha achemon 89-0184 (7861) **Uncommon**

TL 45–55 mm. Pinkish brown FW has darker brown patches at inner median area and apex. HW has large rosy pink basal patch. Thorax has dark brown triangles laterally. **HOSTS:** Grape and peppervine. **NOTE:** Two broods.

VINE SPHINX

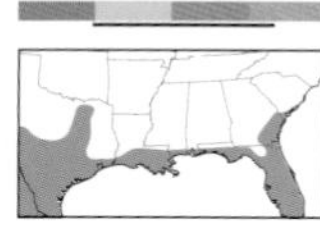

Eumorpha vitis 89-0186 (7864) **Common**

TL 58–62 mm. Dark brown FW has a striking network of thick, pale bands. Three pinkish veins cross inner PM line. Banded HW has a rosy stripe along inner margin. **HOSTS:** Grape and *Cissus* species. **NOTE:** Two broods.

BANDED SPHINX

Eumorpha fasciatus 89-0187 (7865) **Common**

TL 58–62 mm. Similar to Vine Sphinx, but markings are crisper and Banded Sphinx has a light brown costal streak. Rose and black bands of inner part of HW are higher, giving the look of a tiny third wing. **HOSTS:** Evening primrose and grape.

GAUDY SPHINX

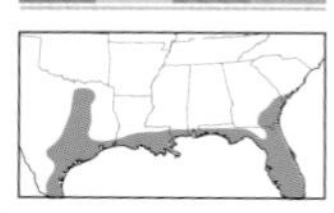

Eumorpha labruscae 89-0188 (7866) **Local**

TL 65–70 mm. Rich green FW has a darker median area. A speckled brown patch marks midpoint of PM line. HW has a black-bordered blue eyespot, reddish stripe along inner margin, and yellow fringe. **HOSTS:** *Cissus* species, including possum vine.

SMALL SPHINX MOTHS

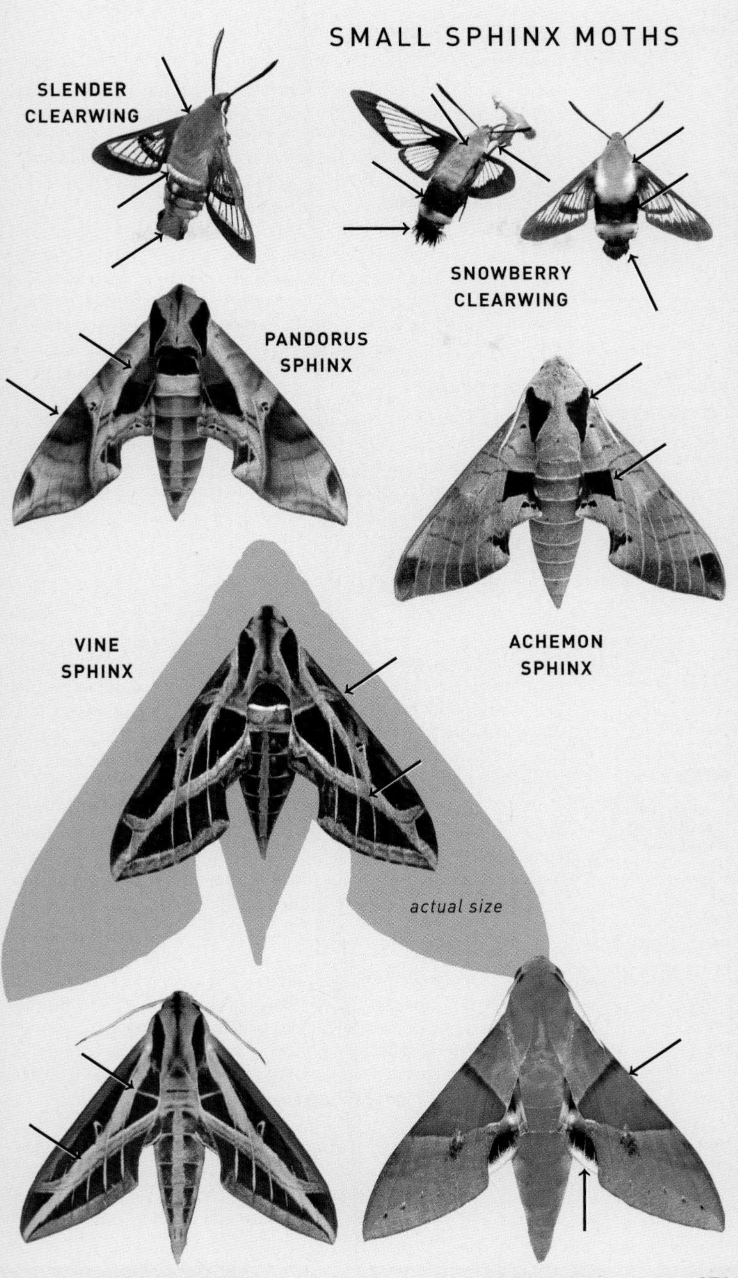

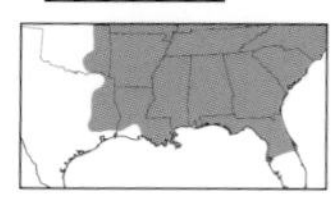

ABBOTT'S SPHINX

Sphecodina abbottii 89-0192 (7870) **Common**

TL 30–40 mm. Violet gray FW has a pattern of narrow black lines, creating appearance of broken bark. Outer margin is deeply scalloped. HW has a broad yellow basal patch. Curls brush-tipped abdomen up at rest. **HOSTS:** Grape and peppervine. **NOTE:** Two broods.

LETTERED SPHINX

Deidamia inscriptum 89-0193 (7871) **Common**

TL 25–38 mm. Brownish gray FW is crossed with thick bands and has tan discal spot and small white spot near apex. Outer margin is strongly scalloped. HW is dull orange with a dusky border. Typically curls abdomen up at rest. **HOSTS:** Grape, peppervine, and Virginia creeper.

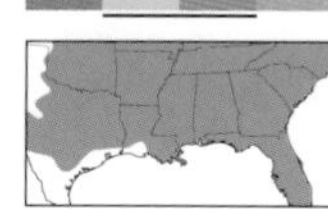

NESSUS SPHINX

Amphion floridensis 89-0194 (7873) **Common**

TL 22–28 mm. Chocolate brown FW has bands of dark brown shading in median area and beyond PM line. HW has a rusty basal area and yellowish fringe. Has one or more (usually two) yellow bands on dark, tufted abdomen. **HOSTS:** Grape, peppervine, and cayenne pepper. **NOTE:** Two broods. Diurnal.

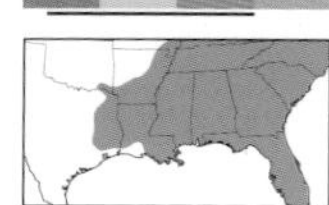

HYDRANGEA SPHINX

Darapsa versicolor 89-0206 (7884) **Local**

TL 35–45 mm. Bronzy green FW is crossed by curved pinkish mauve lines and has a dark spot and a rusty patch at apex. HW is mostly dark orange. Orange-shouldered thorax has a thin white stripe that extends down length of abdomen. **HOSTS:** Smooth hydrangea, buttonbush, and water-willow.

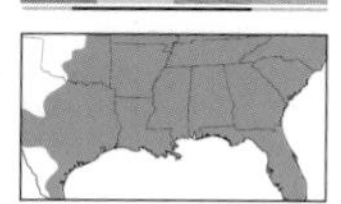

VIRGINIA CREEPER SPHINX

Darapsa myron 89-0207 (7885) **Common**

TL 28–38 mm. FW is variably light brown to pale green with dark AM and PM bands and a dark discal spot. Apex is slightly falcate. HW is mostly salmon to dull orange. **HOSTS:** Virginia creeper, grape, peppervine, and viburnum. **NOTE:** Two broods.

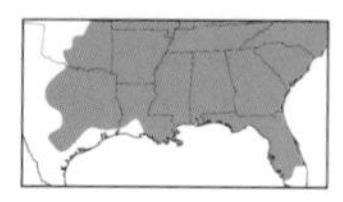

AZALEA SPHINX

Darapsa choerilus 89-0208 (7886) **Common**

TL 30–40 mm. Reddish FW has pinkish mauve shading in median and ST areas. Apex is slightly falcate. HW is mostly dull orange. Thorax has very thin white central stripe. Abdomen is reddish with thin white bands at each segment. **HOSTS:** Azalea, blueberry, black gum, and viburnum. **NOTE:** Two or more broods.

SMALL SPHINX MOTHS
ABBOTT'S SPHINX
NESSUS SPHINX
HYDRANGEA SPHINX
LETTERED SPHINX
actual size
AZALEA SPHINX
VIRGINIA CREEPER SPHINX

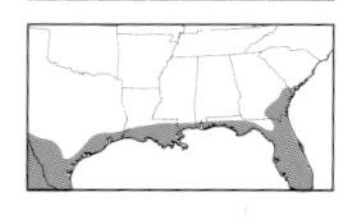

PLUTO SPHINX *Xylophanes pluto* 89-0209 (7887) **Local**

TL 35–45 mm. Mossy green FW has a paler band in upper median area and diffuse white shading near falcate apex. HW has a broad golden yellow median band. Abdomen has tiny white dots at center of each segment. **HOSTS:** Milkberry, firebush, cheese shrub, and others.

TERSA SPHINX *Xylophanes tersa* 89-0211 (7890) **Uncommon**

TL 35–45 mm. Tawny FW is marked with thin longitudinal lines and a broad stripe extending from base to apex. Brown HW is studded with yellowish wedges. Long, sharply tapering abdomen is brown with yellow lateral stripes. **HOSTS:** Madder, starcluster, and Virginia buttonweed.

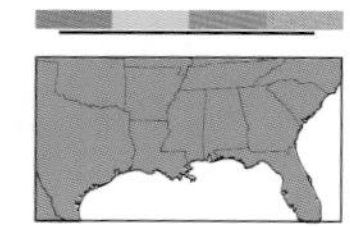

WHITE-LINED SPHINX
Hyles lineata 89-0217 (7894) **Common**

TL 35–50 mm. Dark brown FW has a broad pale stripe from base to apex, crossed by whitish veins. HW has a rosy median band. Thorax has six narrow, pale stripes. Abdomen has pairs of black and white lateral spots and pinkish sides. **HOSTS:** Various trees and plants, including apple, elm, evening primrose, grape, tomato, and purslane. **NOTE:** Two broods.

SCOOPWINGS Family Uraniidae, Subfamily Epipleminae

Flat-winged moths that resemble members of the Geometridae. They habitually crease their HW, creating a gap that lends them a tailed shape. They are nocturnal and will occasionally come to lights.

GRAY SCOOPWING
Callizzia amorata 91-0002 (7650) **Uncommon**

WS 15–22 mm. Gray wings are lightly brindled darker. Rusty-edged AM and PM lines are connected with short bar near inner margin of FW. Brown triangle at midpoint of outer margin is edged black. **HOSTS:** Honeysuckle and snowberry.

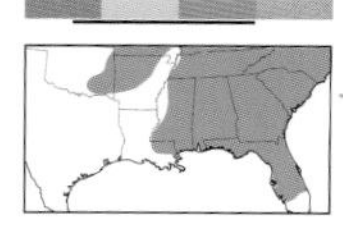

BROWN SCOOPWING
Calledapteryx dryopterata 91-0006 (7653) **Uncommon**

WS 18–22 mm. Orange brown FW has dark edging to scooped-out section of outer margin. AM and PM lines are incomplete. Rectangular patch at midpoint of inner margin is edged blackish. **HOSTS:** Nannyberry and wild-raisin.

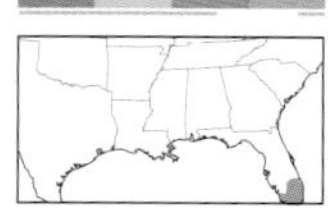

SLOSSON'S SCOOPWING
Philagraula slossoniae 91-0007 (7654) **Local**

WS 23–25 mm. Light brown FW often has dark-edged brown triangles at midpoint of costa and inner margin. Outer margin lacks indentation. HW often has a darker brown median band and small black discal dot. Sometimes rolls FW tightly when resting. **HOSTS:** Unknown.

SMALL SPHINX MOTHS

SCOOPWINGS

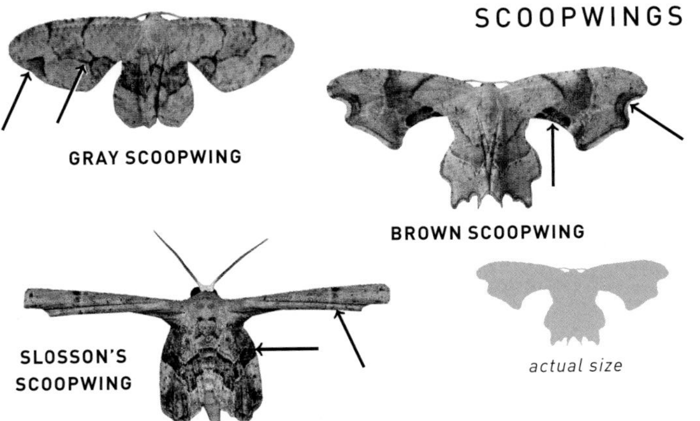

Swallowtail Moths
Family Uraniidae, Subfamily Uraniinae

Large and spectacular moths that resemble swallowtail butterflies. Only one species occurs in the U.S. This common tropical moth undertakes periodic mass migrations every 4–8 years, occasionally reaching our region in late summer and fall. The larval food plant does not grow in North America. Adults are diurnal and can be found taking nectar from flowers (particularly those of leguminous plants) or observed in rapid direct flight.

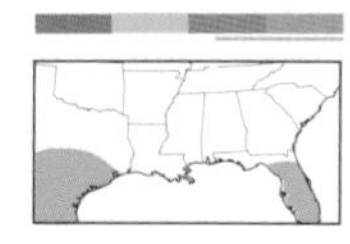

SHINING URANIA *Urania fulgens* 91-0013 (7658) Local

WS 80–90 mm. Unmistakable. Velvety black FW is boldly patterned with shining apple green lines and wide PM band. The green-banded HW has a long white-tipped tail. **HOSTS:** *Omphalea diandra*, a Neotropical liana. **NOTE:** Occurs as an irregular immigrant.

Carpets and Pugs
Family Geometridae, Subfamily Larentiinae

A large group of flimsy, broad-winged moths. Most species adopt a flat position when resting, plastering themselves on tree trunks and branches, where they are extremely cryptic. A few, notably in the genus *Eulithis*, rest with their slender abdomen raised above the level of the wings. Most are woodland species, though some can be found in gardens, even in urban areas. The group is largely nocturnal and will come to lights, though a small number are diurnal and are to be sought along woodland trails or around bogs and fens.

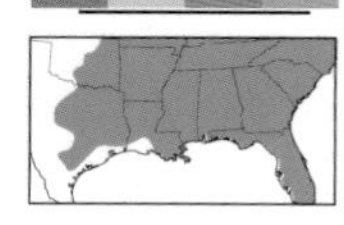

LESSER GRAPEVINE LOOPER

Eulithis diversilineata 91-0031 (7196) Common

WS 28–33 mm. Pale orange FW is crossed by fine brown lines. Inner median area is often tinted lilac. Midpoint of PM line forms a long, outward-pointing spike. Often rests with abdomen curled upward. **HOSTS:** Grape and Virginia creeper. **NOTE:** Two broods.

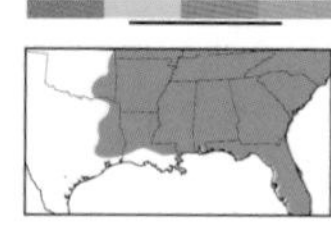

GREATER GRAPEVINE LOOPER

Eulithis gracilineata 91-0032 (7197) Common

WS 35–40 mm. Resembles Lesser Grapevine Looper (many individuals are indistinguishable), but FW has double AM and PM lines that are often filled with brown shading. Median area is uniformly pale. **HOSTS:** Peppervine, grape, hawthorn, and Virginia creeper. **NOTE:** Two broods.

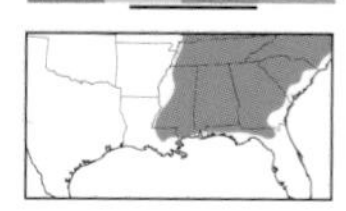

DARK-BANDED GEOMETER

Gandaritis atricolorata 91-0050 (7214) Uncommon

WS 28–32 mm. Dark chocolate brown FW is boldly patterned with strongly angled whitish lines. Longest tooth of AM line usually touches PM line. Irregular whitish ST line is accented with pale veins. **HOSTS:** Unknown.

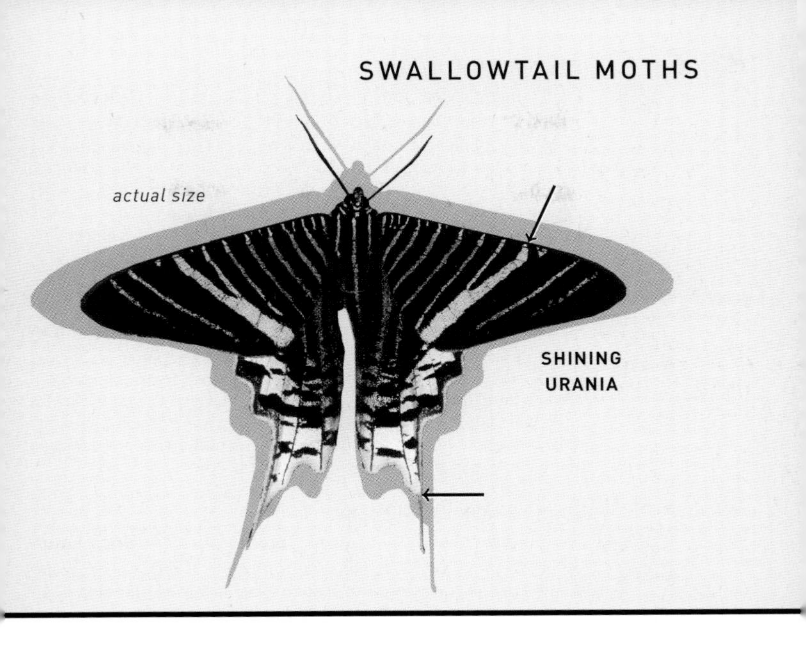
SWALLOWTAIL MOTHS
actual size
SHINING
URANIA

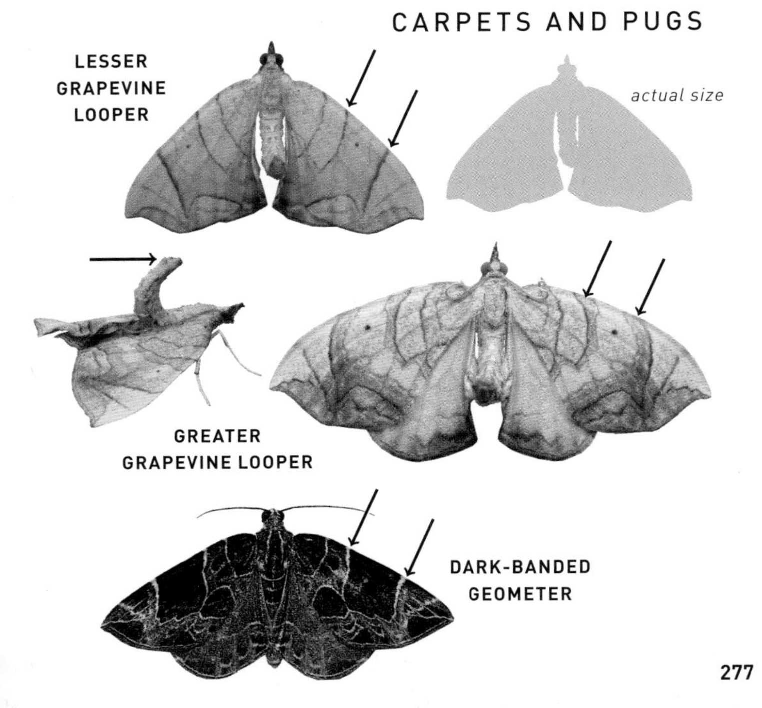
CARPETS AND PUGS
LESSER
GRAPEVINE
LOOPER
actual size
GREATER
GRAPEVINE LOOPER
DARK-BANDED
GEOMETER

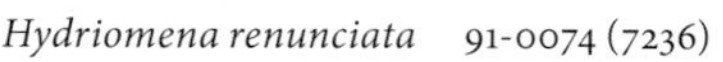

RENOUNCED HYDRIOMENA

Hydriomena renunciata 91-0074 (7236) **Uncommon**

WS 27–30 mm. Pale grayish green FW has many black-bordered bands of varying shades and widths; some may be tinged pink. Pale median band typically is roughly the same width along its length. Paired dots of terminal line are usually distinct. Often has a large, diffuse black patch within ST area at apex. Variation within and overlap among *Hydriomena* species means that most cannot be reliably identified in the field. **HOSTS:** Alder.

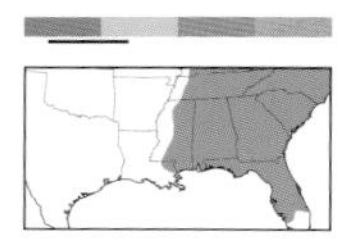

TRANSFIGURED HYDRIOMENA

Hydriomena transfigurata 91-0075 (7237) **Uncommon**

WS 26–33 mm. Difficult to separate from other *Hydriomena* species. Wings are typically narrower than on Renounced Hydriomena. Pale median band gradually widens from inner margin to costa without a noticeable bulge. Paired dots of terminal line are often diffuse. Apex may or may not have diffuse black patch. Sometimes has a dark bar along inner margin of median area. **HOSTS:** Oak.

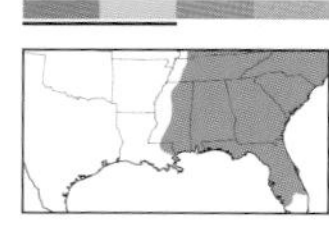

SHARP GREEN HYDRIOMENA

Hydriomena pluviata 91-0077 (7239) **Uncommon**

WS 26–30 mm. Difficult to separate from other *Hydriomena* species. Wings are typically narrower than on Renounced Hydriomena. Pale median band usually bulges in middle. Paired dots of terminal line are distinct. Apex lacks diffuse black patch. Sometimes has a dark bar along inner margin of median area. **HOSTS:** Oak.

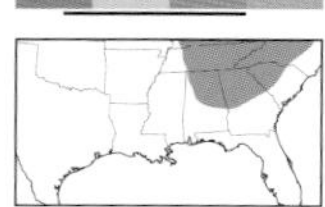

BARBERRY GEOMETER

Coryphista meadii 91-0128 (7290) **Uncommon**

WS 30–36 mm. Variable pale gray or brown FW has a complex pattern of parallel scalloped lines. Outer section of PM line has an outward-pointing tooth. Broad orange bands in basal and ST areas are sometimes present. ST line is edged white near inner margin. HW has a deeply scalloped outer margin. **HOSTS:** Barberry. **NOTE:** Three broods.

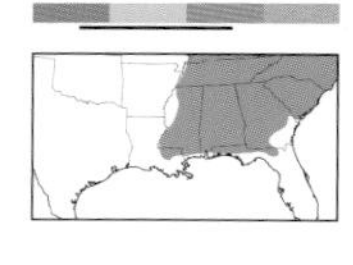

CHERRY SCALLOP SHELL

Rheumaptera prunivorata 91-0130 (7292) **Common**

WS 27–35 mm. Straw-colored wings are overlaid with a dense pattern of undulating lines. A poorly defined paler median band on FW sometimes widens at costa. Jagged white ST line is edged orange. **HOSTS:** Black cherry. **NOTE:** Two broods.

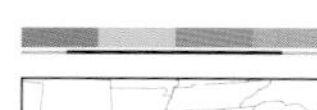

NEW MEXICO CARPET

Archirhoe neomexicana 91-0133 (7295) **Uncommon**

WS 38–42 mm. Light brown FW has many parallel scalloped lines. AM and PM bands are dark brown and join near middle; PM line has a large bulge at midpoint. **HOSTS:** Black cherry. **NOTE:** Two broods.

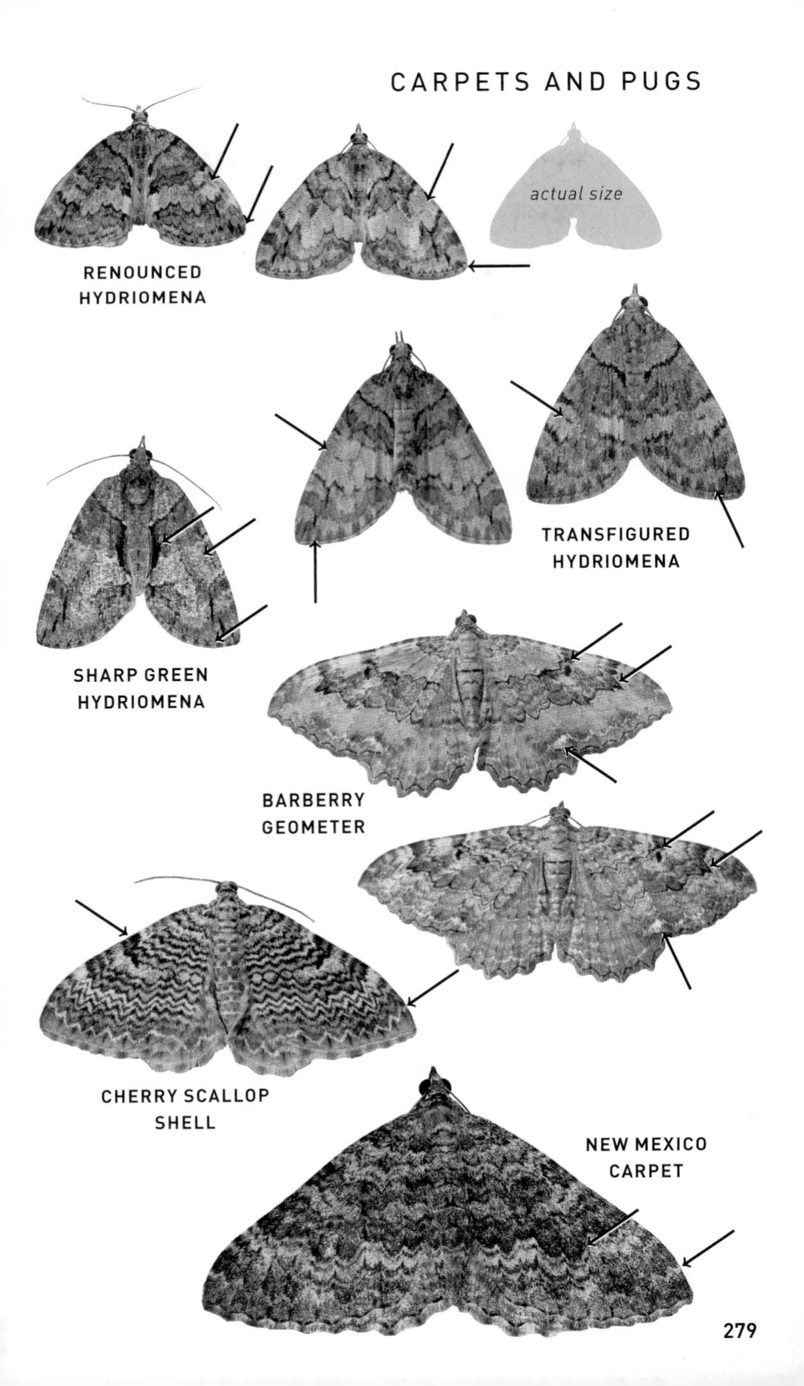
actual size
RENOUNCED HYDRIOMENA
TRANSFIGURED HYDRIOMENA
SHARP GREEN HYDRIOMENA
BARBERRY GEOMETER
CHERRY SCALLOP SHELL
NEW MEXICO CARPET

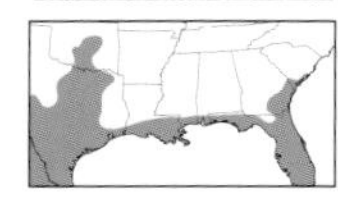

MOSSY CARPET

Hammaptera parinotata 91-0157 (7314) **Local**

WS 18–22 mm. Light brown FW is variably streaked green along costa, through median area, and along ST line. Indistinct blackish lines are sometimes bordered with darker bands, especially near inner margin. Scalloped white ST line has two distinct black dashes near midpoint. **HOSTS:** Unknown.

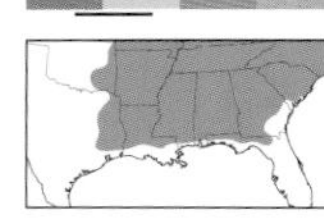

MANY-LINED CARPET

Anticlea multiferata 91-0173 (7330) **Common**

WS 19–25 mm. Dark brown FW is crossed by many parallel, thin yellowish lines, though AM, ST, and terminal lines are white. **HOSTS:** Willow herb and knotweed.

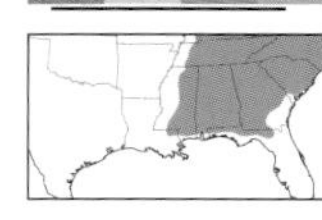

TOOTHED BROWN CARPET

Xanthorhoe lacustrata 91-0234 (7390) **Uncommon**

WS 20–26 mm. Pale gray FW has a warm brown inner basal patch and shading beyond scalloped white ST line. Wide median band is brown with blackish edges and discal spot. Brown subapical patch and ST spots are diffuse. **HOSTS:** Birch, blackberry, hawthorn, and willow. **NOTE:** Two broods.

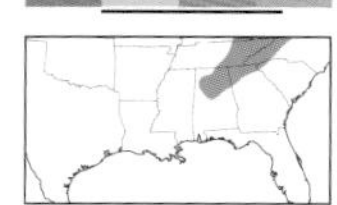

SHARP-ANGLED CARPET

Euphyia intermediata 91-0244 (7399) **Local**

WS 20–27 mm. Resembles Toothed Brown Carpet, but wide dark median band has a distinctly sharp wedge projecting backward from relatively straight PM line. Square blackish subapical patch stands out in otherwise pale ST area. **HOSTS:** Chickweed, elm, impatiens, mustard, and others. **NOTE:** Two broods.

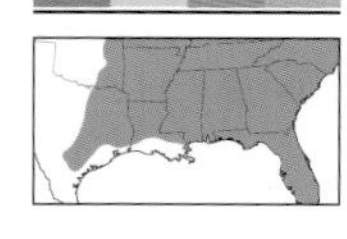

THE GEM *Orthonama obstipata* 91-0258 (7414) **Common**

WS 15–23 mm. Sexually dimorphic. FW of male is light brown with a fragmented dusky median band, blackish apical dash, and dark discal spot. FW of female is maroon with dotted white lines and a white-ringed discal spot. **HOSTS:** Dock, ragwort, and other low plants. **NOTE:** Two or more broods.

BENT-LINE CARPET

Costaconvexa centrostrigaria 91-0260 (7416) **Common**

WS 17–23 mm. Sexually dimorphic. FW of male is pale gray with a blackish AM line and outer section of PM line. Female is similar but has a darker gray median band. Both sexes show a black discal dot. **HOSTS:** Knotweed, smartweed, and other low plants. **NOTE:** Two or more broods.

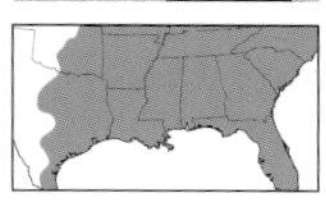

SOMBER CARPET

Disclisioprocta stellata 91-0261 (7417) **Uncommon**

WS 25–33 mm. Straw-colored to dark brown wings are densely patterned with incomplete white edges to scalloped lines. Median area is often slightly darker. Midpoint of PM line has two rounded teeth within the bulge. **HOSTS:** Amaranth and devil's claw.

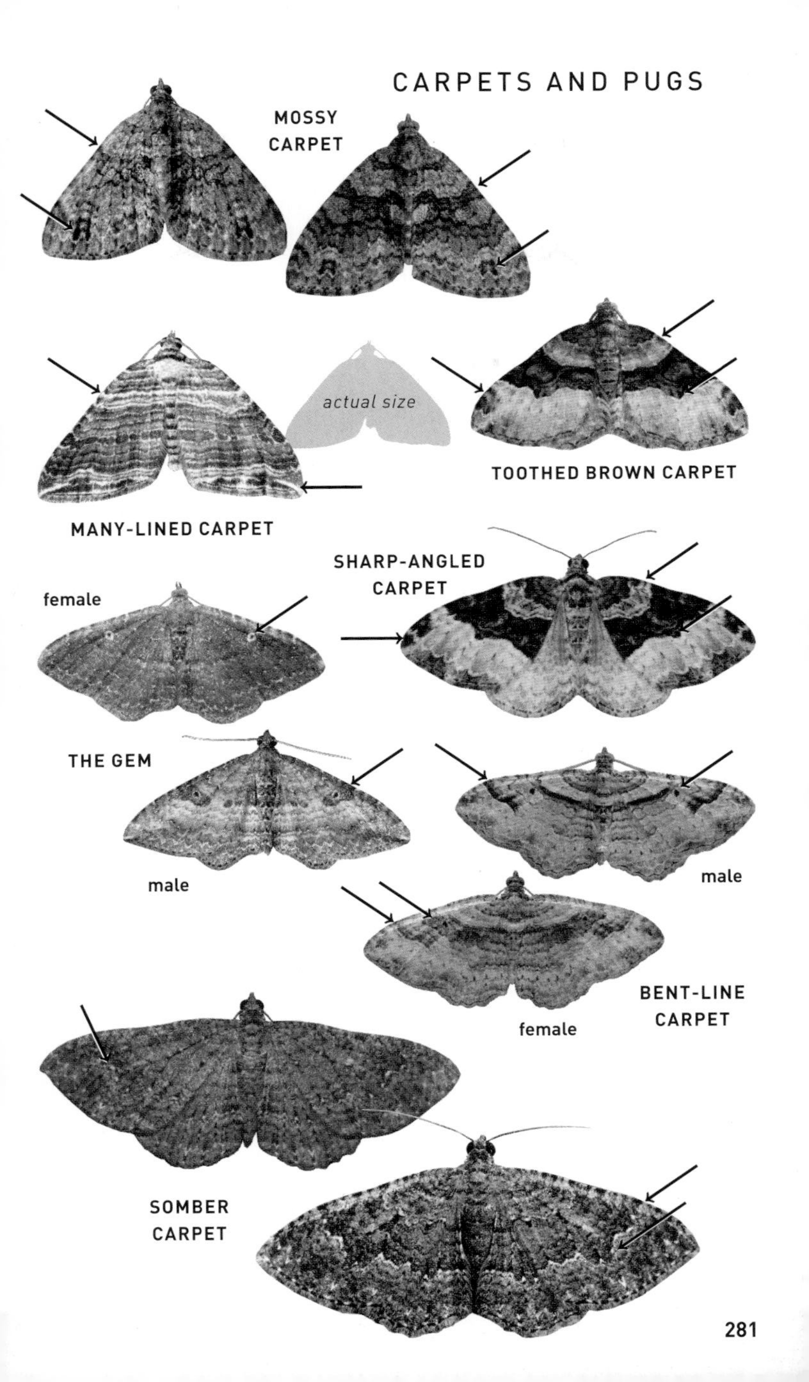
MOSSY
CARPET
actual size
TOOTHED BROWN CARPET
MANY-LINED CARPET
SHARP-ANGLED
CARPET
female
THE GEM
male
male
BENT-LINE
CARPET
female
SOMBER
CARPET

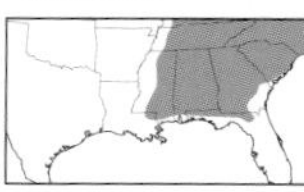

BROWN-SHADED CARPET

Venusia comptaria 91-0272 (7428) **Uncommon**

WS 17–22 mm. Powdery white FW has a faint pattern of wavy brown-edged lines. A patch of brown shading surrounds short black dashes at midpoint of PM line. Terminal line is checkered. **HOSTS:** Alder, beech, and birch.

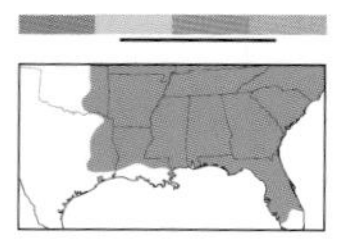

THE BEGGAR *Eubaphe mendica* 91-0286 (7440) **Common**

WS 21–30 mm. Semi-translucent yellow FW has wide purplish gray AM and PM bands fragmented into spots by pale veins. Basal section of costa is suffused orange. **HOSTS:** Violet. **NOTE:** Two or more broods.

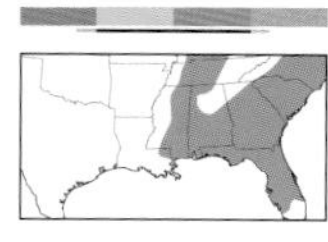

THE LITTLE BEGGAR

Eubaphe meridiana 91-0287 (7441) **Uncommon**

WS 18–25 mm. Resembles The Beggar but is smaller and has deeper orange FW with smaller spots within purplish gray AM and PM bands. **HOSTS:** Unknown.

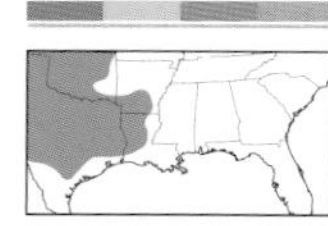

ORANGE BEGGAR

Eubaphe unicolor 91-0291 (7444) **Local**

WS 21–26 mm. Uniformly orange wings and body. Slightly paler veins on wings are the only hint of a pattern. Antennae and legs are contrastingly blackish. **HOSTS:** Violet.

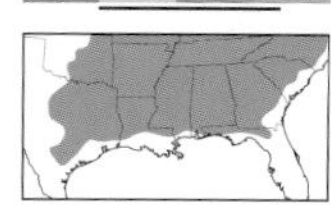

BROWN BARK CARPET

Horisme intestinata 91-0292 (7445) **Common**

WS 21–33 mm. Light brown FW is marked with fine parallel lines. PM and ST lines are deeply scalloped. Broad pale costal streak is almost unmarked. **HOSTS:** Clematis. **NOTE:** Two or more broods.

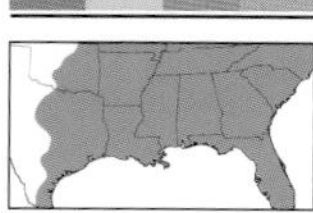

COMMON PUG *Eupithecia miserulata* 91-0324 (7474) **Common**

WS 12–20 mm. Pale gray FW is patterned with numerous faint lines. Black discal spot is conspicuous. Inner ST line is marked with a white spot. Last few segments of abdomen are usually whitish. **HOSTS:** Aster, Canadian horseweed, fleabane, grape, oak, willow, and many others.

SWIFT PUG

Eupithecia jejunata 91-0334 (7486) **Local**

WS 12–20 mm. Gray FW is marked with numerous crisp blackish lines. Blackish discal spot is inconspicuous or absent. Whitish ST line is shallowly scalloped. **HOSTS:** Unknown.

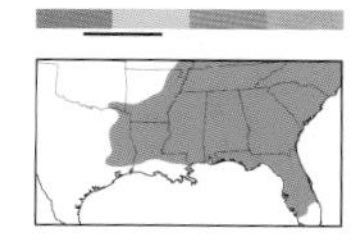

OLIVE-AND-BLACK CARPET

Acasis viridata 91-0476 (7635) **Local**

WS 18–20 mm. Grayish FW has mottled dusky bands broken by black and white dashes along veins. Fresh individuals are streaked mossy green along veins. **HOSTS:** Viburnum.

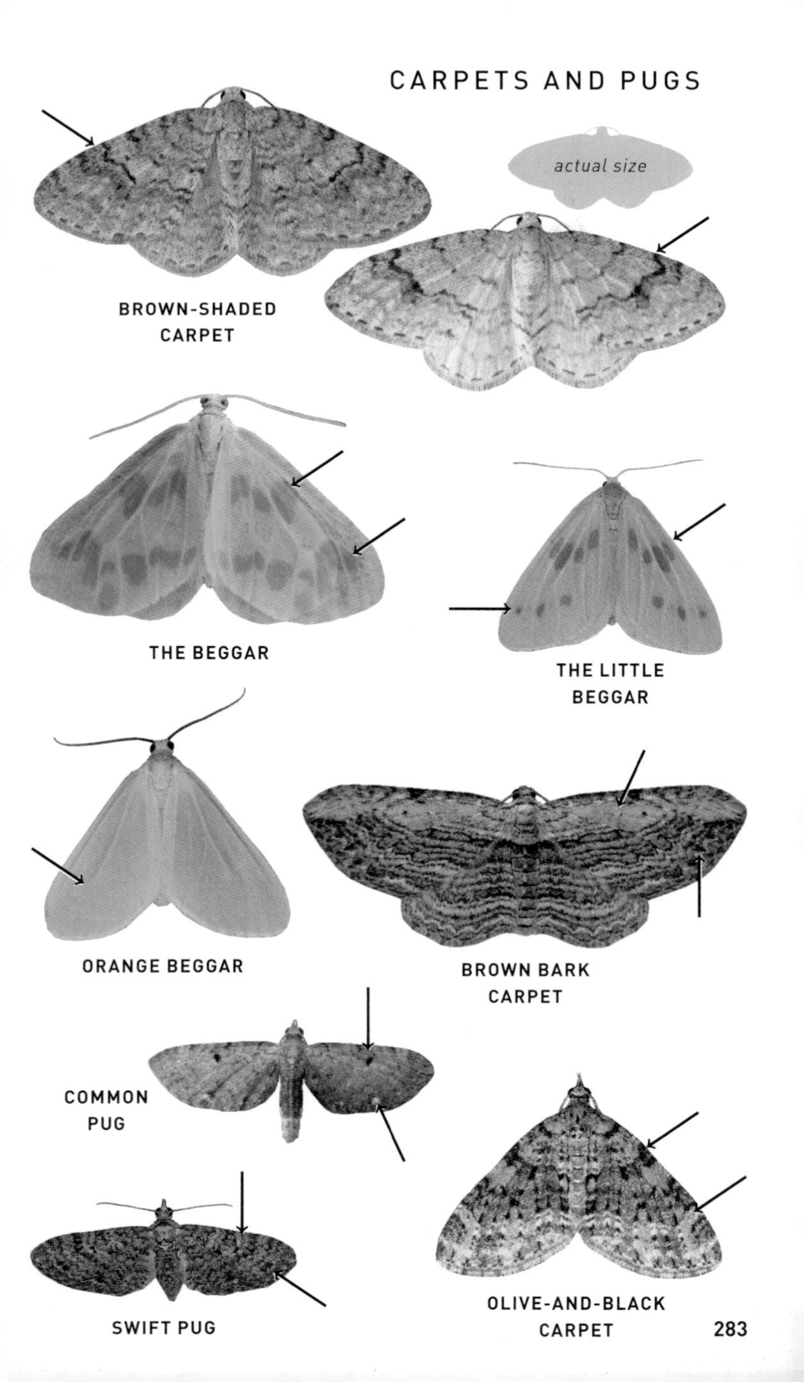
actual size
BROWN-SHADED CARPET
THE BEGGAR
THE LITTLE BEGGAR
ORANGE BEGGAR
BROWN BARK CARPET
COMMON PUG
SWIFT PUG
OLIVE-AND-BLACK CARPET

MOTTLED GRAY CARPET

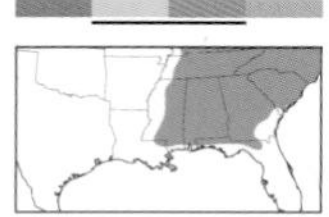

Cladara limitaria 91-0478 (7637) **Common**

WS 21–28 mm. Pale gray FW has wavy lines often edged with fragmented bands of brown, yellow, or green shading. Bulging PM line is broadly edged dark brown. **HOSTS:** Coniferous trees.

THREE-SPOTTED FILLIP

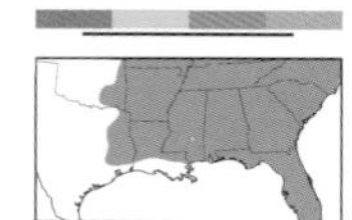

Heterophleps triguttaria 91-0488 (7647) **Local**

WS 18–24 mm. Pale yellowish brown FW is marked with faint AM and PM lines that end as two large chocolate brown patches along costa. A smaller blackish spot usually marks subapical area. **HOSTS:** Maple.

THE BAD-WING

Dyspteris abortivaria 91-0489 (7648) **Common**

WS 20–29 mm. Bluish green FW has faint whitish AM and PM lines that fade before reaching costa. A small white discal spot is noticeable. Rounded HW is much smaller than long FW. **HOSTS:** Grape.

WAVES Family Geometridae, Subfamily Sterrhinae

Small to medium-sized geometrid moths that are often pale or whitish with contrasting darker lines, discal spots, and bands of shading. The males of some species have bipectinate antennae. Most are nocturnal and will come to lights. A few, such as Chickweed Geometer and Cross-lined Wave, are active in daytime.

DRAB BROWN WAVE

Lobocleta ossularia 91-0500 (7094) **Common**

WS 13–19 mm. Pale tan or brown wings are marked with peppery blackish lines. Median line on FW is kinked near inner margin. Terminal line is usually dotted. Small discal dots are black. **HOSTS:** Chickweed, bedstraw, clover, and strawberry. **NOTE:** Two or more broods.

STRAIGHT-LINED WAVE

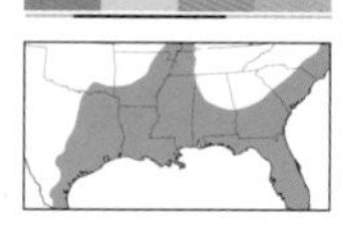

Lobocleta plemyraria 91-0503 (7097) **Common**

WS 14–20 mm. Peppery straw-colored wings are marked with relatively straight faint brown lines. Terminal line is blackish. Black discal dots are usually present. Apex of FW is sharply pointed. **HOSTS:** Unknown.

SPECKLED WAVE

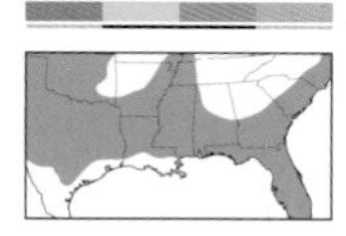

Lobocleta peralbata 91-0506 (7100) **Common**

WS 11–20 mm. Powdery white wings have wavy yellowish brown AM, PM, and ST lines and are variably speckled with dark brown scales. Terminal line is a row of black dashes, not always visible. Tiny black discal dots mark all wings. **HOSTS:** Unknown.

CARPETS AND PUGS

MOTTLED GRAY CARPET

THREE-SPOTTED FILLIP

THE BAD-WING

WAVES

DRAB BROWN WAVE

STRAIGHT-LINED WAVE

SPECKLED WAVE

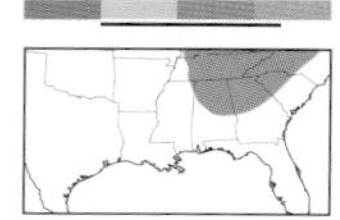

FORTUNATE WAVE *Idaea bonifata* 91-0508 (7102) **Local**

WS 13–15 mm. Shiny straw-colored wings are patterned with parallel brown lines. Orange fringe is dotted brown. Dark discal spots are often diffuse and blend in to median line. **HOSTS:** Decaying leaves and stored grains.

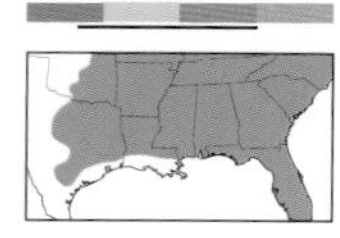

DIMINUTIVE WAVE

Idaea scintillularia 91-0511 (7105) **Common**

WS 11 mm. Silvery gray FW is mostly pale yellow beyond wavy PM line. Pale yellow HW has bold black AM and chestnut PM lines. All wings bear elongated black discal spots. **HOSTS:** Unknown.

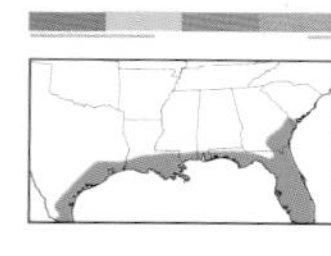

CHESTNUT-BANDED WAVE

Idaea pervertipennis 91-0513 (7107) **Local**

WS 11 mm. Shiny straw-colored wings have chestnut shading in basal area and beyond wavy PM line. Terminal line is a row of black dashes. All wings are marked with black discal dots. **HOSTS:** Unknown.

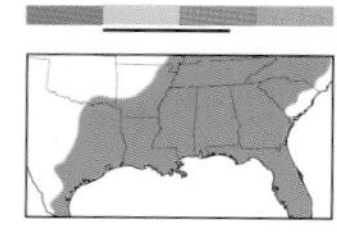

NOTCH-WINGED WAVE

Idaea furciferata 91-0514 (7108) **Common**

WS 14–18 mm. Shiny straw-colored FW is somewhat rectangular, marked with dark wavy AM and PM lines. Inner margin and ST area are broadly banded chestnut. HW of male has a distinctive notch in outer margin that when folded creates a tuft beside abdomen. **HOSTS:** Unknown, but has been reared on clover and dandelion.

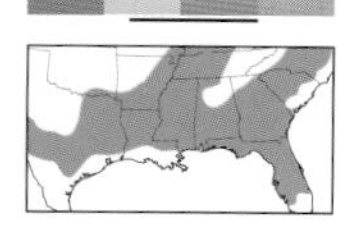

CELTIC WAVE *Idaea celtima* 91-0515 (7109) **Local**

WS 12–13 mm. Peppery yellowish brown wings have a broad chestnut band beyond wavy, dark PM line. AM and median lines, when present, are diffuse. Terminal line is a row of black dashes. **HOSTS:** Unknown.

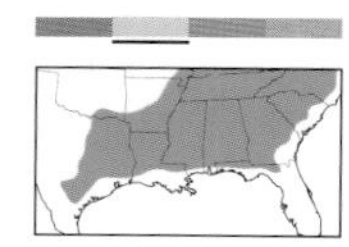

OCHRE WAVE

Idaea productata 91-0521 (7112) **Local**

WS 14–19 mm. Speckled, straw-colored wings have faint AM and median lines, and a stronger, scalloped PM line. Dark discal dots mark all wings. **HOSTS:** Unknown.

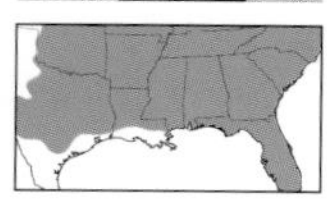

RED-BORDERED WAVE

Idaea demissaria 91-0523 (7114) **Common**

WS 14–19 mm. Shiny straw-colored or pale orange wings have reddish shading in either or both basal area and beyond PM line. FW apex is conspicuously pale on darker individuals. Slightly scalloped lines are reddish brown. **HOSTS:** Unknown.

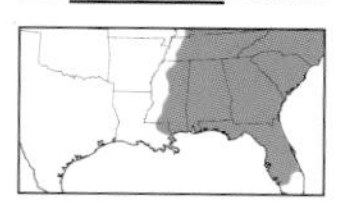

STRAW WAVE *Idaea eremiata* 91-0524 (7115) **Common**

WS 14–19 mm. Uniformly straw-colored wings are often unmarked except for a slightly wavy chestnut PM line. Diffuse median band is sometimes present, as are tiny dark discal dots and narrow terminal line. **HOSTS:** Plant detritus, including dead leaves.

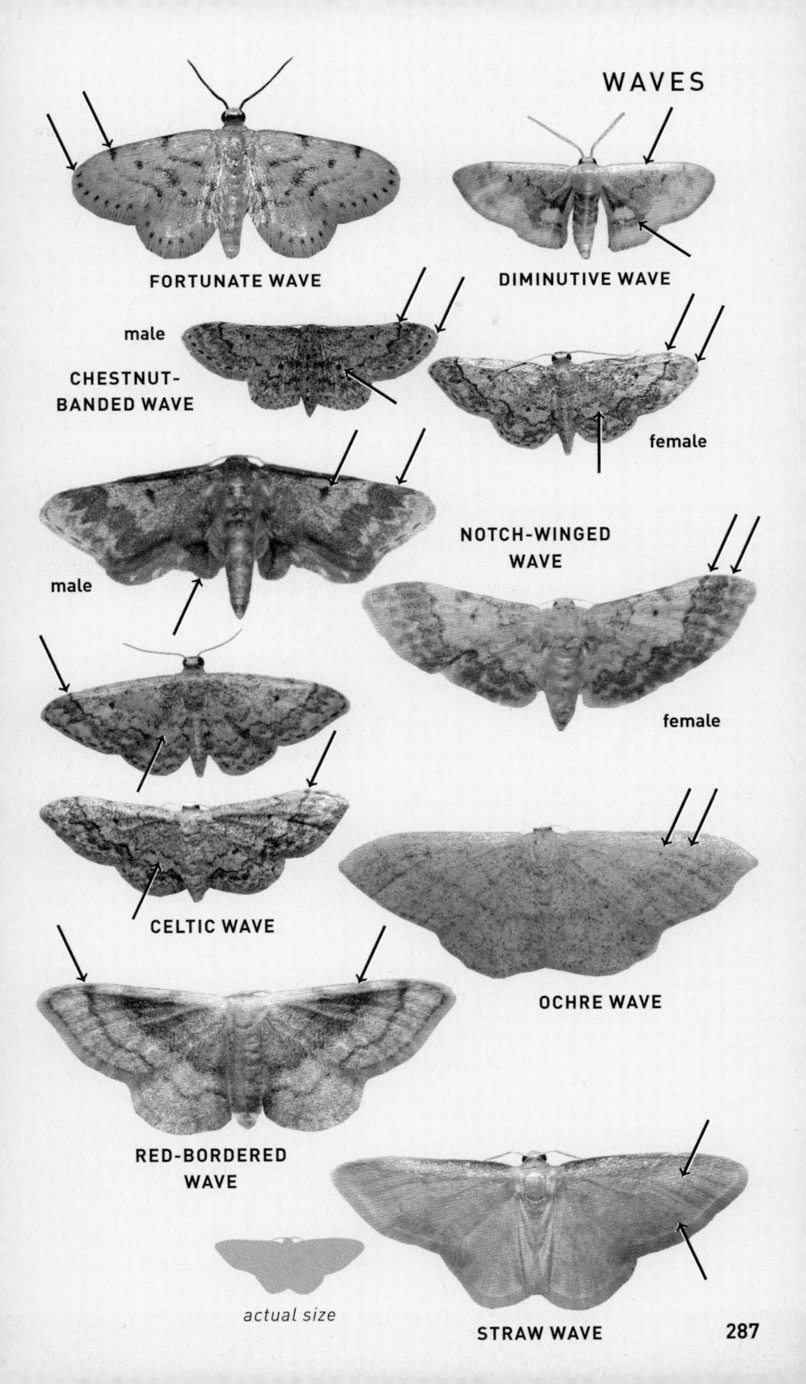
FORTUNATE WAVE
DIMINUTIVE WAVE
male
CHESTNUT-
BANDED WAVE
female
male
NOTCH-WINGED
WAVE
female
CELTIC WAVE
OCHRE WAVE
RED-BORDERED
WAVE
actual size
STRAW WAVE

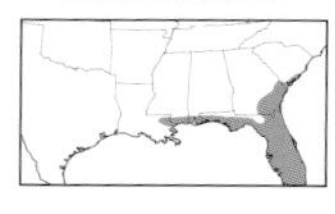

HILL'S WAVE

Idaea hilliata 91-0527 (7118) **Local**

WS 14–19 mm. Light brown wings have a distinctive, meandering black median line. ST area is often shaded darker brown. Terminal line is a row of tiny blackish dots. **HOSTS:** Unknown.

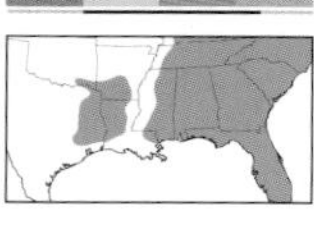

VIOLET WAVE

Idaea violacearia 91-0529 (7120) **Common**

WS 17 mm. Peppery light brown wings are marked with jagged dusky lines that vary in intensity, sometimes appearing dotted. ST area is sometimes tinged warm brown. **HOSTS:** Unknown.

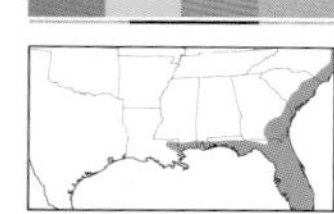

SHOWY WAVE *Idaea ostentaria* 91-0530 (7121) **Local**

WS 16 mm. Peppery whitish wings are marked with scalloped lines that often appear faint or dotted. PM line is often boldest line. Bold black discal dots mark all wings. Terminal line is a row of black dashes. **HOSTS:** Unknown.

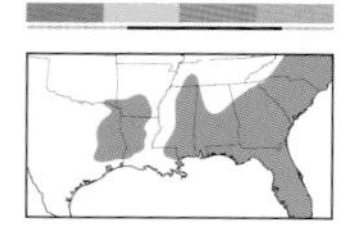

DOT-LINED WAVE *Idaea tacturata* 91-0531 (7122) **Common**

WS 13–21 mm. White wings are marked with finely etched rusty brown zigzag lines. Tiny black discal spots are often lost among brown peppering. ST and terminal lines appear as rows of dusky dashes. **HOSTS:** Unknown, but has been reared on clover.

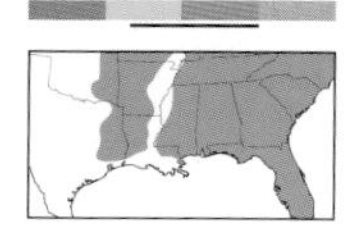

RIPPLED WAVE *Idaea obfusaria* 91-0532 (7123) **Common**

WS 20 mm. White wings are boldly patterned with wavy yellowish brown lines. PM line is often darker. Basal area is often finely speckled brown. Black discal dots are present. Terminal line is a row of black dots when fresh. **HOSTS:** Unknown.

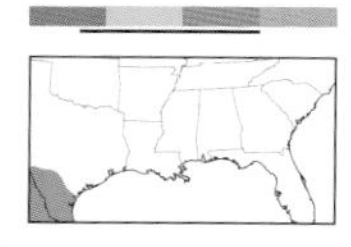

BEAUTIFUL WAVE *Idaea asceta* 91-0536 (7117.1) **Local**

WS 12–14 mm. Shiny pale straw FW is tinged brown along costa. Thick black edging to outer ST line contains metallic blue and chestnut spots. HW is dark in outer half. Thorax is mostly yellow brown. Abdomen is all dark with a narrow white ring and brown tip. **HOSTS:** Unknown.

ORBED WAVE *Odontoptila obrimo* 91-0538 (7130) **Local**

WS 17–24 mm. Wings are strongly banded with pale cream and light orange. Darker median line is straight. Wavy white ST line is bordered above by darker orange. Often rests with wings slightly creased. **HOSTS:** Unknown.

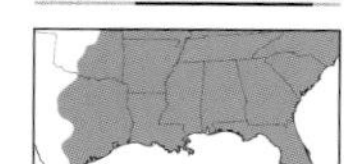

COMMON TAN WAVE

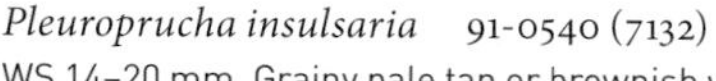

Pleuroprucha insulsaria 91-0540 (7132) **Common**

WS 14–20 mm. Grainy pale tan or brownish wings are faintly marked with scalloped yellow-edged lines. Diffuse, dark median line is often most noticeable marking. Pale ST line is laced with tiny black dots. **HOSTS:** Aster, bedstraw, chestnut, corn, goldenrod, oak, and many other plants. **NOTE:** Two or more broods.

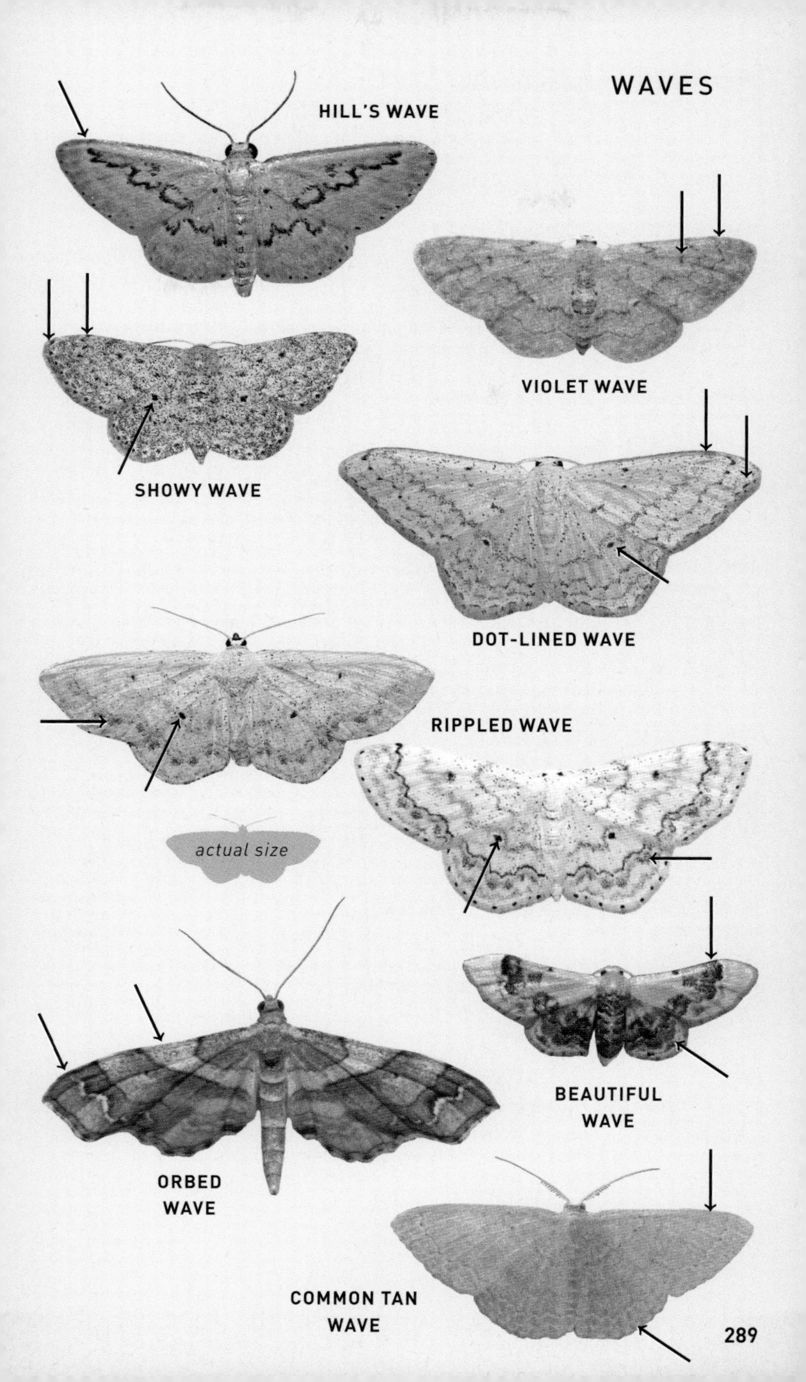

WAVES
HILL'S WAVE
VIOLET WAVE
SHOWY WAVE
DOT-LINED WAVE
RIPPLED WAVE
actual size
BEAUTIFUL WAVE
ORBED WAVE
COMMON TAN WAVE

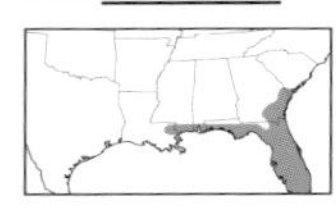

ASTHENE WAVE

Pleuroprucha asthenaria 91-0541 (7133) **Local**

WS 14–18 mm. Similar to Common Tan Wave, but wings are dull olive green. Thin, whitish ST line is often most noticeable marking. **HOSTS:** Mango and other plants.

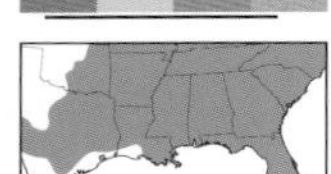

PACKARD'S WAVE

Cyclophora packardi 91-0544 (7136) **Common**

WS 17–23 mm. Pale orange wings are densely peppered reddish. Indistinct AM and PM lines appear dotted. Sometimes has a smudgy dark median band. All wings are marked with black-ringed white discal spots. **HOSTS:** Unknown; possibly oak or sweet fern.

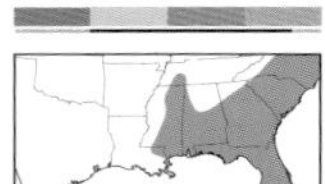

WAXMYRTLE WAVE

Cyclophora myrtaria 91-0545 (7137) **Common**

WS 24–27 mm. Similar to smaller Packard's Wave, but wings are brownish and more densely peppered. All lines are indistinct, often appearing faintly dotted. Black-ringed white discal spots are slightly smaller and less prominent. **HOSTS:** Sweet fern and wax myrtle.

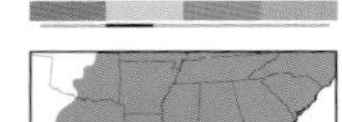

DWARF TAWNY WAVE

Cyclophora nanaria 91-0548 (7140) **Common**

WS 16 mm. Light brown wings are densely peppered darker brown. Blackish AM line has a sharp tooth near costa. Dark median band is often well defined. All wings bear black-ringed white discal spots. **HOSTS:** Unknown.

GOLDEN WAVE

Semaeopus ella 91-0549 (7141) **Local**

WS 25–30 mm. Golden yellow (rarely pale) wings have faint AM and PM lines. ST area sometimes looks lacy with brown blotches. Black rings of discal spots are often so thick that spots appear solid. **HOSTS:** Unknown.

CHICKWEED GEOMETER

Haematopis grataria 91-0554 (7146) **Common**

WS 18–26 mm. Yellow wings have bold pink median and ST lines, and fringe. FW has pink discal dot. Male has bipectinate antennae. **HOSTS:** Chickweed, clover, and other low plants. **NOTE:** Multiple broods.

CROSS-LINED WAVE

Timandra amaturaria 91-0555 (7147) **Common**

WS 20–28 mm. Yellowish wings are finely speckled with brown. Thick reddish PM line crosses thinner ST line near pointed apex of FW. HW is sharply angulate. **HOSTS:** Buckwheat, dock, and knotweed. **NOTE:** Two or more broods.

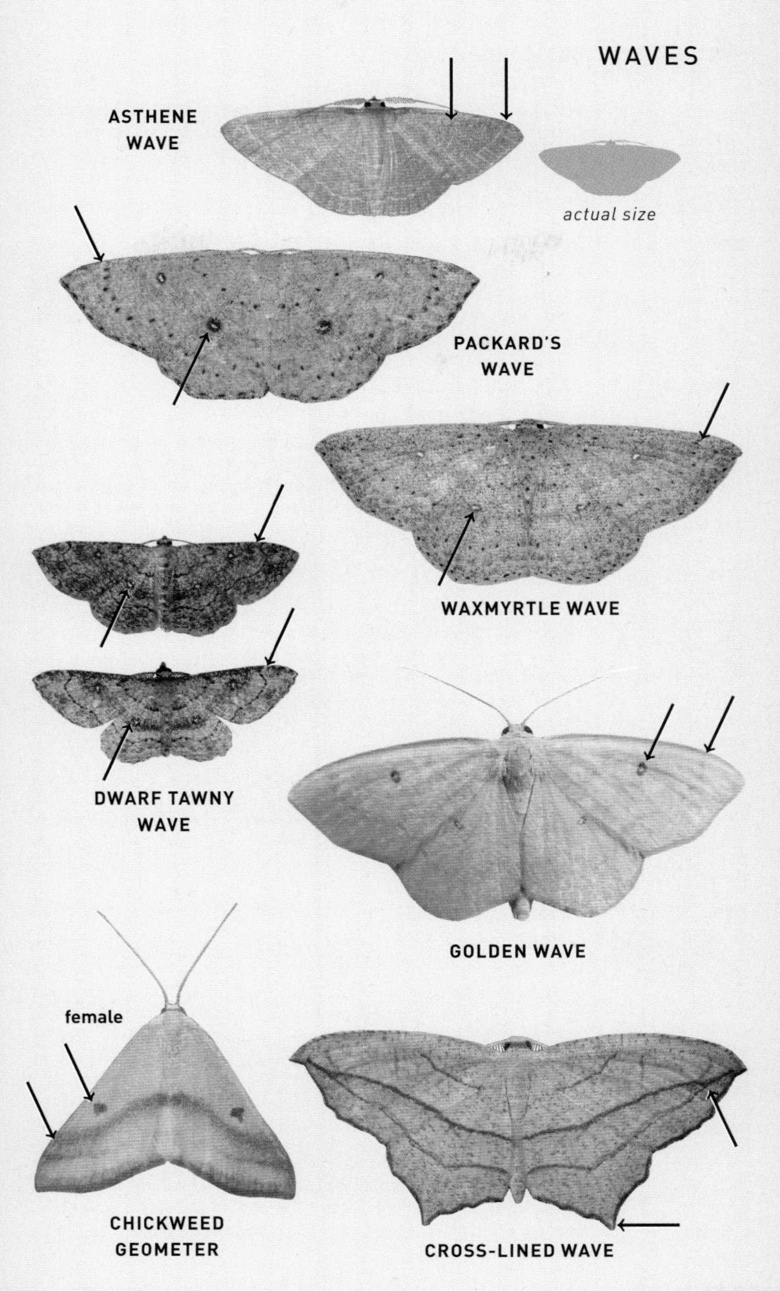
ASTHENE WAVE
actual size
PACKARD'S WAVE
WAXMYRTLE WAVE
DWARF TAWNY WAVE
GOLDEN WAVE
female
CHICKWEED GEOMETER
CROSS-LINED WAVE

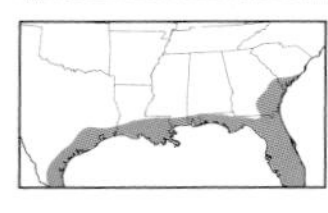

SMALL FROSTED WAVE

Scopula lautaria 91-0557 (7149) **Common**

WS 11–16 mm. Shiny white wings are almost unmarked apart from dark brown shading beyond wavy PM line. Note two blackish blotches at midpoint and inner section of PM line on FW. All wings have tiny black discal dots. **HOSTS:** Unknown, but has been reared on clover.

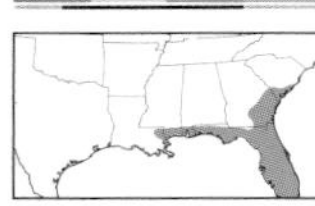

RIVAL WAVE *Scopula aemulata* 91-0559 (7151) **Common**

WS 15 mm. Peppery light brown wings are marked with diffuse brown lines. PM and terminal lines are accented with black dots. All wings have black discal dots. Abdomen has a row of black dorsal spots. **HOSTS:** Beggar tick.

DARK-MARKED WAVE

Scopula compensata 91-0560 (7152) **Common**

WS 15 mm. Peppery whitish or light brown wings are marked with indistinct wavy brown lines that often darken at costa. Some individuals show variably dark blotches at midpoint and inner section of PM line on FW. Small black discal dots are present. HW is bluntly angulate. **HOSTS:** Unknown.

SWAG-LINED WAVE

Scopula umbilicata 91-0564 (7156) **Local**

WS 18 mm. Whitish or straw-colored wings are marked with wavy brown lines. Wings have black discal dots, and a small black spot is present at apex of FW. Terminal line is usually visible as tiny black dots. Fringe is sometimes brownish. **HOSTS:** Unknown.

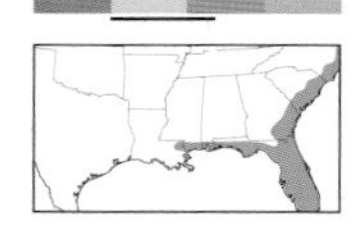

CHALKY WAVE *Scopula purata* 91-0566 (7158) **Local**

WS 17–22 mm. White wings have blotchy blackish shading beyond wavy PM line, which fades before reaching costa. Terminal line is a small chevron. Tiny black discal dots are present. White abdomen has a row of black dorsal spots. **HOSTS:** Unknown, but has been reared on dandelion.

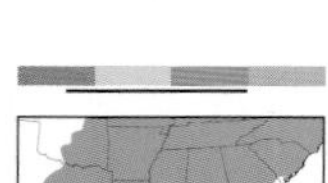

LARGE LACE-BORDER

Scopula limboundata 91-0567 (7159) **Common**

WS 20–31 mm. Whitish or straw-colored wings are marked with wavy light brown lines. ST area is usually filled with either light brown or blackish mottling, creating a lacy effect. Tiny black discal dots mark all wings. **HOSTS:** Apple, bedstraw, blueberry, cherry, dandelion, sweet pepperbush, and many others. **NOTE:** Three or more broods.

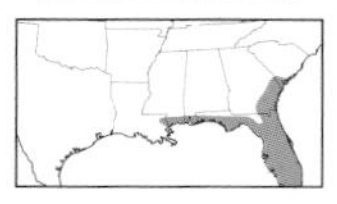

BLACK-PATCHED WAVE

Scopula timandrata 91-0568 (7160) **Local**

WS 24–27 mm. Pale yellowish wings are marked with wavy soft brown lines. Has blackish patches beyond inner section of PM line, and tiny black discal dots on all wings. **HOSTS:** Unknown.

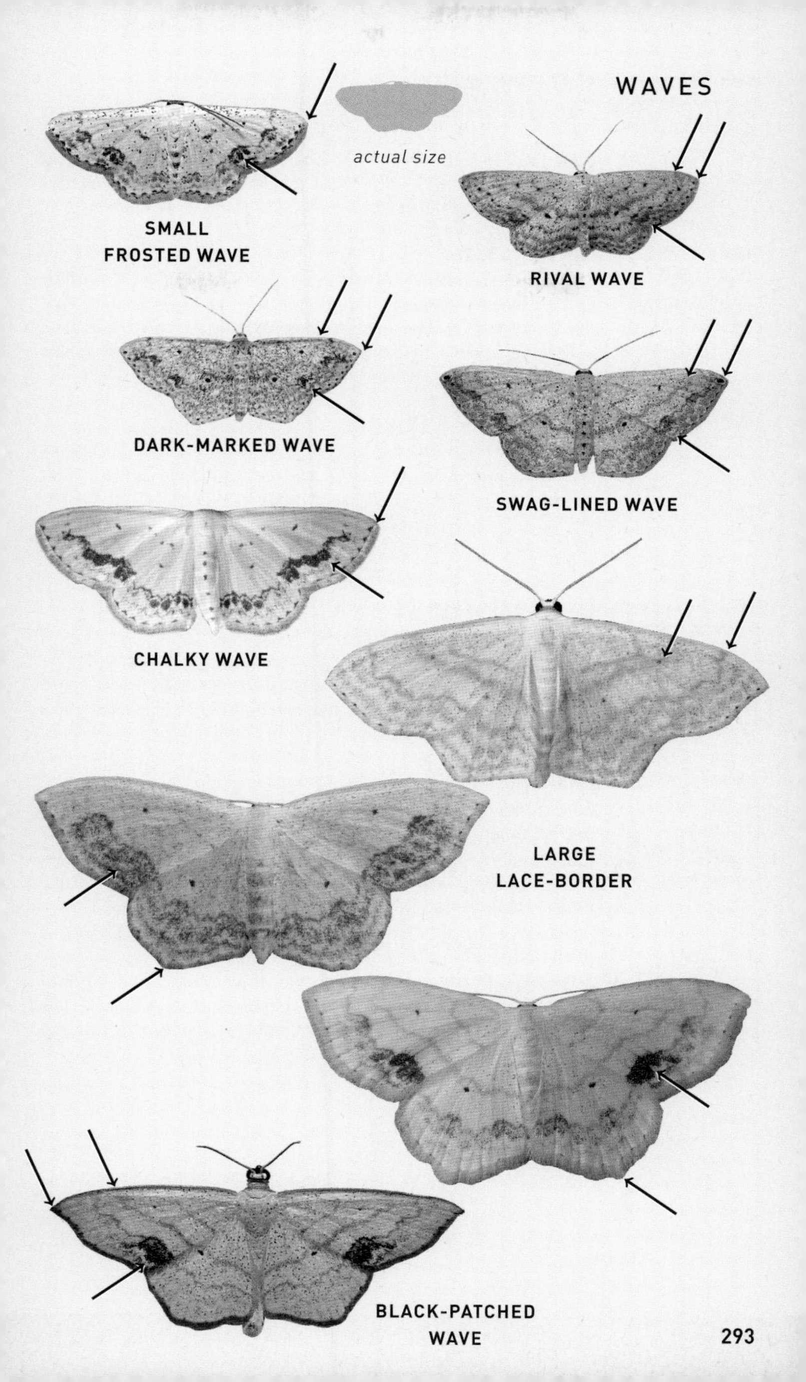
WAVES
actual size
SMALL FROSTED WAVE
RIVAL WAVE
DARK-MARKED WAVE
SWAG-LINED WAVE
CHALKY WAVE
LARGE LACE-BORDER
BLACK-PATCHED WAVE

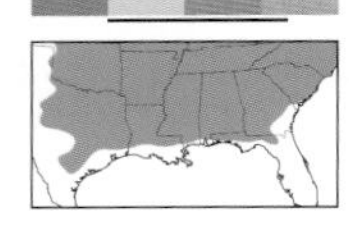

SOFT-LINED WAVE

Scopula inductata 91-0578 (7169) **Common**

WS 17–24 mm. Pale tan wings are lightly peppered brown. Diffuse brownish AM and median lines are relatively straight; PM line is slightly wavy. Has tiny black discal dots on all wings. **HOSTS:** Aster, clover, dandelion, ragweed, sweet clover, and other low plants.

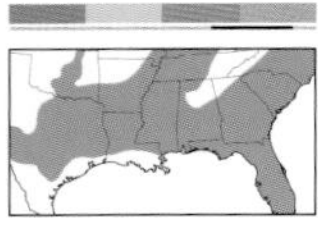

PANNARIA WAVE

Leptostales pannaria 91-0583 (7173) **Common**

WS 14–16 mm. Peppery reddish or (rarely) grayish wings are marked with wavy yellow-edged lines. Costa is yellowish on redder individuals, reddish on paler ones. Fringe is always reddish. Abdomen has a row of yellow dorsal spots. **HOSTS:** Unknown.

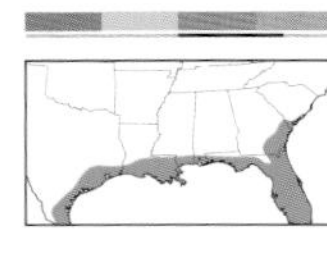

CROSS'S WAVE

Leptostales crossii 91-0584 (7174) **Local**

WS 13–14 mm. Brick red wings are variably patterned with wavy yellow-edged lines. Broad yellow costa continues across thorax. Terminal line is dotted yellow. Reddish abdomen has a row of yellow dorsal bands, sometimes thickening into spots in center. **HOSTS:** Unknown.

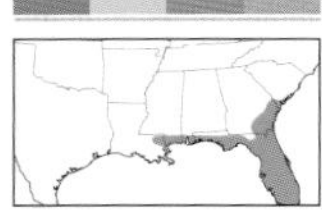

LEMON-BORDERED WAVE

Leptostales hepaticaria 91-0585 (7175) **Local**

WS 13–15 mm. Similar to Cross's Wave, but wings are more orange, especially in median area. Wavy yellow lines on FW blend into yellow costal streak. PM and ST lines are sometimes filled in to form a yellow band. **HOSTS:** Unknown.

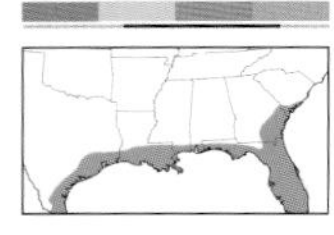

RASPBERRY WAVE

Leptostales laevitaria 91-0587 (7177) **Common**

WS 16–18 mm. Pale yellow wings are brightly patterned with pink lines, broadest in median and ST areas. Note pink costa. Often rests with HW mostly hidden under FW. **HOSTS:** Unknown, but usually found in pine forest.

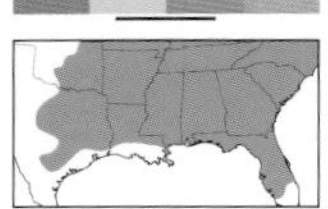

LIGHT-RIBBONED WAVE

Leptostales ferruminaria 91-0590 (7180) **Common**

WS 15–20 mm. Light brown or yellowish wings are marked with two broad bands (one maroon, one orange) that shade area between wavy blackish AM, median, and PM lines. Terminal line is a row of black dashes. **HOSTS:** Unknown.

STAINED LOPHOSIS

Lophosis labeculata 91-0591 (7181) **Common**

WS 12–14 mm. Sexually dimorphic. In male, glossy purple wings have an irregular, sharply defined yellow terminal line and a yellow discal spot on FW. Female has yellow wings marked with a broad purple band below median line and purple in basal area; central portion of yellow body is shaded purple. **HOSTS:** Unknown.

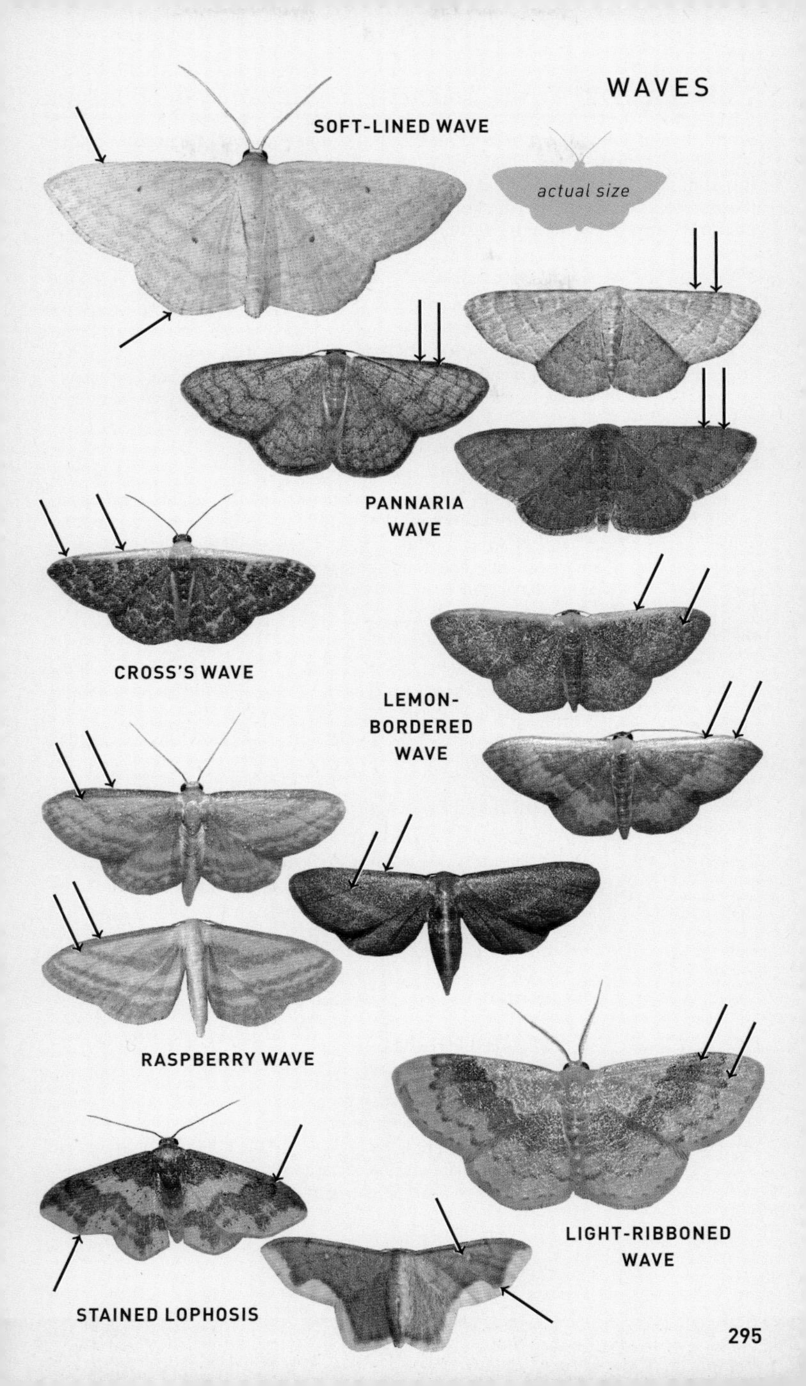
WAVES
SOFT-LINED WAVE
actual size
PANNARIA
WAVE
CROSS'S WAVE
LEMON-
BORDERED
WAVE
RASPBERRY WAVE
LIGHT-RIBBONED
WAVE
STAINED LOPHOSIS

EMERALDS Family Geometridae, Subfamily Geometrinae

Small, broad-winged geometrid moths that are predominantly green, marked with white or pale lines. Some have spots or stripes on the dorsal surface of the abdomen. Tiny dark discal spots are present in about half of species. Females are usually slightly larger than males; males of some species have bipectinate antennae. All are nocturnal and will come to lights in small numbers.

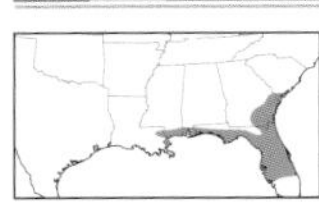

EXTREME EMERALD

Nemoria extremaria 91-0608 (7028) Local

WS 16–20 mm. Bright green wings are crossed by very faint, almost straight whitish AM and PM lines. Discal dots are present. Fringe is pinkish red when fresh. Green abdomen is unmarked. Forehead is white; face and collar are reddish. **HOSTS:** Unknown.

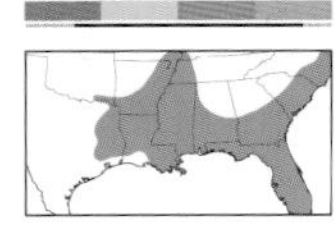

CYPRESS EMERALD

Nemoria elfa 91-0609 (7029) Common

WS 20–26 mm. Green or sometimes brown wings are marked with slightly wrinkly whitish AM and PM lines. Discal dots are present. Fringe is whitish when fresh. Green or brown abdomen is unmarked. Face and collar are same color as body. **HOSTS:** Sweet gum. **NOTE:** Two or more broods.

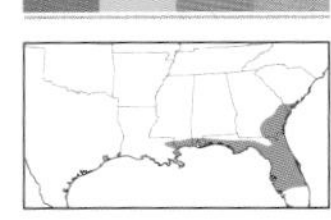

FLORIDA EMERALD

Nemoria catachloa 91-0611 (7031) Local

WS 19–26 mm. Very similar to more common Red-bordered Emerald, but AM and PM lines are more strongly zigzagged, and basal portion of costa usually is washed red. **HOSTS:** Unknown. **NOTE:** Multiple broods.

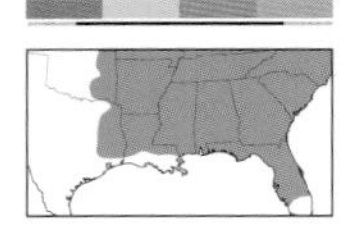

RED-BORDERED EMERALD

Nemoria lixaria 91-0613 (7033) Common

WS 22–32 mm. Green wings are marked with slightly wobbly whitish AM and PM lines. Costa is normally creamy, rarely reddish. Terminal line is red, and pale fringe is boldly checkered red and white. Abdomen has a dorsal line of red-ringed white spots. Discal dots are present. A darker winter form in south has brownish green wings marked with white-edged blackish lines and fringe. **HOSTS:** Oak. **NOTE:** Three or more broods.

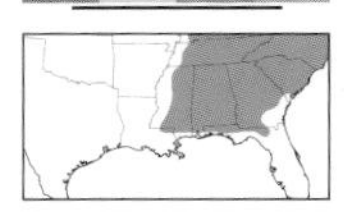

RED-SPOTTED EMERALD

Nemoria saturiba 91-0614 (7034) Common

WS 22–30 mm. Bright green wings are marked with wiggly whitish AM and PM lines and white costa. Discal dots are present but may be faint. Terminal line is red, and pale fringe is boldly checkered red. Abdomen has a dorsal line of dark rusty spots. **HOSTS:** Unknown; has been reared on sweet gum. **NOTE:** Probably two or more broods.

EMERALDS

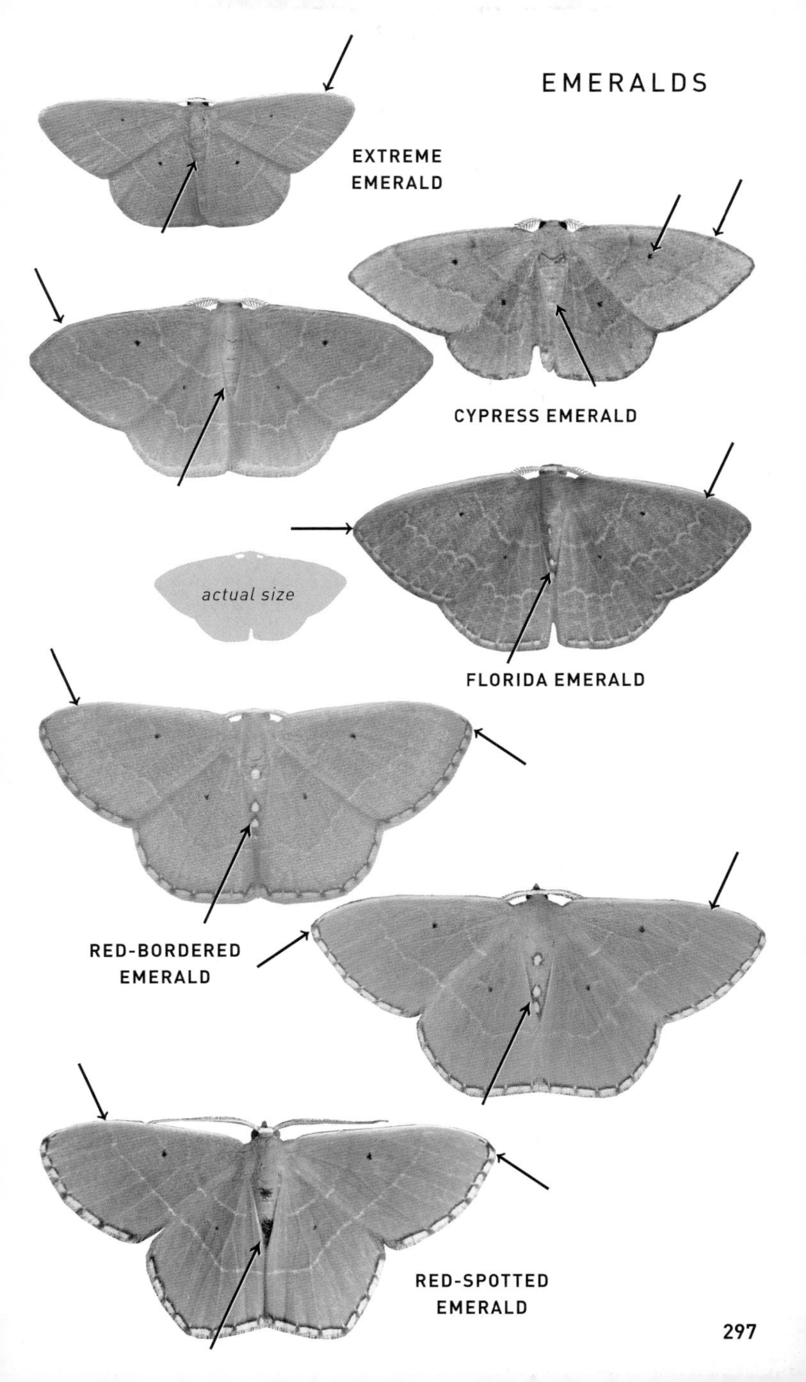

EXTREME EMERALD
CYPRESS EMERALD
actual size
FLORIDA EMERALD
RED-BORDERED EMERALD
RED-SPOTTED EMERALD

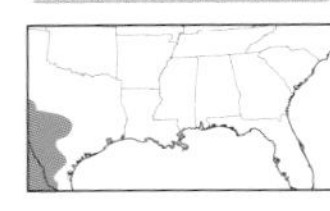

TEXAS EMERALD

Nemoria zygotaria 91-0620 (7040) **Local**

WS 22–28 mm. Bright green wings are marked with narrow, mostly straight, white AM and PM lines. No obvious terminal line. Fringe is whitish when fresh. Green abdomen is unmarked. Discal dots are present but very faint. **HOSTS:** Unknown. **NOTE:** Several broods.

WHITE-BARRED EMERALD

Nemoria bifilata 91-0626 (7045) **Common**

WS 22–28 mm. Pale green or brown (spring brood) wings are marked with white, slightly jagged AM and straight PM lines. Discal dots are present. Terminal line is red. Fringe is uniformly whitish. Abdomen has a pale rusty or whitish dorsal line, sometimes broken into spots. **HOSTS:** Unknown. **NOTE:** Two or more broods.

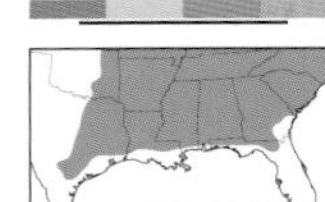

RED-FRINGED EMERALD

Nemoria bistriaria 91-0627 (7046) **Common**

WS 22–25 mm. Green or brown (spring brood) wings are marked with whitish, slightly wavy AM and straight PM lines. Discal dots are faint or absent. Terminal line is red, and pale fringe is boldly checkered red. Forehead and costa are white. Abdomen has a dorsal line of red-ringed cream-colored spots. **HOSTS:** White oak. **NOTE:** Two broods. In late spring some rare individuals may be an intermediate brownish green form.

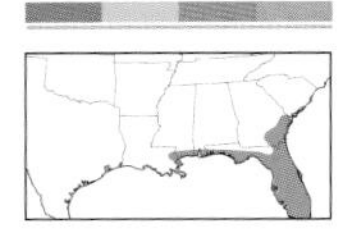

BROWN-SPOTTED EMERALD

Phrudocentra centrifugaria 91-0632 (7051) **Local**

WS 22–28 mm. Sexually dimorphic. Green wings of male have jagged AM and PM lines, usually visible as rows of brown dots. Female (not shown) has large brown blotches (sometimes filled white) in inner ST area of wings. Brown discal dots are present. Terminal line is reddish brown, and fringe is pinkish. Abdomen has a line of tiny white (rarely brown) dorsal dots. **HOSTS:** Unknown.

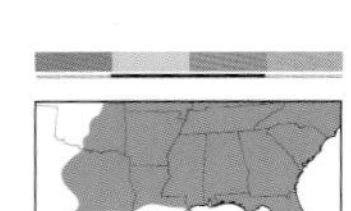

SHOWY EMERALD

Dichorda iridaria 91-0634 (7053) **Common**

WS 24–34 mm. Pea green wings have straight, diffuse white AM (absent on HW) and PM lines. White costa is mottled grayish brown basally. Discal dots are present. **HOSTS:** Sumac and poison ivy. **NOTE:** Two or more broods.

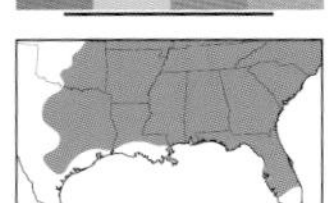

WAVY-LINED EMERALD

Synchlora aerata 91-0639 (7058) **Common**

WS 15–24 mm. Soft green wings have slightly wavy white AM and PM lines and (often) white veins. Terminal line is white dots or scallops, and fringe is pale green. Discal dots are absent. Abdomen has a narrow white dorsal stripe running entire length. **HOSTS:** Trees, shrubs, and low plants, including aster, birch, blackberry, coneflower, goldenrod, and ragweed. **NOTE:** Up to four broods.

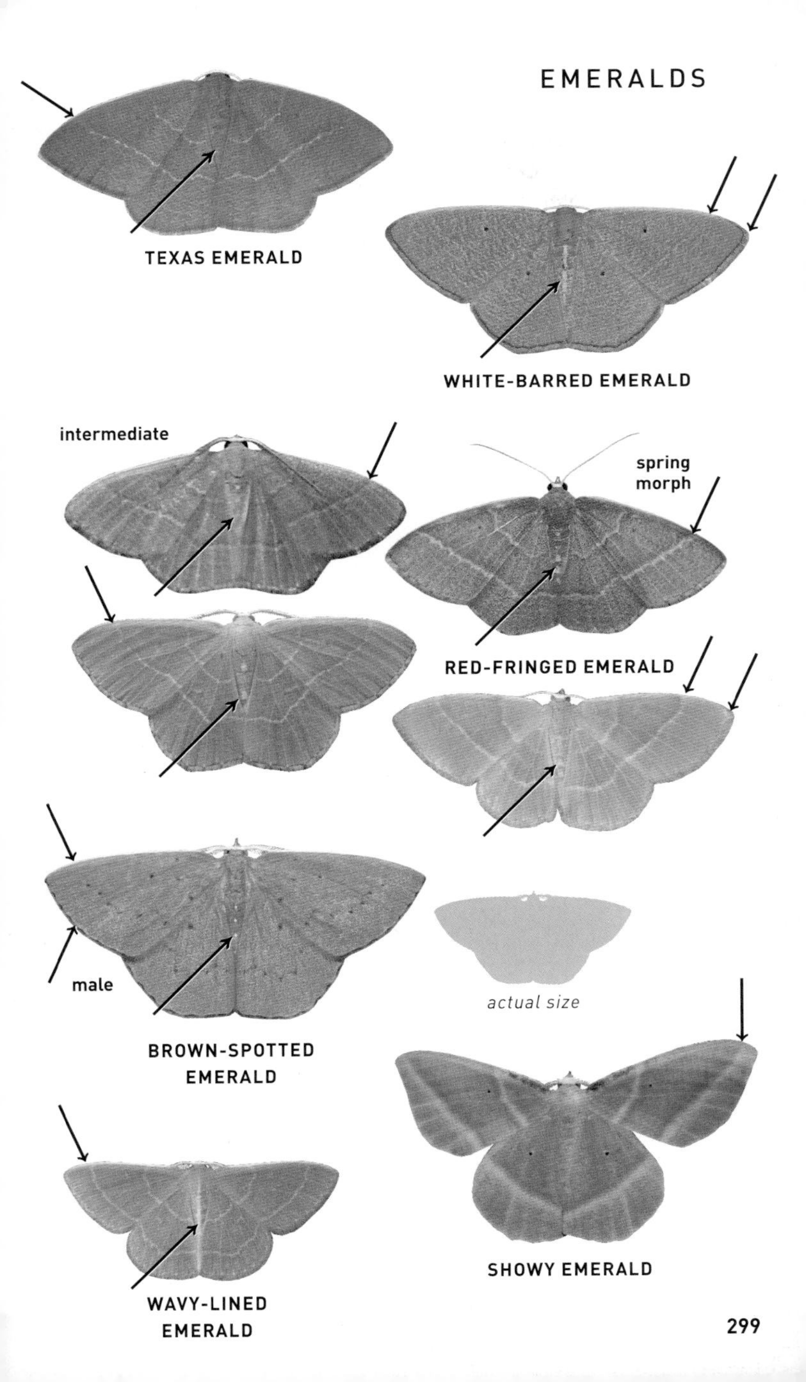
TEXAS EMERALD
WHITE-BARRED EMERALD
intermediate
spring
morph
RED-FRINGED EMERALD
male
BROWN-SPOTTED
EMERALD
actual size
WAVY-LINED
EMERALD
SHOWY EMERALD

SOUTHERN EMERALD

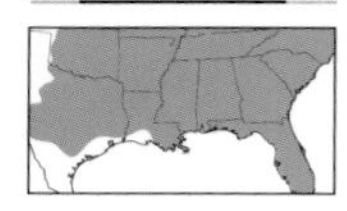

Synchlora frondaria 91-0640 (7059) **Common**

WS 16–20 mm. Resembles Wavy-lined Emerald, but AM and PM lines are more dentate. Faint white veins are only sometimes obvious in median and ST areas. **HOSTS:** Low plants, including blackberry, chrysanthemum, and Spanish needle. **NOTE:** Two or more broods.

SINGED EMERALD

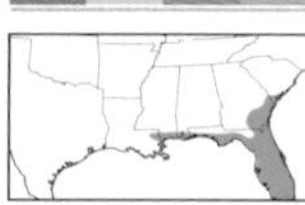

Synchlora xysteraria 91-0642 (7060) **Local**

WS 15–22 mm. Green wings have distinctive brown blotches in ST area. Costa is mostly yellowish. Large hollow discal spots are outlined brown. Fringes are checkered orange and brown. Brown abdomen has three white dorsal spots. **HOSTS:** Reported on mango and lychee.

WHITE-DOTTED EMERALD

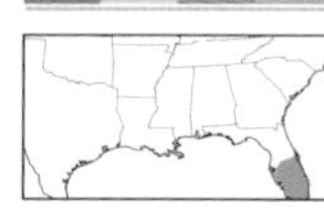

Synchlora herbaria 91-0643 (7061) **Local**

WS 14–20 mm. Green wings have inconspicuous AM and PM lines, usually obvious only as rows of tiny white dots. Costa is creamy, tinged reddish basally. Discal dots are present. Terminal line is usually red, and fringe is checkered red and white, but rare individuals lack any red along outer margin. Green abdomen is marked with three red-bordered white dorsal spots. **HOSTS:** Unknown, but has been found on lantana.

BROWN-BORDERED EMERALD

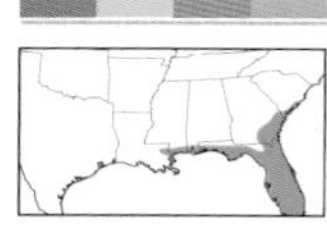

Synchlora cupedinaria 91-0646 (7064) **Local**

WS 15–22 mm. Green wings are bordered with purplish brown costa and fringe. AM and PM lines are pale and indistinct. Brown discal dots are present. Pale abdomen has a wide brown dorsal stripe accented with white spots. **HOSTS:** Unknown.

BLACKBERRY LOOPER

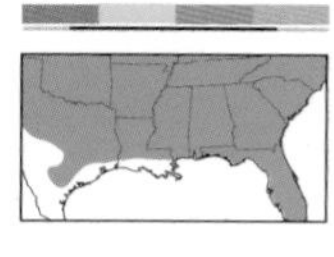

Chlorochlamys chloroleucaria 91-0654 (7071) **Common**

WS 16–23 mm. Grainy bluish green FW has wide cream-colored AM (absent on HW) and PM lines and costal streak. Pale dorsal stripe extends from thorax to tip of abdomen. Discal dots are absent. Male has broadly bipectinate antennae. **HOSTS:** Blackberry and strawberry fruits and Canadian horseweed; also petals of composite flowers. **NOTE:** Two or more broods.

THIN-LINED EMERALD

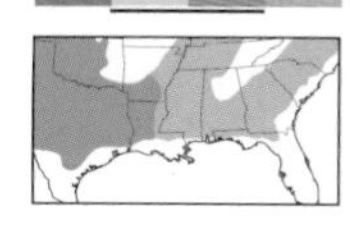

Chlorochlamys phyllinaria 91-0657 (7074) **Uncommon**

WS 16–23 mm. Dull green or (rarely) brownish FW has slightly wavy cream-colored AM (absent on HW) and PM lines and costal streak. Green abdomen is faintly banded whitish. Discal dots are absent. Male has broadly bipectinate antennae tapering to a filiform tip. **HOSTS:** Unknown.

EMERALDS

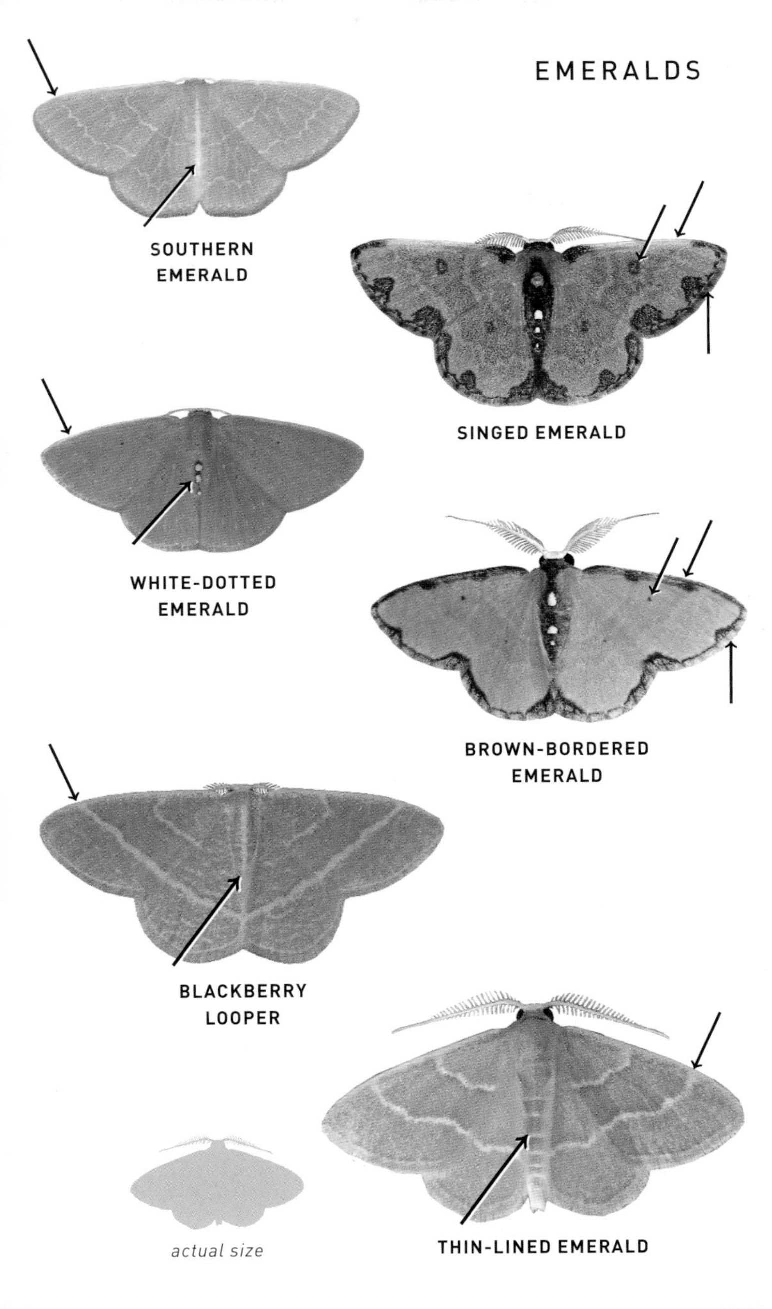

SOUTHERN EMERALD
SINGED EMERALD
WHITE-DOTTED EMERALD
BROWN-BORDERED EMERALD
BLACKBERRY LOOPER
THIN-LINED EMERALD
actual size

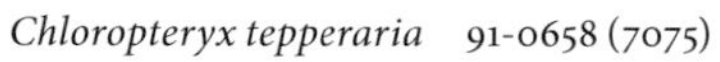

ANGLE-WINGED EMERALD

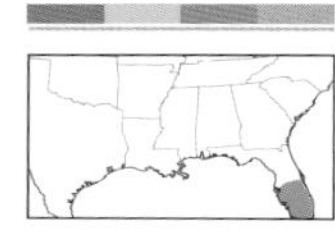

Chloropteryx tepperaria 91-0658 (7075) **Common**

WS 20–26 mm. Grayish green wings have wavy whitish AM and PM lines that usually appear dotted. Terminal line is dark brown. Cream-colored costa and fringe are checkered brown. Green abdomen has creamy second segment marked with a double brown spot. Discal dots are absent. HW is sharply angulate. **HOSTS:** Unknown, but possibly include bald cypress and sumac. **NOTE:** Two or more broods.

FULSOME EMERALD

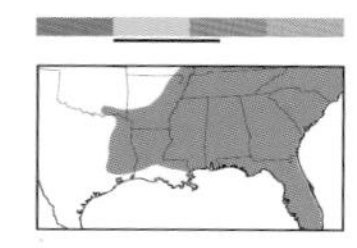

Chloropteryx paularia 91-0660 (7077) **Local**

WS 16–22 mm. Dull grayish green wings have wavy grayish AM and PM lines accented with white dots on veins. Costa is yellowish. Fringes are uniformly pale gray. Green abdomen is unmarked or faintly ringed with white. Discal dots are absent. Male has broadly bipectinate antennae tapering to a filiform tip beyond midpoint. **HOSTS:** Unknown.

PISTACHIO EMERALD

Hethemia pistasciaria 91-0666 (7084) **Common**

WS 16–31 mm. Green, orange green, or orange brown (when worn) wings have fragmented white AM (absent on HW) and PM lines, often faint. Costa and fringes are tinged copper in males. Discal dots are absent. HW is bluntly angulate. **HOSTS:** Basswood, blueberry, ironwood, oak, and other woody plants. **NOTE:** Two broods.

TYPICAL GEOMETERS Family Geometridae, Subfamilies Alsophilinae and Ennominae

Flimsy, broad-winged moths that typically rest with their wings flat and spread, although several species habitually nearly or fully close their wings over their abdomen. Females of some species are wingless, or nearly so. Some groups can be difficult to identify. They are found in a wide variety of habitats, with a few species even venturing into highly urban environments. Most species are nocturnal and will come to lights. A few species may be encountered during the day.

FALL CANKERWORM MOTH

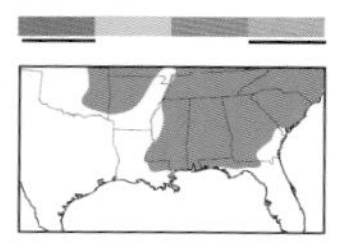

Alsophila pometaria 91-0672 (6258) **Common**

WS 26–32 mm. Gray FW of male has faint white-edged AM and PM lines accented with short blackish dashes. PM line bends at a right angle near costa, forming a white subapical patch. Female is wingless. **HOSTS:** Apple, basswood, elm, maple, oak, and other trees and shrubs.

SEAGRAPE SPANWORM

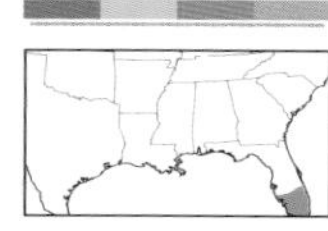

Ametris nitocris 91-0673 (6259) **Local**

WS 38–43 mm. Lime green wings have dotted AM and PM lines and a more solid median line. Paler costa of FW is tinged purple basally. Median area of HW has distinctive double white spots ringed purple. **HOSTS:** Pigeon-plum.

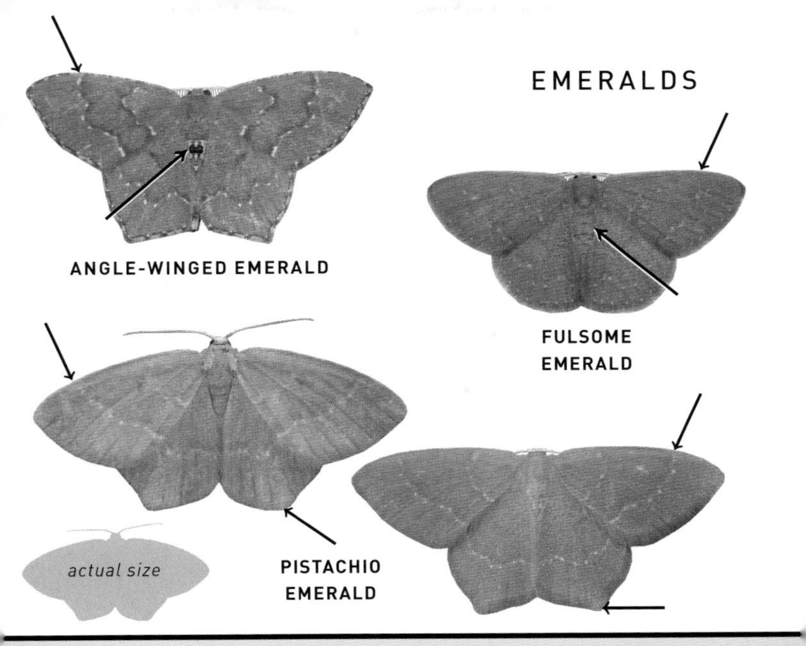
EMERALDS
ANGLE-WINGED EMERALD
FULSOME EMERALD
actual size
PISTACHIO EMERALD

TYPICAL GEOMETERS
actual size
FALL CANKERWORM MOTH
female
SEAGRAPE SPANWORM

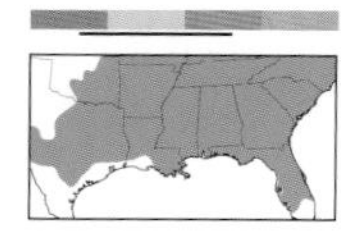

HORNED SPANWORM

Nematocampa resistaria 91-0676 (7010) **Common**

WS 19–25 mm. Speckled yellowish tan wings have a network of brown veins and lines. In most individuals, ST area is extensively shaded brownish on entire HW and inner half of FW. Rarely, this shading is reduced or absent and most distinctive feature is the convergence twice of median and PM lines. **HOSTS:** Alder, ash, birch, maple, oak, strawberry, and many other trees, shrubs, and low plants. **NOTE:** One or two broods.

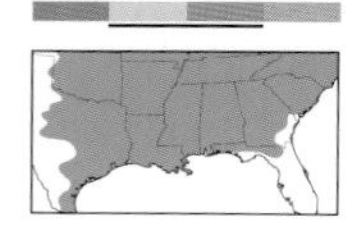

VIRGIN MOTH *Protitame virginalis* 91-0683 (6270) **Local**

WS 20–26 mm. Satin white wings can be unmarked or patterned with brown lines of variable intensity. Wings are often peppered with brown scales, especially along costa of FW. **HOSTS:** Poplar and willow.

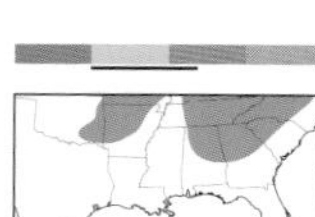

COMMON SPRING MOTH

Heliomata cycladata 91-0686 (6261) **Local**

WS 18–22 mm. Mottled brown FW is marked with large white patches in inner median and subapical areas. White HW has dark border. **HOSTS:** Black locust and honey locust.

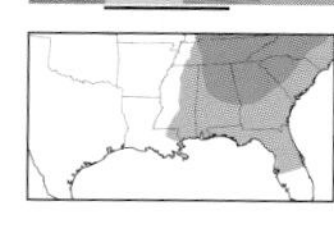

RARE SPRING MOTH

Heliomata infulata 91-0688 (6263) **Local**

WS 26–32 mm. Boldly patterned in black and yellow. Tapered yellow bands in inner median and outer PM areas of FW do not touch. An even yellow band crosses HW. **HOSTS:** Bristly locust.

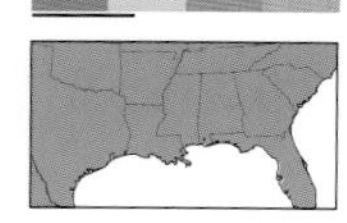

PALE-VEINED ISTURGIA

Isturgia dislocaria 91-0689 (6419) **Common**

WS 30–38 mm. Light brown or grayish FW is finely brindled darker. Blackish AM, median, and PM lines are often bolder toward costa and are overlaid by a network of whitish veins. **HOSTS:** Hackberry.

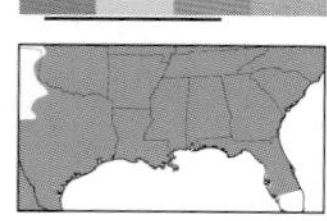

BROWN-BORDERED GEOMETER

Eumacaria madopata 91-0690 (6272) **Uncommon**

WS 20–25 mm. Pale violet gray wings are marked with crisp brown lines. AM and median lines angle sharply inward near costa. Mottled brown ST area is crossed with pale veins. Apex of FW is sharply pointed. **HOSTS:** Apple, cherry, and plum. **NOTE:** Three or more broods.

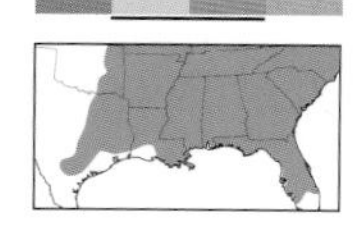

ORANGE WING

Mellilla xanthometata 91-0691 (6271.1) **Common**

WS 20–28 mm. Seasonally and sexually dimorphic. Grayish brown FW of spring brood male is marked with faint lines (wider at costa) and usually has some purplish brown shading beyond almost straight PM line. Summer brood male has FW more uniformly reddish brown. Some individuals have a blackish spot at midpoint of ST line. Larger female is usually paler overall. Orange HW tends to be brighter in males. **HOSTS:** Locust. **NOTE:** Two or more broods.

TYPICAL GEOMETERS

FOUR-SPOTTED ANGLE

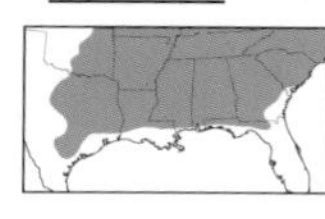

Trigrammia quadrinotaria 91-0692 (6360) **Common**

WS 26–36 mm. Powdery whitish FW is boldly patterned with wavy brown AM line and almost straight median and PM lines. Midpoint of PM line and fragmented ST band is often marked with blackish spots. HW often has a faint blackish discal dot. **HOSTS:** Native buckeyes.

DOT-LINED ANGLE

Psamatodes abydata 91-0697 (6332) **Uncommon**

WS 26–34 mm. Peppery yellowish brown wings have a bold darker ST band beyond dotted PM line. Other lines are indistinct. Blackish discal spots mark all wings but are bolder on angulate HW. **HOSTS:** Sweet acacia, riverhemp, Jerusalem thorn, and soybean.

DARK-SHADED ANGLE

Psamatodes trientata 91-0698 (6332.1) **Local**

WS 22–28 mm. Similar to larger Dot-lined Angle but tends to have a less contrasting wing pattern and less prominent dark spots along inconspicuous ST line. **HOSTS:** Sweet acacia.

SOUTHERN ANGLE

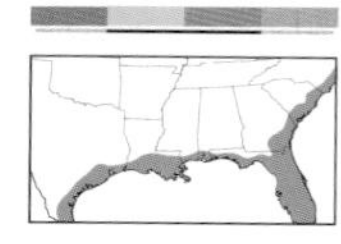

Macaria varadaria 91-0700 (6314) **Uncommon**

WS 20–28 mm. Peppery grayish brown FW has blurry brown-edged lines, often with a paler median band. Faint AM line curves inward near costa. Almost straight PM line is bold at inner margin but fades toward costa. All wings have small black discal dots. **HOSTS:** Groundsel tree.

FOUR-SPOTTED GRANITE

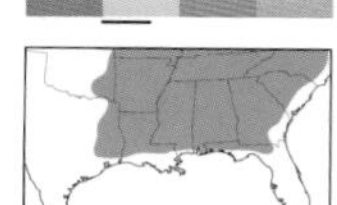

Macaria coortaria 91-0715 (6299) **Uncommon**

WS 22–26 mm. Uniform pale gray FW has four evenly spaced dark brown costal spots. Small blackish discal dot almost touches median costal spot. Unmarked HW is whitish. **HOSTS:** Apple, cherry, hawthorn, pear, and willow.

LESSER MAPLE SPANWORM

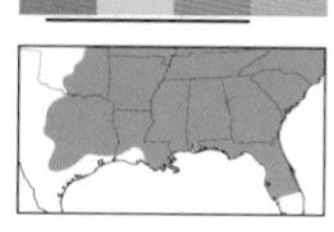

Macaria pustularia 91-0735 (6273) **Common**

WS 18–28 mm. White FW is crossed with four variably fragmented cinnamon brown lines (some may be very faint) that widen to form distinct costal spots. White HW usually has a faint median line. **HOSTS:** Maple.

DECEPTIVE ANGLE

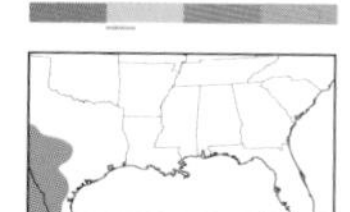

Macaria deceptrix 91-0740 (6312) **Local**

WS 20–26 mm. Similar to Boldly-marked Granite but usually darker and grayer (though some females are whitish). Variably patterned with black and chestnut lines. ST band is often heaviest marking on FW. **HOSTS:** Knifeleaf condalia.

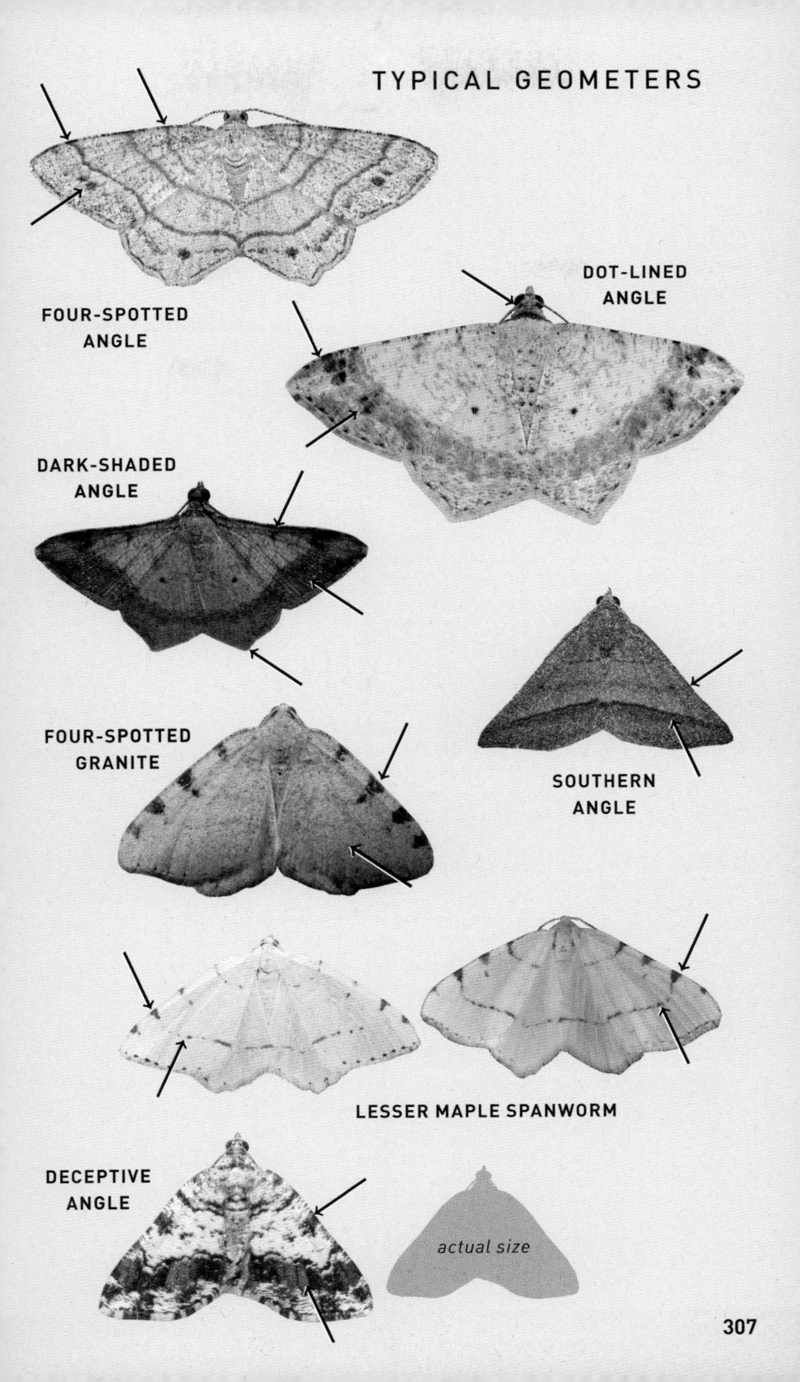
FOUR-SPOTTED ANGLE
DOT-LINED ANGLE
DARK-SHADED ANGLE
SOUTHERN ANGLE
FOUR-SPOTTED GRANITE
LESSER MAPLE SPANWORM
DECEPTIVE ANGLE
actual size

BOLDLY-MARKED GRANITE

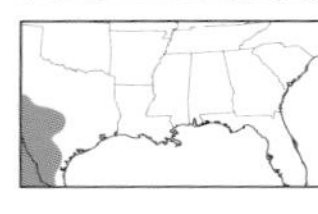

Macaria graphidaria 91-0741 (6311) **Local**

WS 22–28 mm. Boldly marked whitish FW is brindled with blackish dashes, especially in ST area. Black discal spot is often noticeable. Black and brown mottling along ST band is heaviest at inner margin and near bulge of PM line. Fringe is checkered. Pale HW is tinted yellow. **HOSTS:** Saffron plum.

COMMON ANGLE

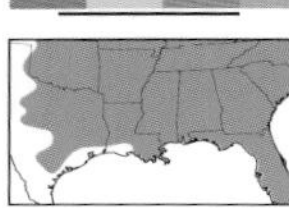

Macaria aemulataria 91-0750 (6326) **Common**

WS 22–28 mm. Cream-colored FW has faint brown lines ending in bold chestnut brown costal spots. Midpoint of ST area is marked with a large blackish paw-print mark. Outer margin has an indentation near apex. Pale HW is sharply angulate. White abdomen has row of double black dorsal spots. **HOSTS:** Maple. **NOTE:** Three broods.

PROMISCUOUS ANGLE

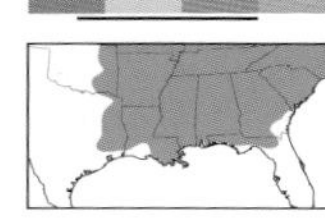

Macaria promiscuata 91-0752 (6331) **Common**

WS 24–30 mm. Similar to slightly smaller Common Angle, but lines tend to be bolder. Fragmented outer ST band almost connects with large black paw-print marking. Bold chestnut PM band is complete on undersides of wings if visible. **HOSTS:** Redbud.

WOODY ANGLE

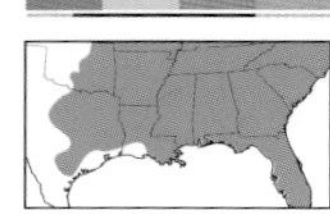

Macaria aequiferaria 91-0755 (6335) **Common**

WS 22–28 mm. Seasonally dimorphic. Darker early spring brood has grayish brown FW with faint brown lines that widen at costa. Midpoints of black PM and ST lines are thickened, sometimes forming an indistinct paw-print marking. Slightly smaller summer broods are paler, often with much reduced dark markings on wings. **HOSTS:** Cypress. **NOTE:** Three or more broods.

RED-HEADED INCHWORM

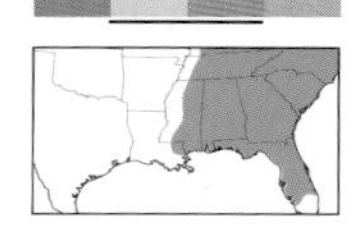

Macaria bisignata 91-0756 (6342) **Common**

WS 26–34 mm. Peppery whitish FW has faint lines obvious only as darker costal spots. Brown ST line is usually marked with a dark spot at midpoint and ends as a large dark brown rectangular spot at costa. Head is orange. **HOSTS:** White pine and other pines. **NOTE:** Three broods.

BICOLORED ANGLE

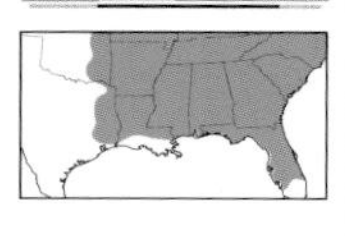

Macaria bicolorata 91-0758 (6341) **Uncommon**

WS 26–36 mm. Resembles Blurry Chocolate Angle but is usually larger and paler. Light purplish gray wings are shaded brown beyond PM line, appearing two-toned. Lines on FW are poorly defined except along costa. Head is orange. **HOSTS:** Pine. **NOTE:** Two or more broods.

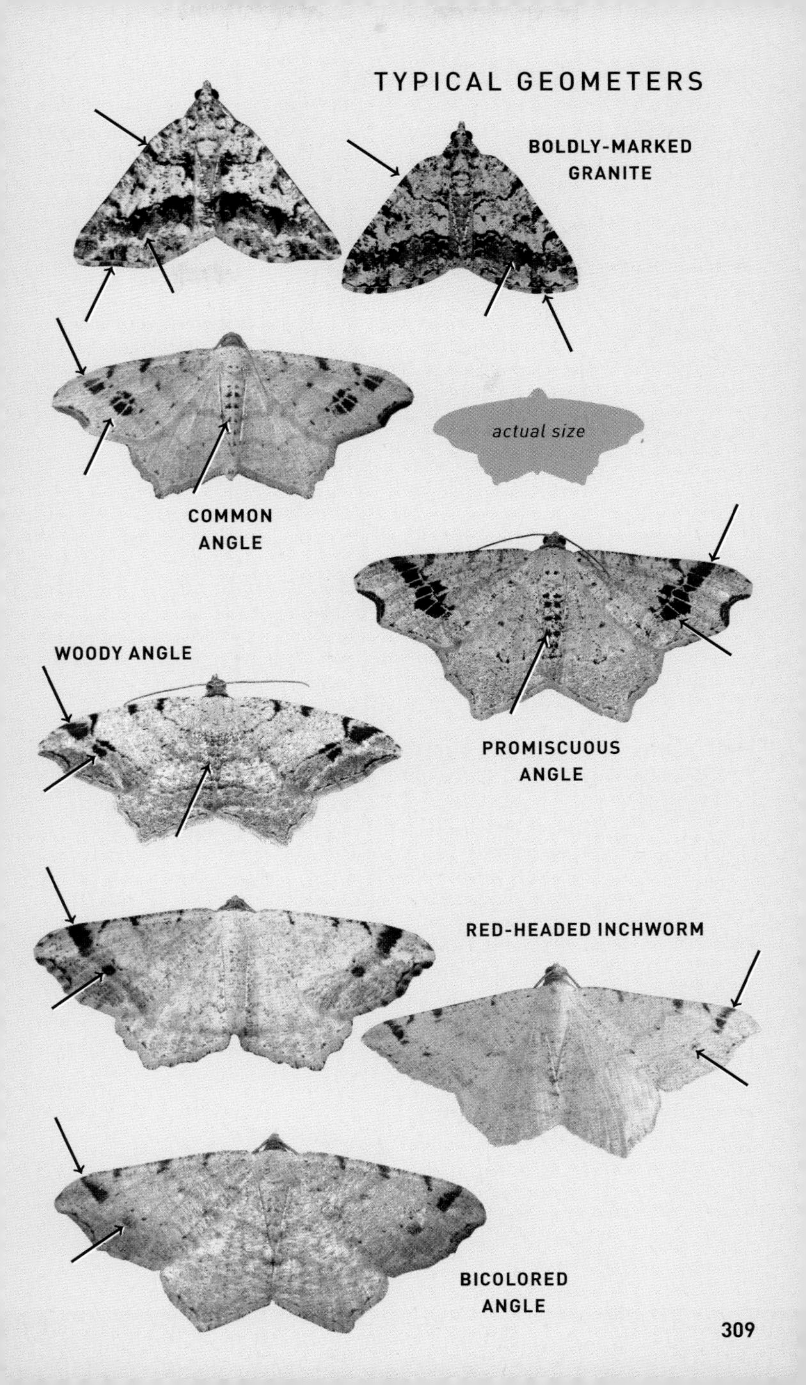
BOLDLY-MARKED GRANITE
actual size
COMMON ANGLE
WOODY ANGLE
PROMISCUOUS ANGLE
RED-HEADED INCHWORM
BICOLORED ANGLE

BLURRY CHOCOLATE ANGLE

Macaria transitaria 91-0761 (6339) **Uncommon**

WS 26–32 mm. Purplish gray FW has faint blackish lines obvious only where they widen at costa. Chocolate brown ST band blends into grayer median area. ST line ends as a thickened darker brown patch near apex. Gray HW has a darker brown ST band. Head is orange. **HOSTS:** Pine. **NOTE:** Two broods.

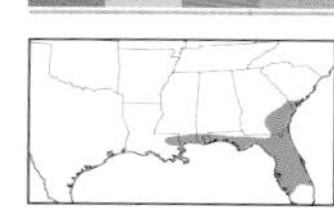

SOUTHERN CHOCOLATE ANGLE

Macaria distribuaria 91-0762 (6336) **Common**

WS 30–40 mm. This large and striking angle has pale violet gray wings brindled darker with sinuous black AM and PM lines. A broad rusty brown band passes through ST area. Head is orange. **HOSTS:** Longleaf and slash pines. **NOTE:** Presumably two or more broods.

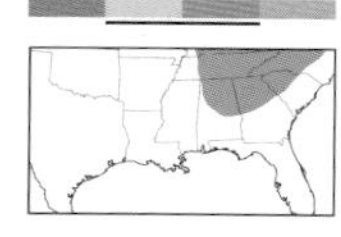

SANFORD'S ANGLE

Macaria sanfordi 91-0763 (6337) **Local**

WS 22–30 mm. Similar to Southern Chocolate Angle but is considerably smaller on average. Moths from s. FL tend to be darker overall with a less contrasting pattern on wings. **HOSTS:** Sand pine.

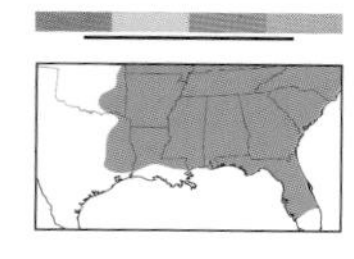

GRANITE ANGLE *Macaria granitata* 91-0771 (6352) **Local**

WS 25–30 mm. Gray FW is variably brindled darker and has bold blackish lines that widen slightly at costa. Outer median area is usually whitish. Note large rusty subapical patch. Darker individuals have dark gray FW with contrasting whitish ST line. **HOSTS:** Pitch pine and perhaps other pines. **NOTE:** Two or more broods.

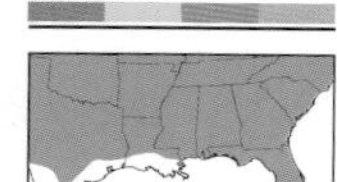

MANY-LINED ANGLE

Macaria multilineata 91-0772 (6353) **Common**

WS 25–30 mm. Pale gray FW is distinctively patterned with almost straight AM and double PM lines and fainter median and ST lines. A warm brown ST band is often present, ending as a chestnut subapical patch. Multilined HW is strongly angulate. Head is pale orange. **HOSTS:** Juniper. **NOTE:** Three broods.

CURVE-LINED ANGLE

Digrammia continuata 91-0789 (6362) **Common**

WS 20–30 mm. Gray FW is variably brindled darker and has bold, gently curved double AM and PM lines. A diffuse median band is sometimes present. Terminal line is a row of black dashes. **HOSTS:** Eastern and southern red cedars. **NOTE:** Two or three broods.

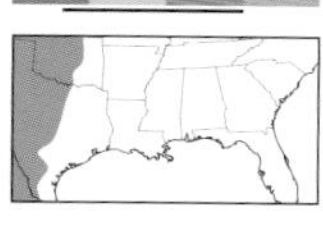

FADED ANGLE

Digrammia pallorata 91-0795 (6363.1) **Local**

WS 22–34 mm. Straw-colored FW is lightly brindled darker. Almost straight AM and slightly kinked PM lines vary in strength. A diffuse median band is sometimes present. Pale HW is marked with a tiny black discal dot. **HOSTS:** One-seed and Pinchot junipers. **NOTE:** Two broods.

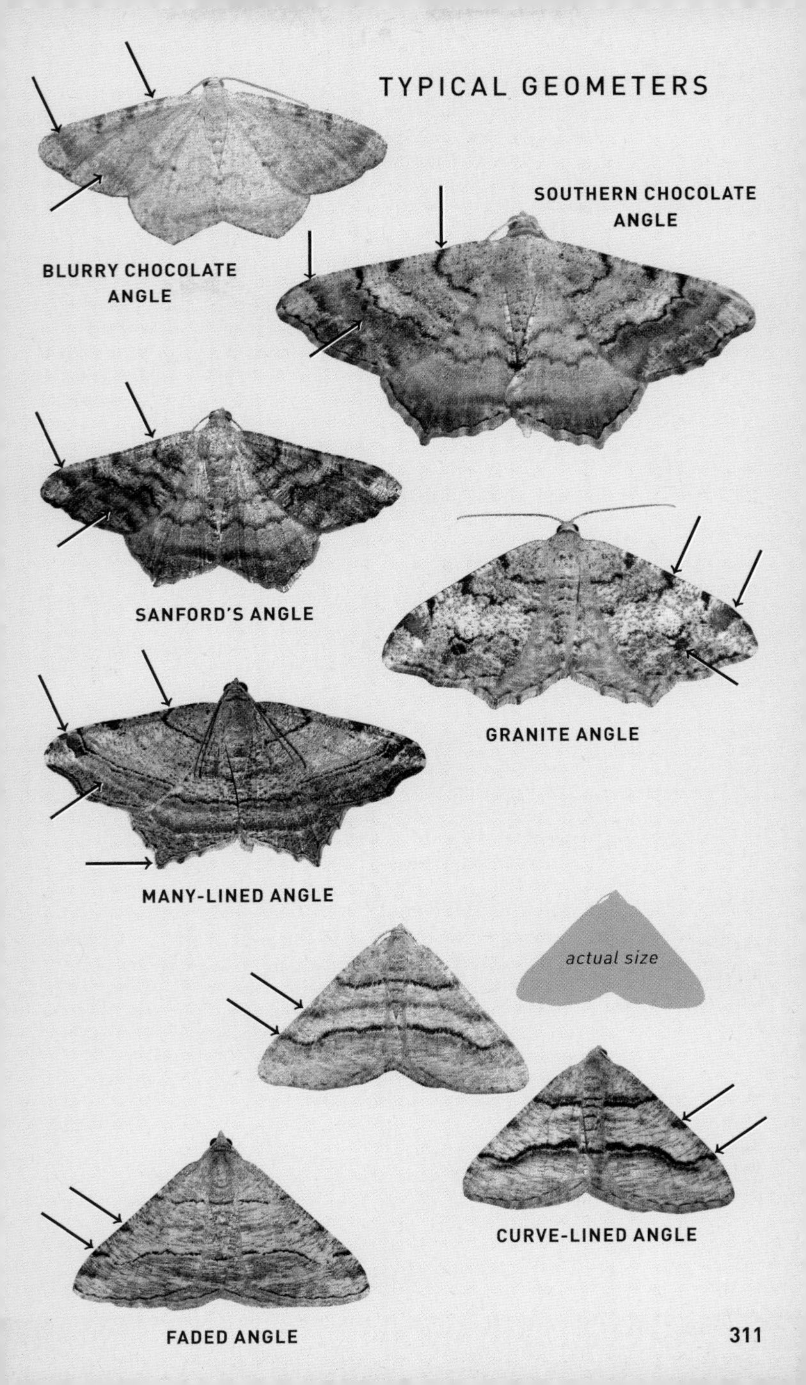
SOUTHERN CHOCOLATE ANGLE
BLURRY CHOCOLATE ANGLE
SANFORD'S ANGLE
GRANITE ANGLE
MANY-LINED ANGLE
actual size
CURVE-LINED ANGLE
FADED ANGLE

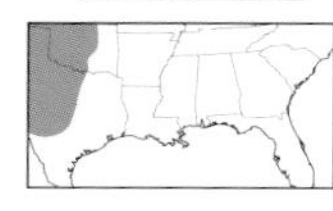

BROAD-LINED ANGLE

Digrammia atrofasciata 91-0798 (6368) **Local**

WS 26–30 mm. Grayish brown FW has an oblique whitish median band within thick double black AM and PM lines. White ST line is edged black toward costa. Terminal line is a row of black dashes. **HOSTS:** Juniper. **NOTE:** Two broods.

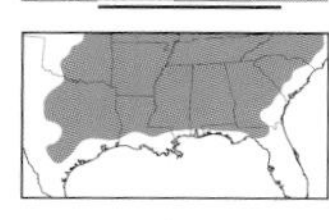

FAINT-SPOTTED ANGLE

Digrammia ocellinata 91-0802 (6386) **Common**

WS 22–30 mm. Peppery gray wings are patterned with blurry, often dotted lines that form bolder costal spots on FW. Darker brown ST band is often accented with short black dashes along veins. **HOSTS:** Black locust and possibly honey locust. **NOTE:** Two or three broods.

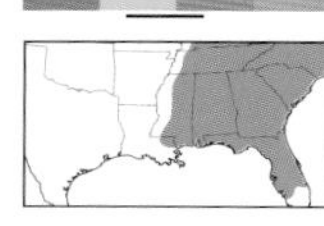

ORDERED ANGLE

Digrammia ordinata 91-0803 (6358) **Local**

WS 24–30 mm. Peppery milky brown wings are faintly patterned with brown AM, median, and PM lines. Slightly scalloped PM line on FW is often accented with tiny black dots. **HOSTS:** False indigo.

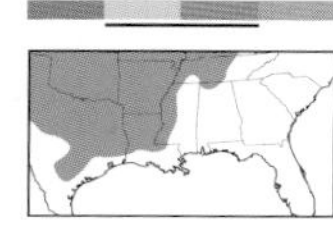

DARK-WAVED ANGLE

Digrammia subminiata 91-0816 (6399) **Uncommon**

WS 20–25 mm. Light gray or reddish brown FW has thick pale-edged black AM and PM lines (sometimes reduced) that fade toward costa. ST band is shaded darker brown. Discal spot is hollow. **HOSTS:** Willow.

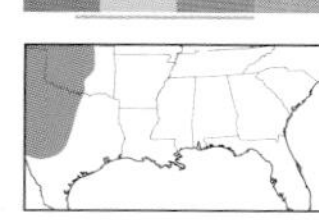

PALE-LINED ANGLE

Digrammia irrorata 91-0818 (6395) **Local**

WS 23–26 mm. Peppery light grayish brown FW is variably patterned with rusty-edged pale yellow AM (often absent) and PM lines. Darker brown median and ST lines often are accented with dusky blotches. **HOSTS:** Poplar and willow. **NOTE:** Two or three broods.

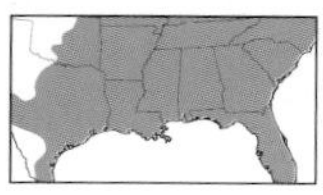

HOLLOW-SPOTTED ANGLE

Digrammia gnophosaria 91-0822 (6405) **Common**

WS 18–24 mm. Peppery grayish brown FW is faintly marked with wavy lines that widen to form blackish costal spots. AM and median lines almost meet at inner margin, sometimes forming a dusky patch. Elliptical blackish reniform spot has a pale center. **HOSTS:** Willow. **NOTE:** Two or three broods.

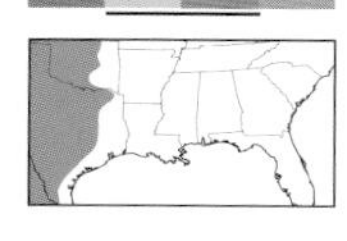

SIGNATE LOOPER

Rindgea s-signata 91-0832 (6414) **Local**

WS 20–30 mm. Whitish to light gray FW is variably brindled darker. Inner PM line is thick and curving, edged below by dark shading. AM and median lines are often faint or nearly absent. ST lines are diffusely darker. **HOSTS:** Mesquite.

TYPICAL GEOMETERS

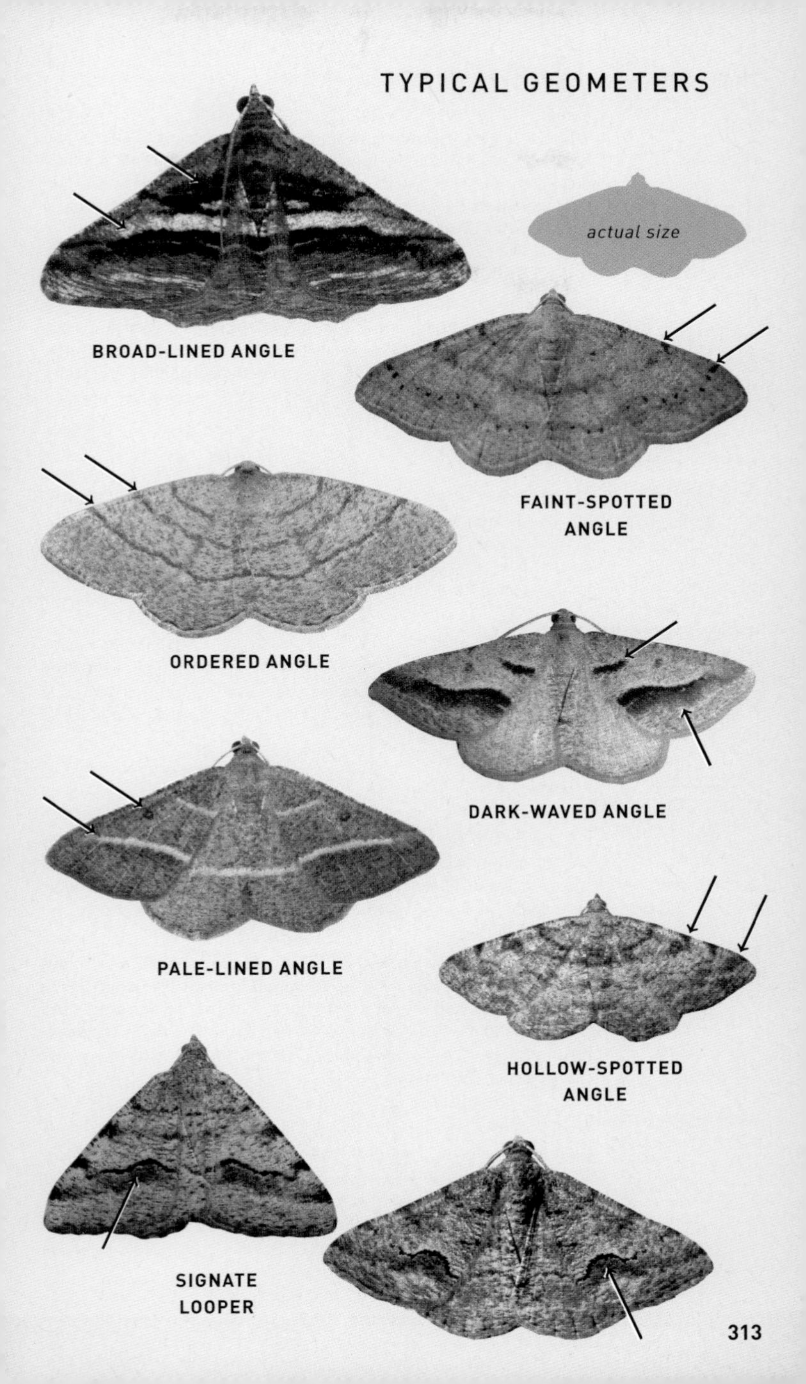

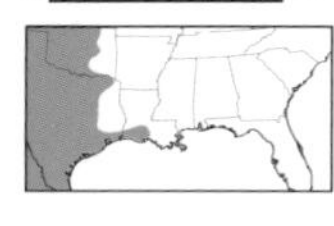

MESQUITE LOOPER

Rindgea cyda 91-0833 (6415) **Local**

WS 20–30 mm. Resembles Signate Looper but generally is less brindled, and PM line has broader pale edging above. AM and median lines are usually visible as diffuse bands. Where range overlaps with Signate Looper, however, some individuals may not be separable in the field. **HOSTS:** Mesquite.

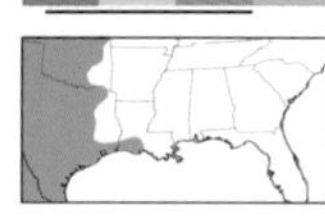

GREEN BROOMWEED LOOPER

Fernaldella fimetaria 91-0837 (6420) **Local**

TL 6–8 mm. A boldly patterned diurnal (and nocturnal) moth that rests with wings tightly closed above abdomen. Undersides of light golden brown wings are boldly marked with paler bands and spots. Brown abdomen is banded white. **HOSTS:** Prairie broomweed. **NOTE:** Two broods.

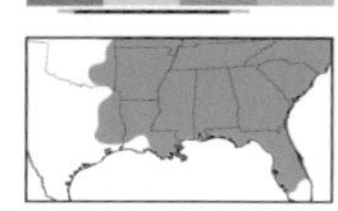

UMBER MOTH

Hypomecis umbrosaria 91-0853 (6439) **Uncommon**

WS 26–40 mm. Peppery light gray wings are patterned with crisp, scalloped lines. All wings are marked with black-ringed elliptical discal spots. Zigzag ST line sometimes is edged with a brownish band. **HOSTS:** Birch and oak. **NOTE:** Two broods.

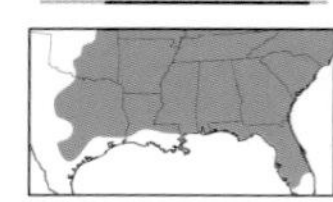

TEXAS GRAY

Glenoides texanaria 91-0858 (6443) **Common**

WS 15–22 mm. Gray FW has indistinct wavy lines that widen at costa. Bold PM line borders a broad rusty brown band. Scalloped ST line is whitish. **HOSTS:** Unknown.

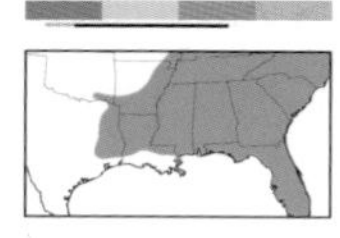

DOTTED GRAY

Glena cribrataria 91-0864 (6449) **Uncommon**

WS 23–31 mm. Whitish or pale gray wings are patterned with evenly spaced dotted or dashed lines. Abdomen has pairs of black dorsal spots on each segment. **HOSTS:** Birch, blueberry, cherry, maple, oak, willow, and other woody plants. **NOTE:** Two or more broods.

BLUEBERRY GRAY *Glena cognataria* 91-0865 (6450) **Local**

WS 20–30 mm. Peppery whitish wings are tinted violet when fresh, and weakly marked with indistinct lines. More noticeable dotted PM line is sometimes edged with a darker band (mostly in females). Terminal line is a row of tiny black dots. **HOSTS:** Blueberry and cherry. **NOTE:** Two broods.

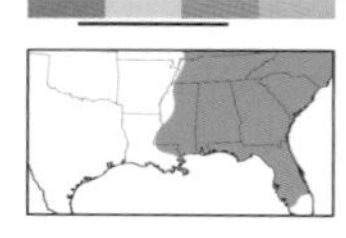

DAINTY GRAY

Glena plumosaria 91-0867 (6452) **Uncommon**

WS 28–32 mm. Pale gray FW is marked with crisp black lines that curve and fade before reaching costa. Basal patch and ST band are rusty brown. Abdomen has narrow black bands on each segment. **HOSTS:** Atlantic white cedar and red cedar. **NOTE:** Two or three broods.

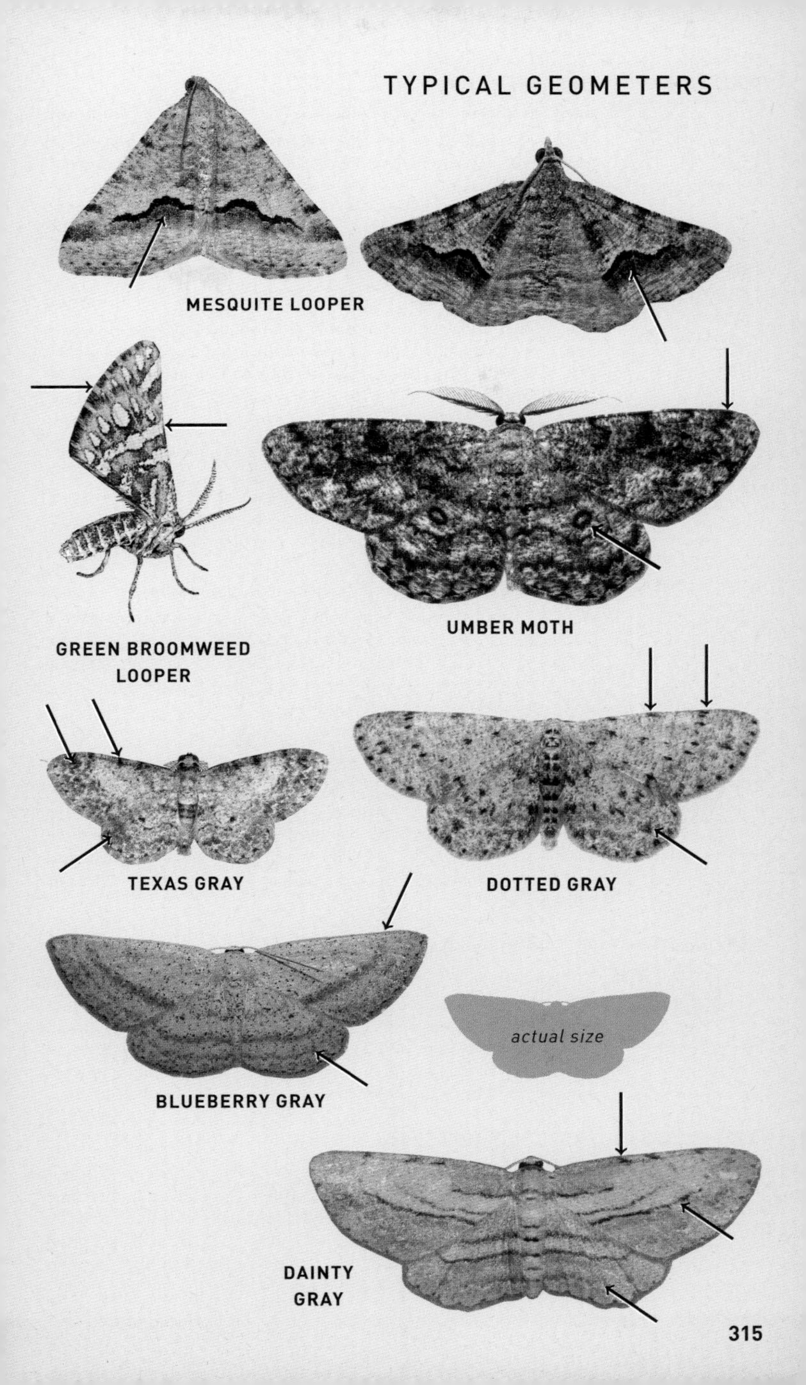
MESQUITE LOOPER
GREEN BROOMWEED LOOPER
UMBER MOTH
TEXAS GRAY
DOTTED GRAY
BLUEBERRY GRAY
actual size
DAINTY GRAY

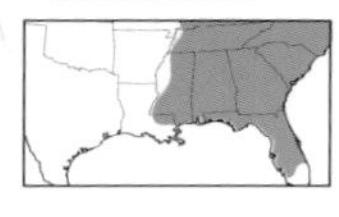

FINE-LINED GRAY

Exelis pyrolaria 91-0893 (6478) **Common**

WS 22–25 mm. Pale to dark gray FW has angulate black AM and PM lines. Indistinct median line passes close to small black discal spot. Zigzag ST line is whitish. Abdomen has a black band near base. Often rests with HW mostly hidden. **HOSTS:** Persimmon and common pipsissewa.

DIMORPHIC GRAY

Tornos scolopacinaria 91-0898 (6486) **Common**

WS 25–28 mm. Sexually dimorphic. Pale yellowish (female) or gray (male) FW has faintly spotted lines and a large blackish discal spot. Often tinged warm brown in basal area and along ST line. Often rests with HW mostly hidden. **HOSTS:** Aster and coreopsis.

FOUR-BARRED GRAY

Aethalura intertexta 91-0988 (6570) **Common**

WS 21–25 mm. Gray wings have four evenly spaced and often fragmented black lines that widen toward costa. Jagged ST line is edged white. **HOSTS:** Alder and birch. **NOTE:** Two broods.

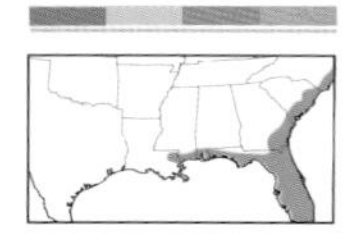

CYPRESS GRAY *Iridopsis pergracilis* 91-0998 (6580) **Local**

WS 26–28 mm. Sexually dimorphic. Whitish wings of male have a distinctive pattern of wide black bands along wavy AM and PM lines. Scalloped white ST line has two black spots near midpoint. Female has yellowish brown basal area and ST band. **HOSTS:** Bald cypress. **NOTE:** Three or more broods.

LARGE PURPLISH GRAY

Iridopsis vellivolata 91-1000 (6582) **Common**

WS 30–36 mm. Purplish gray FW has black AM and PM lines edged with warm brown bands. Median and PM lines converge and often merge at inner margin. HW has an elliptical black-ringed discal spot. **HOSTS:** Pine; also fir, tamarack, and spruce. **NOTE:** Three or more broods.

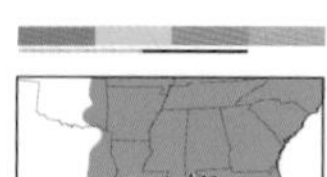

PALE-WINGED GRAY

Iridopsis ephyraria 91-1001 (6583) **Local**

WS 23–28 mm. Pale gray FW is marked with wavy black lines that widen to form costal spots. Sometimes has fragmented yellowish brown bands in basal and ST areas. Elliptical black-ringed discal spots mark all wings. HW often has a dusky median band. **HOSTS:** Ash, birch, cherry, currant, maple, willow, and other woody plants.

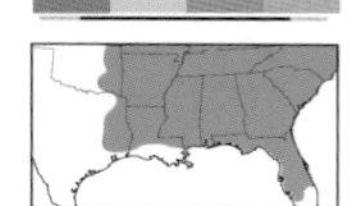

SMALL PURPLISH GRAY

Iridopsis humaria 91-1002 (6584) **Common**

WS 24–28 mm. Resembles Large Purplish Gray but is smaller and has more rounded FW apex. Indistinct median line merges with PM line near inner margin. Reddish brown basal and ST bands are often narrow. **HOSTS:** Birch, clover, hickory, oak, sweet fern, soybean, and many other plants. **NOTE:** Two broods.

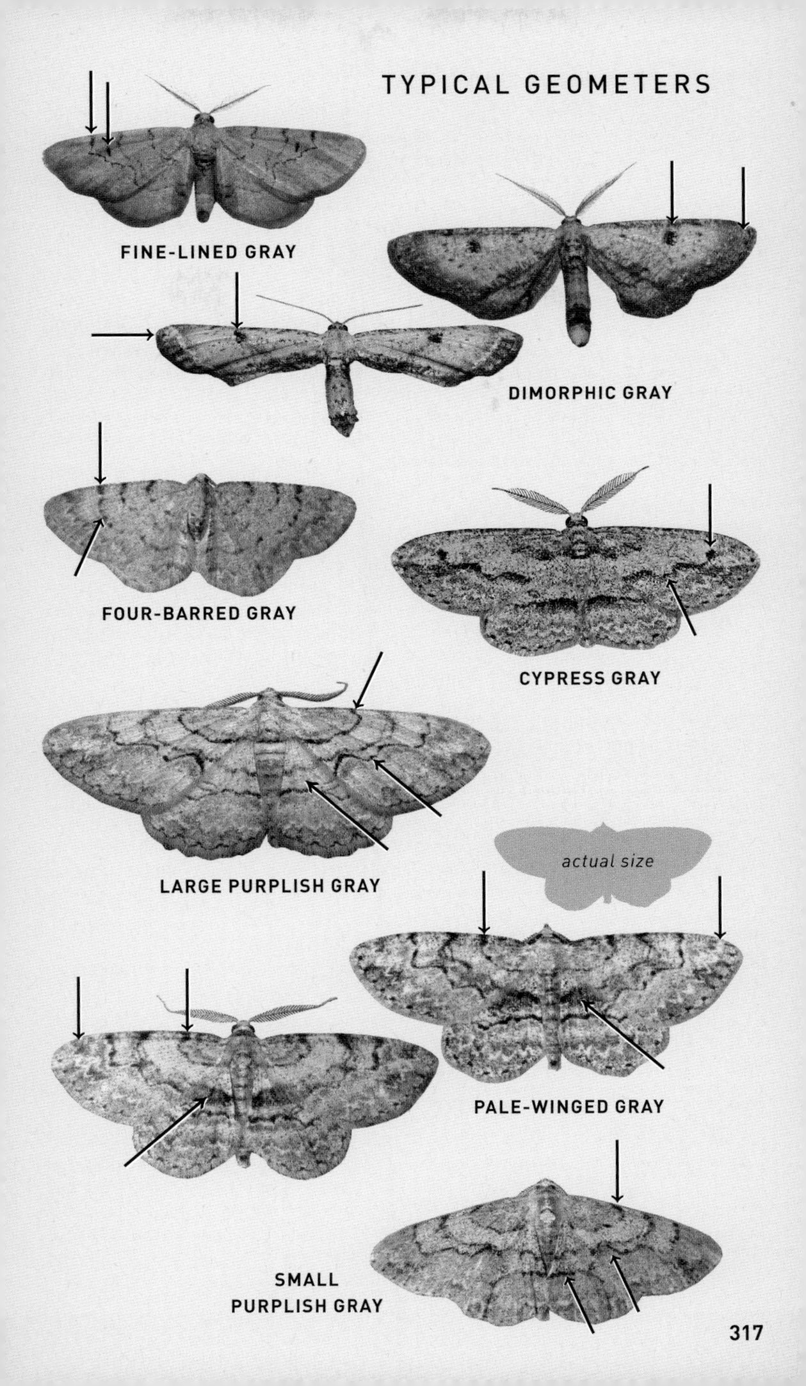
TYPICAL GEOMETERS
FINE-LINED GRAY
DIMORPHIC GRAY
FOUR-BARRED GRAY
CYPRESS GRAY
LARGE PURPLISH GRAY
actual size
PALE-WINGED GRAY
SMALL
PURPLISH GRAY

BROWN-SHADED GRAY

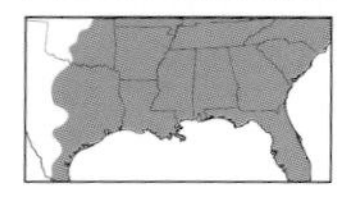

Iridopsis defectaria 91-1004 (6586) **Common**

WS 24–36 mm. Whitish FW is patterned with crisp wavy black lines. Yellowish brown basal and ST bands are usually well defined. Scalloped white ST line has two black spots near midpoint. Elliptical black-ringed discal spots mark all wings. **HOSTS:** Cherry, oak, poplar, walnut, willow, and other woody plants. **NOTE:** Two or more broods.

BENT-LINE GRAY

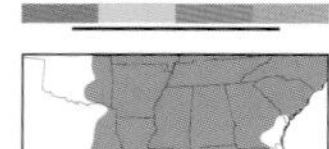

Iridopsis larvaria 91-1007 (6588) **Common**

WS 26–36 mm. Resembles Pale-winged Gray, but wings have sharper, more angulate black lines and outlines to elliptical spots. Often has yellowish brown shading in basal and ST areas. Scalloped whitish ST line has two black dashes near midpoint. **HOSTS:** Alder, apple, birch, cherry, hawthorn, poplar, willow, and many other plants. **NOTE:** Two broods.

COMMON GRAY

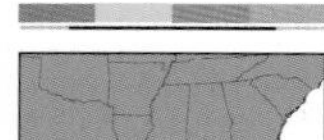

Anavitrinella pampinaria 91-1009 (6590) **Common**

WS 23–34 mm. Peppery light gray FW is marked with crisp black lines that often widen toward inner margin. Median and PM lines converge toward inner margin. Brown basal and ST bands are sometimes present. Abdomen has a white band at base. **HOSTS:** Apple, ash, clover, cotton, fir, tamarack, maple, poplar, and many others. **NOTE:** Two or more broods.

DOUBLE-LINED GRAY

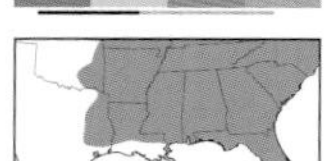

Cleora sublunaria 91-1013 (6594) **Common**

WS 25–30 mm. Whitish or gray FW has double AM and PM lines. Median line does not converge with PM line at inner margin. Midpoint of ST area is often marked with a whitish patch, especially on darker individuals. Elliptical black-ringed discal spots mark all wings. Abdomen has a white-edged black band at base. **HOSTS:** Cherry, chestnut, oak, sweet fern, and others.

PROJECTA GRAY

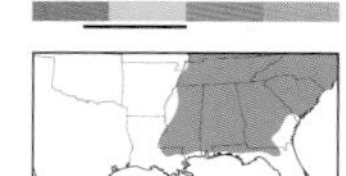

Cleora projecta 91-1014 (6595) **Uncommon**

WS 25–30 mm. Similar to Double-lined Gray but often has darker purplish gray wings and a whitish median band on FW. AM and PM lines are less obviously double. **HOSTS:** Cherry, gale, oak, and probably other woody plants.

SMALL ENGRAILED

Ectropis crepuscularia 91-1016 (6597) **Common**

WS 26–37 mm. Variable. Peppery wings are marked with scalloped black lines, often heaviest on veins. Midpoint of whitish ST line has two black wedges. Pale extremes are whitish with wide dusky PM band. Melanic individuals are sooty with a white ST line. **HOSTS:** Apple, birch, fir, hemlock, tamarack, oak, poplar, spruce, willow, and many others. **NOTE:** Three broods.

TYPICAL GEOMETERS
BENT-LINE GRAY
BROWN-SHADED GRAY
COMMON GRAY
DOUBLE-LINED GRAY
actual size
PROJECTA GRAY
SMALL ENGRAILED

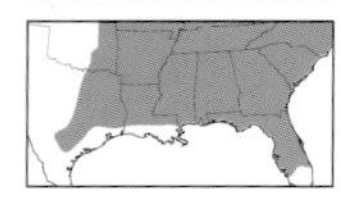

PORCELAIN GRAY

Protoboarmia porcelaria 91-1017 (6598) **Common**

WS 26–35 mm. Peppery gray FW is marked with often indistinct black lines that widen to form costal spots. PM and whitish zigzag ST lines are accented with black wedges on veins. Small black discal spots mark all wings. **HOSTS:** Fir, elm, hemlock, poplar, willow, and many other trees. **NOTE:** Two broods.

TULIP-TREE BEAUTY

Epimecis hortaria 91-1018 (6599) **Common**

WS 43–57 mm. This large geometer has scalloped outer margins to HW. Peppery gray wings have white-edged scalloped lines, sometimes accented with brown bands. Pale extremes are whitish with a broad brown median band. Melanic individuals are sooty with white-edged lines. **HOSTS:** Magnolia, redbay, sassafras, spicebush, and tulip tree. **NOTE:** Two broods.

WHITE-TIPPED BLACK

Melanochroia chephise 91-1034 (6616) **Common**

WS 32–34 mm. Velvety black wings are accented with slightly paler veins. Tip of FW is white. Bright cinnamon thorax contrasts with black abdomen. Diurnal; often found visiting flowers. **HOSTS:** Spurge.

CANADIAN MELANOLOPHIA

Melanolophia canadaria 91-1059 (6620) **Common**

WS 36–39 mm. Peppery light gray FW is marked with often fragmented blackish lines. Wavy PM line is double at midpoint and inner margin, forming two darker patches. Black discal dots mark all wings. **HOSTS:** Alder, ash, birch, elm, locust, tamarack, maple, pine, willow, and other trees and shrubs. **NOTE:** Two broods.

SIGNATE MELANOLOPHIA

Melanolophia signataria 91-1060 (6621) **Common**

WS 30–35 mm. Resembles Canadian Melanolophia, but uniform light brown FW has straighter PM line that is obviously double at inner margin and often less so at midpoint. Darker individuals are difficult to identify. **HOSTS:** Alder, birch, elm, fir, tamarack, maple, oak, spruce, and others.

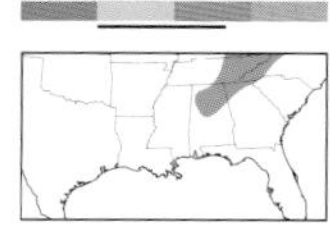

PEPPER-AND-SALT GEOMETER

Biston betularia 91-1062 (6640) **Local**

WS 40–45 mm. Chunky with a distinctively long FW profile. Peppery whitish or light gray wings are marked with wavy black AM and PM lines, often with bands of darker shading. Melanic individuals are sooty black. **HOSTS:** Alder, birch, cherry, dogwood, elm, oak, tamarack, willow, and many other trees. **NOTE:** Two broods.

TYPICAL GEOMETERS
PORCELAIN GRAY
actual size
TULIP-TREE BEAUTY
WHITE-TIPPED BLACK
CANADIAN MELANOLOPHIA
SIGNATE MELANOLOPHIA
typical
PEPPER-AND-SALT GEOMETER
melanic

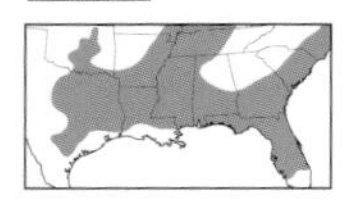

WOOLLY GRAY

Lycia ypsilon 91-1073 (6652) **Common**

WS 30–35 mm. Peppery whitish FW of male has curvy black AM and PM lines that converge at midpoint, forming an hourglass shape. Straight median line converges with PM line at inner margin. Basal area and broad ST band are brown. Blackish discal spots mark all wings. Grayish brown female is almost wingless. **HOSTS:** Apple, cherry, oak, and probably other trees.

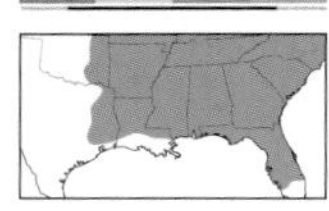

ONE-SPOTTED VARIANT

Hypagyrtis unipunctata 91-1075 (6654) **Common**

WS 20–47 mm. Sexually and seasonally variable. Peppery FW is pale gray or yellowish brown marked with fragmented black lines. Spring brood is generally more variegated. Darker individuals are sooty with brown ST band. Nearly always marked with a bold white spot in outer ST area. Outer margin of wings is scalloped, especially in larger female. **HOSTS:** Alder, birch, elm, hawthorn, ironwood, oak, poplar, willow, and other trees. **NOTE:** Two or three broods.

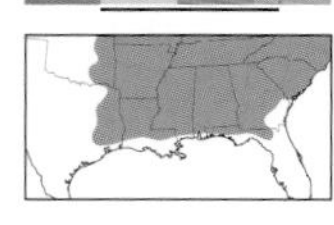

ESTHER MOTH

Hypagyrtis esther 91-1076 (6655) **Common**

WS 25–45 mm. Very similar to One-spotted Variant, but wings are not mottled and average more uniformly purplish brown. Oval-shaped whitish spot in outer ST area is usually well defined. Some individuals may not be safely identified in the field. Perhaps best identified by its preference for pine-barren habitats. **HOSTS:** Pine.

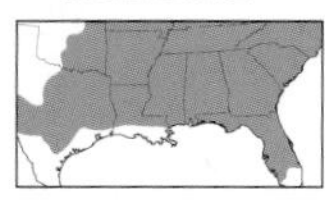

THE HALF-WING

Phigalia titea 91-1079 (6658) **Common**

WS 30–40 mm. Peppery light gray FW of male has wavy double AM and PM lines. Straight median line usually does not touch PM line. Melanic individuals are uniformly sooty with black veins. Grayish female is almost wingless. **HOSTS:** Basswood, blueberry, elm, hickory, maple, oak, and other woody plants.

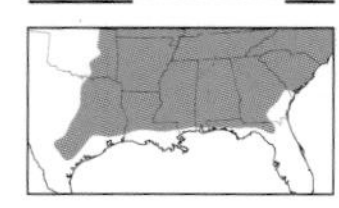

TOOTHED PHIGALIA

Phigalia denticulata 91-1080 (6659) **Common**

WS 30–37 mm. Resembles The Half-Wing, but PM line on FW is jagged, especially near costa. Slightly wavy median line almost touches PM line at inner margin. Flightless female has vestigial wings. **HOSTS:** Unrecorded, but probably a variety of woody plants.

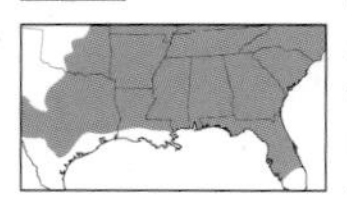

SMALL PHIGALIA

Phigalia strigataria 91-1081 (6660) **Common**

WS 30–38 mm. Resembles Toothed Phigalia, but often-fragmented PM line is straighter and less jagged near inner margin. Flightless female has tiny wings. **HOSTS:** Blueberry, chestnut, elm, oak, willow, and other woody plants.

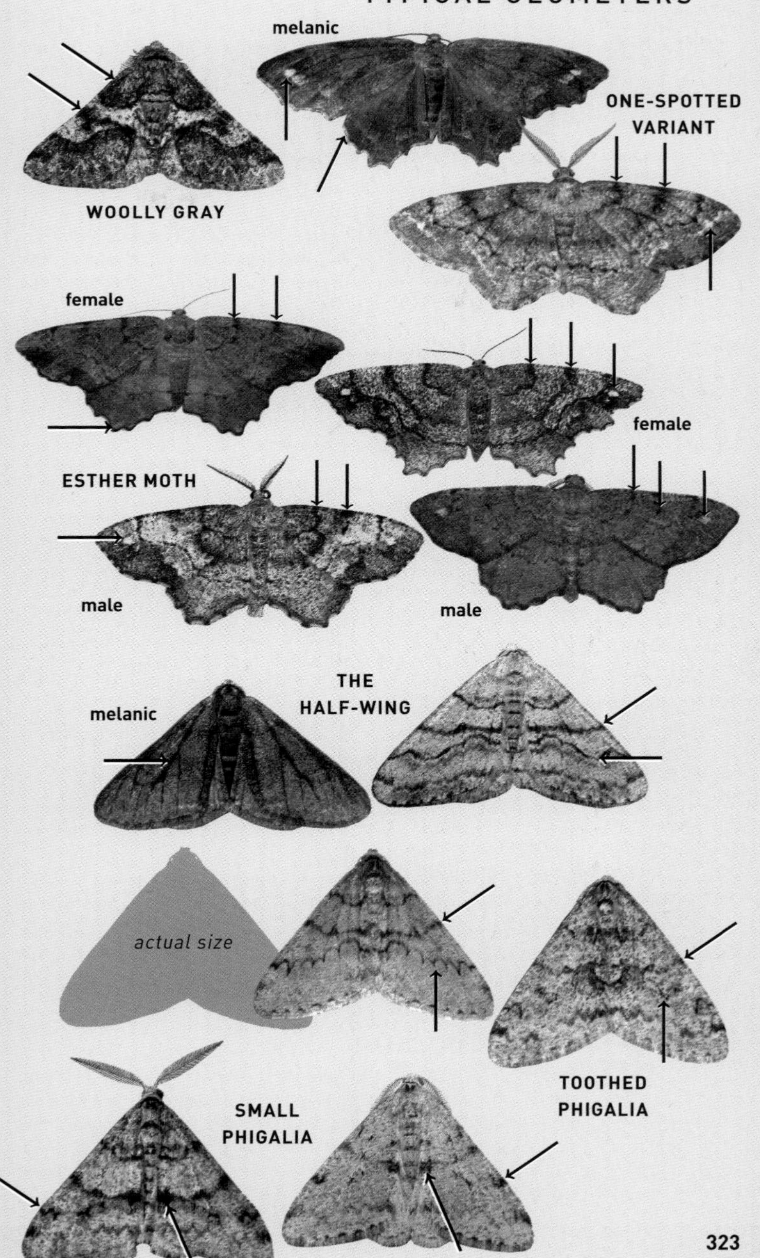
melanic
ONE-SPOTTED VARIANT
WOOLLY GRAY
female
female
ESTHER MOTH
male
male
THE HALF-WING
melanic
actual size
TOOTHED PHIGALIA
SMALL PHIGALIA

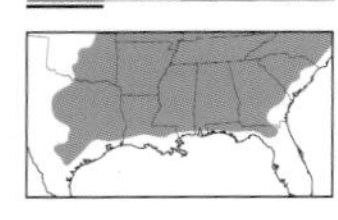

SPRING CANKERWORM

Paleacrita vernata 91-1083 (6662) **Common**

WS 22–35 mm. Silvery gray or brownish FW of male is patterned with often fragmented black lines and veins. Apical dash is connected to outer PM line. Grayish female is wingless. **HOSTS:** Apple, birch, cherry, elm, maple, oak, and other trees.

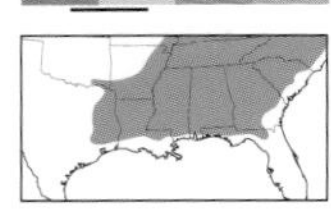

WHITE-SPOTTED CANKERWORM

Paleacrita merriccata 91-1084 (6663) **Uncommon**

WS 28–36 mm. Similar to slightly smaller Spring Cankerworm, but slightly darker FW of male is more uniform, often with indistinct lines. Most obvious feature is black-ringed white discal spot in central median area. Female is wingless. **HOSTS:** Unrecorded, but probably a variety of woody plants.

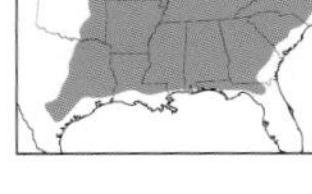

LINDEN LOOPER *Erannis tiliaria* 91-1086 (6665) **Local**

WS 32–42 mm. Finely peppered wings of male are light brown, often with darker brown bands along angulate AM and PM lines. HW is whitish. Wingless female often has pale abdomen blotched black. **HOSTS:** Ash, birch, cherry, elm, maple, oak, poplar, and other deciduous trees.

BLUISH SPRING MOTH

Lomographa semiclarata 91-1088 (6666) **Local**

WS 18–22 mm. Diurnal. Rests with wings closed tight above abdomen. Underside of wings is white with dotted ST and terminal lines and bold black discal spots. Upper surface of FW is powdery bluish gray with diffuse brownish bands. **HOSTS:** Cherry, chokecherry, hawthorn, ninebark, and other members of rose family.

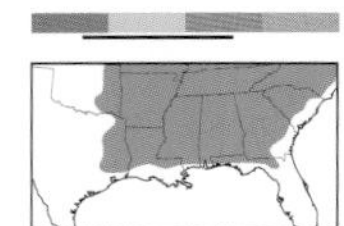

WHITE SPRING MOTH

Lomographa vestaliata 91-1089 (6667) **Common**

WS 15–26 mm. Shiny semi-translucent white wings are without obvious markings. Underside of FW is tinged yellowish brown along costa. **HOSTS:** Apple, cherry, hawthorn, maple, snowberry, and others.

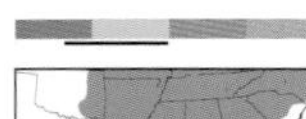

GRAY SPRING MOTH

Lomographa glomeraria 91-1090 (6668) **Uncommon**

WS 22–25 mm. Peppery light gray wings are faintly marked with scalloped brown AM and PM lines and blackish discal spots. Sometimes has brown shading in inner median area. **HOSTS:** Cherry and hawthorn.

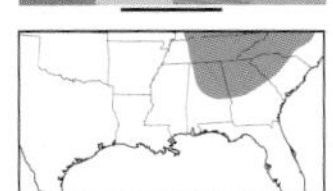

YELLOW-DUSTED CREAM

Cabera erythemaria 91-1098 (6677) **Local**

WS 21–30 mm. Powdery whitish wings are patterned with evenly spaced yellowish AM, median, and PM lines. Some individuals can be densely peppered yellowish brown, especially in basal area. Male has bipectinate antennae. **HOSTS:** Willow; also aspen and poplar. **NOTE:** Two broods.

SPRING CANKERWORM
WHITE-SPOTTED CANKERWORM
LINDEN LOOPER
WHITE SPRING MOTH
BLUISH SPRING MOTH
actual size
GRAY SPRING MOTH
YELLOW-DUSTED CREAM

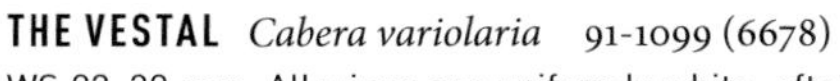

THE VESTAL *Cabera variolaria* 91-1099 (6678) **Local**

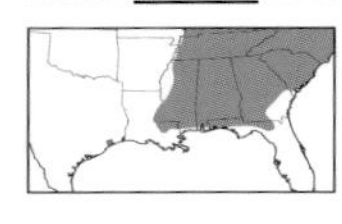

WS 22–32 mm. All wings are uniformly white, often with some fine brown speckling and rarely with traces of faint lines. Front of head and forelegs are tinted yellow or orange. **HOSTS:** Aspen, poplar, and willow. **NOTE:** Two or three broods.

FOUR-LINED CABERA

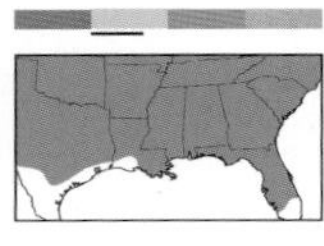

Cabera quadrifasciaria 91-1101 (6680) **Uncommon**

WS 26–32 mm. White wings are boldly patterned with four (three on HW) evenly spaced and parallel yellowish brown lines. **HOSTS:** False indigo.

BICOLORED CHLORASPILATES

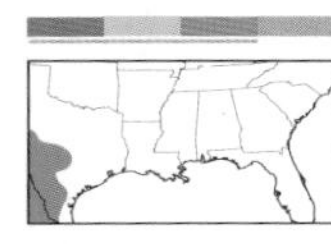

Chloraspilates bicoloraria 91-1121 (6700) **Local**

WS 18–22 mm. Green (occasionally light brown) FW is tinted brown along costa, often with multiple dark blocks. Hollow discal spot is sometimes present. Note dark brown subapical dash. Light rusty brown HW has a darker median line. **HOSTS:** Unknown. **NOTE:** Up to three broods.

BROAD-LINED ERASTRIA

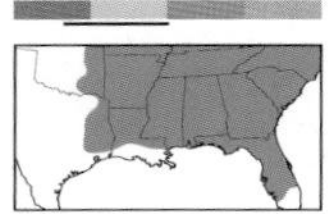

Erastria coloraria 91-1125 (6704) **Uncommon**

WS 27–37 mm. Largely diurnal. Rests with wings closed tight above abdomen. Seasonally dimorphic. Grayish brown wings of spring brood have broad dark chestnut AM and PM lines visible on undersides. Pale yellow wings of summer brood have less obvious lines, and undersides of wings are shaded pink. **HOSTS:** New Jersey tea. **NOTE:** Two broods, but single-brooded in south.

THIN-LINED ERASTRIA

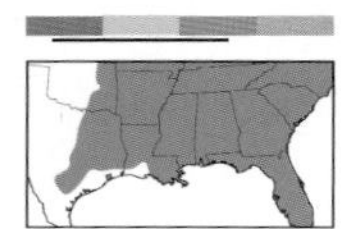

Erastria cruentaria 91-1126 (6705) **Common**

WS 27–37 mm. Largely diurnal. Rests with wings flat, or mostly so. Seasonally dimorphic. Light brown FW has curved AM and straight PM lines finely etched in brown. A faint median line is sometimes present. FW of spring brood has black spots at midpoint and inner section of reddish-shaded ST area. **HOSTS:** Blackberry. **NOTE:** Two broods.

BLACK-DOTTED RUDDY

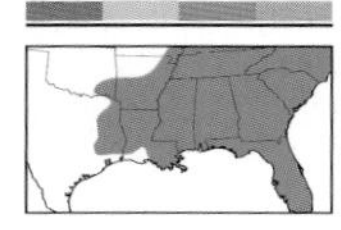

Ilexia intractata 91-1132 (6711) **Common**

WS 22–25 mm. Peppery wings are variably ochre to reddish brown and patterned with three evenly spaced darker brown lines. Black discal dots mark all wings. **HOSTS:** Holly. **NOTE:** Three or more broods.

SOLITARY RUDDY

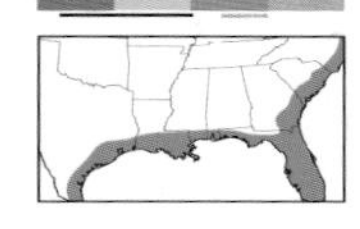

Episemasia solitaria 91-1137 (6713) **Uncommon**

WS 26–30 mm. Light brown wings are faintly patterned with dotted lines. Some individuals have extensive dark brown shading in inner basal area and beyond wavy PM line. Tiny black discal dots are inconspicuous. **HOSTS:** Holly.

TYPICAL GEOMETERS

FAWN RUDDY

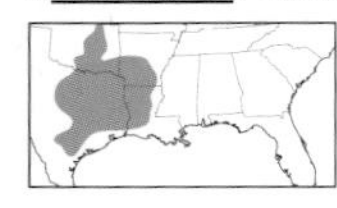

Episemasia cervinaria 91-1138 (6714) **Local**

WS 24–28 mm. Peppery light brown wings are patterned with gently curving pale-edged AM (absent on HW) and PM lines. Black discal dots mark all wings. **HOSTS:** Unknown.

COMMON LYTROSIS

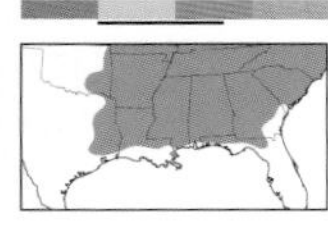

Lytrosis unitaria 91-1145 (6720) **Common**

WS 45–57 mm. This large moth has pale gray or yellowish wings with rust and black shading beyond mostly straight PM line. Fine black lines create a barklike pattern. Larger females are generally grayer than males. HW has scalloped outer margin. **HOSTS:** Hawthorn, rose, and serviceberry; also maple and oak.

SINUOUS LYTROSIS

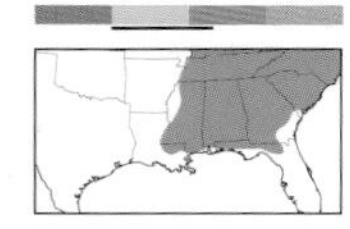

Lytrosis sinuosa 91-1146 (6721) **Uncommon**

WS 48–60 mm. Similar to Common Lytrosis, but mostly brown FW has a contrasting wide, pale yellow median area. Inner section of PM line is wavy. Inner terminal area of HW is white. **HOSTS:** Unknown, but oak and box elder in captivity.

GRAY LYTROSIS

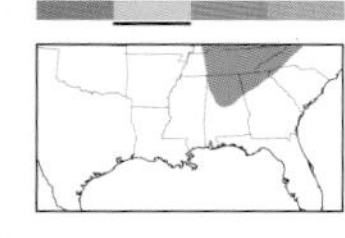

Lytrosis permagnaria 91-1148 (6723) **Local**

WS 50–75 mm. Light gray wings are faintly brindled darker. Black PM line on FW curves inward before reaching costa. AM line is only obvious as a black spot at costa. HW has a single black median line and scalloped outer margin. **HOSTS:** Unknown, but red oak and hickory in captivity.

THE SAW-WING

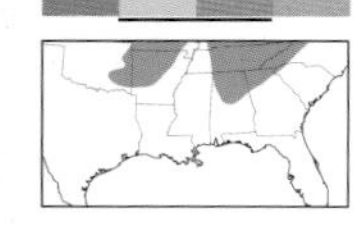

Euchlaena serrata 91-1149 (6724) **Local**

WS 40–53 mm. Pale yellow wings are shaded chocolate brown beyond straight PM line. Incomplete AM line and small discal spots are present on all wings. Outer margin of HW is deeply scalloped. **HOSTS:** Apple, blueberry, and maple.

MUZARIA EUCHLAENA

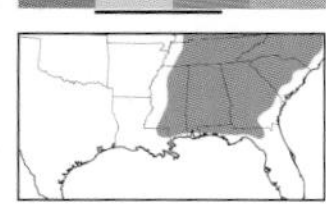

Euchlaena muzaria 91-1150 (6725) **Uncommon**

WS 27–48 mm. Peppery tawny or purplish brown wings are often slightly darker beyond curved yellow-spotted PM line. Short blackish apical dash is usually not spotted. Tiny black discal spots mark all wings. Apex of FW is sharply pointed. Outer margin of HW is deeply scalloped. **HOSTS:** Black cherry and chokecherry.

OBTUSE EUCHLAENA

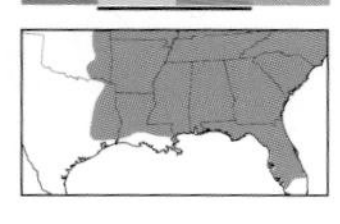

Euchlaena obtusaria 91-1151 (6726) **Common**

WS 27–48 mm. Similar to Muzaria Euchlaena, and wing color is likewise variable. Faint apical dash is accented with two or more bold black dots. **HOSTS:** Rose and impatiens; also birch and black cherry. **NOTE:** Two broods.

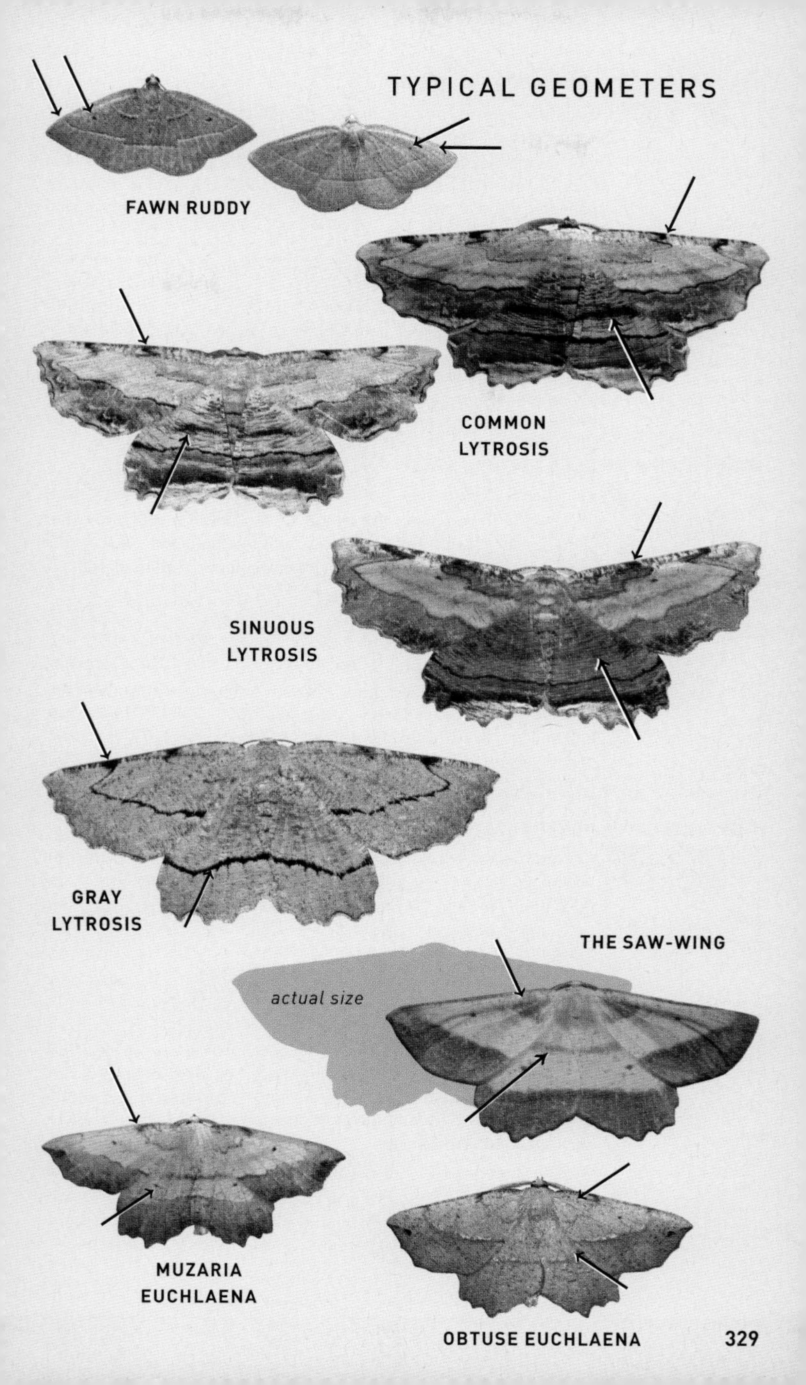
TYPICAL GEOMETERS
FAWN RUDDY
COMMON LYTROSIS
SINUOUS LYTROSIS
GRAY LYTROSIS
THE SAW-WING
actual size
MUZARIA EUCHLAENA
OBTUSE EUCHLAENA

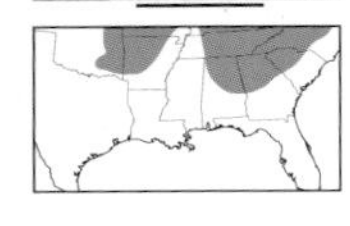

JOHNSON'S EUCHLAENA

Euchlaena johnsonaria 91-1154 (6729) **Local**

WS 26–37 mm. Uniform tawny or chocolate brown wings are marked with crisp black lines and faint dark veins. ST line often has blackish blotches at inner margin and near apex. Outer margins of wings are deeply scalloped. **HOSTS:** Ash, birch, elm, hawthorn, willow, and others. **NOTE:** Two broods.

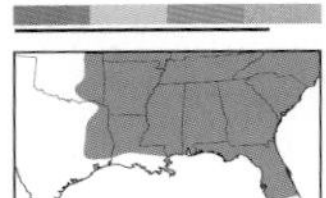

SCRUB EUCHLAENA

Euchlaena madusaria 91-1156 (6731) **Common**

WS 26–37 mm. Peppery pale yellow wings are usually boldly patterned with brown lines and veins. Brown shading beyond curved PM line on FW is often fragmented. Short apical dash on FW is edged pale yellow. Outer margin of HW is barely scalloped. **HOSTS:** Unknown, but reported on buffalo berry and conifers in north.

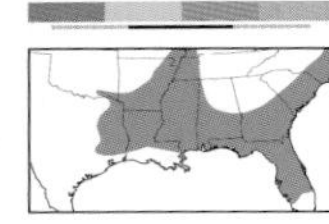

BANDED EUCHLAENA

Euchlaena deplanaria 91-1157 (6732) **Local**

WS 26–37 mm. Peppery tan wings are variably shaded brown beyond curved PM line. Pale apical dash on FW. Bold black discal dots mark all wings. Outer margin of HW is shallowly scalloped. **HOSTS:** Unknown.

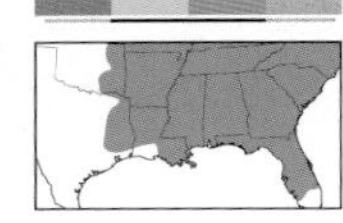

DEEP YELLOW EUCHLAENA

Euchlaena amoenaria 91-1158 (6733) **Common**

WS 30–50 mm. Peppery yellow or pale orange wings are variably shaded warm brown in basal area and beyond curved PM line. Pale apical dash on FW. Black discal dots mark all wings. Outer margin of HW often has a double point. **HOSTS:** Unknown. **NOTE:** Two broods.

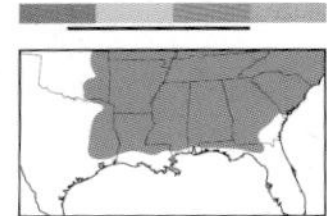

OCHRE EUCHLAENA

Euchlaena marginaria 91-1159 (6734) **Common**

WS 30–49 mm. Straw-colored wings have curved dark brown AM (absent on HW) and PM lines. Warm brown ST band is often present. Pale apical dash on FW. Black discal dots mark all wings. Outer margin of HW is weakly scalloped. **HOSTS:** Alder, ash, birch, elm, maple, viburnum, willow, and other woody plants.

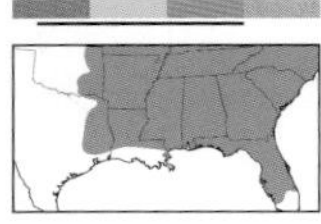

FORKED EUCHLAENA

Euchlaena pectinaria 91-1160 (6735) **Common**

WS 30–49 mm. Pale yellowish wings are heavily peppered darker and often shaded warm brown in basal area and beyond curved PM line. Small black discal dots mark all wings. Outer margin of HW is scalloped. **HOSTS:** Cherry.

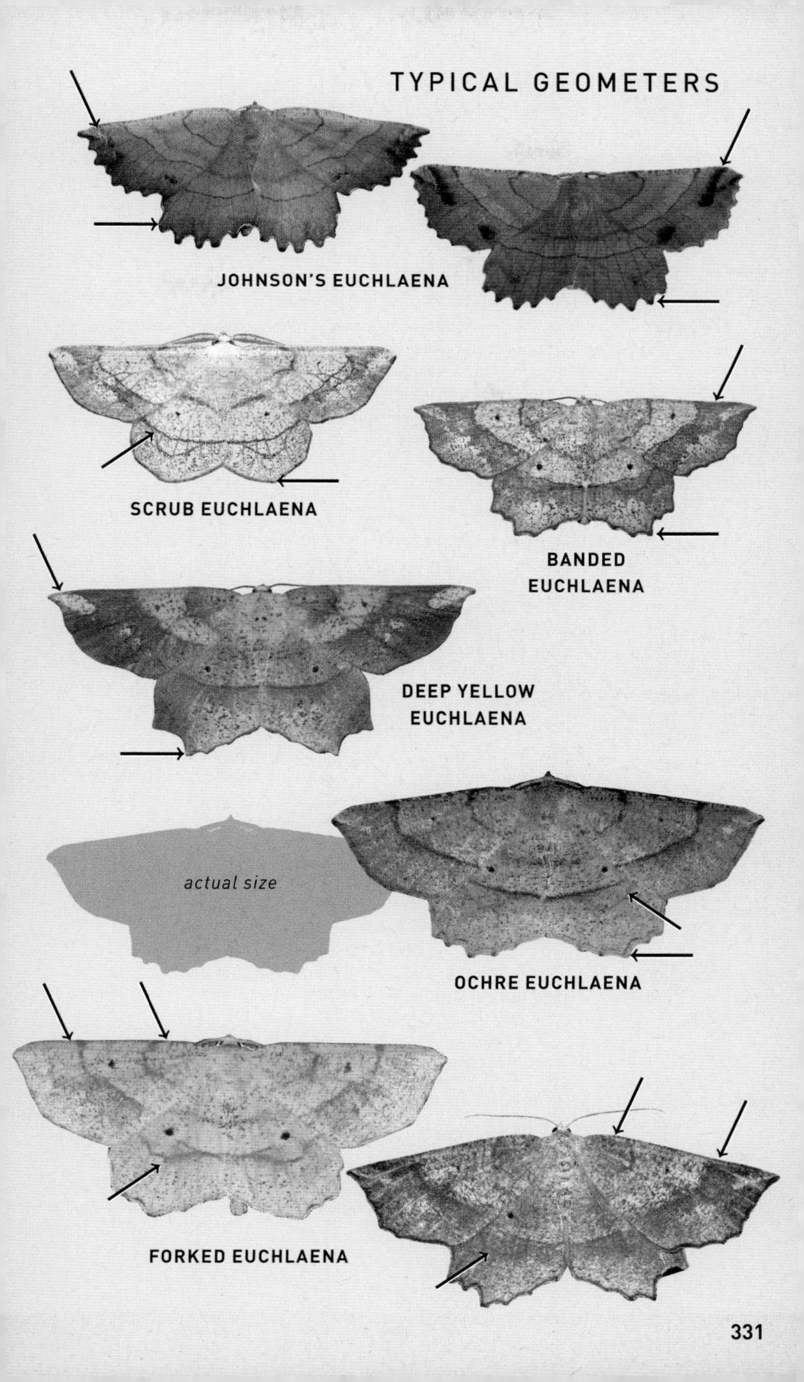
JOHNSON'S EUCHLAENA
SCRUB EUCHLAENA
BANDED EUCHLAENA
DEEP YELLOW EUCHLAENA
actual size
OCHRE EUCHLAENA
FORKED EUCHLAENA

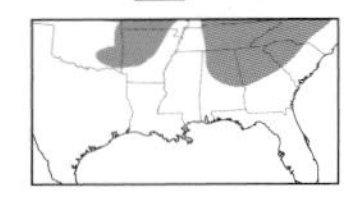

LEAST-MARKED EUCHLAENA

Euchlaena irraria 91-1164 (6739) **Local**

WS 37–48 mm. Uniform straw-colored wings have curved AM (absent on HW) and PM lines. Often has diffuse brown shading beyond inner section of PM line on FW. Outer margin of HW is weakly or not scalloped. **HOSTS:** Birch, dogwood, maple, oak, poplar, viburnum, and other woody plants.

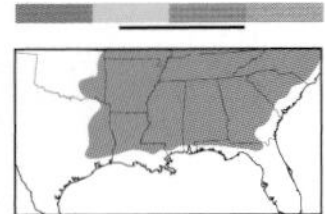

FALSE CROCUS GEOMETER

Xanthotype urticaria 91-1165 (6740) **Common**

WS 30–46 mm. Similar to larger Crocus Geometer. Bright yellow wings are variably speckled and blotched purplish brown, sometimes with almost complete AM and PM bands. Males are typically more heavily patterned than larger females. **HOSTS:** Azalea, blueberry, currant, dogwood, goldenrod, rose, sweet fern, and many others. **NOTE:** Two broods.

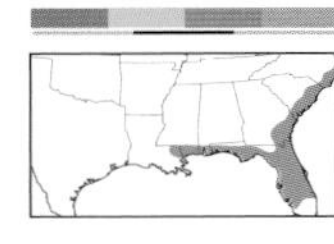

RUFOUS GEOMETER

Xanthotype rufaria 91-1167 (6742) **Local**

WS 30–50 mm. Deep yellow wings are lightly speckled and blotched purplish brown. PM line is boldly marked to midpoint of FW. Males are typically more heavily patterned than larger, paler females. Similar to other two *Xanthotype* species, but AM and PM bands are intermediate in boldness and extent. The only *Xanthotype* in much of its range. **HOSTS:** Unknown.

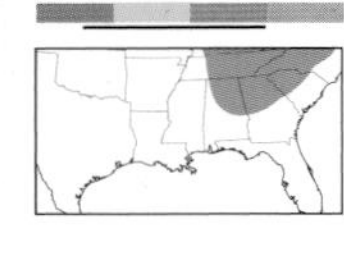

CROCUS GEOMETER

Xanthotype sospeta 91-1168 (6743) **Local**

WS 35–50 mm. Similar to slightly smaller False Crocus Geometer but is typically paler yellow with sparser brown speckles and narrower AM and PM bands. Many individuals of both species are intermediate and not reliably identified. **HOSTS:** Basswood, blueberry, cherry, currant, elm, maple, viburnum, and many others. **NOTE:** Two broods.

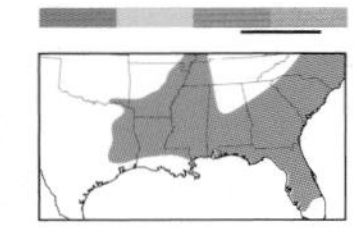

GIANT GRAY

Cymatophora approximaria 91-1170 (6745) **Uncommon**

WS 34–51 mm. Gray wings are finely brindled darker and often blotched brown. Black AM and strongly curved PM lines converge at inner margin. Inner median and PM lines are often shaded brown, forming a dark patch. Outer margin of HW is scalloped. **HOSTS:** Fringed meadowberry, greenbrier, and oak.

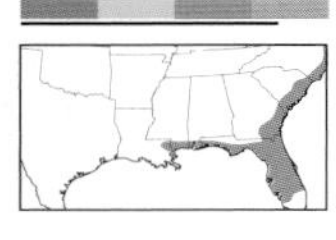

FALSE PERO

Stenaspilatodes antidiscaria 91-1171 (6746) **Local**

WS 28 mm. Resembles *Pero* species in pattern and wing posture. Peppery warm brown wings are shaded light gray in basal area and beyond almost-straight white-edged PM line. FW has black discal dots. **HOSTS:** Unknown.

TYPICAL GEOMETERS

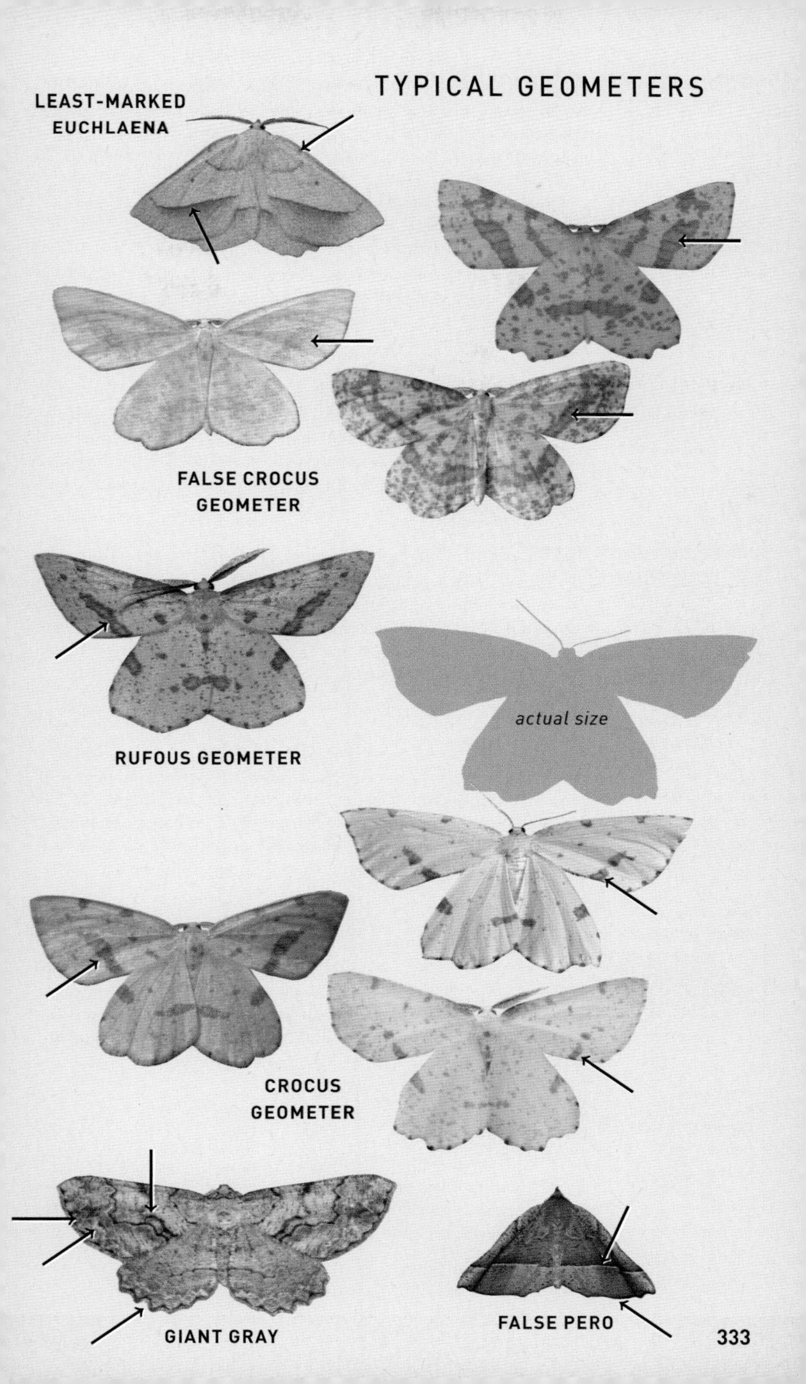

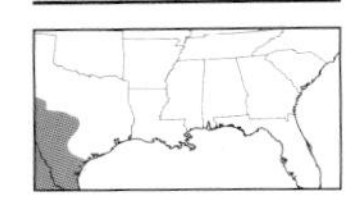

MESKE'S PERO *Pero meskaria* 91-1172 (6747) Common

WS 28 mm. This and other *Pero* species have outer part of FW creased at rest. Variable. Gray to tan, sometimes peppery, FW is usually darker through median area. Inner PM line is heavily edged with dark shading. ST area is often tan. Large white discal spot is roundish. **HOSTS:** Clematis.

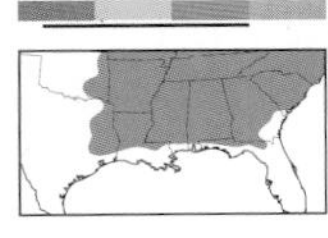

COASTAL PERO *Pero zalissaria* 91-1178 (6752) Local

WS 28–30 mm. Similar to other *Pero* species, but pale-edged PM line is almost straight. Basal and median areas are plain warm brown. Costa is pale. Small black-ringed discal dot is present on FW. **HOSTS:** Unknown.

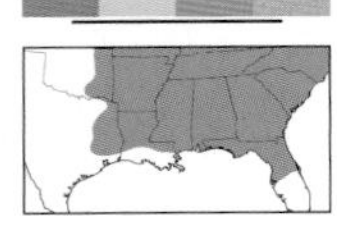

HUBNER'S PERO *Pero ancetaria* 91-1179 (6748) Common

WS 30–36 mm. Peppery dark brown FW is lighter in basal area and beyond wavy white-edged PM line. Inner PM line is obviously bulged. FW is often shaded beyond ST line. A white crescent-shaped discal spot often marks FW. **HOSTS:** Alder, birch, and willow; perhaps conifers. **NOTE:** Two broods.

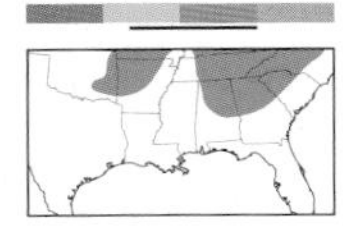

HONEST PERO *Pero honestaria* 91-1182 (6753) Common

WS 34–36 mm. Sexually dimorphic. FW is dark gray (male) or violet brown (female) and unspeckled. Basal area contrasts only slightly with median area. PM line is weakly curved. A tiny white discal spot is rarely present. **HOSTS:** Unclear because of potential confusion with other *Pero* species. **NOTE:** Two broods.

MORRISON'S PERO

Pero morrisonaria 91-1183 (6755) Common

WS 34–40 mm. Light brown FW is heavily peppered. Median area contrasts with lighter basal and PM areas. Costa is paler. PM line has a shallow bulge near inner margin. ST area is usually shaded, often bounded by a jagged, pale ST line. Small white discal spot is normally present. **HOSTS:** Fir, tamarack, pine, and spruce. **NOTE:** Two broods.

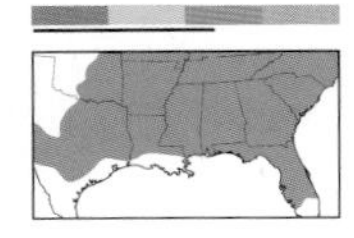

OAK BEAUTY

Phaeoura quernaria 91-1191 (6763) Common

WS 37–56 mm. Charcoal gray wings are variably mottled brown and white along jagged lines. Females are larger and tend to show wider white bands. Dark extremes are sooty with white costal spots. **HOSTS:** Ash, basswood, birch, cherry, elm, oak, poplar, willow, and other trees. **NOTE:** One or two broods.

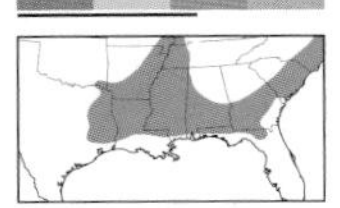

HORNED CERATONYX

Ceratonyx satanaria 91-1208 (6780) Uncommon

TL 22–24 mm. Rests with wings closed and abdomen raised above head. Grayish brown FW has white costal streak interrupted by gray median band. Curvy AM and PM lines are black. **HOSTS:** Sweet gum.

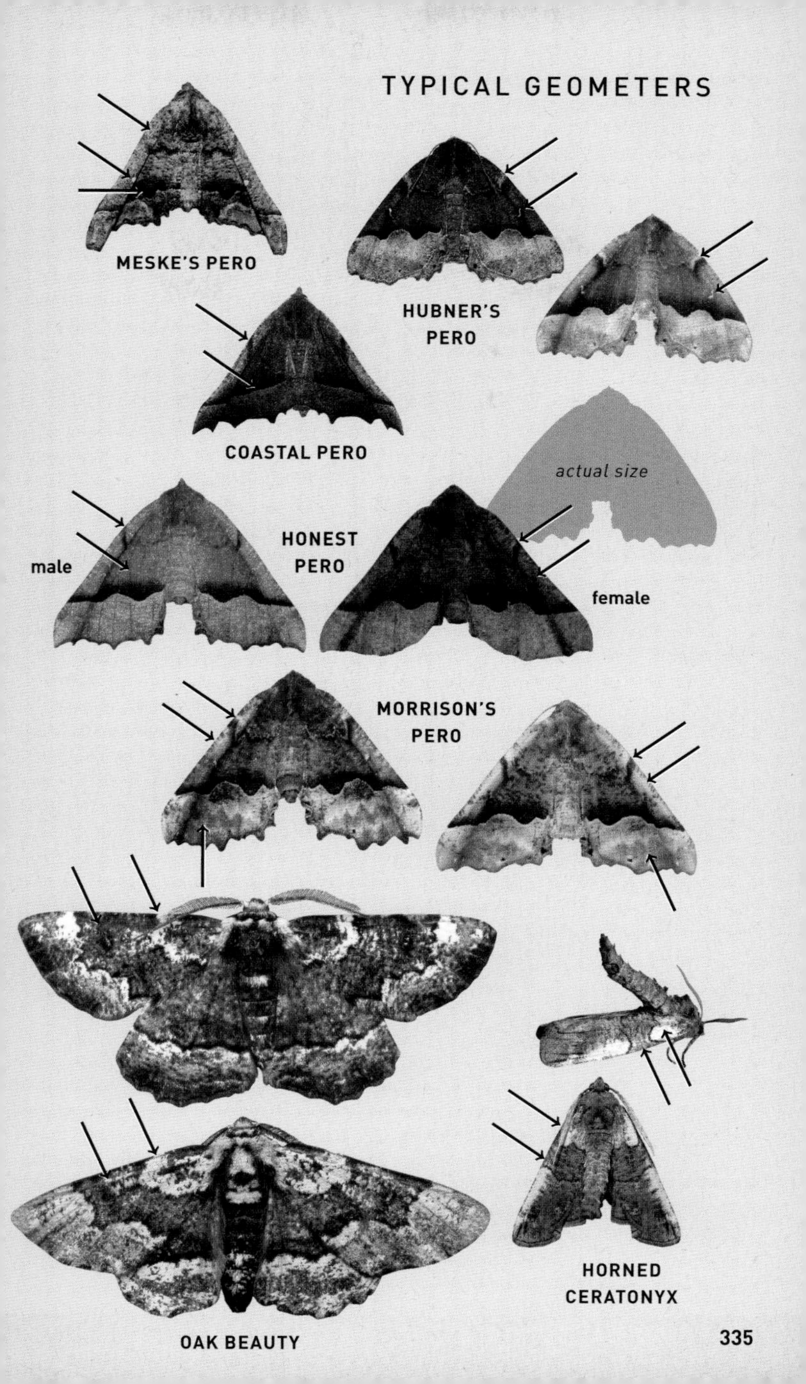
MESKE'S PERO
HUBNER'S PERO
COASTAL PERO
actual size
male
HONEST PERO
female
MORRISON'S PERO
HORNED CERATONYX
OAK BEAUTY

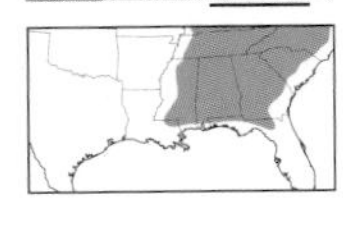

MAPLE SPANWORM

Ennomos magnaria 91-1227 (6797) **Common**

WS 43–60 mm. Pale orange wings are variably speckled and blotched brown. Indistinct lines are obvious only as spots along costa. Outer margins of wings are unevenly scalloped. Rests with wings in a dihedral. **HOSTS:** Alder, basswood, maple, oak, poplar, willow, and other trees.

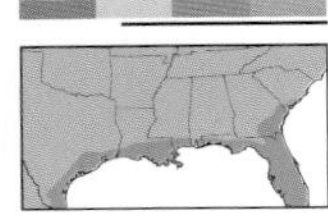

ELM SPANWORM

Ennomos subsignaria 91-1229 (6798) **Common**

WS 35–40 mm. Costa of all-white FW is often tinged green. Underside of HW has a small black discal spot. Bipectinate antennae are lime green. Sometimes rests with wings held together above abdomen. **HOSTS:** Apple, birch, elm, hickory, maple, oak, and other woody plants.

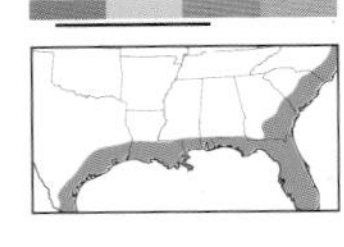

WOUNDED EPAULETTE

Sphacelodes vulneraria 91-1232 (6800) **Local**

WS 34–36 mm. Sexually dimorphic. Dark brown wings are patterned with evenly spaced AM, median, and PM lines. Male has a bold creamy or orange shallow triangle at midpoint of costa. Female is often lighter colored. HW has a tiny white discal dot. **HOSTS:** Unknown, but *Zizyphus guatemalensis* in Costa Rica.

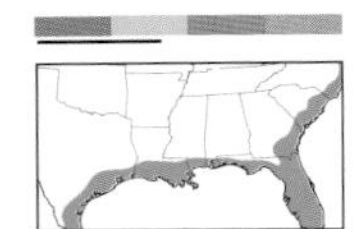

COMMON PETROPHORA

Petrophora divisata 91-1235 (6803) **Local**

WS 25–32 mm. Light violet brown FW has diffuse AM line edged brown toward inner margin. Conspicuous pale-edged brown PM line curves gently toward apical area. Pale HW has a faint PM line. **HOSTS:** Ferns.

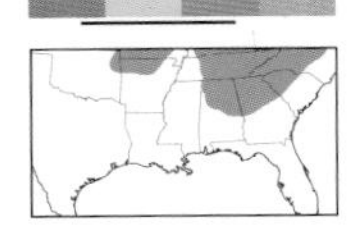

SOUTHERN PALE ALDER

Tacparia zalissaria 91-1237 (6805) **Common**

WS 28–30 mm. Light grayish or brownish wings have white-dotted PM line suffused with reddish brown shading. Sparsely dotted AM line and tiny black discal dot give FW a speckled appearance. **HOSTS:** Unknown; possibly sweet fern and wax myrtle.

KENT'S GEOMETER *Selenia kentaria* 91-1250 (6818) **Local**

WS 33–52 mm. Rests with wings tightly compressed over abdomen. Underside of HW has broad black median line and whitish band beyond PM line. Tiny white crescent is present on HW. Rarely seen upper surface of FW typically has extensive white along costa and wide white band beyond PM line. **HOSTS:** Basswood, beech, birch, elm, maple, oak, and other woody plants. **NOTE:** One or two broods.

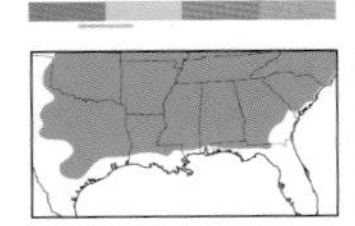

PALE METANEMA *Metanema inatomaria* 91-1251 (6819) **Local**

WS 26–36 mm. Dove gray wings are marked with paler veins and rust-edged yellowish lines. Dark crescent curves along outer margin from apex. Dark-edged discal spot on FW is often filled orange. **HOSTS:** Poplar and willow.

MAPLE SPANWORM
ELM SPANWORM
female
WOUNDED
EPAULETTE
male
COMMON PETROPHORA
actual size
SOUTHERN PALE ALDER
KENT'S
GEOMETER
PALE METANEMA

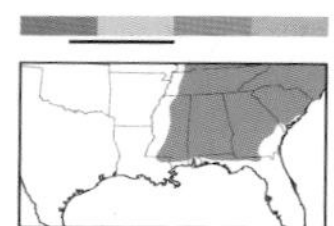

RUDDY METARRANTHIS

Metarranthis duaria 91-1254 (6822) **Local**

WS 35–40 mm. Light brown wings are tinted purplish or reddish, especially in median area. Diffuse AM and PM lines are blackish. Short black apical dash is not always present. Black discal dots mark all wings. **HOSTS:** Birch, blueberry, and cherry in captivity.

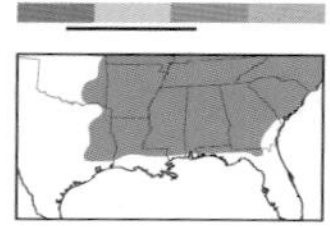

SCALLOPED METARRANTHIS COMPLEX *Metarranthis* spp.

91-1257/91-1258/91-1259 (6825/6826/6827) **Common**

WS 31–44 mm. These three species overlap greatly in appearance. Brown to straw-colored wings have sharply pointed PM line. Pale Metarranthis (*M. indeclinata*) averages darker shading along PM line and dark discal spot on FW. Refracted Metarranthis (*M. refractaria*) is very similar but usually has shading in basal area and a strong kink in the outer AM line. Scalloped Metarranthis (*M. hypochraria*) often lacks strong shading along PM line, FW discal spot may be indistinct, and outer margin is often strongly scalloped. **HOSTS:** Includes apple, blueberry, cherry, persimmon, and sassafras.

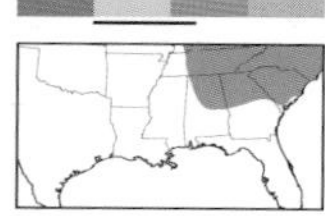

ANGLED METARRANTHIS

Metarranthis angularia 91-1255 (6823) **Common**

WS 31–44 mm. Speckled brown wings are typically paler in median area and show dark discal spots. AM and PM lines are slightly scalloped and sometimes edged darker distally. **HOSTS:** Cherry.

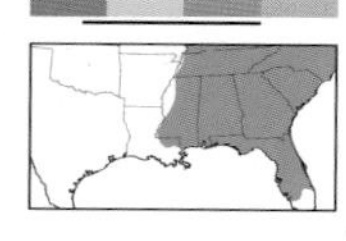

PURPLISH METARRANTHIS

Metarranthis homuraria 91-1261 (6828) **Common**

WS 26–32 mm. Pinkish brown wings are boldly patterned with wide rusty bands along angled AM and PM lines. Pale apical dash is bordered with rust. Small black discal dots mark all wings. Outer margin of HW is distinctly scalloped. **HOSTS:** Unknown.

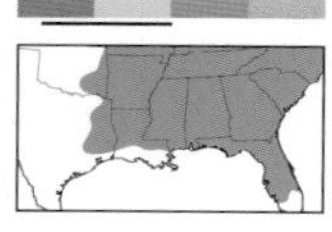

YELLOW-WASHED METARRANTHIS

Metarranthis obfirmaria 91-1265 (6832) **Common**

WS 26–36 mm. Peppery reddish brown wings usually have contrasting paler median area. Straight rusty AM and PM lines are yellow-edged. Costal half of HW is strongly tinged yellow. Black discal spots are more obvious on HW. **HOSTS:** Bearberry and blueberry.

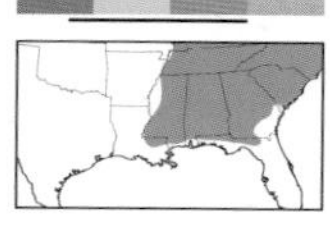

DARK SCALLOP MOTH

Cepphis decoloraria 91-1267 (6834) **Uncommon**

WS 26–33 mm. Dark bluish gray (fresh) or brownish (worn) wings have black-edged AM and PM lines. Black scallops edge indented sections of outer margins of all wings. **HOSTS:** Birch, blackberry, and cherry. **NOTE:** One or two broods.

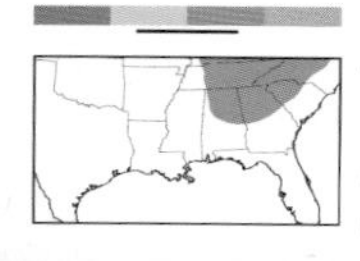

SCALLOP MOTH *Cepphis armataria* 91-1268 (6835) **Local**

WS 26–33 mm. Tawny FW has a paler costal streak and slightly wavy parallel brown lines. PM line is double. All wings have brown-edged scalloped indentations along outer margin. **HOSTS:** Apple, birch, currant, holly, maple, oak, and others. **NOTE:** One or two broods.

TYPICAL GEOMETERS

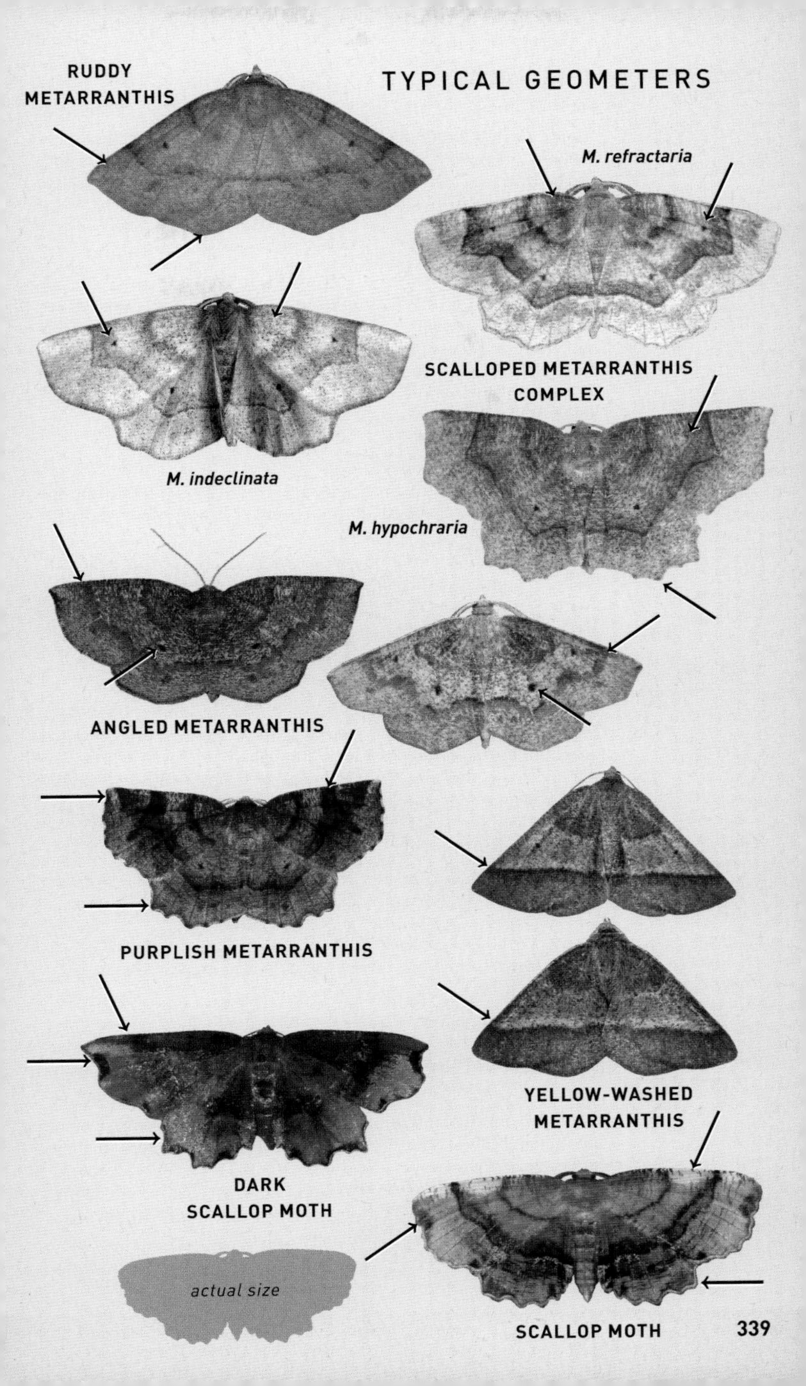

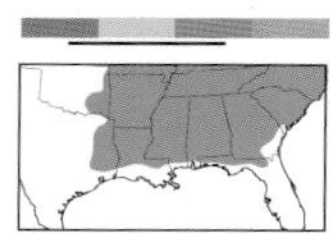

ALIEN PROBOLE

Probole alienaria 91-1269 (6837) **Common**

WS 23–34 mm. Variable. Straw-colored wings are brindled with brown streaks, and veins are indistinctly marked. Midpoint of PM line on FW is usually sharply pointed. **HOSTS:** Basswood, birch, dogwood, hawthorn, maple, and other deciduous trees. **NOTE:** Two broods.

FRIENDLY PROBOLE

Probole amicaria 91-1270 (6838) **Common**

WS 23–34 mm. Strongly resembles Alien Probole, and many individuals may not be identifiable to species. Wings may be more consistently shaded strongly purple or brown beyond PM line, and veins more distinctly marked. Tooth of PM line is typically blunted, not sharp. **HOSTS:** Primarily dogwood. **NOTE:** Two broods.

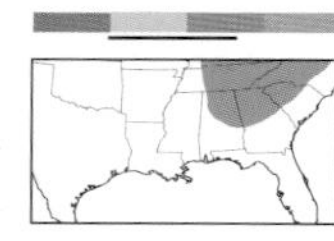

STRAIGHT-LINED PLAGODIS

Plagodis phlogosaria 91-1274 (6842) **Common**

WS 21–33 mm. Seasonally dimorphic. FW of spring brood is straw-colored, suffused with purple shading along lines. Summer brood is more uniformly orange, often with a blackish blotch in inner ST area. FW sometimes has blackish discal dots. Paler HW has a fragmented PM line. **HOSTS:** Alder, apple, basswood, birch, cherry, oak, poplar, and other trees. **NOTE:** Two broods.

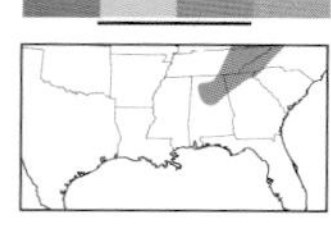

AMERICAN BARRED UMBER

Plagodis pulveraria 91-1275 (6836) **Local**

WS 26–32 mm. Light brown FW has darker brown or maroon median area between straight AM and bulging PM lines. Light brown HW has a thin PM line. **HOSTS:** Alder, birch, cherry, fir, hemlock, spruce, willow, and others. **NOTE:** Two broods.

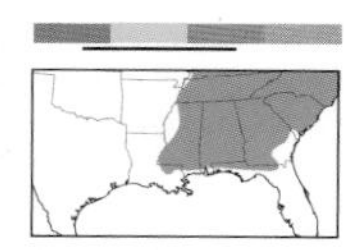

FERVID PLAGODIS

Plagodis fervidaria 91-1276 (6843) **Common**

WS 23–31 mm. Seasonally dimorphic. FW of spring brood is yellowish, strongly brindled brown, and with purple shading along inner PM line. Summer brood is more uniformly light brown with fragmented PM line. Summer brood often has black discal dots on FW. Yellowish HW has a fragmented PM line. **HOSTS:** Ash, birch, maple, oak, spruce, and others. **NOTE:** Two broods.

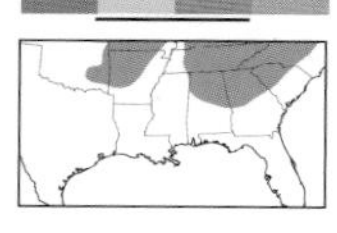

HOLLOW-SPOTTED PLAGODIS

Plagodis alcoolaria 91-1277 (6844) **Common**

WS 26–35 mm. Resembles Straight-lined Plagodis but is not as seasonally dimorphic. Bold blackish PM line curves outward slightly toward apex. Discal spots on FW are usually hollow. Median area is often strongly brindled brown. **HOSTS:** Alder, basswood, birch, chestnut, maple, oak, and other deciduous trees. **NOTE:** Two broods.

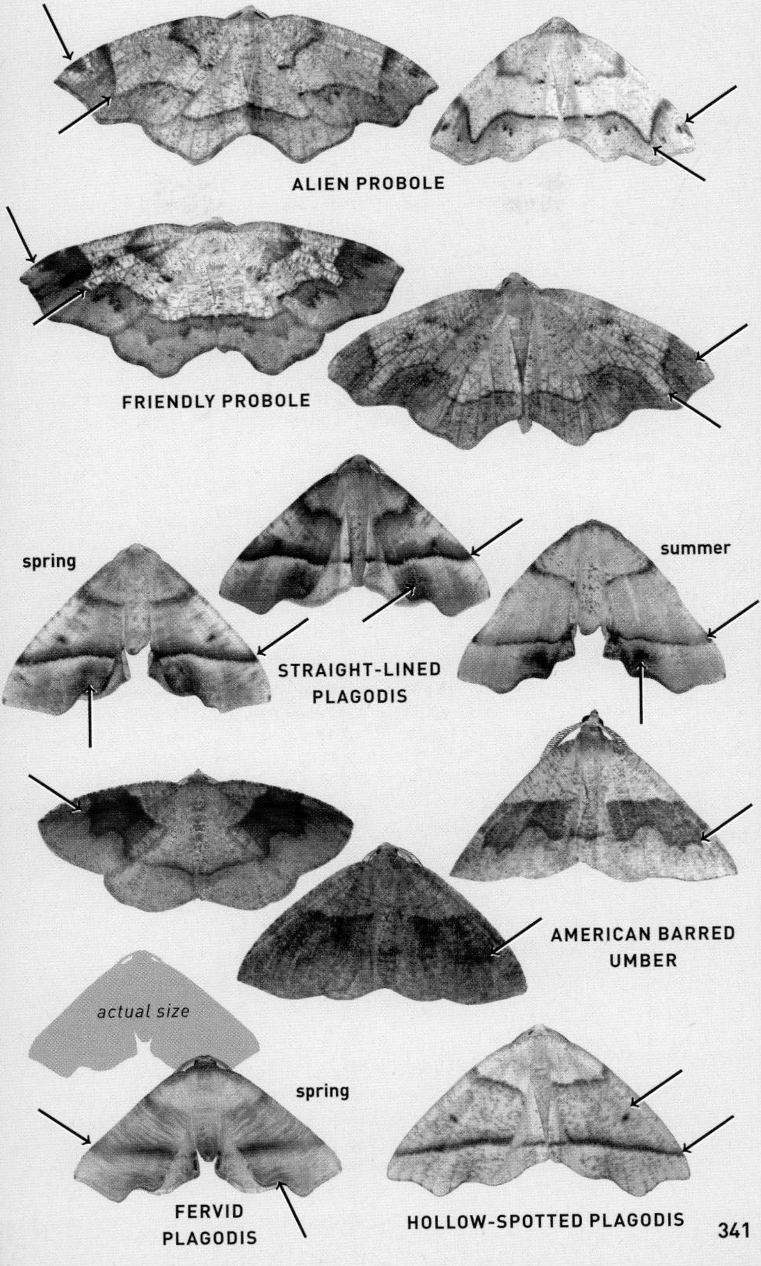

ALIEN PROBOLE

FRIENDLY PROBOLE

spring

summer

STRAIGHT-LINED PLAGODIS

AMERICAN BARRED UMBER

actual size

spring

FERVID PLAGODIS

HOLLOW-SPOTTED PLAGODIS

SPECKLED LAMPLIGHTER

Lychnosea intermicata 91-1291 (6858) **Common**

WS 30–32 mm. Straw-colored wings are evenly speckled brown, sometimes profusely on females. Broad pale-edged brown PM line on FW fades before reaching apex. **HOSTS:** Unknown.

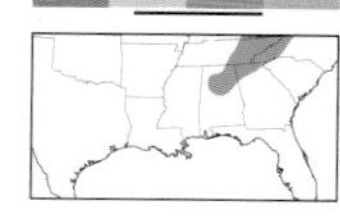

GRAY SPRUCE LOOPER

Caripeta divisata 91-1296 (6863) **Local**

WS 27–38 mm. Grayish brown FW is lightly speckled darker. Dark brown median area (sometimes fragmented) is bordered with jagged white-edged AM and PM lines. Large white discal spot is conspicuous. **HOSTS:** Fir, hemlock, tamarack, pine, and spruce.

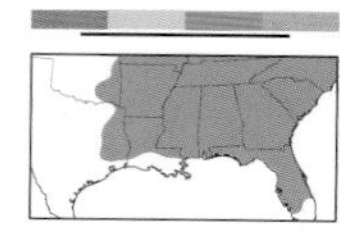

SOUTHERN PINE LOOPER

Caripeta aretaria 91-1303 (6869) **Uncommon**

WS 34–40 mm. Orange brown FW has peppery light gray basal area and ST band. Pale orange costal streak ends at PM line. Angulate AM and wavy PM lines are broadly edged white. Fringe is checkered when fresh. **HOSTS:** Reported on red pine.

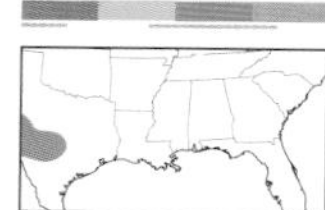

ANGLED GIRDLE

Caripeta triangulata 91-1309 (6957) **Local**

WS 30–40 mm. Light brown FW has black AM and PM lines (sometimes fragmented) angled near costa. Median area is sometimes shaded reddish brown. Note peppery gray triangular subapical patch. **HOSTS:** Unknown.

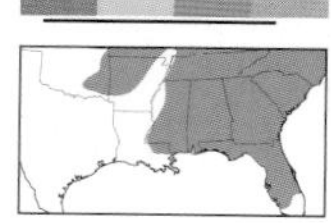

OAK BESMA

Besma quercivoraria 91-1324 (6885) **Common**

WS 27–41 mm. Variable and sexually dimorphic. Females are paler and less boldly patterned than males. Pale FW is often finely peppered brown. AM and PM lines are often well defined; in some individuals these bend to meet at midpoint of FW. Other individuals may have prominent orange or brown veins, or brown shading in ST area. Tiny black discal dots mark all wings. **HOSTS:** Alder, birch, elm, maple, oak, willow, and other trees. **NOTE:** Two broods.

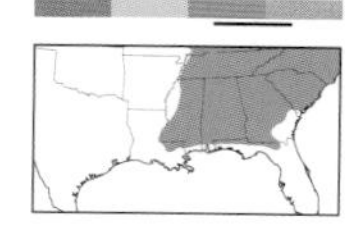

HEMLOCK LOOPER

Lambdina fiscellaria 91-1327 (6888) **Common**

WS 29–45 mm. Peppery semi-translucent grayish brown wings have bold yellow-edged AM (absent on HW) and PM lines. PM line is kinked at midpoint on all wings. Median area is sometimes darker. **HOSTS:** Apple, birch, cherry, fir, hemlock, maple, oak, spruce, and many other trees.

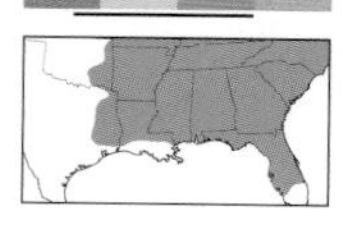

YELLOW-HEADED LOOPER

Lambdina pellucidaria 91-1331 (6892) **Common**

WS 29–40 mm. Resembles Hemlock Looper, but wide dusky AM and PM lines are slightly curved, not angulate. Median area is sometimes accented with darker veins. Head is pale yellow. **HOSTS:** Pine, including loblolly, pitch, shortleaf, and Virginia pines.

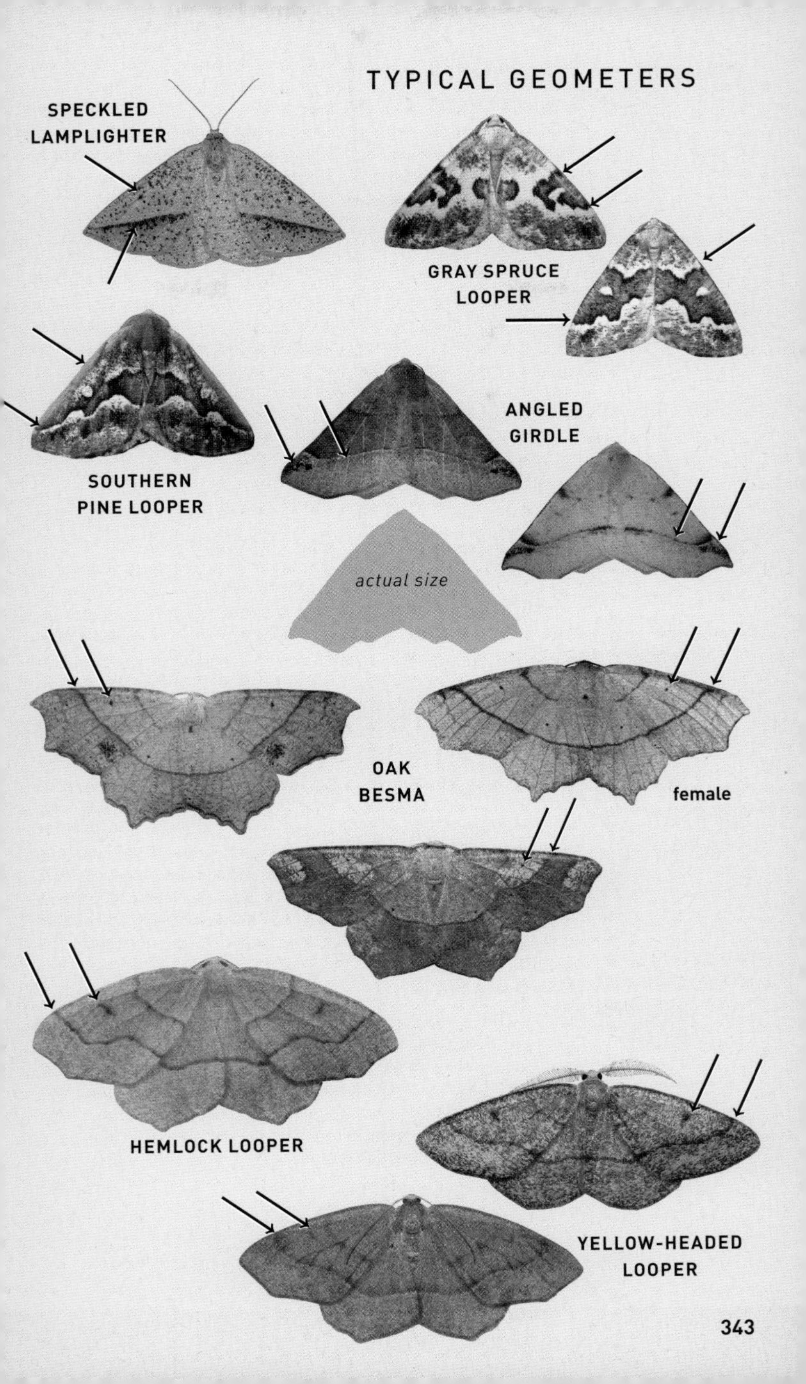
TYPICAL GEOMETERS
SPECKLED LAMPLIGHTER
GRAY SPRUCE LOOPER
SOUTHERN PINE LOOPER
ANGLED GIRDLE
actual size
OAK BESMA
female
HEMLOCK LOOPER
YELLOW-HEADED LOOPER

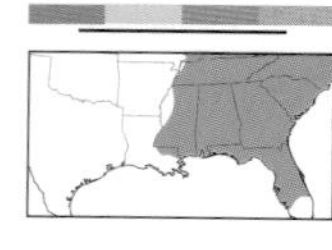

CURVE-LINED LOOPER

Lambdina fervidaria 91-1333 (6894) **Common**

WS 25–37 mm. Resembles Yellow-headed Looper, but PM line is usually angled at midpoint. Median area usually lacks darker veins and often has a blackish discal spot. **HOSTS:** Hemlock; also birch, chestnut, hickory, maple, oak, and many other trees.

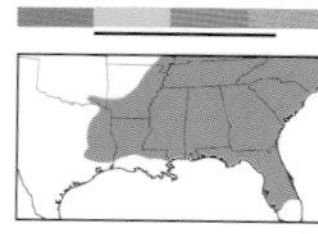

PINE CONELET LOOPER

Nepytia semiclusaria 91-1348 (6908) **Common**

WS 36 mm. Peppery gray FW has slightly darker gray shading in median area. Bold white-edged AM and PM lines are scalloped along veins. Paler HW has a single PM line. All wings are marked with bold discal spots. **HOSTS:** Pine.

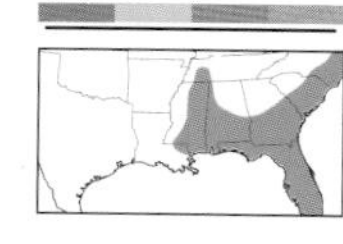

DARK-EDGED EUSARCA

Eusarca fundaria 91-1376 (6933) **Common**

WS 35 mm. Yellow wings have warm brown shading in basal area and beyond straight PM line. AM line on FW is strongly curved. Darker individuals are grayish brown with contrasting yellow-edged PM line. All wings have tiny black discal dots. **HOSTS:** Eastern baccharis.

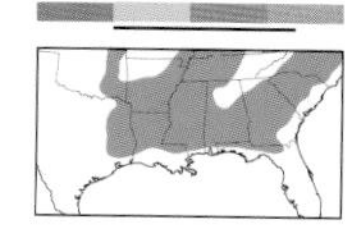

PACKARD'S EUSARCA

Eusarca packardaria 91-1379 (6936) **Uncommon**

WS 23–37 mm. Peppery grayish brown wings often have blackish blotches in ST area. Crisp brown AM and PM lines on FW are strongly angulate and widen at costa. Median and ST lines are obvious only as streaks along costa. Apex of FW is slightly falcate. **HOSTS:** Unknown.

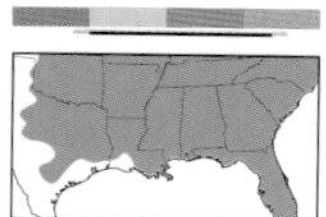

CONFUSED EUSARCA

Eusarca confusaria 91-1384 (6941) **Common**

WS 30–40 mm. Straw-colored wings are marked with yellow-edged AM (absent on HW) and PM lines. Less obvious AM line is rounded. Straight PM line fades before reaching apex. Some individuals have blackish blotches in ST area. All wings have tiny black discal dots. **HOSTS:** Composites, including aster, dandelion, and goldenrod.

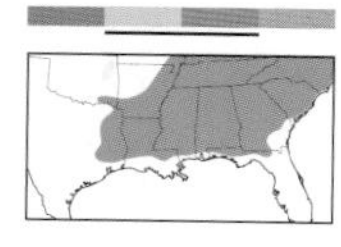

YELLOW SLANT-LINE

Tetracis crocallata 91-1400 (6963) **Common**

WS 25–45 mm. Creamy yellow FW is sometimes lightly peppered with brown scales. Thick brown PM line slants from midpoint of inner margin to apex. HW has a brown median line in second brood. Tiny black discal dots mark all wings. **HOSTS:** Alder, birch, chestnut, sumac, willow, and other trees. **NOTE:** Two broods.

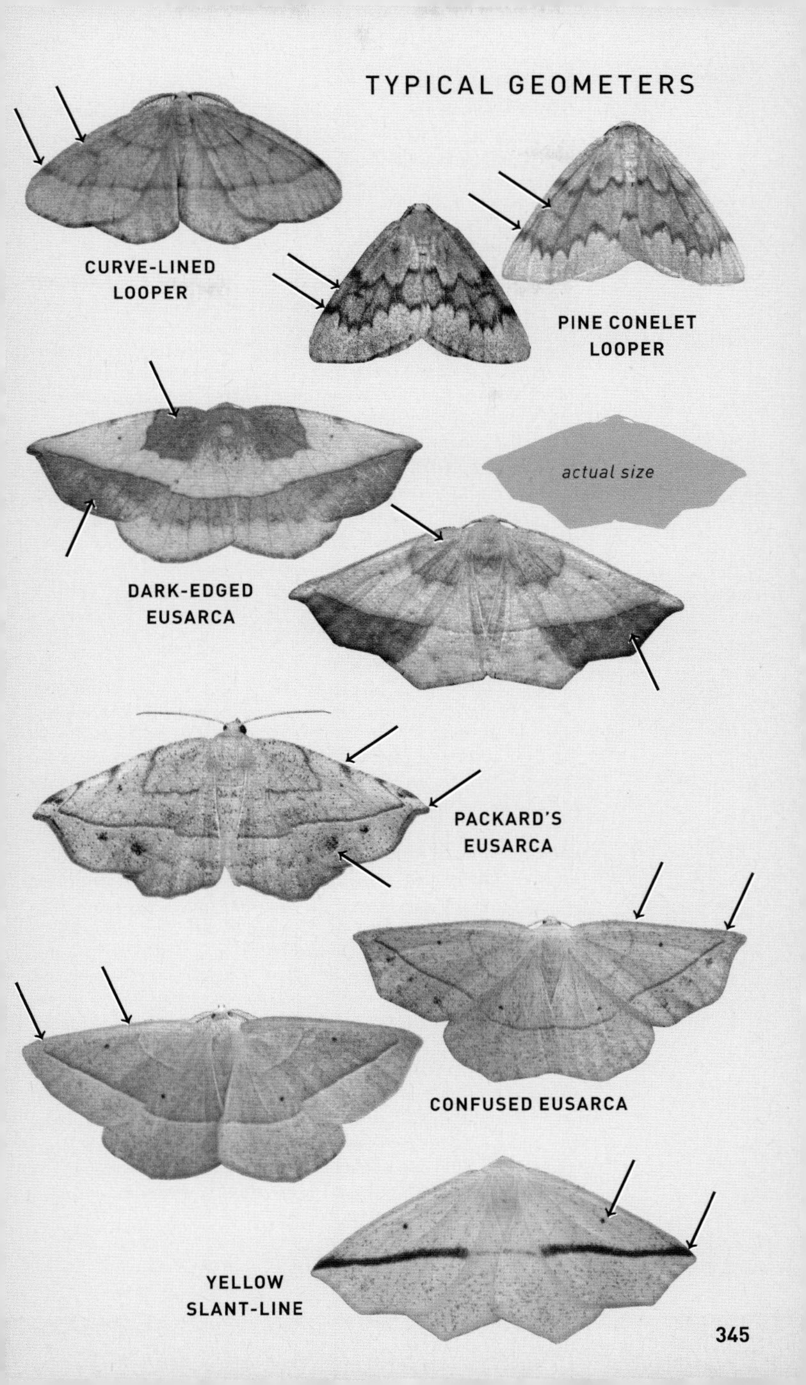
CURVE-LINED LOOPER
PINE CONELET LOOPER
actual size
DARK-EDGED EUSARCA
PACKARD'S EUSARCA
CONFUSED EUSARCA
YELLOW SLANT-LINE

WHITE SLANT-LINE

Tetracis cachexiata 91-1401 (6964) **Common**

WS 34–50 mm. White FW has a bold orange brown PM line that slants from midpoint of inner margin to apex. HW is uniformly white. **HOSTS:** Alder, ash, birch, cherry, maple, spruce, willow, and many other trees.

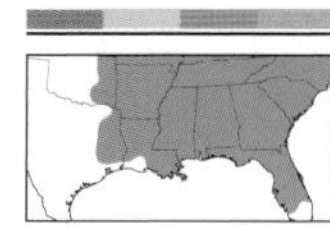

CURVE-TOOTHED GEOMETER

Eutrapela clemataria 91-1414 (6966) **Common**

WS 38–56 mm. Light brown to dark purplish brown wings are often shaded darker in median and ST areas. Yellow-edged PM line on FW jags sharply inward near costa. Apex of FW is falcate. Wiggly, dark ST line is sometimes present. Outer margin of HW is slightly scalloped. **HOSTS:** Ash, basswood, birch, cherry, elm, maple, poplar, and many other trees. **NOTE:** One or two broods.

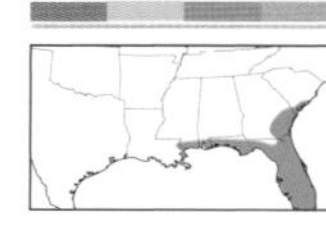

SPURGE SPANWORM

Oxydia vesulia 91-1415 (6967) **Local**

WS 55 mm. Light brown wings are variably shaded darker in median and ST areas. FW has inconspicuous wavy AM line and straight yellow-edged PM line angling inward near costa. Apex of FW is falcate. HW has a distinct black blotch beyond outer section of PM line. **HOSTS:** Citrus.

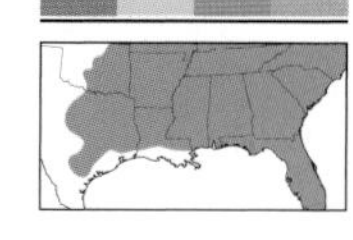

JUNIPER-TWIG GEOMETER

Patalene olyzonaria 91-1423 (6974) **Common**

WS 32–39 mm. Light brown or orange wings have faint AM line (absent on HW) and bold pale-edged PM line that kinks inward before reaching costa on FW. Often shows blackish blotches in ST area. Apex of FW is usually falcate, especially in female. **HOSTS:** Atlantic white cedar and juniper. **NOTE:** Two or more broods.

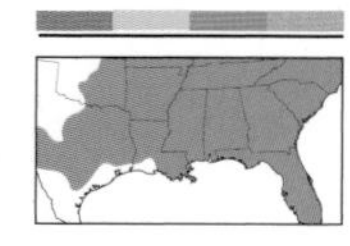

LARGE MAPLE SPANWORM

Prochoerodes lineola 91-1432 (6982) **Common**

WS 43–46 mm. Variable. Wings range from pale yellow to purplish brown. Bold, straight PM line jags just before costa. AM line is sometimes indistinct. Some individuals have a dark zigzag ST line. Tiny black discal dots mark all wings. Wingtips are rarely falcate. **HOSTS:** Birch, blueberry, cherry, gale, hemlock, oak, poplar, willow, and many other woody plants, grass, and crops. **NOTE:** Two broods.

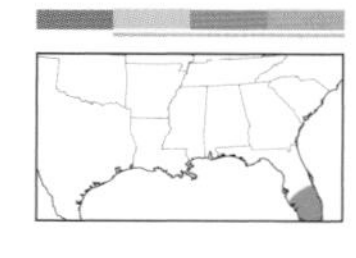

SPOTTED SWALLOW-TAILED MOTH

Nepheloleuca politia 91-1435 (6985) **Local**

WS 42–44 mm. Bright yellow wings are variably speckled brown. FW has brown-edged blotches in subapical and inner ST areas. HW has broad brown ST line and shading at anal angle. Outer margin of HW is strongly angulate, forming short tail. **HOSTS:** Unknown.

TYPICAL GEOMETERS

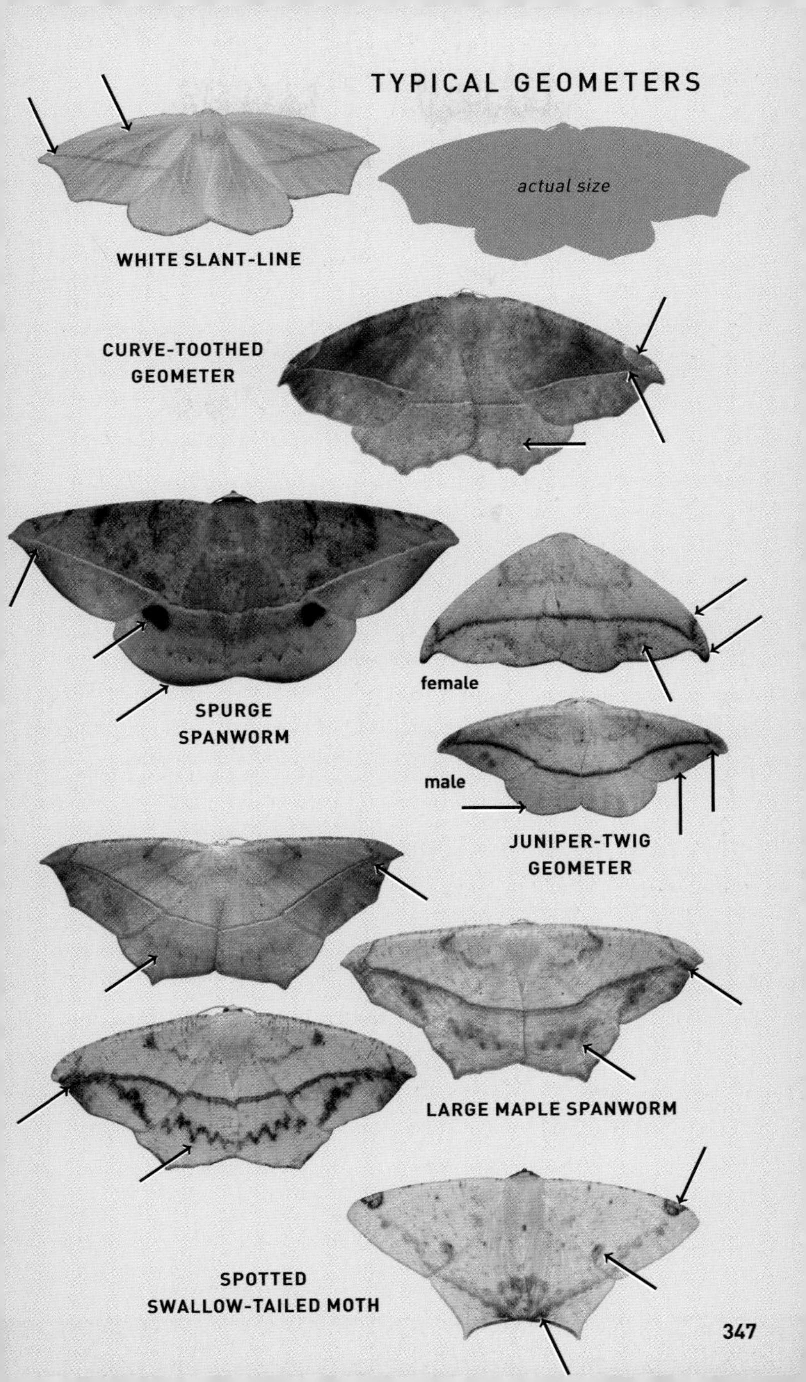

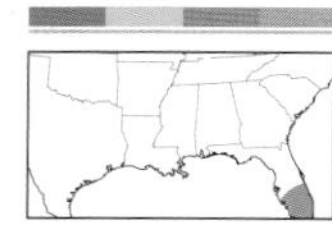

PALE SWALLOW-TAILED MOTH

Nephelolecua floridata 91-1436 (6986) **Local**

WS 42–44 mm. Very similar to Spotted Swallow-tailed Moth, and many intermediate individuals may not be safely identified in the field. Subapical patches of FW are usually indistinct, and ST line on HW is narrower, with a smaller brown patch at inner margin. **HOSTS:** Unknown.

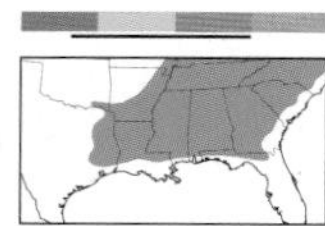

VARIABLE ANTEPIONE

Antepione thisoaria 91-1437 (6987) **Common**

WS 27–40 mm. Seasonally and sexually dimorphic. Both sexes of spring brood have brown FW marked with dark triangular subapical patch and a dusky smudge along inner PM line. In summer brood, wings are instead yellow and sexes are dimorphic; male has reddish brown shading beyond PM line. **HOSTS:** Apple, bittersweet, cherry, maple, sumac, and other woody plants. **NOTE:** Two broods.

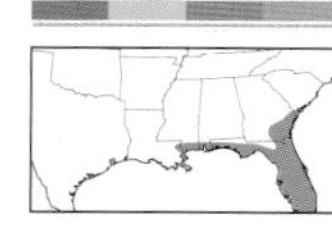

VIRGIN SILKWING

Sericoptera virginaria 91-1441 (6994) **Local**

WS 39–42 mm. Silky white wings are largely unmarked. PM line is typically barely visible as a thin line with a dark spot at costa, but more boldly marked individuals can have large chestnut olive blotches in subapical and inner ST areas of FW and a lacy gray pattern along ST line of HW. Outer margin of HW is strongly angulate, sometimes forming short tail. Head is black. **HOSTS:** Unknown.

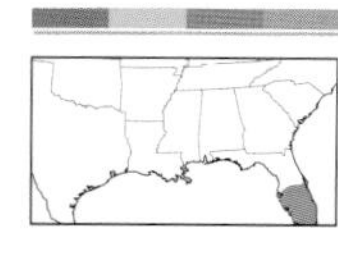

GOLDEN-WINGED PALYAS

Phrygionis auriferaria 91-1456 (6670) **Local**

WS 28–30 mm. Dimorphic. Bright golden yellow or grayish olive wings are lightly brindled brown and patterned with metallic lines. Broad brown ST band on FW fades before costa. HW has a double median line and a distinct black and red spot at midpoint of partial red terminal line. **HOSTS:** Unknown; possibly marlberry.

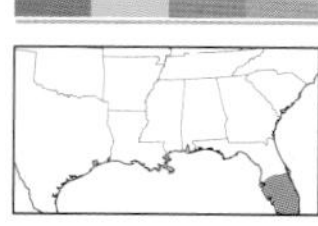

JEWELED SATYR MOTH

Phrygionis paradoxata 91-1457 (6671) **Local**

WS 24–28 mm. Dove gray wings are boldly patterned with metallic-edged yellow AM and PM bands. Apex of FW is often tinged with rust. Orange and white terminal area of bluntly angulate HW is accented with two silver-studded red spots. **HOSTS:** Marlberry.

TYPICAL GEOMETERS

Superfamily NOCTUOIDEA

The Noctuoidea are a massive assemblage of stout-bodied, often hairy moths, considered taxonomically to be the most evolutionarily recent moth lineage. Members of this superfamily are generally medium small to large, but a few are very tiny and could be confused for micromoths. Many fold their wings over their abdomen when at rest, while others rest with their wings spread like members of the Geometridae. A large number are nondescript or even somber in appearance, clad in shades of gray or brown, often with intricate and contrasting patterns on the FW. Some, notably the underwing moths in the genus *Catocala*, have vividly colorful wing patterns. Most are nocturnal and likely to be found only at lights or sugar bait, though a small number are habitual day-fliers.

Prominents Family Notodontidae

Stout, often beautifully patterned noctuid moths. Many are furry; others have a short thoracic crest or tufts of hairlike scales along the costa of the HW. Some are twig mimics: blunt-headed, resembling snapped-off twigs, or resting with their wings rolled into a tube and held elevated from the substrate. Most are found in mature woodland, though many also occur in well-established gardens. Adults do not feed. All are nocturnal and will visit lights in small numbers.

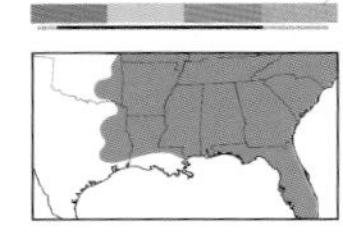

ANGLE-LINED PROMINENT

Clostera inclusa 93-0004 (7896) **Common**

TL 17–19 mm. Gray FW is crisscrossed by thin white lines. Median line angles to form a shaded triangular median area. Wavy PM line cuts through a variably golden orange to chestnut subapical patch. Tufted abdomen usually protrudes from wings. **HOSTS:** Aspen, poplar, and willow. **NOTE:** Two broods.

GEORGIAN PROMINENT

Hyperaeschra georgica 93-0010 (7917) **Common**

TL 19–24 mm. Grayish brown FW has toothed, white-edged AM and PM lines. Inner basal area and outer half of thorax have buffy oval patches. ST area is marked with black dashes at apex and anal angle. **HOSTS:** Oak. **NOTE:** Two or more broods.

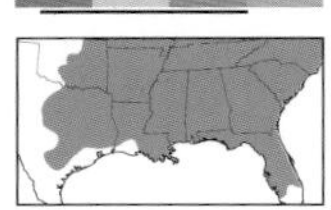

DOUBLE-TOOTHED PROMINENT

Nerice bidentata 93-0018 (7929) **Common**

TL 20–21 mm. FW is bisected by a dark, double-toothed longitudinal streak from base to outer margin. Costal side of FW is warm brown; inner half is light gray. **HOSTS:** Elm. **NOTE:** Two or more broods.

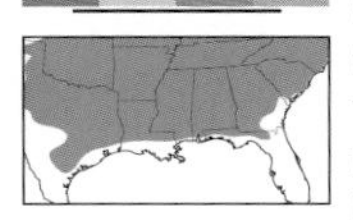

COMMON GLUPHISIA

Gluphisia septentrionis 93-0019 (7931) **Common**

TL 14–17 mm. Peppery gray FW has warm brown patches in inner median and basal areas. Slightly wavy AM and PM lines are roughly parallel. Thorax and legs are densely hairy. Darker individuals have blackish median area. **HOSTS:** Poplar. **NOTE:** Two or three broods.

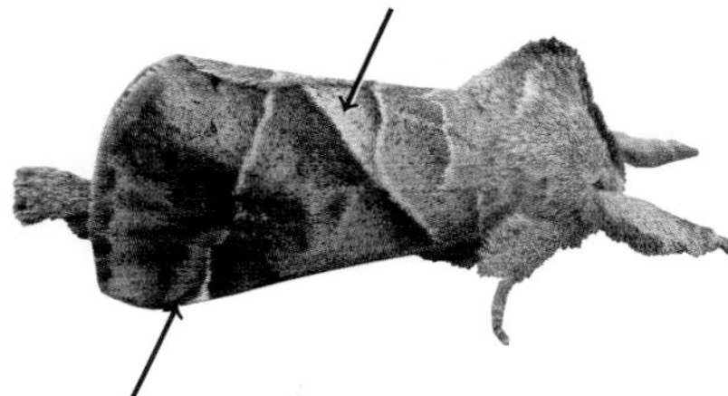

ANGLE-LINED PROMINENT

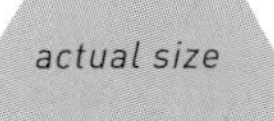

GEORGIAN PROMINENT

DOUBLE-TOOTHED PROMINENT

COMMON GLUPHISIA

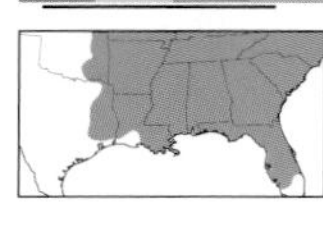

WHITE FURCULA

Furcula borealis 93-0024 (7936) **Common**

TL 18–23 mm. White FW has peppery dark gray median band and subapical patch. AM, PM, and ST lines are accented with orange and small metallic blue spots. Thorax has a topknot of gray, orange, and metallic blue scales. **HOSTS:** Cherry. **NOTE:** Two broods.

GRAY FURCULA

Furcula cinerea 93-0025 (7937) **Common**

TL 17–22 mm. Similar to White Furcula, but FW is pearly gray. Median band is often broken at midpoint. Markings are usually faded or diffuse. Thorax has horizontal bands of black, orange, and metallic blue. **HOSTS:** Aspen, poplar, and willow. **NOTE:** Two or more broods.

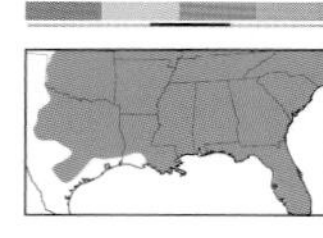

BLACK-ETCHED PROMINENT

Tecmessa scitiscripta 93-0030 (7942) **Uncommon**

TL 15–22 mm. White FW is crossed with finely etched black lines. Double AM and ST lines are sometimes filled gray. Crescent-shaped reniform spot is outlined black. Fluffy thorax is marked with black lines and spots. **HOSTS:** Poplar and willow. **NOTE:** Two or more broods.

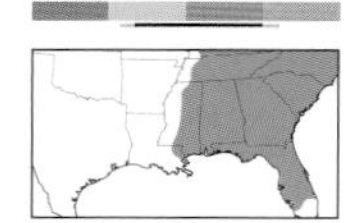

YELLOW-NECKED CATERPILLAR MOTH

Datana ministra 93-0033 (7902) **Common**

TL 22–26 mm. Tawny brown FW has four straight brown lines. PM line is indistinct. Orbicular spot is faint or absent. Outer margin is deeply scalloped when fresh. Front of thorax is chestnut. **HOSTS:** Trees and shrubs, including apple, oak, birch, and willow. **NOTE:** Two broods.

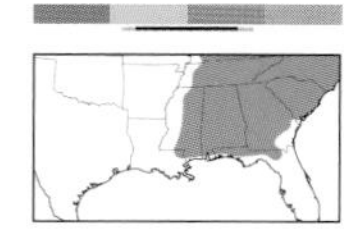

ANGUS' DATANA

Datana angusii 93-0034 (7903) **Uncommon**

TL 20–25 mm. Coppery brown FW has five straight dark brown lines. ST line is double; inner is often very close and/or fainter. Orbicular and reniform spots are sometimes present as indistinct dark blotches. Outer margin is slightly scalloped. Front of thorax is dark brown. **HOSTS:** Butternut, hickory, and walnut. **NOTE:** Two broods.

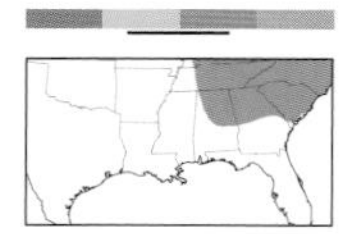

DREXEL'S DATANA

Datana drexelii 93-0035 (7904) **Uncommon**

TL 22–27 mm. Tawny brown FW has four or five brown lines; PM line is often indistinct, and ST line is sometimes double. Blackish orbicular and reniform blotches are usually relatively distinct. Black apical dash is noticeable. Outer margin is slightly scalloped. Front of thorax is chestnut. **HOSTS:** Blueberry and witch hazel. **NOTE:** Two or more broods.

PROMINENTS

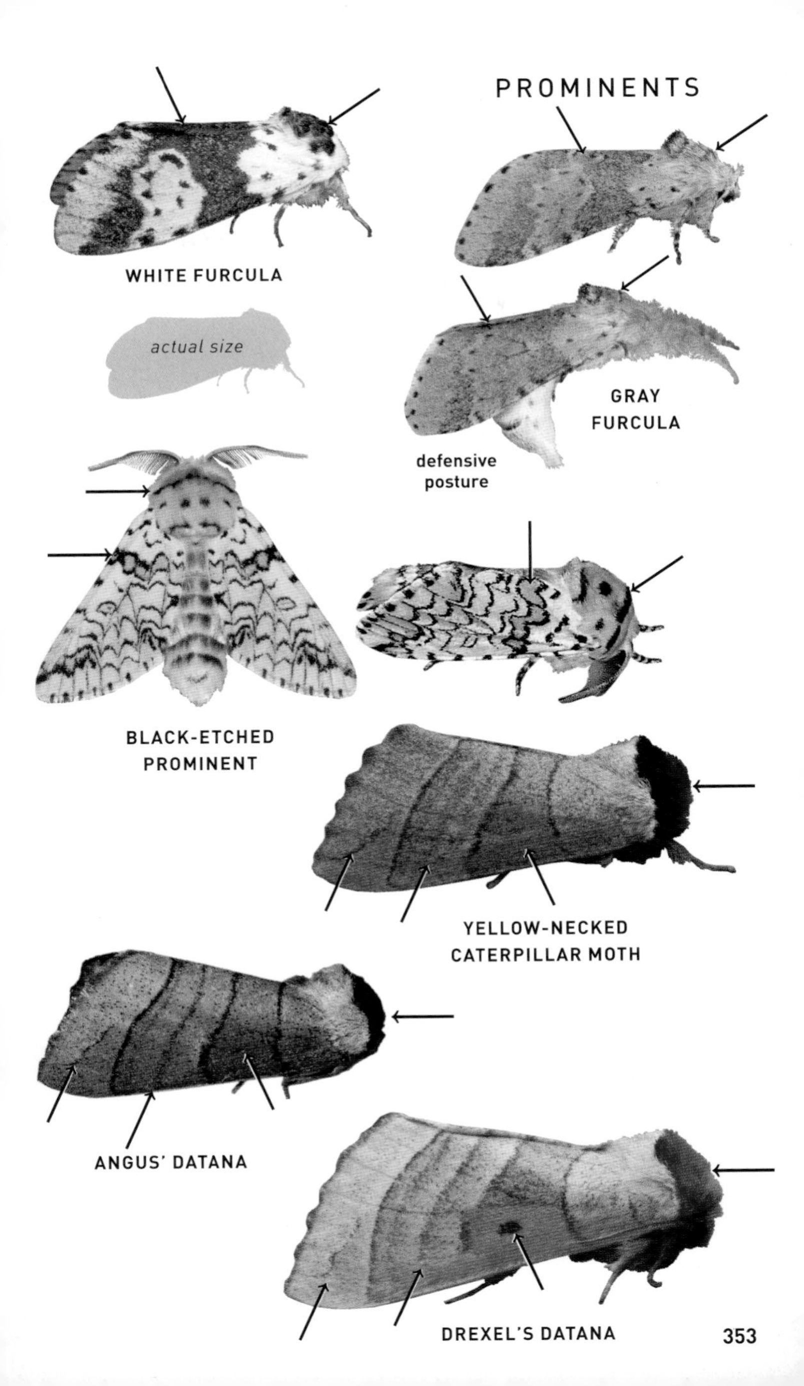

WHITE FURCULA
actual size
GRAY FURCULA
defensive posture
BLACK-ETCHED PROMINENT
YELLOW-NECKED CATERPILLAR MOTH
ANGUS' DATANA
DREXEL'S DATANA

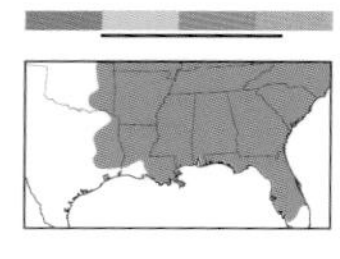

MAJOR DATANA *Datana major* 93-0036 (7905) Common

TL 21–27 mm. Peppery brown FW has four or five yellow-edged, often indistinct, lines; ST line is sometimes double. Median area may have darker brown shading between lines. Orbicular and reniform blotches are indistinct. Outer margin is straight. Front of thorax is rusty. **HOSTS:** Azalea, leatherleaf, staggerbush, and other heaths. **NOTE:** Two or more broods.

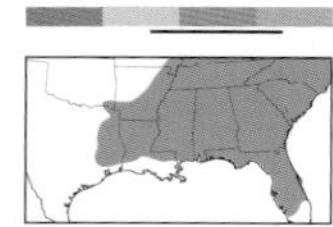

CONTRACTED DATANA

Datana contracta 93-0037 (7906) Common

TL 21–27 mm. Tawny FW has four yellow-edged, dusky lines; ST line rarely is noticeably double. Orbicular and reniform blotches are not always distinguishable. Outer margin is straight. Front of thorax is brown. **HOSTS:** Chestnut and oak; perhaps also blueberry, hickory, and witch hazel. **NOTE:** Two or more broods.

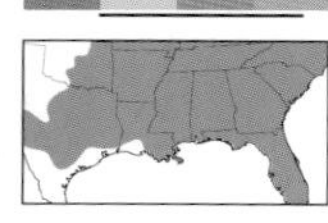

WALNUT CATERPILLAR MOTH

Datana integerrima 93-0038 (7907) Common

TL 21–28 mm. Virtually identical to Contracted Datana, and many individuals are not safely identified in the field. Yellowish edging to lines may be bolder. Apical dash is usually distinct. Rarely shows orbicular and reniform blotches. Front of thorax is dark brown to rusty. **HOSTS:** Butternut, hickory, pecan, and walnut. **NOTE:** Two broods.

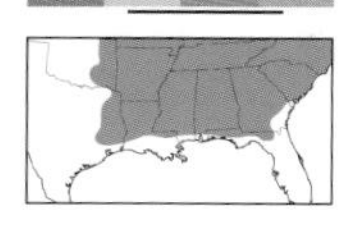

SPOTTED DATANA

Datana perspicua 93-0039 (7908) Common

TL 25–30 mm. Straw-colored or tawny FW has five reddish brown lines; ST line is double, and inner may be faint. Brown orbicular and reniform spots are conspicuous. Brown apical dash is usually distinct. Front of thorax is brown. Outer margin may be slightly scalloped. **HOSTS:** Sumac and smoke tree. **NOTE:** Two broods.

ROBUST DATANA

Datana robusta 93-0040 (7909) Local

TL 21–27 mm. Similar to Spotted Datana but is slightly shaded in median area. Apical dash reaches ST line. **HOSTS:** Unknown. **NOTE:** Two broods.

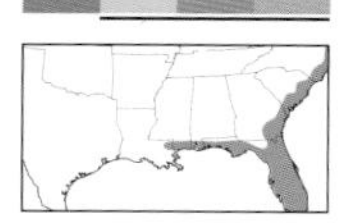

POST-BURN DATANA

Datana ranaeceps 93-0042 (7911) Local

TL 21–27 mm. Peppery reddish brown FW appears uniform, with very faint pattern of darker lines. Dusky orbicular and reniform blotches are sometimes present. **HOSTS:** Staggerbush; also fetterbush and maleberry. **NOTE:** Two broods.

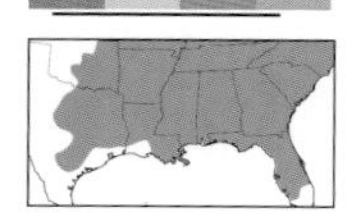

WHITE-DOTTED PROMINENT

Nadata gibbosa 93-0046 (7915) Common

TL 20–30 mm. Peppered tawny FW has straight AM and PM lines. Two small, round white spots mark median area. Outer margin is scalloped and often falcate. Thorax has pointed crest. **HOSTS:** Oak; also birch, cherry, maple, and others. **NOTE:** Two or three broods.

PROMINENTS

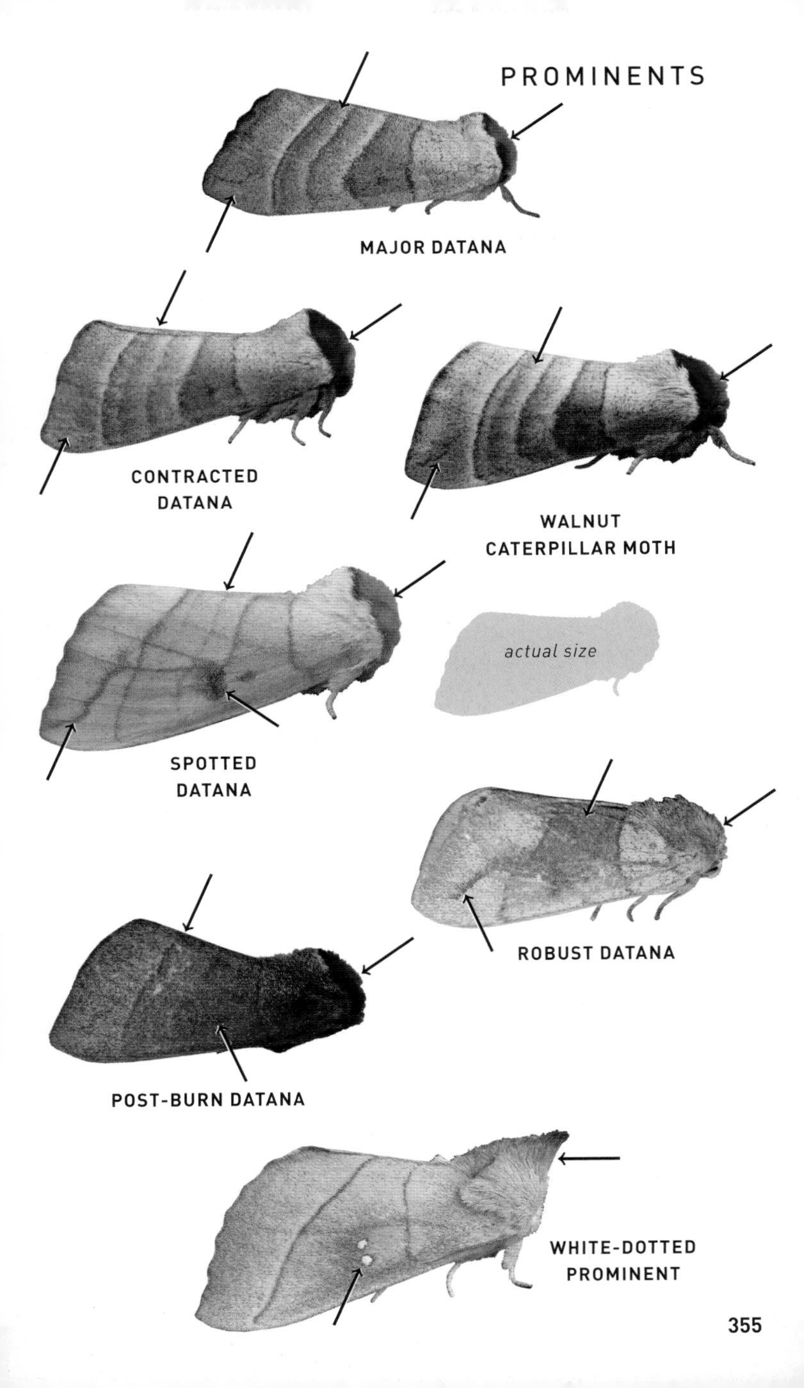
MAJOR DATANA
CONTRACTED DATANA
WALNUT CATERPILLAR MOTH
SPOTTED DATANA
actual size
ROBUST DATANA
POST-BURN DATANA
WHITE-DOTTED PROMINENT

OVAL-BASED PROMINENT

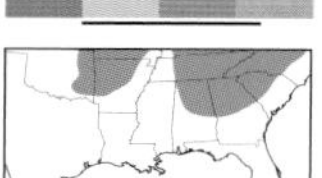

Peridea basitriens 93-0048 (7919) **Common**

TL 19–26 mm. Peppery light gray FW has a brown, ovate inner basal patch. Teeth of scalloped PM line point to dark veins in ST area. Costa of HW is hairy and often projects from under FW. **HOSTS:** Maple. **NOTE:** Two or more broods.

ANGULOSE PROMINENT

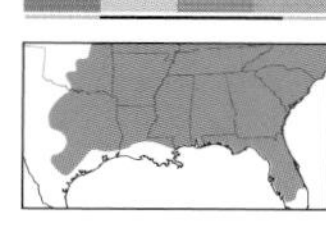

Peridea angulosa 93-0049 (7920) **Common**

TL 19–26 mm. Dark gray FW has jagged double AM and PM lines variably filled orange. Often has an incomplete whitish patch in outer median area. ST area is crossed with thin black veins. Costa of HW is hairy and often projects from under FW. **HOSTS:** Oak. **NOTE:** Two or more broods.

CHOCOLATE PROMINENT

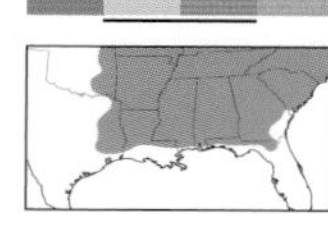

Peridea ferruginea 93-0050 (7921) **Common**

TL 20–27 mm. Grayish brown FW has chocolate brown shading, especially in basal and median areas. Double AM and PM lines are filled white. Reniform spot is white. Costa of HW is hairy and often projects from under FW. **HOSTS:** Birch. **NOTE:** Two or more broods.

LINDEN PROMINENT

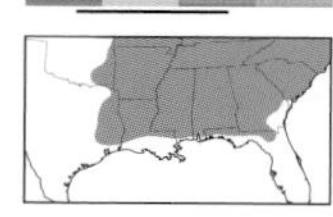

Ellida caniplaga 93-0051 (7930) **Uncommon**

TL 19–24 mm. Silvery gray FW has three thin, parallel black lines forming outer section of AM line. White-edged reniform spot is visible as short black line. ST line is a row of black dashes. Brown shading in subapical area. Usually appears tubular and blunt-fronted at rest. **HOSTS:** Basswood. **NOTE:** Two or more broods.

ORANGE-BANDED PROMINENT

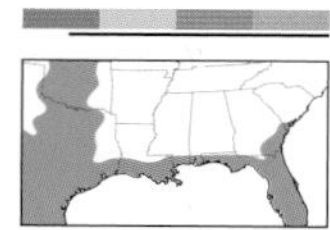

Litodonta hydromeli 93-0060 (7968) **Local**

TL 17 mm. Peppery gray or pale green FW has scalloped, double AM and PM lines filled white. ST line is a row of black scallops. Area below median line is usually shaded with gray, sometimes darkly. Basal and terminal areas are frequently partly or fully shaded orange. **HOSTS:** Gum bumelia.

DRAB PROMINENT

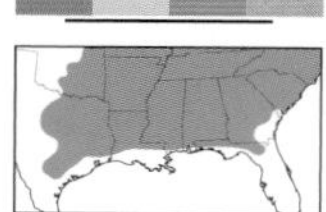

Misogada unicolor 93-0066 (7974) **Common**

TL 24–26 mm. Pale grayish brown FW has a greenish tint when fresh. All markings are indistinct, but faintly scalloped PM line is accented with tiny white dots. Thorax has short orange crest. Abdomen has orange band near base. Costa of HW is hairy and often projects from under FW. **HOSTS:** Sycamore and perhaps cottonwood. **NOTE:** Two or more broods.

PROMINENTS

OVAL-BASED PROMINENT

ANGULOSE PROMINENT

actual size

CHOCOLATE PROMINENT

LINDEN PROMINENT

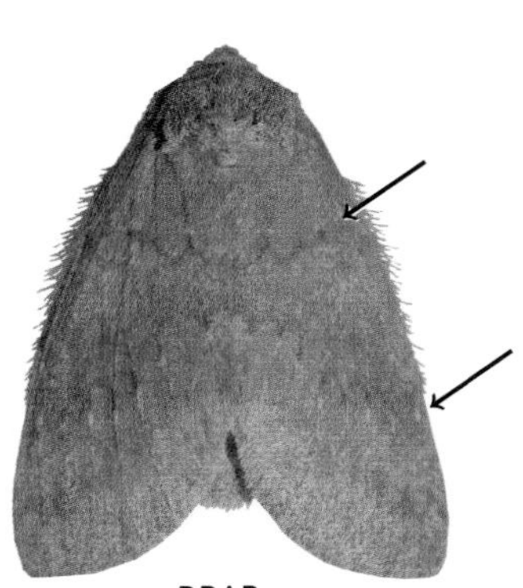

DRAB PROMINENT

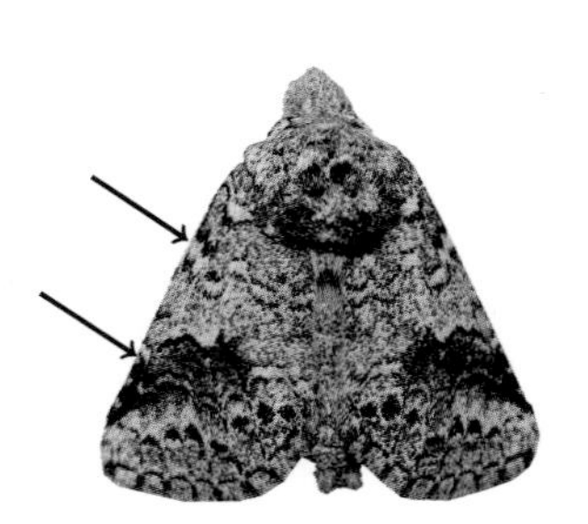

ORANGE-BANDED PROMINENT

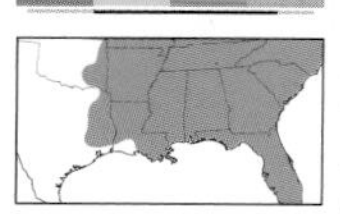

MOTTLED PROMINENT

Macrurocampa marthesia 93-0067 (7975) **Common**

TL 24–28 mm. Gray FW is often washed green when fresh. Scalloped AM and PM lines are double. Basal and ST areas are usually strongly shaded gray. Costa of HW is hairy and often projects from under FW. **HOSTS:** Oak; perhaps also poplar and maple. **NOTE:** Two or more broods.

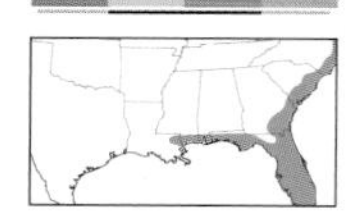

PALE GREEN HETEROCAMPA

Heterocampa astarte 93-0069 (7977) **Local**

TL 24–26 mm. Pale green FW is usually shaded brown in basal and ST areas. ST line is heavily bordered white above. Scalloped, double AM and PM lines are often inconspicuous. **HOSTS:** Unknown.

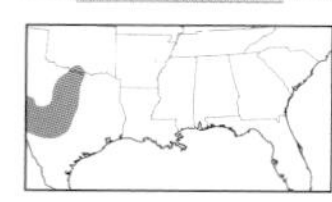

TEXAS HETEROCAMPA

Heterocampa astartoides 93-0070 (7978) **Local**

TL 26–28 mm. Light gray FW is tinged pale green when fresh. Basal area is usually shaded brownish or gray. ST line is heavily bordered white above. Wavy double AM and PM lines are inconspicuous. **HOSTS:** Unknown.

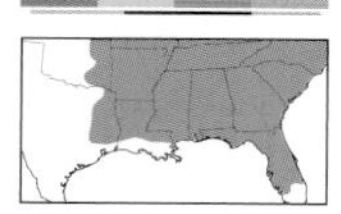

OBLIQUE HETEROCAMPA

Heterocampa obliqua 93-0075 (7983) **Common**

TL 20–28 mm. Gray or mossy FW has large dark gray or brownish patch in outer half and shading in basal area. Whitish patch in subapical area is tipped rusty in female. **HOSTS:** Oak. **NOTE:** Two broods.

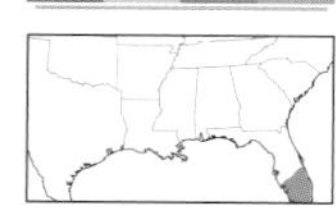

CUBAN HETEROCAMPA

Heterocampa cubana 93-0076 (7984) **Local**

TL 20–28 mm. Golden brown FW is faintly tinged yellowish green along costa. Thin black lines trace a semicircle across outer-middle of FW, which is often shaded darker, especially in females. Costa of HW is hairy and often projects from under FW. **HOSTS:** Unknown.

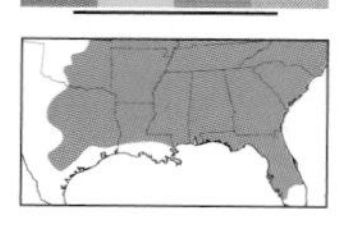

SMALL HETEROCAMPA

Heterocampa subrotata 93-0077 (7985) **Common**

TL 16–20 mm. Grayish brown FW is often washed greenish, especially through median area. Scalloped, double AM and PM lines are filled orange. Thin black crescent in reniform area creates border to a rusty patch sometimes present in outer part of FW. Oblique white subapical patch is marked with black streaks along veins. **HOSTS:** Hackberry. **NOTE:** Two or more broods.

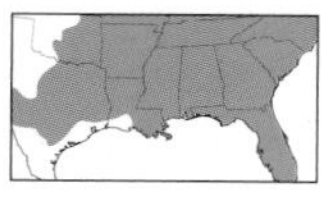

WHITE-BLOTCHED HETEROCAMPA

Heterocampa umbrata 93-0082 (7990) **Common**

TL 23–33 mm. Gray FW is strongly washed mossy green. Curved black bar extending from top of reniform crescent to inner ST line usually borders a large pale or whitish patch. Outer ST line is accented with a row of bold black triangles. Costa of HW is hairy and often projects from under FW. **HOSTS:** Oak. **NOTE:** Multiple broods.

PROMINENTS

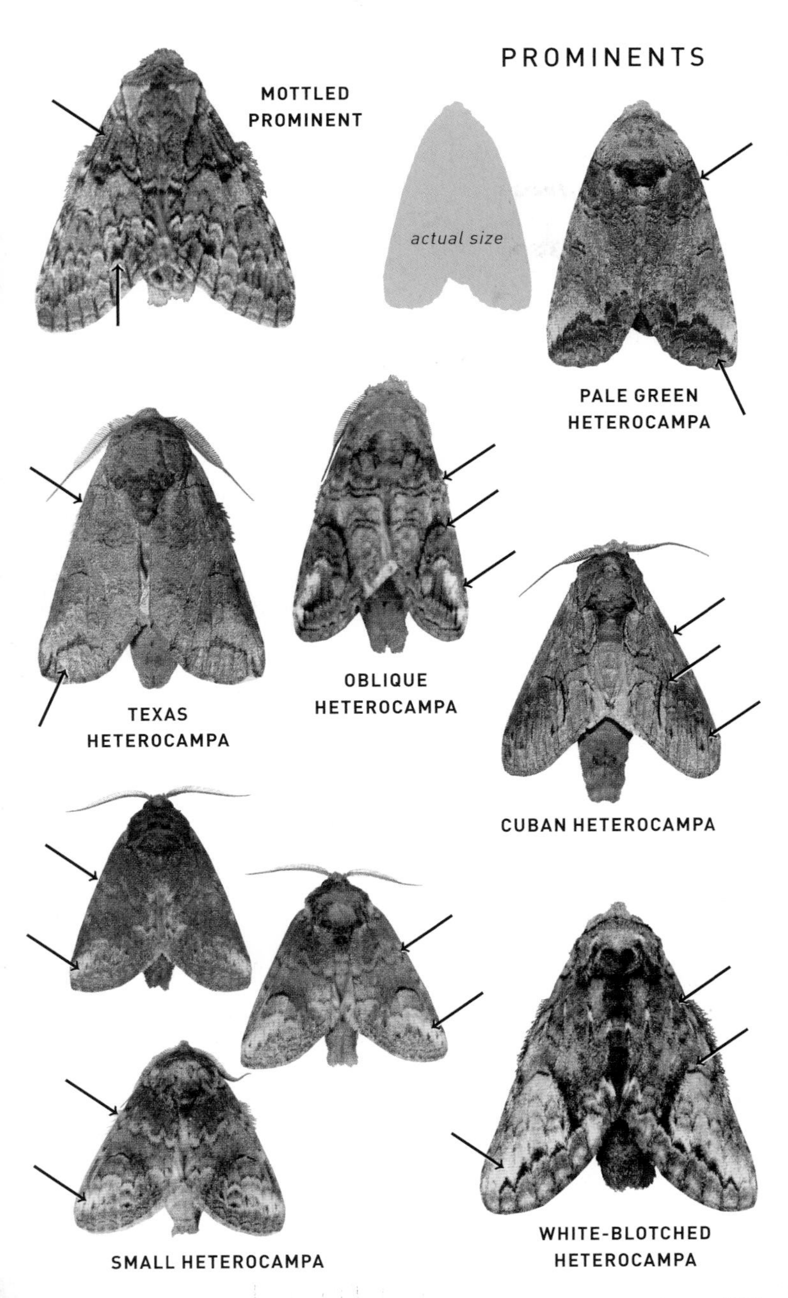
MOTTLED PROMINENT
actual size
PALE GREEN HETEROCAMPA
TEXAS HETEROCAMPA
OBLIQUE HETEROCAMPA
CUBAN HETEROCAMPA
SMALL HETEROCAMPA
WHITE-BLOTCHED HETEROCAMPA

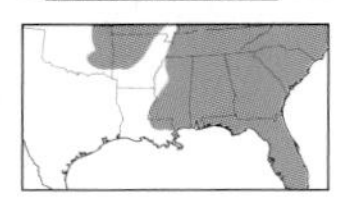

SADDLED PROMINENT

Heterocampa guttivitta 93-0086 (7994) **Common**

TL 23–25 mm. Variable. FW is gray to yellowish green; sometimes mint green. Scalloped AM and PM lines may be inconspicuous but are marked with tiny white dashes at veins. ST line is usually a series of dark spots, and terminal line is tiny white dots. In some individuals, ST area is often shaded whitish, blending into white-edged reniform spot. Costa of HW is hairy and often projects from under FW. **HOSTS:** Apple, birch, maple, oak, sumac, and many other woody plants. **NOTE:** Two broods.

WAVY-LINED HETEROCAMPA

Heterocampa biundata 93-0087 (7995) **Common**

TL 25–30 mm. Grayish brown or mossy green FW has scalloped, double AM and PM lines filled orange. ST line is a row of black wedges. Dark shading is usually present at inner reniform area. Some individuals have whitish subapical patch. Costa of HW is hairy and often projects from under FW. **HOSTS:** Basswood, birch, hickory, maple, oak, and many other woody plants. **NOTE:** Two or more broods.

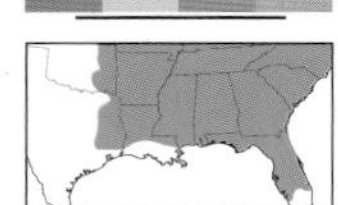

VARIABLE OAKLEAF CATERPILLAR MOTH

Lochmaeus manteo 93-0090 (7998) **Common**

TL 20–27 mm. Gray FW has scalloped, double AM line filled pale gray. Zigzag PM line is partly double and often accented with tiny white dashes at veins. Reniform spot is usually a thin black crescent edged with white but can be mostly black or whitish. Terminal line is small black dots. **HOSTS:** Beech, chestnut, and oak; perhaps also birch, elm, and walnut. **NOTE:** Two broods.

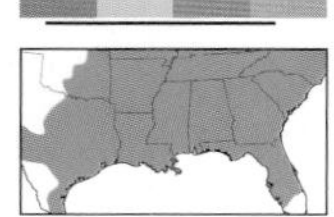

DOUBLE-LINED PROMINENT

Lochmaeus bilineata 93-0091 (7999) **Common**

TL 18–23 mm. Smooth gray FW has wavy double AM and PM lines filled whitish. Diffuse whitish ST line is often present. Reniform spot is a thin black crescent edged whitish. **HOSTS:** Elm and basswood; perhaps also beech, birch, oak, and walnut. **NOTE:** Two broods.

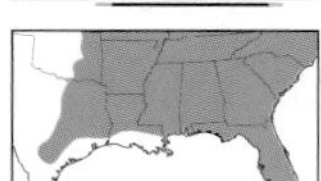

MORNING-GLORY PROMINENT

Schizura ipomoeae 93-0098 (8005) **Common**

TL 20–25 mm. Grayish to dark brown FW is paler along costa. Jagged double AM and PM lines are visible on paler individuals. Reniform spot is a thin black crescent outlined white. Fringe is checkered. **HOSTS:** Basswood, beech, birch, maple, oak, and many other woody plants. **NOTE:** Two or more broods.

PROMINENTS

SADDLED PROMINENT

WAVY-LINED HETEROCAMPA

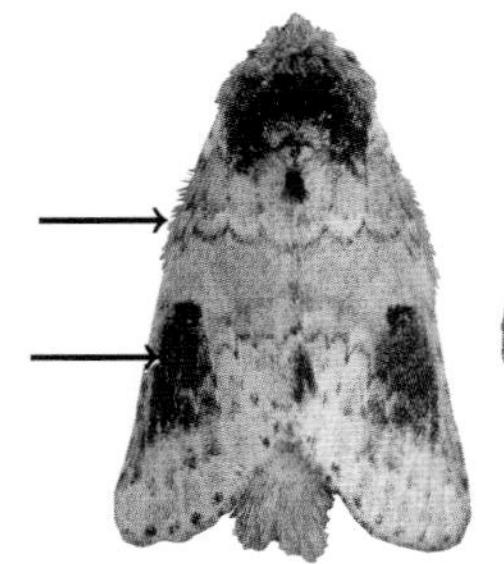

VARIABLE OAKLEAF CATERPILLAR MOTH

actual size

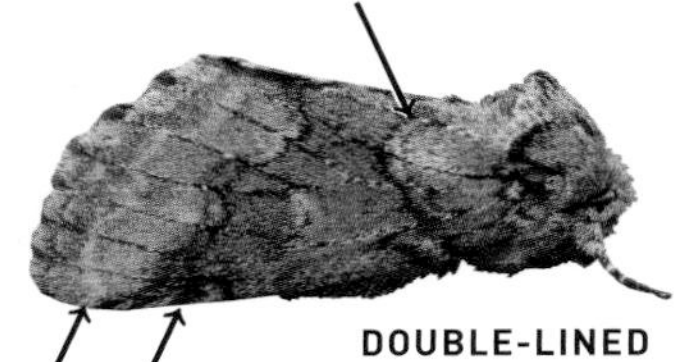

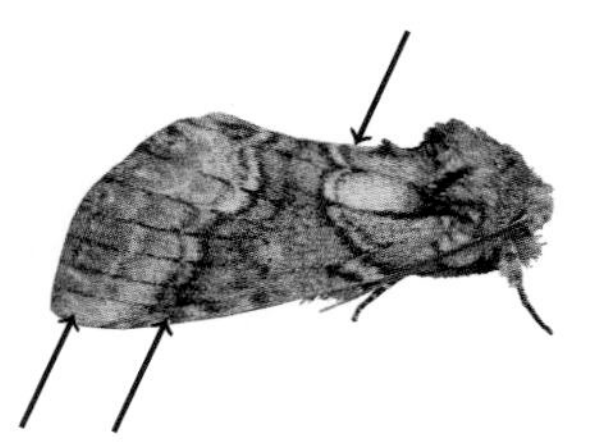

DOUBLE-LINED PROMINENT

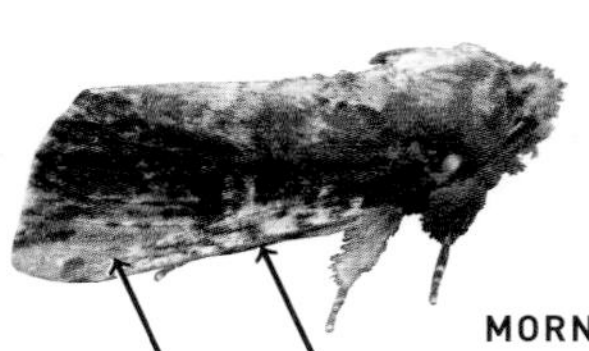

MORNING-GLORY PROMINENT

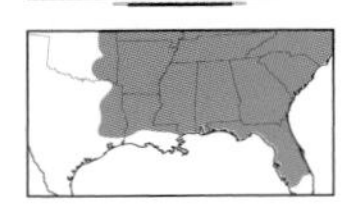

CHESTNUT SCHIZURA

Schizura badia 93-0099 (8006) **Uncommon**

TL 17–20 mm. Straw-colored FW is washed reddish in basal and ST areas. Dark streaks extend from crescent-shaped reniform spot to outer margin. Thorax is dark. **HOSTS:** Wild-raisin and other viburnum species. **NOTE:** Two or more broods.

UNICORN PROMINENT

Schizura unicornis 93-0100 (8007) **Common**

TL 18–25 mm. Multicolored FW has scalloped, double AM and PM lines. A large pale brownish patch in ST area is marked with black dashes at each side and a short white dash at anal angle. Inconspicuous black reniform dash is edged with a whitish patch. Basal area is pale green. **HOSTS:** Hickory, maple, oak, willow, and other deciduous trees. **NOTE:** Two or more broods.

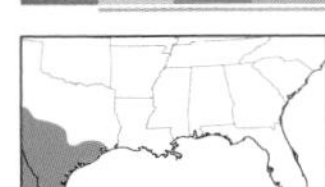

RAINBOW PROMINENT

Schizura errucata 93-0101 (8008) **Local**

TL 18–25 mm. Similar to Unicorn Prominent but grayer overall and less boldly marked. Patch in ST area is not well defined, and median area often blends into rest of wing. Black dashes are often present across entire ST area, with outermost edged white. **HOSTS:** Unknown.

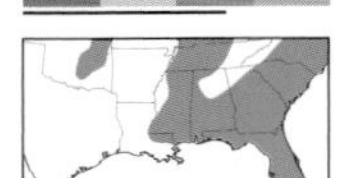

PLAIN SCHIZURA

Schizura apicalis 93-0102 (8009) **Uncommon**

TL 18–25 mm. Grayish to brown FW has scalloped, double AM and PM lines. Apex and costal median area are whitish, and inner median area is often golden brown. Reniform spot is a bold black crescent. Rarely rests with wings rolled like other *Schizura* species. **HOSTS:** Bayberry, blueberry, sweet fern, and wax myrtle.

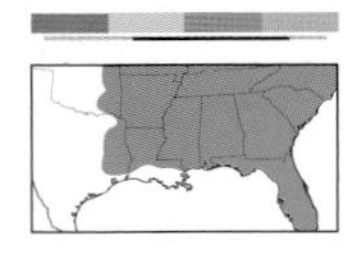

RED-HUMPED CATERPILLAR MOTH

Schizura concinna 93-0103 (8010) **Common**

TL 17–21 mm. Brown to straw-colored FW is shaded reddish brown along inner margin and grayish along costa. A tiny black reniform dot and small black streaks at outer ST line are only obvious markings. Very diffuse median band is often visible. **HOSTS:** Apple, blueberry, elm, hickory, maple, oak, and many other woody plants. **NOTE:** Two or more broods.

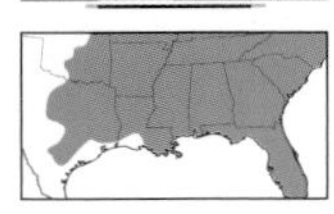

BLACK-BLOTCHED SCHIZURA

Schizura leptinoides 93-0104 (8011) **Common**

TL 22–26 mm. Gray FW has double AM and PM lines. A large dark blotch is present below small black reniform crescent. Small white patch at apex. A black basal dash is usually visible. **HOSTS:** Hickory and walnut; also apple, birch, oak, poplar, and others. **NOTE:** Two or more broods.

PROMINENTS

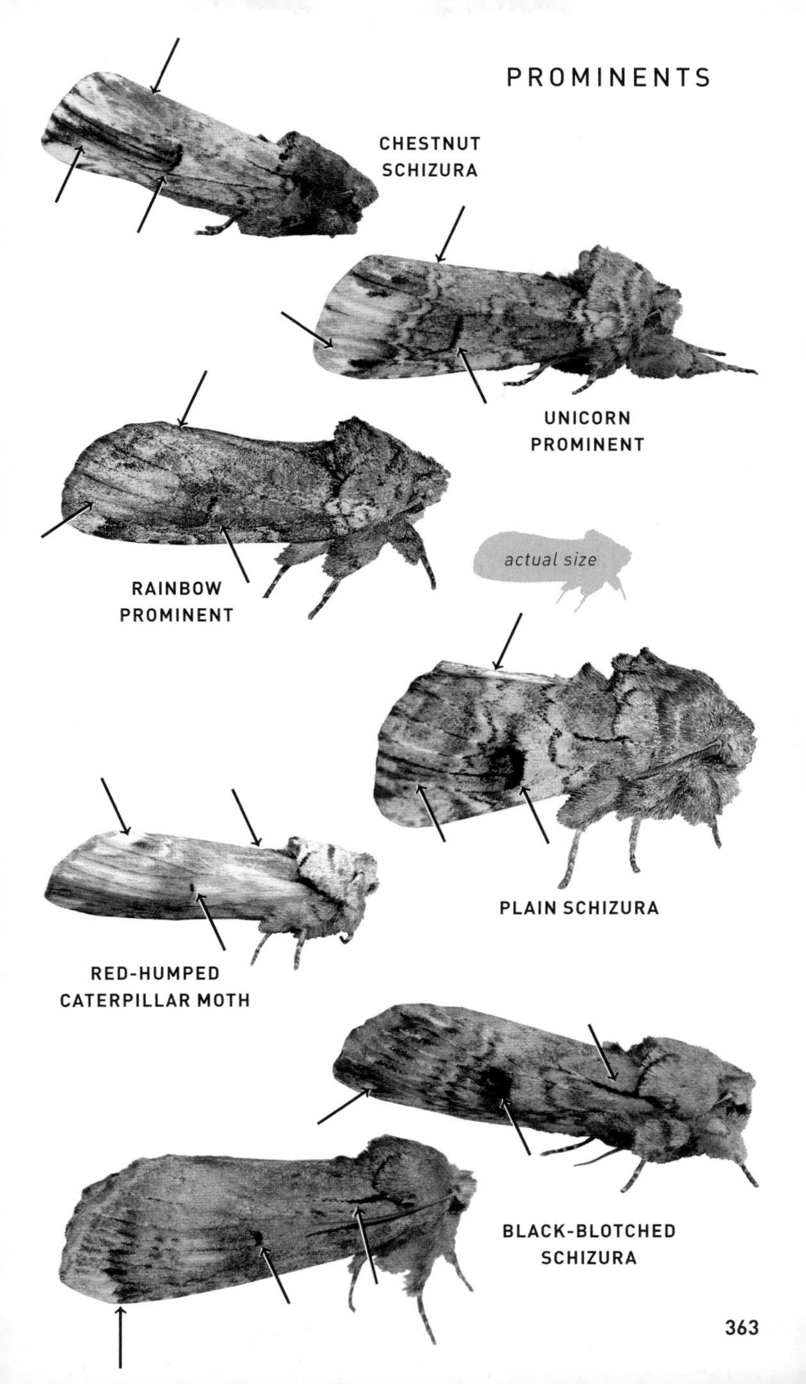

CHESTNUT SCHIZURA
UNICORN PROMINENT
RAINBOW PROMINENT
actual size
PLAIN SCHIZURA
RED-HUMPED CATERPILLAR MOTH
BLACK-BLOTCHED SCHIZURA

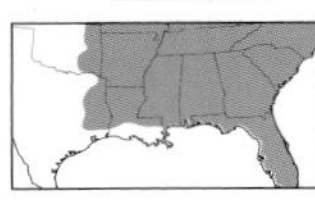

RED-WASHED PROMINENT

Oligocentria semirufescens 93-0105 (8012) **Common**

TL 24–25 mm. Straw-colored FW is usually washed reddish along inner margin and sometimes grayish along costa. Inner half of double PM line is often only visible line. Reniform spot is a small black dot, or sometimes crescent, bordered by a dark patch. Whitish spot is often present at apex. **HOSTS:** Poplar and willow; also alder, birch, oak, and rose. **NOTE:** Two broods.

WHITE-STREAKED PROMINENT

Oligocentria lignicolor 93-0110 (8017) **Common**

TL 25–26 mm. Gray FW is usually paler toward costa, sometimes forming a thin, indistinct whitish streak. Reniform spot is a tiny black dot, sometimes with a dark streak beneath. Black basal dash often is also noticeable. Black streaks in ST area are edged white. Thorax has a dark central stripe. **HOSTS:** Beech, chestnut, and oak. **NOTE:** Two or more broods.

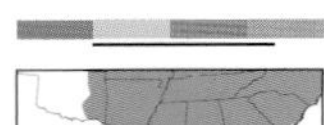

PINK PROMINENT

Hyparpax aurora 93-0115 (8022) **Uncommon**

TL 17–22 mm. Pale yellow FW has powdery pink basal and ST areas. Pink costa often bleeds into median area. Reniform spot is a thin pink crescent. Yellow thorax is variably shaded pink. Strongly resembles Rosy Maple Moth, a silkworm moth. **HOSTS:** Scrub oak and perhaps viburnum. **NOTE:** Two broods.

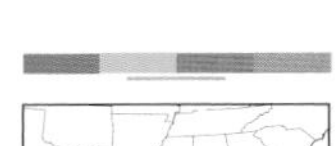

GOLD-STRIPED PROMINENT

Hyparpax aurostriata 93-0118 (8025) **Local**

TL 17–22 mm. Straw-colored FW is lightly peppered pink to rusty. Pattern of lines resembles that of Pink Prominent, but basal and ST areas are not shaded. **HOSTS:** Unknown.

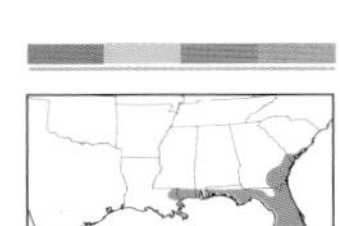

BROWN-STRIPED PROMINENT

Hyparpax perophoroides 93-0119 (8026) **Local**

TL 17–19 mm. Resembles Gold-striped Prominent, but FW has a narrow rusty brown border. Light speckling across wings is also brown, as are AM and PM lines. Reniform crescent touches midpoint of AM line. **HOSTS:** Oak.

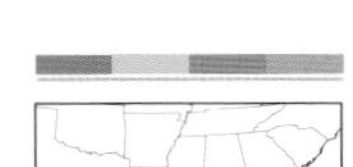

FLORIDA PROMINENT

Nystalea eutalanta 93-0123 (7947) **Local**

TL 24–26 mm. Brown FW is shaded darker along costa, especially in outer PM area. Has pale patches at midpoint of costa and near apex. Median line is a series of pale-edged dark spots, the most obvious of which is at inner margin. **HOSTS:** Unknown.

PROMINENTS

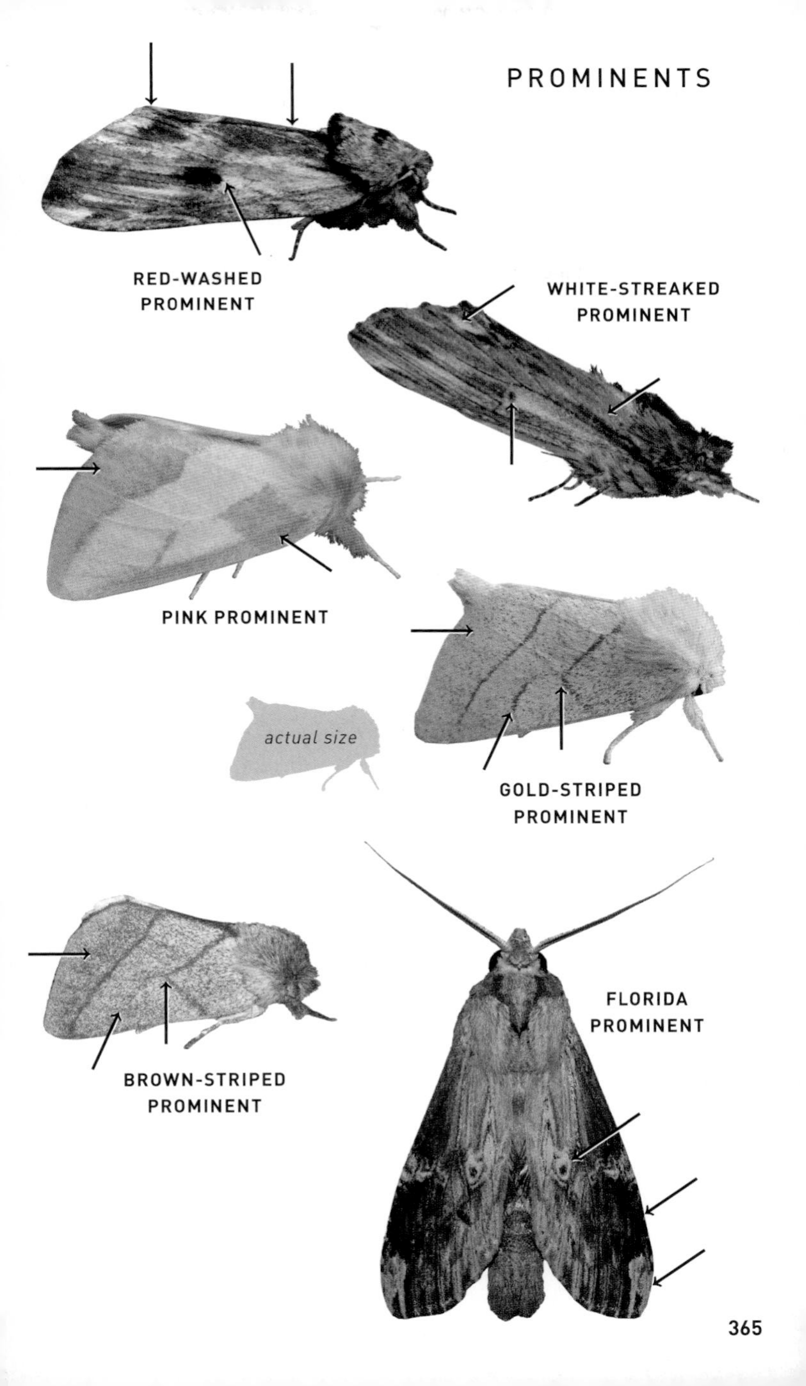

RED-WASHED PROMINENT
WHITE-STREAKED PROMINENT
PINK PROMINENT
actual size
GOLD-STRIPED PROMINENT
BROWN-STRIPED PROMINENT
FLORIDA PROMINENT

PACKARD'S PROMINENT

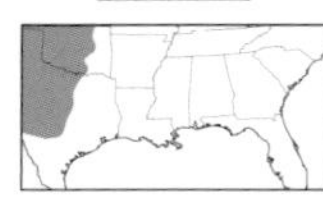

Elasmia packardii 93-0125 (7949) **Local**

TL 14–18 mm. Light to medium gray FW is washed pinkish brown, especially along costa and in ST area. Jagged double AM and PM lines are often fragmented or indistinct. Apex is divided diagonally: pale toward costa, dark toward outer margin. Thin black reniform crescent is edged with brown, occasionally surrounded by diffuse shading. Center of thorax is dark in male, brown in female. **HOSTS:** Western soapberry and Mexican buckeye.

MANDELA'S PROMINENT

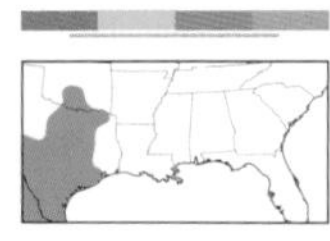

Elasmia mandela 93-0126.1 (7950.1) **Local**

TL 18–22 mm. Similar to slightly smaller Packard's Prominent, and some lighter individuals may not be safely identifiable to species. FW is brownish to light gray with darker shading in outer median and ST areas. Dark blotch behind reniform spot. Whitish half of apex usually stands out more than darker side toward outer margin. Center of thorax is dark in male, brown in female. **HOSTS:** Mexican buckeye.

WHITE-HEADED PROMINENT

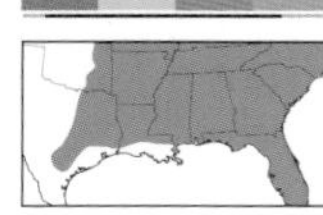

Symmerista albifrons 93-0127 (7951) **Common**

TL 18–22 mm. Light gray FW is marked with indistinct scalloped lines. White costal bar extending from AM line to apex has a blunt spur jutting into brownish median area. Has blunt front to buffy thorax, and head looks like a snapped twig. **HOSTS:** Beech, chestnut, and oak. **NOTE:** Two broods.

BLACK-SPOTTED PROMINENT

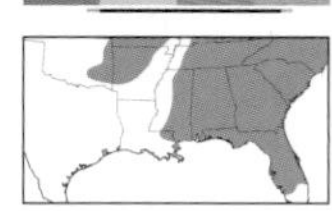

Dasylophia anguina 93-0133 (7957) **Common**

TL 24 mm. Sexually dimorphic. Gray FW has gray basal area with strong black-edged vein in male, and tawny basal area with lightly brown-edged vein in female. Dark PM area contrasts with whitish PM line and pale ST area. Two black spots in inner ST area. With blunt front to thorax and long labial palps, resembles a snapped twig. **HOSTS:** Legumes, including bush clover, leadplant, locust, and wild indigo. **NOTE:** Two or more broods.

GRAY-PATCHED PROMINENT

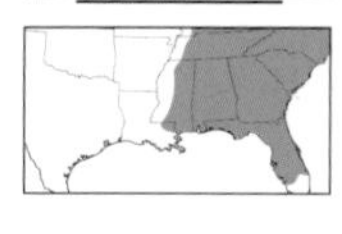

Dasylophia thyatiroides 93-0134 (7958) **Local**

TL 24 mm. Similar to Black-spotted Prominent but lacks pale ST area and dark-edged central vein. Basal area is orange in female, gray in male. White PM line is edged strongly black at inner margin. **HOSTS:** Beech. **NOTE:** Two broods.

SILVERED PROMINENT

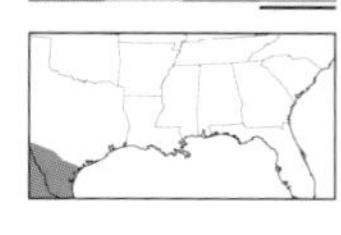

Didugua argentilinea 93-0137 (7961) **Local**

TL 17–19 mm. Tan FW has two flattened white triangles parallel to costa, broadly edged with dark brown shading. **HOSTS:** *Serjania*, *Urvillea*, Turk's-cap, and mesquite seeds; also hackberry, ash, bluewood condalia, and other woody plants.

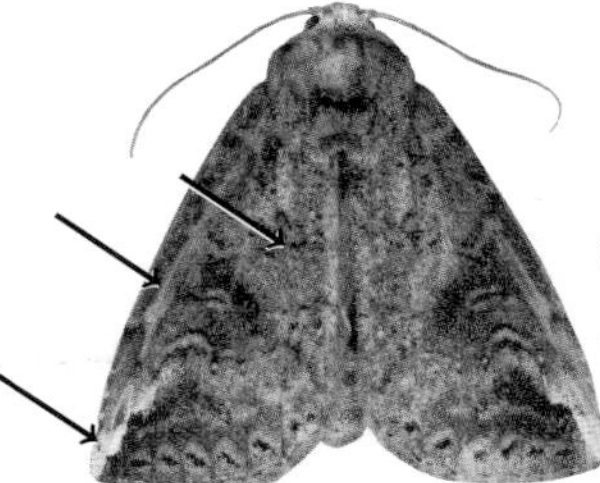

PACKARD'S PROMINENT

MANDELA'S PROMINENT

WHITE-HEADED PROMINENT

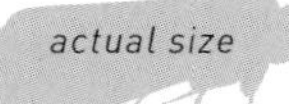

BLACK-SPOTTED PROMINENT

SILVERED PROMINENT

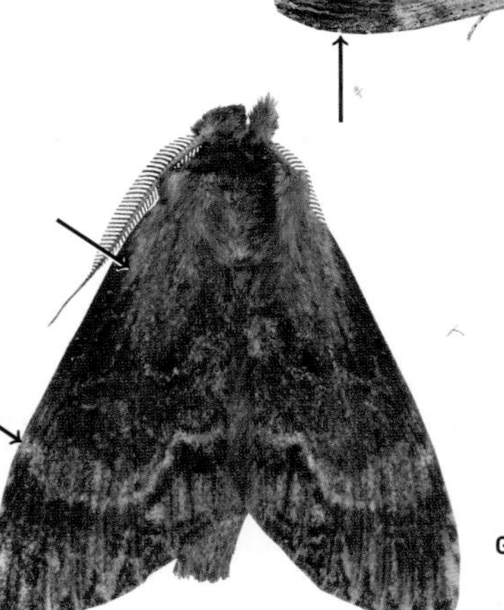

GRAY-PATCHED PROMINENT

Tussock Moths

Family Erebidae, Subfamily Lymantriinae

Extremely hairy delta-shaped moths that rest with their long hairy forelegs sprawled out in front. They often have variable and confusing FW patterns that make specific identification a challenge. They inhabit woodlands and larger gardens. Females tend to be larger than males. Adults do not feed. They are nocturnal and will come to light in small numbers.

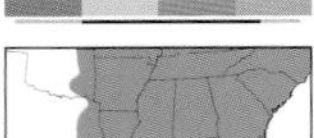

TEPHRA TUSSOCK MOTH

Dasychira tephra 93-0144 (8292) **Common**

TL 17–27 mm. Variable. FW brown to brownish gray, typically marked with a thick black streak extending from base to scalloped PM line, which is usually bold. In some individuals black streak may be absent; these individuals often have brown AM and PM lines. ST line is absent. **HOSTS:** Oak. **NOTE:** Two broods.

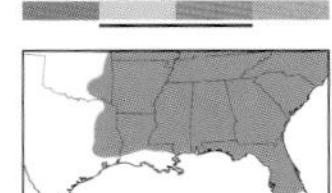

YELLOW-BASED TUSSOCK MOTH

Dasychira basiflava 93-0148 (8296) **Common**

TL 17–28 mm. Variable. Grayish FW usually has warm brown to brownish gray basal and PM areas. Area around reniform spot is usually whitish. Rare individuals, mostly females, show a narrow black streak from base to PM line. ST line is whitish and often faint; usually has a small white spot near inner margin. **HOSTS:** Oak; also hickory. **NOTE:** Two broods.

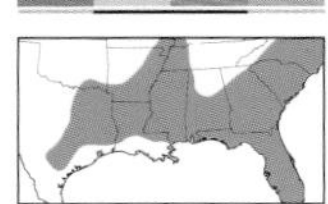

SOUTHERN TUSSOCK MOTH

Dasychira meridionalis 93-0150 (8298) **Common**

TL 17–28 mm. Variable. FW gray to brownish, usually with warm brown shading in basal and ST areas. Area around reniform spot is usually whitish. PM line is typically bold and relatively smooth. ST line is absent. Some individuals have a black streak from base to PM line. S. FL form is uniformly light brown, often with paler median area. **HOSTS:** Oak. **NOTE:** Two broods.

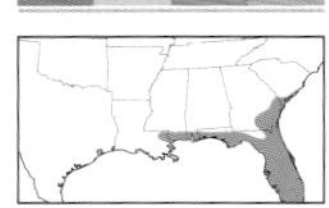

PEPPERED TUSSOCK MOTH

Dasychira leucophaea 93-0153 (8301) **Local**

TL 17–28 mm. Sexually dimorphic and variable. Female is whitish with rich brown in basal and PM areas. Terminal line is bold black dashes. Male is peppery gray with dark basal and PM areas. Area around reniform spot is whitish. Dashes of terminal line are edged white. Base of thorax in both sexes has two black-edged orange circles. **HOSTS:** Oak. **NOTE:** Two broods.

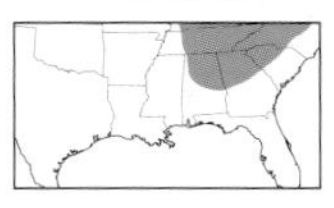

STREAKED TUSSOCK MOTH

Dasychira obliquata 93-0154 (8302) **Local**

TL 18–30 mm. Light gray FW has distinct rusty brown bands behind scalloped black AM and PM lines and two thin black streaks running from base to ST line. Reniform spot creates white patch in median area. White ST line is diffuse. Black terminal line is thin and continuous. **HOSTS:** Oak; also beech, birch, elm, and hickory.

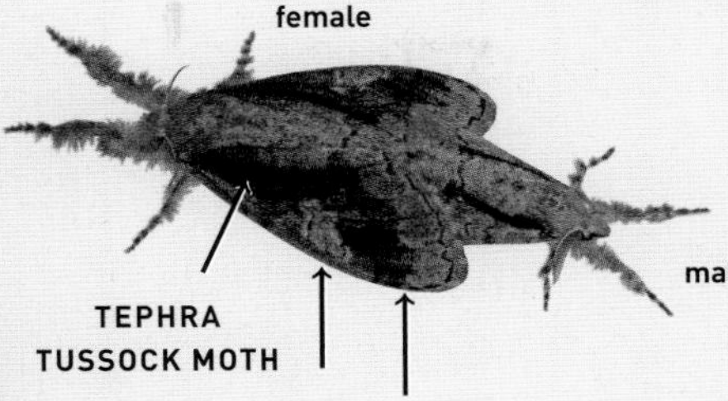

female

male

TEPHRA TUSSOCK MOTH

YELLOW-BASED TUSSOCK MOTH

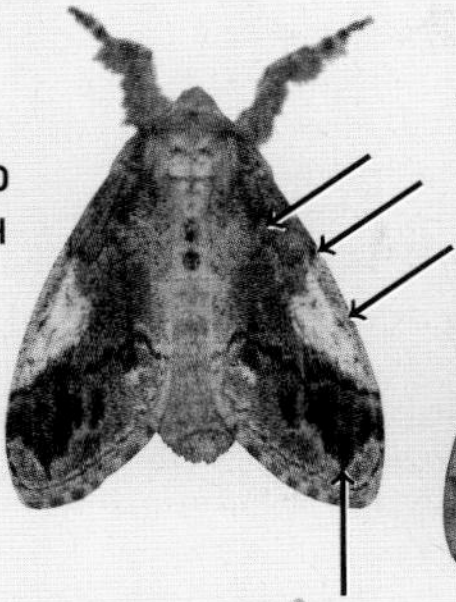

SOUTHERN TUSSOCK MOTH

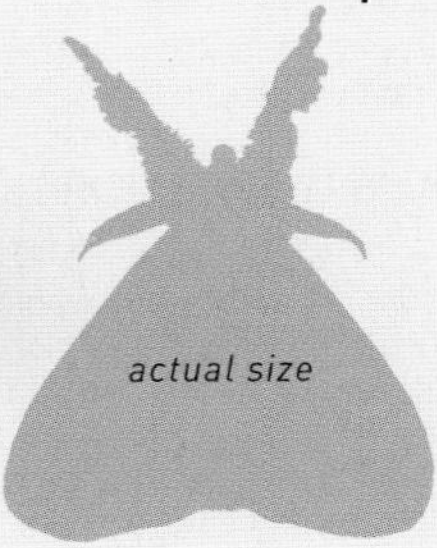

actual size

PEPPERED TUSSOCK MOTH

STREAKED TUSSOCK MOTH

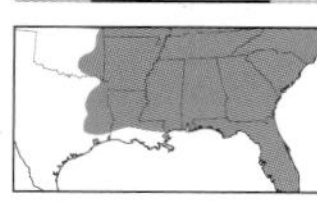

MANTO TUSSOCK MOTH

Dasychira manto 93-0159 (8307) **Common**

TL 17–26 mm. Variable. Grayish to brown FW has chestnut brown basal and PM areas. Slightly wavy black AM and PM lines are bold. Reniform spot creates white patch in median area. Diffuse white ST line is sometimes faint; usually has white spot at inner margin. Rarely, some individuals show a black basal patch. **HOSTS:** Pine. **NOTE:** Two or more broods.

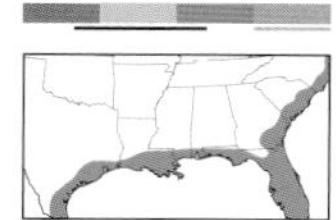

LIVE OAK TUSSOCK MOTH

Orgyia detrita 93-0165 (8313) **Common**

TL 15–17 mm. Sexually dimorphic and variable. Rounded brown FW of male has thin scalloped PM line; AM line is indistinct. Fragmented reniform spot is often filled white. White spot in inner ST area is usually small or faint. Whitish female (not shown) is wingless. **HOSTS:** Live oak and cypress. **NOTE:** Also called Fir Tussock Moth, a misnomer.

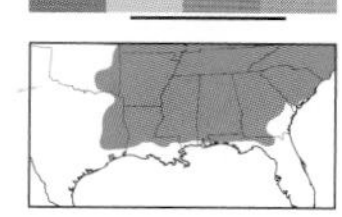

DEFINITE TUSSOCK MOTH

Orgyia definita 93-0166 (8314) **Common**

TL 12–18 mm. Sexually dimorphic and variable. Grayish to brown FW of male has dark shading in a semicircle around strongly white-edged reniform spot. Costa is whitish in median area and at end of ST line. Bold PM line is relatively smooth; AM line is usually diffuse. Often has a purplish sheen in median area when fresh. Inner ST area has a bold white spot. Whitish female (not shown) is wingless. **HOSTS:** Basswood, birch, oak, maple, willow, and other deciduous trees. **NOTE:** Two broods.

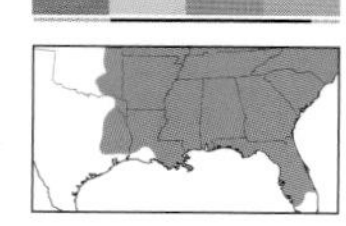

WHITE-MARKED TUSSOCK MOTH

Orgyia leucostigma 93-0168 (8316) **Common**

TL 15–22 mm. Sexually dimorphic and variable. Male resembles Definite Tussock Moth but has more uniformly gray FW with little to no shading around thinly white-edged reniform spot. Usually has no white along costa. Bold PM line is lightly scalloped; AM line is diffuse. White spot in inner ST area is variable, and may be nearly absent on some individuals. Whitish female (not shown) is wingless. **HOSTS:** Apple, birch, cherry, elm, fir, hemlock, rose, willow, and many other woody plants. **NOTE:** Two or more broods.

MANTO
TUSSOCK
MOTH
LIVE OAK
TUSSOCK MOTH
actual size
WHITE-MARKED
TUSSOCK MOTH
DEFINITE
TUSSOCK MOTH

Lichen Moths

Family Erebidae, Subfamily Arctiinae, Tribe Lithosiini

Small, often strikingly colorful moths whose larvae feed mostly on lichen in wooded areas. Most species are nocturnal and will visit lights in small numbers. A small number, such as Black-and-yellow Lichen Moth, are primarily diurnal and can be found resting among vegetation and flower heads.

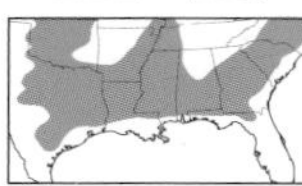

ONE-BANDED LICHEN MOTH

Cisthene unifascia 93-0177 (8060) **Local**

TL 8–10 mm. Dark gray FW has a yellow-orange median band of variable width that usually narrows at midpoint before extending along basal section of inner margin, often widening toward thorax. Pink HW has a small black apical patch. Yellow-orange thorax has a black dorsal patch. **HOSTS:** Presumably lichens.

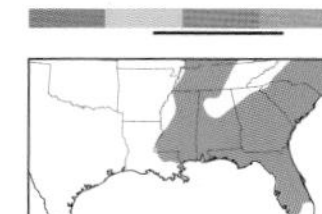

KENTUCKY LICHEN MOTH

Cisthene kentuckiensis 93-0178 (8061) **Common**

TL 8–10 mm. Blackish FW has a wide yellow-orange median band that narrows slightly at midpoint before extending along basal section of inner margin. Pink HW has a black apical patch. Yellow-orange thorax has a black dorsal patch. **HOSTS:** Lichens.

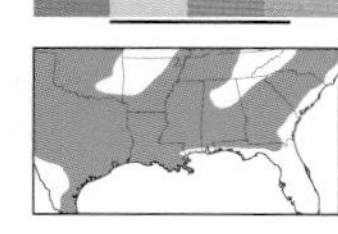

THIN-BANDED LICHEN MOTH

Cisthene tenuifascia 93-0183 (8066) **Common**

TL 8–10 mm. Blackish FW has a narrow yellow-orange median band that narrows slightly at midpoint. Yellow-orange patch at base of inner margin tapers to a point and does not usually meet median band. Reddish HW has a narrow black border. Yellow-orange thorax has a black dorsal patch. **HOSTS:** Lichens and algae.

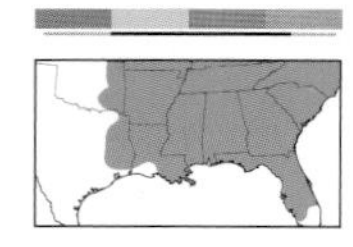

LEAD-COLORED LICHEN MOTH

Cisthene plumbea 93-0184 (8067) **Common**

TL 9–11 mm. Dark gray FW has yellow-orange streaks along costa and inner margin. A yellow-orange triangle extends into inner PM area. Pink HW has a narrow black border. Pinkish orange thorax has a grayish dorsal patch. **HOSTS:** Lichens. **NOTE:** Two broods.

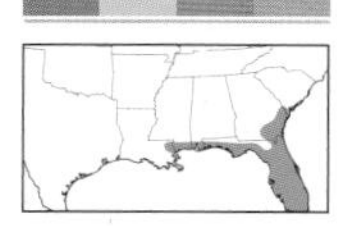

STRIATED LICHEN MOTH

Cisthene striata 93-0185 (8068) **Local**

TL 8–10 mm. Light gray FW is boldly striated blackish along veins. Has a narrow pinkish orange costal streak and patch at inner PM area. Pink HW is sometimes gray at apex. Gray thorax has pink lateral stripes. **HOSTS:** Presumably lichens.

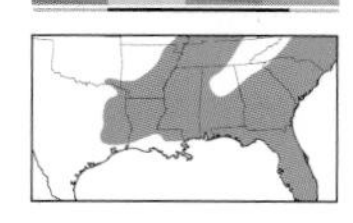

SUBJECT LICHEN MOTH

Cisthene subjecta 93-0188 (8071) **Common**

TL 7–9 mm. Dark gray FW has a rosy triangle at outer PM area. A yellow-orange stripe parallel to inner margin joins pink patch at inner PM area. Reddish HW has black border that widens toward apex. **HOSTS:** Lichens.

LICHEN MOTHS

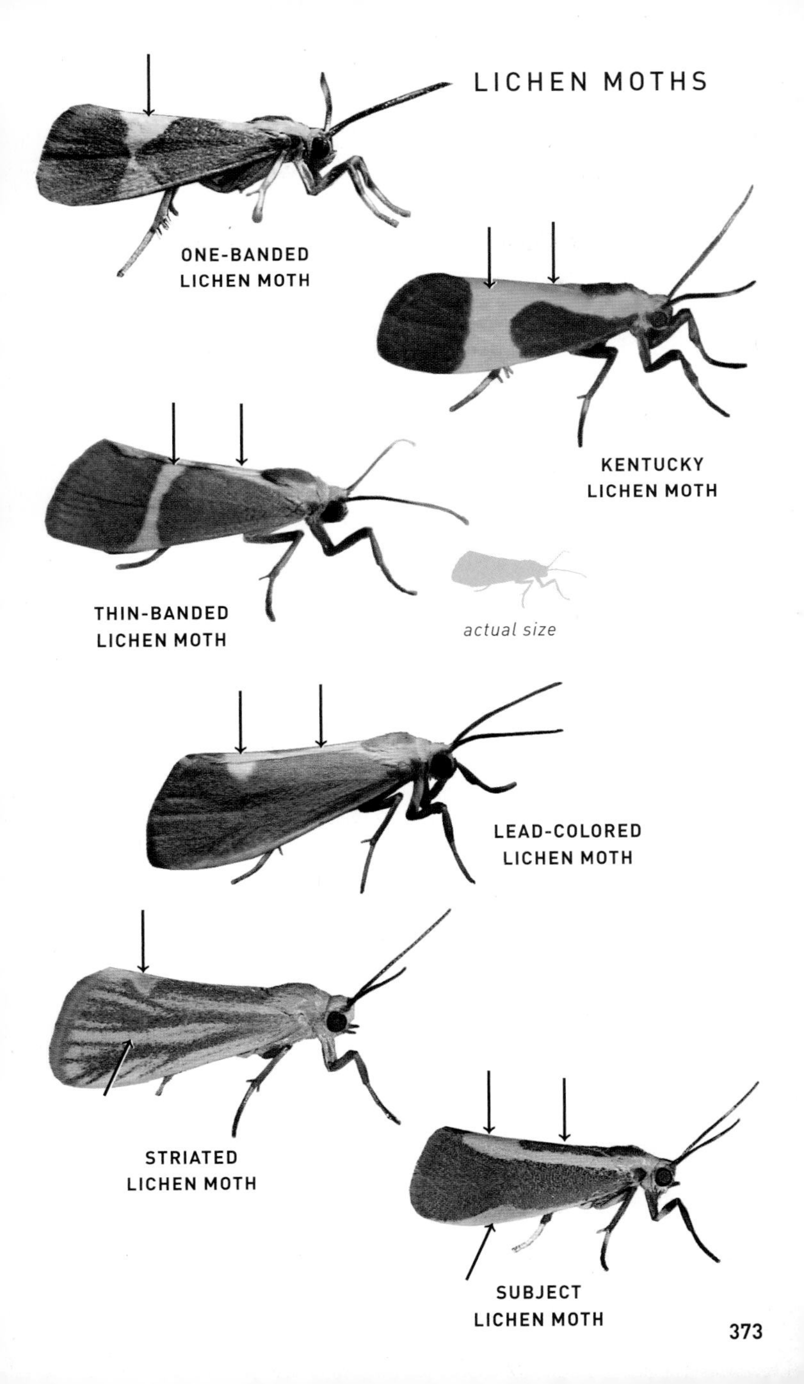

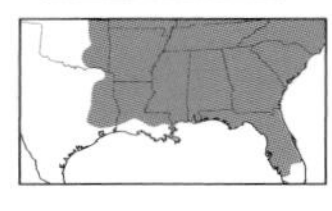

PACKARD'S LICHEN MOTH

Cisthene packardii 93-0189 (8072) **Common**

TL 8–10 mm. Resembles Subject Lichen Moth, but rosy patch at outer PM area is rounded, and patch at inner PM area is pink only along inner margin. Reddish HW has black border and extensive apical patch. **HOSTS:** Lichens and algae. **NOTE:** Two broods.

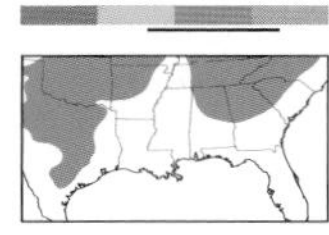

BLACK-AND-YELLOW LICHEN MOTH

Lycomorpha pholus 93-0201 (8087) **Common**

TL 13–17 mm. Yellow-orange FW has a bluish black outer half. Head and thorax are black. **HOSTS:** Lichens. **NOTE:** Resembles net-winged beetles in the genus *Calopteron*.

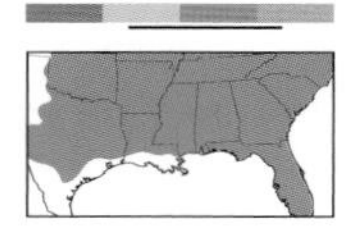

SCARLET-WINGED LICHEN MOTH

Hypoprepia miniata 93-0204 (8089) **Common**

TL 15–21 mm. Dark gray FW has wide scarlet border and shading along central veins. Black HW has red basal patch. Red thorax sometimes has a black dorsal spot. **HOSTS:** Lichens and blue-green algae. **NOTE:** Two or more broods.

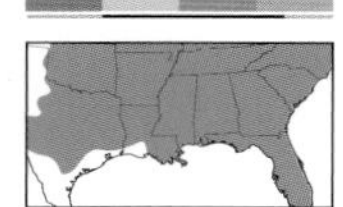

PAINTED LICHEN MOTH

Hypoprepia fucosa 93-0205 (8090) **Common**

TL 14–18 mm. Resembles Scarlet-winged Lichen Moth, but FW has wide yellow-orange border and shading along central veins, tinged red in ST area. Yellowish pink thorax has a gray dorsal spot. **HOSTS:** Lichens, algae, and moss.

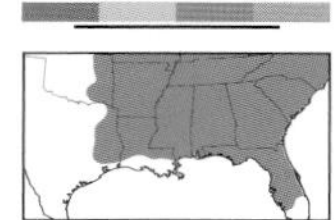

LITTLE WHITE LICHEN MOTH

Clemensia albata 93-0215 (8098) **Common**

TL 9–13 mm. Peppery white FW has black or dark brownish AM, median, and PM lines that are sometimes fragmented. Black patch is usually present at inner margin of PM line. Wide pale band above median line contains small black discal spot. Reniform spot is a black crescent. Terminal line is checkered. **HOSTS:** Lichens. **NOTE:** Two or more broods.

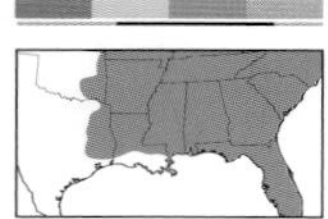

DARK GRAY LICHEN MOTH

Crambidia lithosioides 93-0218 (8045) **Common**

TL 8–10 mm. Sexually dimorphic. Brownish gray FW of male is evenly striated with thin, light brown veins. The vein that meets upper midpoint of horizontal bar is usually only faintly shaded. Darker female (not shown) has a narrow, pale costal streak and less obvious veins. Male has bipectinate antennae. **HOSTS:** Lichens.

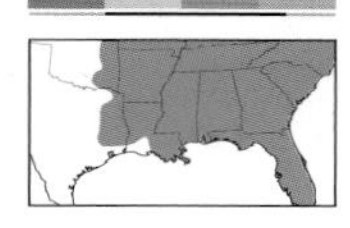

PALE LICHEN MOTH

Crambidia pallida 93-0219 (8045.1) **Common**

TL 10–13 mm. Resembles slightly smaller Dark Gray Lichen Moth, but veins are generally more heavily lined, and horizontal bar in middle of wing is more strongly noticeable. The vein that meets upper midpoint of horizontal bar is usually strongly shaded. Both sexes have filiform antennae. **HOSTS:** Lichens. **NOTE:** Two or more broods.

LICHEN MOTHS

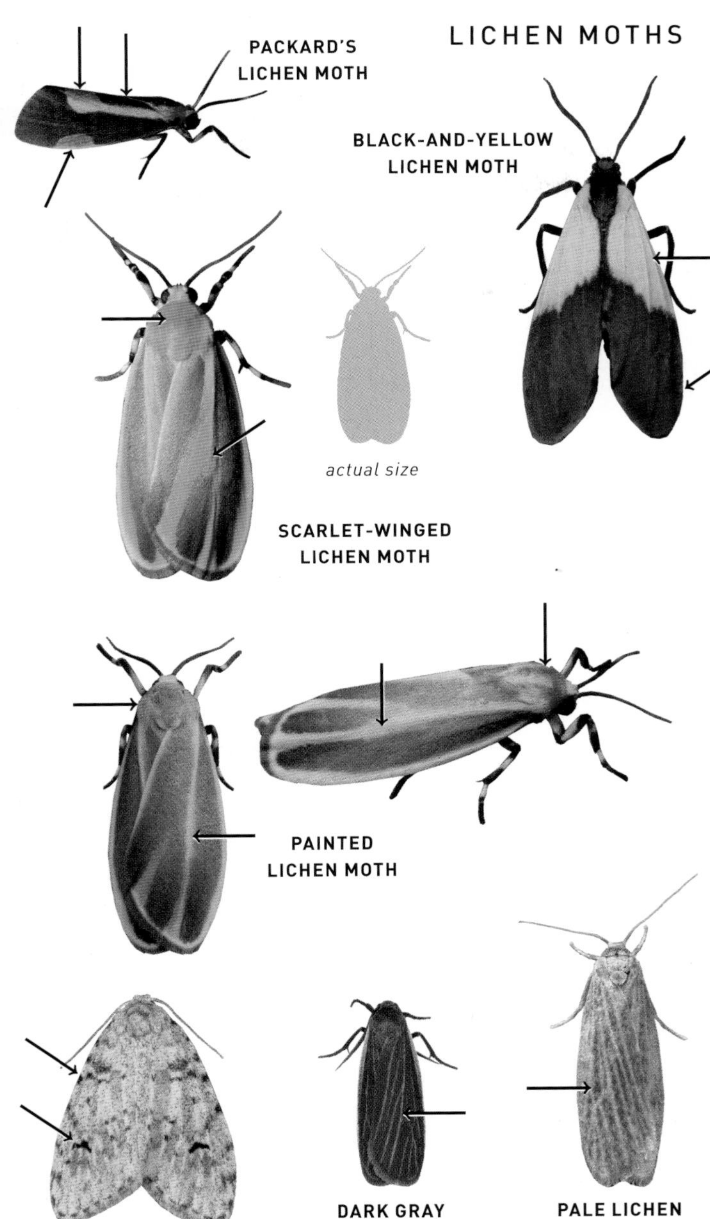

UNIFORM LICHEN MOTH

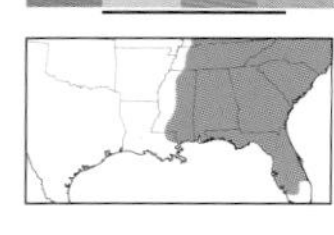

Crambidia uniformis 93-0220 (8046) **Common**

TL 7–10 mm. Resembles Dark Gray Lichen Moth, but FW is usually darker, with strong gray or straw-colored veins. The shading along the vein that meets upper midpoint of horizontal bar usually peters out before reaching bar. **HOSTS:** Lichens.

TIGER MOTHS

Family Erebidae, Subfamily Arctiinae, Tribe Arctiini

A varied group of strikingly attractive moths found in woodlands, fields, and gardens. The *Haploa* and *Cycnia* moths are broad-winged and flimsy-looking. The smaller *Virbia* moths are both nocturnal and diurnal and rest with their wings tightly closed. Yellow-collared Scape Moth and the colorful wasp moths are primarily day-fliers. The other species are more robust and hairy, often with striped, banded, or spotted patterns. Some species are more familiar as larvae, such as the Woolly Bear caterpillar of Isabella Tiger Moth. Many species do not feed as adults. Most species will visit lights in varying numbers.

ARGE MOTH *Grammia arge* 93-0240 (8199) **Common**

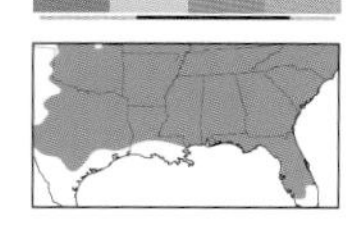

TL 20–26 mm. FW is heavily marked with thick creamy to pinkish white bands along veins, creating a fractured pattern of variable black triangles. Pale HW has a pink flush at base and small black spots along ST line. **HOSTS:** Various low plants and vines, including sunflower, cotton, and grape. **NOTE:** Two or more broods.

DORIS TIGER MOTH

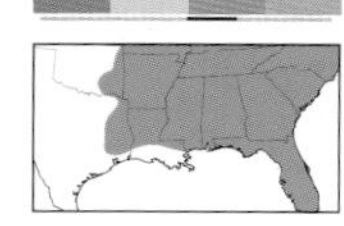

Grammia doris 93-0241 (8198) **Uncommon**

TL 20–26 mm. FW is virtually identical to that of Arge Moth but not usually pink-tinged. Deep pink HW has large blackish spots in ST area. **HOSTS:** Low plants, including dandelion and lettuce.

PHYLLIRA TIGER MOTH

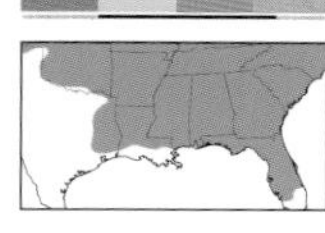

Grammia phyllira 93-0242 (8194) **Uncommon**

TL 17–21 mm. Bold creamy bands mark major wing veins of FW, forming a W in ST area. Minor wing veins are thin lines. AM and PM lines create bold cross-bars across FW; AM line connects to major vertical band in inner wing. Major vertical band does not fork at outer margin. Points of W do not touch fringe. Salmon pink HW has a fragmented black terminal line. **HOSTS:** Corn, lupine, tobacco, and other low plants.

PLACENTIA TIGER MOTH

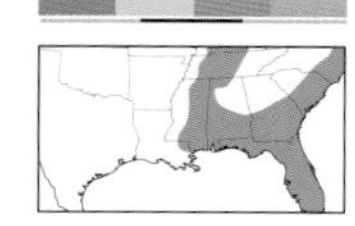

Grammia placentia 93-0243 (8191) **Common**

TL 18–22 mm. Sexually dimorphic. Black FW of male is thinly edged whitish. Major bands are similar to those of Phyllira Tiger Moth but lack thin vein markings. AM line joins vertical median band to costa. Apical triangle of W is closed by pale costa. Female is distinct; FW is black with whitish spots in AM and PM areas. Deep pink HW has an irregular black terminal line. **HOSTS:** Low plants, including plantain.

LICHEN MOTHS

TIGER MOTHS

VIRGIN TIGER MOTH

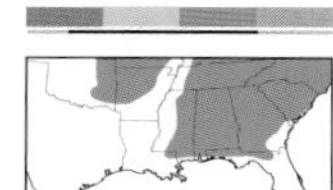

Grammia virgo 93-0244 (8197) **Common**

TL 24–37 mm. Very similar to Phyllira Tiger Moth, but major bands are slightly narrower and often tinged orange. AM line does not meet major vertical band, instead forming a tooth that points toward inner margin; on rare individuals, AM band is absent. PM line is bent at midpoint. Major vertical band forks at outer margin. Points of W rarely or barely touch fringe. Red HW is boldly marked with black spots in median and terminal areas. **HOSTS:** Bedstraw, clover, plantain, and other low plants.

ANNA TIGER MOTH

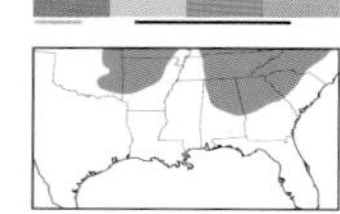

Grammia anna 93-0245 (8176) **Common**

TL 23–28 mm. Very similar to Virgin Tiger Moth, but major bands are tinged yellowish. AM line almost never touches major vertical band; on rare individuals, AM band is absent. Thin veins often are barely marked. Yellow HW typically has a wide black terminal line and a single black spot near costa, but can be all black in some females. **HOSTS:** Dandelion, plantain, and other low plants.

PARTHENICE TIGER MOTH

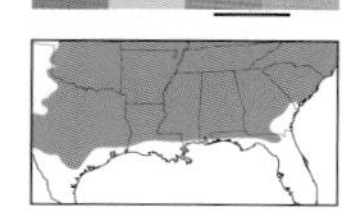

Grammia parthenice 93-0246 (8196) **Common**

TL 18–28 mm. Very similar to Virgin Tiger Moth, but FW bands have a slight pinkish orange tint. AM line virtually always connects to major vertical band. PM band usually is almost straight. Points of W typically touch fringe. Salmon pink HW has fragmented black terminal line. **HOSTS:** Dandelion, ironweed, thistle, and other low plants.

FIGURED TIGER MOTH

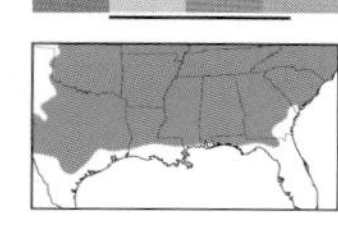

Grammia figurata 93-0253 (8188) **Common**

TL 16–22 mm. Similar to male Placentia Tiger Moth but lacks creamy edging to FW, and W marking is usually incomplete or absent; when present, apical triangle is open at costa. Red or yellow HW typically has an irregular black terminal line and one or more black spots in median area. Rarely, HW is black with a single red median spot. **HOSTS:** Alfalfa, plantain, and other low plants. **NOTE:** Two broods.

HARNESSED TIGER MOTH

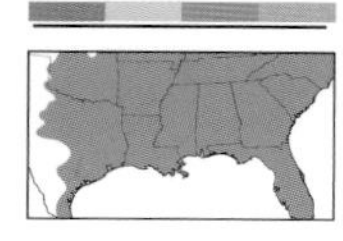

Apantesis phalerata 93-0278 (8169) **Common**

TL 16–22 mm. Variable. Black FW is edged creamy orange. Central vertical line is usually crossed in PM area to form an X. In some individuals (especially females) this is reduced or absent, resembling Banded Tiger Moth. AM line is absent. Red or yellow HW has a fragmented black terminal line. Thoracic collar is marked with two black spots. **HOSTS:** Clover, corn, dandelion, plantain, and other low plants. **NOTE:** Two broods.

TIGER MOTHS

VIRGIN TIGER MOTH

actual size

ANNA TIGER MOTH

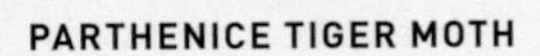

PARTHENICE TIGER MOTH

FIGURED TIGER MOTH

HARNESSED TIGER MOTH

BANDED TIGER MOTH

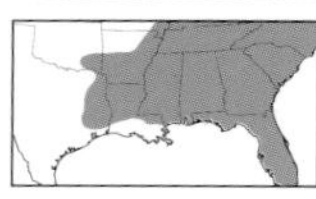

Apantesis vittata 93-0279 (8170) **Common**

TL 16–22 mm. Variable. Black FW is edged creamy orange. Central vertical line usually splits halfway down and ends in PM area. Outer arm of split bends at 90 degrees to connect to costa. Red or rosy (never yellow) HW has a broad black terminal line. Thoracic collar has two black spots, sometimes reduced. **HOSTS:** Dandelion and other low plants.

NAIS TIGER MOTH

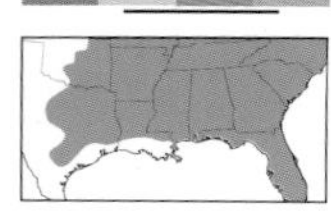

Apantesis nais 93-0280 (8171) **Common**

TL 16–22 mm. Variable. FW is virtually identical to that of Harnessed Tiger Moth. Lacks spots on thoracic collar. Pale yellow HW (sometimes with a reddish wash at base) has a wide black terminal line that is sometimes fragmented into spots. **HOSTS:** Clover, grass, plantain, violet, and other low plants. **NOTE:** Two broods.

JOYFUL VIRBIA

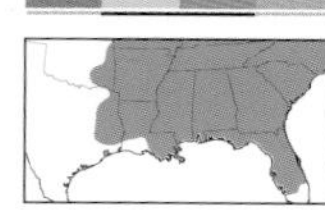

Virbia laeta 93-0294 (8114) **Common**

TL 11–13 mm. Uniform dark gray FW usually has a thin pink costal streak. Rosy HW has a broad black border. Narrow thoracic collar is pink. **HOSTS:** Dandelion and plantain.

TAWNY VIRBIA

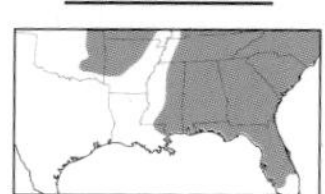

Virbia opella 93-0297 (8118) **Common**

TL 12–16 mm. Yellowish to light brown FW sometimes has slightly darker veins. An inconspicuous discal crescent is sometimes visible. Orange HW of female has a dark discal spot; HW of male is uniformly dark. **HOSTS:** Low plants, including dandelion.

ORANGE VIRBIA

Virbia aurantiaca 93-0299 (8121) **Common**

TL 10–14 mm. Orange to brown FW sometimes has an indistinct dusky discal spot and shading along ST line. Some individuals have whitish spots in median area. Orange HW usually has an almost complete black ST band and discal spot. **HOSTS:** Corn, dandelion, pigweed, plantain, and other low plants and crops.

RUDDY VIRBIA

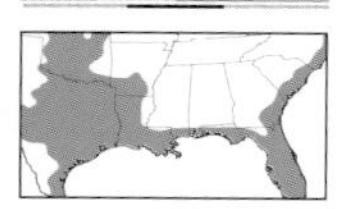

Virbia rubicundaria 93-0300 (8122) **Common**

TL 10–12 mm. Orange to pinkish brown FW has no obvious markings. HW is uniformly bright pinkish orange. **HOSTS:** Unknown, but probably a variety of low plants.

BANDED
TIGER MOTH
NAIS TIGER
MOTH
JOYFUL
VIRBIA
TAWNY VIRBIA
actual size
ORANGE VIRBIA
RUDDY VIRBIA

ECHO MOTH

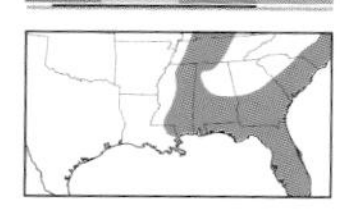

Seirarctia echo 93-0308 (8130) **Local**

TL 27–30 mm. White FW has a striking pattern of olive brown shading along pale yellow veins. White HW has a faintly spotted ST band. Legs are banded black and yellow. **HOSTS:** Coontie, cabbage palmetto, cotton, lupine, oak, persimmon, and other woody plants. **NOTE:** Multiple broods.

AGREEABLE TIGER MOTH

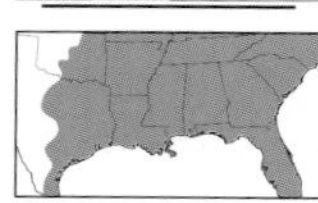

Spilosoma congrua 93-0309 (8134) **Common**

TL 19–24 mm. White FW is variably spotted brown along AM, PM, and ST lines; rare individuals are unmarked. HW and abdomen are plain white. Coxa and femur of forelegs are orange yellow. **HOSTS:** Dandelion, plantain, pigweed, and other low plants.

DUBIOUS TIGER MOTH

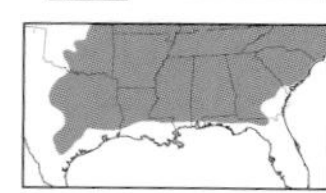

Spilosoma dubia 93-0310 (8136) **Uncommon**

TL 19–24 mm. Similar to Agreeable Tiger Moth, but white FW usually has more obvious brown spotting along AM, PM, and ST lines. HW has a spotted ST line. Abdomen is marked with yellow and black spotting. Coxa and femur of forelegs are partly orange yellow. **HOSTS:** Dandelion, plantain, pigweed, and other low plants.

VIRGINIAN TIGER MOTH

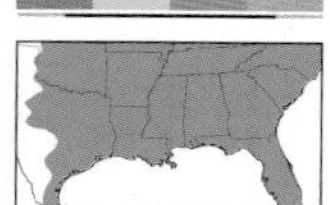

Spilosoma virginica 93-0316 (8137) **Common**

TL 17–26 mm. White FW usually has a tiny black dot in center. White HW has a few indistinct black spots in ST area. Undersides of wings have a bold black spot. Abdomen is marked with yellow patches and rows of black spots. Coxa and femur of forelegs are orange yellow. **HOSTS:** Birch, cabbage, maple, tobacco, walnut, willow, and many other trees, shrubs, and low plants. **NOTE:** Two broods.

SALT MARSH MOTH

Estigmene acrea 93-0317 (8131) **Common**

TL 24–35 mm. White FW is marked with fragmented lines of black dots, often boldest along costa. HW and undersides of all wings are orange in male, white in female, with small black blotches. Orange abdomen has a white tip and rows of black dots on each segment. **HOSTS:** Apple, cabbage, corn, potato, tobacco, and other trees and plants. **NOTE:** Two broods.

FALL WEBWORM MOTH

Hyphantria cunea 93-0319 (8140) **Common**

TL 14–19 mm. White FW can be uniform or marked with brown spots along lines, sometimes heavily. Dark extremes have mostly brown FW with white veins. HW is white, sometimes with a fragmented brown ST band in more strongly marked individuals. White abdomen often has brown bands near tip. Femora and tibiae of forelegs are orange and black. **HOSTS:** Ash, hickory, maple, oak, walnut, and many other woody plants. **NOTE:** Up to four broods.

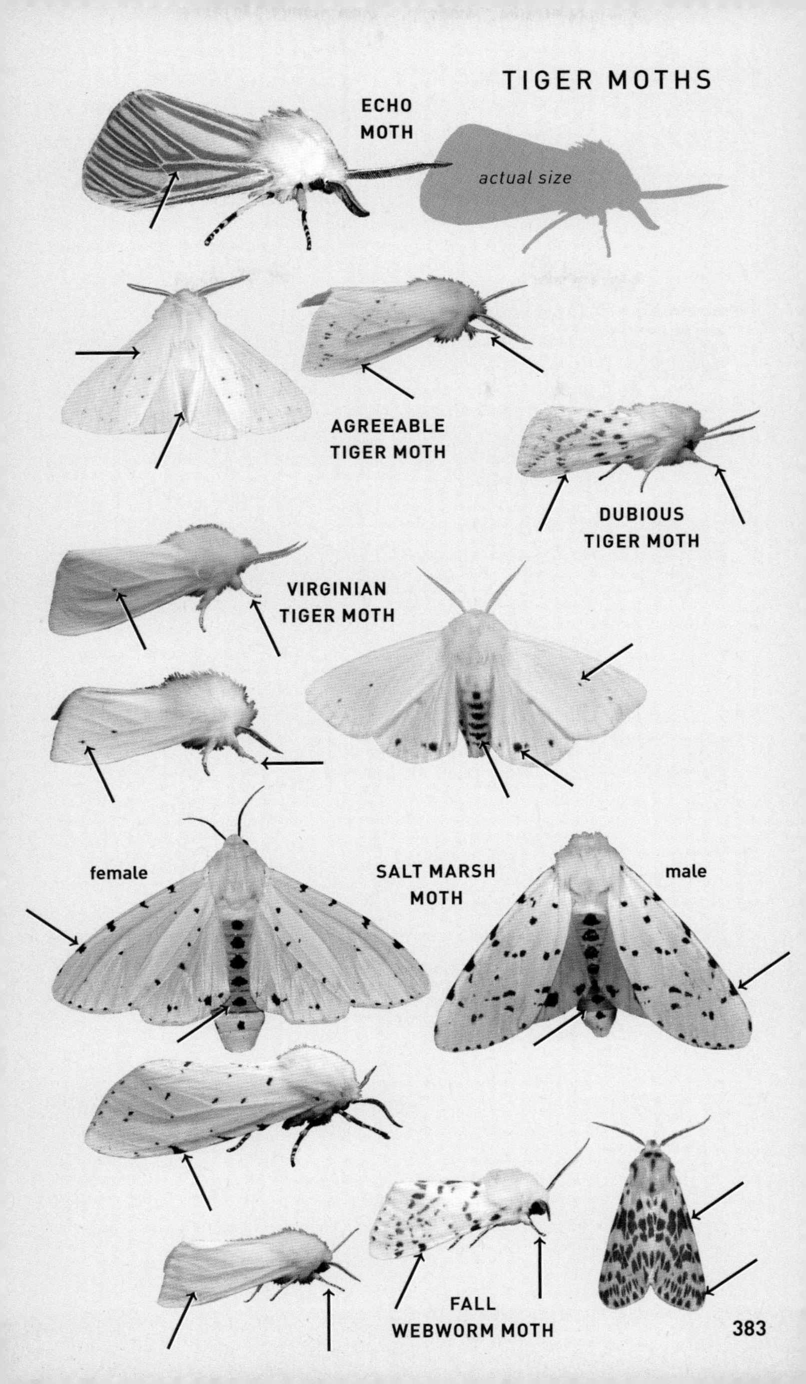
TIGER MOTHS
ECHO MOTH
actual size
AGREEABLE TIGER MOTH
DUBIOUS TIGER MOTH
VIRGINIAN TIGER MOTH
female
SALT MARSH MOTH
male
FALL WEBWORM MOTH

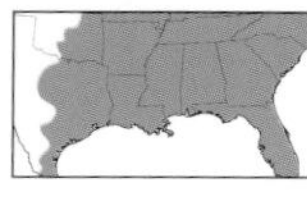

GIANT LEOPARD MOTH

Hypercompe scribonia 93-0323 (8146) **Common**

TL 33–45 mm. White FW is boldly marked with glossy blue-black spots that are mostly hollow toward inner margin. Thorax is patterned with blue-black spots. In some individuals, nearly all spots are hollow, filled light gray. Blue-black abdomen has orange lateral stripes and segment edges. **HOSTS:** Cabbage, cherry, maple, sunflower, willow, and many other trees and plants. **NOTE:** Two or more broods.

ISABELLA TIGER MOTH

Pyrrharctia isabella 93-0335 (8129) **Common**

TL 24–33 mm. Pale orange FW is marked with faint brown lines and spots of varying intensity. Rosy (female) or tan (male) HW has a few blackish spots. Orange abdomen has a row of black dorsal spots. **HOSTS:** Aster, dandelion, grass, lettuce, meadowsweet, and other low plants. **NOTE:** Two or more broods. Larva is the familiar Woolly Bear caterpillar.

CLYMENE MOTH

Haploa clymene 93-0341 (8107) **Common**

TL 22–28 mm. Creamy white (rarely pale orange) FW has a broad black border, broken only at apex and anal angle. Inner margin has a barlike extension; folded wings create a swordlike shape. Orange HW has one or two black spots in inner ST area. **HOSTS:** *Eupatorium* species; also oak, peach, and willow.

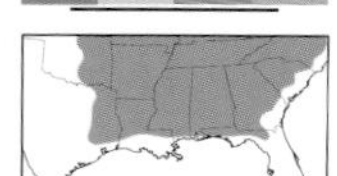

REVERSED HAPLOA

Haploa reversa 93-0343 (8109) **Uncommon**

TL 22–28 mm. White FW has a narrow dark brown border, broken at apex. An oblique brown band extends from midpoint of costa to anal angle. Apical side of oblique line is broken into four or five white spots. HW is yellow with a single dark spot in ST area. **HOSTS:** Deciduous trees and shrubs, including apple, ash, and hackberry.

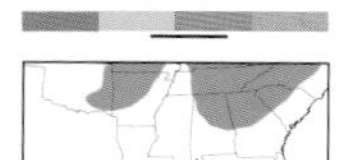

THE NEIGHBOR

Haploa contigua 93-0344 (8110) **Local**

TL 19–26 mm. White FW has a blackish border, broken at apex. Lines in lower half of wing form a T and are sometimes fragmented. Head is orange. **HOSTS:** Probably *Eupatorium* species and other low plants.

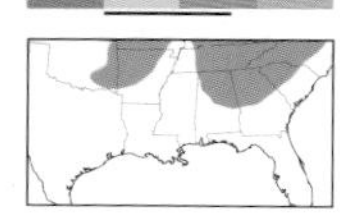

LECONTE'S HAPLOA

Haploa lecontei 93-0345 (8111) **Common**

TL 19–26 mm. Variable. White FW has a brown to blackish border, broken at costa and anal angle. An oblique black line reaches from inner PM area to apex. AM, PM, and ST lines are variably present, sometimes incomplete. Well-marked individuals look spotted. **HOSTS:** Apple, thoroughwort, willow, and other trees and low plants.

TIGER MOTHS
GIANT LEOPARD MOTH
ISABELLA TIGER MOTH
CLYMENE MOTH
THE NEIGHBOR
actual size
REVERSED HAPLOA
LECONTE'S HAPLOA

ORNATE MOTH

Utetheisa ornatrix 93-0348 (8105) **Common**

TL 16–24 mm. Rosy to golden orange FW typically has several rows of black-studded white bands. Pale individuals are mostly pale pink with just a few black spots. Pink HW has an uneven black border and white fringe. **HOSTS:** Legumes, such as rattlebox; also various trees and low plants. **NOTE:** Two or more broods.

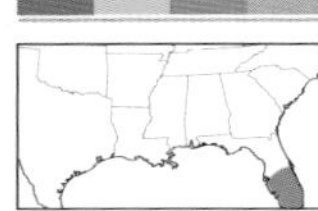

FAITHFUL BEAUTY

Composia fidelissima 93-0349 (8038) **Local**

WS 48–64 mm. Diurnal. White-spotted black wings have metallic blue accents and three red spots along inner costa. HW is often blue through basal area, with white spots along ST line. **HOSTS:** Spurge; also devil's potato, bay bean, leafless cynanchum, and oleander.

BANDED TUSSOCK MOTH

Halysidota tessellaris 93-0360 (8203) **Common**

TL 22–25 mm. Straw-colored FW has an irregular pattern of finely edged brownish bands. Whitish HW has a yellowish flush along inner margin. Thorax has turquoise and yellow dorsal stripes. **HOSTS:** Alder, ash, birch, elm, oak, willow, and many other woody plants. **NOTE:** Two broods. Florida Tussock Moth (*H. cinctipes*) is very similar but has a third turquoise stripe, down center of thorax. Found in s. FL.

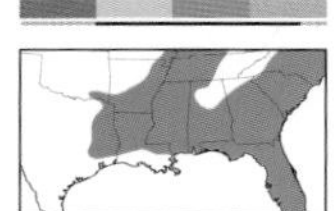

LONG-STREAKED TUSSOCK MOTH

Leucanopsis longa 93-0376 (8217) **Common**

TL 22–25 mm. Pale orange FW is densely peppered dark brown. A diffuse brown streak in median area fades into spotted apical dash. Whitish HW is semi-translucent. **HOSTS:** Marsh grass in freshwater wetlands.

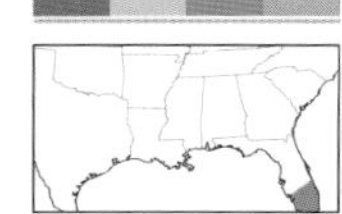

STREAKED CALIDOTA

Calidota laqueata 93-0384 (8224) **Local**

TL 22–28 mm. Semi-translucent FW is finely streaked black and white along veins. Basal section of inner margin is tinted pinkish orange. Black thorax has whitish double lateral stripes. Abdomen is bright pink. **HOSTS:** Hammock velvetseed.

YELLOW-WINGED PAREUCHAETES

Pareuchaetes insulata 93-0387 (8227) **Local**

TL 15–20 mm. Creamy yellow FW lacks any markings. Pale HW is semi-translucent toward base. Yellow abdomen has a row of black dorsal spots. **HOSTS:** *Eupatorium* species.

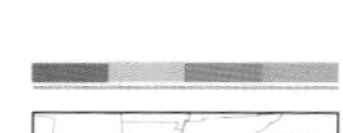

SNOWY EUPSEUDOSOMA

Eupseudosoma involuta 93-0389 (8257) **Local**

TL 18–22 mm. Satin white FW is semi-translucent with white veins; costa is sometimes edged blackish. Deep pink abdomen has a row of white dorsal spots and white tip. HW is much reduced in size. **HOSTS:** *Eucalyptus* species.

TIGER MOTHS
ORNATE MOTH
FAITHFUL BEAUTY
male
actual size
BANDED TUSSOCK MOTH
FLORIDA TUSSOCK MOTH
LONG-STREAKED TUSSOCK MOTH
STREAKED CALIDOTA
SNOWY EUPSEUDOSOMA
YELLOW-WINGED PAREUCHAETES

MOUSE-COLORED LICHEN MOTH

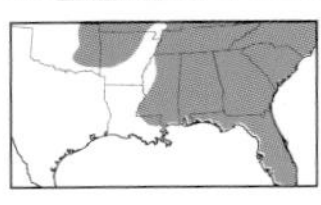

Pagara simplex 93-0396 (8099) **Uncommon**

TL 12–14 mm. Smoky gray semi-translucent wings are unmarked apart from slightly paler veins. FW has a small yellow tinge at base. Pale yellow abdomen has a row of brownish dorsal spots. Male has bipectinate antennae. **HOSTS:** Unknown, but has been reared on dandelion and lettuce.

RED-TAILED SPECTER

Euerythra phasma 93-0398 (8141) **Uncommon**

TL 16–18 mm. Snow-white FW has a broad peppery gray stripe from base to outer margin, crossed by partial AM and oblique PM bands; bands are accented with reddish to golden veins. Abdomen is strongly banded bright red. **HOSTS:** Unknown.

THREE-SPOTTED SPECTER

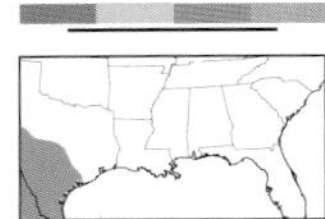

Euerythra trimaculata 93-0399 (8142) **Local**

TL 14–16 mm. Snow-white FW has patches of peppery golden brown spots marked with orange or red veins at outer AM, inner PM, and subapical areas. Abdomen is narrowly banded dull red toward tip. **HOSTS:** Unknown.

COLLARED CYCNIA

Cycnia collaris 93-0402 (8229) **Common**

TL 15–19 mm. White FW is sometimes shaded pale gray with white veins. Costa is yellow-orange in basal area, leading into yellow-orange head and collar. Pale yellow abdomen is marked with rows of black dorsal spots. **HOSTS:** Butterfly weed. **NOTE:** Two broods.

DELICATE CYCNIA

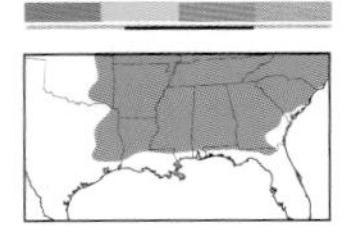

Cycnia tenera 93-0404 (8230) **Common**

TL 16–22 mm. Similar to slightly smaller Collared Cycnia, but yellow-orange costal streak is more diffuse and almost reaches apex. **HOSTS:** Indian hemp and milkweed. **NOTE:** Two or more broods.

MILKWEED TUSSOCK MOTH

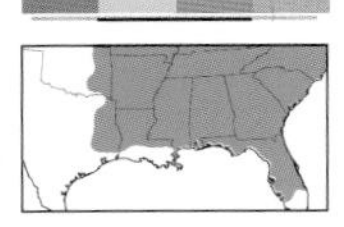

Euchaetes egle 93-0412 (8238) **Common**

TL 17–22 mm. Mousy gray FW is often lightly flecked whitish along veins. HW is gray. Head often has some yellow around eyes. Yellow-orange abdomen has rows of black spots. **HOSTS:** Milkweed. **NOTE:** Two broods.

FRECKLED TIGER MOTH

Ectypia bivittata 93-0420 (8247) **Local**

TL 23–25 mm. White FW is patterned with short black dashes. White thorax has black lateral lines edged orange. Abdomen has yellow dorsal surface and lines of black dorsal and lateral dots. Black legs have wide white bands. **HOSTS:** Unknown.

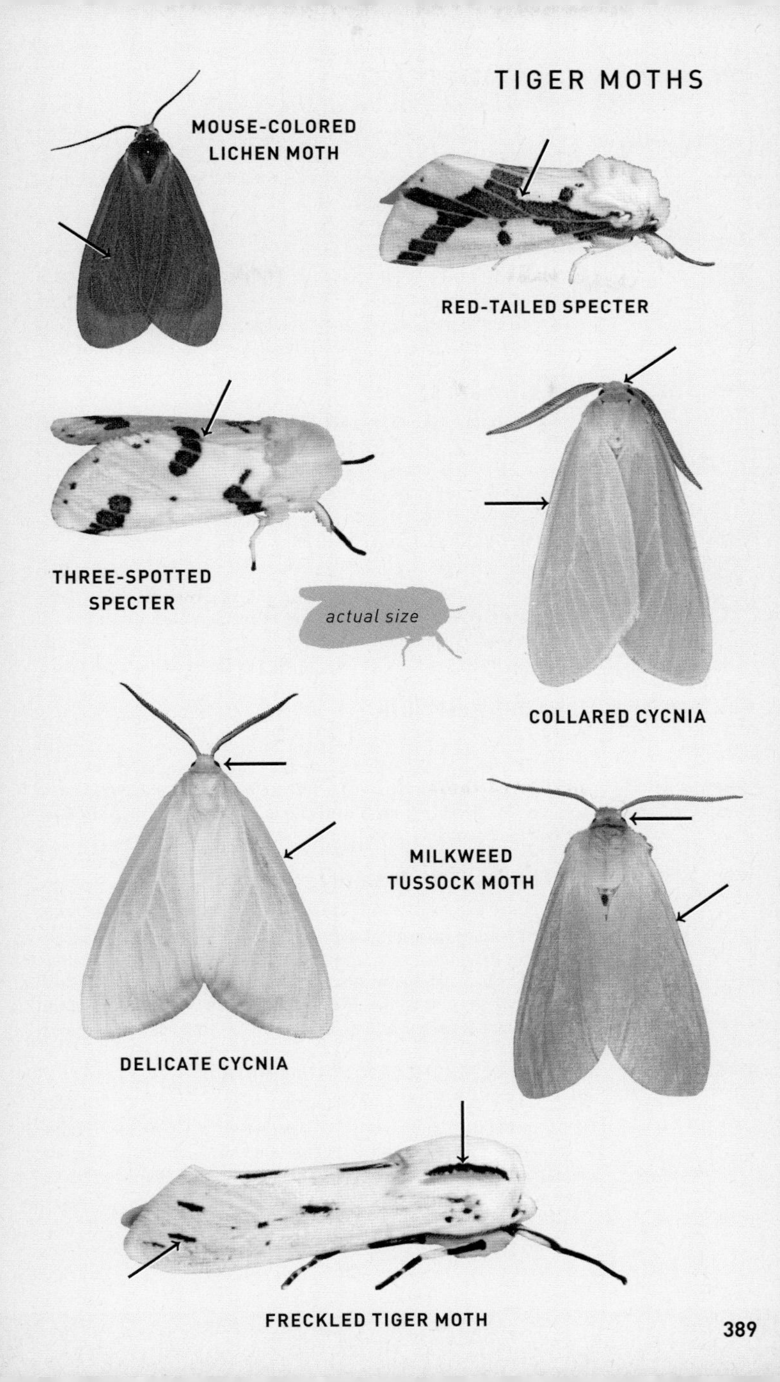
MOUSE-COLORED LICHEN MOTH
RED-TAILED SPECTER
THREE-SPOTTED SPECTER
actual size
COLLARED CYCNIA
DELICATE CYCNIA
MILKWEED TUSSOCK MOTH
FRECKLED TIGER MOTH

YELLOW-EDGED PYGARCTIA

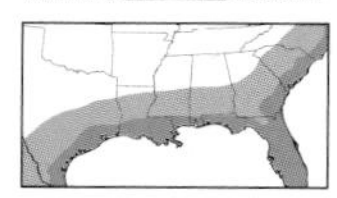

Pygarctia abdominalis 93-0430 (8255) **Local**

TL 18–24 mm. Dove gray FW has yellow streaks along costa and inner margin, tinted orange at base. Gray thorax has reddish lateral stripes. Pale orange abdomen has a row of tiny black dorsal dots. **HOSTS:** Spurge and dogbane.

VEINED CTENUCHA

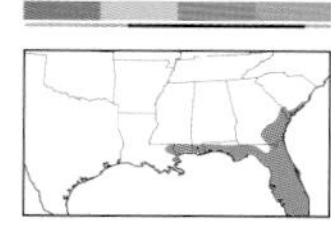

Ctenucha venosa 93-0433 (8260) **Local**

TL 15–17 mm. Dark leaden gray FW is boldly patterned with three yellow streaks along veins. Apex is edged white. Metallic ink blue thorax has yellow lateral stripes. Head is bright red. **HOSTS:** Grass.

BLACK-WINGED DAHANA

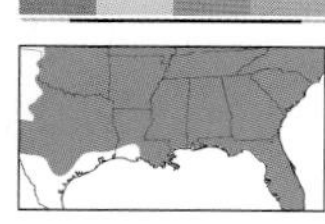

Dahana atripennis 93-0439 (8266) **Local**

TL 18–22 mm. Crepuscular. Narrow, blackish FW is marked with a pale streak at anal angle. Bright orange abdomen has metallic blue patches at base. Head is orange. **HOSTS:** Spanish moss.

YELLOW-COLLARED SCAPE MOTH

Cisseps fulvicollis 93-0440 (8267) **Common**

TL 16–20 mm. Narrow, sooty brown FW has thin yellowish edge along costa. HW is semi-translucent in median area. Dark thorax, sometimes with a metallic blue sheen, has a contrasting golden yellow collar. **HOSTS:** Grass and sedge. **NOTE:** Three or more broods.

EDWARD'S WASP MOTH

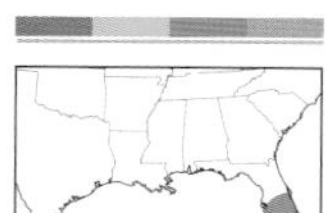

Lymire edwardsii 93-0443 (8270) **Local**

TL 19–22 mm. Narrow, light to dove gray FW has slightly paler costa toward apex. Abdomen is mostly dull metallic blue, and thorax has a blue sheen. Head and base of forelegs are bright orange. **HOSTS:** Fig and lancewood.

LITTLE CAROL'S WASP MOTH

Nelphe carolina 93-0446 (8271) **Local**

TL 16–19 mm. Heavy yellow-edged gray veins partly obscure black-and-white banded pattern of FW. Has thick black band in median area, with black patches in AM and ST areas. Semi-translucent HW has dusky veins and border. Blackish abdomen often has a blue tint and has pairs of yellow-orange spots near tip. **HOSTS:** Swallowwort.

SPOTTED OLEANDER CATERPILLAR MOTH

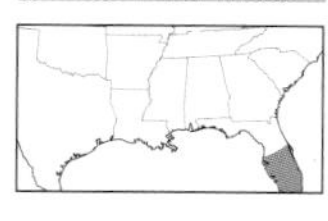

Empyreuma affinis 93-0448 (8272) **Local**

TL 22 mm. Long, sunset red FW is edged dark brown along outer margin. Orange HW has a narrow blackish border. Metallic blue abdomen and thorax have pairs of white dots. **HOSTS:** Oleander. **NOTE:** Previously classified as *E. pugione*.

TIGER MOTHS
YELLOW-EDGED PYGARCTIA
VEINED
CTENUCHA
actual size
BLACK-WINGED
DAHANA
EDWARD'S
WASP MOTH
YELLOW-COLLARED
SCAPE MOTH
LITTLE CAROL'S
WASP MOTH
SPOTTED OLEANDER
CATERPILLAR MOTH

SCARLET-BODIED WASP MOTH

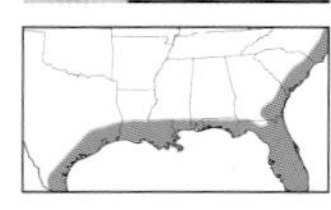

Cosmosoma myrodora 93-0458 (8280) **Common**

WS 30–35 mm. Strikingly bright red body; abdomen is tipped black with metallic blue spots. Translucent wings have black borders and veins. Antennae are tipped white. **HOSTS:** Climbing hempvine.

DOUBLE-TUFTED WASP MOTH

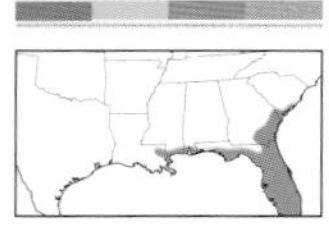

Didasys belae 93-0459 (8281) **Local**

WS 22–24 mm. Translucent wings have black borders and veins. FW has ST band of reddish orange to orange spots and a red to reddish orange discal spot. Red abdomen is tipped with two wispy black and red tufts. Black thorax is boldly striped white. **HOSTS:** Unknown.

YELLOW-BANDED WASP MOTH

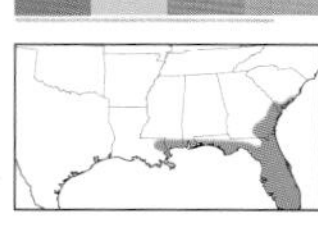

Syntomeida ipomoeae 93-0460 (8282) **Local**

WS 35–42 mm. Long, blackish blue FW is marked with small whitish spots in PM, and sometimes AM, area. HW is transparent at base, but this is rarely visible when at rest. Golden orange abdomen is boldly banded black, often with a metallic blue sheen. **HOSTS:** Morning glory.

POLKA-DOT WASP MOTH

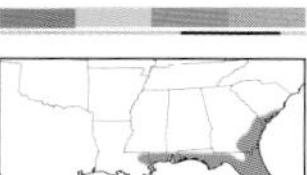

Syntomeida epilais 93-0462 (8284) **Local**

WS 42–52 mm. Wings and body are blackish with metallic blue sheen, marked with splashy white spots. Tip of abdomen is bright reddish orange. Antennae have a white band near tip. **HOSTS:** Oleander and devil's potato.

LESSER WASP MOTH

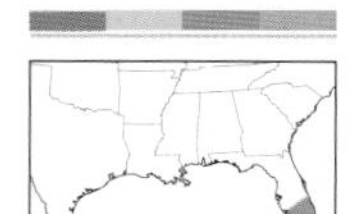

Pseudocharis minima 93-0463 (8286) **Local**

WS 30–35 mm. Similar to Polka-dot Wasp Moth, but abdomen is entirely black. Antennae are rusty orange. Hindlegs have distinctive tufts and white tips. **HOSTS:** Christmas berry.

TEXAS WASP MOTH

Horama panthalon 93-0465 (8287) **Local**

WS 32–34 mm. Long, narrow FW is dull rusty brown. Blackish abdomen has metallic purplish sheen and is boldly banded yellow. Thorax has four yellowish spots bordered by two lateral stripes. Antennae have pale tips. In male, hindlegs have distinctive black-and-yellow–flanged tips. **HOSTS:** Unknown.

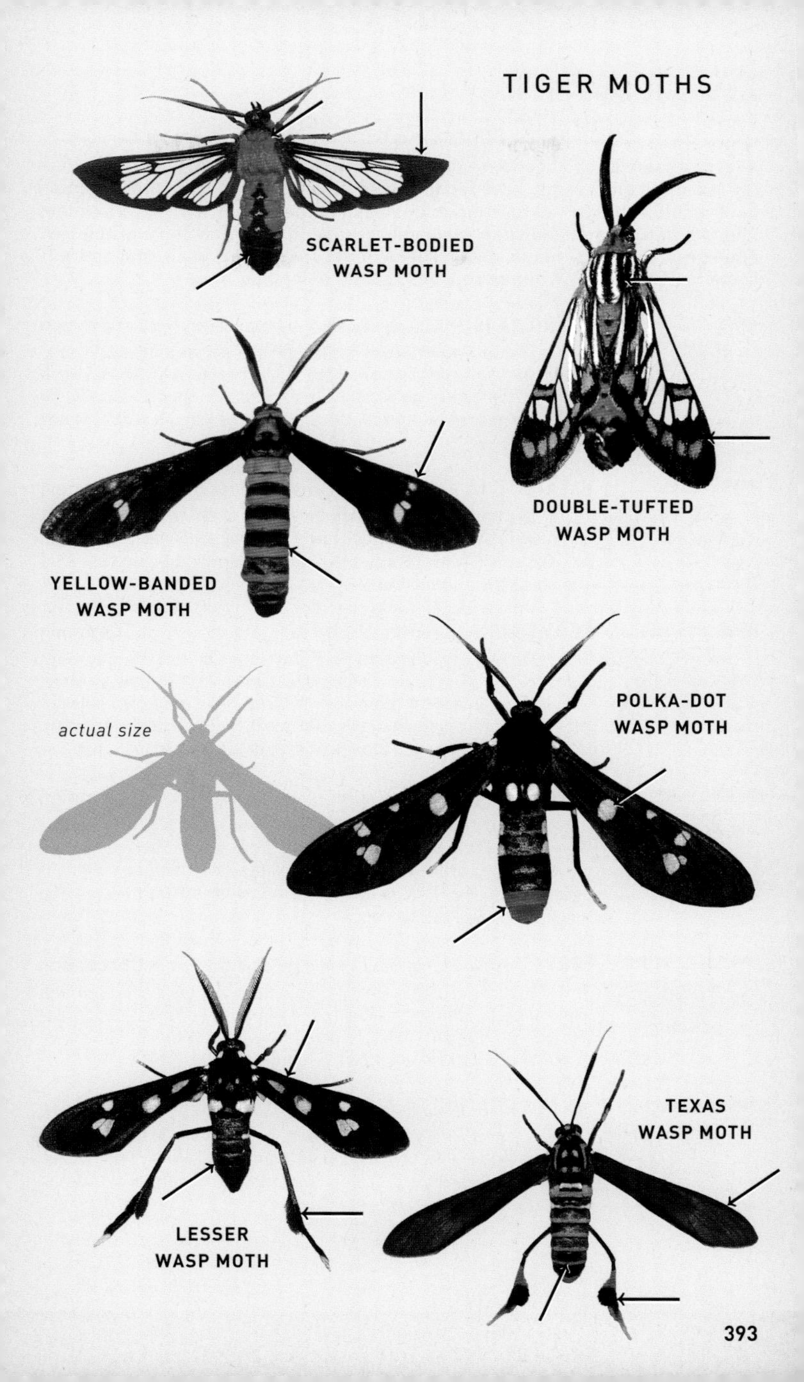
SCARLET-BODIED
WASP MOTH
DOUBLE-TUFTED
WASP MOTH
YELLOW-BANDED
WASP MOTH
actual size
POLKA-DOT
WASP MOTH
TEXAS
WASP MOTH
LESSER
WASP MOTH

Litter Moths

Family Erebidae, Subfamily Herminiinae

A large assemblage of delta-winged, flat-looking moths. Some species, notably fan-foots and renias, have long, often upturned, labial palps. Male renias have a noticeable tuft of hairlike scales beyond the midpoint of the antennae. The two palthis moths display a curious crease in the outer third of the FW. Found in woodlands, fields, and gardens, most species are nocturnal and will come to lights in small numbers. Idias readily visit sugar bait.

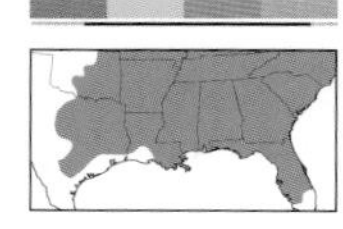

AMERICAN IDIA *Idia americalis* 93-0469 (8322) **Common**

TL 13–14 mm. Pale gray FW has jagged black lines boldest at costa. Orbicular and reniform spots are rusty brown. A wash of warm brown shading often colors interior of wing. Terminal line is slightly thickened near apex. **HOSTS:** Dead leaves and lichens. **NOTE:** Multiple broods.

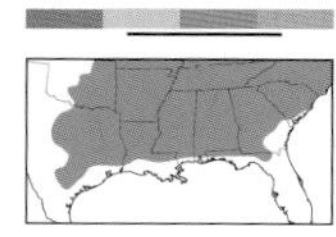

GREATER IDIA *Idia majoralis* 93-0470 (8324) **Uncommon**

TL 16–18 mm. Similar to darker individuals of American Idia but has contrasting paler costal streak interrupted by wide black squares at AM, PM, and ST lines. Terminal line is uniformly thin. **HOSTS:** Presumably debris in woodrat nests.

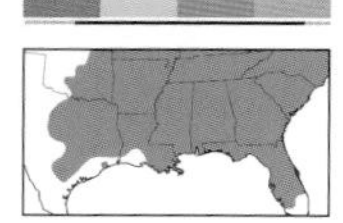

COMMON IDIA *Idia aemula* 93-0471 (8323) **Common**

TL 11–16 mm. FW is variably pale grayish brown to dark gray, peppered with dusky scales. Jagged lines widen only slightly at costa. Diffuse median band is often bold. Orbicular and reniform spots are conspicuously pale yellowish. Terminal line is uniformly thin. **HOSTS:** Presumably leaf litter. **NOTE:** Multiple broods.

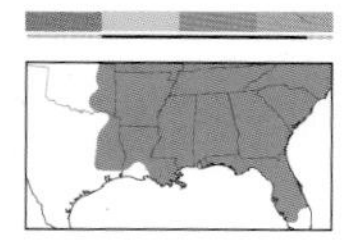

ROTUND IDIA *Idia rotundalis* 93-0474 (8326) **Common**

TL 11–13 mm. Shiny, sooty brown FW is peppered with gray scales. Wavy black AM and PM lines are indistinct, sometimes edged dark gray. Pale gray orbicular and reniform spots are small and often inconspicuous. Pale HW is lightly barred. **HOSTS:** Dead leaves and fungi. **NOTE:** Multiple broods.

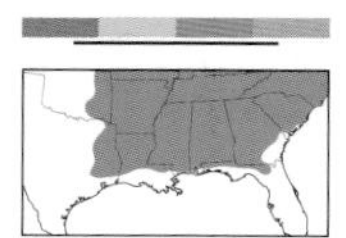

FORBES' IDIA *Idia forbesii* 93-0475 (8327) **Uncommon**

TL 11–13 mm. Dark gray FW has indistinct black lines accented with white at costa. Blackish median band is diffuse and most clearly visible at inner margin. Wavy ST line is highlighted with tiny white dots. Gray HW is boldly barred. **HOSTS:** Unknown.

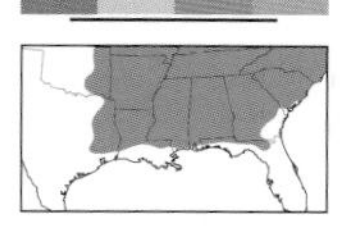

JULIA'S IDIA *Idia julia* 93-0476 (8328) **Common**

TL 11–13 mm. Similar to Rotund Idia, but whitish AM and PM lines are more noticeable, widening slightly at costa to form white spots. Small whitish reniform spot is sometimes conspicuous. **HOSTS:** Dead leaves. **NOTE:** Multiple broods.

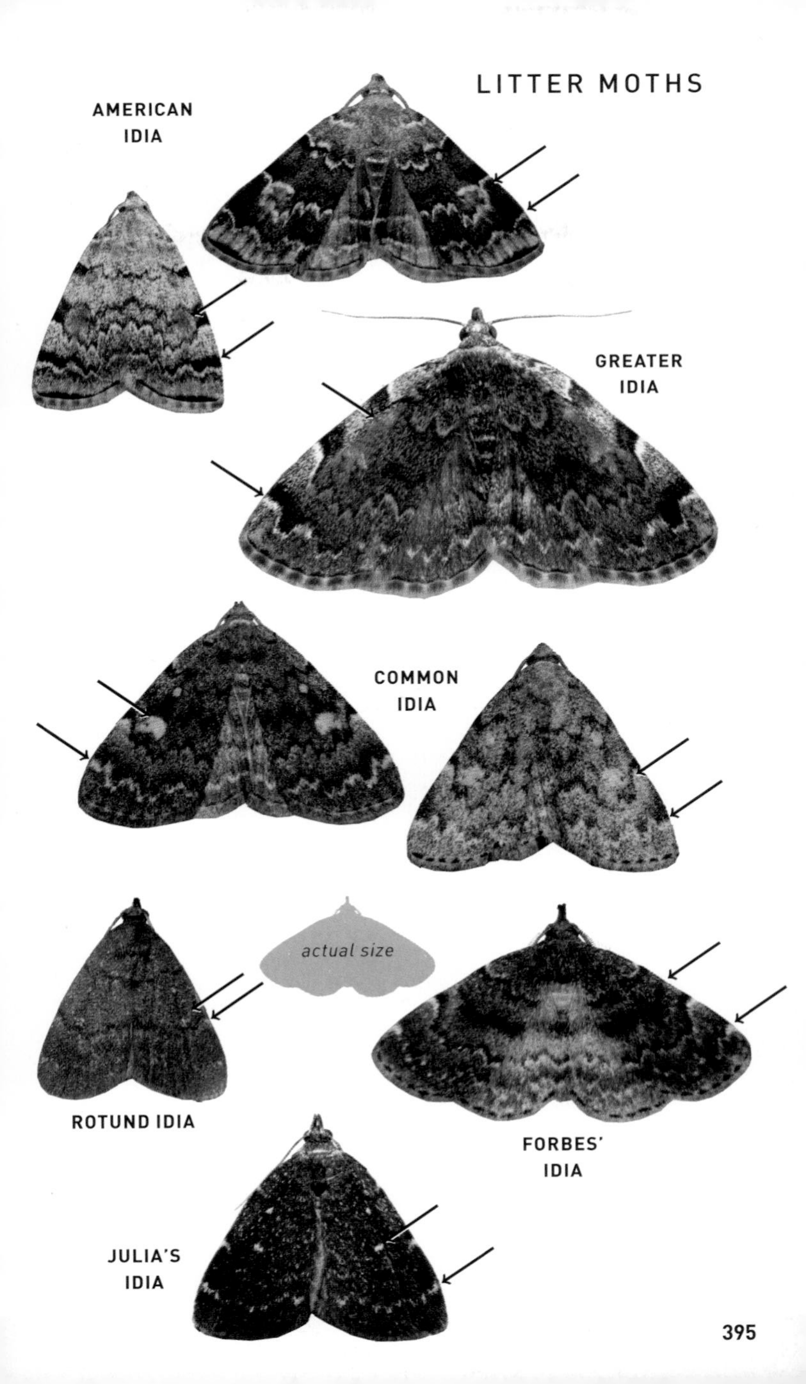
LITTER MOTHS
AMERICAN IDIA
GREATER IDIA
COMMON IDIA
actual size
ROTUND IDIA
FORBES' IDIA
JULIA'S IDIA

ORANGE-SPOTTED IDIA

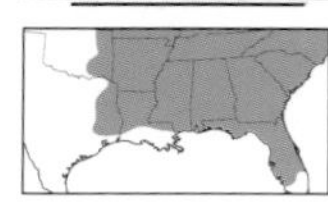

Idia diminuendis 93-0477 (8329) **Common**

TL 9–11 mm. FW is variably light gray to blackish with inconspicuous jagged black AM and PM lines; dark individuals resemble Rotund Idia. Small pale orange orbicular and reniform spots are usually conspicuous. **HOSTS:** Dead leaves. **NOTE:** Multiple broods.

GLOSSY BLACK IDIA

Idia lubricalis 93-0482 (8334) **Common**

TL 18–21 mm. Considerably larger than other idias. Shiny, sooty brown to blackish FW has toothed, yellow-edged AM, PM, and ST lines that widen slightly at costa. Reniform spot is outlined pale yellow. **HOSTS:** Fungi and lichens. **NOTE:** Multiple broods.

DARK-BANDED OWLET

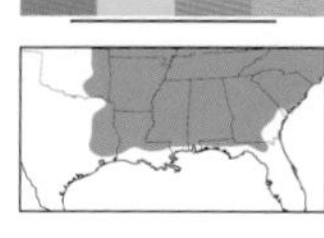

Phalaenophana pyramusalis 93-0487 (8338) **Common**

TL 12–14 mm. Lilac gray FW has pale, straight AM and slightly wavy ST lines bordered with wide, dark brown bands. Reniform spot contains two black dots. **HOSTS:** Dead leaves. **NOTE:** Two or more broods.

LETTERED FAN-FOOT

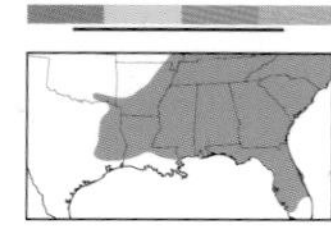

Zanclognatha lituralis 93-0489 (8340) **Common**

TL 12–15 mm. Pale gray FW has dotted lines that end as bold black spots along costa. Reniform spot is a thin black crescent. Jagged ST line is accented with white triangles. **HOSTS:** Dead leaves. **NOTE:** One or two broods.

FLAGGED FAN-FOOT

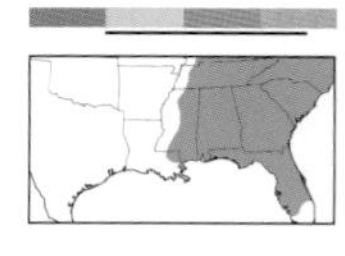

Zanclognatha theralis 93-0490 (8341) **Common**

TL 12–15 mm. Pale grayish brown FW has crisp jagged lines that end as bold black spots along costa. Straight PM line forms vertical line near costa. ST line is white-edged. Reniform spot is a thin black crescent, often indistinct. **HOSTS:** Presumably dead leaves.

BLACK-LINED FAN-FOOT

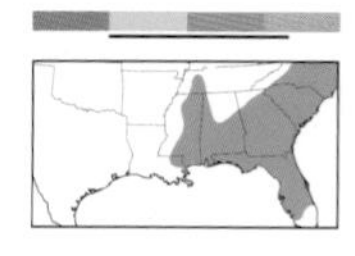

Zanclognatha atrilineella 93-0493 (8346) **Uncommon**

TL 12–15 mm. Light brown FW is peppered with reddish scales. AM, PM, and ST lines are broadly edged black. Black reniform spot is boldly marked. **HOSTS:** Presumably dead leaves.

DARK FAN-FOOT

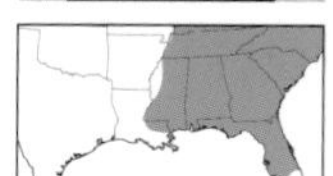

Zanclognatha obscuripennis 93-0494 (8347) **Common**

TL 12–15 mm. Yellowish to brown FW has straight AM line that turns gently up at costa, or curves evenly. PM line bulges around reniform crescent, then angles away from AM line at inner margin. Straight ST line is edged yellow. **HOSTS:** Dead leaves. **NOTE:** Multiple broods.

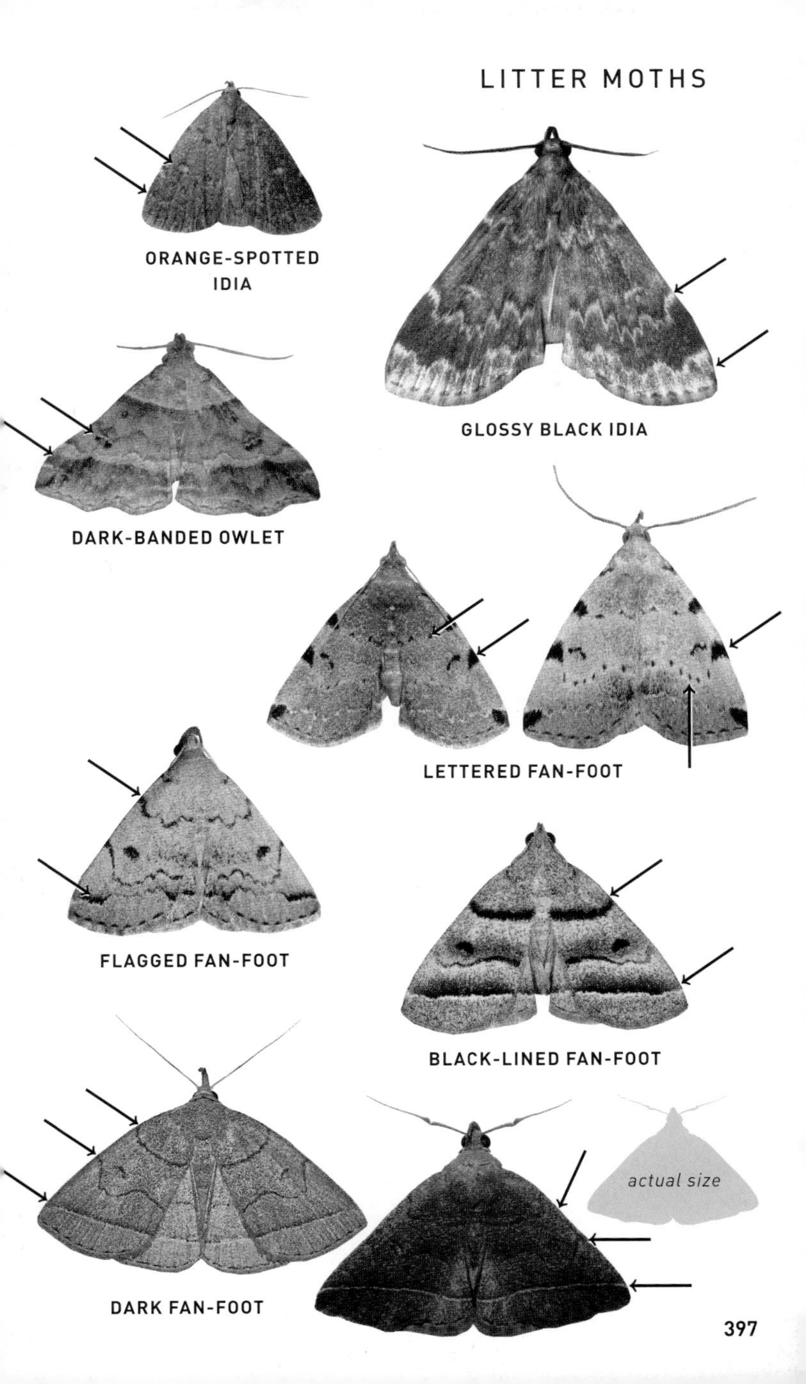

ORANGE-SPOTTED IDIA

GLOSSY BLACK IDIA

DARK-BANDED OWLET

LETTERED FAN-FOOT

FLAGGED FAN-FOOT

BLACK-LINED FAN-FOOT

DARK FAN-FOOT

COMPLEX FAN-FOOT

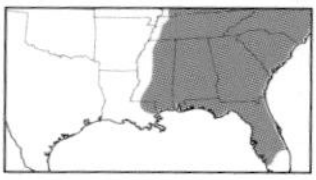

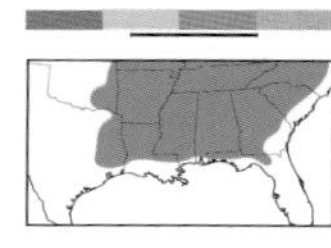

Zanclognatha protumnusalis 93-0496 (8349) **Common**

TL 12–13 mm. Resembles Flagged Fan-Foot, but FW is yellowish tan to brown with a noticeable dark reniform spot. Jagged AM and PM lines are slightly thicker at costa. PM line curves around reniform spot. Straight ST line is sometimes edged yellow. **HOSTS:** Coniferous trees. **NOTE:** Multiple broods.

PINE BARRENS FAN-FOOT

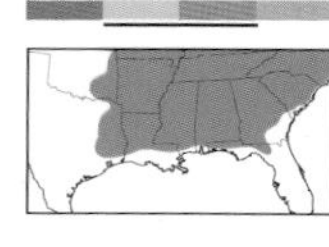

Zanclognatha martha 93-0497 (8350) **Local**

TL 13–16 mm. Resembles Complex Fan-Foot, but FW is dusky brown and lines are indistinct. PM line is almost straight, and rises sharply to meet costa. Reniform spot is dark. Straight ST line is weakly edged pale yellow. **HOSTS:** Pitch pine.

EARLY FAN-FOOT

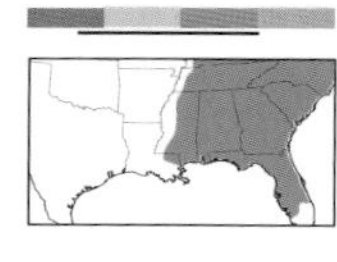

Zanclognatha cruralis 93-0498 (8351) **Common**

TL 15–17 mm. Resembles Dark Fan-Foot, and some individuals may not be safely identified. FW is pale yellowish tan (darker in summer brood). AM line bends sharply to meet costa. PM line meets inner margin parallel to AM line. **HOSTS:** Deciduous and coniferous trees, including beech, hazel, hemlock, maple, and others. **NOTE:** Two broods.

YELLOWISH FAN-FOOT

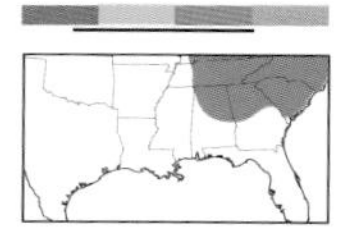

Zanclognatha marcidilinea 93-0499 (8352) **Common**

TL 17–18 mm. Resembles Early Fan-Foot, but yellowish to brown FW is marked with indistinct wavy AM and PM lines. AM line is nearly straight to costa, or just slightly curved. Reniform spot is dark. ST line is boldly edged pale yellow. **HOSTS:** Probably dead leaves. **NOTE:** One or two broods.

WAVY-LINED FAN-FOOT

Zanclognatha jacchusalis 93-0500 (8353) **Common**

TL 17–18 mm. Resembles Yellowish Fan-Foot, but wavy AM and PM lines are distinct. AM line bends strongly to meet costa. Straight ST line is edged pale yellow and often bordered with darker brown shading. Reniform spot is usually hollow. **HOSTS:** Probably dead leaves. **NOTE:** One or two broods. Revision of the *Zanclognatha* genus resulted in reassigning the species name *Z. jacchusalis* to the species that was formerly identified as *Z. ochreipennis*, which is now synonymous. The species previously known as *Z. jacchusalis* now goes by the name *Z. marcidilinea* (Yellowish Fan-Foot).

MORBID OWLET

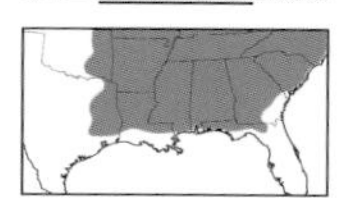

Chytolita morbidalis 93-0502 (8355) **Common**

TL 16–19 mm. Straw-colored FW is finely peppered with brown scales. Wavy AM line and bulging PM line are pale brown. ST line is accented with a row of outward-pointing dusky wedges, bordered white. Labial palps are long, as in *Renia* species. **HOSTS:** Dead leaves.

LITTER MOTHS

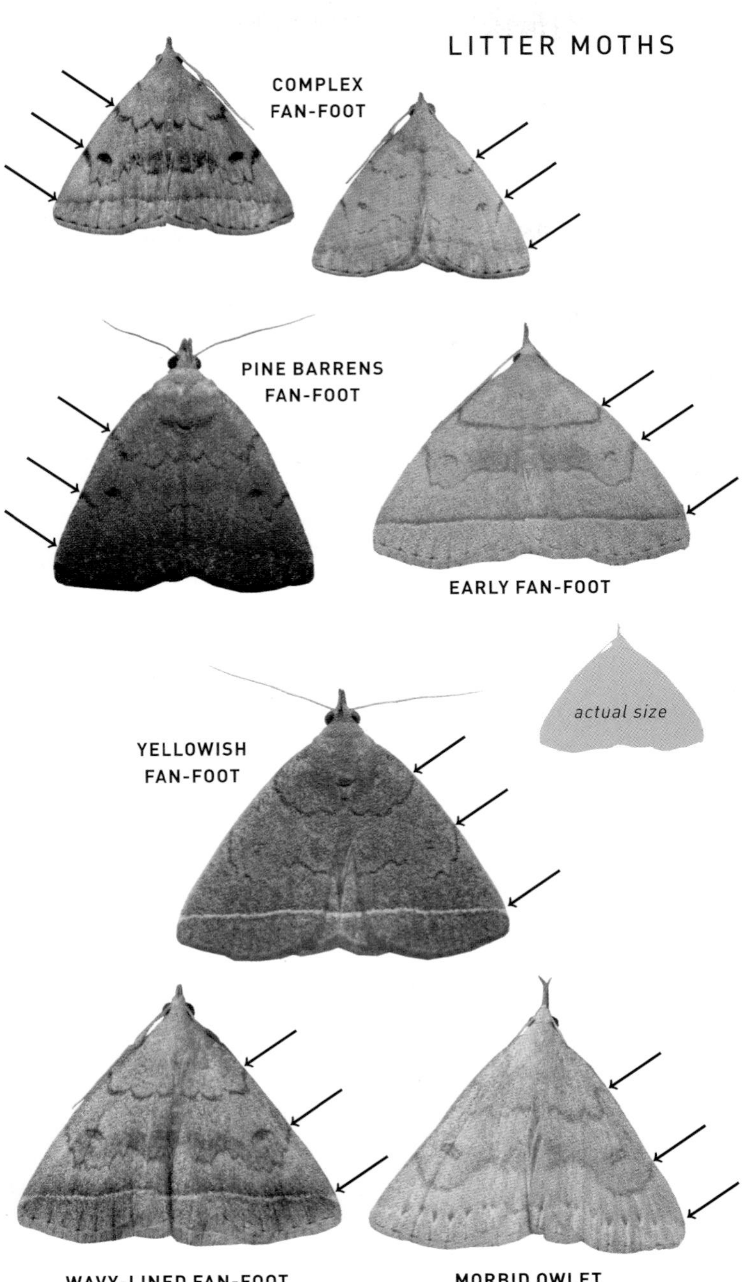

MAGAS FRUIT-BORER

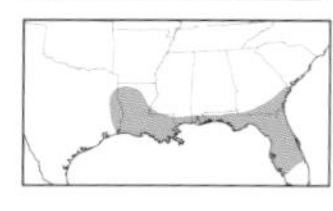

Simplicia cornicalis 93-0504 (8339.1) **Local**

TL 12–14 mm. Narrow, brownish olive FW has indistinct, slightly curved AM and PM lines. Straight pale yellow ST line is boldest marking. **HOSTS:** Dead leaves and plant debris, particularly palm thatch. **NOTE:** A recently introduced Asian species that seems likely to expand its range.

LOUISIANA OWLET

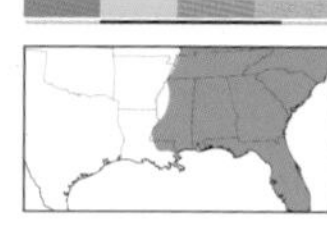

Macrochilo louisiana 93-0505 (8361) **Common**

TL 14–15 mm. Lightly peppered brown FW has curving PM line usually bordered by brown shading. Four tiny black dots mark orbicular and reniform spots. **HOSTS:** Unknown.

TWIN-DOTTED OWLET

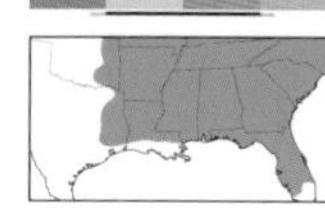

Macrochilo hypocritalis 93-0509 (8357.1) **Common**

TL 10–12 mm. Peppery straw-colored FW has an oblique, broadly shaded brown PM line that strongly angles upward before reaching costa. Reniform spot contains two tiny black dots. ST line is edged pale, most obvious on darker individuals. **HOSTS:** Unknown. **NOTE:** Presumably two or more broods.

TWO-LINED OWLET

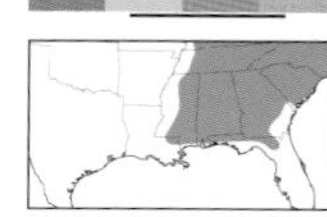

Macrochilo litophora 93-0510 (8358) **Common**

TL 12–14 mm. Light brown FW has parallel brown AM and PM lines that are strongly angled before reaching costa. ST line is indistinct. **HOSTS:** Live and dead grass and clover.

BRONZY MACROCHILO

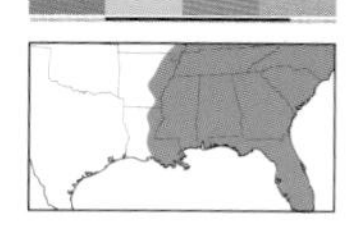

Macrochilo orciferalis 93-0511 (8360) **Common**

TL 14–15 mm. Narrow bronzy to tan FW has a blackish streak passing through central median area that joins a slanting apical dash. Whitish reniform spot contains two tiny black dots. Some individuals have dark shading along veins. **HOSTS:** Possibly grass. **NOTE:** Two broods.

BLACK-BANDED OWLET

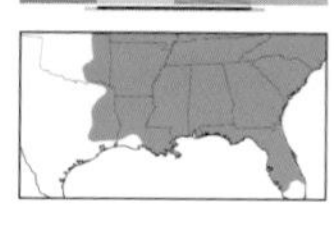

Phalaenostola larentioides 93-0514 (8364) **Common**

TL 10–13 mm. Violet gray FW has wavy AM and PM lines. Wide bands of black shading pass through median area and along wavy whitish ST line. **HOSTS:** Dead and living grass, leaves, and clover. **NOTE:** Two or more broods.

SMOKY TETANOLITA

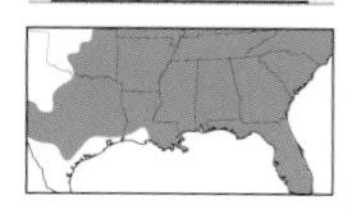

Tetanolita mynesalis 93-0516 (8366) **Common**

TL 11–13 mm. Dark gray FW has inconspicuous scalloped AM and PM lines. Pale yellow reniform spot is most obvious feature. Whitish ST line is fragmented into a row of dots. **HOSTS:** Probably dead leaves. **NOTE:** Probably two or more broods.

LITTER MOTHS

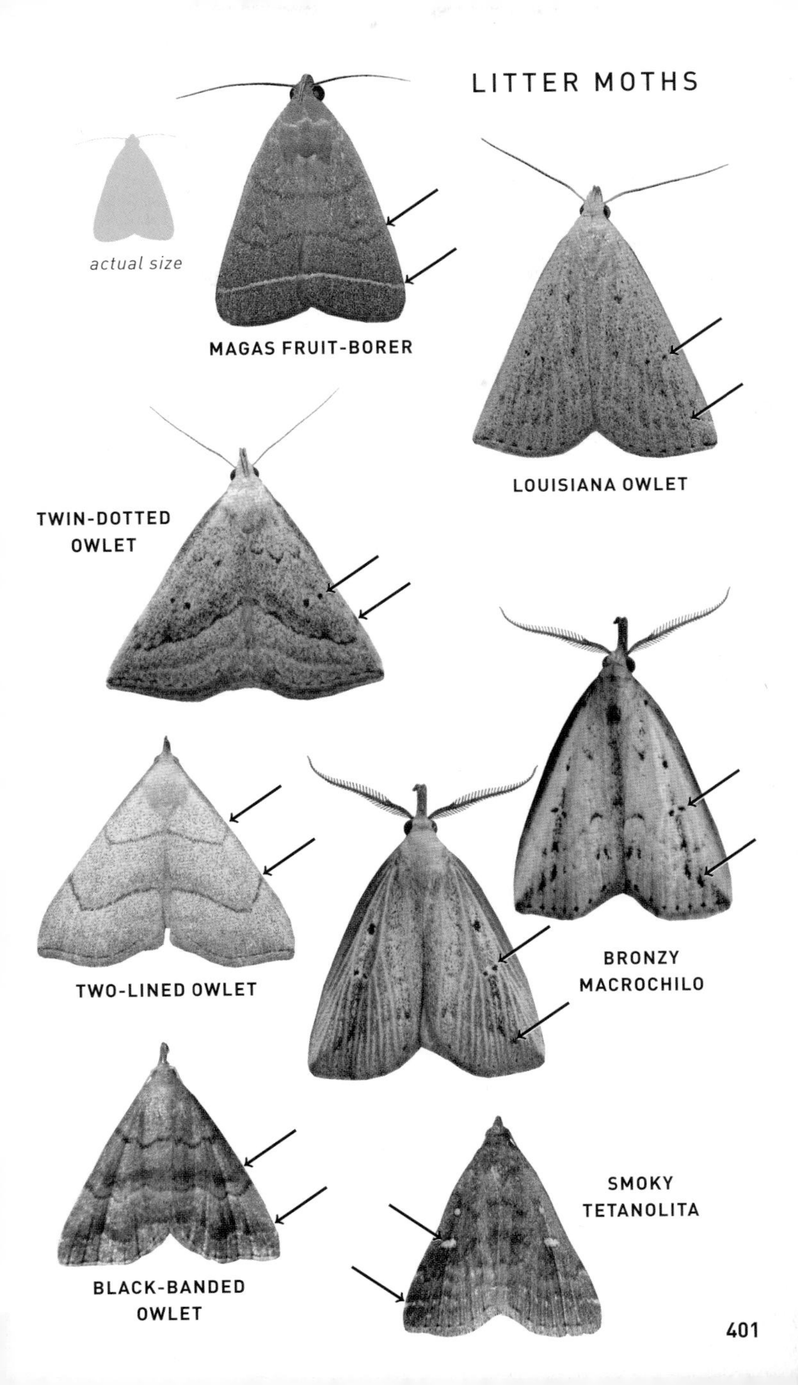

FLORIDA TETANOLITA

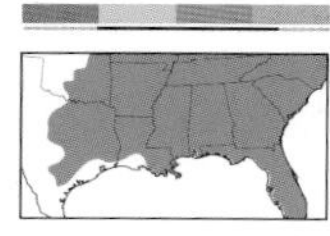

Tetanolita floridana 93-0518 (8368) **Common**

TL 11–13 mm. Pale grayish brown FW has inconspicuous jagged AM and PM lines; PM line is sometimes edged pale. Median and ST areas have diffuse dark bands. Rectangular reniform spot is pale yellow. Whitish ST line is often fragmented into a row of dots. **HOSTS:** Probably dead leaves. **NOTE:** Two or more broods.

BENT-WINGED OWLET

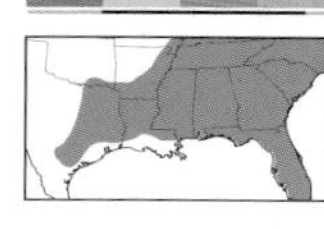

Bleptina caradrinalis 93-0520 (8370) **Common**

TL 12–17 mm. Light brown to violet gray FW has inconspicuous jagged AM and PM lines. A bold median band is usually present. ST line is pale and distinct, often bordered by dark shading. Reniform spot can be orange or black. Male has concave costal edge. **HOSTS:** Dead leaves. **NOTE:** Multiple broods.

INFERIOR OWLET

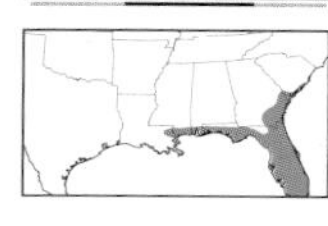

Bleptina inferior 93-0522 (8371) **Common**

TL 12–14 mm. Lightly peppered gray FW has jagged AM and PM lines. Pale ST line is thin and often edged with dark shading. Orbicular and reniform spots are boldly outlined brown. **HOSTS:** Dead leaves. **NOTE:** One or two broods.

LONG-HORNED OWLET

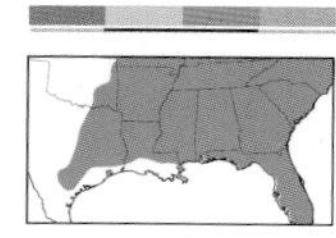

Hypenula cacuminalis 93-0527 (8376) **Local**

TL 15–16 mm. Dark brown FW has indistinct jagged AM and PM lines that are sometimes thinly edged whitish. Orbicular and reniform spots are orange; reniform spot contains white dots at each end. **HOSTS:** Dead leaves. **NOTE:** Multiple broods.

DOTTED RENIA

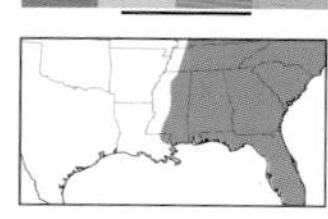

Renia salusalis 93-0529 (8378) **Common**

TL 14 mm. Yellowish brown FW has AM and PM lines that are most boldly marked at veins and so sometimes appear dotted. A diffuse brown median band is sometimes present. Indistinct reniform spot can be pale orange with a tiny black dot at each end, or all black. Irregular ST line is accented with tiny white dots. **HOSTS:** Dead leaves. **NOTE:** Two to three broods.

SOCIABLE RENIA

Renia factiosalis 93-0530 (8379) **Common**

TL 14 mm. Resembles Dotted Renia, but dark brown FW has distinctly jagged, often yellow-edged, AM and PM lines. Orbicular and two-dotted reniform spots are dark orange to brown. ST line is often bordered by dark shading and/or accented with short pale dashes. **HOSTS:** Dead leaves. **NOTE:** Two broods.

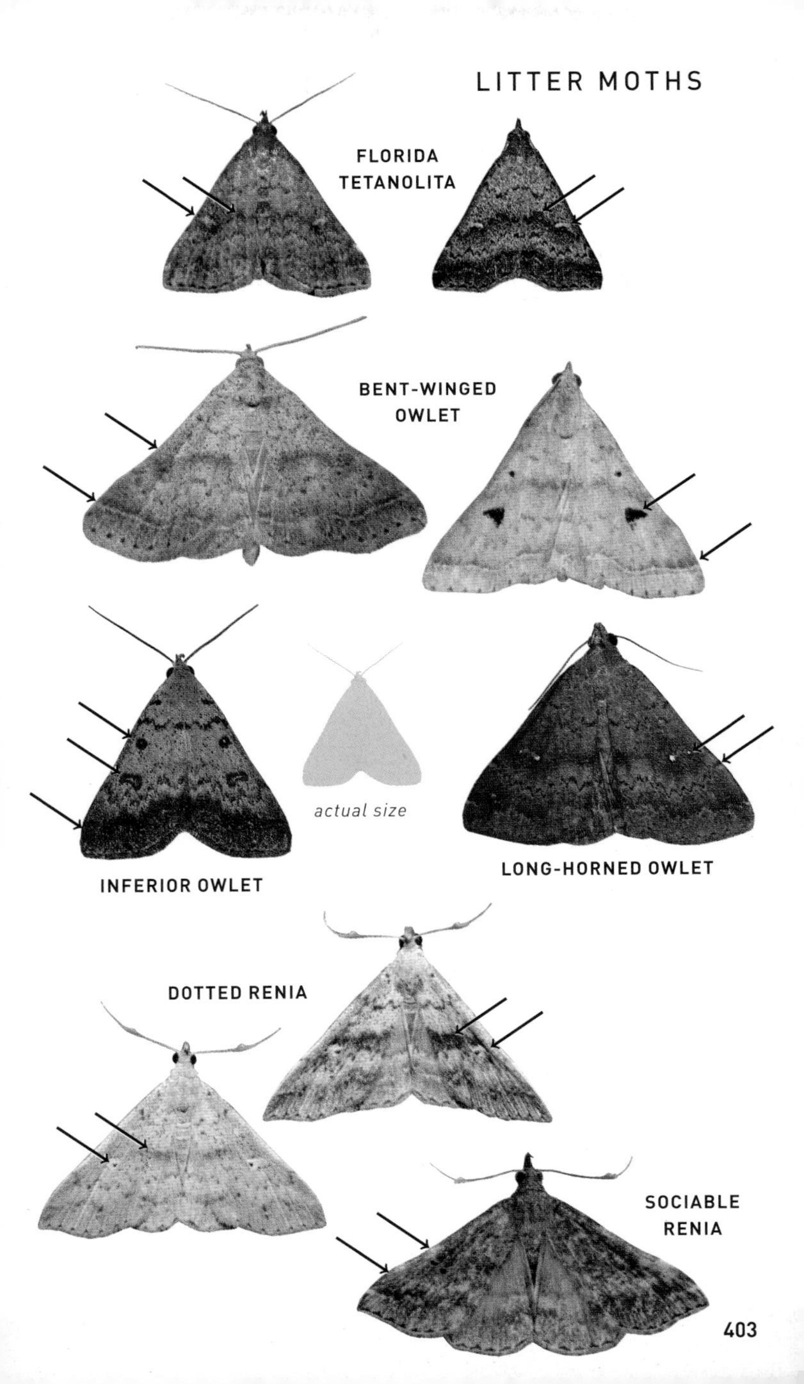
FLORIDA
TETANOLITA
BENT-WINGED
OWLET
actual size
INFERIOR OWLET
LONG-HORNED OWLET
DOTTED RENIA
SOCIABLE
RENIA

CHOCOLATE RENIA

Renia nemoralis 93-0531 (8380) **Common**

TL 14 mm. Resembles Sociable Renia, but yellowish to brown FW has a straight AM line. Indistinct PM line is jagged and faintly edged pale. ST line is a row of pale dots and is edged with dark shading. Orbicular and two-dotted reniform spots can be dark orange or all black. **HOSTS:** Dead leaves.

DISCOLORED RENIA

Renia discoloralis 93-0532 (8381) **Common**

TL 19–24 mm. Variable. FW can be yellowish, brown, or dark chocolate brown. Jagged AM and PM lines are faint or nearly absent. Brown median band is often present; on dark individuals, entire median area may be shaded. ST line is pale, bordered by darker shading. Orbicular and reniform spots are usually pale orange but may be dark. Two dots in reniform spot are often connected by a thin bar. **HOSTS:** Dead leaves.

YELLOW-SPOTTED RENIA

Renia flavipunctalis 93-0536 (8384.1) **Common**

TL 16–17 mm. Yellowish to violet gray FW has straight, pale-edged AM and PM lines; the latter curves smoothly upward at costa. Orbicular and two-dotted reniform spots are orange, sometimes pale. Indistinct ST line is bordered with a band of brown shading above. **HOSTS:** Dead leaves.

FRATERNAL RENIA

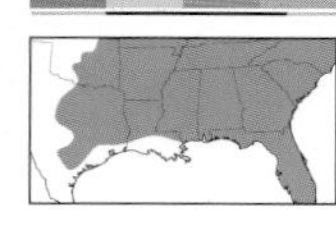

Renia fraternalis 93-0537 (8385) **Common**

TL 14–16 mm. Resembles pale morphs of Discolored Renia, but warm brown FW has a row of pale dots along ST line, sometimes lightly edged with dark shading. Darker median band is narrow and most obvious toward inner margin. Orbicular spot is always pale orange, but two-dotted reniform spot can be pale or black. Long labial palps usually diverge at tip. **HOSTS:** Probably dead leaves.

SPECKLED RENIA

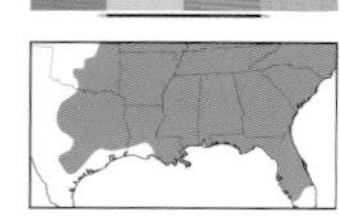

Renia adspergillus 93-0538 (8386) **Common**

TL 16–17 mm. Resembles well-marked individuals of Discolored Renia, but straw-colored to brown FW has obvious scalloped AM and PM lines. A darker median band is usually present. Pale orbicular spot is usually outlined brown. Dark shading that borders indistinct ST line is fragmented. **HOSTS:** Dead leaves. **NOTE:** Multiple broods.

SOBER RENIA *Renia sobrialis* 93-0539 (8387) **Common**

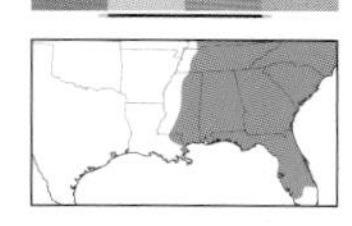

TL 13–15 mm. Resembles Chocolate Renia, but purplish gray FW has jagged AM and PM lines. A blackish median band is often present. Orbicular and two-dotted reniform spots are orange. Irregular ST line is often reduced to a row of whitish dots bordered by darker shading. **HOSTS:** Dead leaves.

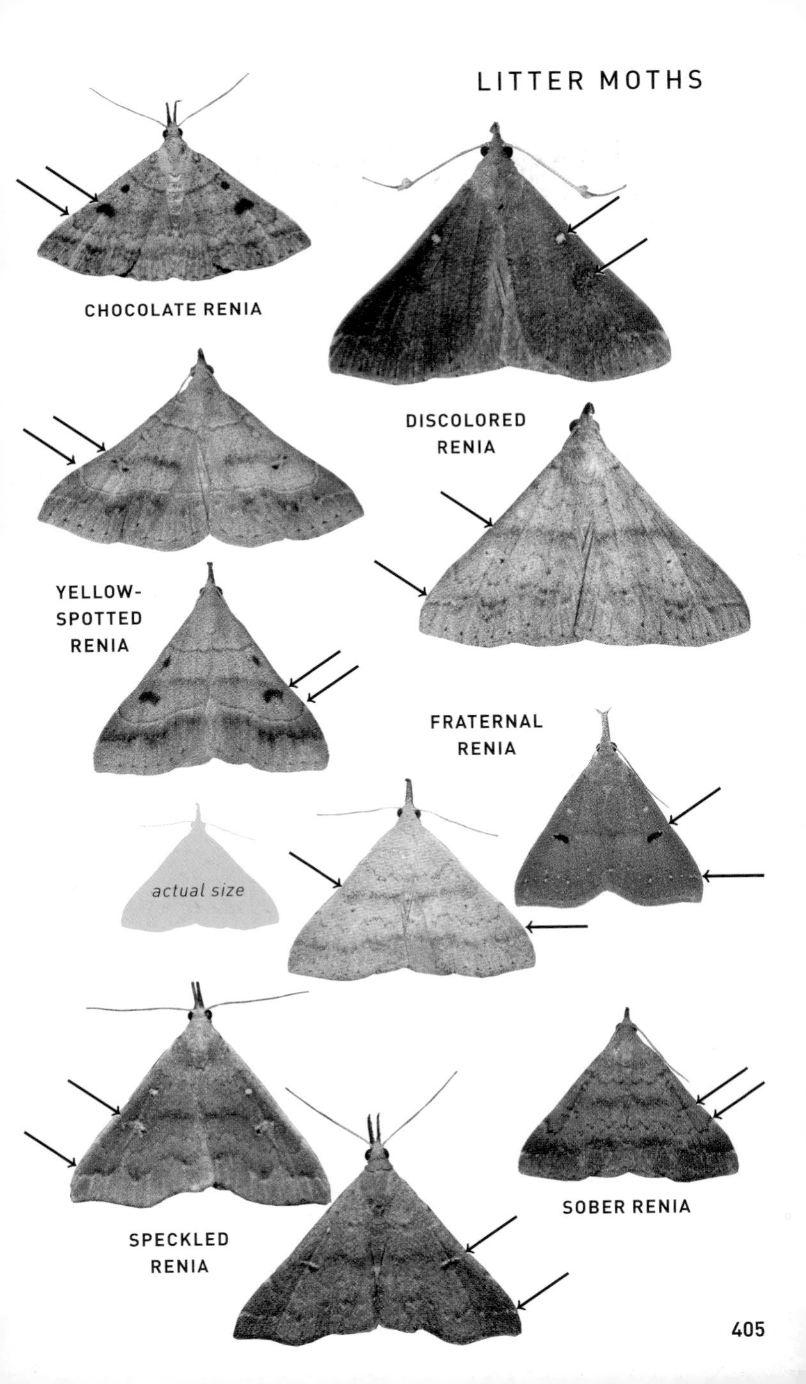
CHOCOLATE RENIA
DISCOLORED RENIA
YELLOW-SPOTTED RENIA
FRATERNAL RENIA
actual size
SPECKLED RENIA
SOBER RENIA

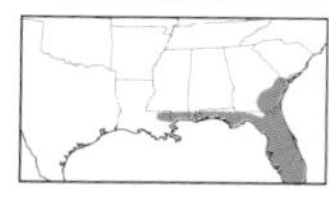

WHITE-BORDERED OWLET

Physula albipunctilla 93-0543 (8390.2) **Local**

TL 15–16 mm. Chocolate brown FW has a contrasting light gray band beyond irregular white ST line. Jagged AM and PM lines are inconspicuous. Tiny white dots mark orbicular and reniform spots. Dusky HW has partial white ST line. **HOSTS:** Unknown.

SKIRTED OWLET

Phlyctaina irrigualis 93-0546 (8392) **Local**

TL 13–15 mm. Golden brown FW bulges outward at midpoint of costa. Wavy AM and pale ST lines are bordered with dark shading. Two-dotted reniform spot is bordered by a large, dusky blotch. **HOSTS:** Unknown.

AMBIGUOUS MOTH

Lascoria ambigualis 93-0547 (8393) **Common**

TL 12–14 mm. Gray to brown FW has almost straight AM line broadly edged diffusely black. Median and PM lines are irregular. Faint pale ST line is bordered above by dark shading. Male has notch at midpoint of outer margin. Reniform spot is an oblique white crescent, often faint in female. **HOSTS:** Chrysanthemum, ragweed, and horseradish. **NOTE:** Two broods.

ENIGMATIC MOTH

Lascoria orneodalis 93-0550 (8396) **Common**

TL 12–14 mm. Similar to Ambiguous Moth, but FW has wavy AM line. Reniform spot is two tiny white dots; sometimes one is faint. ST line is usually very faint or absent. Male has a noticeable tuft at midpoint of costa and a notch at midpoint of outer margin. **HOSTS:** Unknown.

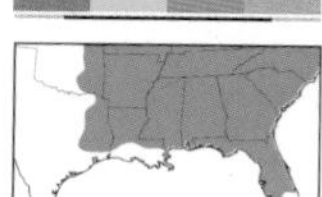

DARK-SPOTTED PALTHIS

Palthis angulalis 93-0551 (8397) **Common**

TL 12–18 mm. Pinkish tan FW is usually held slightly creased when at rest. An oblique brown median band stops short of costa. Reniform spot is a slanting brown crescent. Apical patch is rusty brown. Male has long, upturned labial palps. **HOSTS:** Coniferous and deciduous trees, including ash, alder, maple, spruce, and willow. **NOTE:** Two or three broods.

FAINT-SPOTTED PALTHIS

Palthis asopialis 93-0552 (8398) **Common**

TL 12–16 mm. Resembles Dark-spotted Palthis, but grayish brown FW has a conspicuous black patch near apex. AM line and dark border are straight. Reniform spot is bisected by thin whitish line. Has a distinctly humpbacked profile. Male has long, upturned labial palps. **HOSTS:** Aster, bean, coralberry, corn, Spanish needle, and wild indigo. **NOTE:** Multiple broods.

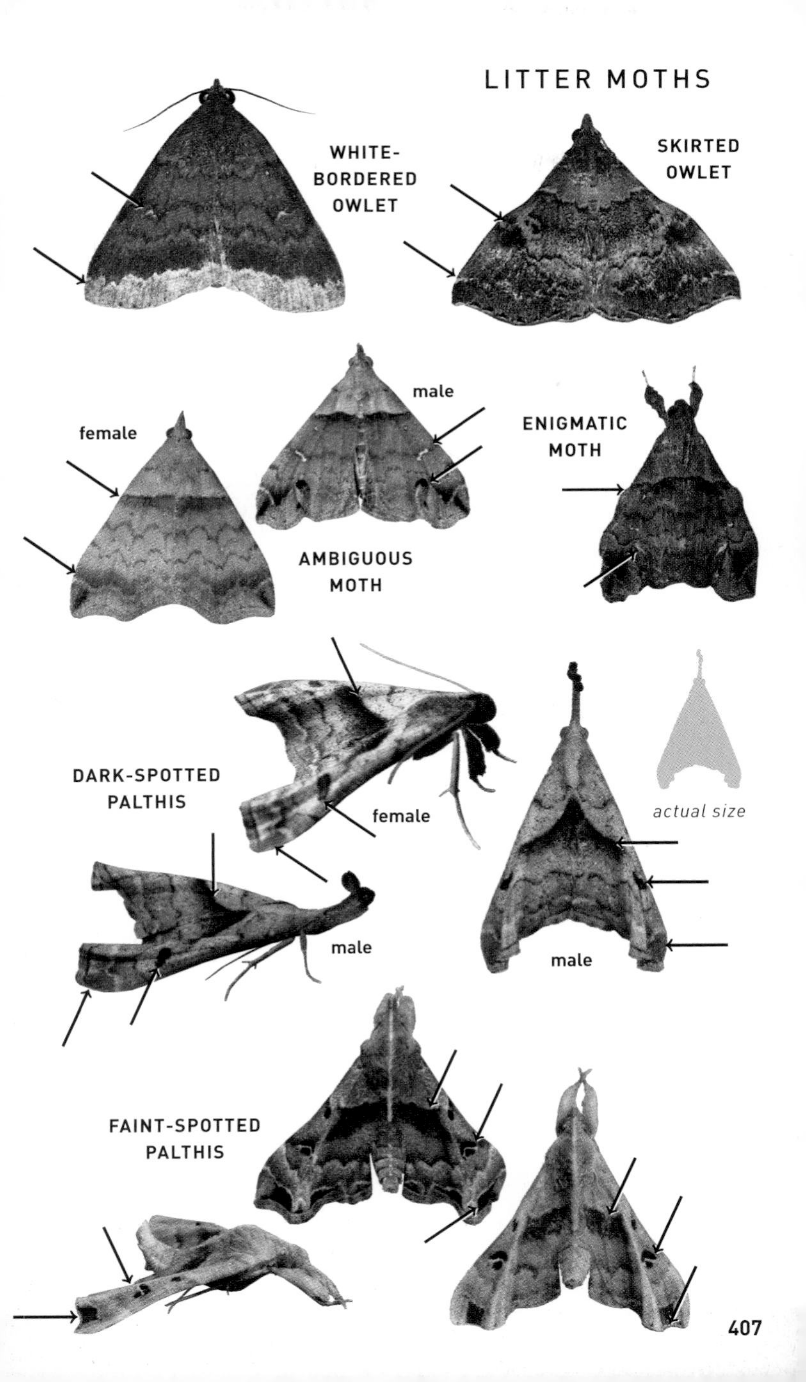
WHITE-BORDERED OWLET
SKIRTED OWLET
female
male
ENIGMATIC MOTH
AMBIGUOUS MOTH
DARK-SPOTTED PALTHIS
female
actual size
male
male
FAINT-SPOTTED PALTHIS

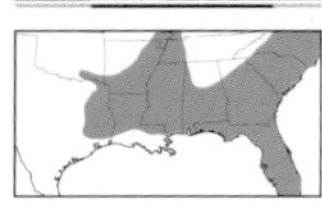

PYGMY REDECTIS

Redectis pygmaea 93-0554 (8400) **Common**

TL 15 mm. Grayish tan FW has scalloped AM and PM lines, sometimes indistinct. Orange reniform spot is bisected by dark line. Small white patch where PM line meets costa. Outer margin is pointed at apex and midpoint. **HOSTS:** Unknown.

WHITE-SPOTTED REDECTIS

Redectis vitrea 93-0555 (8401) **Common**

TL 15 mm. Resembles Pygmy Redectis, but reniform spot consists of a white spot bordered by two indistinct orangish spots. Brownish to violet gray FW has golden brown patch in outer median area. Male has long, upturned labial palps. **HOSTS:** Crabgrass.

PANGRAPTINE OWLETS

Family Erebidae, Subfamily Pangraptinae

Broad-winged moths. Both species are common in damp woodlands and heathy areas. They are mainly nocturnal and attracted to lights, though Decorated Owlet is also often encountered during daytime.

DECORATED OWLET

Pangrapta decoralis 93-0559 (8490) **Uncommon**

TL 11–15 mm. Pale violet gray to brown FW is peppered with warm brown. Wavy PM line is bordered with a diffuse rusty band above and whitish or grayish shading below. Triangular black discal spots are bordered by orange on FW and white on HW. Usually holds wings in a dihedral above body when at rest. **HOSTS:** Heaths, including blueberry, huckleberry, sourwood, and staggerbush. **NOTE:** Two or more broods.

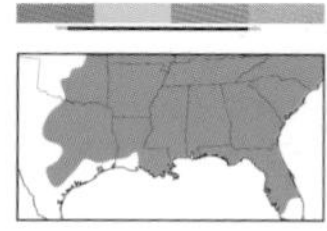

LOST OWLET

Ledaea perditalis 93-0560 (8491) **Common**

TL 13–16 mm. Pale gray FW has a smoothly curving PM line bordered above by dark brown at inner margin. Diffuse brown band crosses below PM line. AM line is wavy. Orbicular and reniform spots are small black dots. **HOSTS:** Buttonbush. **NOTE:** Two or three broods.

LITTER MOTHS

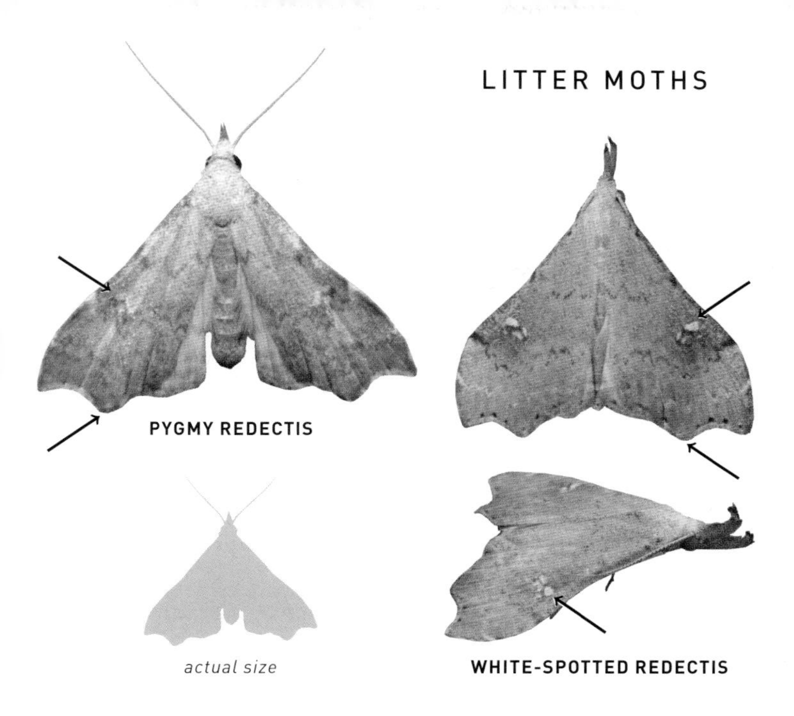

PANGRAPTINE OWLETS

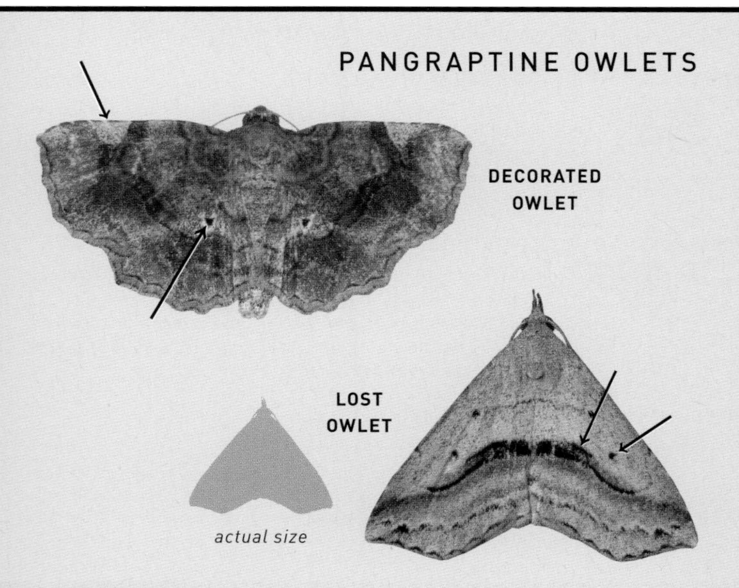

Snouts and Allies

Family Erebidae, Subfamily Hypeninae

Delta-shaped noctuids characterized by their ample, strikingly patterned FW. Long labial palps give them a snoutlike appearance. Most species are nocturnal and will visit lights in small numbers. Green Cloverworm is commonly encountered in grassy areas during daylight hours, as well as at sugar bait. Most species are found in wooded areas, although some are present in waste areas and gardens, even in big cities.

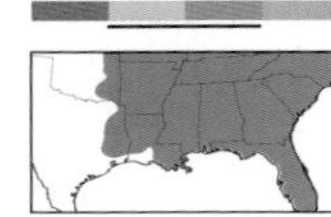

FLOWING-LINE SNOUT

Hypena manalis 93-0561 (8441) **Common**

TL 13–16 mm. Grayish brown FW has a rounded dark brown patch covering most of median area. Black apical dash points to finger on median patch. **HOSTS:** False nettle. **NOTE:** Two or more broods.

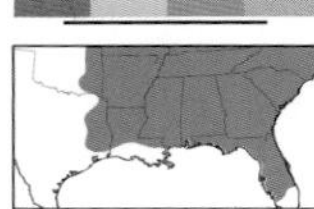

BALTIMORE SNOUT

Hypena baltimoralis 93-0562 (8442) **Common**

TL 16–18 mm. Resembles Flowing-line Snout, but median patch has a wavy edge and runs parallel to inner margin. Females average paler. **HOSTS:** Maple. **NOTE:** Two or more broods.

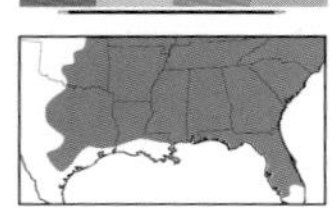

DIMORPHIC SNOUT

Hypena bijugalis 93-0564 (8443) **Common**

TL 14–17 mm. Sexually dimorphic. Female resembles Baltimore Snout, but lower side of brown patch runs along median line parallel to outer margin, and is edged by diffuse whitish shading. Male has sooty black FW with a square white patch where PM line meets inner margin. **HOSTS:** Dogwood. **NOTE:** Two or more broods.

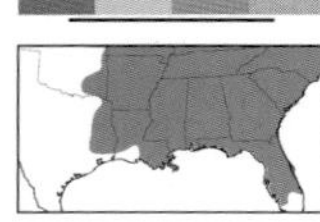

MOTTLED SNOUT *Hypena palparia* 93-0565 (8444) **Common**

TL 15–19 mm. Resembles Baltimore Snout, but AM line defines upper side of brown patch. Grayish brown FW has a mottled appearance, especially in basal and ST areas. Males average darker than females. **HOSTS:** Eastern hornbeam, ironwood, and hazel. **NOTE:** Two or more broods.

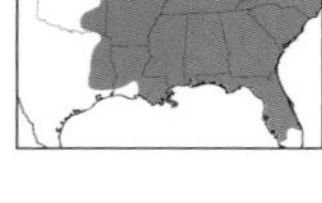

WHITE-LINED SNOUT

Hypena abalienalis 93-0566 (8445) **Common**

TL 14–18 mm. Sexually dimorphic. Resembles Mottled Snout, but chocolate FW has lacy gray ST area beyond curved white PM line. Basal area of female is also lacy; that of male is dark. **HOSTS:** Slippery elm. **NOTE:** Two or more broods.

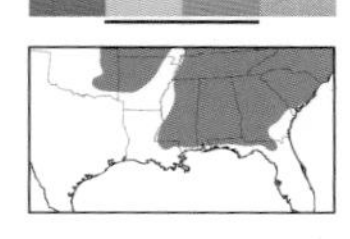

GRAY-EDGED SNOUT

Hypena madefactalis 93-0568 (8447) **Common**

TL 19 mm. Sexually dimorphic. FW of male is sooty with indistinctly darker median area. Female is paler with rusty brown median area edged below wavy PM line with diffuse gray. Apical dash is not well defined. **HOSTS:** Walnut.

SNOUTS AND ALLIES

SOOTY SNOUT

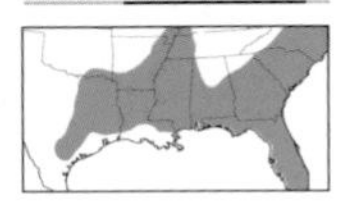

Hypena minualis 93-0580 (8457) **Common**

TL 10–12 mm. Mottled grayish brown FW is narrower than in other snouts. Rusty PM line curves around dark gray patch in outer median area. A whitish semicircle borders a gray subapical patch. Often rests upside down. **HOSTS:** Globe mallow, false mallow, and sida. **NOTE:** Multiple broods. Migrates north in late fall.

YELLOW-LINED SNOUT

Hypena degesalis 93-0582 (8459) **Common**

TL 12–14 mm. Narrow grayish brown FW has an oblique, yellow-edged rusty PM line, bordered by diffuse grayish shading. Rusty AM and ST lines are indistinct. Often rests upside down. **HOSTS:** Unknown.

GREEN CLOVERWORM

Hypena scabra 93-0588 (8465) **Common**

TL 15–21 mm. Variable. Resembles Sooty Snout, but black PM line is straight and noticeably thickens in inner half of FW. Gray patch in outer half of median area is sometimes absent. Whitish subapical patch lacks gray center. **HOSTS:** Low plants and crops, including alfalfa, bean, clover, ragweed, raspberry, and strawberry. **NOTE:** Multiple broods.

YELLOW-LINED OWLET

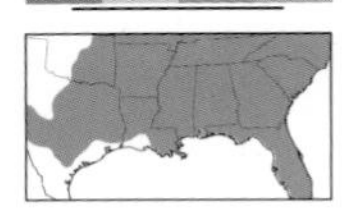

Colobochyla interpuncta 93-0590 (8411) **Uncommon**

TL 11–13 mm. Violet gray FW has parallel, yellow-edged AM, median, and PM lines. Orbicular and reniform spots are tiny black dots. **HOSTS:** Willow. **NOTE:** Two or more broods.

GOLD-LINED MELANOMMA

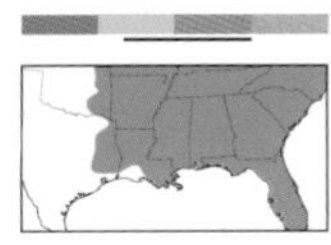

Melanomma auricinctaria 93-0591 (8412) **Uncommon**

TL 8–10 mm. Gray FW has diffuse, dark median band. Wavy AM line is edged whitish. Large reniform spot resembles an eye; black spot is bordered by thin pale yellow ring and has metallic silvery dots in middle. Outer ST line is studded with metallic silvery dots. Male has broadly bipectinate antennae. **HOSTS:** Huckleberry. **NOTE:** Two or more broods.

SOOTY SNOUT

YELLOW-LINED SNOUT

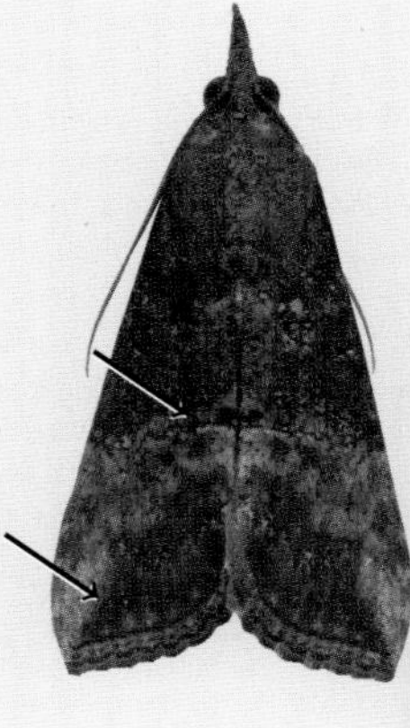

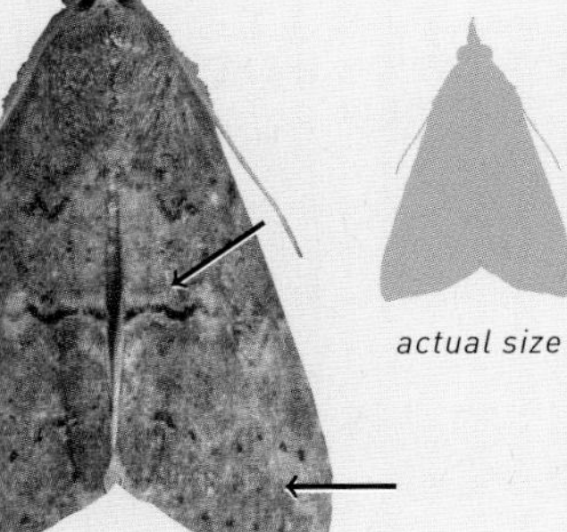

actual size

GREEN CLOVERWORM

YELLOW-LINED OWLET

GOLD-LINED MELANOMMA

Assorted Owlets

Family Erebidae, Subfamilies Rivulinae, Scoliopteryginae, Calpinae, Hypocalinae, Scolecocampinae, Hypenodinae, Boletobiinae, Eublemminae, Aventiinae, and Phytometrinae

A diverse assemblage of mostly broad-winged noctuids. Most species are somberly clad in shades of tan, gray, or brown, often with complex patterns on the FW. A few, such as The Herald and Moonseed Moth, are strikingly colorful. Most are delta-shaped, but a few—such as the *Eupithecia*-look-alike *Sigela* species—hold their wings spread open. Most of the smaller species (*Ablemma, Eublemma,* and *Hyperstrotia*) adopt a more tentlike posture. These are nocturnal moths that will come to lights in small numbers. Some species, notably The Herald, are also frequent visitors to sugar bait.

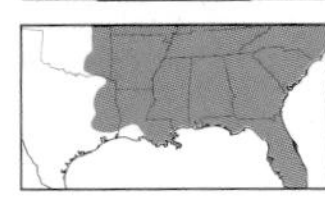

SPOTTED GRASS MOTH

Rivula propinqualis 93-0592 (8404) **Common**

TL 10–11 mm. Straw-colored FW has a purplish gray reniform spot with two black dots. AM and PM lines are warm brown and are parallel near inner margin. **HOSTS:** Grass. **NOTE:** Two or more broods.

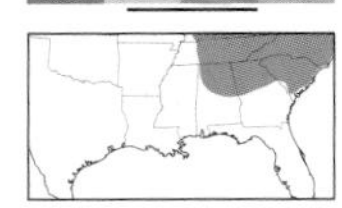

THE HERALD

Scoliopteryx libatrix 93-0601 (8555) **Common**

TL 23 mm. Rusty brown to brownish gray FW has wide fiery orange streak through basal and median areas. Whitish PM line is double. Basal and orbicular spots are white. **HOSTS:** Poplar and willow. **NOTE:** Often visits sugar bait.

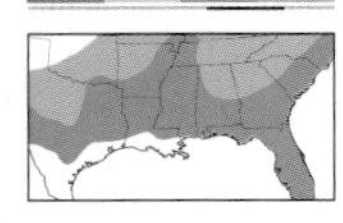

YELLOW SCALLOP MOTH

Anomis erosa 93-0602 (8545) **Common**

TL 15–19 mm. Lightly peppered yellow to orange FW has rusty AM and median lines that join at inner margin. PM line joins reniform spot and loops back before reaching costa. Orbicular spot is white, bordered by diffuse brown. PM and ST areas have diffuse grayish shading. **HOSTS:** Mallow.

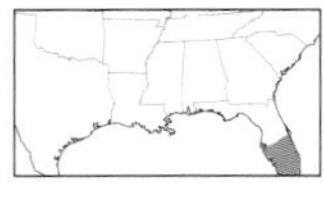

CITRUS FRUIT-PIERCER

Gonodonta nutrix 93-0617 (8540) **Local**

TL 15–19 mm. Unmistakable! Dark brown FW has a contrasting creamy stripe slanting from central base along costa almost to apex. ST area is violet beyond zigzag ST line. Orange HW has a broad blackish ST band and stripe along inner margin. Head is largely white. **HOSTS:** Pond apple and related *Annona* species.

UNICA FRUIT-PIERCER

Gonodonta unica 93-0618 (8541) **Local**

TL 15–19 mm. Shiny brown FW has a paler lilac gray ST band beyond brown-edged, saw-toothed ST line. Blackish streak along concave inner margin has a pale border. Orange HW has a spotted ST band. Head is white. **HOSTS:** Unknown.

ASSORTED OWLETS

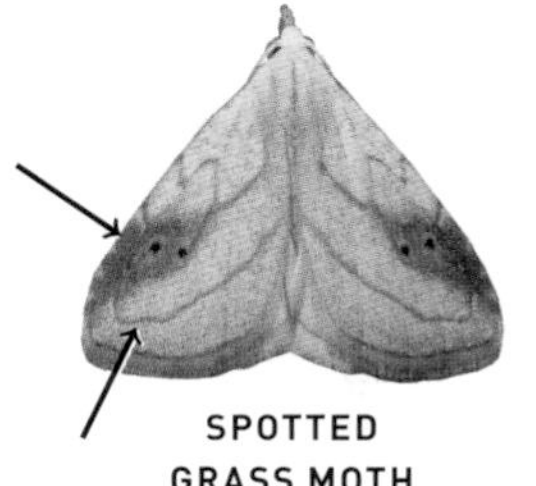

SPOTTED
GRASS MOTH

THE HERALD

YELLOW
SCALLOP MOTH

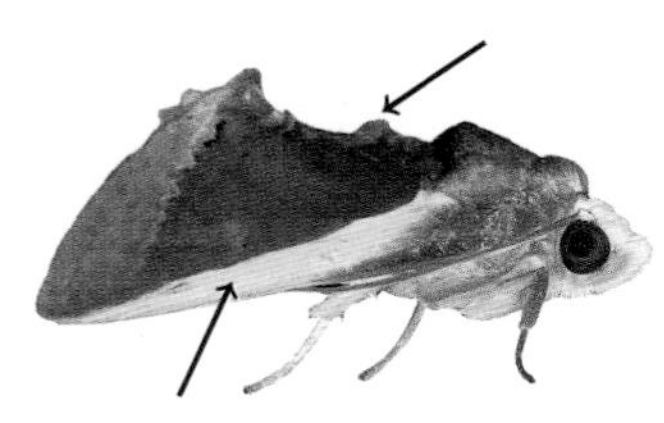

CITRUS FRUIT-PIERCER

actual size

UNICA
FRUIT-PIERCER

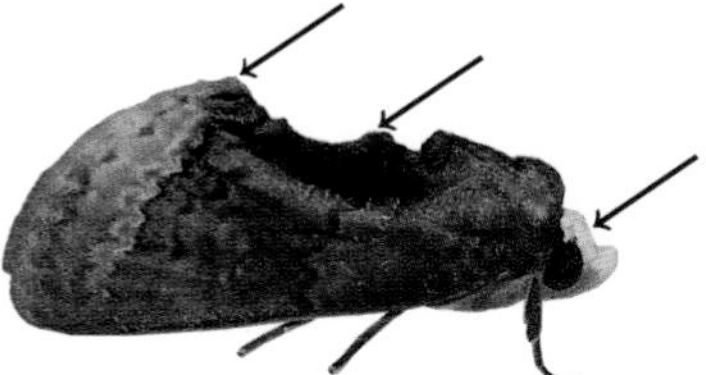

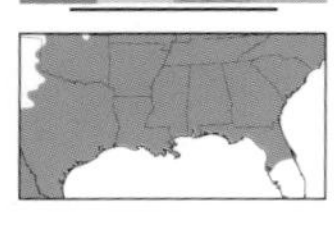

MOONSEED MOTH

Plusiodonta compressipalpis 93-0622 (8534) **Uncommon**

TL 14–18 mm. Fawn-colored FW has brown median band that tapers to a tuft of hairlike scales protruding from inner margin. AM and PM lines are edged with lilac and gray. Basal area is pale with gold edging. Has large pale-edged spot with small scale tuft near anal angle. **HOSTS:** Moonseed and snailseed. **NOTE:** Two broods.

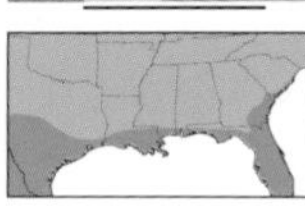

HYPOCALA MOTH

Hypocala andremona 93-0627 (8642) **Common**

TL 14–18 mm. Warm brownish gray FW is relatively plain. Inner ST line is orange, edged dark. Sometimes has a large hollow blackish reniform spot. Golden yellow HW has an irregular black ST band and stripes along margins. **HOSTS:** Persimmon.

LARGE NECKLACE MOTH

Hypsoropha monilis 93-0628 (8527) **Common**

TL 17–22 mm. Uniform tan to brown FW has a string of four white spots that decrease in size toward inner margin. ST line is weakly accented white. Sometimes has purplish shading below PM line. Antennae are bipectinate in both sexes. **HOSTS:** Persimmon.

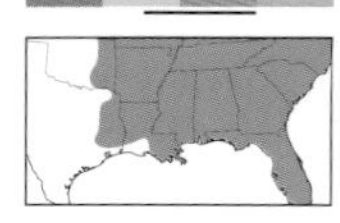

SMALL NECKLACE MOTH

Hypsoropha hormos 93-0629 (8528) **Common**

TL 14–18 mm. Pale gray to brown FW has a string of three or four white spots that increase in size toward inner margin. Diffuse AM and PM lines are often visible. Antennae are filiform. **HOSTS:** Persimmon. **NOTE:** Two or more broods.

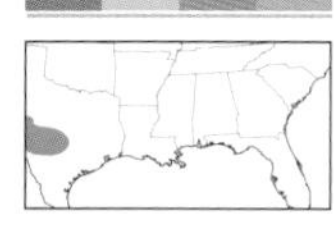

AON MOTH

Aon noctuiformis 93-0632 (8954) **Common**

TL 16–18 mm. Pale gray FW has sparse blackish vermiculation and is variably shaded brown in inner basal area. Reniform spot is black, sometimes indistinct within wing pattern. Tufted brown thorax has narrow black collar. Typically rests with wings rolled into a cylinder. **HOSTS:** Unknown.

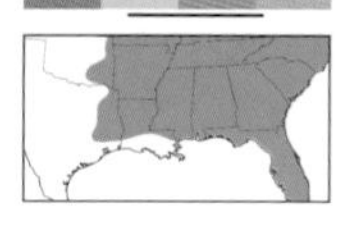

COMMON ARUGISA

Arugisa lutea 93-0634 (8509) **Common**

TL 9–10 mm. Shiny tan FW has a blotch of dusky shading at midpoint of AM line. Curved PM line is fragmented or dotted. Orbicular and reniform spots are small black dots. Jagged ST line is edged diffuse brown. **HOSTS:** Live and dead grass and blue-green algae.

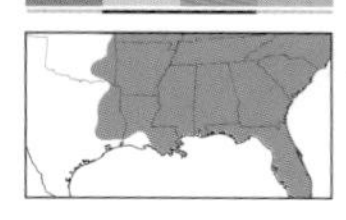

WATSON'S ARUGISA

Arugisa latiorella 93-0635 (8510) **Common**

TL 9–12 mm. Resembles Common Arugisa, but AM and PM lines are usually more complete, and shading along ST line is reduced. **HOSTS:** Live and dead bluegrass.

ASSORTED OWLETS

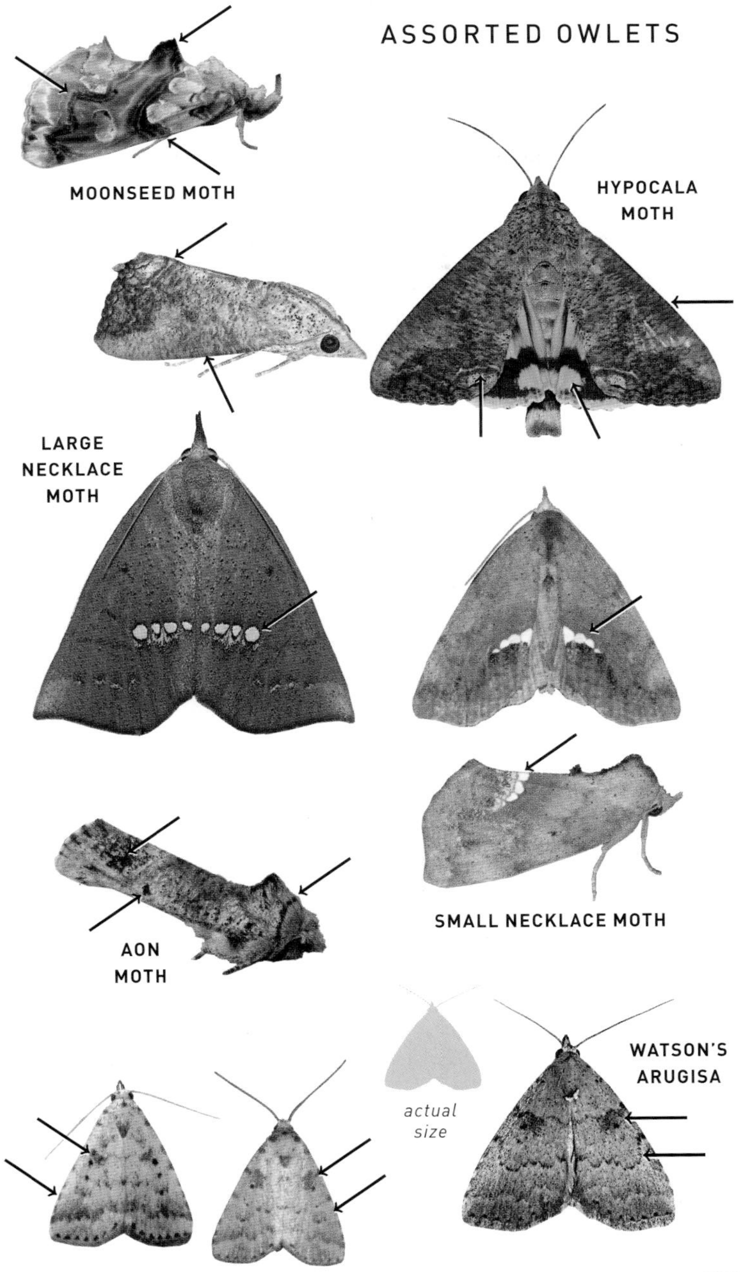

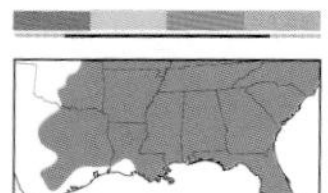

DEAD-WOOD BORER MOTH

Scolecocampa liburna 93-0637 (8514) **Common**

TL 19–24 mm. Straw-colored FW has row of dark brown dots along PM, ST, and terminal lines. Reniform spot is outlined dark brown (rarely solid brown). Has dusky blotch at midpoint of ST area. **HOSTS:** Larvae bore into decaying logs and stumps of deciduous trees. **NOTE:** Two or more broods.

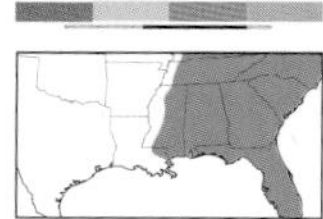

STREAKED GABARA

Gabara distema 93-0644 (8523) **Uncommon**

TL 12–14 mm. Straw-colored FW is variably peppered dark brown. A dark streak (sometimes reduced or absent) passes through central FW, fading before outer margin. Orbicular and reniform spots are tiny black dots. Short labial palps are pointed. **HOSTS:** Grass.

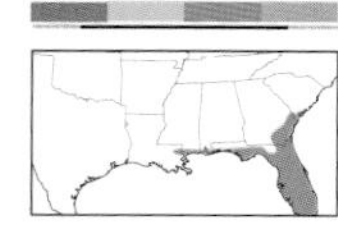

PALE PALPIDIA

Palpidia pallidior 93-0646 (8517) **Local**

TL 18–20 mm. Narrow straw-colored FW is streaked with peppery gray shading between veins. HW is white. **HOSTS:** Unknown.

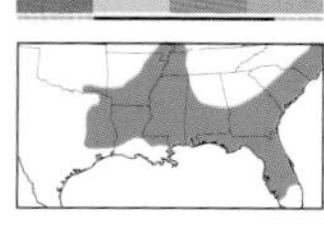

BRIMLEY'S ABABLEMMA

Ablemma brimleyana 93-0651 (8437) **Common**

TL 8–9 mm. Finely vermiculated FW has pale double AM and PM lines and large black reniform spot. Abdomen is tipped with metallic bronze scales. **HOSTS:** Green algae and perhaps lichens. **NOTE:** Two or more broods.

THIN-WINGED OWLET

Nigetia formosalis 93-0655 (8440) **Common**

TL 12 mm. Narrow, sharply pointed FW is white with dark costa and a dark, curved median bar. White ST line is bordered with diffuse brown shading above and gray below. **HOSTS:** Green algae and perhaps lichens. **NOTE:** Two or more broods.

NOTE: These next five species often rest with wings outspread, resembling geometer moths in the genus *Eupithecia,* but head is stouter and broader.

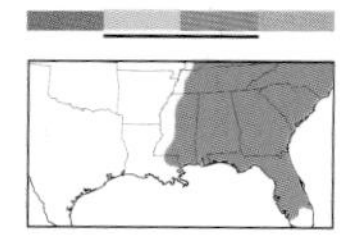

BROWN FALSE PUG

Sigela brauneata 93-0657 (8432) **Common**

WS 11–12 mm. Variable. Wings are whitish, tan, or gray with indistinct wavy lines, often bolder on HW, sometimes creating a dark median band. Narrow black reniform spot is diffusely edged with brown. **HOSTS:** Green algae in captivity. **NOTE:** Two or more broods.

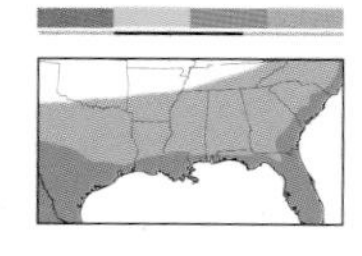

STIPPLED FALSE PUG

Sigela penumbrata 93-0658 (8433) **Local**

WS 11–12 mm. Wings are variably whitish to brownish gray. Dotted lines form a spotted pattern on well-marked individuals. Black orbicular dot is usually visible. Dotted terminal line is often strongest marking. **HOSTS:** Unknown. **NOTE:** Likely two or more broods.

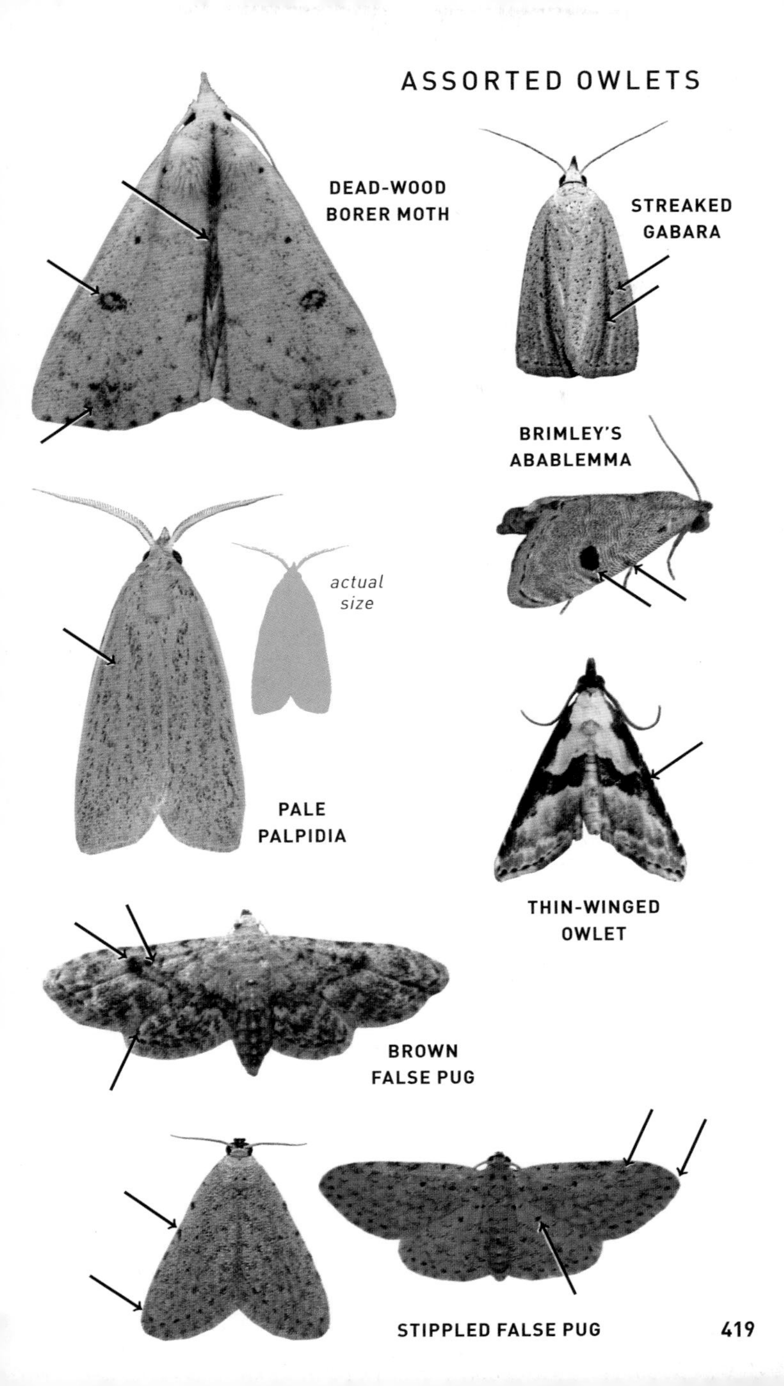
ASSORTED OWLETS
DEAD-WOOD BORER MOTH
STREAKED GABARA
BRIMLEY'S ABABLEMMA
actual size
PALE PALPIDIA
THIN-WINGED OWLET
BROWN FALSE PUG
STIPPLED FALSE PUG

WAVY FALSE PUG

Sigela basipunctaria 93-0659 (8434) **Local**

WS 11–12 mm. Pale straw-colored wings are weakly patterned. PM lines are wavy when visible. Black orbicular and reniform dots are usually distinct. **HOSTS:** Unknown.

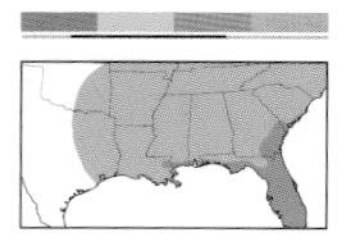

TAWNY FALSE PUG *Sigela eoides* 93-0660 (8435) **Local**

WS 11–12 mm. Tawny brown wings are lightly dusted with reddish orange scales. Costa is checkered and sometimes shaded gray. Wavy lines are often indistinct. Diffuse, blackish, jagged ST line is fragmented. A tiny black orbicular spot is noticeable. **HOSTS:** Unknown. **NOTE:** Likely two or more broods.

UNKNOWN

FLORIDA FALSE PUG

New genus nr. *Sigela* 93-0660.97 (8435.97) **Local**

WS 11–12 mm. Resembles scoopwings at rest. Brown FW is weakly patterned with zigzag white lines, strongest at inner PM line. Central median area is shaded black. Narrow HW is speckled green at base and is held slightly apart from FW, creating a small gap. **HOSTS:** Unknown. **NOTE:** This new species has not yet been formally described and is still awaiting a scientific name and official numerical assignment.

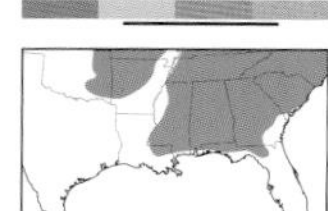

BROKEN-LINE HYPENODES

Hypenodes fractilinea 93-0662 (8421) **Common**

TL 6–8 mm. Dark reniform spot and broken spotted inner PM line form a line of dots across middle of gray FW. Orbicular spot is two dark dots. Indistinct pale ST line is bordered by diffuse dark shading. **HOSTS:** Unknown. **NOTE:** Two or more broods.

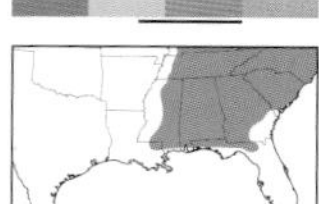

MASKED PARAHYPENODES

Parahypenodes quadralis 93-0667 (8430) **Uncommon**

TL 9–10 mm. Pale gray to tan FW is crossed by broad shaded bands. A large blackish reniform spot is most noticeable feature. Jagged AM and PM lines are pale-edged. **HOSTS:** Unknown.

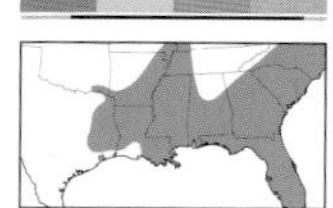

BLACK-SPOTTED SCHRANKIA

Schrankia macula 93-0668 (8431) **Common**

TL 8–10 mm. Narrow, grayish brown FW has a pale, straight PM line that divides pale PM area from dark median area. Inner median area is often washed pale. Basal area is frequently tan. Black reniform spot is not obvious. **HOSTS:** Bracket fungus. **NOTE:** Likely two or more broods.

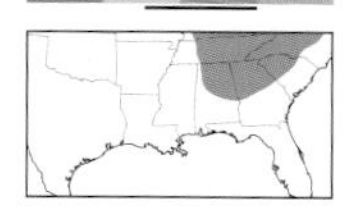

SPOT-EDGED DYSPYRALIS

Dyspyralis puncticosta 93-0670 (8427) **Local**

TL 8–9 mm. Pale gray FW is peppered with black scales. AM and PM lines are wavy, but often indistinct except at costa, where they widen into dark blocks. Reniform spot is a thin black crescent. Black dashes of terminal line curve around apex and carry on up costa. **HOSTS:** Unknown.

ASSORTED OWLETS
WAVY FALSE PUG
TAWNY FALSE PUG
FLORIDA FALSE PUG
actual size
BROKEN-LINE HYPENODES
MASKED PARAHYPENODES
BLACK-SPOTTED SCHRANKIA
SPOT-EDGED DYSPYRALIS

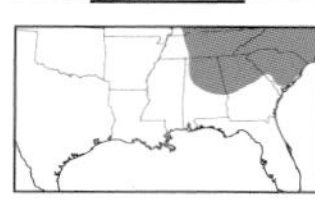

NIGELLA DYSPYRALIS

Dyspyralis nigellus 93-0671 (8428) **Local**

TL 9–10 mm. Slate gray FW has a white reniform crescent and five white dots along lower costa. AM and PM lines are often indistinct; PM line is sometimes edged pale. **HOSTS:** Unknown.

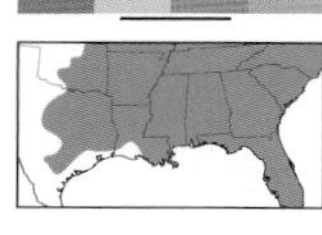

COMMON FUNGUS MOTH

Metalectra discalis 93-0679 (8499) **Common**

WS 20–27 mm. Dark FW is heavily mottled black and brown. Whitish AM line is most distinct line; PM line is a row of tiny white dots, ending in a whitish patch at costa. Median area is usually shaded darker. On paler individuals, dark reniform spot may be obvious. **HOSTS:** Mostly bracket fungi. **NOTE:** Two or more broods.

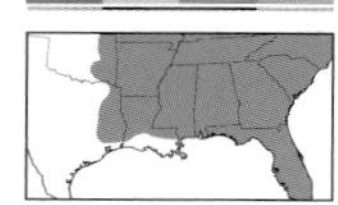

FOUR-SPOTTED FUNGUS MOTH

Metalectra quadrisignata 93-0680 (8500) **Common**

WS 18–35 mm. Dark brown to tan FW shows dark patches along costa at AM, median, and PM lines. PM line is a row of tiny white dots, usually ending in a pale spot at costa. Often has a dark median band. Black reniform spot is obvious on lighter individuals. **HOSTS:** Mostly bracket fungi. **NOTE:** Two or more broods.

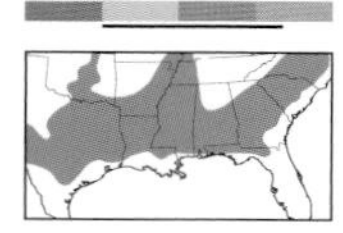

BLACK FUNGUS MOTH

Metalectra tantillus 93-0682 (8502) **Common**

WS 20–25 mm. Sooty black wings are speckled white. Has small white spots along costa at basal, AM, PM, and ST lines. Terminal line is tiny white dots. **HOSTS:** Algae and lichens. **NOTE:** Two or more broods.

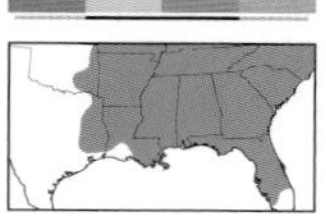

DIABOLICAL FUNGUS MOTH

Metalectra diabolica 93-0683 (8503) **Local**

WS 18–23 mm. FW is mottled tan and dusky brown. Diffuse, dark brown median band is double, bordered below by a tan patch. AM and PM lines are wavy and pale, most clearly marked by a pale yellow spot at costa. Terminal line is accented with pale dots. **HOSTS:** Fungi.

RICHARDS' FUNGUS MOTH

Metalectra richardsi 93-0685 (8505) **Common**

WS 15–17 mm. Resembles larger Diabolical Fungus Moth, but median line is a single thick blackish line. FW has a smoother, less mottled appearance, and pale AM and ST lines are usually more apparent. Abdomen has a pale band near tip. Pale dots of terminal line are faint or almost absent. **HOSTS:** Mostly bracket fungi. **NOTE:** Likely two or more broods.

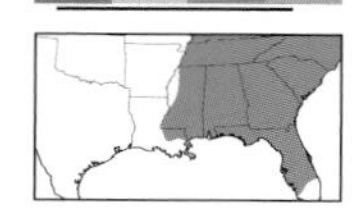

THE WHITE EDGE

Oruza albocostaliata 93-0692.5 (9025) **Common**

WS 18–21 mm. Chocolate brown FW has a broad whitish costal streak that extends across thorax as a whitish collar. Straight AM and PM lines are pale. **HOSTS:** Unknown.

ASSORTED OWLETS

NIGELLA DYSPYRALIS

COMMON FUNGUS MOTH

FOUR-SPOTTED FUNGUS MOTH

BLACK FUNGUS MOTH

DIABOLICAL FUNGUS MOTH

RICHARDS' FUNGUS MOTH

THE WHITE EDGE

EVERLASTING EUBLEMMA

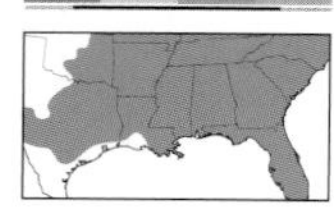

Eublemma minima 93-0693 (9076) **Common**

TL 8–9 mm. Whitish FW has broad bands of brown shading in median and ST areas. Wide median band extends as a tuft at midpoint of inner margin. Indistinct reniform spot is two tiny black dots. ST line is a series of sometimes indistinct black dots. **HOSTS:** Cudweed. **NOTE:** Likely has two or more broods.

CINNAMON EUBLEMMA

Eublemma cinnamomea 93-0694 (9077) **Local**

TL 8–9 mm. Light cinnamon brown FW has a bold dark-edged median line that angles upward at costa. Slightly wavy AM and PM lines are less obvious. ST line often has a series of blackish spots along outer half. Terminal area is shaded darker brown. **HOSTS:** Unknown.

STRAIGHT-LINED EUBLEMMA

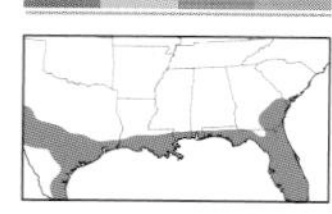

Eublemma recta 93-0695 (9078) **Local**

TL 8–10 mm. Light brown FW has a straight whitish median line that slants toward apex. ST area has a line of tiny black dots, often faint or absent. ST line is often reddish orange, especially near apex, contrasting with pale fringe. **HOSTS:** Seeds of bindweed, morning glory, and sweet potato.

FLORIDA PROBLEM MOTH

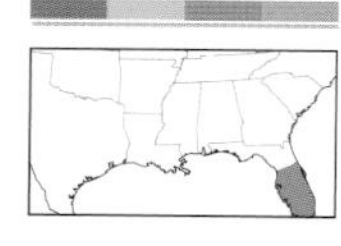

Proroblemma testa 93-0697 (9080) **Local**

TL 7–9 mm. Lightly speckled brown FW has a strongly angled PM line (sometimes dotted). PM area is variably shaded rusty brown in inner half. Broad, rounded FW is often slightly falcate at apex. Pointed labial palps impart a shrewlike appearance. **HOSTS:** Unknown.

VARIABLE TROPIC

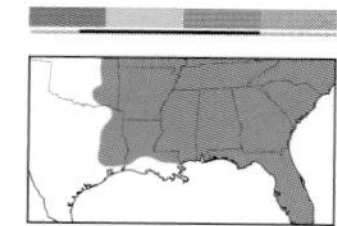

Hemeroplanis scopulepes 93-0700 (8467) **Common**

WS 20–25 mm. Variable. Yellow to warm brown FW has diffuse brownish PM line that runs in a straight line from apex to apex when wings are open at rest. PM and ST area are shaded darker, sometimes boldly. Often has blackish blotches just below PM line near inner margin. Dark wedges mark costa at AM, median, and PM lines. Reniform spot may be hollow or filled brownish. **HOSTS:** Clover. **NOTE:** Likely has two or more broods.

BLACK-DOTTED TROPIC

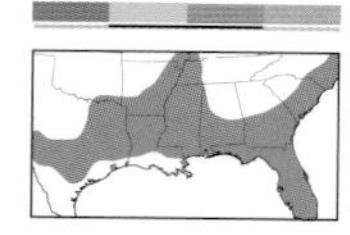

Hemeroplanis habitalis 93-0704 (8471) **Common**

WS 20–25 mm. Rusty brown FW has parallel yellowish AM and PM lines. Large reniform spot is black. ST line is a row of black-edged white dots. Long labial palps resemble those of renias or fan-foots. **HOSTS:** Unknown. **NOTE:** Likely has two or more broods.

ASSORTED OWLETS

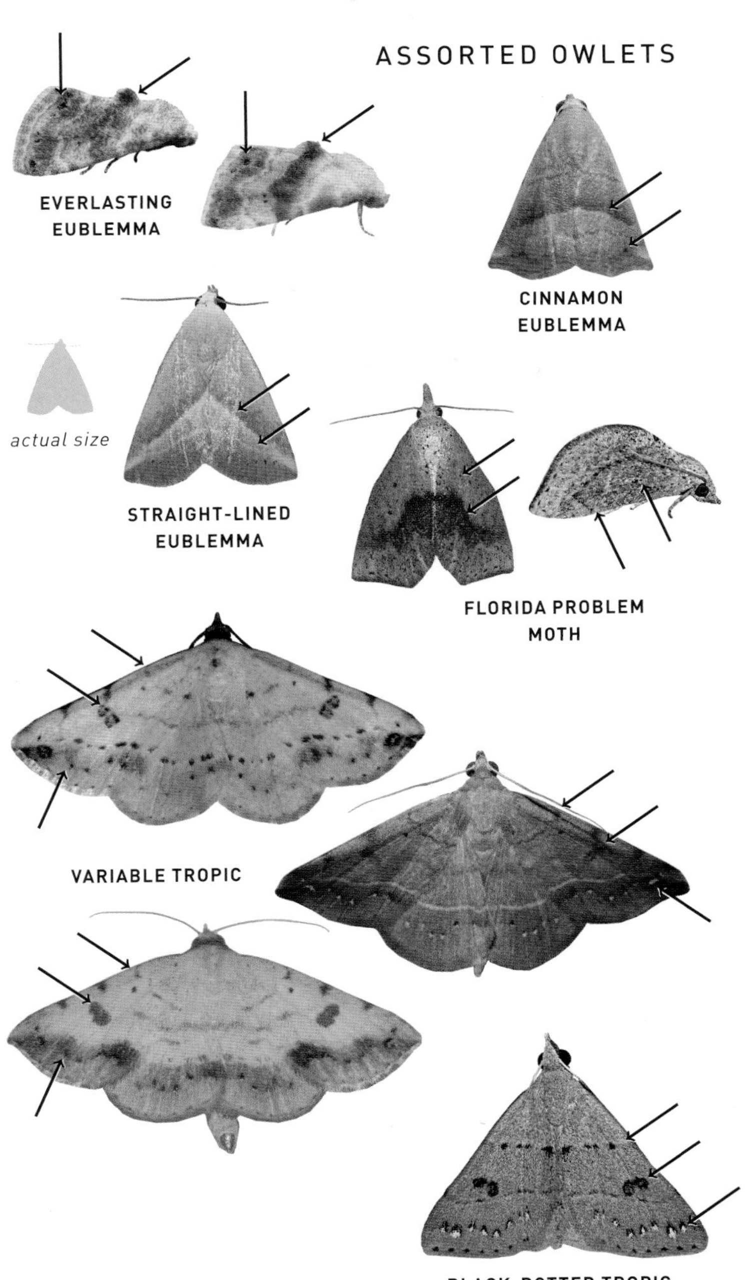

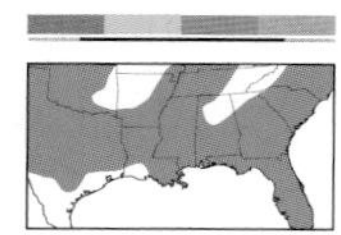

ERNESTINE'S MOTH

Phytometra ernestinana 93-0716 (8480) **Common**

TL 10–11 mm. Pale yellow FW has a slanting pink PM line. Basal half of costa and terminal line are also pink. Tiny black reniform dot touches PM line. ST area is often lightly shaded pink. **HOSTS:** Unknown. **NOTE:** Likely has two or more broods.

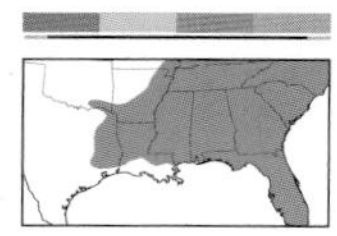

PINK-BORDERED YELLOW

Phytometra rhodarialis 93-0717 (8481) **Common**

TL 10–11 mm. Resembles Ernestine's Moth, but bright yellow FW has strong pink shading below PM line, crossed by wavy yellow lines of variable thickness. Small claviform spot is also pink. **HOSTS:** Milkwort. **NOTE:** Two or more broods.

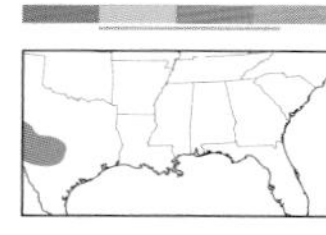

LEMON YELLOW *Phytometra orgiae* 93-0720 (8484) **Local**

TL 9–11 mm. Bright yellow FW has a fragmented PM line that usually widens into a rusty patch at inner margin. Orbicular and reniform spots are often inconspicuous. ST area is sometimes shaded pink in outer half. **HOSTS:** Unknown.

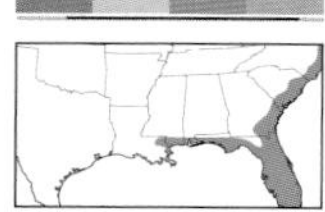

DOUBLE-LINED BROWN

Hormoschista latipalpis 93-0724 (8488) **Common**

TL 10–12 mm. Violet gray FW has parallel chestnut AM and PM lines, sometimes accented with tiny white dots. Median band and outer ST area are often shaded darker. Reniform spot is usually black but sometimes indistinct. **HOSTS:** Unknown.

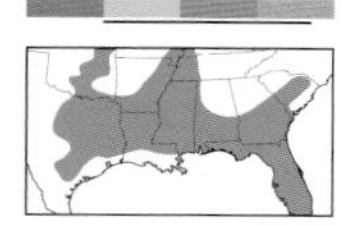

OMMATOCHILA MOTH

Ommatochila mundula 93-0725 (8489) **Local**

TL 7–9 mm. Violet gray FW has roughly parallel, angled AM and PM lines; PM line is sometimes edged whitish. Black dot at apex is edged by whitish terminal line. Median area is often shaded dark. Reniform spot may be black or indistinct. **HOSTS:** Unknown.

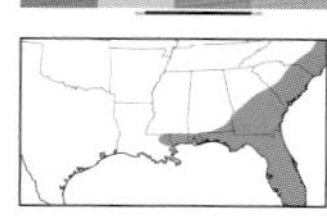

WHITE-LINED GRAYLET

Hyperstrotia nana 93-0727 (9035) **Uncommon**

TL 8–10 mm. Pale gray FW has faint white-edged AM and PM lines. Brown patches are present in inner median area, and along costa at AM and ST lines. Reniform spot consists of two tiny black dots. **HOSTS:** Chestnut, oak, and elm. **NOTE:** Two or more broods. The species that previously bore the name White-lined Graylet (*H. villificans*) is now considered synonymous with this species.

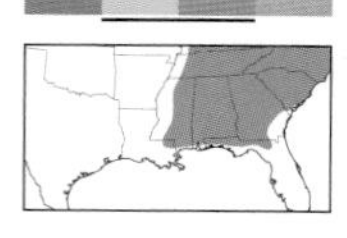

DOTTED GRAYLET

Hyperstrotia pervertens 93-0729 (9037) **Common**

TL 8–10 mm. Pale gray FW is crossed by faint brownish lines. PM line has whitish edging that widens at inner margin, bordered by brownish shading. Has brownish patches along costa at AM, median, and ST lines. Reniform spot contains two tiny black dots. **HOSTS:** Beech and oak. **NOTE:** Two or more broods.

ASSORTED OWLETS

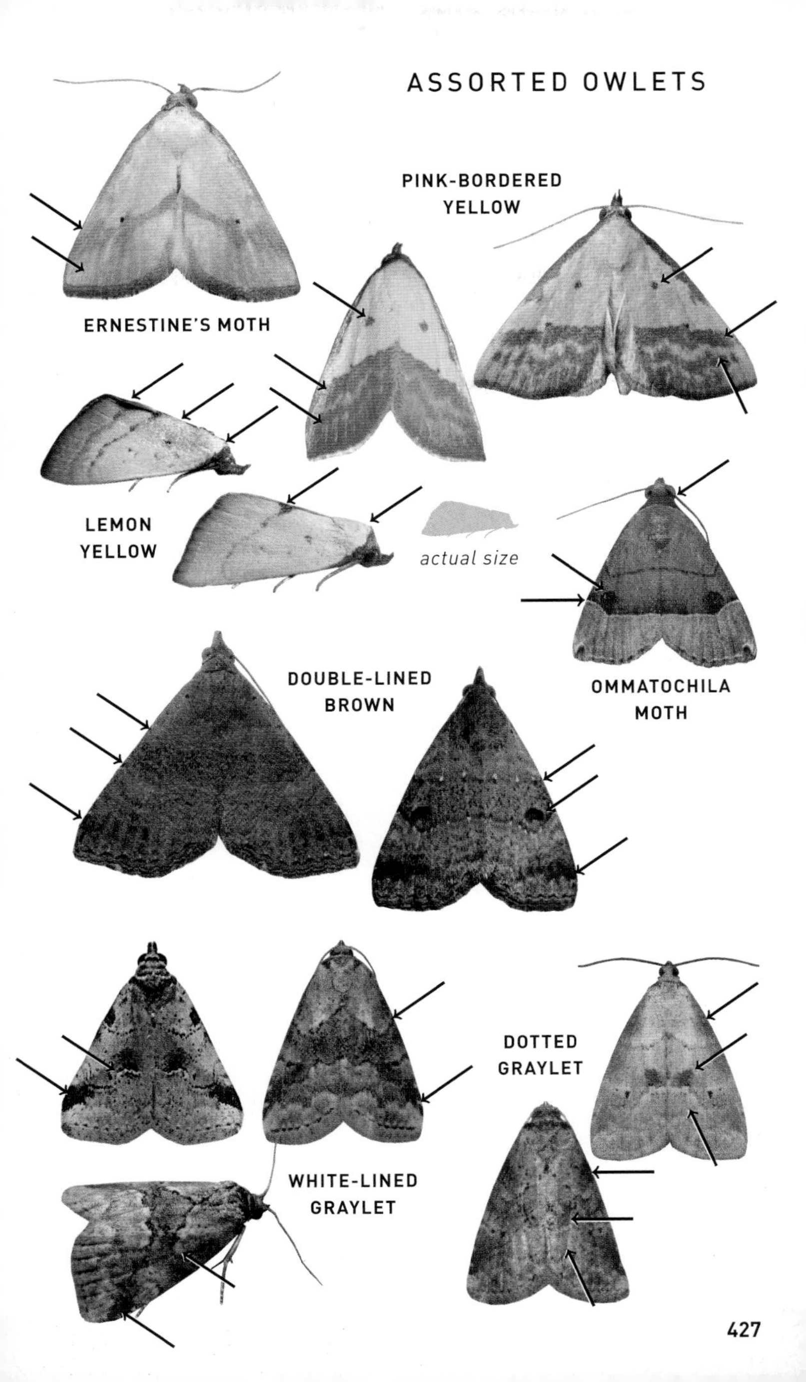

YELLOW-SPOTTED GRAYLET

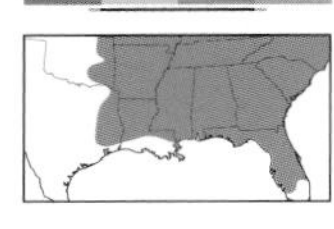

Hyperstrontia flaviguttata 93-0731 (9039) **Common**

TL 8–9 mm. Silvery gray FW has indistinct or almost absent AM and PM lines. Rusty median band is often reduced to patches at inner margin, accented with pale yellow spots. Diffuse, pale reniform spot contains two small black dots. **HOSTS:** Red and white oaks. **NOTE:** Two broods.

BLACK-PATCHED GRAYLET

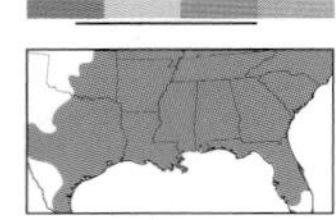

Hyperstrotia secta 93-0732 (9040) **Common**

TL 8–9 mm. Resembles Yellow-spotted Graylet, but inner median patch is blackish. PM line is edged strongly whitish at inner margin. ST line is often edged brownish. Has dark patches along costa at AM, median, and ST lines. **HOSTS:** Oak, especially white oak. **NOTE:** Two broods.

THIN-LINED OWLET *Isogona tenuis* 93-0734 (8493) **Common**

TL 13–16 mm. Pale grayish brown FW has pale yellow veins. Yellow-edged AM line curves up at costa. Strongly angled PM line is connected to yellow apical dash. Large, hollow reniform spot is outlined pale yellow. Thorax has dark collar. **HOSTS:** Hackberry.

UNDERWINGS, ZALES, AND RELATED OWLETS Family Erebidae, Subfamily Erebinae

A large and diverse assemblage of broad-winged noctuids. Most species are somberly clad in shades of gray or brown, often with complex barklike patterns on the FW. The familiar underwing moths in the genus *Catocala* have dramatically vibrant HW patterns that are usually kept hidden. Many species in this group, especially zales and underwings, are frequent visitors to sugar bait. With a few exceptions, they are nocturnal and will come to lights in small numbers.

LEVANT BLACKNECK

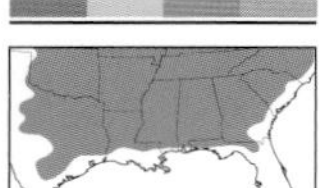

Tathorhynchus exsiccata 93-0754 (8466) **Uncommon**

TL 14–16 mm. Narrow grayish brown FW has dark streaks behind pale orbicular and two-dotted reniform spots. Veins are often noticeably pale. ST area is shaded dark brown. Dark brown basal dashes meet lateral stripes on thorax. **HOSTS:** Alfalfa and Spanish broom. **NOTE:** Introduced; native to eastern half of Southern Hemisphere.

TRANSVERSE HEMEROBLEMMA

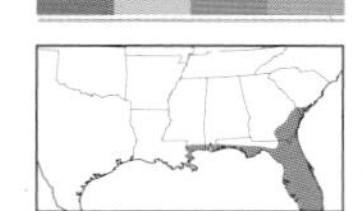

Hemeroblemma opigena 93-0755 (8645) **Local**

WS 58–72 mm. Sexually dimorphic. Male has boldly patterned wings; a broad whitish streak crosses FW from apex to apex, and maroon median area is crisply bordered by a brownish ST area. Smaller female has light to dark brown FW with paired black spots in outer and central ST areas; PM line is a series of dark crescents bordered with pale gray. **HOSTS:** Unknown.

ASSORTED OWLETS

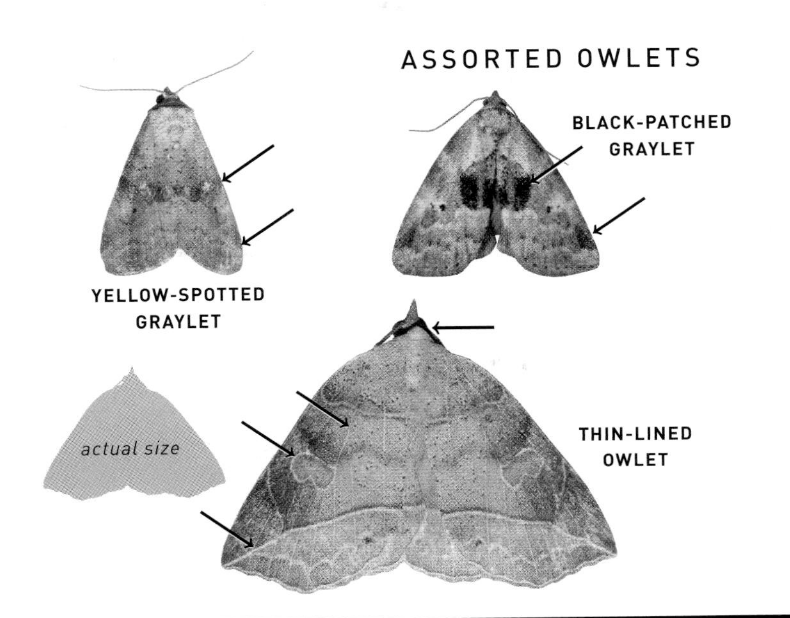

UNDERWINGS, ZALES, AND RELATED OWLETS

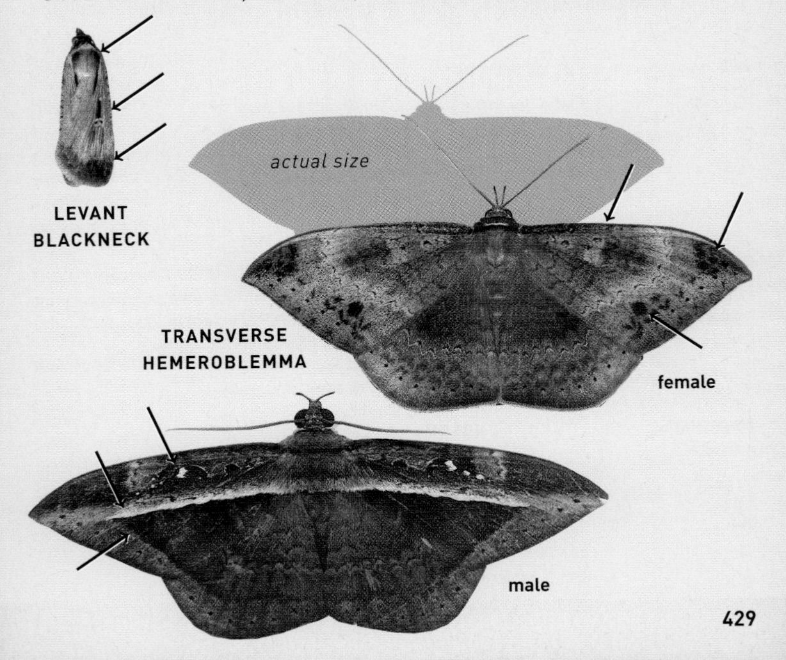

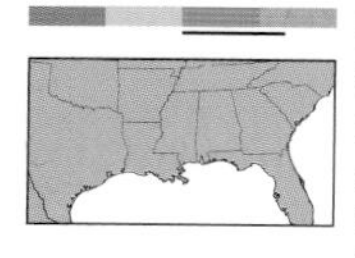

OWL MOTH *Thysania zenobia* 93-0758 (8647) Uncommon

WS 100–150 mm. Unmistakable. Sexually dimorphic. Pale gray to brown wings have coarse barklike pattern. Two dark brown triangles mark costa at median and AM lines. ST line forms straight line across HWs. Male has strong black line that runs from apex to apex. **HOSTS:** Senna and cassia. **NOTE:** A tropical species that is a casual visitor to our region.

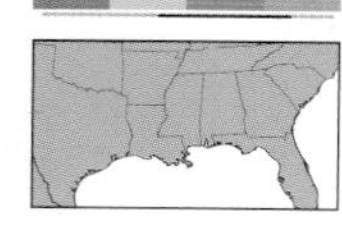

BLACK WITCH

Ascalapha odorata 93-0759 (8649) Uncommon

WS 110–150 mm. Unmistakable. Sexually dimorphic. Brown FW has large comma-shaped, metallic blue reniform eyespot. HW has paired eyespots at anal angle. Yellow-dotted ST line on FW is edged rusty and metallic blue along inner half. Female has whitish median band that forms a straight line across wings when at rest. **HOSTS:** Leguminous trees, including cassia and mesquite.

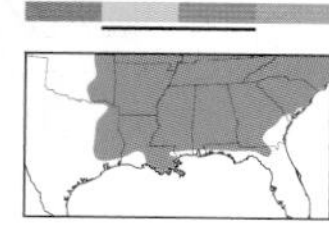

MOON-LINED MOTH

Spiloloma lunilinea 93-0760 (8769) Common

TL 22–27 mm. Brownish gray FW has faint, dotted lines that widen into bold dark patches along costa. Dark double median line is visible at inner margin of FW and across HW. Thorax has black collar. **HOSTS:** Honey locust. **NOTE:** Two broods.

NOTE: The *Catocala* underwings are a large assemblage of cryptically patterned moths. In addition to HW color (which may not be visible), size can be useful in helping with identification. We indicate relative size at the beginning of each account, using six categories: small, medium-small, medium-sized, medium-large, large, very large.

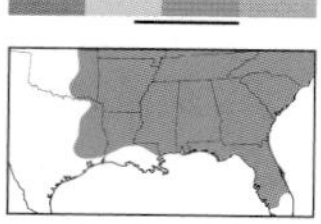

THE BETROTHED

Catocala innubens 93-0761 (8770) Common

TL 30–38 mm. Medium-large. Mottled brown FW has a dusky band from base to apex, often noticeable only where it interrupts whitish ST line and lighter ST area. Small subreniform spot is whitish. Dark form is uniformly chocolate brown above ST line. Orange HW has black median and ST bands and pale fringe. **HOSTS:** Honey locust.

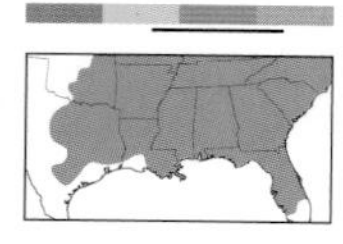

THE PENITENT *Catocala piatrix* 93-0762 (8771) Common

TL 36–44 mm. Large. Grayish FW has a wide pale band bordering black AM line. Brownish shading is often present in inner basal and PM areas. Jagged black PM line has two long teeth forming a W at midpoint. Golden orange HW has black median and ST bands and pale fringe. **HOSTS:** Pecan, hickory, and walnut.

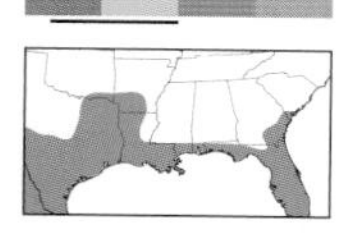

CONSORT UNDERWING

Catocala consors 93-0763 (8772) Uncommon

TL 32–37 mm. Medium-large. Purplish gray FW has strong, wavy black AM and PM lines. Bands of warm brown and whitish gray border PM line. Warm brown reniform spot touches whitish subreniform spot. Golden orange HW has wavy black median and ST bands and checkered fringe. **HOSTS:** Hickory.

UNDERWINGS, ZALES, AND RELATED OWLETS

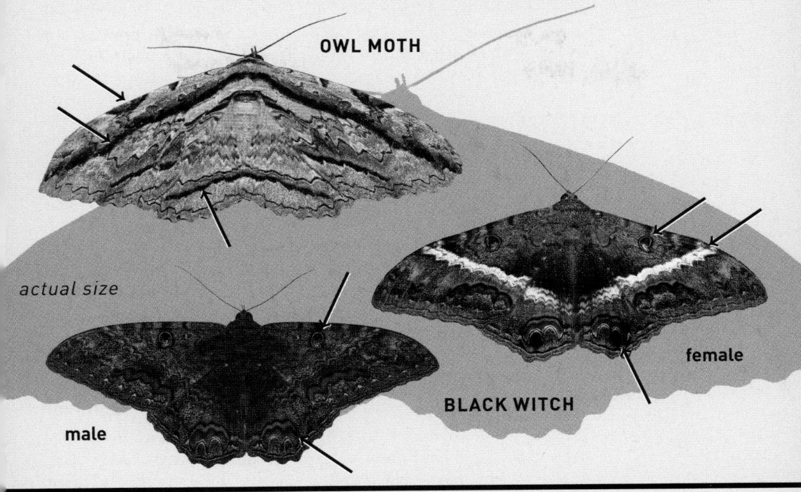

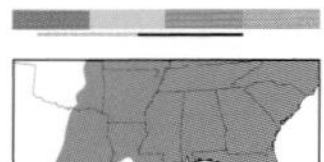

EPIONE UNDERWING

Catocala epione 93-0764 (8773) **Common**

TL 29–34 mm. Medium-sized. Resembles Consort Underwing, but HW is entirely black with a white fringe. AM line is also bordered by brown shading. **HOSTS:** Hickory.

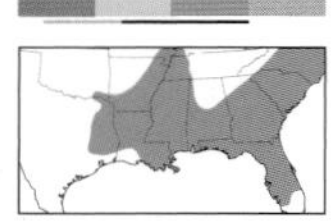

LITTLE WIFE UNDERWING

Catocala muliercula 93-0765 (8774) **Common**

TL 29–37 mm. Medium-large. Dark reddish brown FW has dark brown basal and, often, ST areas. Median area may be light, or dark enough to blend into basal area. Inner median area is often washed blackish. Some individuals have a paler band bordering black AM line, ending in a golden brown subreniform spot. Jagged black PM line has two long teeth forming a W at midpoint. Orange HW has black median and ST bands, and bicolored fringe that is blackish along inner half. **HOSTS:** Wax myrtle and bayberry.

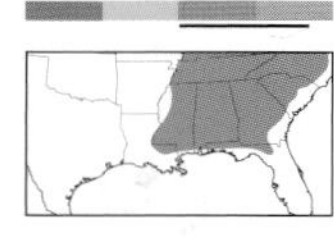

ROBINSON'S UNDERWING

Catocala robinsonii 93-0769 (8780) **Uncommon**

TL 37–42 mm. Large. Pale gray FW has a diffuse whitish ST line bordered by blackish below and warm brown above. Jagged AM line is edged pale and bordered with diffuse brown. PM line has two teeth forming an uneven W at midpoint. Reniform spot is often filled warm brown. Black HW has a narrow white fringe. **HOSTS:** Hickory.

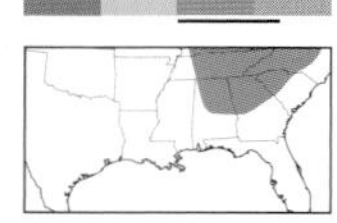

ANGUS' UNDERWING

Catocala angusi 93-0770 (8783) **Local**

TL 32–39 mm. Medium-large. Two forms. Typical form has gray FW with a vertical black dash at inner PM line that is shaded with brown. AM line is weakly defined at costa. AM and PM lines may be variably edged with warm brown. Reniform spot is brown but sometimes indistinct. Diffuse ST line is whitish. Form "lucetta" has broad black streak from base to apex, broken by reniform and subreniform spots. Black HW has a bicolored fringe, gray along inner half. **HOSTS:** Hickory.

MOURNING UNDERWING

Catocala flebilis 93-0775 (8782) **Common**

TL 30–36 mm. Medium-large. Resembles "lucetta" form of Angus' Underwing, but HW fringe is completely white. AM line is boldly marked at costa. Reniform spot is usually large and strongly filled with brown. **HOSTS:** Hickory and walnut.

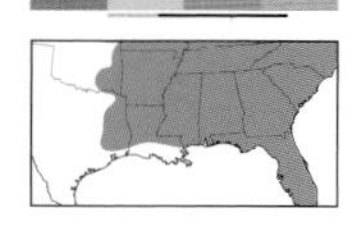

SAPPHO UNDERWING

Catocala sappho 93-0776 (8786) **Common**

TL 33–38 mm. Medium-large. White FW has large black spots along costa at AM, median, and PM lines; the lines themselves are indistinct. Bold chestnut reniform spot has a double black outline. Black HW has a narrow white fringe. **HOSTS:** Hickory.

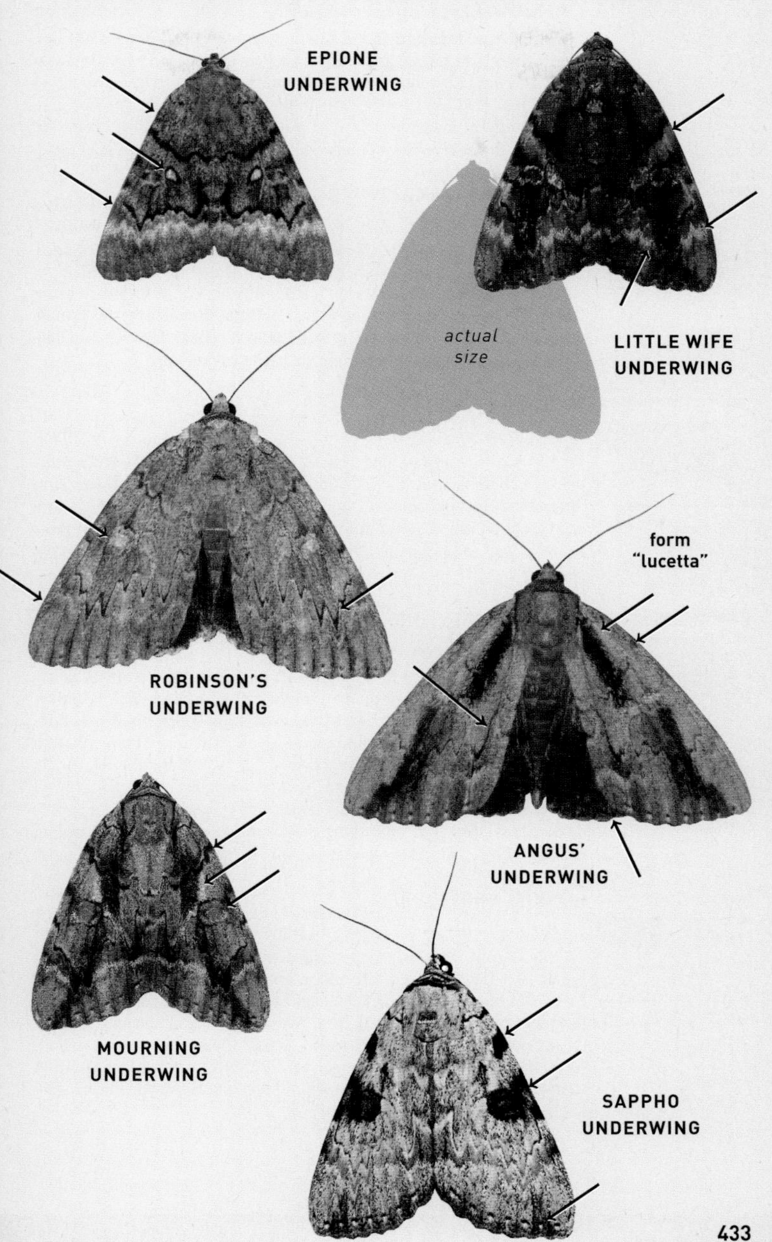
EPIONE
UNDERWING
actual
size
LITTLE WIFE
UNDERWING
form
"lucetta"
ROBINSON'S
UNDERWING
ANGUS'
UNDERWING
MOURNING
UNDERWING
SAPPHO
UNDERWING

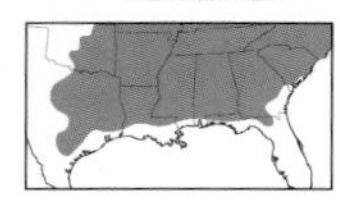

DEJECTED UNDERWING

Catocala dejecta 93-0780 (8790) **Uncommon**

TL 30–38 mm. Medium-large. Medium gray FW has double AM line filled whitish, boldly edged black along outer half. Inner PM line is a vertical, bold black bar. Brown shading is sometimes present along both lines. Two long teeth form a black W at midpoint of PM line, surrounding a tiny white triangle, beneath which ST area is shaded blackish. Black HW has a checkered white fringe. **HOSTS:** Hickory.

INCONSOLABLE UNDERWING

Catocala insolabilis 93-0781 (8791) **Common**

TL 35–40 mm. Large. Peppery light gray FW has thick black bands along length of inner margin, heaviest in basal and inner median areas, contrasting noticeably with a paler streak in inner median area. Black AM and PM lines are edged with muted bands of brown shading. Midpoint of ST area is shaded blackish. Black HW has a bicolored fringe, blackish along inner half. **HOSTS:** Hickory.

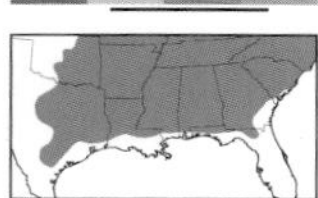

WIDOW UNDERWING

Catocala vidua 93-0782 (8792) **Common**

TL 37–44 mm. Large. Light gray FW has a black crescent from midpoint of costa to midpoint of outer margin. Black bands reach from inner base to anal angle. Double AM line is filled with white and boldly edged black along outer half. Diffuse white ST line is sometimes edged brownish. Black HW has a broad white fringe. **HOSTS:** Hickory.

TEARFUL UNDERWING

Catocala lacrymosa 93-0783 (8794) **Common**

TL 32–43 mm. Large. Light to dark gray FW has diffuse black bands along inner margin of median area that isolate a short, boldly white-edged section of PM (and sometimes AM) line. Black AM line is bordered by diffuse black shading above and a pale area below. Whitish ST line is edged black below and brown above. Midpoint of ST area is strongly shaded blackish. Form "paulina" has brownish basal area and dusky green lower median area. Form "evelina" has a boldly dark basal area that continues down inner margin to anal angle. Black HW has a checkered white fringe. **HOSTS:** Hickory.

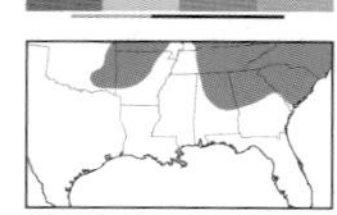

OLDWIFE UNDERWING

Catocala palaeogama 93-0784 (8795) **Local**

TL 32–37 mm. Medium-large. Light to medium gray FW has bold black AM line that is often edged white at inner margin. Jagged black PM line forms a thick vertical bar near inner margin, and two long teeth create a W at midpoint. Whitish ST line is bordered blackish below and warm brown above, most obvious along inner half. Reniform spot is filled brown or rarely black; subreniform spot is usually pale. Orange HW has black median and ST bands and checkered orangish fringe. **HOSTS:** Hickory.

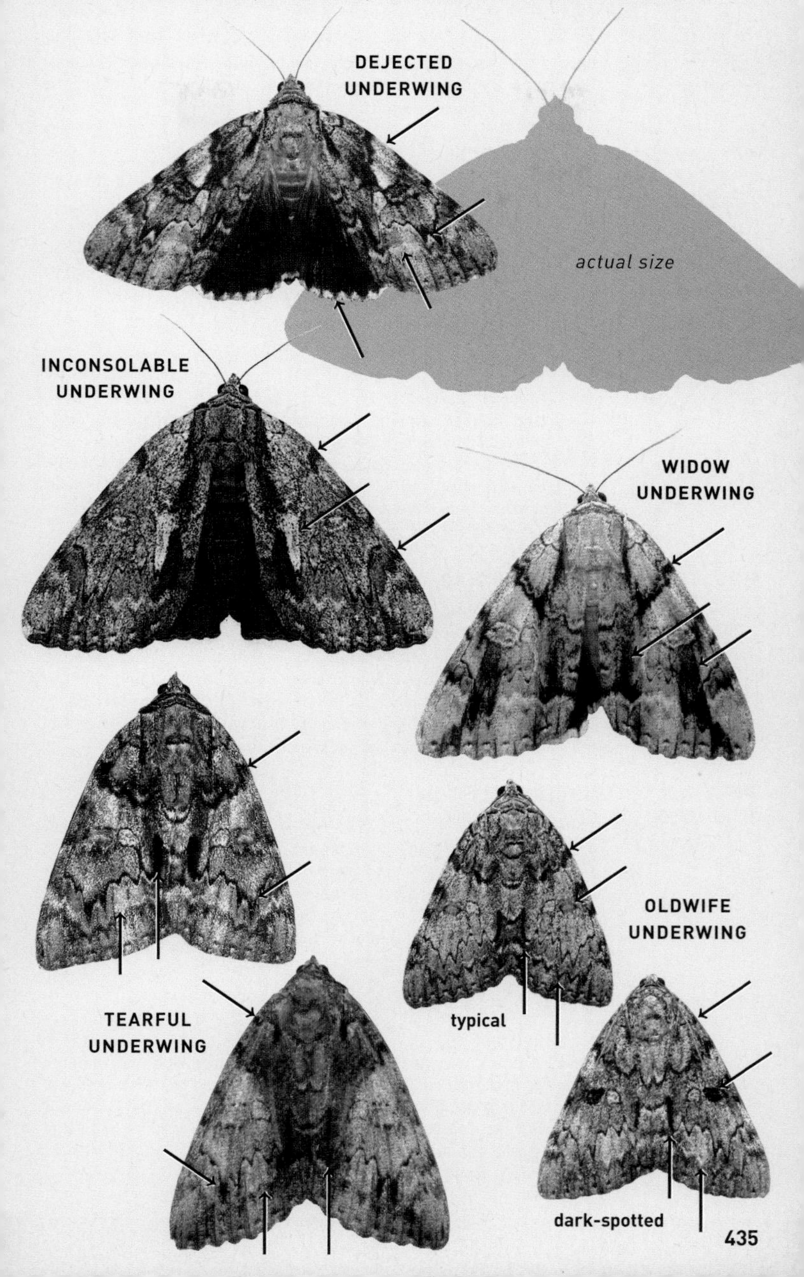
DEJECTED
UNDERWING
actual size
INCONSOLABLE
UNDERWING
WIDOW
UNDERWING
OLDWIFE
UNDERWING
typical
TEARFUL
UNDERWING
dark-spotted

AGRIPPINA UNDERWING

Catocala agrippina 93-0785 (8787) **Common**

TL 35–45 mm. Large. Resembles Mourning Underwing, but dark band from base to apex is much weaker, sometimes nearly absent. Brown to brownish gray FW has brown shading in basal area. Black HW has a checkered white fringe. **HOSTS:** Pecan.

SAD UNDERWING

Catocala maestosa 93-0789 (8793) **Common**

TL 40–50 mm. Very large. Light gray FW has dark crescent curving from midpoint of costa to apex of outer margin, touching edge of warm brown reniform spot. Indistinct AM and PM lines are bordered with muted light brown and whitish bands. Black HW has a checkered white fringe. **HOSTS:** Pecan and walnut.

THE BRIDE

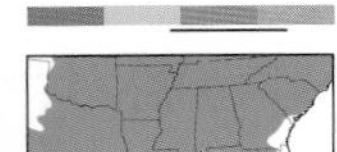

Catocala neogama 93-0790 (8798) **Common**

TL 37–44 mm. Large. Variable. Light individuals resemble Robinson's Underwing, but pale ST line and bordering shading are indistinct, and W at midpoint of PM line is rounder and usually more well marked. Darker individuals resemble Dejected Underwing but lack white triangle in W, and subreniform spot is usually shaded brownish. Golden orange HW has black median and ST bands and checkered, pale yellow fringe. **HOSTS:** Walnut and butternut.

ILIA UNDERWING

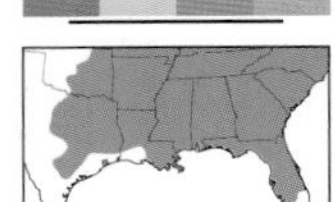

Catocala ilia 93-0792 (8801) **Common**

TL 34–45 mm. Large. Variable. Gray to brownish FW usually has a bold black AM line and a white or white-outlined reniform spot. Some individuals have a paler median band. Dark shading typically runs diagonally from apex. Blackish basal dash often connects to horizontal bar on thorax. Reddish orange HW has black median and ST bands and checkered, pale orange fringe. **HOSTS:** Oak.

UMBER UNDERWING

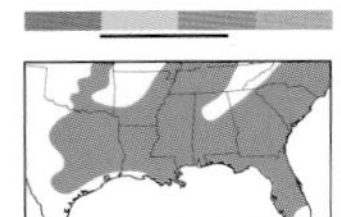

Catocala umbrosa 93-0793 (8801.1) **Local**

TL 36–40 mm. Large. Resembles the Penitent, but wavy brown AM line is double, and PM line does not form vertical bar near inner margin. ST line lacks brownish border, and ST area does not contain dark shading. Reniform spot is usually pale. Reddish orange HW has black median band that often tapers toward inner margin, black ST band, and checkered, pale fringe. **HOSTS:** Oak.

MARBLED UNDERWING

Catocala marmorata 93-0796 (8804) **Local**

TL 47–50 mm. Very large. Resembles Sad Underwing, but medium gray FW has dark reniform spot that blends into blackish crescent band. Double AM line is filled pale gray. Diffuse whitish ST line is relatively even. Reddish to orange HW has black median band that tapers toward inner margin, black ST band, and irregularly edged white fringe. **HOSTS:** Poplar and probably willow.

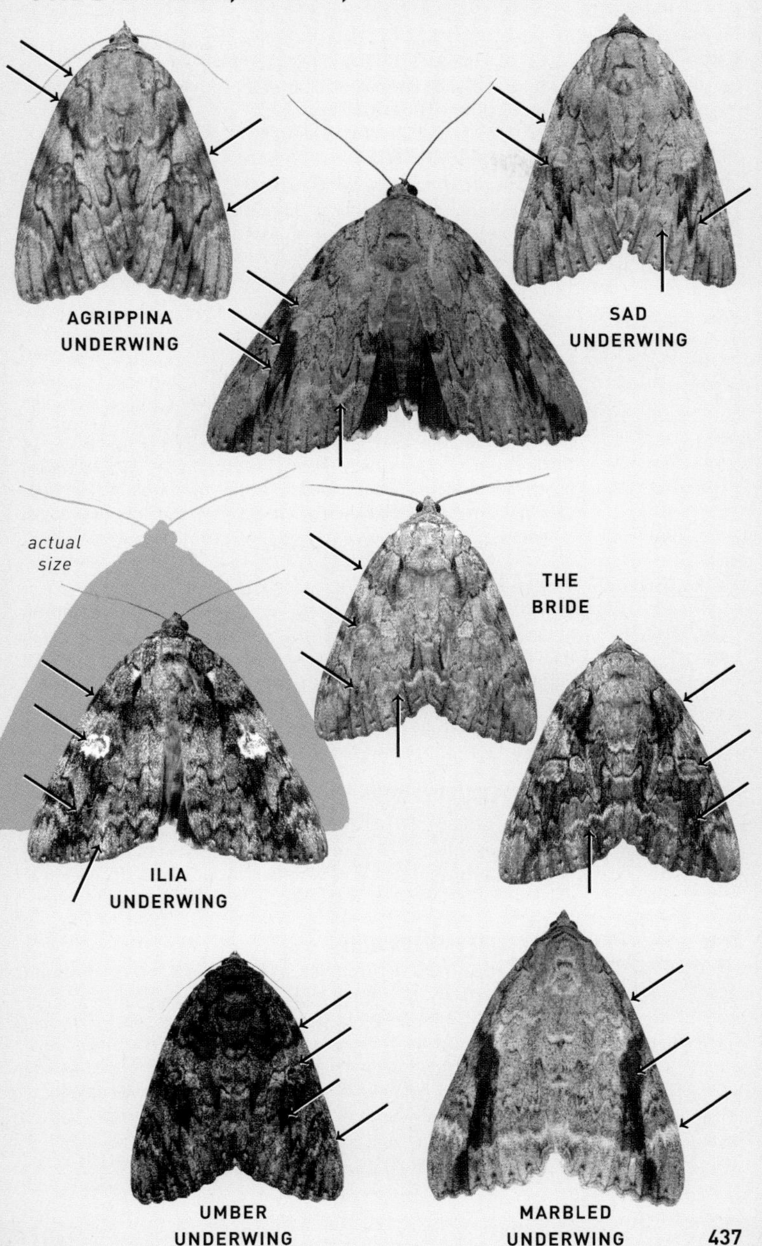
AGRIPPINA UNDERWING
SAD UNDERWING
actual size
THE BRIDE
ILIA UNDERWING
UMBER UNDERWING
MARBLED UNDERWING

DARLING UNDERWING

Catocala cara 93-0812 (8832) **Uncommon**

TL 37–44 mm. Large. Violet brown FW is peppered with green scales along lines. Fine black AM and PM lines are most noticeable toward costa. Bright pink HW has black median and ST bands and checkered white fringe. **HOSTS:** Willow.

CARISSIMA UNDERWING

Catocala carissima 93-0813 (8832.1) **Uncommon**

TL 37–44 mm. Large. Similar to Darling Underwing but usually has large pale blotch at apex. FW is washed rusty, especially along costa, and peppering tends to be orangish. AM and PM lines are bold along entire length. **HOSTS:** Willow.

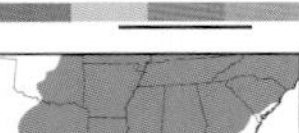

THE SWEETHEART

Catocala amatrix 93-0815 (8834) **Uncommon**

TL 37–44 mm. Large. Resembles Mourning Underwing, but HW is reddish pink with a white, sometimes checkered, fringe. Brownish gray FW is more uniformly colored. PM line lacks dark vertical bar near inner margin. Faint ST line is barely shaded brownish above. Reniform spot is indistinct. Form "selecta" lacks dark diagonal bands; subreniform spot is joined to AM line by thin black line. Reddish pink HW has black bands and checkered white fringe. **HOSTS:** Poplar.

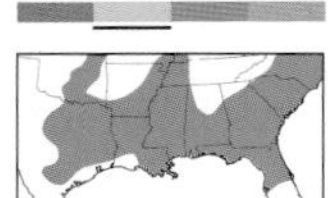

DELILAH UNDERWING

Catocala delilah 93-0816 (8835) **Uncommon**

TL 30–33 mm. Medium-large. Brown FW has thick black AM line that is doubled above and diffusely brown. Inner PM line forms bold black upward projection parallel to inner margin. Basal and PM areas are often slightly paler. Three dark wedges mark middle of ST area. Golden orange HW has black median and ST bands and checkered yellowish fringe. **HOSTS:** Oak.

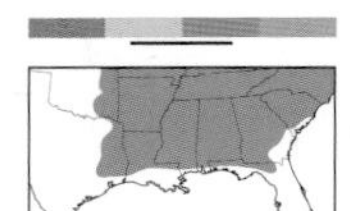

MAGDALEN UNDERWING

Catocala illecta 93-0826 (8840) **Uncommon**

TL 39–48 mm. Large. Resembles form "selecta" of the Sweetheart, but HW is golden orange and subreniform spot is not connected to AM line. Outline of reniform spot is double. **HOSTS:** Honey locust.

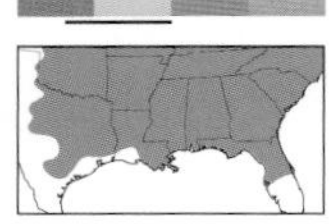

THREE-STAFF UNDERWING

Catocala amestris 93-0830 (8844) **Uncommon**

TL 20–24 mm. Medium-small. Gray FW has dusky shading in outer median area. AM and PM lines are double, filled brownish, and thicken toward costa. Narrow apical dash reaches W at midpoint of PM line. Golden orange HW has black median and ST bands and checkered pale fringe. **HOSTS:** Desert false indigo and leadplant.

UNDERWINGS, ZALES, AND RELATED OWLETS

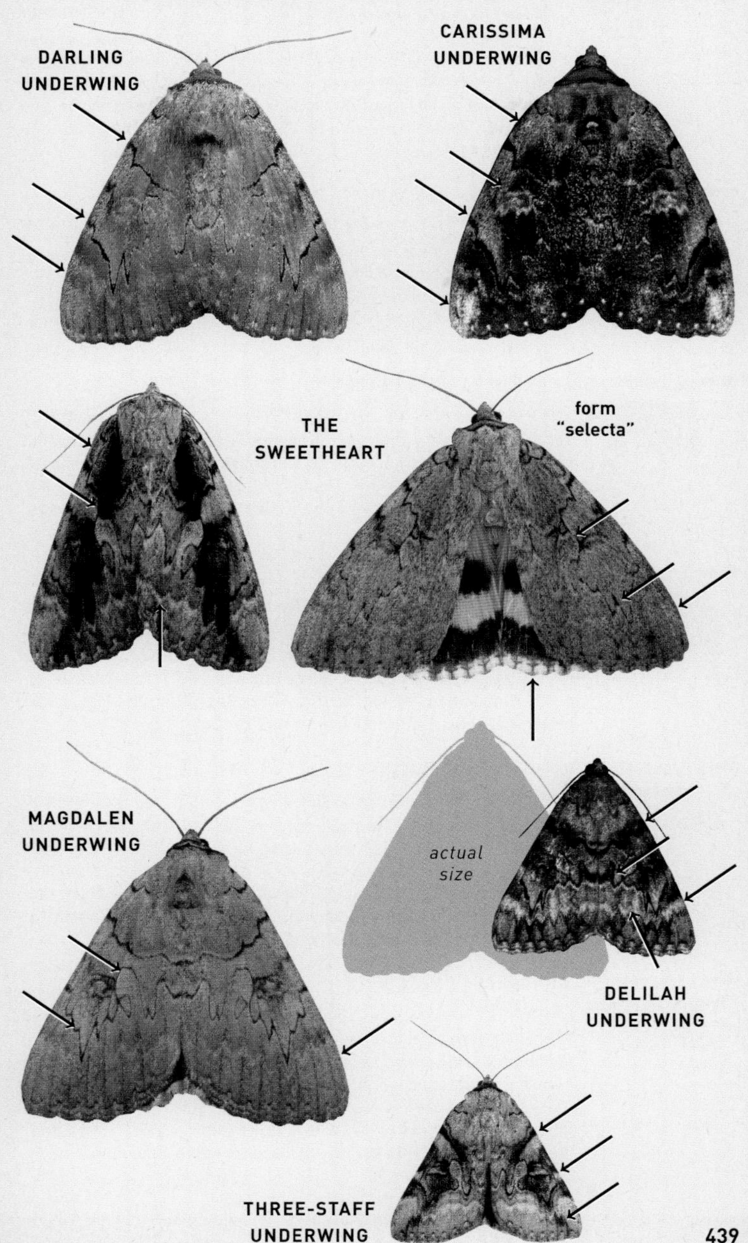

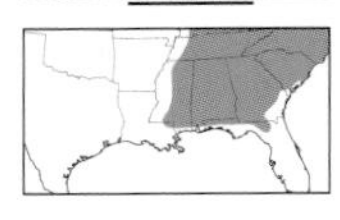

SORDID UNDERWING

Catocala sordida 93-0832 (8846) **Uncommon**

TL 20–24 mm. Medium-small. Peppery gray FW has wavy, sometimes fragmented AM and PM lines that widen at costa. PM line joins AM line near inner margin in a bold hook. Gray reniform spot is diffusely outlined whitish. Zigzag pale ST line is edged gray. Golden orange HW has black median and ST bands and checkered orange fringe. **HOSTS:** Blueberry.

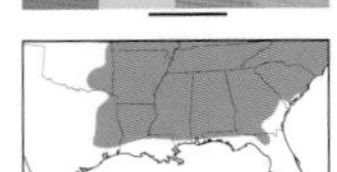

GRACEFUL UNDERWING

Catocala gracilis 93-0833 (8847) **Uncommon**

TL 20–24 mm. Medium-small. Resembles Sordid Underwing, but FW has dark shading along inner margin. Golden orange HW has black median and ST bands and checkered orange fringe (sometimes all black in inner half). **HOSTS:** Blueberry and swamp doghobble.

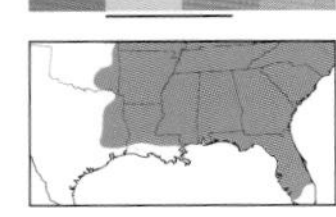

ANDROMEDA UNDERWING

Catocala andromedae 93-0835 (8849) **Common**

TL 22–24 mm. Medium-small. Resembles Graceful Underwing, but PM line is edged strongly whitish and subreniform spot is filled pale, with a black blotch above. Reniform spot is filled dark gray and edged blackish. Black HW has a black fringe, white at apex. **HOSTS:** Blueberry and pink azalea.

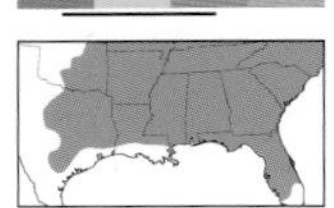

SCARLET UNDERWING

Catocala coccinata 93-0837 (8851) **Uncommon**

TL 30–37 mm. Medium-large. Streaky gray FW usually has a pale brownish patch just inside W of PM line. Pale yellowish green reniform spot is usually edged whitish. Blackish basal dash crosses inner part of double AM line. Bright pink HW has black median and ST bands and checkered white fringe. **HOSTS:** Oak.

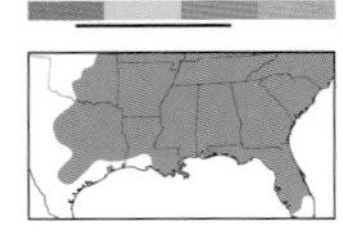

ULTRONIA UNDERWING

Catocala ultronia 93-0841 (8857) **Common**

TL 25–33 mm. Medium-sized. Bluish gray to brownish FW has blackish shading in inner basal area, sometimes extending along inner margin. Pale brown apex is bordered by brown subapical patch and blackish shading in ST area. Reddish HW has black median and ST bands and a mostly gray fringe, white at apex. **HOSTS:** Cherry and plum; also apple and hawthorn.

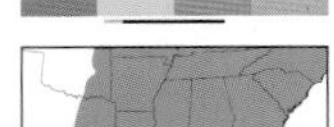

WONDERFUL UNDERWING

Catocala mira 93-0844 (8863) **Common**

TL 22–27 mm. Medium-sized. Resembles Sordid Underwing, but AM and PM lines are bordered by brownish shading, and PM line has a pronounced point (sometimes double point) at midpoint. Upper half of median area is usually pale. Subreniform spot is whitish. Golden orange HW has wavy black median and ST bands and checkered yellow fringe. **HOSTS:** Hawthorn, crab apple, and plum.

UNDERWINGS, ZALES, AND RELATED OWLETS

WOODY UNDERWING

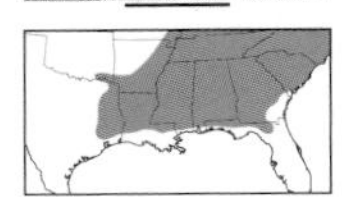

Catocala grynea 93-0845 (8864) **Common**

TL 22–27 mm. Medium-sized. Smooth gray FW is chestnut along inner margin. Wavy AM and PM lines are most obvious as thick black dashes at costa. Golden orange HW has black median and ST bands and checkered pale yellow fringe. **HOSTS:** Apple, crab apple, and plum.

HAWTHORN UNDERWING

Catocala crataegi 93-0846 (8858) **Uncommon**

TL 22–27 mm. Medium-sized. Resembles Wonderful Underwing but usually has heavy brown shading through basal and inner median areas. Central and outer median area is contrastingly evenly pale. Golden orange HW has wavy black median and ST bands and checkered pale orange fringe. **HOSTS:** Apple and hawthorn.

PRAECLARA UNDERWING

Catocala praeclara 93-0847 (8865) **Uncommon**

TL 22–27 mm. Medium-sized. Resembles Woody Underwing but has brown shading in inner ST area. Pale gray FW often has a greenish tinge when fresh. Thin double AM line is edged chestnut. Golden orange HW has black median and ST bands and checkered pale fringe. **HOSTS:** Chokeberry, serviceberry, and hawthorn.

ALABAMA UNDERWING

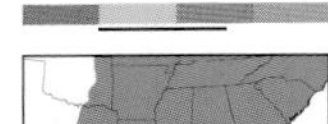

Catocala alabamae 93-0848 (8869) **Common**

TL 17–22 mm. Small. Resembles Woody Underwing, but FW is paler and less well marked; brown shading along inner margin is usually faint. AM line is straight. Some individuals have a band of strong brown shading in PM area; others have a large patch of dark brown shading through inner median area. Golden orange HW has black median and ST bands that are often broken, and a checkered yellow fringe. **HOSTS:** Hawthorn.

CLINTON'S UNDERWING

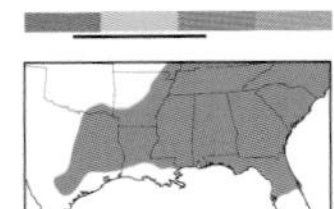

Catocala clintonii 93-0853 (8872) **Common**

TL 25–29 mm. Medium-sized. Pale gray FW has fine white-edged AM and PM lines. Dark basal dash and thick vertical bar of PM line are often most conspicuous markings. Costa is slightly paler. Golden orange HW has black median and ST bands (often broken near inner margin) and pale fringe. **HOSTS:** Apple, cherry, hawthorn, and plum.

SIMILAR UNDERWING

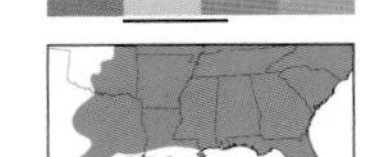

Catocala similis 93-0855 (8873) **Common**

TL 20–25 mm. Medium-small. Variable. Light to medium gray FW has white-edged reniform spot that bleeds toward costa and large pale subapical spot. Median area is often shaded darker. AM and PM lines are edged brown. PM line is relatively straight. Golden orange HW has black median and ST bands (sometimes broken) and a checkered or gray fringe. **HOSTS:** Oak.

UNDERWINGS, ZALES, AND RELATED OWLETS

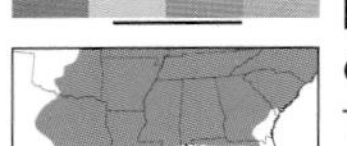

LITTLE UNDERWING

Catocala minuta 93-0856 (8874) **Common**

TL 19–24 mm. Medium-small. Brown to brownish gray FW has a white ST line that widens dramatically at costa. A wide blackish band of shading along AM line may or may not be present. Some individuals have a large black patch in inner median area. Golden orange HW has black median and ST bands and checkered pale orange fringe. **HOSTS:** Honey locust.

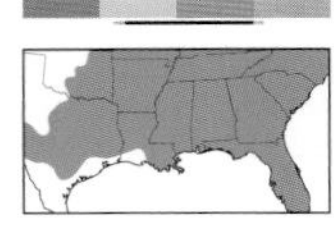

THE LITTLE NYMPH

Catocala micronympha 93-0857 (8876) **Common**

TL 17–26 mm. Small. Variable. Brown to grayish FW has a dark crescent curving from apex to midpoint of costa; faint on some individuals. Diffuse ST line is whitish. Wavy AM and PM lines are often bordered brownish. Many individuals have a dark median area contrasting with a pale ST area. Golden orange HW has black median and ST bands, often broken, and checkered pale fringe. **HOSTS:** Oak.

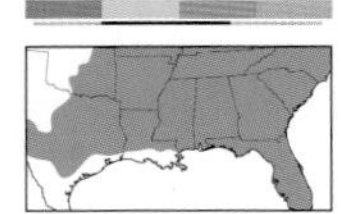

CONNUBIAL UNDERWING

Catocala connubialis 93-0858 (8877) **Common**

TL 20–25 mm. Medium-small. Pale FW has dark basal and ST areas, and dark patch along costa beside reniform spot. AM and PM lines may be crisply black or indistinct. PM line is often edged with warm brown. Golden orange HW has broken black ST and median bands and checkered or gray fringe. **HOSTS:** Oak.

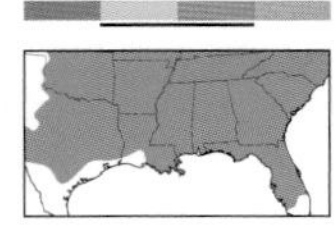

GIRLFRIEND UNDERWING

Catocala amica 93-0859 (8878) **Common**

TL 17–24 mm. Small. Peppery gray FW is often shaded yellowish when fresh. Fragmented AM and PM lines are usually edged faintly brownish. Median line forms black patch at costa. Some individuals have a partial or full blackish crescent, similar to the Little Nymph but with oblique black median bar broken. Golden orange HW has a single black ST band, often incomplete toward anal angle. **HOSTS:** Oak.

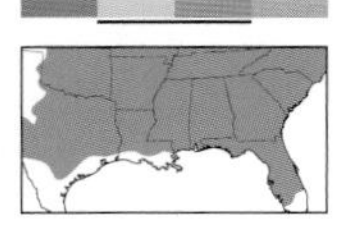

LITTLE LINED UNDERWING

Catocala lineella 93-0860 (8878.1) **Common**

TL 22–24 mm. Medium-small. Resembles Girlfriend Underwing and was formerly considered a subspecies; some individuals may not be identifiable in the field. AM and PM lines are bold and unfragmented. Pale or brownish subreniform spot is prominently outlined black. Dark crescent is usually indistinct when present and lacks oblique median bar. **HOSTS:** Oak.

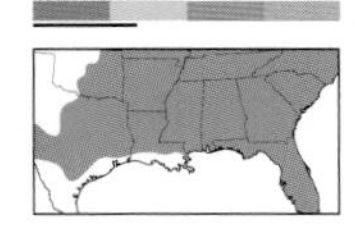

COMMON OAK MOTH

Phoberia atomaris 93-0862 (8591) **Common**

TL 18–23 mm. Grayish tan to brown FW has gently wavy yellow-edged AM and PM lines. ST area is shaded with darker brown. Reniform spot is brown, edged pale yellow. Antennae are always filiform. **HOSTS:** Oak.

UNDERWINGS, ZALES, AND RELATED OWLETS

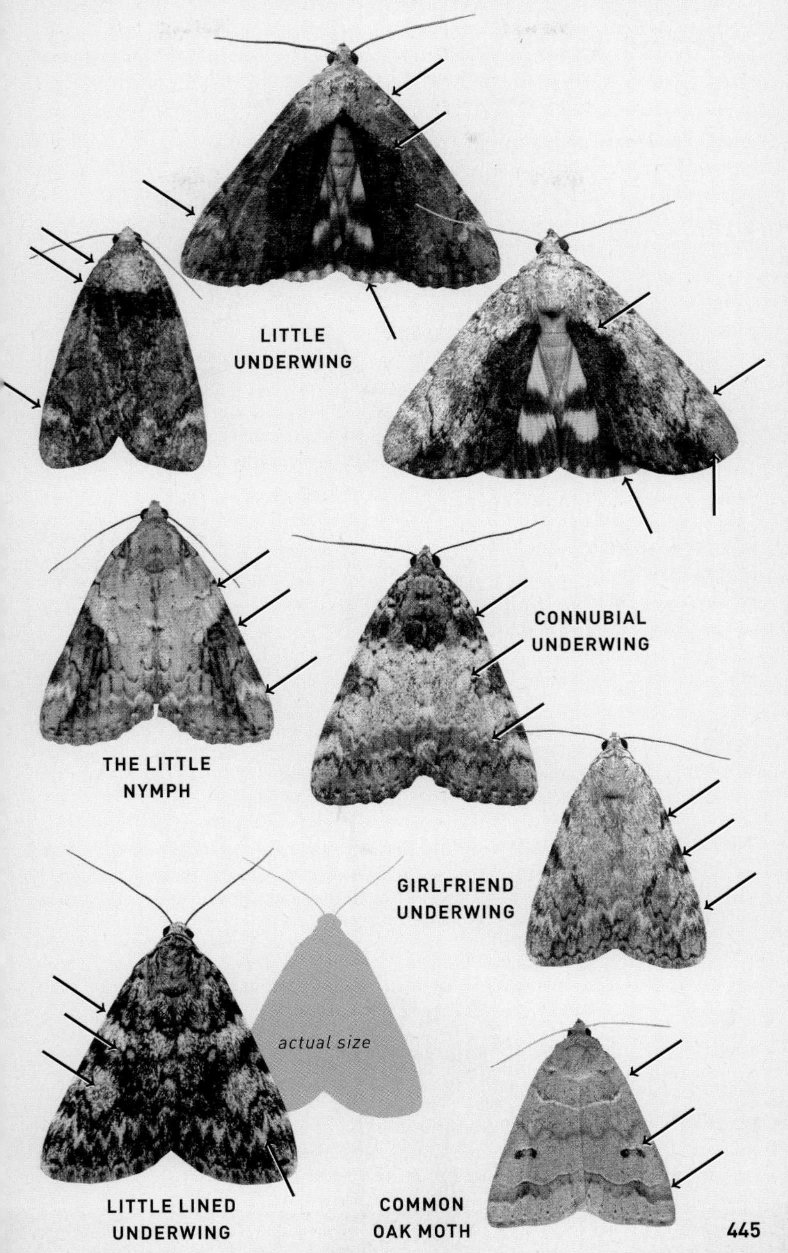

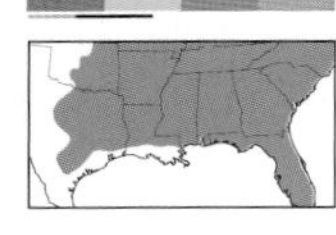

BLACK-DOTTED BROWN

Cissusa spadix 93-0864 (8592) **Common**

TL 17–22 mm. Reddish brown FW has almost parallel, white-edged AM and median lines. Curvy PM line fades before reaching inner margin. Fragmented ST line is accented with black dots and two large spots near apex. **HOSTS:** Oak, sometimes hickory.

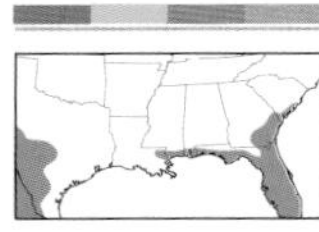

PERPENDICULAR GRAPHIC

Melipotis perpendicularis 93-0869 (8598) **Local**

TL 20–22 mm. Sexually dimorphic. Gray to brownish FW has a thick, slanting pale AM band. A large pale spot is framed by sinuous, smooth PM line. Inner PM line forms a squarish dark wedge. ST area is often paler. Basal area of male is dark; that of female blends into AM band. **HOSTS:** Hopbush.

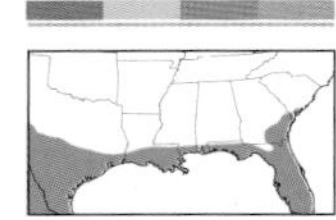

FASCIOLATED GRAPHIC

Melipotis fasciolaris 93-0870 (8599) **Local**

TL 19–23 mm. Sexually dimorphic. Gray to buffy FW has a dark brown, triangular median area and rounded subapical patch. Male has chocolate brown basal area with slanting pale AM band; that of female is uniformly buffy. **HOSTS:** Mesquite; east of TX, unknown.

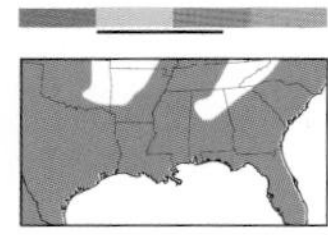

INDOMITABLE GRAPHIC

Melipotis indomita 93-0871 (8600) **Common**

TL 21–26 mm. Sexually dimorphic. Resembles Perpendicular Graphic, but large pale spot is toothed. Wedge formed by inner PM line has a rounded lip. **HOSTS:** Mesquite; east of TX, unknown. **NOTE:** Multiple broods.

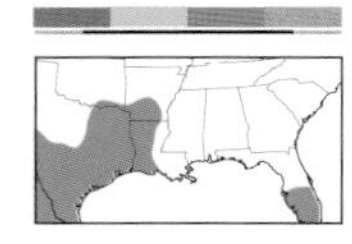

CELLAR GRAPHIC *Melipotis cellaris* 93-0872 (8601) **Local**

TL 21–26 mm. Sexually dimorphic. Light brown FW has a narrow, horizontal pale AM band. Sinuous PM line forms a dark rounded dome near inner margin. Male has a large dark triangle near inner margin; basal area of female is pale with a reduced triangle. **HOSTS:** Acacia and related legumes. **NOTE:** Multiple broods.

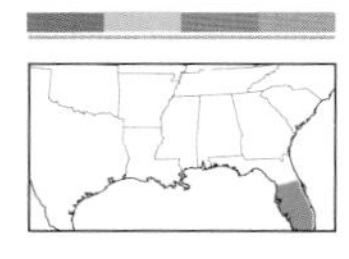

JANUARY GRAPHIC *Melipotis januaris* 93-0874 (8603) **Local**

TL 17–22 mm. Sexually dimorphic. Variable. Purplish to brown FW of male has a pale AM band that sometimes bleeds into median area. Pale spot bordered by sinuous PM line is weakly highlighted. Dark wedge formed by PM line near inner margin is rounded but may be absent in some individuals. Female is similar, but basal area blends into AM band, and overall pattern is often muted or faint. **HOSTS:** Sacky sac bean. **NOTE:** Multiple broods.

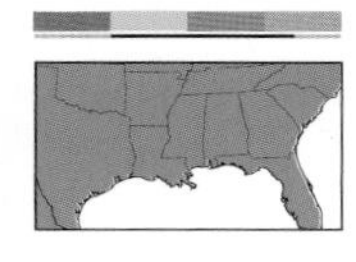

MERRY GRAPHIC *Melipotis jucunda* 93-0878 (8607) **Common**

TL 21–26 mm. Streaky gray FW has a pale AM band that connects to a large pale spot, both of which are often bordered by bold black dashes along inner margin. Some individuals are uniformly streaky gray with a brownish apical dash. **HOSTS:** Catclaw blackbeard, oak, and willow. **NOTE:** Multiple broods.

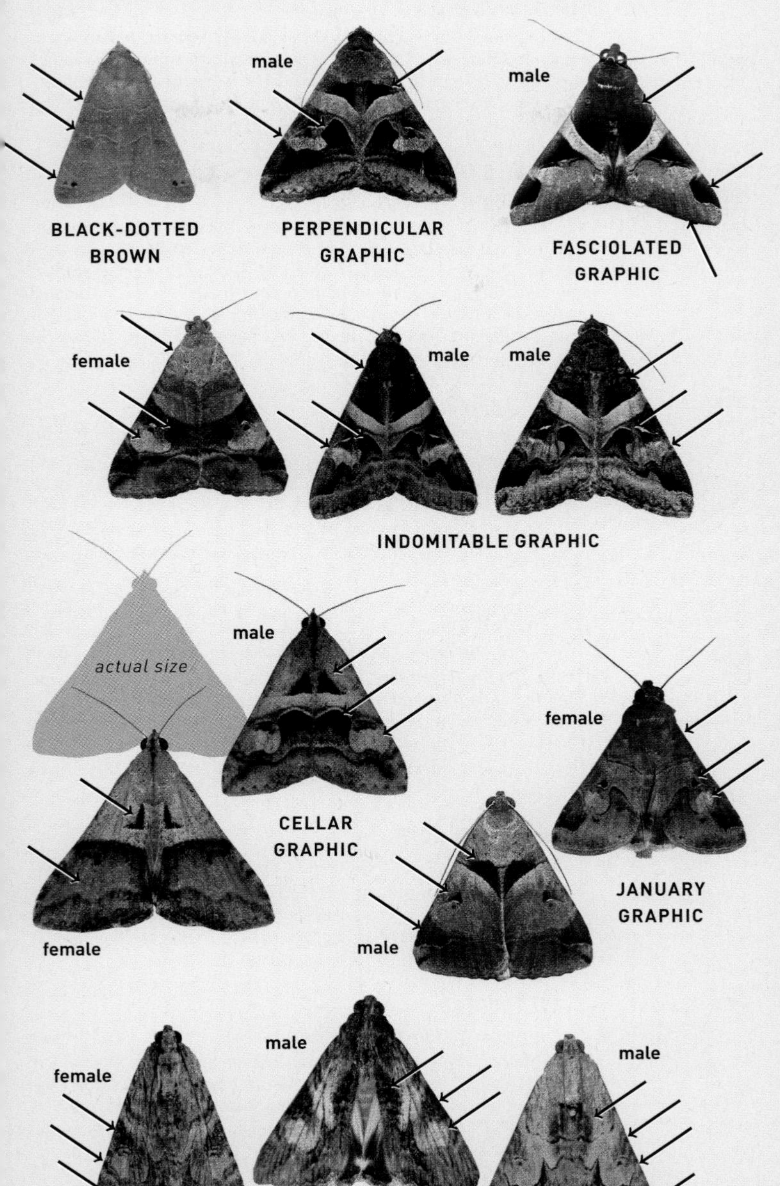

BLACK-DOTTED BROWN

PERPENDICULAR GRAPHIC

FASCIOLATED GRAPHIC

INDOMITABLE GRAPHIC

CELLAR GRAPHIC

JANUARY GRAPHIC

MERRY GRAPHIC

ROYAL POINCIANA GRAPHIC

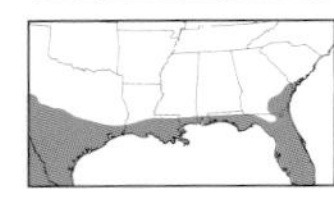

Melipotis acontioides 93-0881 (8610) **Local**

TL 22–24 mm. Narrow, streaky gray FW is variably shaded orange in central median and subapical areas. Double AM and median lines are often visible only at costa. Thin PM line forms a vertical black dash near inner margin. Typically rests with forelegs stretched out. **HOSTS:** Royal poinciana. **NOTE:** Multiple broods.

DEDUCED GRAPHIC

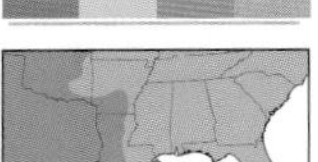

Bulia deducta 93-0885 (8614) **Uncommon**

TL 16–20 mm. Sexually dimorphic. Male resembles Perpendicular Graphic, but pale AM band is horizontal and bordered with dark shading above. Dark wedge framed by PM line at inner margin is rounded. Upper basal area and thorax are grayish. Some individuals may have opaque orange shading through median and ST areas. Females are plain, with just an indistinct dark reniform spot, pale ST line, and small black apical spot. **HOSTS:** Mesquite.

COASTAL GRAPHIC

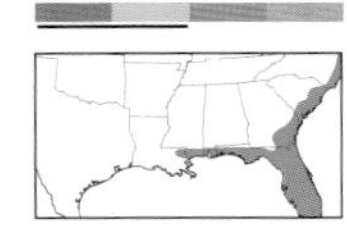

Drasteria graphica 93-0891 (8618) **Uncommon**

TL 18–20 mm. Pale AM band curves up near inner margin and merges with lower median area. Basal area may be dark or pale. Sinuous PM line forms a dark shark-fin–shaped wedge in inner half. ST line is a row of pale dots. **HOSTS:** Mountain goldenheather and woolly beachheather. **NOTE:** Two broods. Also known as Graphic Moth.

PALER GRAPHIC

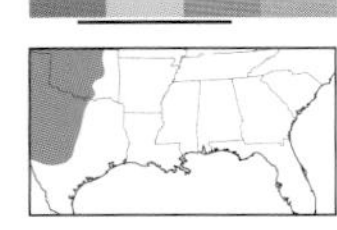

Drasteria pallescens 93-0900 (8628) **Local**

TL 16–20 mm. Resembles Deduced Graphic, but ST area is grayish. Slanting buffy AM band merges with cinnamon median band. Sinuous PM line forms triangular wedge beside large pale tan patch. Costa is checkered with diffuse gray and blackish spots. Upper basal area and thorax are cinnamon. **HOSTS:** Unknown. **NOTE:** Two broods.

FIGURE-SEVEN MOTH

Drasteria grandirena 93-0915 (8641) **Uncommon**

TL 18–20 mm. Resembles Perpendicular Graphic, but buffy AM band is narrow and bordered by cinnamon median band. Sinuous PM line forms a sharply pointed dark wedge. Thorax has two narrow brown stripes. **HOSTS:** Witch hazel. **NOTE:** Two broods.

CLOVER LOOPER

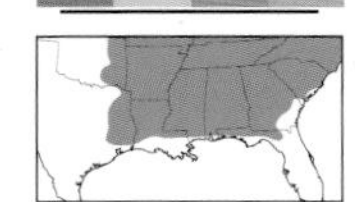

Caenurgina crassiuscula 93-0923 (8738) **Common**

TL 17–22 mm. Pale gray to brownish FW has broad dark AM and PM bands that converge near inner margin and may touch in some individuals. AM band reaches inner margin. Two black dots mark end of indistinct ST line. **HOSTS:** Mostly legumes and grass. **NOTE:** Multiple broods.

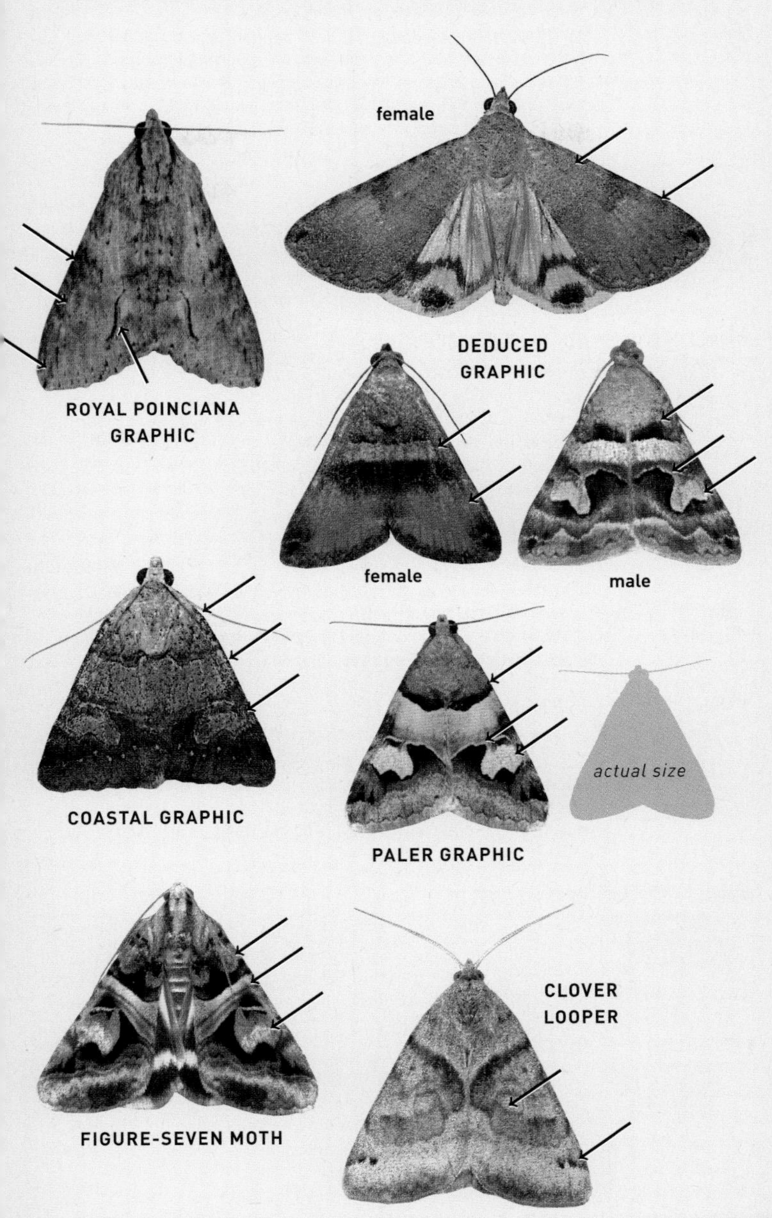

ROYAL POINCIANA GRAPHIC

DEDUCED GRAPHIC

COASTAL GRAPHIC

PALER GRAPHIC

FIGURE-SEVEN MOTH

CLOVER LOOPER

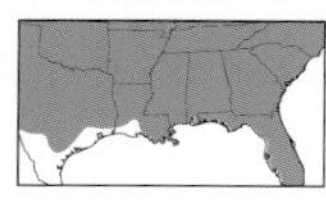

FORAGE LOOPER

Caenurgina erechtea 93-0924 (8739) **Common**

TL 17–23 mm. Resembles Clover Looper, but AM and PM bands never connect, and AM band doesn't touch inner margin. Sometimes bands are simply traced outlines. ST line is usually a series of dark dots. Some individuals may not be identifiable in the field. **HOSTS:** Clover and grass. **NOTE:** Multiple broods.

DOUBLE-LINED DORYODES

Doryodes bistrialis 93-0925 (8765) **Local**

TL 17–20 mm. Narrow, pointed FW is straw-colored with brownish to grayish shading along costa. Two thin white streaks bordered by dark brown shading run from base to apex. **HOSTS:** Wiregrass in coastal environs. **NOTE:** Multiple broods.

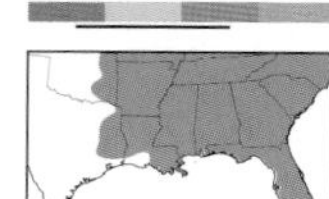

TOOTHED SOMBERWING

Euclidia cuspidea 93-0929 (8731) **Common**

TL 23 mm. Gray to brownish FW has a dark brown AM band with a pointed tooth that almost touches a blackish spot near inner margin. PM line has a sharp, dark brown tooth at midpoint. Indistinct ST line ends in jagged subapical patch. **HOSTS:** Clover and other legumes. **NOTE:** One or more broods.

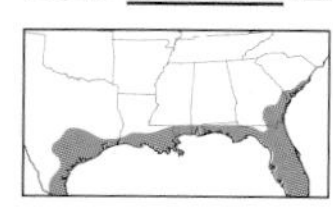

BLACK-TIPPED PTICHODIS

Ptichodis vinculum 93-0931 (8749) **Common**

TL 23 mm. Pale gray FW has parallel yellow AM and PM lines edged brown. Inconspicuous reniform spot is dark at both ends and outlined whitish. ST line is a row of small black dots ending as a bold blackish spot at apex. **HOSTS:** Unknown.

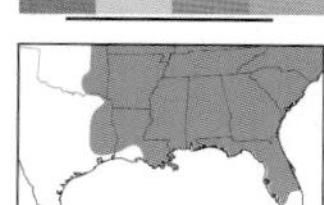

COMMON PTICHODIS

Ptichodis herbarum 93-0932 (8750) **Common**

TL 15–18 mm. Sexually dimorphic. Resembles Black-tipped Ptichodis, but apex lacks blackish spot and reniform spot is uniform. Male has darker shading to AM and PM lines and a small black spot in inner basal area. **HOSTS:** Bush clover. **NOTE:** Two or more broods.

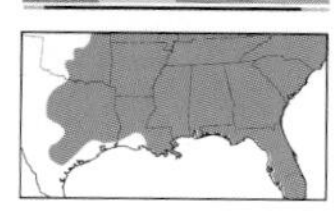

VETCH LOOPER

Caenurgia chloropha 93-0938 (8733) **Common**

TL 15–18 mm. Tan to orange brown FW usually has a dusky blotch just beyond midpoint of thin PM line. ST line is a row of tiny dark dots. **HOSTS:** Vetch and other legumes. **NOTE:** Two or more broods.

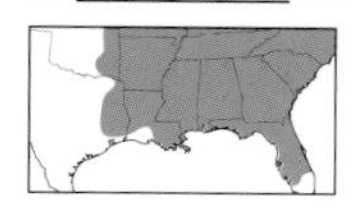

BLACK BIT MOTH

Celiptera frustulum 93-0940 (8747) **Common**

TL 19–21 mm. Grayish brown FW has a bold black triangle near inner margin of basal area. Pale AM line and double PM line are bordered with dark shading. Orbicular spot is a tiny white dot. **HOSTS:** Black locust. **NOTE:** Two broods.

UNDERWINGS, ZALES, AND RELATED OWLETS

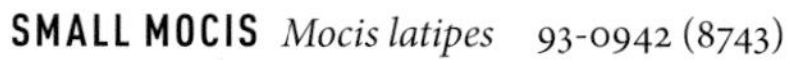

SMALL MOCIS *Mocis latipes* 93-0942 (8743) Common

TL 19–23 mm. Yellowish to tan FW has relatively straight, pale-edged brown AM and PM lines. Circular, brown-outlined subreniform spot abuts kidney-shaped reniform spot. Tiny white orbicular dot is often present. Usually has dark brown shading in PM area. Some individuals have a black spot where AM line meets inner margin. **HOSTS:** Grass, including corn, rice, and sorghum. **NOTE:** Multiple broods.

WITHERED MOCIS

Mocis marcida 93-0943 (8744) Common

TL 24–26 mm. Very similar to Small Mocis, but PM line fades before reaching inner margin. A double median band is usually visible as two dark dashes at inner margin. Black spot, when present, does not touch AM line. **HOSTS:** Grass. **NOTE:** Multiple broods.

TEXAS MOCIS

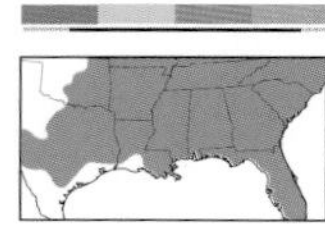

Mocis texana 93-0944 (8745) Common

TL 22–26 mm. Very similar to Small Mocis but lacks subreniform spot. Thin dusky median line touches reniform spot. Black dot, when present, is tiny. **HOSTS:** Grass. **NOTE:** Multiple broods.

YELLOW MOCIS

Mocis disseverans 93-0945 (8746) Local

TL 22–26 mm. Resembles Small Mocis, but AM line and subreniform spot are faint. Often has a row of dark dots along middle of PM line. **HOSTS:** Grass. **NOTE:** Multiple broods.

YELLOW-LINED CHOCOLATE

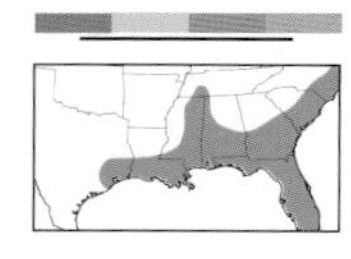

Argyrostrotis flavistriaria 93-0951 (8759) Common

TL 15–17 mm. Grayish brown FW has angled yellow PM line, fading toward apex and usually edged dark brown. Thin brown AM line is sometimes visible. Basal area is often slightly brownish. **HOSTS:** Leatherwood. **NOTE:** Multiple broods.

WOODLAND CHOCOLATE

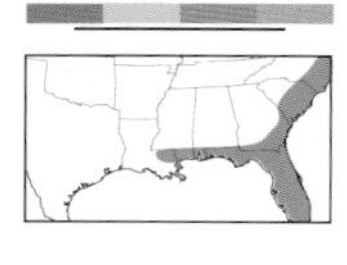

Argyrostrotis sylvarum 93-0952 (8760) Local

TL 15–17 mm. Speckled grayish brown FW has soft brown AM and PM lines that converge slightly toward inner margin; pale yellowish edging is noticeable on darker individuals. PM area is usually dark brown. Irregular ST line is edged with a brown band. **HOSTS:** Unknown.

FOUR-LINED CHOCOLATE

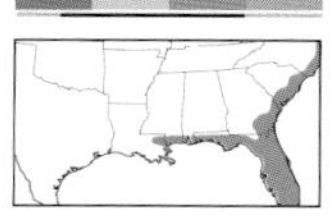

Argyrostrotis quadrifilaris 93-0954 (8762) Local

TL 15–18 mm. Chocolate brown FW has straight white (rarely dark) AM and PM lines that converge slightly toward inner margin. White fringe is grayish at anal angle. **HOSTS:** Maleberry, and perhaps other *Lyonia* species. **NOTE:** One or more broods.

UNDERWINGS, ZALES, AND RELATED OWLETS

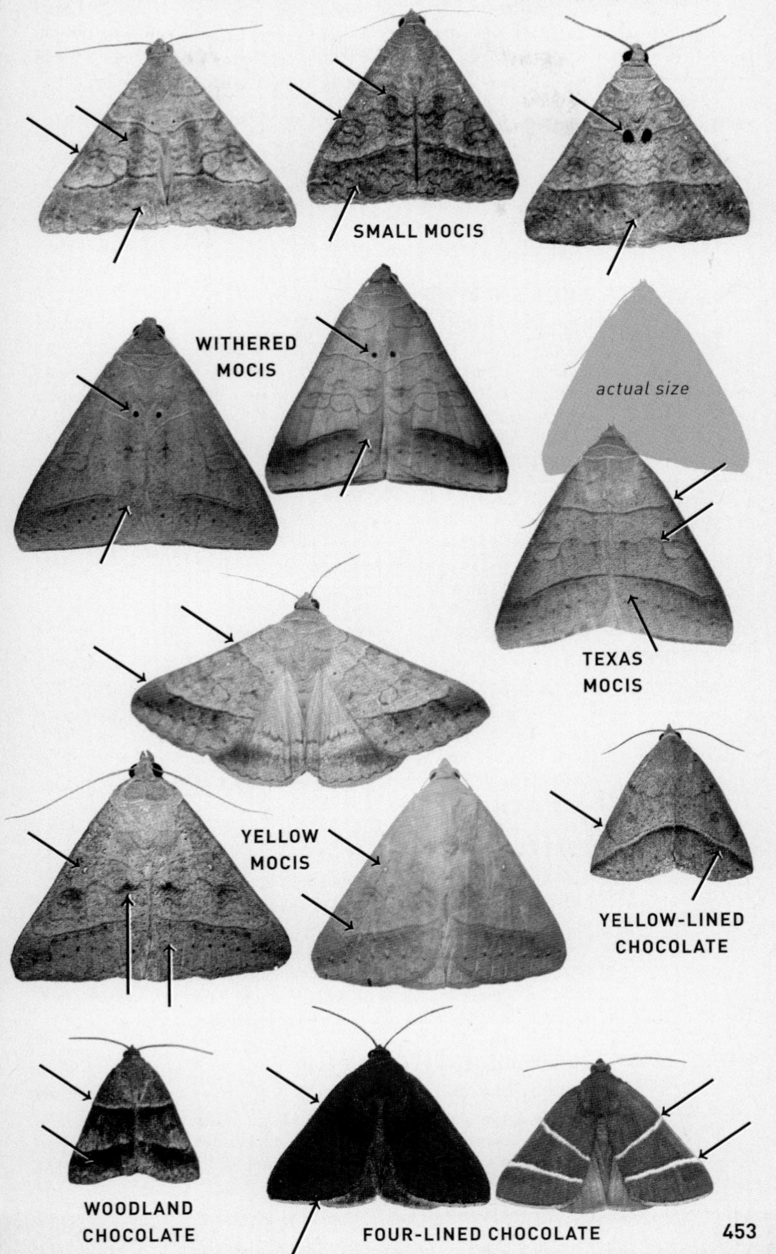

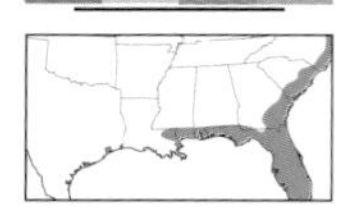

RUINED CHOCOLATE

Argyrostrotis deleta 93-0955 (8763) **Local**

TL 12–14 mm. Chocolate brown FW has wavy dark brown AM and PM lines edged warm brown. Some individuals have a violet tint to FW. ST area is grayish. Antennae are white. **HOSTS:** Unknown.

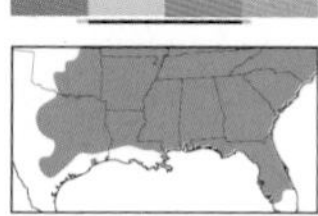

SHORT-LINED CHOCOLATE

Argyrostrotis anilis 93-0956 (8764) **Common**

TL 15–20 mm. Similar to Four-lined Chocolate, but PM line is incomplete. Three tiny white dots mark costa of subapical area. Whitish fringe wraps around anal angle to meet AM line. **HOSTS:** Plum, crab apple, and hawthorn. **NOTE:** Two broods.

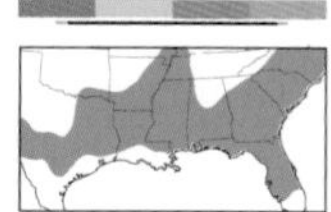

SMITH'S DARKWING

Gondysia smithii 93-0959 (8726) **Common**

TL 20–22 mm. Grayish brown FW has a slightly wavy AM line and double-toothed PM line that are broadly edged with dark brown shading. Inconspicuous ST line ends as a black wedge at apex. **HOSTS:** Unknown.

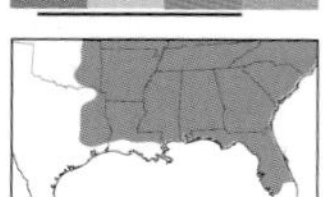

MAPLE LOOPER MOTH

Parallelia bistriaris 93-0961 (8727) **Common**

TL 20–24 mm. Brown FW has a slightly darker median area bordered with parallel yellowish AM and PM lines. ST area is pale gray. Pale apical dash is edged with darker brown shading. **HOSTS:** Maple. **NOTE:** Two or more broods.

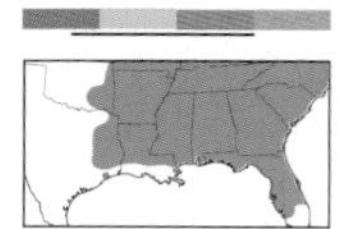

FALSE UNDERWING

Allotria elonympha 93-0962 (8721) **Common**

TL 18–24 mm. Pale gray FW has scalloped AM and PM lines. Some individuals may have heavily shaded basal area, rarely also median area. Reniform spot is hollow. Pale ST line is variably shaded with dark brown, most consistently at apex. Yellow-orange HW has a wide blackish ST area and checkered fringe. **HOSTS:** Tupelo, including black gum. **NOTE:** Two or more broods.

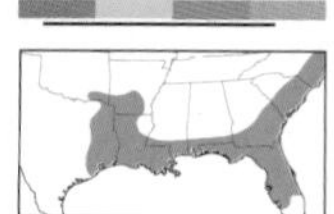

GRAY CYPRESS OWLET

Cutina albopunctella 93-0963 (8728) **Local**

TL 13–16 mm. Grayish brown FW always has three tiny white dots along costa below PM line. Typical individuals have diffuse white streaks along veins, and a thin dark streak through center. Pale ST line is edged broadly dark brown. Some individuals have a uniformly brown FW paler below thin, jagged PM line, and a whitish reniform crescent. **HOSTS:** Bald cypress. **NOTE:** Multiple broods.

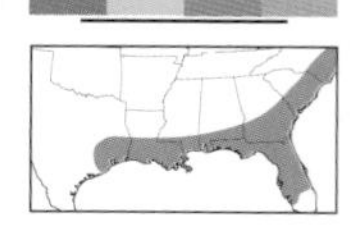

DISTINGUISHED CYPRESS OWLET

Cutina distincta 93-0964 (8729) **Common**

TL 12–14 mm. Light brown FW has a broad whitish band below dark AM line. Basal area is often shaded darker. White wedge at inner margin of PM line is set against a large dark brown patch. **HOSTS:** Cypress. **NOTE:** Multiple broods.

RUINED CHOCOLATE
SHORT-LINED CHOCOLATE
actual size
SMITH'S DARKWING
MAPLE LOOPER MOTH
FALSE UNDERWING
GRAY CYPRESS OWLET
DISTINGUISHED CYPRESS OWLET

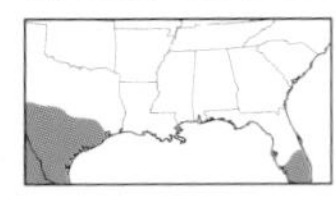

MULTI-LINED OWLET

Tyrissa multilinea 93-0969 (8650) **Local**

WS 22–24 mm. Light brown FW is patterned with many parallel lines across wings. Usually has patches of slightly darker shading in inner median and central ST areas of FW. Typically rests with wings spread open. **HOSTS:** Unknown.

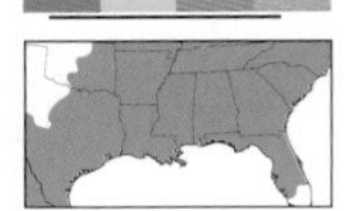

DETRACTED OWLET

Lesmone detrahens 93-0970 (8651) **Common**

WS 27–31 mm. Brown to violet gray FW has a pale band that runs in a straight line apex-to-apex when at rest. Straight PM line angles sharply at costa. Tiny white reniform crescent stands out against diffuse dark median line. Wavy ST line is bordered above with dark shading. **HOSTS:** Cassia. **NOTE:** Multiple broods.

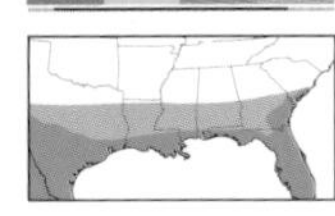

LIFTED OWLET

Lesmone hinna 93-0972 (8653) **Local**

WS 27–31 mm. Light gray FW has scalloped AM and PM lines. Indistinct ST line is bordered by diffuse dark shading. Small reniform spot is often white. Often has a small black apical spot. **HOSTS:** Unknown; possibly cassia. **NOTE:** Multiple broods.

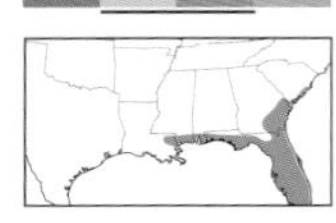

ROUND OWLET

Epidromia rotundata 93-0977 (8585.3) **Common**

WS 40–45 mm. Warm brown FW has a straight yellow PM line that curves upward at costa. Median area is shaded blackish along inner PM line. AM line is scalloped. Reniform spot is gray. Apex is often slightly falcate. **HOSTS:** Unknown.

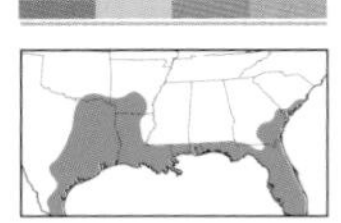

PALE-EDGED SELENISA

Selenisa sueroides 93-0981 (8658) **Common**

WS 25–32 mm. Dark brown FW has a broad yellowish costal band that continues across thorax. Wavy ST line is accented yellowish. Some individuals have a diffuse yellowish median band on HW. **HOSTS:** Woody legumes, including acacia, blackbead, and senna. **NOTE:** Multiple broods.

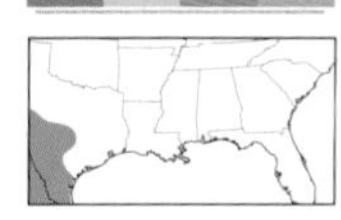

THE MIME *Heteranassa mima* 93-0982 (8659) **Local**

TL 14–18 mm. Sooty brown (rarely gray) FW is marbled with warm brown. AM line is double. Thin bold PM line curves around lower edge of kidney-shaped reniform spot. Reniform spot is sometimes partially or all white. **HOSTS:** Mesquite and acacia.

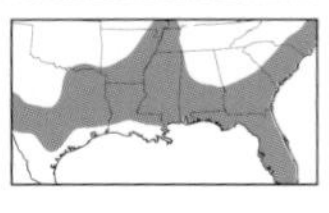

LIVE OAK METRIA

Metria amella 93-0992 (8666) **Common**

TL 20–22 mm. Brown FW has a pale band below double, dark median line. Yellowish reniform spot blends into pale median area. Thin AM line is bordered by pale yellow. Some individuals have brown shading through inner median area. **HOSTS:** Live and turkey oaks. **NOTE:** Multiple broods.

UNDERWINGS, ZALES, AND RELATED OWLETS

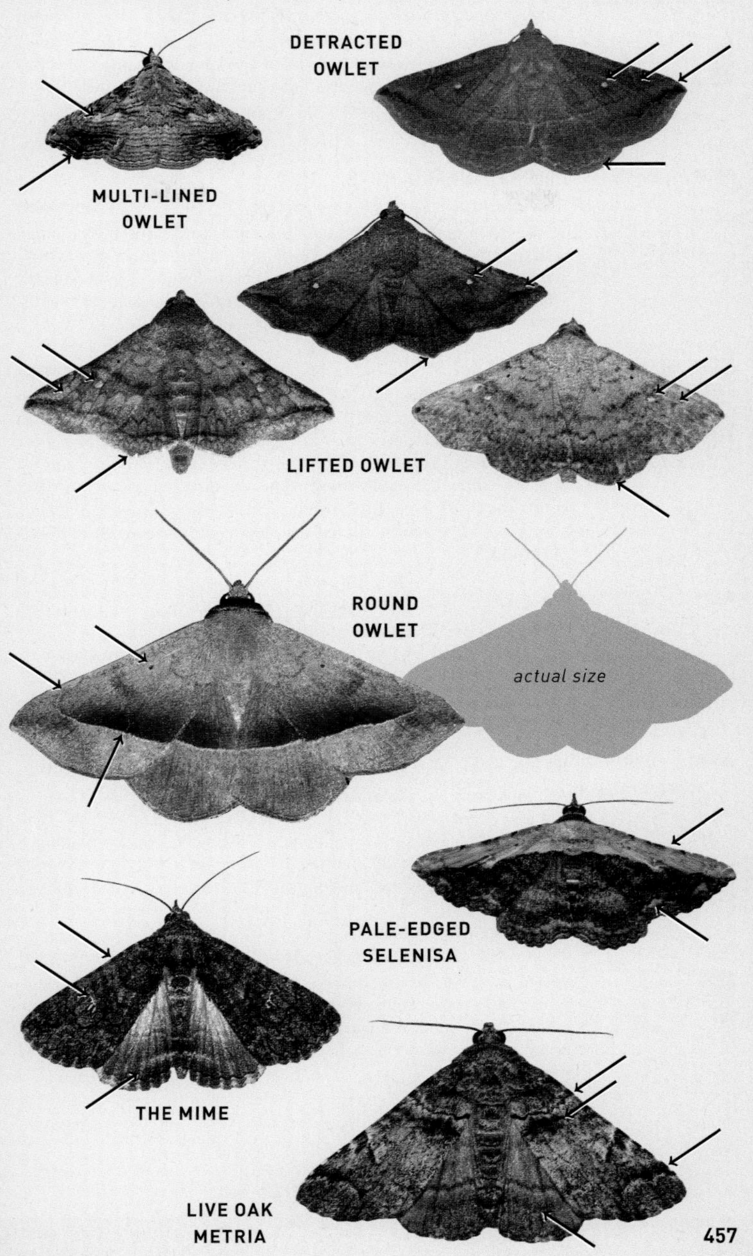

CRUEL TOXONPRUCHA

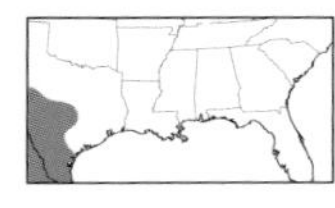

Toxonprucha crudelis 93-1001 (8674) **Local**

WS 20–24 mm. Brown wings are crossed by slightly scalloped lines. Dark median band is double. Thin PM line is usually edged with yellow-orange. ST line is usually a series of tiny white dots, becoming a thin white line edged with dark brown at apex. **HOSTS:** Acacia. **NOTE:** Multiple broods.

PSEUDANTHRACIA MOTH

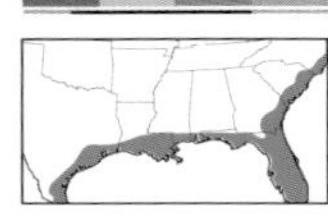

Pseudanthracia coracias 93-1015 (8683) **Common**

WS 34–35 mm. Resembles zales. Dark gray wings are finely banded with parallel black lines, edged with white dots at costa. Some individuals have a whitish band in median area. Reniform spot is partly edged white. Terminal line is a series of white dashes. **HOSTS:** Unknown.

LUNATE ZALE *Zale lunata* 93-1023 (8689) **Common**

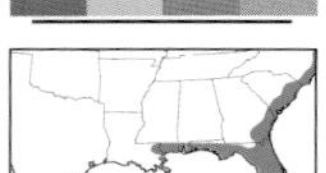

WS 40–55 mm. Variable. Brown FW has a barklike pattern of thin wavy black lines. Double PM line is boldest along inner half, with shading curving to meet outer margin at midpoint. Often has darker patches along costa at PM, median, and AM lines. Reniform spot is obscure. Some individuals have whitish edging to ST lines, or whitish bands in ST area. **HOSTS:** Various deciduous trees and woody plants, including apple, cherry, oak, plum, and willow. **NOTE:** Multiple broods.

DIXIE ZALE *Zale declarans* 93-1025 (8691) **Local**

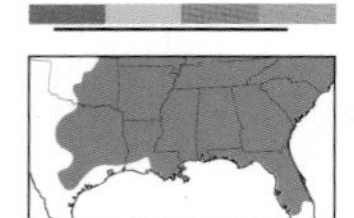

WS 35–40 mm. Brown FW usually has a paler terminal line beyond thin, pale-edged ST line. Dark reniform spot is incompletely outlined light brown. Orbicular spot is a small black dot. Some individuals are finely brindled dusky brown. **HOSTS:** Live oak. **NOTE:** Multiple broods.

MAPLE ZALE *Zale galbanata* 93-1026 (8692) **Common**

WS 35–41 mm. Variable. Resembles Lunate Zale, but dark kidney-shaped reniform spot is usually obvious on FW, and often lacks dark shading along inner PM line. AM line is strongly double and usually bordered below by paler median area. Double PM line clearly widens at costa into brown wedge. **HOSTS:** Box elder. **NOTE:** Two or more broods.

GOAT ZALE *Zale edusina* 93-1027 (8693) **Local**

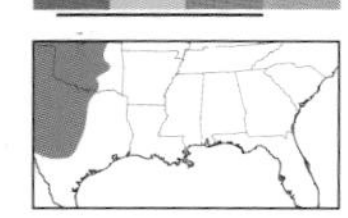

WS 30–40 mm. Barklike brown wings have a subtle orange brown wash along major veins. Pale orange reniform spot is often partly edged white. Upper median area and ST area are usually paler. Terminal line is a row of tiny pale dots. **HOSTS:** Unknown. **NOTE:** Two or more broods.

UNDERWINGS, ZALES, AND RELATED OWLETS

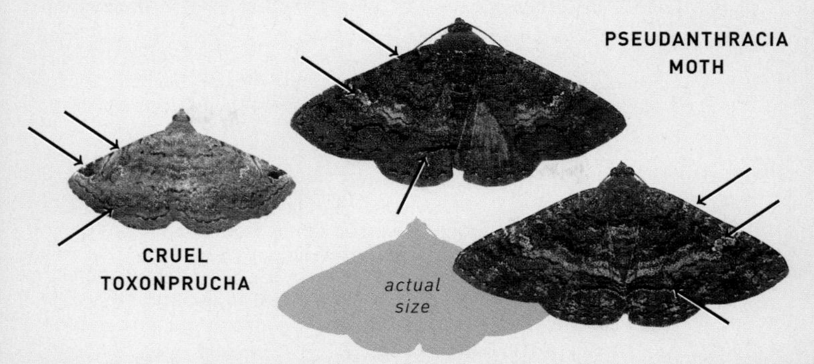

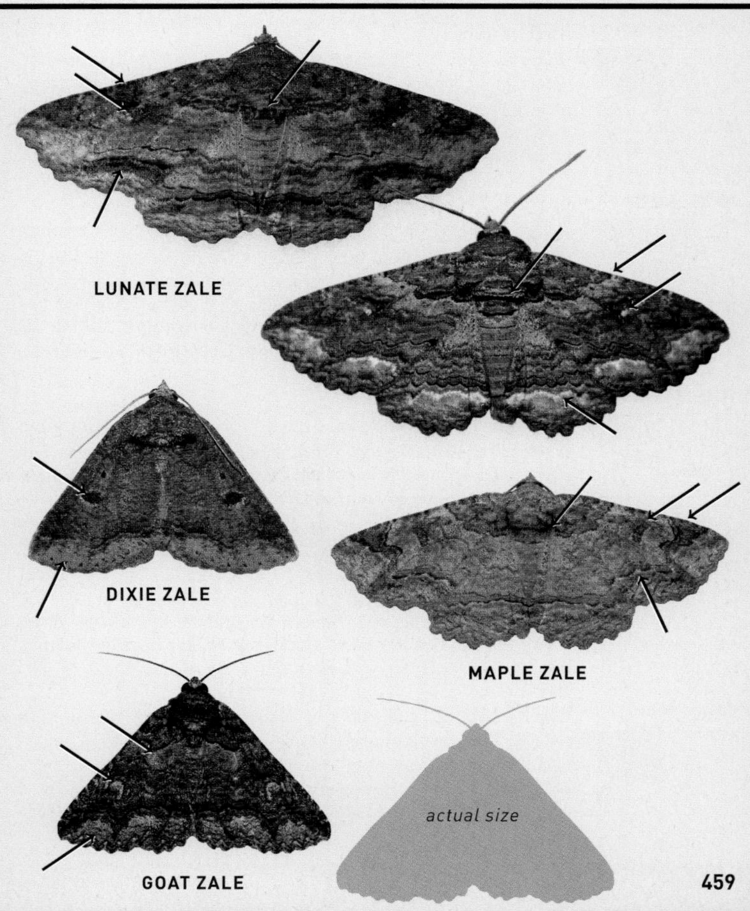

GREEN-DUSTED ZALE

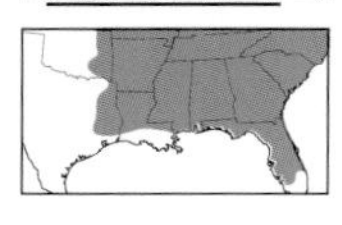

Zale aeruginosa 93-1029 (8694) **Uncommon**

WS 35–42 mm. Variable. Resembles Lunate Zale, but sooty gray wings have mint green edging along AM and ST lines. Many individuals also have violet shading along inner PM line. Reniform spot is sometimes a thin white crescent. **HOSTS:** Primarily oak; also blueberry. **NOTE:** Two or more broods.

COLORFUL ZALE *Zale minerea* 93-1032 (8697) **Common**

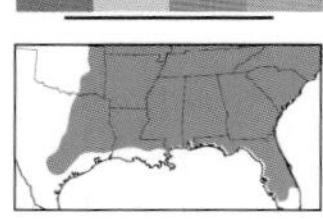

WS 37–50 mm. Variable. Resembles Lunate Zale, but inner median area is usually strongly washed tan to brown, and dark kidney-shaped reniform spot is usually obvious on FW. Basal area is usually noticeably darker. Costa has dark triangular wedge at PM line. Some individuals have a darker lower median area and a pale band beneath AM line. **HOSTS:** Chestnut, hawthorn, oak, and probably others. **NOTE:** One or two broods.

HAZEL ZALE

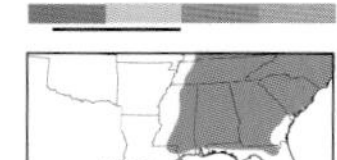

Zale phaeocapna 93-1033 (8698) **Uncommon**

WS 37–50 mm. Resembles Maple Zale, but wings are typically more uniformly colored, and there is no pale band below AM line. Double median line is usually shaded darker. Median area lacks warm brown wash. **HOSTS:** Hazel.

OBLIQUE ZALE

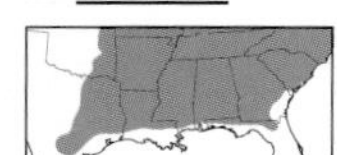

Zale obliqua 93-1034 (8699) **Common**

WS 35–40 mm. Grayish brown FW has warm brown patch below brown reniform spot. Major veins are subtly shaded warm brown. Usually has white patch in ST area near anal angle. Double PM line is filled dark brown toward inner margin on HW. **HOSTS:** Pine, including loblolly, longleaf, pitch, and pond pines. **NOTE:** Two broods.

BROWN-SPOTTED ZALE

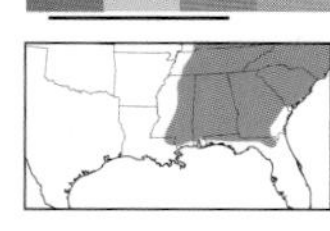

Zale helata 93-1039 (8704) **Common**

WS 35–41 mm. Resembles Oblique Zale, but wings are paler with more contrasting pattern. AM and PM lines are crisp black, strongly edged warm brown. Reniform spot is blackish. **HOSTS:** Pine.

WASHED-OUT ZALE

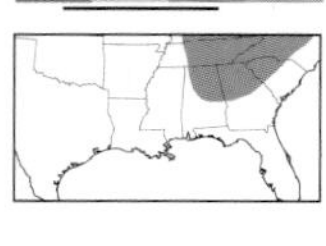

Zale metatoides 93-1042 (8707) **Local**

WS 32–40 mm. Resembles Oblique Zale, but AM and PM lines are faint, except for thickened innermost section of PM line. Warm brown PM band on HW lacks blackish shading near inner margin. **HOSTS:** Pine, including jack, pitch, red, and Virginia pines.

INTENT ZALE *Zale intenta* 93-1049 (8713.1) **Uncommon**

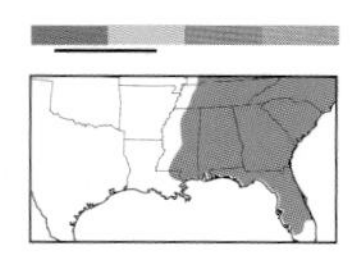

WS 40–45 mm. Finely vermiculated, dark gray FW has broad pale edging below AM line. Pale ST area is strongly vermiculated. Thick AM line is curved or gently angled. Warm brown reniform spot is edged pale yellow. Orbicular spot is reduced or absent. **HOSTS:** Cherry and plum.

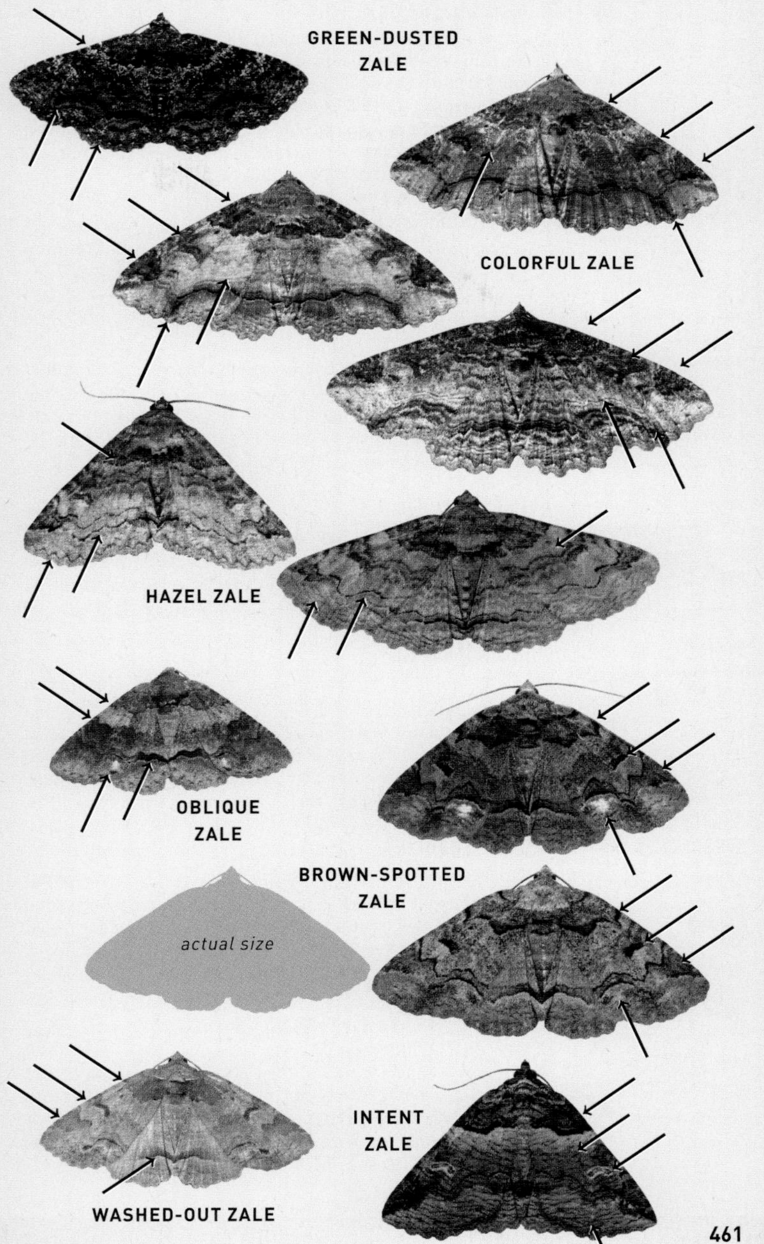
GREEN-DUSTED ZALE
COLORFUL ZALE
HAZEL ZALE
OBLIQUE ZALE
BROWN-SPOTTED ZALE
actual size
INTENT ZALE
WASHED-OUT ZALE

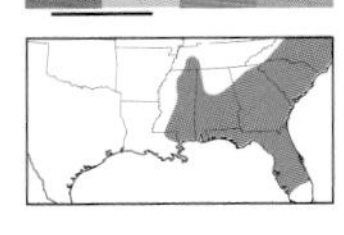

BOLD-BASED ZALE

Zale lunifera 93-1048 (8713) **Common**

WS 37–45 mm. Very similar to Intent Zale, and used to be considered the same species; many individuals may be unidentifiable in the field. Smaller than Intent Zale. Thick black AM is strongly angled. Weakly vermiculated ST area does not contrast markedly with median area. Small black orbicular spot is present. **HOSTS:** New leaves of scrub oak. **NOTE:** Restricted to east and south of Appalachian Mts., but may overlap with Intent Zale in this area.

DOUBLE-BANDED ZALE

Zale calycanthata 93-1050 (8714) **Common**

WS 42–45 mm. Resembles Intent Zale, but ST area is whitish and weakly vermiculated. Area beneath dark reniform spot is usually paler. **HOSTS:** Oak.

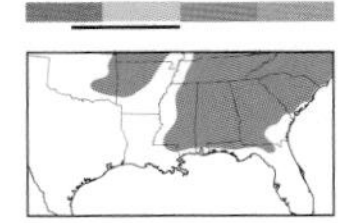

ONE-LINED ZALE *Zale unilineata* 93-1052 (8716) **Local**

WS 40–50 mm. Light brown FW has a wavy yellowish PM line that fades into dark brown subapical patch. AM and median lines are faint. Brownish reniform spot is indistinct. HW has a bold black PM line. **HOSTS:** Black locust.

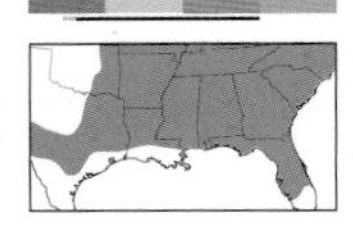

HORRID ZALE *Zale horrida* 93-1053 (8717) **Common**

WS 35–40 mm. Slate gray FW has broad, vermiculated yellowish band along outer margin. Wavy black AM and PM lines are edged yellowish at costa. Thorax has distinct cinnamon tufts. **HOSTS:** Viburnum, including arrowwood. **NOTE:** Two or more broods.

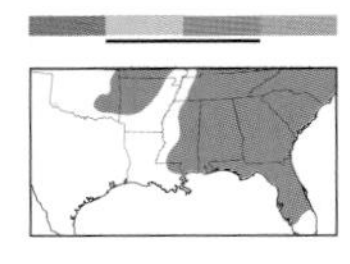

LOCUST UNDERWING

Euparthenos nubilis 93-1055 (8719) **Common**

TL 30–37 mm. Sexually dimorphic. Gray FW has double PM line that is filled whitish toward costa, sometimes bleeding into median area. Male has paler band below AM line and in ST area; female does not. Golden orange HW has three wavy black bands and pale fringe. **HOSTS:** Primarily black locust. **NOTE:** Two broods.

IMPARTIAL EUBOLINA

Eubolina impartialis 93-1056 (8720) **Local**

TL 14–16 mm. Light gray FW is tinged warm brown in ST area and along thin, sometimes fragmented, AM and PM lines. Reniform spot is boldly outlined black and edged whitish along bottom. ST line is bordered by a broad dusky band. **HOSTS:** Unknown.

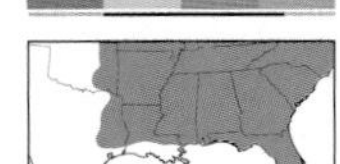

FEEBLE GRASS MOTH

Amolita fessa 93-1060 (9818) **Common**

TL 13–17 mm. Peppery straw-colored FW is marked with a bold brown streak that gently curves from base to apex. A second, sometimes lighter, line curves from outer margin to midpoint of inner margin. Orbicular and reniform spots are tiny black dots. **HOSTS:** Grass. **NOTE:** One or two broods.

UNDERWINGS, ZALES, AND RELATED OWLETS

OBLIQUE GRASS MOTH

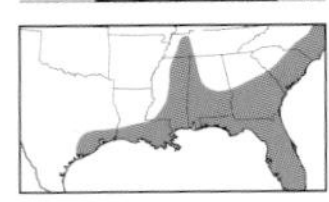

Amolita obliqua 93-1061 (9819) **Common**

TL 9–11 mm. Resembles Feeble Grass Moth, but outer brown streak is typically Y-shaped instead of curved. Both streaks are often indistinct. **HOSTS:** Unknown.

ROSY GRASS MOTH

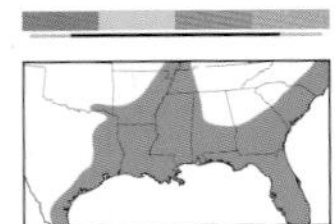

Amolita roseola 93-1063 (9821) **Common**

TL 9–10 mm. Resembles Feeble Grass Moth, but curving lines are faint and usually rosy or purplish. Black orbicular, reniform, and subreniform spots are often bisected by pale veins; may be faint on some individuals. **HOSTS:** Unknown.

EULEPIDOTINE OWLETS

Family Erebidae, Subfamily Eulepidotinae

Medium-sized owlet moths, some of which spread their ample wings open when at rest, like geometer moths. Most are clad in shades of brown, often with oblique lines that cut across all wings. With a few exceptions they are nocturnal and will come to lights in small numbers.

VARIABLE METALLATA

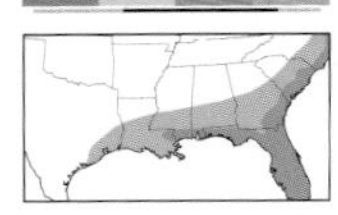

Metallata absumens 93-1075 (8573) **Common**

WS 33–40 mm. Tan to brown wings have a straight, brown-edged, yellowish PM line that runs apex-to-apex when at rest. Large reniform spot is usually white or blackish but sometimes blends into FW; spot is connected to costa by a short dark dash. Tiny black orbicular dot is edged white. Discal spot on HW is all black. ST line is a row of tiny white dots; often dark on individuals with black reniform spots. **HOSTS:** Unknown. **NOTE:** Presumably two or more broods.

VELVETBEAN CATERPILLAR MOTH

Anticarsia gemmatalis 93-1077 (8574) **Common**

WS 33–40 mm. Resembles Variable Metallata but lacks discal spot on HW, and orbicular spot is not edged white. Many individuals have a diffuse median line and strong mottling across wings. Some individuals have median area entirely shaded blackish, with a pale subapical patch. **HOSTS:** Low plants and crops, including alfalfa, peanut, soybean, and velvetbean. **NOTE:** Multiple broods in south.

HELPFUL ATHYRMA

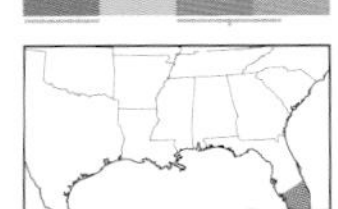

Athyrma adjutrix 93-1084 (8583) **Common Local**

TL 18–20 mm. Shiny light brown FW is boldly patterned with a large blackish triangle in inner basal area and an irregular black patch of variable size in central median area. Rests in a delta shape. **HOSTS:** Unknown. **NOTE:** Probably two broods.

UNDERWINGS, ZALES, AND RELATED OWLETS

OBLIQUE GRASS MOTH

ROSY GRASS MOTH

EULEPIDOTINE OWLETS

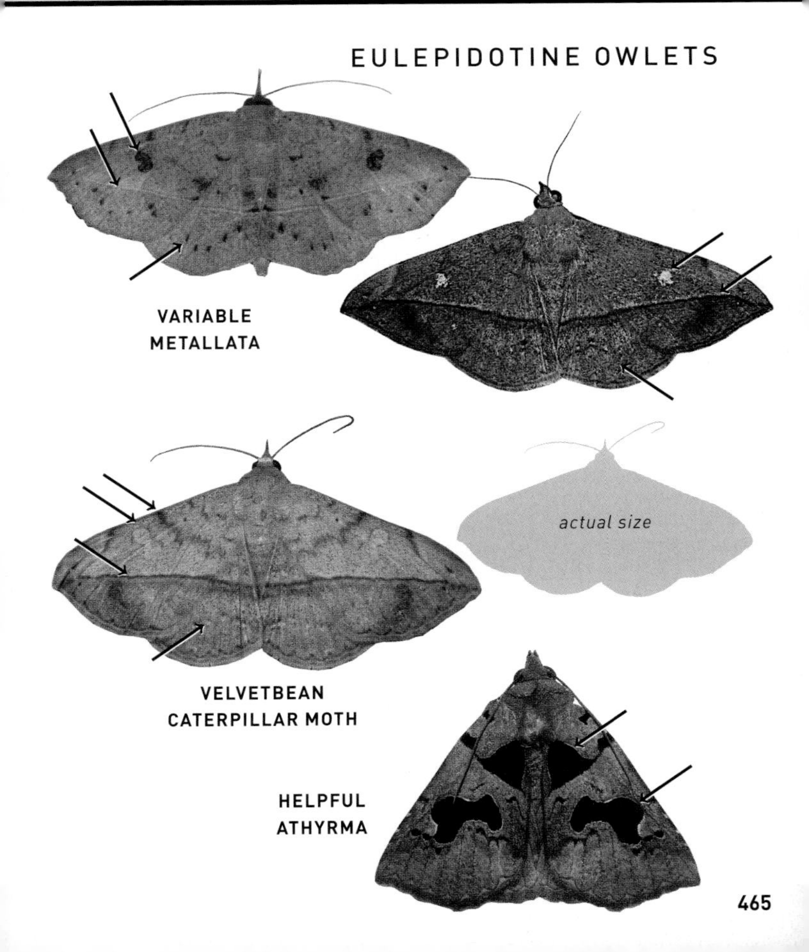

VARIABLE METALLATA

VELVETBEAN CATERPILLAR MOTH

HELPFUL ATHYRMA

CARIBBEAN LEAF MOTH

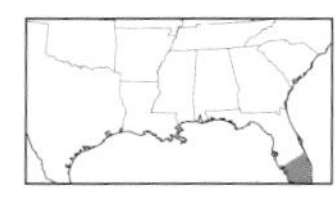

Syllectra erycata 93-1086 (8584) **Local**

WS 20 mm. Reddish brown FW has a pale subapical patch and small spot at midpoint of ST area. Dark brown AM and PM lines kink sharply upward at costa. Outer margin of HW is irregularly scalloped. Rests with spread wings slightly raised. **HOSTS:** Unknown.

RED-LINED PANOPODA

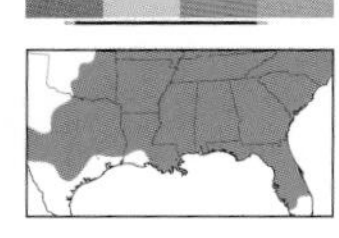

Panopoda rufimargo 93-1089 (8587) **Common**

TL 22–25 mm. Tan to brownish FW has almost parallel reddish brown AM and PM lines, edged yellowish. Reniform spot is typically yellowish but can be black. ST line is small white dots set in diffuse blackish spots. Head and collar are reddish orange to chestnut. **HOSTS:** Beech and oak. **NOTE:** Two or more broods.

BROWN PANOPODA

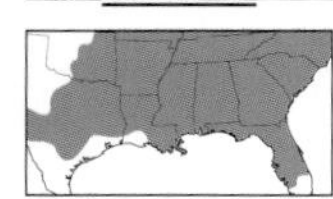

Panopoda carneicosta 93-1090 (8588) **Common**

TL 21–25 mm. Grayish brown FW has slightly wavy brown AM and PM lines, thinly edged yellowish, and a diffuse median band. Black reniform spot is L-shaped; on some individuals, it is reduced to a thin crescent. Orbicular spot is a black dot. Head and collar are brown. **HOSTS:** Hickory. **NOTE:** Two or more broods.

ORANGE PANOPODA

Panopoda repanda 93-1091 (8589) **Common**

TL 21–25 mm. Orange brown FW has dotted AM and PM lines, sometimes faint. Reniform spot is pale and often inconspicuous. Some individuals have a black-edged gray spot near inner margin of median line. **HOSTS:** Live oak. **NOTE:** Two or more broods.

TWO-BARRED ANTIBLEMMA

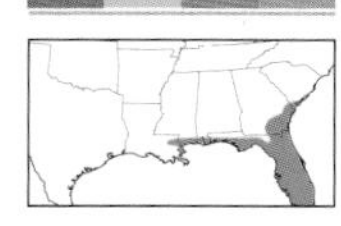

Antiblemma concinnula 93-1095 (8579) **Local**

WS 18–20 mm. Peppery light to dark brown FW usually has white bars along costa at AM and PM lines. Hourglass-shaped reniform spot is indistinct and outlined brown. **HOSTS:** Unknown.

CURVE-LINED OWLET

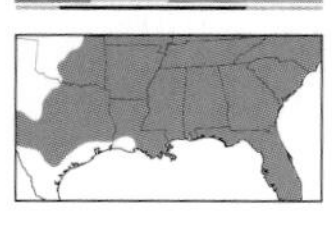

Phyprosopus callitrichoides 93-1101 (8525) **Common**

TL 16–18 mm. Brown FW has a nearly straight, white-edged, brown PM line that runs from apex to midpoint of inner margin. Whitish AM line is often indistinct. Reniform spot contains two small black dots. Pointed labial palps give moth a shrewlike appearance. **HOSTS:** Greenbriar. **NOTE:** Two or more broods.

EULEPIDOTINE OWLETS

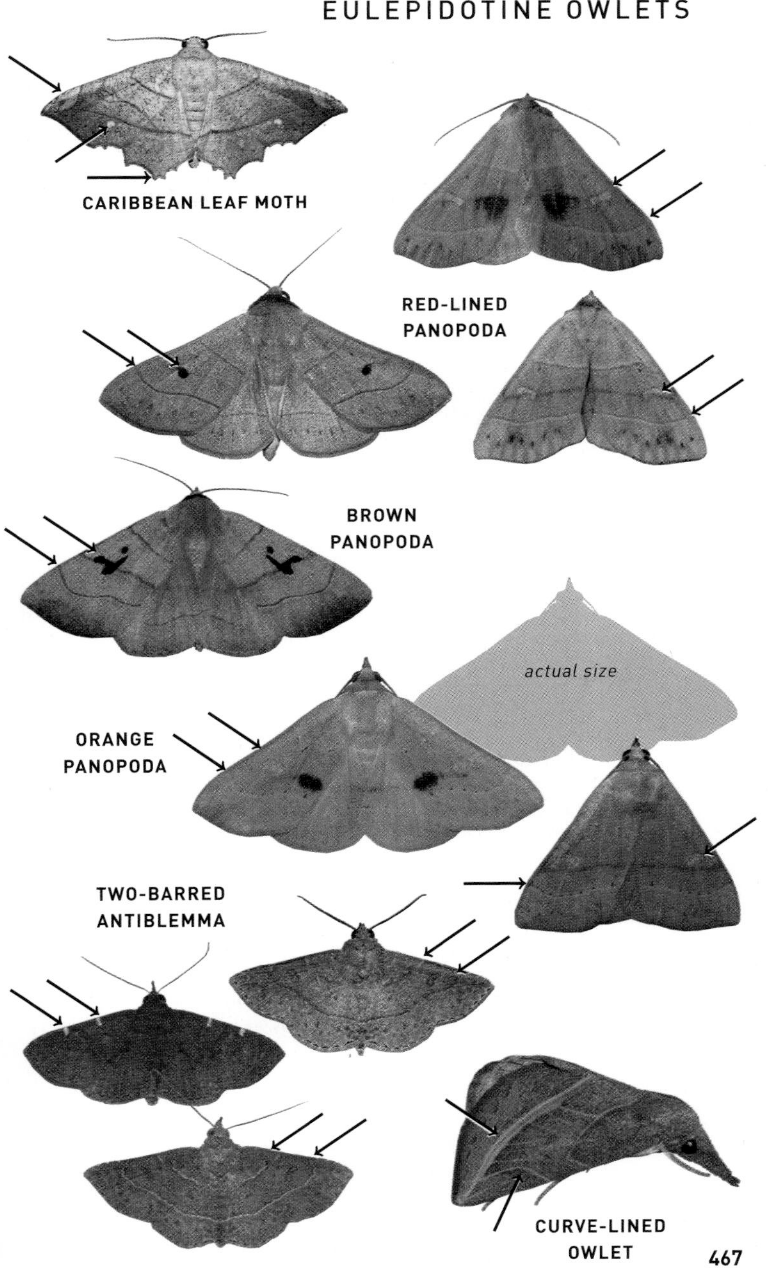

Marathyssas and Paectes

Family Euteliidae, Subfamily Euteliinae

A diverse group of small to medium-sized, often colorful, noctuids. Some, notably the two *Marathyssa* species, have eye-catchingly acrobatic resting positions. The *Paectes* species rest with their wings flat, but raise the tip of their abdomen above the level of the wings. Both groups occur in a variety of habitats, even in highly urban areas. All are nocturnal and will come to lights in small numbers.

DARK MARATHYSSA

Marathyssa inficita 93-1103 (8955) **Common**

TL 14–17 mm. Rests with wings tightly rolled. Gray to brownish FW has broad grayish median area and rusty bands in basal and ST areas. AM and PM lines are blackish, sometimes indistinct. Base of abdomen has a narrow black band bordered grayish. Male has serrate antennae. **HOSTS:** Sumac. **NOTE:** Two or more broods.

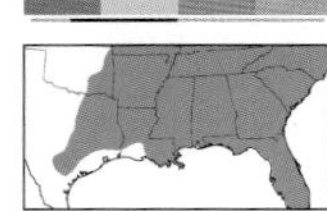

LIGHT MARATHYSSA

Marathyssa basalis 93-1104 (8956) **Common**

TL 15–18 mm. Resembles Dark Marathyssa, but paler FW has a broad straw-colored streak extending from base to apex. Broad, dark AM line is edged grayish. Abdomen has paired thin white lines down center. Male has bipectinate antennae. **HOSTS:** Poison ivy. **NOTE:** One or two broods.

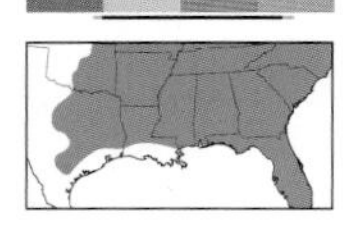

EYED PAECTES *Paectes oculatrix* 93-1106 (8957) **Common**

TL 14–16 mm. Grayish brown FW has an eyelike pattern in ST area bordered by curved black PM line. Straw-colored basal patch is edged with a thick black line. **HOSTS:** Poison ivy and poison sumac. **NOTE:** Two or more broods.

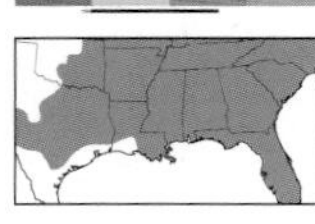

PYGMY PAECTES *Paectes pygmaea* 93-1107 (8959) **Common**

TL 11–12 mm. Grayish brown FW has an oval tan to dark gray basal patch edged by curved double AM line. Smoothly curved double PM line is sharply kinked before reaching costa. Thin, scalloped median line is most noticeable near inner margin. **HOSTS:** Winged sumac. **NOTE:** Two or more broods.

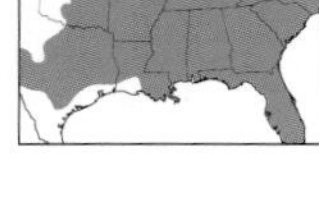

BARRENS PAECTES

Paectes abrostolella 93-1108 (8959.1) **Uncommon**

TL 12–14 mm. Resembles Pygmy Paectes, and some individuals may not be identifiable in the field. FW is paler with pale gray to whitish basal patch. Apex is usually pale gray to whitish. **HOSTS:** Unknown; possibly sumac.

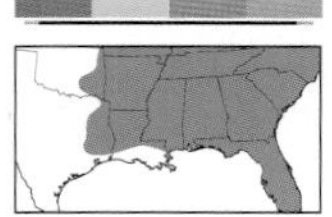

LARGE PAECTES

Paectes abrostoloides 93-1111 (8962) **Common**

TL 15–17 mm. Resembles smaller Pygmy Paectes but usually has a pale band through median area. Tan patch in basal area is reduced to a crescent or entirely absent. Double PM line has a second, small tooth at midpoint. **HOSTS:** Sweet gum. **NOTE:** Two or more broods.

MARATHYSSAS AND PAECTES

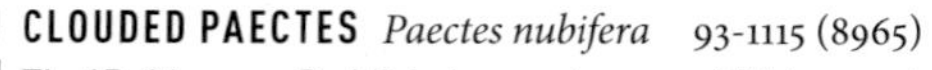

CLOUDED PAECTES *Paectes nubifera* 93-1115 (8965) Local

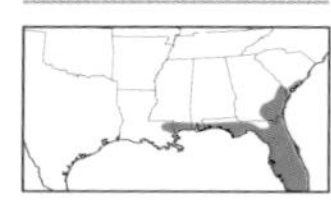

TL 15–16 mm. Reddish brown to gray FW has pale gray band through lower median area. Dark swirls in inner ST area create an indistinct eyespot. Scalloped AM and PM lines are most noticeable at inner margin. **HOSTS:** Live oak. **NOTE:** Multiple broods.

BEAUTIFUL EUTELIA

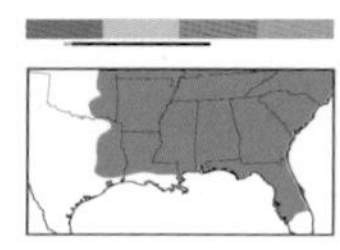

Eutelia pulcherrimus 93-1118 (8968) Uncommon

TL 14–20 mm. FW is held partly rolled and away from body at rest. Broad orange costa meets and continues across orange head and collar. Inner basal area of FW is maroon. A white-edged maroon spot is prominent near inner margin. Tip of orange abdomen has pincerlike tufts in male. **HOSTS:** Poison sumac and poison ivy.

HIEROGLYTHIC MOTH
Family Nolidae, Subfamily Diphtherinae

A distinctive species whose crisp pattern calls to mind Egyptian artwork. Generally common in its mostly coastal range. It will occasionally come to light.

HIEROGLYPHIC MOTH

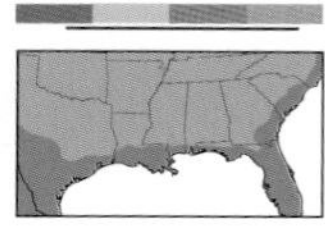

Diphthera festiva 93-1120.1 (8560) Common

TL 20–24 mm. Pale yellow-orange FW has a unique pattern of metallic ink blue lines and dots. Blackish HW has a white fringe. Rests with rounded wings tightly compressed. **HOSTS:** Chocolate weed, pyramid flower, soybean, clover, mimosa, and mallow.

NOLAS Family Nolidae, Subfamily Nolinae

Small delta-shaped moths that are predominantly gray or white with patterns of dotted or broken lines. Many species have raised tufts of hairlike scales on the FW. *Meganola* species are found mostly in oak woodlands, whereas some *Nola* species inhabit more riparian or swampy habitats. They are nocturnal and are attracted to lights in small numbers.

CONFUSED MEGANOLA

Meganola minuscula 93-1121 (8983) Common

TL 8–12 mm. Gray FW has dotted PM line, often double, which bulges around circular, grayish reniform spot. Median area is often slightly shaded. Three tufts of raised scales make FW appear lumpy. **HOSTS:** Oak. **NOTE:** Two broods.

COASTAL PLAIN MEGANOLA

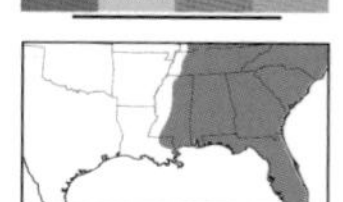

Meganola phylla 93-1122 (8983.1) Common

TL 8–12 mm. Resembles Confused Meganola but has dark gray blotches along costa in basal and median areas. Double PM line is often most obvious near inner margin. AM line is indistinct or nearly absent. Reniform spot is often inconspicuous. **HOSTS:** Beech and oak. **NOTE:** One or two broods.

MARATHYSSAS AND PAECTES

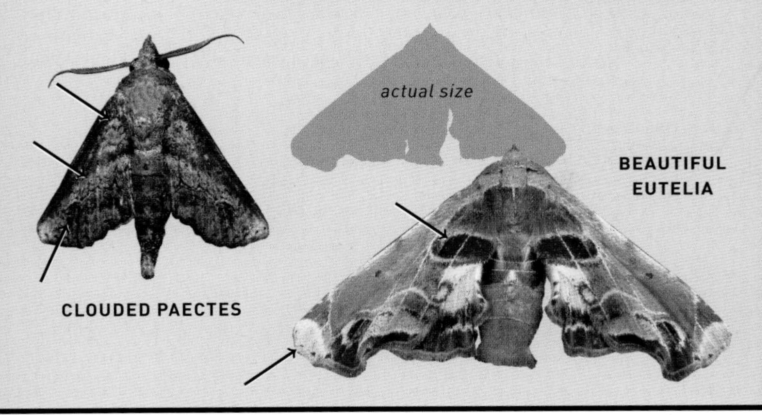

CLOUDED PAECTES

BEAUTIFUL EUTELIA

HIEROGLYTHIC MOTH

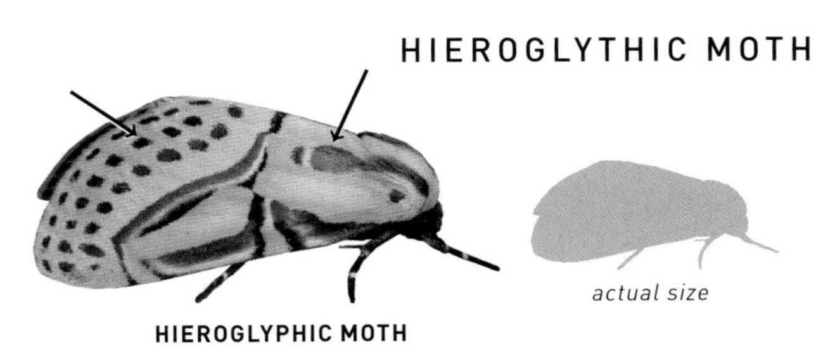

HIEROGLYPHIC MOTH

NOLAS

CONFUSED MEGANOLA

COASTAL PLAIN MEGANOLA

ASHY MEGANOLA

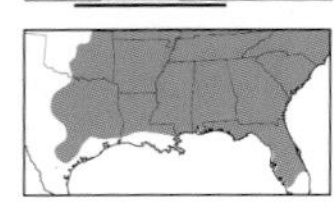

Meganola spodia 93-1123 (8983.2) **Uncommon**

TL 8–13 mm. Resembles Coastal Plain Meganola, but PM line is strongly double and AM line is distinct. Reniform spot is usually noticeable. **HOSTS:** Red and white oaks.

SHARP-BLOTCHED NOLA

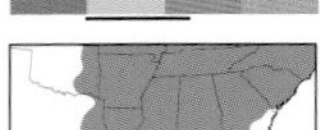

Meganola pustulata 93-1129 (8989) **Uncommon**

TL 8–13 mm. White FW has a sharply demarcated brown median band that widens toward costa. Orbicular and reniform spots are filled with raised tufts of silvery scales. **HOSTS:** Maleberry.

BLURRY-PATCHED NOLA

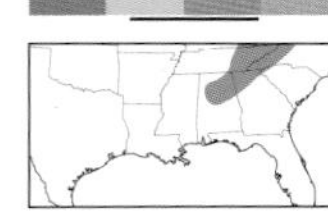

Nola cilicoides 93-1130 (8990) **Common**

TL 9–10 mm. White FW has blurry fawn-colored median band that passes through raised black and silver reniform spot. Two raised tufts of white scales are present in basal and median areas. **HOSTS:** Fringed loosestrife. **NOTE:** Up to three broods.

SORGHUM WEBWORM

Nola cereella 93-1131 (8991) **Common**

TL 7–10 mm. Creamy white FW has fawn median band and PM line. Pale grayish patch in ST area has two blackish wedges. Up to five tufts of scales run along costa. **HOSTS:** Grass and sorghum seedheads. **NOTE:** Multiple broods.

THREE-SPOTTED NOLA

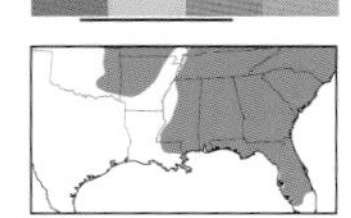

Nola triquetrana 93-1132 (8992) **Common**

TL 9–10 mm. Gray FW has blackish marks along costa at basal, AM, and median lines; mark at AM line is strongly kinked, forming a V-shape. Toothed PM and ST lines are faint or nearly absent. Three tufts of scales are present along costa. **HOSTS:** Witch hazel.

WOOLLY NOLA *Nola ovilla* 93-1135 (8995) **Rare**

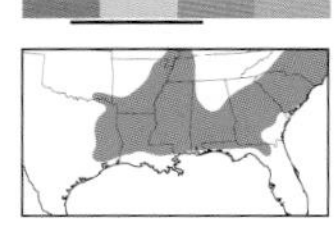

TL 9–10 mm. Peppery gray FW has toothed AM and PM lines that turn sharply at costa; AM line often has small black wedge at angle. Lower median area is often lightly shaded. Terminal line is dotted when fresh. Males have fasciculate antennae. **HOSTS:** Oak.

SWEET PEPPERBUSH NOLA

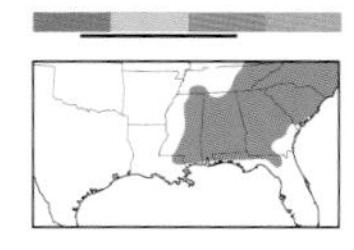

Nola clethrae 93-1136 (8996) **Uncommon**

TL 9–10 mm. Dimorphic. Light individuals resemble Woolly Nola, but males have pectinate antennae; females are probably not safely identifiable in the field. May have on average less well defined PM and ST lines and shading. Some individuals have a darker inner median area with blackish veins. **HOSTS:** Coastal and mountain sweet pepperbush. **NOTE:** Two broods. More common than Woolly Nola.

ASHY MEGANOLA

SHARP-BLOTCHED NOLA

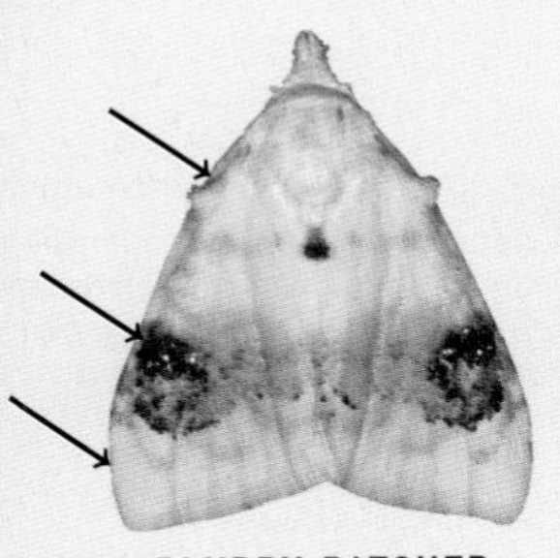

BLURRY-PATCHED NOLA

actual size

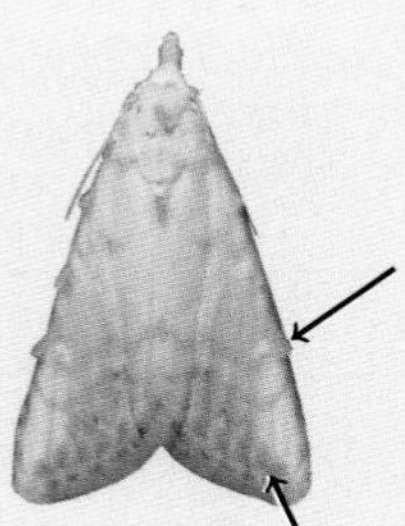

SORGHUM WEBWORM

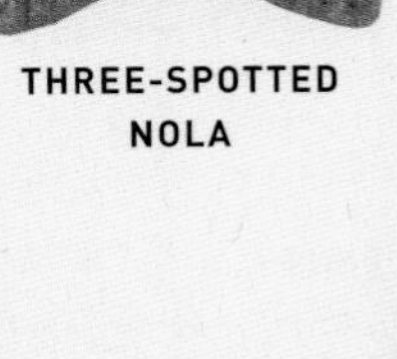

THREE-SPOTTED NOLA

WOOLLY NOLA

SWEET PEPPERBUSH NOLA

Baileyas, Nycteolas, and Allies

Family Nolidae, Subfamilies Chloephorinae, Risobinae, Collomeninae, and Afridinae

Predominately gray, delta-shaped noctuids. The *Baileya* species sometimes curl the tip of their abdomen above the level of the wings. Most species are common in wooded areas and will visit lights in small numbers.

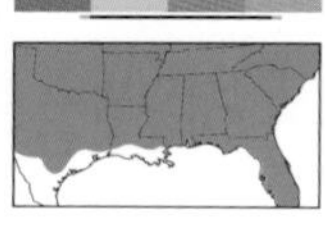

SMALL GARELLA *Garella nilotica* 93-1141 (8974) **Local**

TL 8–9 mm. Easily mistaken for a member of the Tortricidae. Highly variable. Nearly always has a small white spot adjacent to a short, thin black dash near inner margin of median area. FW is usually grayish brown with a pale AM band of variable width. Often has a dark brown basal area and rusty brown in median area, with a faint, broad, dark apical dash. Some individuals have two bold blackish streaks through median area. **HOSTS:** Black almond and willow. **NOTE:** Two broods.

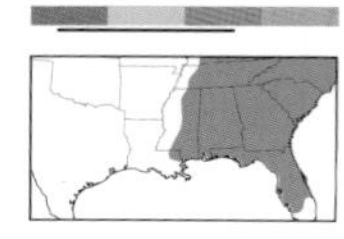

FORGOTTEN FRIGID OWLET

Nycteola metaspilella 93-1145 (8978) **Uncommon**

TL 13–16 mm. Pale gray FW has zigzag double AM line ending as a two-pronged black wedge near inner margin. Outer median area is usually shaded dark gray. Reniform spot is orange or brown. Irregular ST line is a row of blackish dots. Thorax has two black-outlined circles. **HOSTS:** Poplar and willow. **NOTE:** Two broods.

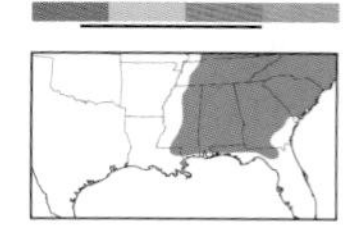

DOUBLEDAY'S BAILEYA

Baileya doubledayi 93-1148 (8969) **Uncommon**

TL 14–17 mm. Gray FW has white PM and ST lines broadly edged with dark brown. Thorax is bordered by bold white basal line strongly edged dark brown. Kidney-shaped black reniform spot is outlined white. Has small blackish wedge near inner margin of ST line. **HOSTS:** Alder. **NOTE:** Two or three broods.

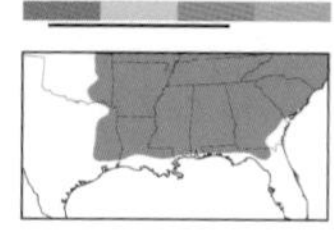

EYED BAILEYA

Baileya ophthalmica 93-1149 (8970) **Common**

TL 13–18 mm. Sexually dimorphic. Peppery grayish brown FW has narrow buff basal area. Outer median area and apex are pale gray. Large reniform spot is outlined blackish with dot in center. Jagged white ST line is edged black, with broad blackish apical patch at costa. Male has buffy thorax; female's is grayish. **HOSTS:** Ironwood, eastern hornbeam, and hazel. **NOTE:** Two broods.

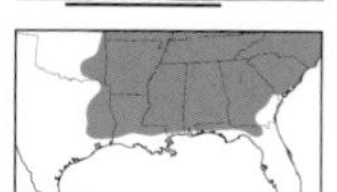

SLEEPING BAILEYA

Baileya dormitans 93-1150 (8971) **Uncommon**

TL 14–18 mm. Resembles Eyed Baileya, but reniform spot is a tiny black dot and PM line is faint or absent. ST line is weakly defined at midpoint. Basal area and thorax are usually grayish. **HOSTS:** Hickory and walnut. **NOTE:** Two broods.

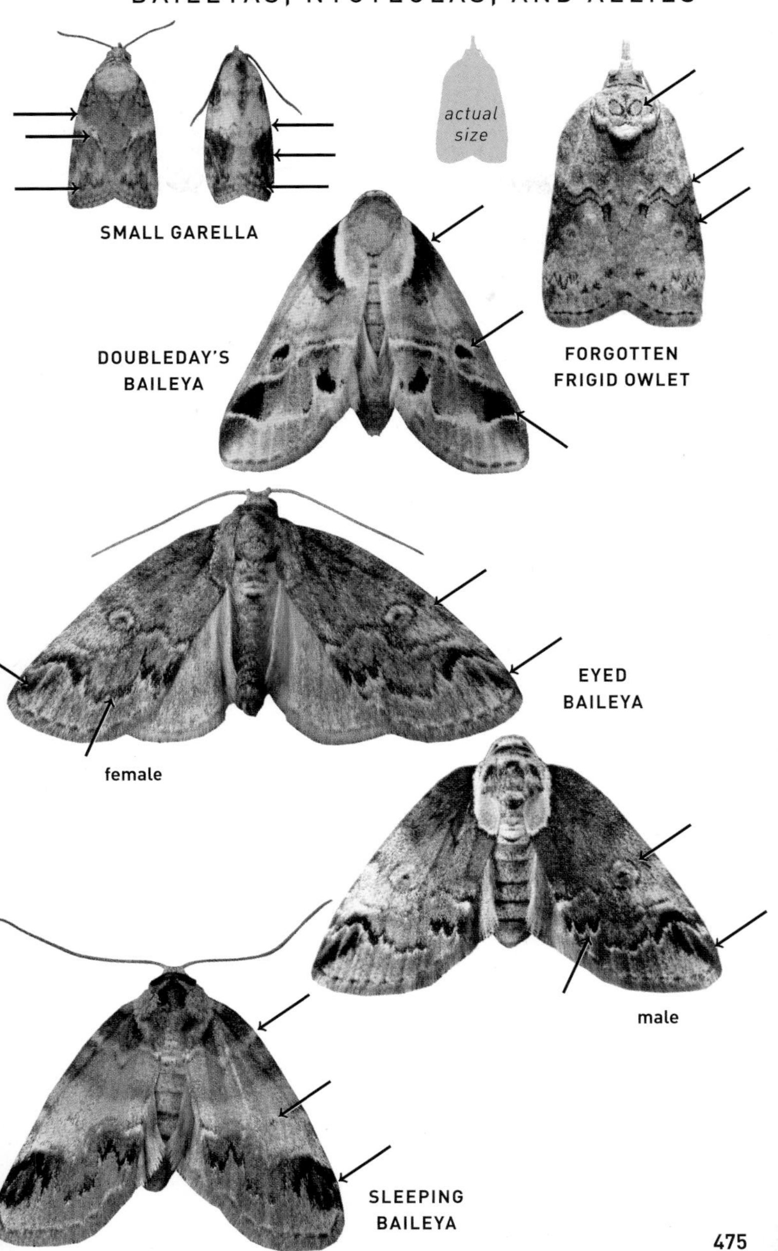
actual
size
SMALL GARELLA
DOUBLEDAY'S
BAILEYA
FORGOTTEN
FRIGID OWLET
EYED
BAILEYA
female
male
SLEEPING
BAILEYA

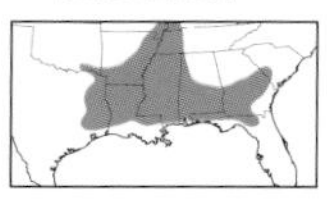

ACADIAN BAILEYA

Baileya acadiana 93-1151 (8971.1) **Common**

TL 14–18 mm. Similar to Sleeping Baileya but has wavy black PM line. ST line is well defined. **HOSTS:** Unknown. **NOTE:** Three broods.

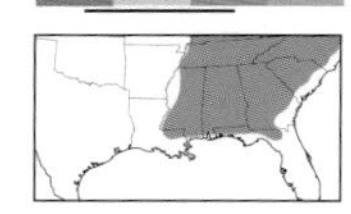

PALE BAILEYA

Baileya levitans 93-1152 (8972) **Uncommon**

TL 14–18 mm. Resembles Eyed Baileya, but basal area is gray. Reniform spot is usually joined to costa by broad dark bar of outer PM line. **HOSTS:** Hickory and walnut. **NOTE:** Two broods.

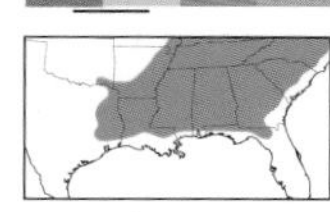

DARK BAILEYA

Baileya ellessyoo 93-1153 (8972.1) **Uncommon**

TL 15–16 mm. Resembles darker individuals of Pale Baileya, but reniform spot is a single black spot or a small black-outlined circle without dot in center. Median area is usually only slightly paler. **HOSTS:** Unknown.

SMALL BAILEYA

Baileya australis 93-1154 (8973) **Common**

TL 12–16 mm. Resembles Eyed Baileya, but basal area is tan, edged buff. Reniform spot is often indistinct. PM line is noticeably edged white at inner margin. **HOSTS:** Black walnut. **NOTE:** Three broods.

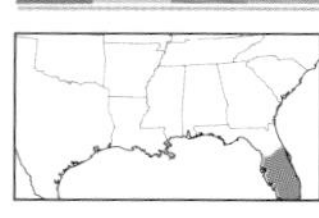

NORTHERN MOTYA

Motya abseuzalis 93-1155 (8981) **Local**

TL 13–15 mm. Grayish brown FW is weakly patterned with fragmented jagged black lines. ST line is a row of black spots, bolder near inner margin. Thorax has tufts of raised black scales. Typically rests with blunt-tipped abdomen raised erectly above head. **HOSTS:** Button mangrove.

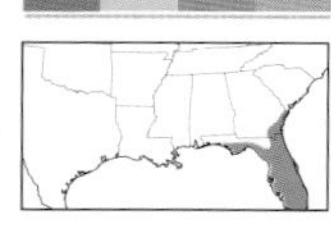

NORTHERN COLLOMENA

Collomena filifera 93-1156 (8982) **Local**

TL 14–16 mm. Pearly gray FW is sharply marked with jagged black basal, AM, and PM lines. Reniform spot is crisply outlined black. PM line is bordered below by diffuse whitish band. Thorax has a double black collar behind head. Blunt-tipped abdomen has paired black spots near end; typically rests with it raised above wings. **HOSTS:** Unknown.

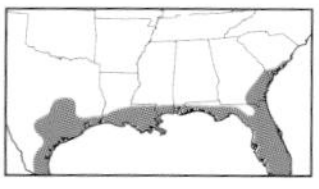

DYAR'S LICHEN MOTH

Afrida ydatodes 93-1158 (8102) **Uncommon**

TL 6–7 mm. Peppery mint green FW has a brownish gray median band and ST area. Jagged black basal, AM, and PM lines are sharply defined. Antennae are bipectinate. Typically rests with head curled under, giving it a pug-faced look. **HOSTS:** Lichens.

BAILEYAS, NYCTEOLAS, AND ALLIES

PALM BORERS Family Noctuidae, Subfamily Dyopsinae

A group containing a single genus of brown to tan moths bearing eyespots on the HW. This species may be a minor household pest, as larvae build their cocoons using natural fibers. Adults will come to lights in small numbers.

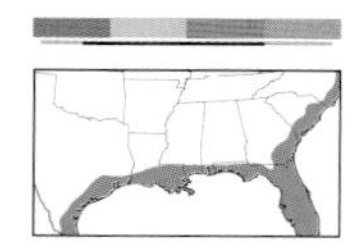

PALMETTO BORER

Litoprosopus futilis 93-1160.1 (8556) **Local**

TL 22–25 mm. Grayish brown or tawny FW has fragmented double AM and PM lines. Orbicular spot is a black spot. Hollow reniform spot is incompletely outlined black. Brown HW has a black and metallic blue, pale-edged eyespot near anal angle. **HOSTS:** Saw palmetto, cabbage palmetto, and related palms.

LOOPERS Family Noctuidae, Subfamily Plusiinae

A distinctive group of sleek-looking noctuids that typically rest with wings tented over the abdomen. Many display metallic sheens and silvery stigmas on the FW. Several species also have distinctive thoracic crests and tufts of scales at the anal angle of the FW, creating a flared appearance. Some are migratory, appearing well north of their normal ranges late in the fall, often during drought years or periods of favorable winds. The larvae of some widespread species can be serious pests on commercial crops. A few species are active during daylight hours, notably Celery Looper, though most are nocturnal and will usually be encountered at lights.

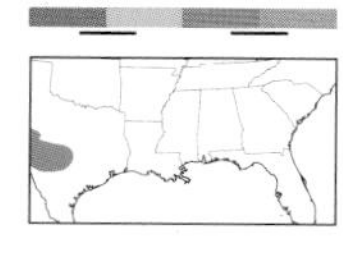

MINUTE OVAL ABROSTOLA

Abrostola microvalis 93-1164 (8883) **Local**

TL 11–13 mm. Brownish gray FW has a paler basal patch edged warm brown along double, curved AM line. Large pale gray or brownish claviform and orbicular spots are fused (or almost so) into an hourglass shape. Thorax has a pair of black-outlined circles on collar. **HOSTS:** Possibly live oak. **NOTE:** Adults may overwinter.

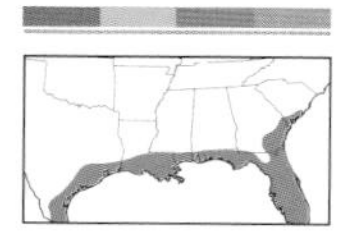

MOCK ABROSTOLA

Mouralia tinctoides 93-1165 (8884) **Uncommon**

TL 17–20 mm. Resembles Minute Oval Abrostola, but basal area is gray and AM and PM lines are single. Claviform, orbicular, and reniform spots are not noticeably paler. Thorax has a long, forward-pointing crest. **HOSTS:** Spiderwort.

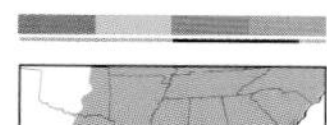

GOLDEN LOOPER

Argyrogramma verruca 93-1166 (8885) **Common**

TL 16–19 mm. Golden brown FW is peppered with rusty brown scales. Toothed AM and PM lines are edged with bands of lilac shading. Golden stigma has a hollow inner part and a smaller, solid outer spot. Stigma of females is FW color. **HOSTS:** Low plants, including arrowhead, curled dock, and tobacco. **NOTE:** Multiple broods.

PALM BORERS
PALMETTO
BORER
actual size

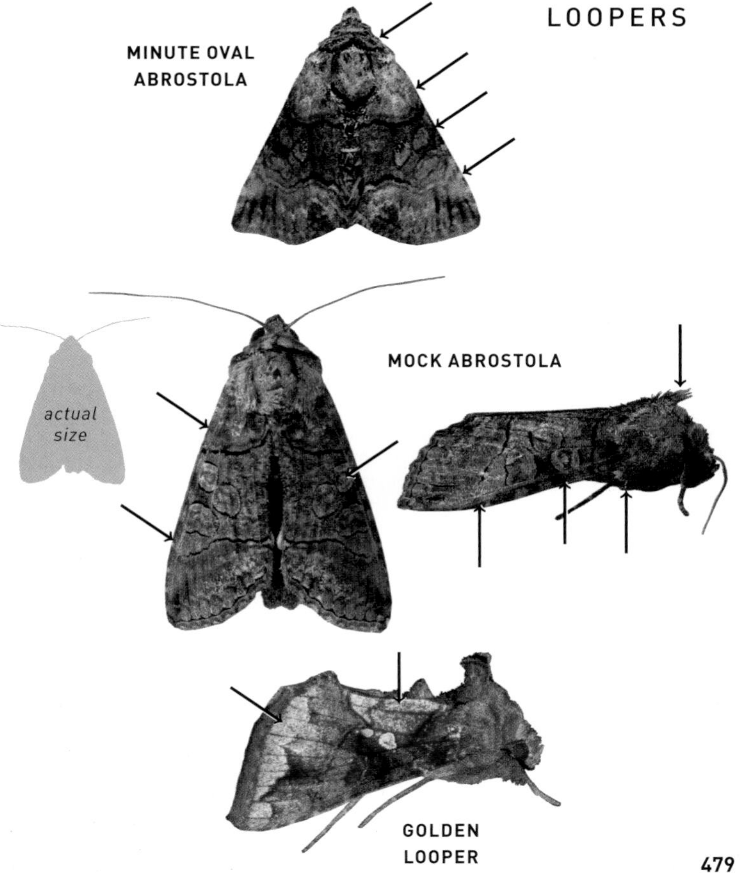
LOOPERS
MINUTE OVAL
ABROSTOLA
actual
size
MOCK ABROSTOLA
GOLDEN
LOOPER

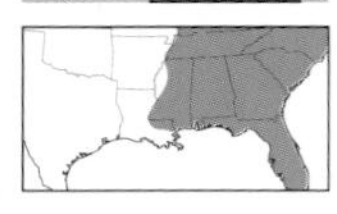

PINK-WASHED LOOPER

Enigmogramma basigera 93-1167 (8886) **Uncommon**

TL 16–19 mm. Pinkish brown FW has a contrasting patch of warm brown shading in inner median area, forming a "saddle." Silvery stigma has an inverted U-shaped inner part and a solid outer spot. **HOSTS:** Umbellate water pennywort. **NOTE:** Multiple broods.

NI MOTH *Trichoplusia ni* 93-1168 (8887) **Common**

TL 17–19 mm. Mottled grayish brown FW is peppered with whitish scales along indistinct jagged lines. Silvery stigma has an inverted U-shaped inner part and a solid outer spot; on some individuals, the two spots are joined. **HOSTS:** Low plants and crops, including asparagus, cabbage, corn, tobacco, and watermelon. **NOTE:** Three or more broods. Also known as Cabbage Looper.

SHARP-STIGMA LOOPER

Ctenoplusia oxygramma 93-1169 (8889) **Common**

TL 19–22 mm. Gray FW is peppered with golden scales along inconspicuous jagged lines. Blackish median patch is pierced by an oblique, gently pointed, silver-edged stigma. **HOSTS:** Aster, goldenrod, horseweed, and tobacco. **NOTE:** Multiple broods in south.

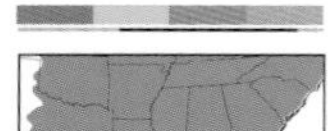

SOYBEAN LOOPER

Chrysodeixis includens 93-1170 (8890) **Common**

TL 16–20 mm. Gray-brown FW has a distinctly bronzy sheen, especially at inner median area. Broad blackish patch runs through middle of FW behind stigma and PM line. Silvery stigma has an inverted U-shaped inner part and a solid outer spot. **HOSTS:** Low plants and crops, including goldenrod, lettuce, soybean, and tobacco. **NOTE:** Multiple broods.

GRAY LOOPER

Rachiplusia ou 93-1176 (8895) **Common**

TL 17–22 mm. Ash gray FW has brown shading along indistinct AM and PM lines. Median area is often shaded darker behind stigma. Silvery stigma has an inverted V-shaped inner part that extends to costa, and a solid outer spot, often joined. **HOSTS:** Low plants, including clover, mint, corn, and cosmos. **NOTE:** Multiple broods in south.

UNSPOTTED LOOPER

Allagrapha aerea 93-1177 (8898) **Local**

TL 16–22 mm. Pinkish brown FW has diffuse bands of orange brown shading along wavy AM and PM lines. Inconspicuous reniform spot is outlined brown. Thorax and head are orange. **HOSTS:** Low plants, including aster, dandelion, and stinging nettle. **NOTE:** Three or more broods in south.

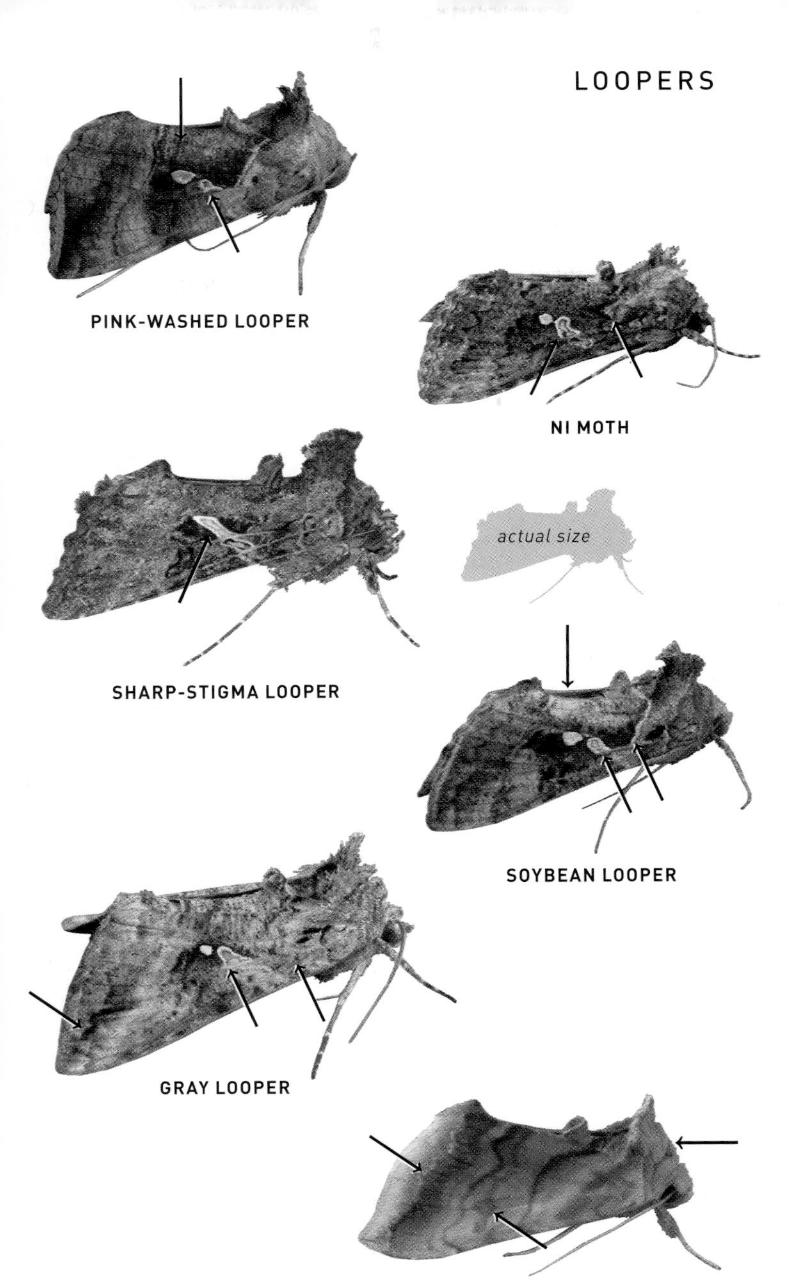

PINK-WASHED LOOPER

NI MOTH

SHARP-STIGMA LOOPER

SOYBEAN LOOPER

GRAY LOOPER

UNSPOTTED LOOPER

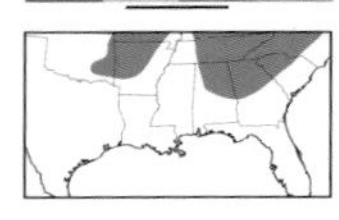

FORMOSA LOOPER

Chrysanympha formosa 93-1186 (8904) **Local**

TL 16–19 mm. FW has a bold, pale pinkish basal patch that extends along costa into warm brown median area. ST area is banded with evenly curving pale pinkish, brown, and black bands. **HOSTS:** Blueberry and dwarf huckleberry.

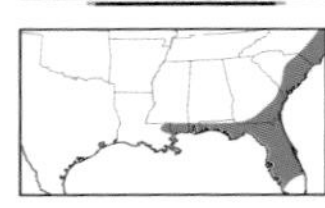

PITCHERPLANT-MINING LOOPER

Exyra semicrocea 93-1189 (9024) **Local**

TL 16–19 mm. Basal half of FW is pale yellow; distal half is grayish brown. Front of thorax and head are blackish. **HOSTS:** Hooded pitcher plant and other pitcher plants. **NOTE:** Two or more broods.

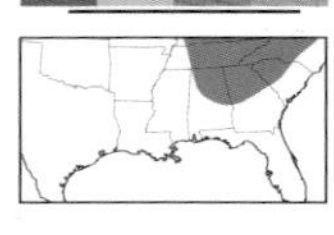

COMMON LOOPER

Autographa precationis 93-1191 (8908) **Common**

TL 18–20 mm. Grayish brown FW has a bronze and lilac sheen. Inner median area is often shaded darker, especially behind stigma. Inner part of stigma is a slanting inverted V-shape and outer part is a solid circle; typically joined, but sometimes separated. Thorax has a thin reddish collar immediately behind head. **HOSTS:** Low plants, including bean, cabbage, dandelion, and plantain. **NOTE:** Three or more broods.

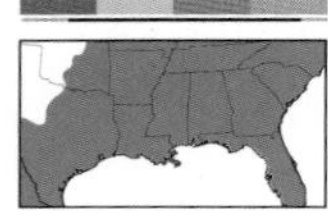

BILOBED LOOPER

Megalographa biloba 93-1209 (8907) **Common**

TL 19–20 mm. Brown FW has bronze shading in median area. Large, satin white bilobed stigma covers most of central median area. Reniform spot is partly outlined satin white. **HOSTS:** Low plants, including alfalfa, cabbage, and tobacco. **NOTE:** Two or more broods.

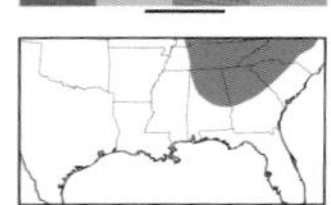

SALT-AND-PEPPER LOOPER

Syngrapha rectangula 93-1227 (8942) **Local**

TL 17–19 mm. Black FW has boldly contrasting white AM and PM lines, and ST area, creating a pied look. Crisp white stigma extends from AM line at costa and splits into two rounded lobes; sometimes one lobe is broken into a separate white dot. Thorax occasionally has a large rusty patch. **HOSTS:** Coniferous trees, including fir, hemlock, pine, and spruce.

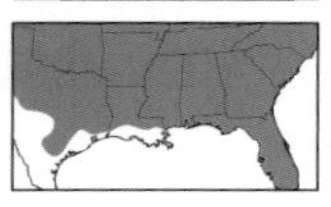

CELERY LOOPER

Anagrapha falcifera 93-1234 (8924) **Common**

TL 18–22 mm. Light brown FW has a darker brown wash across inner median area. Slanting silvery white stigma has a notch at costal end; one side touches whitish AM line. **HOSTS:** Low plants, including beet, celery, clover, corn, and dandelion. **NOTE:** Three or more broods.

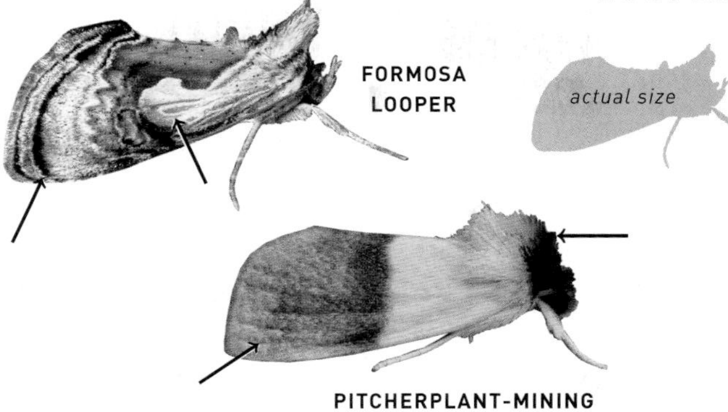

FORMOSA LOOPER

PITCHERPLANT-MINING LOOPER

COMMON LOOPER

BILOBED LOOPER

SALT-AND-PEPPER LOOPER

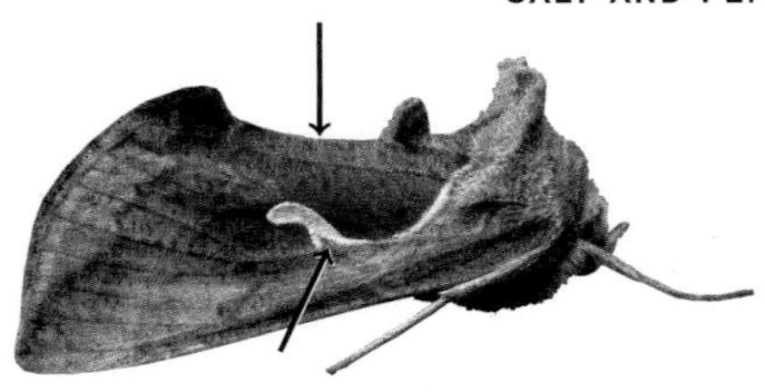

CELERY LOOPER

GLYPHS Family Noctuidae, Subfamilies Bagisarinae, Cydosiinae, and Eustrotiinae

Often colorful, small to medium-sized noctuids, many with cryptic lichenlike markings. All are nocturnal and will come to lights. A few, such as Bog Deltote, can be flushed from vegetation during daytime.

WAVY-LINED MALLOW MOTH

Bagisara repanda 93-1240 (9168) **Common**

TL 12–14 mm. Light brown FW has parallel, pale AM, PM, and ST lines that are thinly edged brownish; all curve upward at costa. Reniform spot has two smudgy dark spots. Outer margin is slightly angulate. **HOSTS:** Sida and false mallow.

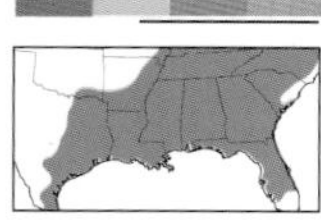

STRAIGHT-LINED MALLOW MOTH

Bagisara rectifascia 93-1241 (9169) **Uncommon**

TL 14–16 mm. Similar to Wavy-lined Mallow Moth, but white lines on FW lack any noticeable brown edging. Reniform spot is only slightly darker than rest of wing. **HOSTS:** Hibiscus and Turk's-cap.

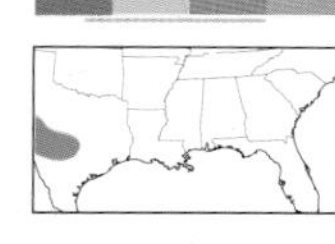

PALE YELLOW MALLOW MOTH

Bagisara buxea 93-1244 (9172) **Local**

TL 12–14 mm. The inverse of Straight-lined Mallow Moth: FW is light yellowish brown with brown lines. Reniform spot is a thin crescent line. **HOSTS:** Globe mallow.

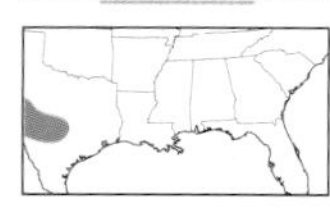

RAPID MALLOW MOTH

Bagisara praecelsa 93-1248 (9175.1) **Local**

TL 14–17 mm. Shiny light brown FW has bronzy shading in ST area. Whitish ST line is sharply pointed near apex. Soft, whitish AM and PM lines curve upward at costa. **HOSTS:** Unknown.

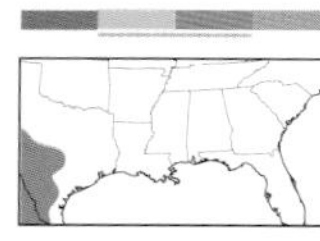

SAD MALLOW MOTH *Bagisara tristicta* 93-1250 (9176) **Local**

TL 13–15 mm. Resembles Pale Yellow Mallow Moth, but peppery FW has two rusty streaks running from pale three-parted reniform spot, through rusty PM and ST lines, to outer margin. Terminal line is a row of black dots. **HOSTS:** Unknown.

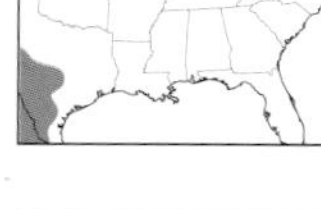

HOOK-TIPPED AMYNA

Amyna bullula 93-1252 (9069) **Common**

TL 12–14 mm. Reddish brown FW has indistinct lines that are lightly speckled white along costa. Reniform spot is figure eight–shaped and partly outlined white, rarely filled. Fringe is whitish. Angulate outer margin is slightly hook-tipped. **HOSTS:** Unknown.

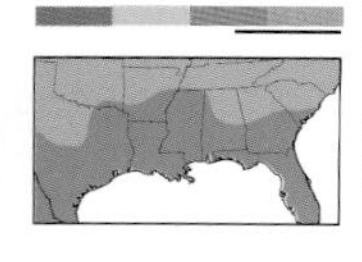

EIGHT-SPOT *Amyna stricta* 93-1253 (9070) **Common**

TL 12–14 mm. Resembles Hook-tipped Amyna, but FW is brown and inner half of reniform spot is white or pale orange. Pale apical patch is reduced or nearly absent. **HOSTS:** Amaranth and croton.

WAVY-LINED MALLOW MOTH
STRAIGHT-LINED MALLOW MOTH
RAPID MALLOW MOTH
PALE YELLOW MALLOW MOTH
actual size
SAD MALLOW MOTH
HOOK-TIPPED AMYNA
EIGHT-SPOT

STRAIGHT-LINED CYDOSIA

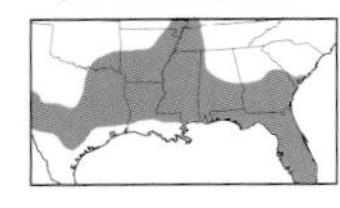

Cydosia aurivitta 93-1254 (8999) **Local**

TL 11–13 mm. Sexually dimorphic. Narrow bluish black FW is boldly marked with wide reddish orange AM and PM bands and spot in central median area. Female has white spots in black areas; male does not. **HOSTS:** Unknown.

CURVE-LINED CYDOSIA

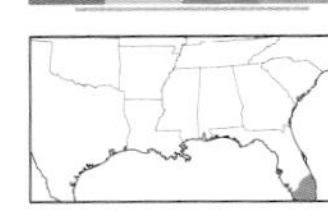

Cydosia nobilitella 93-1255 (9000) **Local**

TL 11–13 mm. Similar to female Straight-lined Cydosia, but white spots on FW are larger and crowd the narrow red lines. Red spot in median area is triangular. HW is mostly white in male, black in female. **HOSTS:** West Indian pinkroot.

HARP-WINGED TRIPUDIA

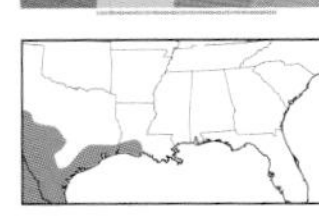

Tripudia quadrifera 93-1260 (9003) **Common**

TL 7–9 mm. Light grayish brown FW has a bold, dark brown saddle in inner median area; pale reniform spot creates concave outer edge. Wavy AM and PM lines are indistinct. ST area is often shaded dark brown, especially toward costa. **HOSTS:** Unknown.

RECTANGULAR TRIPUDIA

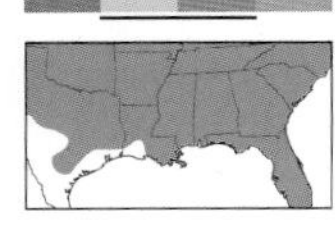

Tripudia rectangula 93-1261 (9003.1) **Uncommon**

TL 7–9 mm. Similar to Harp-winged Tripudia, but outer edge of blackish saddle is nearly straight. FW is darker overall, and dark brown shading in ST area is more uniform. Terminal line is edged warm brown. **HOSTS:** Seeds of wild petunia. **NOTE:** Multiple broods.

FLORIDA TRIPUDIA

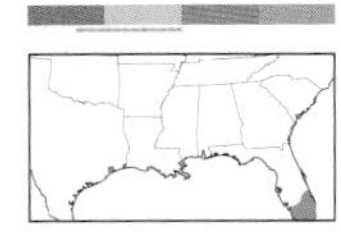

Tripudia grapholithoides 93-1262 (9004) **Local**

TL 7–9 mm. Strongly resembles Rectangular Tripudia, but FW is generally darker; reniform spot blends in to dark median patch, and terminal line is edged darkish. **HOSTS:** Unknown.

BELTED TRIPUDIA

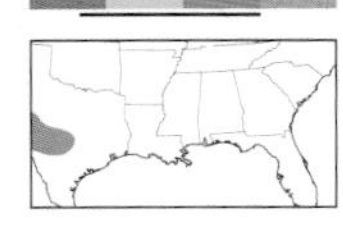

Tripudia balteata 93-1264 (9005) **Local**

TL 6–8 mm. Median area of FW is tan to yellow brown in upper half and chestnut brown in lower half. Basal area is blackish speckled with white. ST area is mottled black and chestnut brown. **HOSTS:** Unknown.

LUXURIOUS TRIPUDIA

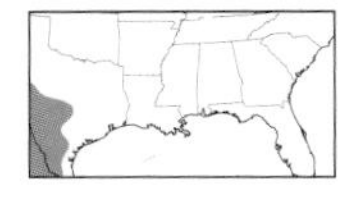

Tripudia luxuriosa 93-1267 (9008) **Local**

TL 7–8 mm. Brown FW has a pale AM band that passes above whitish orbicular spot. Basal area is often grayish; distal half of wing is mottled brown and black. Long, whitish fringe has blackish patch near midpoint. **HOSTS:** Unknown.

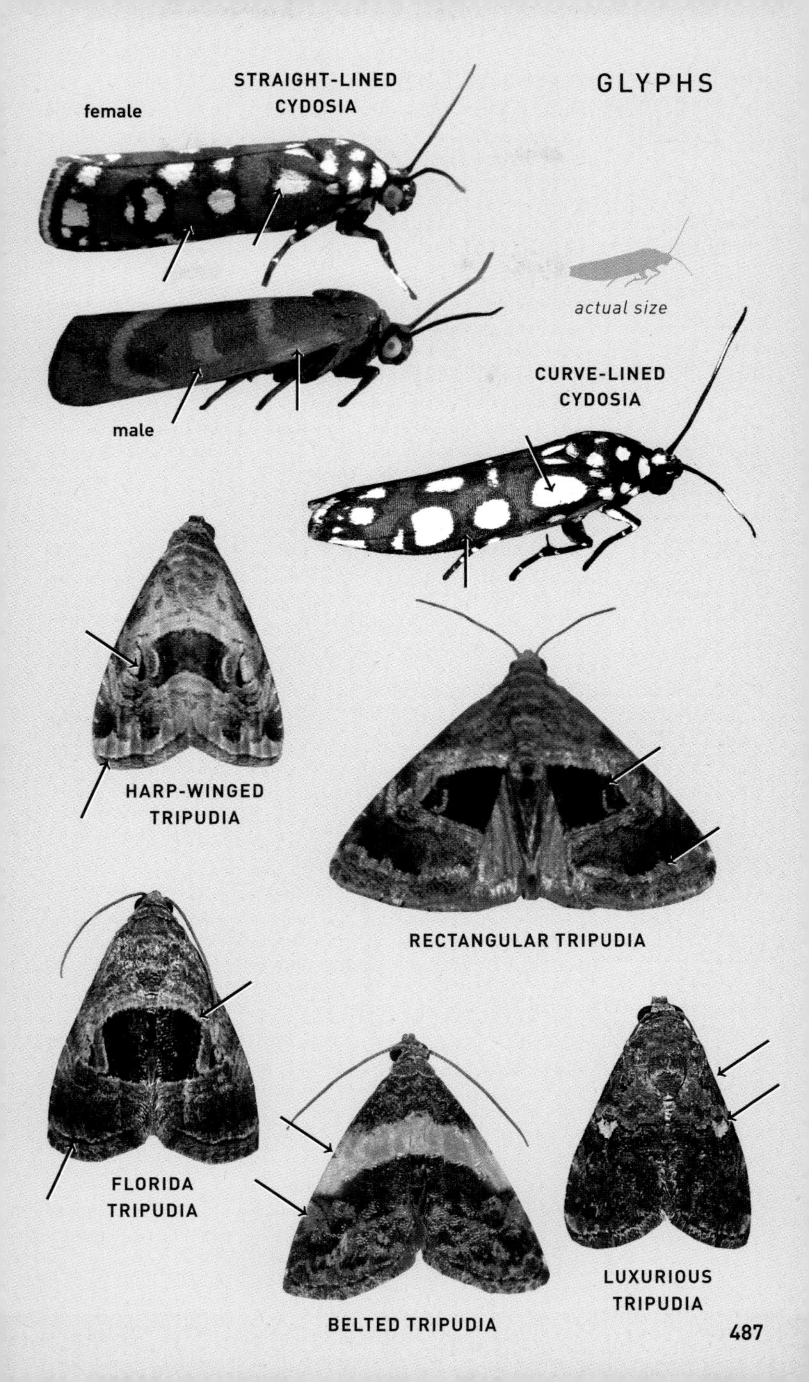
STRAIGHT-LINED CYDOSIA
female
male
actual size
CURVE-LINED CYDOSIA
HARP-WINGED TRIPUDIA
RECTANGULAR TRIPUDIA
FLORIDA TRIPUDIA
BELTED TRIPUDIA
LUXURIOUS TRIPUDIA

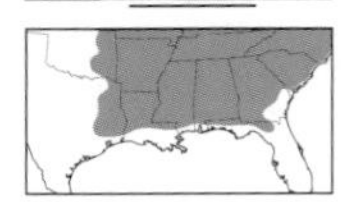

ORANGE-BANDED TRIPUDIA

Tripudia flavofasciata 93-1268 (9009) **Common**

TL 7–9 mm. Resembles Belted Tripudia, but median area is bright orange and red. Basal area has minimal speckling. ST area is mottled black and reddish orange. **HOSTS:** Unknown.

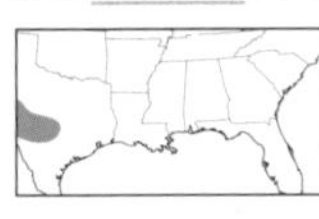

DARK-BANDED COBUBATHA

Cobubatha lixiva 93-1275 (9014) **Local**

TL 7–8 mm. Bicolored FW has a gray base and pale orange brown distal half, separated by a wide, white-edged, blackish AM band. Costa has a blackish patch at ST line. **HOSTS:** Unknown.

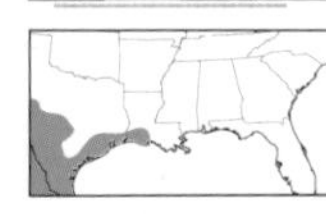

SHARP-BANDED COBUBATHA

Cobubatha orthozona 93-1276 (9017) **Local**

TL 9–11 mm. Resembles slightly smaller Dark-banded Cobubatha, but narrow blackish median band has a pointed tooth jutting into light gray basal area. **HOSTS:** Unknown.

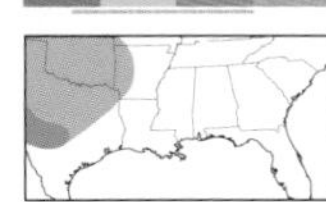

DIVIDED COBUBATHA

Cobubatha dividua 93-1277 (9018) **Local**

TL 7–9 mm. Bicolored FW has a brown base and pale tan distal half. Straight, black AM line is usually edged basally with an irregular band of dusky shading. Costa has a blackish patch at ST line. **HOSTS:** Shrimp plant.

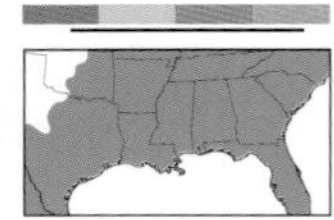

BLACK-BORDERED LEMON

Marimatha nigrofimbria 93-1284 (9044) **Common**

TL 10–12 mm. Lemon yellow FW has tiny blackish claviform and reniform dots. Terminal line is black. Fringe is purplish gray. Rarely has faint orangish, curving PM line. **HOSTS:** Grass, including crabgrass. **NOTE:** Two broods.

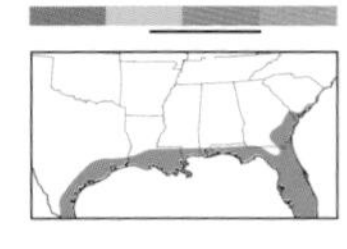

THREE-SPOTTED LEMON

Marimatha tripuncta 93-1287 (9045) **Local**

TL 10–12 mm. Resembles Black-bordered Lemon, but FW has an additional black orbicular dot. Faint brown PM line is strongly angulate. Fringe is dark brown. **HOSTS:** Crabgrass and morning glory.

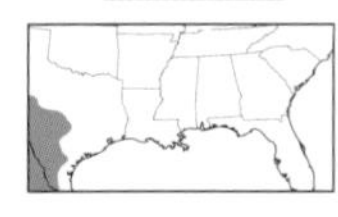

BROWN-LINED LEMON MOTH

Marimatha piscimala 93-1288 (9045.1) **Local**

TL 8–9 mm. Resembles Three-spotted Lemon but is slightly smaller, and diffuse brown PM line is distinct. **HOSTS:** Unknown.

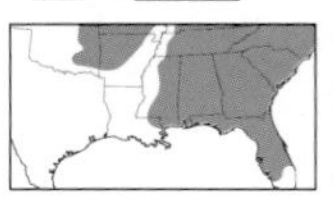

BOG DELTOTE *Deltote bellicula* 93-1289 (9046) **Local**

TL 10–12 mm. Violet gray to brownish FW has a distinctive white M formed by PM line when wings are folded at rest. Median and ST areas are rusty brown. Orbicular and reniform spots are partly edged white and connected with a pale orange bar. **HOSTS:** Probably sedge.

ORANGE-BANDED TRIPUDIA

DARK-BANDED COBUBATHA

SHARP-BANDED COBUBATHA

DIVIDED COBUBATHA

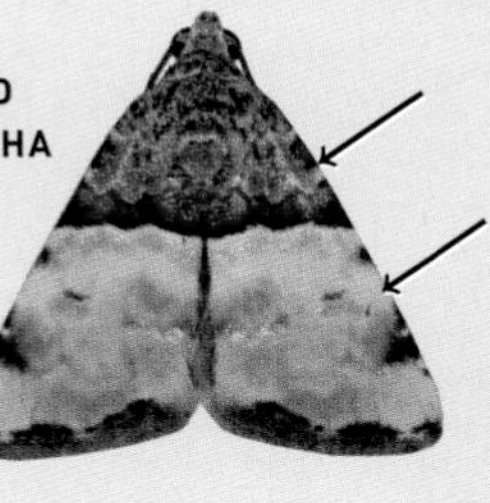

actual size

BLACK-BORDERED LEMON

THREE-SPOTTED LEMON

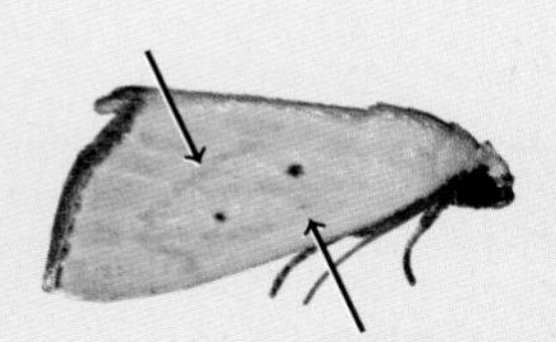

BROWN-LINED LEMON MOTH

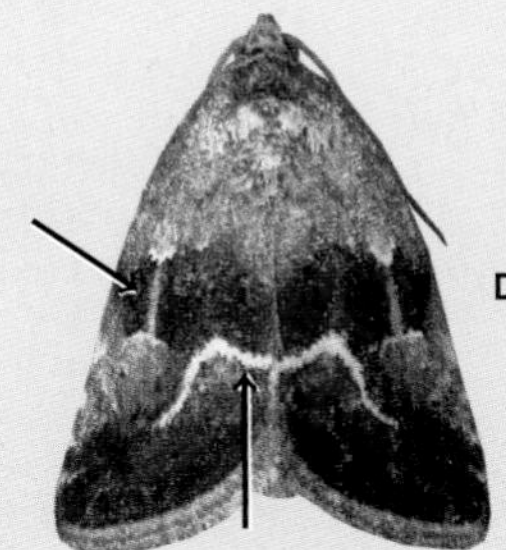

BOG DELTOTE

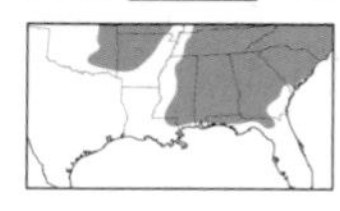

LARGE MOSSY GLYPH

Protodeltote muscosula 93-1290 (9047) **Common**

TL 12–14 mm. Mottled, mossy olive green FW has a narrow white patch where double PM line meets inner margin. Reniform, claviform, and orbicular spots are edged (sometimes partially filled) whitish. **HOSTS:** Grass, including sawgrass. **NOTE:** Two broods.

SMALL MOSSY GLYPH

Lithacodia musta 93-1292 (9051) **Common**

TL 8–10 mm. Green FW has narrow, double black AM and PM lines filled with pale orange. Inner PM line is broadly edged white. Black claviform spot is partly outlined white. Black orbicular spot is outlined orange. Large white-edged reniform spot is partly filled orange. Green color often fades with wear. **HOSTS:** Unknown.

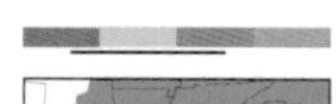

BLACK-DOTTED GLYPH

Maliattha synochitis 93-1295 (9049) **Common**

TL 9–11 mm. White FW has triangular, olive green saddle in inner median area. Basal and ST areas are shaded green and pink. Black apical dashes often blend into ST shading. Orbicular spot is sharply black. White-outlined reniform spot is gray. **HOSTS:** Grass, including crabgrass.

RED-SPOTTED GLYPH

Maliattha concinnimacula 93-1296 (9050) **Uncommon**

TL 10–12 mm. Peppery mint green FW has bold, white-edged red claviform and reniform spots and black orbicular spot. White AM and PM lines are often edged black. ST line ends at black subapical patch. **HOSTS:** Unknown.

FORKED HALTER MOTH

Argillophora furcilla 93-1299 (9060) **Uncommon**

TL 12–13 mm. Light brown FW has a bent white line running from base to outer ST area, backed by diffuse brown shading that extends down center of FW to apex. **HOSTS:** Cane.

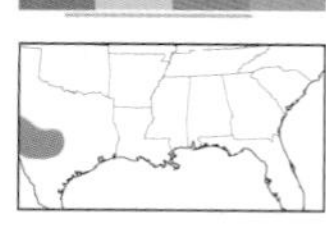

ETCHED CERATHOSIA

Cerathosia tricolor 93-1301 (9064) **Local**

TL 15 mm. Narrow white FW is crisply patterned with incomplete black lines and outlines to spots. Dotted ST line ends at a black smudge along costa. HW and abdomen are pale yellow. **HOSTS:** Unknown.

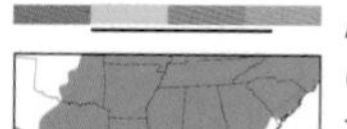

AERIAL BROWN

Ozarba aeria 93-1302 (9030) **Common**

TL 10–12 mm. Light grayish to brown FW has broad blackish patch along distal half of costa, interrupted by pale reniform spot and thin ST line. Double AM and PM lines are often inconspicuous. ST area is pale. **HOSTS:** Clammy cuphea and *Jacobinia* species.

LARGE MOSSY GLYPH

SMALL MOSSY GLYPH

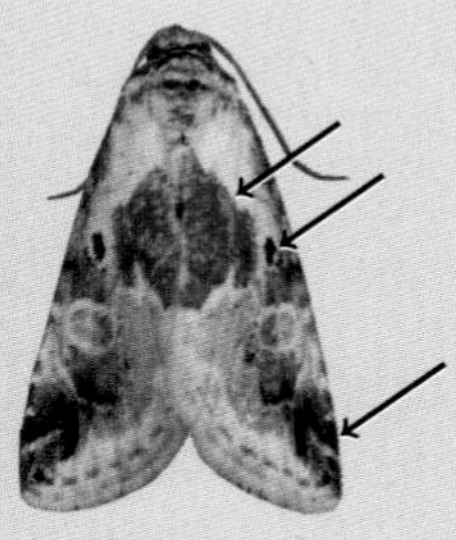

BLACK-DOTTED GLYPH

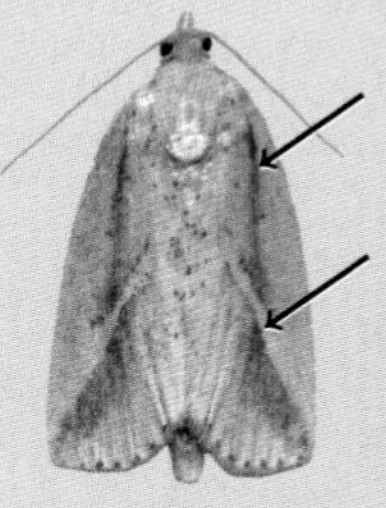

FORKED HALTER MOTH

RED-SPOTTED GLYPH

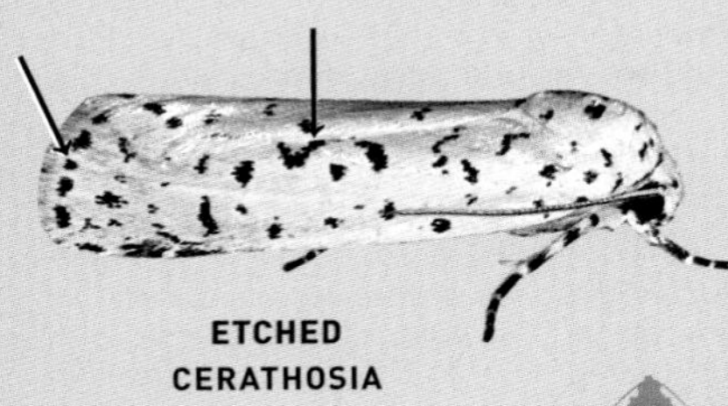

ETCHED CERATHOSIA

actual size

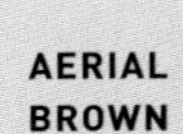

AERIAL BROWN

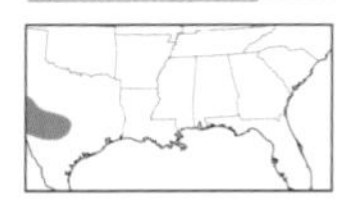

CATALINA BROWN

Ozarba catilina 93-1304 (9032) **Local**

TL 8–10 mm. Bicolored FW is dark chocolate brown with contrasting straw-colored ST area and reniform spot. Inner ST line is edged dark brown. **HOSTS:** Unknown.

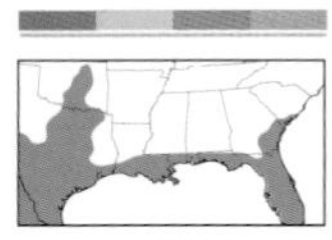

NEBULA BROWN

Ozarba nebula 93-1305 (9033) **Local**

TL 8–9 mm. Light grayish to brown FW has dark brown shading in lower median and outer ST areas. Double PM line curves around messy pale reniform spot. Has two black dashes in central ST area. **HOSTS:** Unknown.

Bird-Dropping Moths

Family Noctuidae, Subfamily Acontiinae

Small moths, many of which are remarkable bird-dropping mimics. Commonly encountered at woodland edges, roadsides, dunes, and old-fields in midsummer. Most are nocturnal and will come to lights, though some, such as Olive-shaded Bird-dropping Moth, may be commonly flushed from vegetation during the daytime.

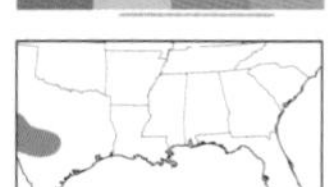

HALF-YELLOW MOTH

Ponometia semiflava 93-1308 (9085) **Common**

TL 9–13 mm. Bicolored FW is chocolate brown with a sharply demarcated yellow basal area. **HOSTS:** Various composites, including Maryland goldenaster. **NOTE:** Two or more broods.

PRETTY BIRD-DROPPING MOTH

Ponometia venustula 93-1311 (9087) **Local**

TL 9–10 mm. Sexually dimorphic. Male (not shown) has light orange FW with a narrow, dark band across median area, forming a crescent at rest. Female is more strongly orange, with a wide, white-edged, peppery gray median band. **HOSTS:** Unknown.

VIRGIN BIRD-DROPPING MOTH

Ponometia virginalis 93-1312 (9088) **Local**

TL 9–10 mm. Satin white FW has diffuse brownish median band and slanting PM band that meet at a silvery patch at inner margin; costal half of median line is sometimes indistinct. Large, dark reniform spot sits at middle. **HOSTS:** Unknown.

PRAIRIE BIRD-DROPPING MOTH

Ponometia binocula 93-1313 (9089) **Local**

TL 9–11 mm. Resembles Virgin Bird-dropping Moth, but PM band is dark olive brown and silvery patch is reduced or absent. ST area often has gray shading. **HOSTS:** Unknown.

GLYPHS

CATALINA BROWN

actual size

NEBULA BROWN

BIRD-DROPPING MOTHS

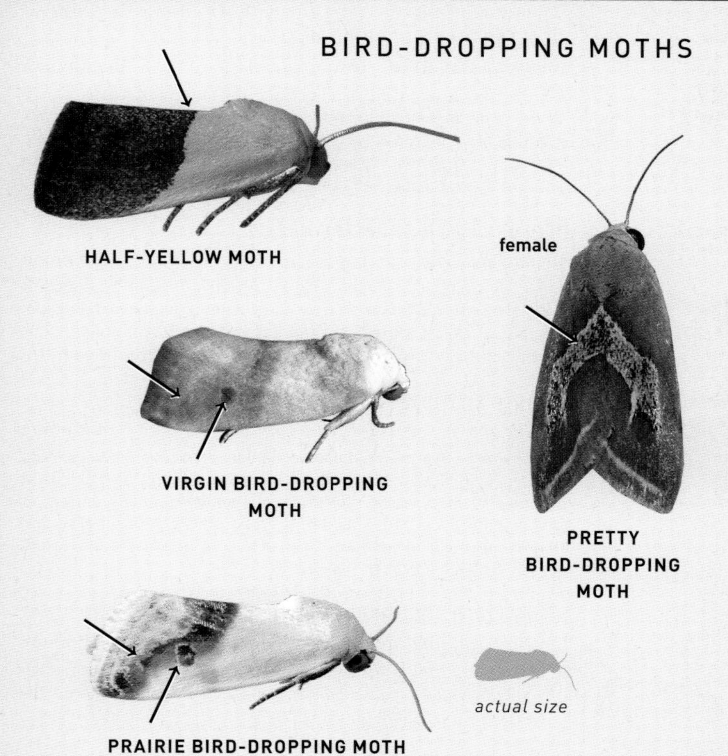

HALF-YELLOW MOTH

female

VIRGIN BIRD-DROPPING MOTH

PRETTY BIRD-DROPPING MOTH

PRAIRIE BIRD-DROPPING MOTH

actual size

OLIVE-SHADED BIRD-DROPPING MOTH

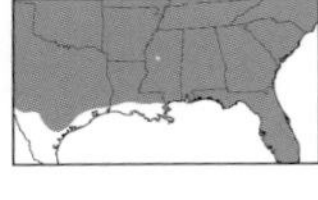

Ponometia candefacta 93-1314 (9090) **Common**

TL 12 mm. White FW has dark olive brown median band that is often lighter in outer half, resulting in a dark saddle. Slanting, yellowish olive PM band meets median band at a grayish patch at inner margin. White-outlined gray reniform spot sits at middle. ST area is usually shaded gray. **HOSTS:** Ragweed. **NOTE:** Two or more broods.

SMALL BIRD-DROPPING MOTH

Ponometia erastrioides 93-1319 (9095) **Common**

TL 9–10 mm. White FW has a dark olive brown saddle, formed by median, PM, and ST lines that slant toward apex. Inner half of PM line is black edged with dark gray. **HOSTS:** Ragweed. **NOTE:** Two or more broods.

OPEN-SPOTTED BIRD-DROPPING MOTH

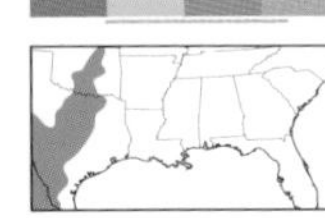

Ponometia phecolisca 93-1322 (9098) **Local**

TL 9–11 mm. White FW has mottled brown shading along inner margin and in ST area. Inner PM line is bordered by two black patches in ST area. Reniform spot is outlined black. **HOSTS:** Unknown.

SPECKLED BIRD-DROPPING MOTH

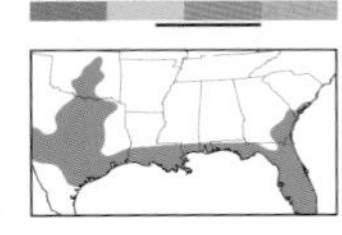

Ponometia fasciatella 93-1329 (9102) **Uncommon**

TL 9–10 mm. Sexually dimorphic. Male (not shown) has pale orange FW with indistinct banding; diffuse terminal line and fringe are dark brown. Female is speckled grayish brown with whitish costa; pale PM line is sometimes obvious at inner margin, and dark reniform spot is often noticeable. **HOSTS:** Unknown.

BICOLORED BIRD-DROPPING MOTH

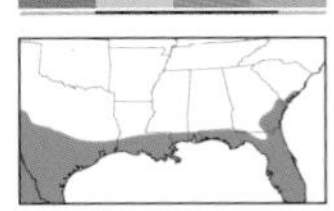

Ponometia exigua 93-1335 (9115) **Common**

TL 12–13 mm. Sexually dimorphic. Striking female has a dark chocolate brown FW with sharply contrasting creamy costal stripe. Male has a similar pattern but not demarcated: warm brown inner margin diffuses into orange costa. **HOSTS:** American burnweed.

EXPOSED BIRD-DROPPING MOTH

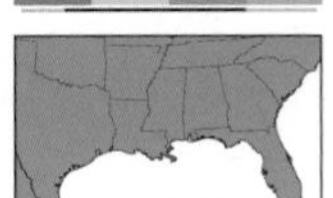

Tarache aprica 93-1343 (9136) **Common**

TL 9–15 mm. Sexually dimorphic. Male resembles Open-spotted Bird-dropping Moth but has large dark brown patches along costa at AM and PM lines, and reniform spot is solid. Female's pattern is similar but grayer, with shading extending through basal area and connecting to costal patches. Orbicular spot is a small black dot. **HOSTS:** Hollyhock. **NOTE:** Two broods.

CURVE-LINED BIRD-DROPPING MOTH

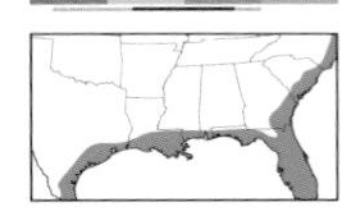

Tarache terminimaculata 93-1351 (9145) **Common**

TL 11–13 mm. Light gray FW has warm brown patch in inner median area between white median and PM lines, curving to bleed into diffuse brown ST area. AM band is diffuse brown in inner half. **HOSTS:** Basswood.

BIRD-DROPPING MOTHS

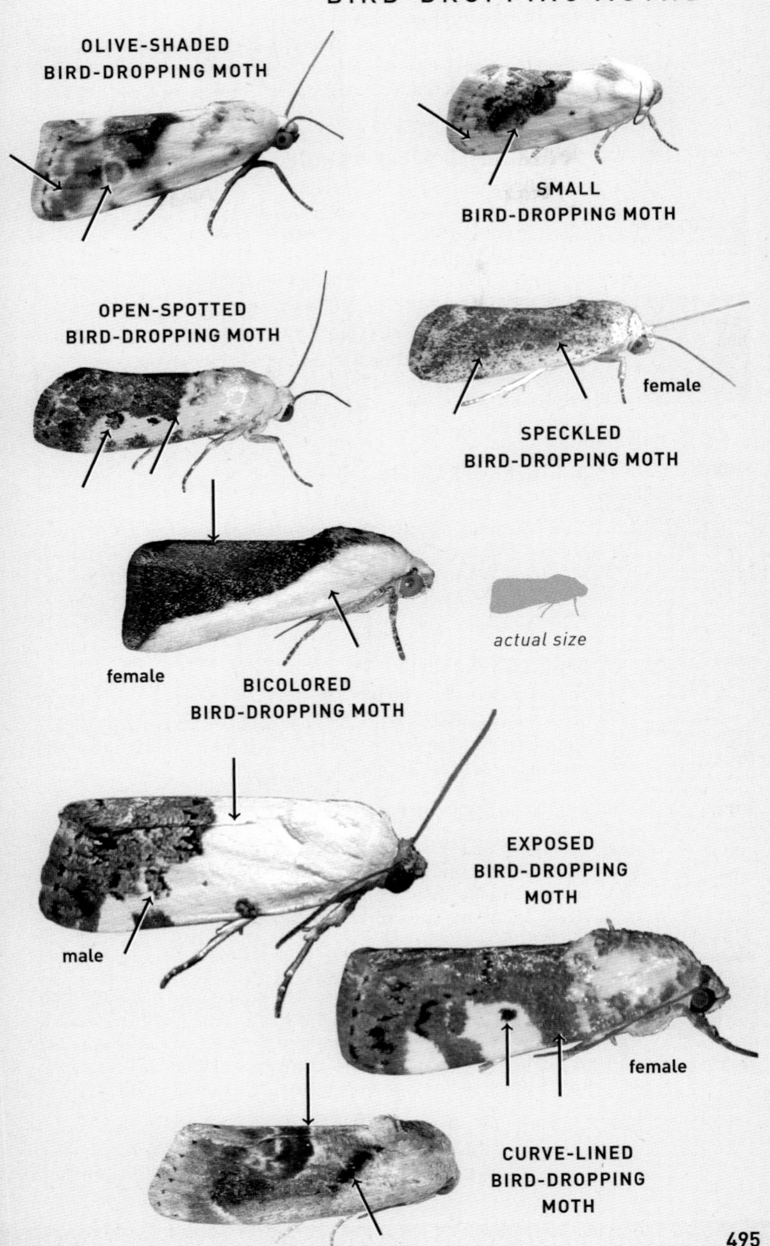

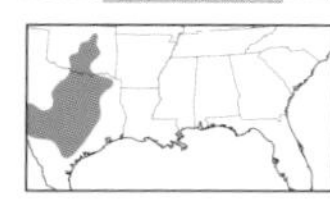

FOUR-PATCHED BIRD-DROPPING MOTH

Tarache quadriplaga 93-1353 (9142) **Local**

TL 10–13 mm. Sexually dimorphic. Resembles Exposed Bird-dropping Moth but lacks orbicular dot. Male's brown costal patches are reduced (not shown). Female's pattern is more uniformly colored. **HOSTS:** Unknown.

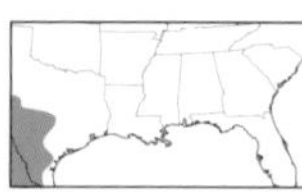

POLISHED BIRD-DROPPING MOTH

Tarache expolita 93-1363 (9149) **Local**

TL 9–11 mm. Shiny, dark grayish brown FW has a broad white band along costa and a large whitish patch, often tinged with brown, in inner ST area. **HOSTS:** Unknown.

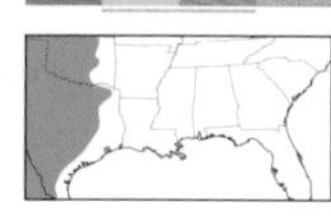

CHALKY BIRD-DROPPING MOTH

Acontia cretata 93-1375 (9161) **Local**

TL 12–14 mm. Satin white FW has soft olive brown patch in ST and inner median area, traced through by thin white ST line. Basal area is gray, extending along costa to midpoint. Fringe is shaded in outer half. **HOSTS:** Unknown.

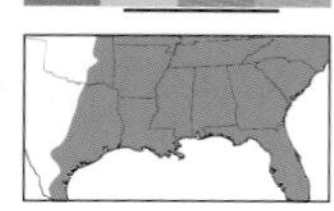

SOUTHERN SPRAGUEIA

Spragueia dama 93-1382 (9122) **Common**

TL 9–12 mm. Sexually dimorphic. Orange-and-black FW has a creamy U-shaped patch at midpoint of costa, and rectangular patches at AM and PM lines. FW in between creamy patches is orange in male, black in female. **HOSTS:** Bindweed. **NOTE:** Two or more broods.

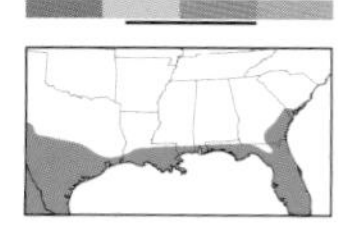

SPOTTED SPRAGUEIA

Spragueia guttata 93-1385 (9125) **Local**

TL 7–9 mm. Creamy FW is boldly patterned with a crisp network of black lines. Orange patches mark ST and inner median areas and front of thorax. **HOSTS:** Feverfew.

BLACK-DOTTED SPRAGUEIA

Spragueia onagrus 93-1386 (9126) **Local**

TL 7–10 mm. Orange-and-black FW has zebra-stripe pattern of creamy and black stripes in basal area and along costa. **HOSTS:** Mostly annual ragweed. **NOTE:** Two or more broods.

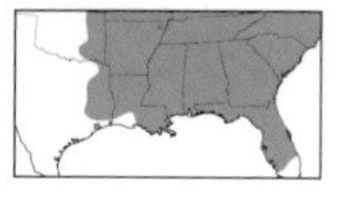

COMMON SPRAGUEIA

Spragueia leo 93-1387 (9127) **Common**

TL 7–10 mm. Orange-and-black FW has a broad black stripe down center that is often broken into three parts by narrow orange AM and PM lines. Creamy patches mark inner FW as well as costa. **HOSTS:** Ragweed; also bindweed. **NOTE:** Two or more broods.

BIRD-DROPPING MOTHS

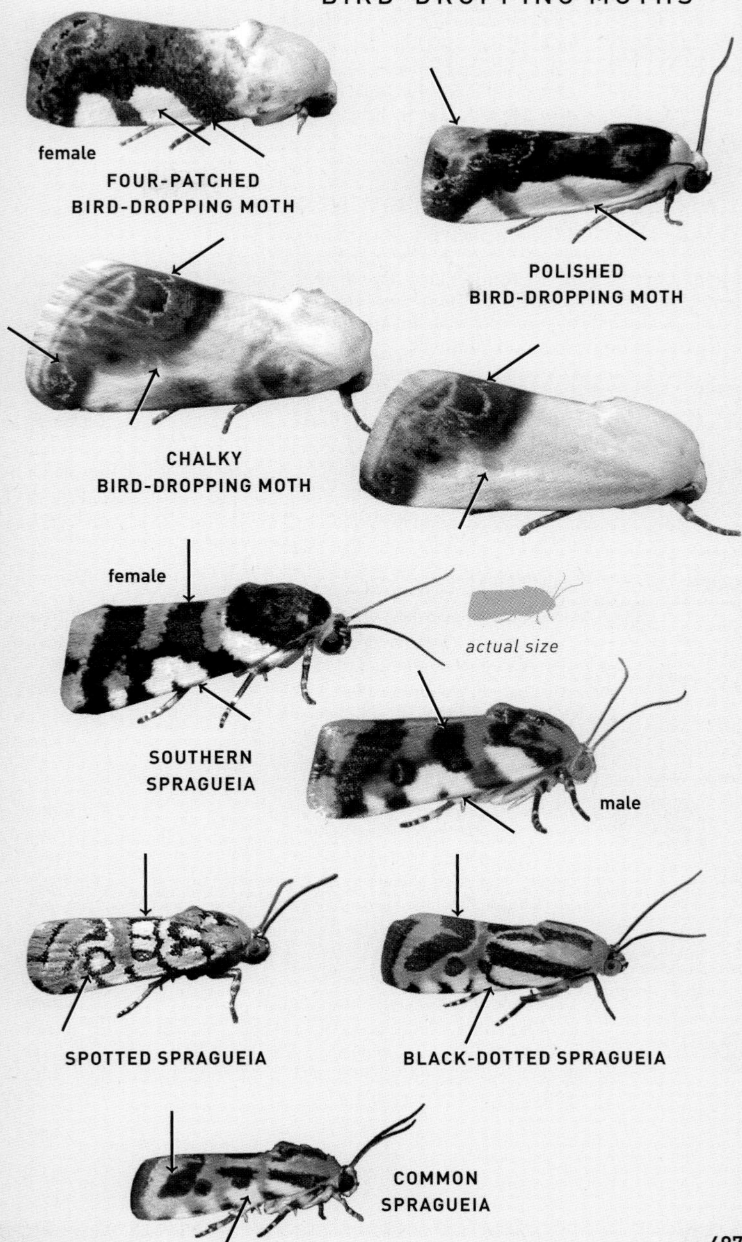

YELLOW SPRAGUEIA

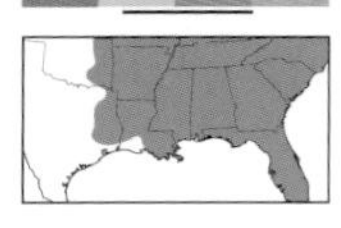

Spragueia apicalis 93-1391 (9131) **Common**

TL 7–10 mm. Sexually dimorphic. Rusty orange thorax has a thick crest. Male has straw-colored FW with olive brown shading down center and in ST area; anal angle is blackish. Female has peppery gray FW with creamy basal patch and apical area. **HOSTS:** Sida. **NOTE:** Two or more broods.

PANTHEAS AND YELLOWHORNS

Family Noctuidae, Subfamily Pantheinae

Chunky moths with a slightly hairy appearance. The yellowhorns take their name from their yellowish pectinate antennae. All are common in mixed woodlands and large gardens, mostly from spring to late summer. These moths are attracted to lights in small numbers.

EASTERN PANTHEA

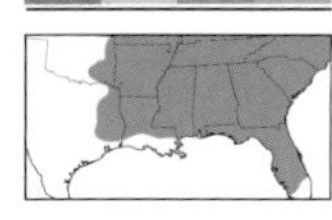

Panthea furcilla 93-1396 (9182) **Common**

TL 20–25 mm. Peppery FW is variably whitish to dark gray with crisp black lines. AM and median lines are almost straight; PM line bends sharply to touch median line before reaching inner margin. Orbicular and reniform spots are absent. **HOSTS:** Coniferous trees, including tamarack, pine, and spruce. **NOTE:** Two or more broods.

SADDLED YELLOWHORN

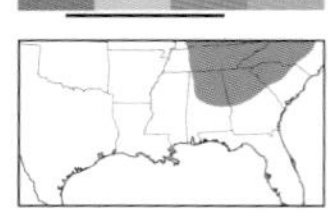

Colocasia flavicornis 93-1400 (9184) **Common**

TL 18–22 mm. Gray FW has large blackish patch covering inner median area. Kinked AM and PM lines are connected by black bar in central median area. White orbicular spot is outlined black. Antennae are yellow. **HOSTS:** Deciduous trees, including beech, elm, ironwood, oak, and maple. **NOTE:** Two broods.

THE LAUGHER

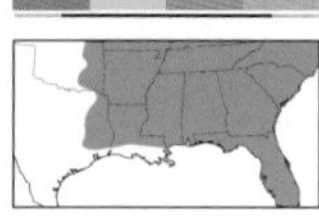

Charadra deridens 93-1406 (9189) **Common**

TL 20–24 mm. Resembles Saddled Yellowhorn, but median area lacks gray patch and orbicular spot has a black dot in center. Diffuse median line often shades median area. Antennae are grayish to olive. **HOSTS:** Deciduous trees, including beech, birch, elm, maple, and oak. **NOTE:** Two broods.

THE DEBATER

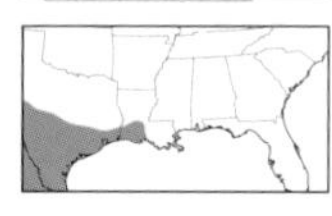

Charadra displusa 93-1408 (9190) **Local**

TL 17–19 mm. Resembles The Laugher, but AM line is almost straight and does not connect to PM line. FW is lighter, with less distinct median and ST lines. Dot in orbicular spot is often faint. **HOSTS:** Unknown; possibly oak.

BIRD-DROPPING MOTHS

PANTHEAS AND YELLOWHORNS

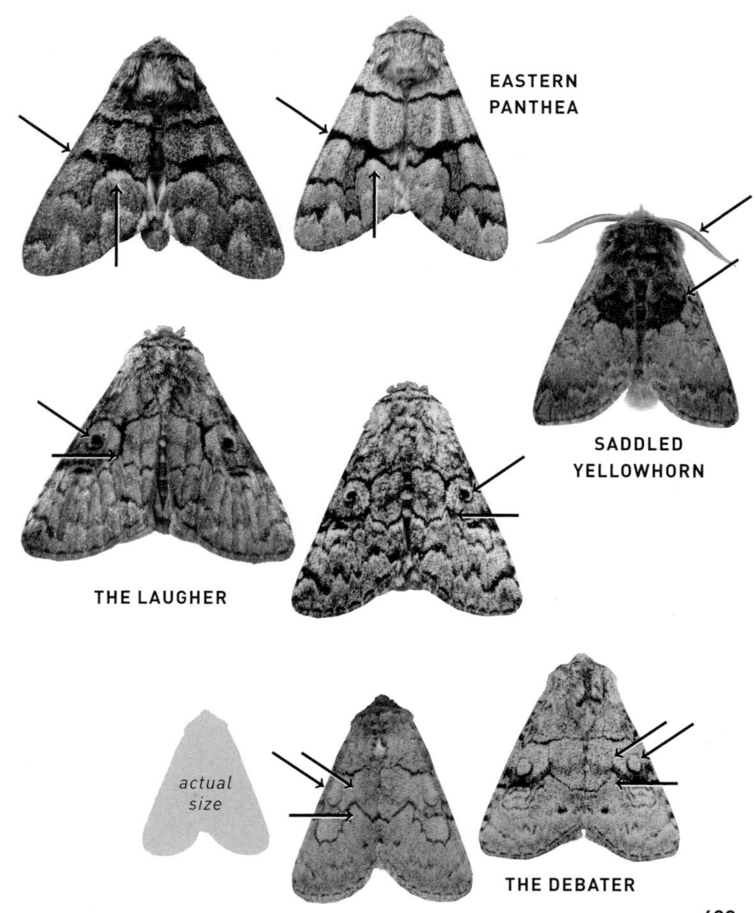

BROTHERS Family Noctuidae, Subfamily Raphiinae

This chunky, slightly hairy noctuid resembles the yellowhorns in size and shape. It is a common inhabitant of mixed woodlands and will visit lights in small numbers.

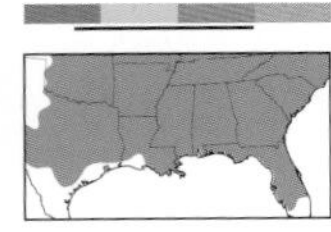

THE BROTHER *Raphia frater* 93-1412 (9193) **Common**

TL 17–19 mm. Gray FW has curved, slightly wavy AM and PM lines, usually connected by a thin vertical bar parallel to inner margin. Orbicular and reniform spots are narrowly outlined black and sometimes tinged brownish. Veins in ST area are traced darker. **HOSTS:** Unknown. **NOTE:** Two broods. Abrupt Brother (*R. f. abrupta*) is now considered a subspecies.

BALSAS Family Noctuidae, Subfamily Balsinae

Small, predominantly gray moths that are patterned with fine black lines and streaks. They are commonly encountered in woodlands and gardens in the summer months. All three will come to lights in small numbers.

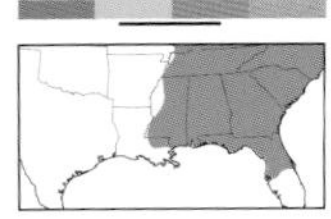

MANY-DOTTED BALSA

Balsa malana 93-1417 (9662) **Common**

TL 13–16 mm. Peppery ash gray FW has a bold, sharply pointed V at midpoint of pale gray costa. Thin, wavy AM line and jagged ST line are black. Veins in ST area are lightly traced blackish. **HOSTS:** Apple, cherry, pear, and plum. **NOTE:** Two broods.

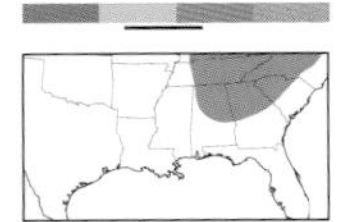

THREE-LINED BALSA

Balsa tristrigella 93-1418 (9663) **Common**

TL 11–12 mm. Resembles Many-dotted Balsa, but bold V is broken at point, and streaking along veins extends into median area. AM line thickens into bold dash at costa, which is not noticeably paler than rest of FW. Often shaded slightly brownish through ST area. **HOSTS:** Apple and hawthorn. **NOTE:** Two broods.

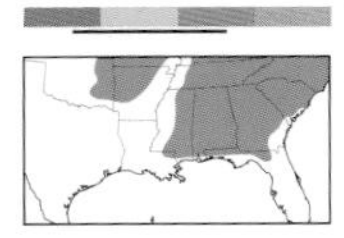

WHITE-BLOTCHED BALSA

Balsa labecula 93-1419 (9664) **Common**

TL 13–16 mm. Gray FW has a narrow bold Y at midpoint of costa, stem of which crosses pale reniform spot then turns to join jagged ST line. Orbicular spot is a noticeable white circle. Streaking along veins is reduced to midpoint of ST area. **HOSTS:** Chokeberry and hawthorn.

BROTHERS

BALSAS

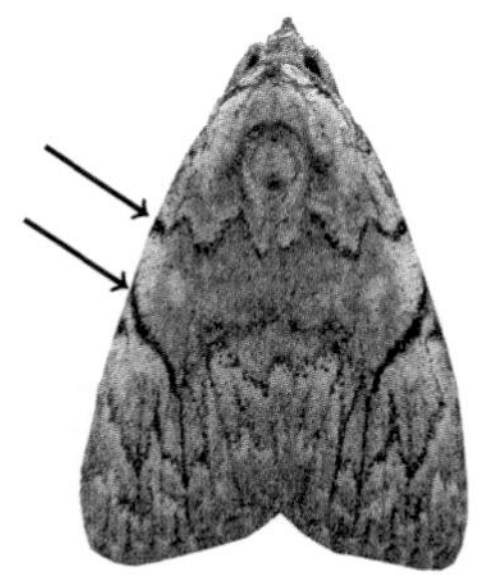

MANY-DOTTED BALSA

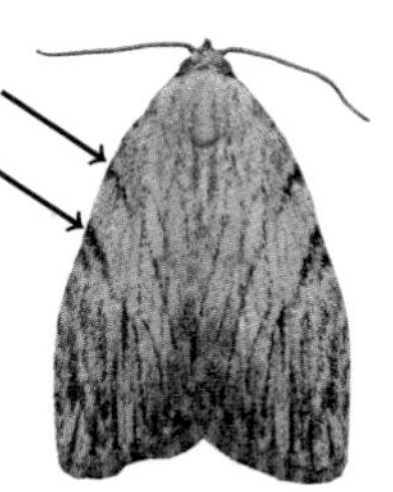

THREE-LINED BALSA

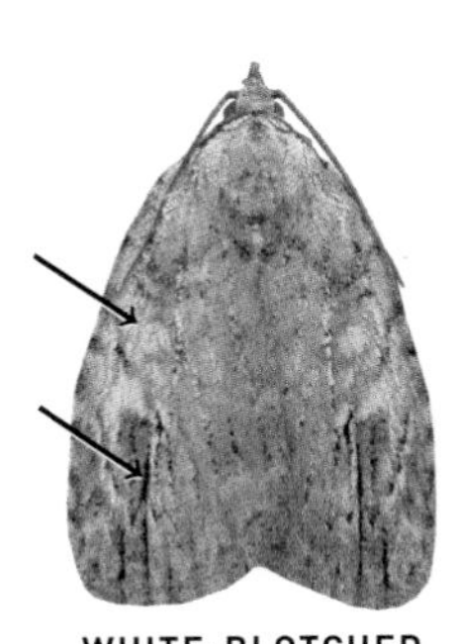

WHITE-BLOTCHED BALSA

DAGGERS Family Noctuidae, Subfamily Acronictinae

A large group of predominantly gray noctuids that commonly have bold black dashes as their most distinctive markings, giving the group its name. Most can be found in wooded habitat, though a few species occur in highly urban areas, even in large cities. They are nocturnal and will visit lights in small numbers.

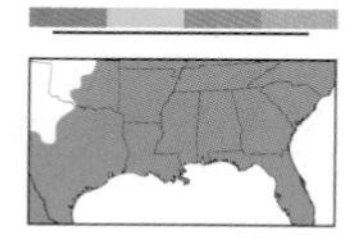

RUDDY DAGGER

Acronicta rubricoma 93-1420 (9199) **Common**

TL 20–24 mm. Peppery light gray FW has double AM and PM lines edged with dark gray shading; whitish filling often is more noticeable on PM line. Basal area is lightly shaded dark gray. Orbicular spot is outlined black. Large reniform spot typically has gray center. Anal and especially basal dashes are thin and often indistinct. **HOSTS:** Hackberry and elm. **NOTE:** Two broods.

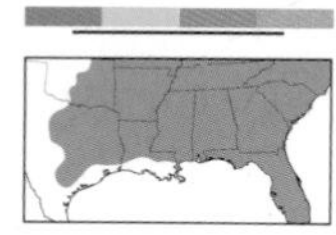

AMERICAN DAGGER

Acronicta americana 93-1421 (9200) **Common**

TL 27–38 mm. The largest of our daggers. Resembles Ruddy Dagger, but AM line is usually indistinct, and PM line and basal area lack strong shading. Thin anal dash cuts through PM line. **HOSTS:** Trees and woody plants, including alder, ash, basswood, elm, chestnut, and hickory. **NOTE:** Two or three broods.

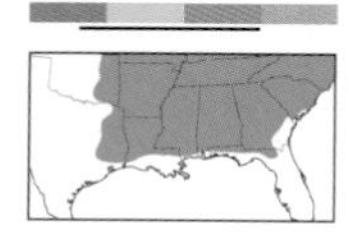

BIRCH DAGGER *Acronicta betulae* 93-1429 (9208) **Common**

TL 18–22 mm. Pale yellowish gray FW has jagged, dark brown PM line edged thinly white above and with diffuse warm brown shading below. Reniform spot is a dark brown crescent edged below with warm brown shading. Basal area is sometimes shaded brown. **HOSTS:** River birch. **NOTE:** Two or three broods.

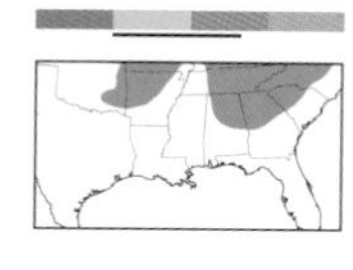

RADCLIFFE'S DAGGER

Acronicta radcliffei 93-1430 (9209) **Uncommon**

TL 18–20 mm. Gray FW has bold double AM and PM lines filled whitish. Long basal and anal dashes cut through lines. Black-edged orbicular spot is often pale. **HOSTS:** Apple, cherry, chokeberry, hawthorn, and others. **NOTE:** Two broods.

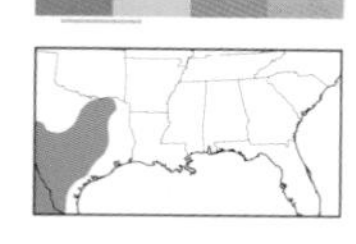

COMPLETE DAGGER *Acronicta tota* 93-1431 (9210) **Local**

TL 16–19 mm. Resembles Ruddy Dagger, but basal dash connecting to double AM line contrasts against paler basal area. White-edged black PM line is smoothly scalloped and lacks noticeable shading. Anal dash is often reduced or absent. **HOSTS:** Unknown.

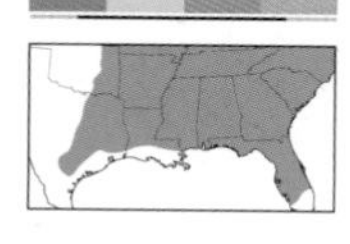

TRITON DAGGER *Acronicta tritona* 93-1432 (9211) **Common**

TL 18–20 mm. Purplish gray FW has fragmented, dark brown PM line edged below with diffuse brown shading. Dark brown reniform crescent is edged with warm brown, and connected to costa by diffuse blackish bar. Anal dash is strongly black and cuts through PM line. Basal dash barely reaches faint AM line. **HOSTS:** Azalea and blueberry. **NOTE:** Two or more broods.

RUDDY DAGGER

AMERICAN DAGGER

BIRCH DAGGER

actual size

RADCLIFFE'S DAGGER

COMPLETE DAGGER

TRITON DAGGER

FALCON DAGGER

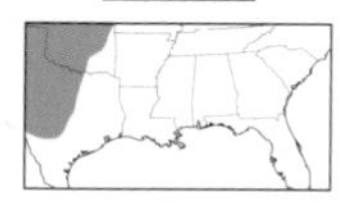

Acronicta falcula 93-1434 (9214) **Local**

TL 16–18 mm. Medium gray FW has bold, slightly scalloped black PM line edged whitish; anal dash just touches near inner margin. Thick black basal dash connects to sharp tooth of sometimes-indistinct double AM line. Basal area is usually noticeably pale. Soft whitish ST line diffuses into ST area. Commonly, some individuals have a thick black dash in inner median area, and heavy shading behind all dashes creates a broad blackish streak parallel to inner margin. **HOSTS:** *Prunus* species.

CONNECTED DAGGER

Acronicta connecta 93-1436 (9219) **Common**

TL 19–20 mm. Gray to brownish FW has a conspicuous blackish streak connecting basal and anal dashes. Pale orbicular spot is separated from brown-edged reniform crescent by a patch of dusky shading. Double PM line is filled whitish. **HOSTS:** Willow. **NOTE:** Two broods.

FUNERARY DAGGER

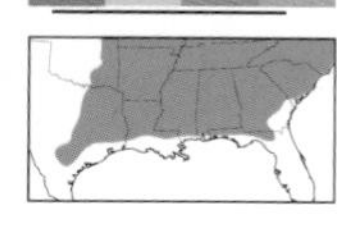

Acronicta funeralis 93-1438 (9221) **Uncommon**

TL 17–21 mm. Resembles black-streaked Falcon Dagger but has bold black patch at midpoint of costa. Shading along inner margin is strongly black. PM line is usually fragmented. Basal area is not noticeably paler. **HOSTS:** Deciduous trees, including apple, birch, blueberry, cottonwood, elm, maple, oak, and others. **NOTE:** Two or more broods.

ASHY DAGGER

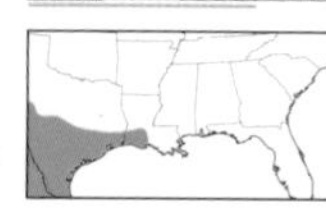

Acronicta lepetita 93-1440 (9223) **Local**

TL 20–24 mm. Resembles Delightful Dagger, but FW is generally paler and midpoint of ST area has faint subapical dash instead of shading. Orbicular and reniform spots are joined by a short, thin bar. Double AM line is often indistinct. **HOSTS:** Unknown.

DELIGHTFUL DAGGER

Acronicta vinnula 93-1442 (9225) **Common**

TL 15–17 mm. Whitish to gray FW has conspicuous white orbicular spot outlined black. Black PM line is edged white above and bordered by gray shading below. Anal dash barely reaches beyond PM line. Basal dash widens or forks where it touches gray, double AM line. ST area has diffuse blackish patch at midpoint. **HOSTS:** Elm. **NOTE:** Multiple broods.

GREEN MARVEL

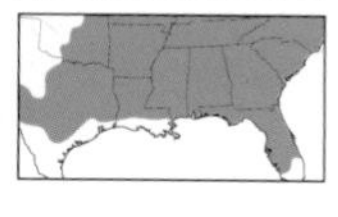

Acronicta fallax 93-1442.1 (9281) **Common**

TL 17–20 mm. Pale green FW has chunky black patches along AM, median, and PM lines. Fringe is checkered. Green color rapidly fades, and worn moths often appear whitish. **HOSTS:** Viburnum. **NOTE:** Two broods. Previously in genus *Agriopodes*.

FALCON DAGGER

CONNECTED DAGGER

FUNERARY DAGGER

actual size

ASHY DAGGER

DELIGHTFUL DAGGER

GREEN MARVEL

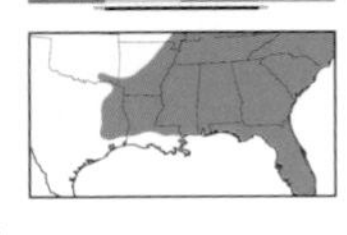

PLEASANT DAGGER

Acronicta laetifica 93-1444 (9227) **Common**

TL 20–22 mm. Whitish FW has thick black basal dash that usually forks at faint double AM line. Black anal dash cuts through white-edged, scalloped PM line. Thin subapical dash is often faint or absent. Whitish reniform spot is narrowly edged black above. **HOSTS:** Ironwood and eastern hornbeam. **NOTE:** Two broods.

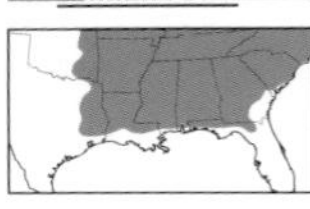

SPEARED DAGGER

Acronicta hasta 93-1445 (9229) **Common**

TL 22–23 mm. Resembles Delightful Dagger, but orbicular and reniform spots are joined by bold black bar and partial edging. Basal and anal dashes are often thick. Subapical dash may be strong or indistinct. White-edged PM line is scalloped. **HOSTS:** Cherry and plum. **NOTE:** Two or three broods.

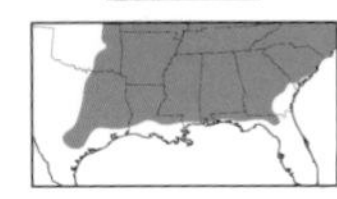

NONDESCRIPT DAGGER

Acronicta spinigera 93-1452 (9235) **Common**

TL 21–25 mm. Resembles Ruddy Dagger, but basal area is not shaded darker, and veins in ST area are unmarked. Dashes are usually distinct; basal dash stops at top of double AM line. **HOSTS:** Basswood, apple, and elm. **NOTE:** Two broods.

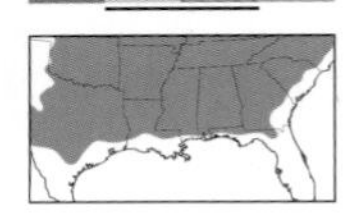

OCHRE DAGGER

Acronicta morula 93-1453 (9236) **Common**

TL 22–30 mm. Pale gray FW has yellowish brown shading along lines and in spots. Basal, anal, and subapical dashes are bold. Thorax usually has a distinctive ochre dorsal stripe. **HOSTS:** Elm. **NOTE:** Two or more broods.

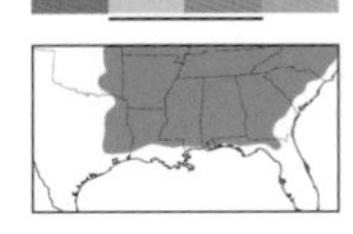

INTERRUPTED DAGGER

Acronicta interrupta 93-1454 (9237) **Common**

TL 21 mm. Resembles Ochre Dagger, but thorax is gray and FW lacks brown shading except at reniform spot. Subapical dash is faint or absent. **HOSTS:** Deciduous trees, including apple, cherry, plum, and serviceberry. **NOTE:** Two broods.

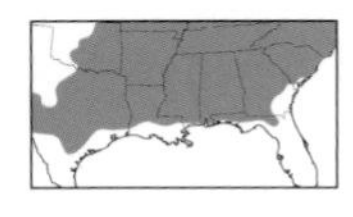

GREAT OAK DAGGER

Acronicta lobeliae 93-1455 (9238) **Common**

TL 21–32 mm. Resembles Speared Dagger, but longer FW is paler and without mottled shading. AM line is weakly double. Thin subapical dash is always present. **HOSTS:** Oak. **NOTE:** Two or three broods.

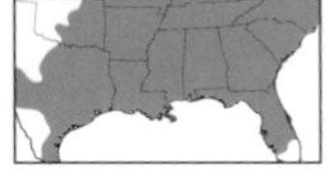

CLEAR DAGGER

Acronicta clarescens 93-1460 (9246) **Common**

TL 20–22 mm. Resembles Speared Dagger, but bold basal dash crosses through AM line, and subapical dash is not well defined. Reniform spot is often tinged brown. **HOSTS:** Apple, cherry, hawthorn, plum, and others. **NOTE:** Two broods.

PLEASANT DAGGER

NONDESCRIPT DAGGER

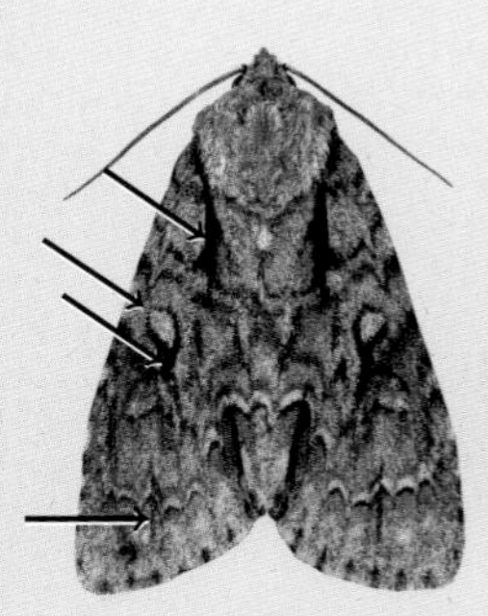

SPEARED DAGGER

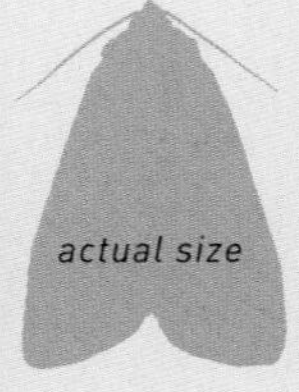

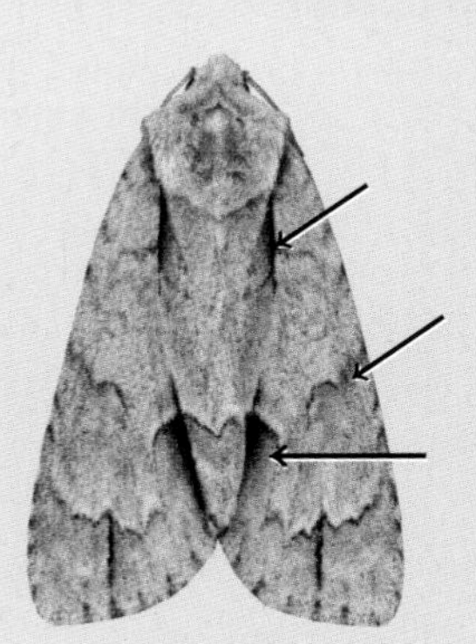

OCHRE DAGGER

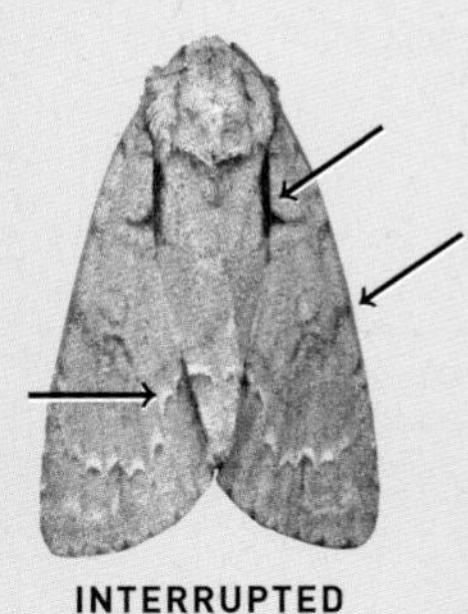

INTERRUPTED DAGGER

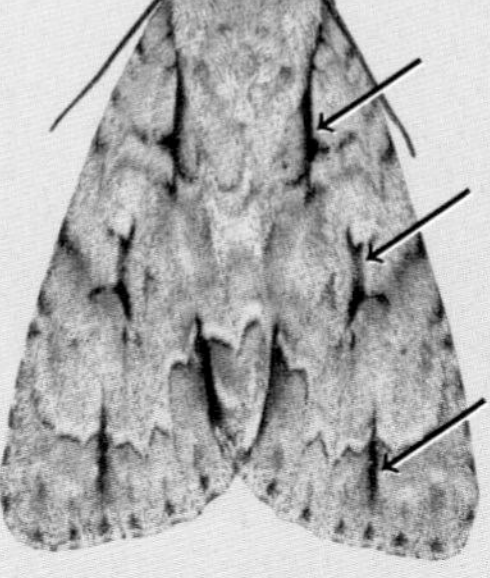

GREAT OAK DAGGER

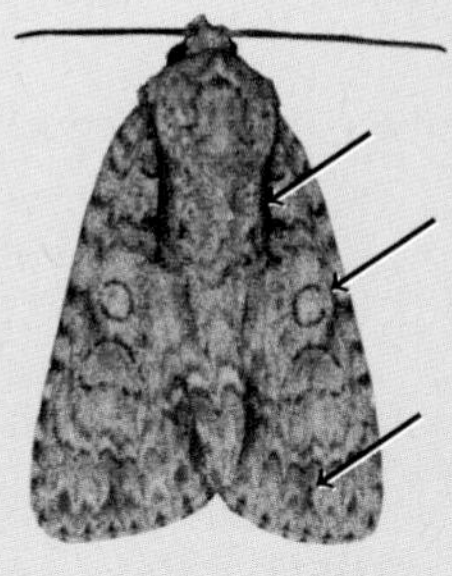

CLEAR DAGGER

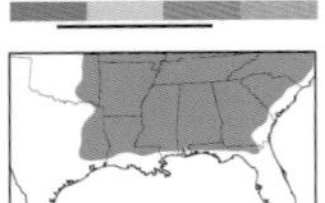

OVATE DAGGER *Acronicta ovata* 93-1463 (9243) Common

TL 15–18 mm. Pale gray to olive brown FW has thick basal dash connecting to double AM line; inner half of AM line is filled blackish, isolating pale basal area. Base of basal dash has small tawny patch. Reniform spot is washed brownish. Area between anal and subapical dashes is often lightly shaded blackish. Some individuals lack shading in AM line. **HOSTS:** White and red oaks.

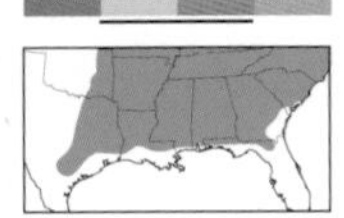

EXILED DAGGER *Acronicta exilis* 93-1464 (9242) Common

TL 13–16 mm. Gray to tan FW has thin basal dash connecting to strongly double, gray-filled AM line. Basal and upper median areas are subtly paler. Orbicular and reniform spots are shaded brownish. **HOSTS:** White oak. **NOTE:** Two broods.

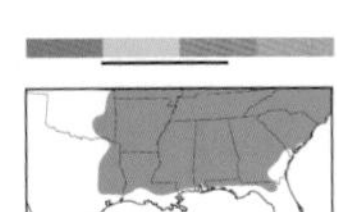

MEDIUM DAGGER

Acronicta modica 93-1465 (9244) Common

TL 18–22 mm. Resembles Exiled Dagger, but orbicular and reniform spots are gray, ringed in white, occasionally with a hint of brown; dark edging is faint or fragmented. Basal and median areas are commonly uniform in color with rest of FW. Subapical dash is indistinct, often reduced to a dark patch in ST area. **HOSTS:** Oak. **NOTE:** Two broods.

HESITANT DAGGER

Acronicta haesitata 93-1466 (9245) Common

TL 24 mm. Resembles Medium Dagger, but double AM line is usually filled dark gray. Upper edge of reniform spot is outlined black. Subapical dash is reduced to a small dark patch bordering terminal line. **HOSTS:** Oak. **NOTE:** Two broods.

SMALL OAK DAGGER

Acronicta increta 93-1467 (9249) Common

TL 15–20 mm. Resembles Ovate Dagger, but double AM line is completely filled blackish, to costa. Anal and subapical dashes rarely show shading in between. Median area typically is less contrastingly pale. **HOSTS:** Chestnut, oak, and beech. **NOTE:** Two broods. Also commonly known as Raspberry Bud Dagger.

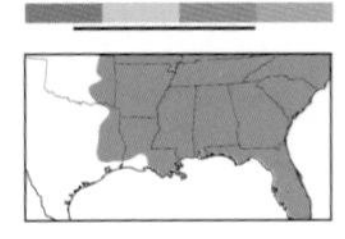

RETARDED DAGGER

Acronicta retardata 93-1470 (9251) Common

TL 14–16 mm. Gray FW has a small black triangle where thin basal dash meets double AM line. Scalloped, double PM line is filled white. Inner median area and below reniform spot are clear of mottling and often contrastingly pale. Anal dash is often present as a long, dark smudge. **HOSTS:** Maple. **NOTE:** Multiple broods.

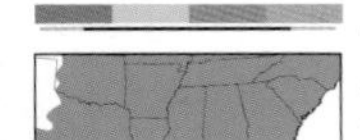

AFFLICTED DAGGER

Acronicta afflicta 93-1471 (9254) Common

TL 18–22 mm. Dark, sooty FW has contrastingly pale orbicular spot filled with gray. ST and terminal lines are a series of white Vs. Center of thorax is pale. **HOSTS:** Oak. **NOTE:** Two or three broods.

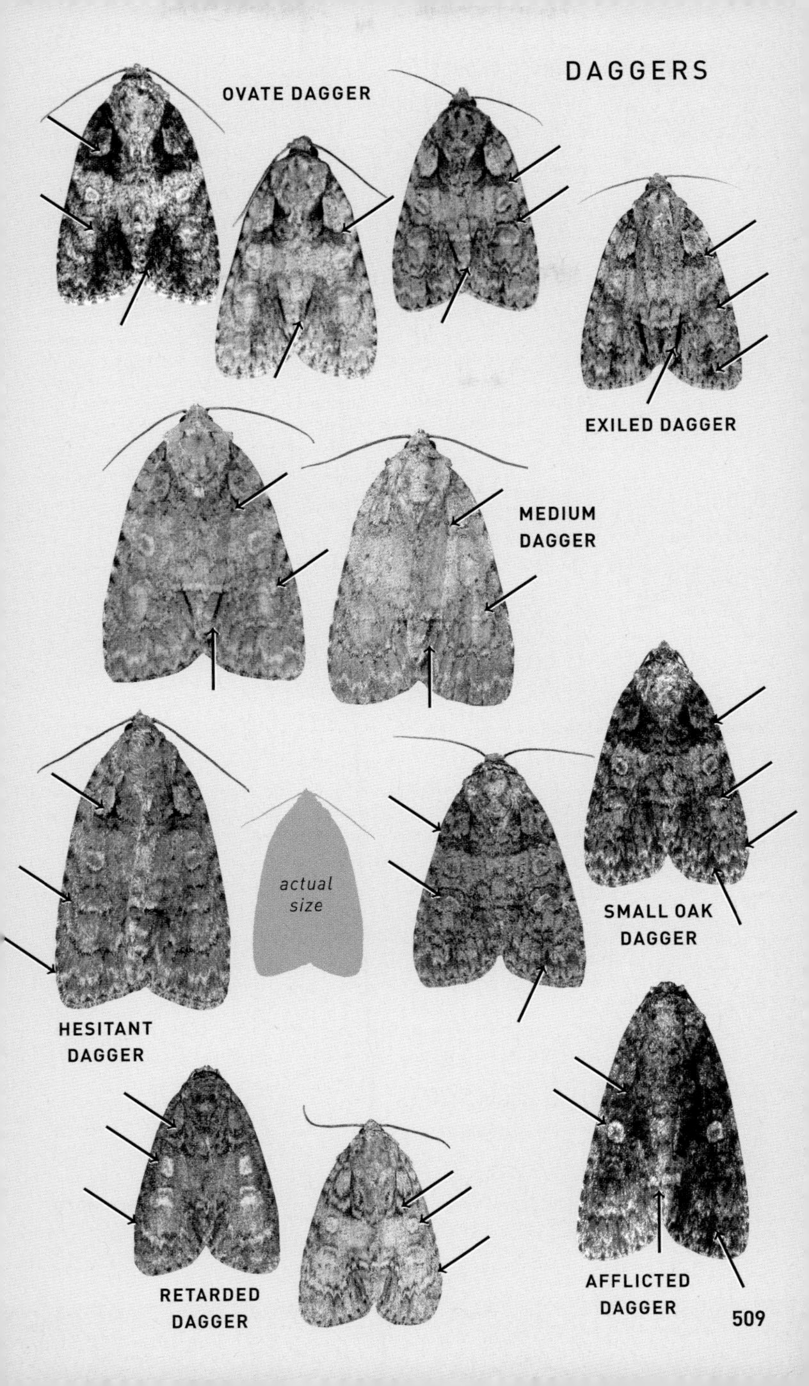
DAGGERS
OVATE DAGGER
EXILED DAGGER
MEDIUM DAGGER
actual size
SMALL OAK DAGGER
HESITANT DAGGER
RETARDED DAGGER
AFFLICTED DAGGER

CHARRED DAGGER

Acronicta brumosa 93-1472 (9255) **Uncommon**

TL 18–20 mm. Gray FW has heavily shaded basal dash, reaching into inner median area. Anal dash and area below reniform spot are also usually strongly shaded black. Subapical dash is lightly shaded. Double AM line is filled dark gray; double PM line is filled whitish. Orbicular spot is often contrastingly pale, with gray center. **HOSTS:** Blackjack oak.

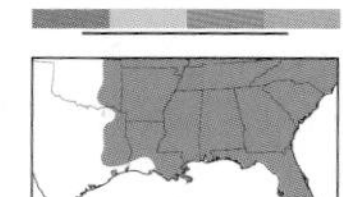

YELLOW-HAIRED DAGGER

Acronicta impleta 93-1474 (9257) **Uncommon**

TL 21–27 mm. Peppery gray FW has narrow black outlines and dark centers to orbicular and reniform spots. Angled blackish bar of outer median line connects to center of reniform spot. Strongly scalloped, white-edged PM line appears dashed. All dashes are indistinct smudges. Darker individuals are sooty with contrasting white-filled PM line. **HOSTS:** Hickory and walnut; also alder, ash, elm, and others. **NOTE:** Multiple broods. Common name refers to distinctive caterpillar.

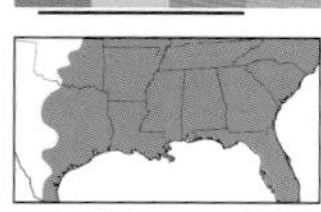

LONG-WINGED DAGGER

Acronicta longa 93-1478 (9264) **Common**

TL 18–24 mm. Pale gray FW has black-outlined orbicular and reniform spots. AM and median lines are most obvious as dark bars at costa. Strongly toothed, thin PM line is cut through by long black anal dash. **HOSTS:** Deciduous trees, including alder, birch, cherry, oak, and others. **NOTE:** Two or more broods.

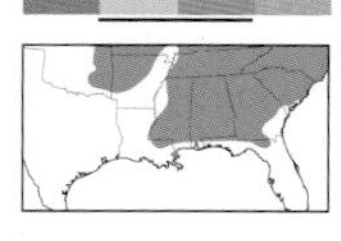

STREAKED DAGGER

Acronicta lithospila 93-1480 (9266) **Uncommon**

TL 18-23 mm. Gray FW has white and blackish streaking along veins. Whitish orbicular spot and black-outlined reniform spot are indistinct, hidden beneath streaking. **HOSTS:** Primarily oak and chestnut; also hickory and walnut. **NOTE:** Two broods.

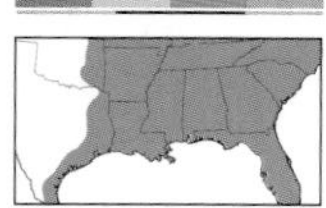

SMEARED DAGGER

Acronicta oblinita 93-1485 (9272) **Uncommon**

TL 20–28 mm. Peppery whitish FW has smudgy dark reniform spot and broken orbicular spot; often has a round pale patch between spots. Strongly toothed PM line is very thin and lightly edged below with grayish shading. **HOSTS:** Trees, shrubs, and low plants, including apple, corn, elm, pine, and willow. **NOTE:** Two or more broods.

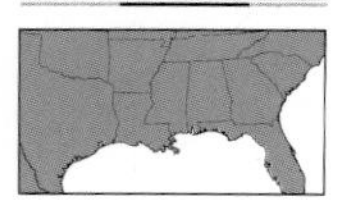

HENRY'S MARSH MOTH

Acronicta insularis 93-1493 (9280) **Local**

TL 20–22 mm. Pale tan FW has contrasting whitish veins. Three streaks of dark brown shading can sometimes be indistinct. **HOSTS:** Cattail, grass, sedge, smartweed, poplar, willow, and others. **NOTE:** Two or more broods.

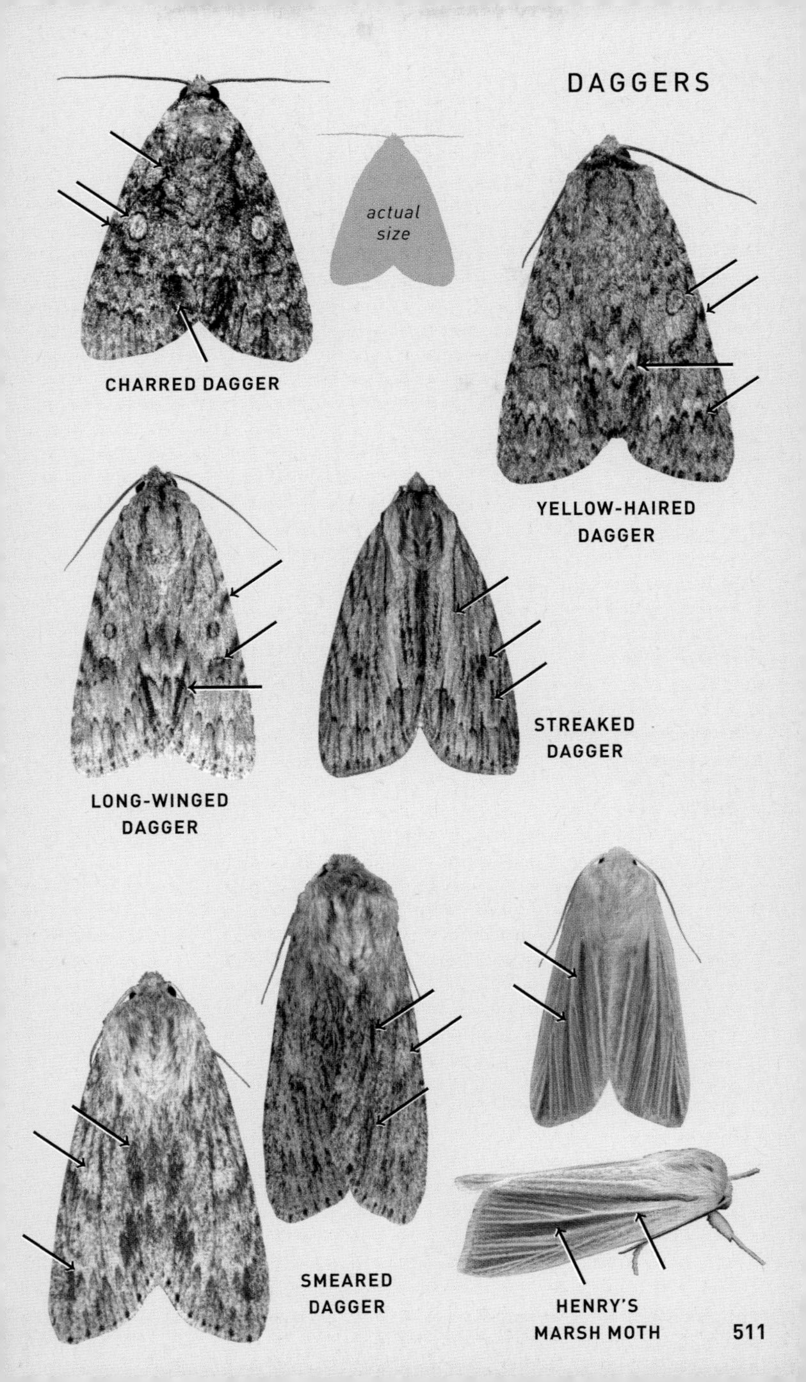
DAGGERS
actual size
CHARRED DAGGER
YELLOW-HAIRED DAGGER
LONG-WINGED DAGGER
STREAKED DAGGER
SMEARED DAGGER
HENRY'S MARSH MOTH

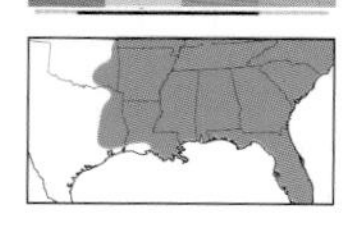

THE HEBREW

Polygrammate hebraeicum 93-1497 (9285) **Common**

TL 13–15 mm. White FW has crisp black markings. AM line is broken at midpoint. Partially outlined reniform spot usually touches jagged median line directly below orbicular dot. **HOSTS:** Tupelo. **NOTE:** Two or three broods.

HARRIS'S THREE-SPOT

Harrisimemna trisignata 93-1498 (9286) **Common**

TL 17–20 mm. White FW has a distinctive pattern of three large, warm brown spots set amid crisp black lines. Abdomen has a tuft of brown hairlike scales at midpoint. Thorax is velvety brown. **HOSTS:** Trees and woody plants, including apple, blueberry, cherry, honeysuckle, and willow. **NOTE:** Two broods.

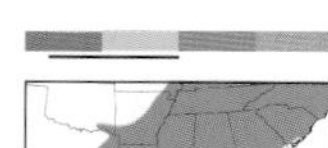

CADBURY'S LICHEN MOTH

Comachara cadburyi 93-1499 (8104) **Uncommon**

TL 10–12 mm. Light gray FW has dark gray median area. PM line is diffusely edged whitish below. Dotted ST ends as a series of black dashes at inner margin. **HOSTS:** Tupelo. **NOTE:** One or two broods.

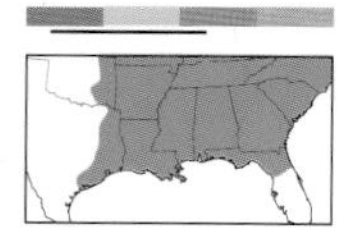

OWL-EYED BIRD-DROPPING MOTH

Cerma cora 93-1500 (9061) **Uncommon**

TL 15–17 mm. FW is olive in basal and ST areas and whitish through median area, crossed by thick, jagged black lines. White orbicular spot is boldly outlined black. **HOSTS:** Pin cherry and hawthorn.

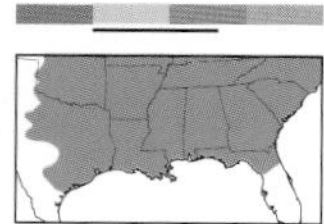

TUFTED BIRD-DROPPING MOTH

Cerma cerintha 93-1501 (9062) **Common**

TL 15–17 mm. White median area of FW has a diffuse gray, double median line. Basal and ST areas are a mosaic of brown, black, white, bluish gray, and moss green. White orbicular and reniform spots are narrowly outlined black. **HOSTS:** Cherry and hawthorn; also apple, cotoneaster, and plum. **NOTE:** Two broods.

THE HEBREW

HARRIS'S THREE-SPOT

CADBURY'S LICHEN MOTH

OWL-EYED BIRD-DROPPING MOTH

TUFTED BIRD-DROPPING MOTH

actual size

Hooded Owlets

Family Noctuidae, Subfamily Cuculliinae

A distinctive group of sleek-looking moths with a thick thoracic crest that curls forward over the head to create a "hooded" appearance. These moths are called "sharks" in Europe, and more recently, "paints" in the U.S., to reflect the outlandishly colorful larvae. Generally uncommon, these moths are found in woodlands, old-fields, and large gardens. They are nocturnal and will visit lights in small numbers, but are more commonly encountered feeding on flowers.

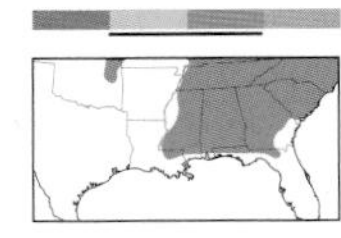

GOLDENROD HOODED OWLET

Cucullia asteroides 93-1504 (10200) **Uncommon**

TL 26–29 mm. Ash gray to tan FW has an inverted brown U edged by a black spot at anal angle. Pale orbicular and reniform spots sit against dark brown costal streak. Upper abdomen has gray tufts. **HOSTS:** Aster and goldenrod. **NOTE:** Two broods. Also known as The Asteroid.

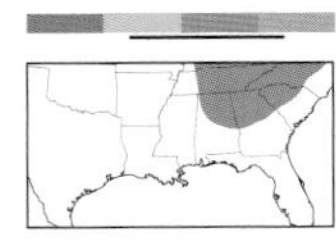

BROWN HOODED OWLET

Cucullia convexipennis 93-1513 (10202) **Uncommon**

TL 22–27 mm. Tan FW has wide brown bands along lower half of costa and inner margin. Outer margin is edged rusty brown. Narrow black streak runs vertically through center of wing. Upper abdomen has brown tufts. **HOSTS:** Aster and goldenrod. **NOTE:** Two broods. Also known as Brown-bordered Cucullia.

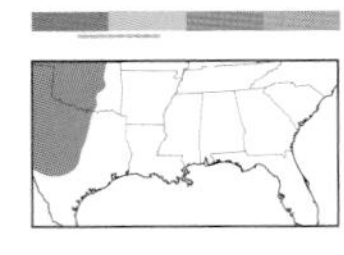

RABBITBUSH HOODED OWLET

Cucullia laetifica 93-1519 (10191) **Local**

TL 20–23 mm. Blue gray FW has warm brown shading in outer median and inner basal areas. Bold anal dash is capped by a small crescent. Fine lines vertically trace wing. Upper abdomen has small gray tufts. **HOSTS:** Rabbitbush and spiderling.

Amphipyrine Sallows

Family Noctuidae, Subfamily Amphipyrinae

These small to medium-sized noctuids are most common in spring and fall. Most rest with their wings folded tentlike over their back. They are largely nocturnal, and many will visit lights in good numbers. Copper Underwing is also frequent at sugar bait.

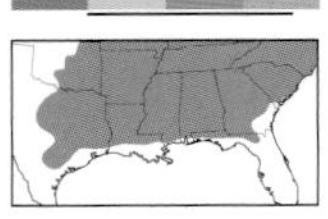

COPPER UNDERWING

Amphipyra pyramidoides 93-1544 (9638) **Common**

TL 23–28 mm. Broad, grayish brown FW has a paler ST area. Double AM line is fragmented into individual scallops. Double PM line is filled whitish. Pale orbicular spot has a dark spot in center. **HOSTS:** Trees and vines, including birch, elm, oak, Virginia creeper, and willow.

HOODED OWLETS

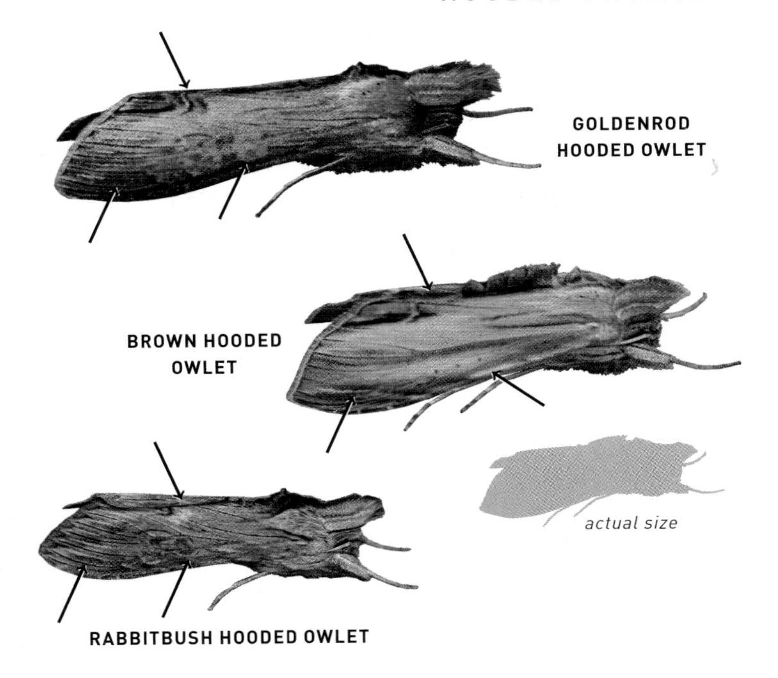

AMPHIPYRINE SALLOWS

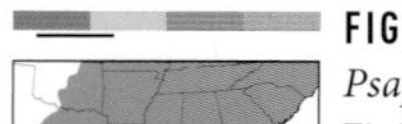

FIGURE-EIGHT SALLOW

Psaphida resumens 93-1548 (10019) **Common**

TL 20–21 mm. Medium gray FW has large whitish claviform and orbicular spots fused to form a figure eight. Reniform spot and ST area are also pale. Thorax is uniformly gray. **HOSTS:** Oak.

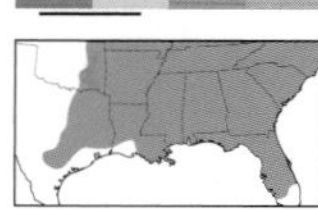

ROLAND'S SALLOW

Psaphida rolandi 93-1550 (10014) **Common**

TL 17–21 mm. Medium gray FW has gray orbicular and reniform spots circled thinly with speckled white and edged in black. ST line is speckled white. Thorax has paired oblique whitish stripes. **HOSTS:** Oak.

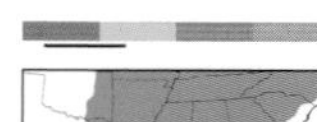

GRAY SALLOW

Psaphida grandis 93-1551 (10013) **Uncommon**

TL 18–21 mm. Slate gray FW has a narrow whitish patch bordering inner section of double AM line. Median area is often slightly darker. Dark-outlined reniform spot is inconspicuous. ST area is pale. **HOSTS:** Unknown, but probably oak.

CHOSEN SALLOW

Psaphida electilis 93-1552 (10012) **Uncommon**

TL 21–22 mm. Variable. Slate gray FW has hooked black basal dash typically edged warm brown. Black-edged orbicular and reniform spots, and white-edged ST line, are usually shaded brown. Some individuals lack brown shading but are otherwise similar in pattern. Others have a pale basal area, and broad black basal dash extending to inner ST line. **HOSTS:** Hickory and walnut.

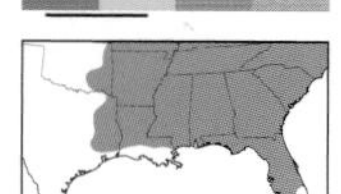

FAWN SALLOW

Psaphida styracis 93-1553 (10016) **Uncommon**

TL 17–21 mm. Peppery fawn-colored FW has a darker median area between rusty brown AM and PM lines. Pale orbicular and reniform spots are incompletely outlined rusty brown. Thorax is densely hairy. **HOSTS:** Oak.

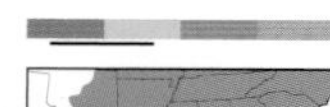

GROTE'S SALLOW

Copivaleria grotei 93-1557 (10021) **Common**

TL 21 mm. Resembles Figure-eight Sallow, but dark gray FW has mossy green accents. Claviform spot and ST area are gray. Whitish orbicular and reniform spots have dark centers. Thorax has a whitish band down middle. **HOSTS:** Ash.

MAJOR SALLOW

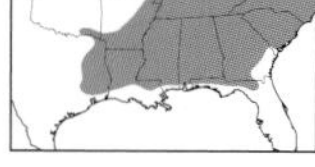

Feralia major 93-1563 (10007) **Common**

TL 20–22 mm. Dimorphic. Typical form is pale green with heavy black peppering through median area. Light form has jagged white-edged black lines and a black-edged square at midpoint of costa, between partially outlined orbicular and reniform spots. **HOSTS:** Pine and spruce.

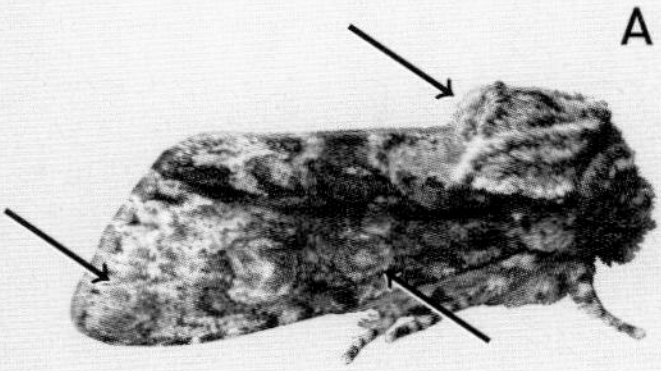

AMPHIPYRINE SALLOWS

FIGURE-EIGHT SALLOW

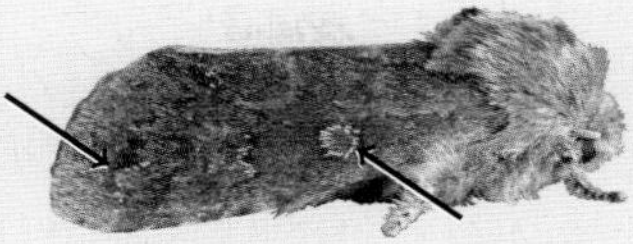

ROLAND'S SALLOW

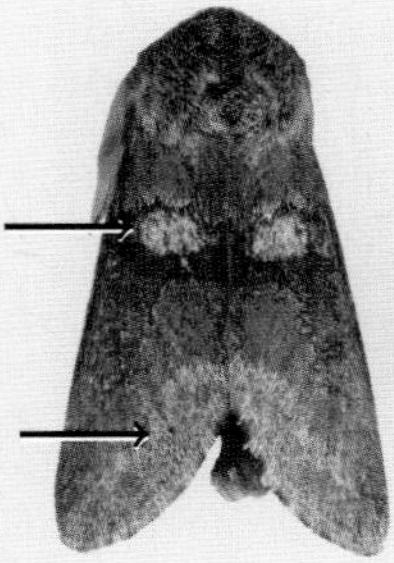

GRAY SALLOW

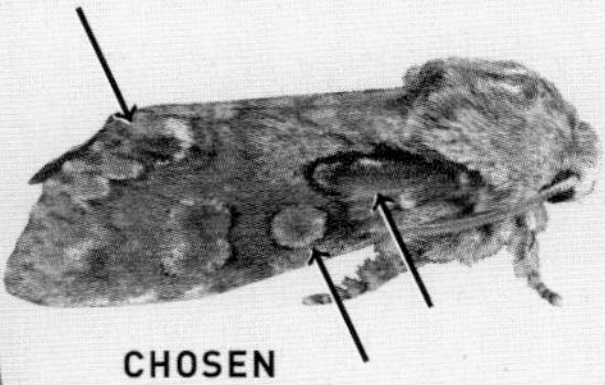

actual size

CHOSEN SALLOW

FAWN SALLOW

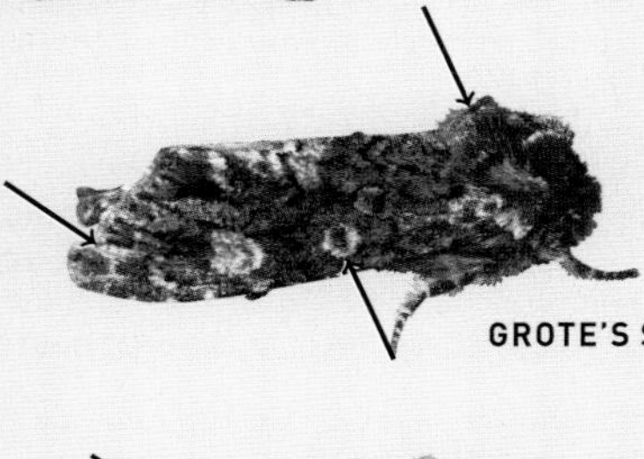

GROTE'S SALLOW

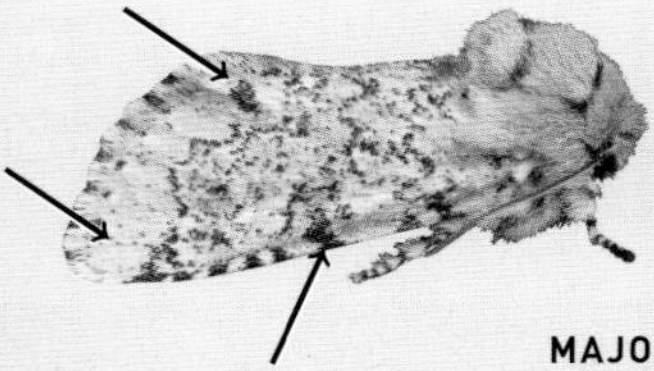

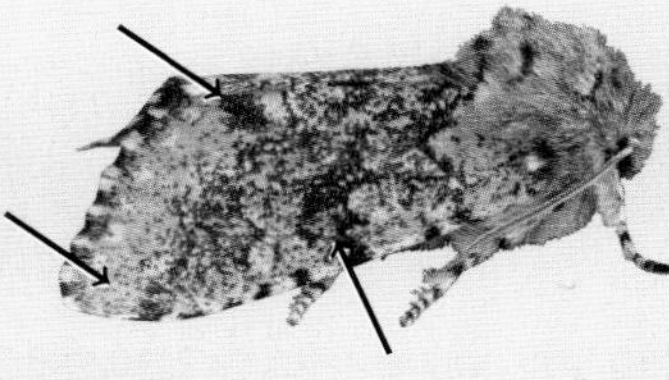

MAJOR SALLOW

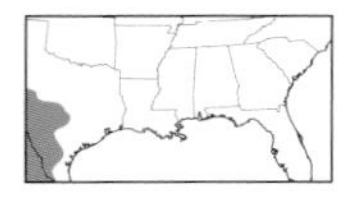

GREEN OSLARIA

Oslaria viridifera 93-1575 (9791) **Local**

TL 14–16 mm. Light green FW has darker olive green patches at inner median area and along costa in median and ST areas. Dark green AM and PM lines converge toward inner margin. Thorax is tinged olive. **HOSTS:** Unknown.

BELOVED EMARGINEA

Emarginea percara 93-1606 (9718) **Common**

TL 12–14 mm. Greenish FW has blackish shading to outer median area, with a large pale patch at costa. Lines are edged diffusely whitish. Black edge of thorax is blackish. **HOSTS:** Mistletoe.

INCA SALLOW *Aleptina inca* 93-1633 (9071) **Local**

TL 9–11 mm. Medium gray FW has a pale basal area tinged orangish. Pale orbicular spot has dark center and black outline. Whitish shading inside reniform spot bleeds to costa. **HOSTS:** Unknown.

ROGENHOFER'S SALLOW

Metaponpneumata rogenhoferi 93-1640 (9074) **Local**

TL 11–13 mm. Grayish to dark brown FW often has a paler basal area. Kidney-shaped reniform spot has a brownish center and is partly edged black. Wavy black AM and PM lines are often edged brownish. **HOSTS:** Herbaceous plants, including corn and sorghum.

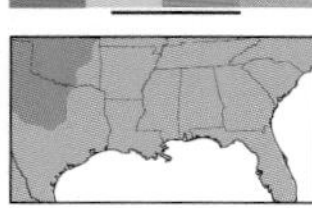

FROTHY MOTH

Plagiomimicus spumosum 93-1651 (9748) **Uncommon**

TL 14–18 mm. Light brown FW has a darker brown median area and is heavily speckled with whitish scales. AM and ST lines are inconspicuous. Straight white PM line is sharply angled and widens at costa. Apex is pointed. **HOSTS:** Seeds of sunflower.

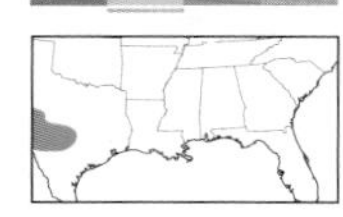

NAVY MOTH *Plagiomimicus navia* 93-1658 (9751) **Local**

TL 10–12 mm. Resembles Frothy Moth, but PM line is curved and uniform in width, and thin, pale AM line is often visible. Sometimes inconspicuous hourglass-shaped reniform spot contains two tiny blackish dots. **HOSTS:** Unknown.

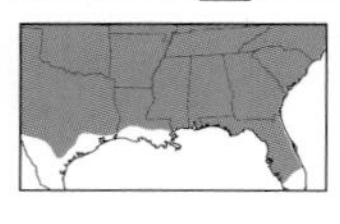

BLACK-BARRED BROWN

Plagiomimicus pityochromus 93-1661 (9754) **Uncommon**

TL 14–18 mm. Peppery grayish brown FW has a large, dark brown subapical patch. Blackish claviform and orbicular spots are fused into an hourglass-shaped bar. AM and PM lines are pale. **HOSTS:** Great ragweed.

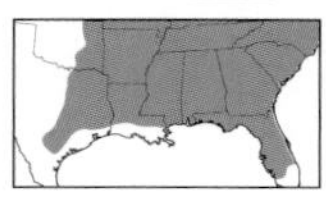

GOLD MOTH *Basilodes pepita* 93-1676 (9781) **Common**

TL 18–24 mm. Metallic gold FW has strongly angled AM and PM lines that are finely etched in brown. Round orbicular and reniform spots are narrowly outlined brown. Reniform spot contains a tiny black dot near center. **HOSTS:** Crownbeard.

AMPHIPYRINE SALLOWS

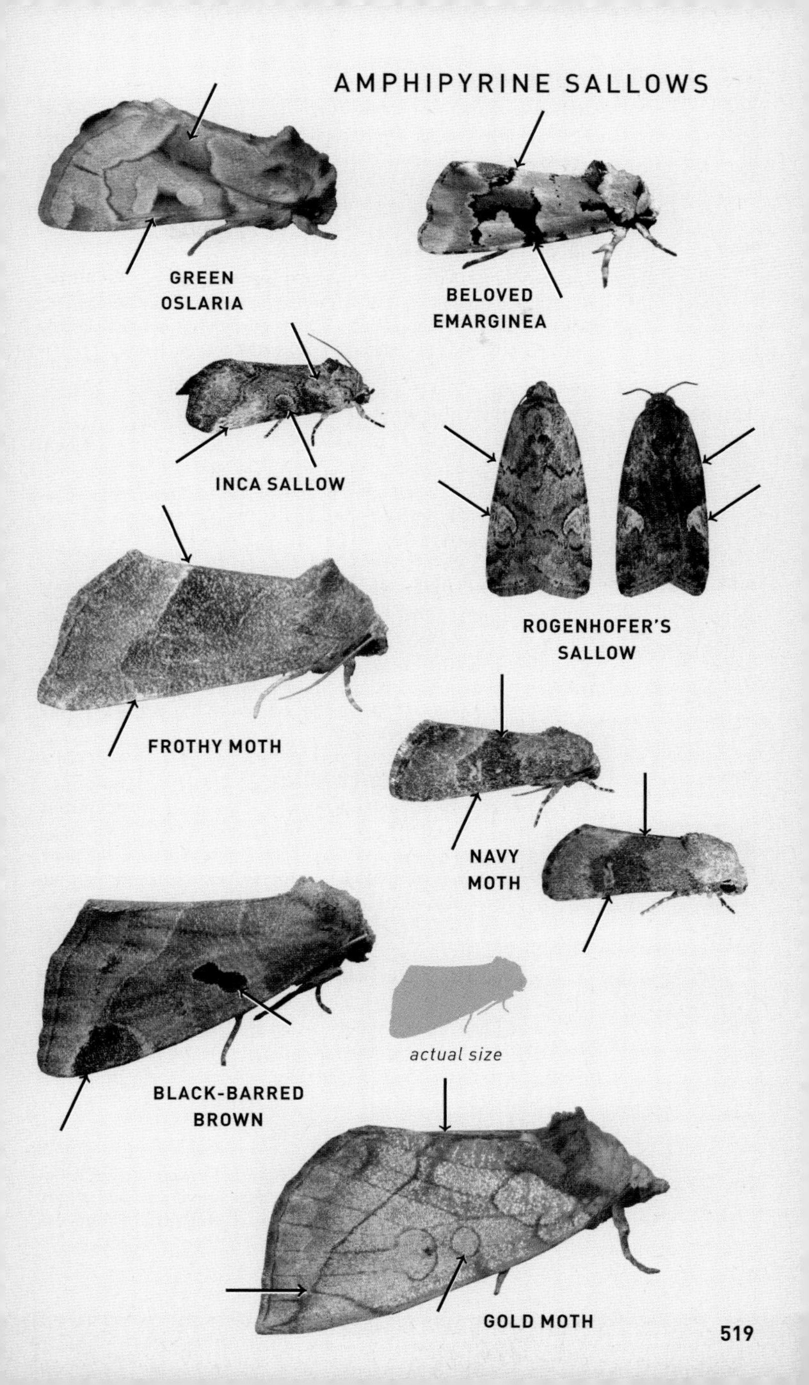

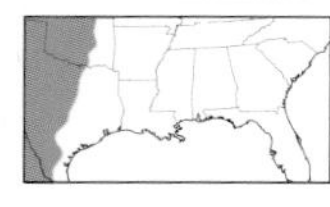

BRONZE MOTH

Basilodes chrysopis 93-1678 (9780) **Local**

TL 16–18 mm. Golden FW has a darker, bronzy median area. Wavy brown AM and PM lines converge toward inner margin. Rounded golden orbicular and reniform spots are outlined brown. Outer half of ST area is shaded grayish. **HOSTS:** Golden crownbeard.

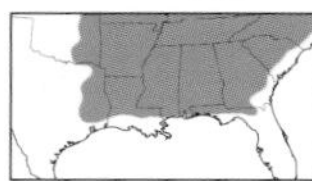

GOLDENROD STOWAWAY

Cirrhophanus triangulifer 93-1681 (9766) **Common**

TL 20–21 mm. Shining yellow FW has orange streaks along veins above sinuous PM line. Broad, diffuse ST band grows faint in pale inner ST area. **HOSTS:** Spanish needle. **NOTE:** Often found on yellow flowers (especially goldenrod) during daylight.

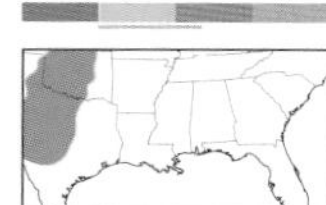

EXPENSIVE STOWAWAY

Cirrophanus pretiosa 93-1682 (9766.1) **Local**

TL 14–16 mm. Resembles slightly larger Goldenrod Stowaway, but ST line is narrower and more defined. Inner ST area is often contrastingly whitish. Flight period and range are easiest means of identification. **HOSTS:** Unknown.

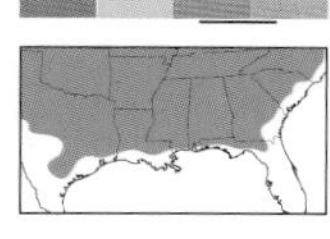

YELLOW SUNFLOWER MOTH

Stiria rugifrons 93-1688 (9785) **Uncommon**

TL 18–20 mm. Yellow FW has contrasting chocolate brown saddle along inner margin. Brown fringe spills into midpoint of ST area, forming a two-humped bulge. Scalloped lines are weakly defined. Inconspicuous yellow reniform spot contains a tiny brown dot. Thorax is brown. **HOSTS:** Sunflower.

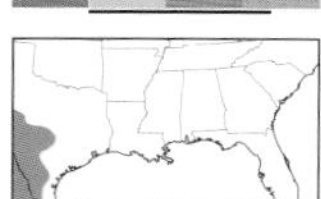

INTERMINGLED SUNFLOWER MOTH

Stiria intermixta 93-1689 (9785.1) **Uncommon**

TL 18–21 mm. Resembles Yellow Sunflower Moth, but brown shading in ST area is typically more rounded and tapers evenly to apex. The only sunflower moth flying in spring in its range; some individuals may not be identifiable in the field in the fall. **HOSTS:** Sunflower.

OBTUSE YELLOW

Azenia obtusa 93-1724 (9725) **Common**

TL 12–14 mm. Yellow FW has wide median line broken into three patches. ST line is a row of dots ending in a square subapical patch. AM and PM lines are reduced to scattered dots. **HOSTS:** Great ragweed.

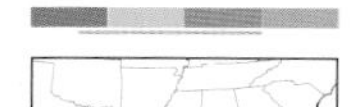

TEXAS YELLOW

Azenia perflava 93-1726 (9727) **Local**

TL 14–16 mm. Resembles Obtuse Yellow, but brown patches at inner margin and costa are smaller, and lacks third median patch in center. Head and thorax are yellowish brown. **HOSTS:** Unknown.

AMPHIPYRINE SALLOWS

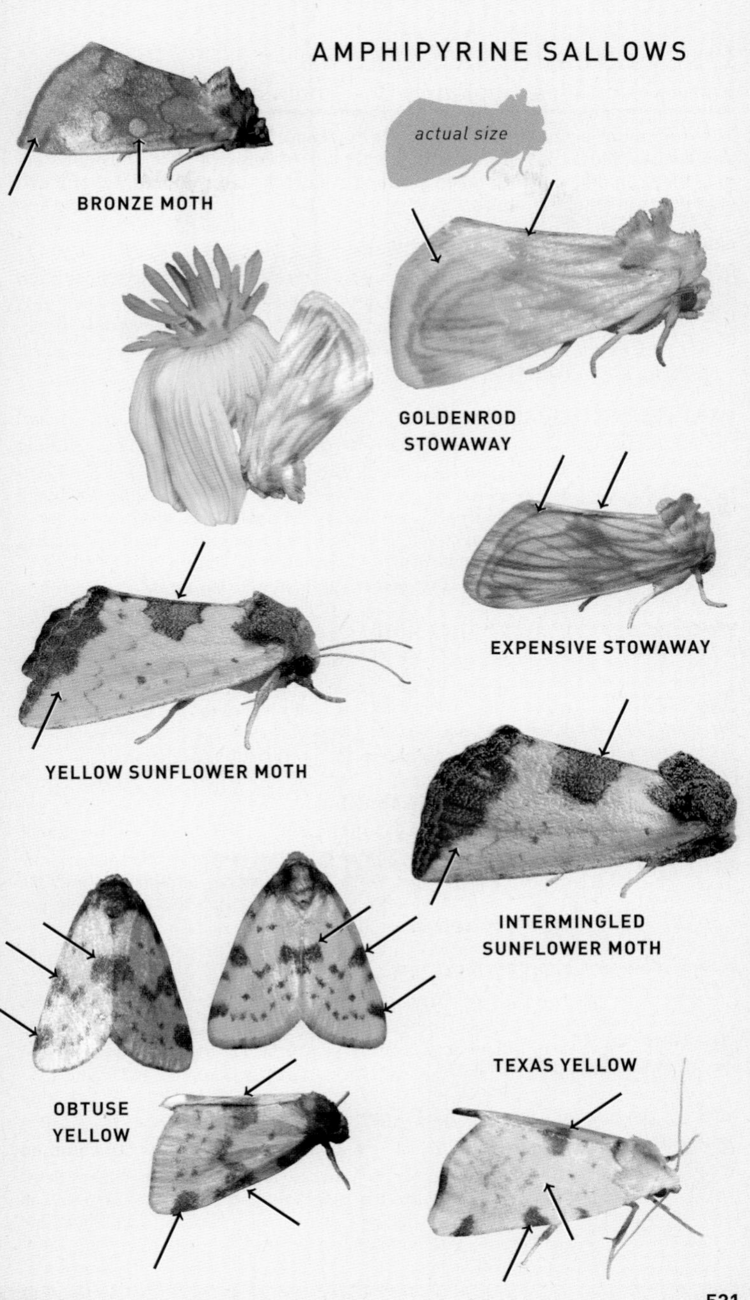

Oncocnemidine Sallows

Family Noctuidae, Subfamily Oncocnemidinae

These medium-sized grayish moths are boldly patterned with crisp black lines and dashes. Most rest with their wings tented. Mainly dwellers of wooded areas, they can also occur in large gardens. They are strictly nocturnal and will visit lights in small numbers.

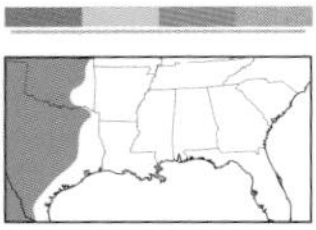

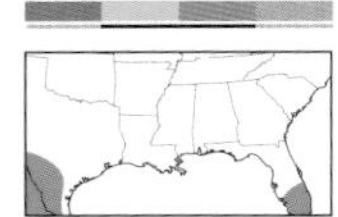

FINE-LINED SALLOW

Catabena lineolata 93-1765 (10033) **Uncommon**

TL 13–14 mm. Ash gray FW has black veins finely edged white. Thin blackish PM line is strongly jagged and frequently indistinct. Orbicular spot is often present as a long, narrow, black-edged pale dash. **HOSTS:** Goldenrod and hoary vervain.

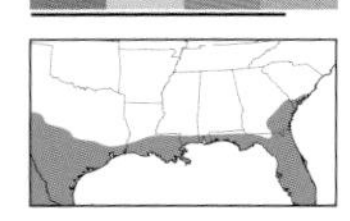

GLASS SALLOW

Catabenoides vitrina 93-1767 (10036) **Local**

TL 12–14 mm. Sexually dimorphic. Male has light grayish brown FW with blackish blotches between veins of ST area. Black veins are thinly edged pale, particularly in ST area. Diffusely brown PM line is often indistinct and is bordered by a pale patch containing small black dots at inner margin. Female is similar but has a thick black streak from base to outer margin. **HOSTS:** Unknown.

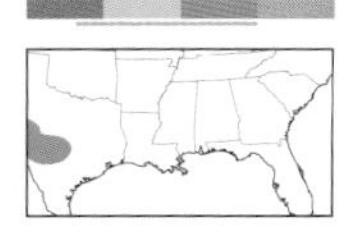

LANTANA STICK MOTH

Neogalea sunia 93-1770 (10032) **Uncommon**

TL 16–18 mm. Peppery grayish brown FW is streaked blackish between veins. Jagged PM line is inconspicuous but is edged whitish at inner margin and costa. Orbicular spot is a small white dot. Thorax has a blunt, forward-pointing crest. **HOSTS:** Lantana.

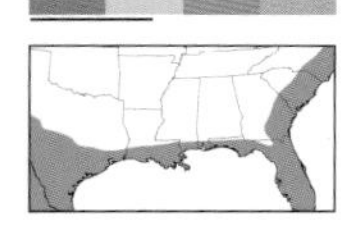

GOLD-WINGED COPANARTA

Copanarta aurea 93-1781 (10169) **Local**

TL 10–12 mm. Gray FW often has a darker median band above large, whitish reniform spot. White ST line touches terminal line multiple times, and is most noticeable near anal angle. Bright orange HW has a thick black border. **HOSTS:** Unknown.

SCRIBBLED SALLOW

Sympistis perscripta 93-1797 (10154) **Uncommon**

TL 16–17 mm. Light gray FW has bold, evenly wavy black AM and PM lines, usually edged warm brown. Orbicular, claviform, and reniform spots are thinly outlined black. **HOSTS:** Snapdragon and toadflax.

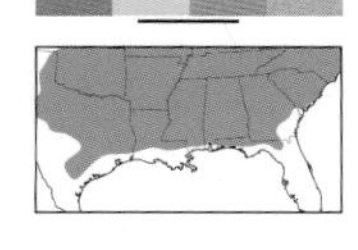

BROWN-LINED SALLOW

Sympistis badistriga 93-1821 (10059) **Uncommon**

TL 17–18 mm. Peppery gray to brownish FW has a thick black basal dash and a subapical dash extending up to AM line. ST area has streaks of brown shading between veins. Narrow white-edged AM and PM lines are strongly curved. **HOSTS:** Honeysuckle.

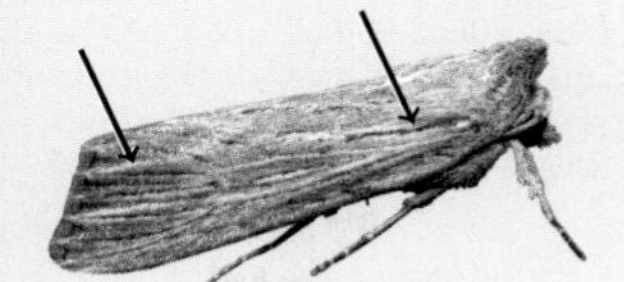

FINE-LINED SALLOW

GLASS SALLOW

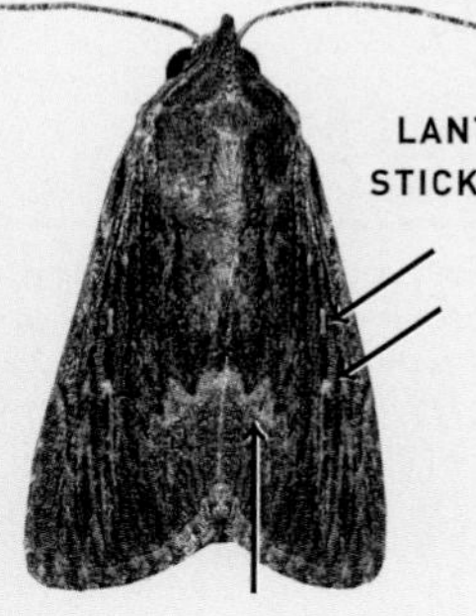

LANTANA STICK MOTH

GOLD-WINGED COPANARTA

SCRIBBLED SALLOW

actual size

BROWN-LINED SALLOW

BROAD-LINED SALLOW

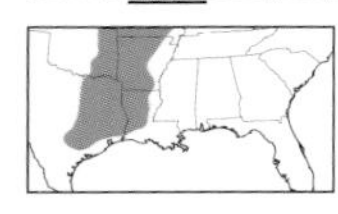

Sympistis dinalda 93-1823 (10066.1) **Uncommon**

TL 17–18 mm. Resembles Brown-lined Sallow, but subapical dash extends only to middle of median area, and basal dash widens at PM line. Orbicular and reniform spots are sometimes outlined pale. Thorax has whitish yoke. **HOSTS:** Snowberry.

FRINGE-TREE SALLOW

Sympistis chionanthi 93-1906 (10067) **Uncommon**

TL 18–22 mm. Ash gray FW has bold, smoothly curving AM line that connects to V-shaped claviform spot. PM line runs along inner edge of reniform spot. Circular orbicular spot is edged black and sometimes shaded gray in center. Thick black anal dash cuts through jagged white ST line. **HOSTS:** Ash and fringe tree.

WOOD-NYMPHS AND FORESTERS

Family Noctuidae, Subfamily Agaristinae

This subfamily contains two very different groups of moths. The nocturnal, woodland-dwelling wood-nymphs have broad white wings that are held tented, and long forelegs covered in downy tufts that are outstretched at rest. They freely come to lights. The foresters are day-fliers that generally have a white-spotted, black FW and either similar or bright orange HW; they are best sought in flowery meadows and open woodlands.

WILSON'S WOOD-NYMPH

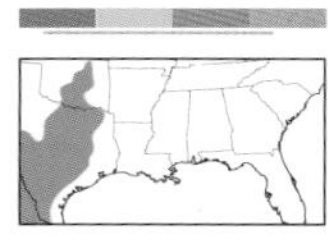

Xerociris wilsonii 93-1963 (9298) **Local**

TL 18–20 mm. Satin white FW has a gray ST area edged by a wavy olive ST line that bleeds into a gray band along inner margin and inner AM area. Olive orbicular and reniform spots are connected by a gray costal patch. Thorax has a gray-speckled black band down center. **HOSTS:** Sorrelvine.

PEARLY WOOD-NYMPH

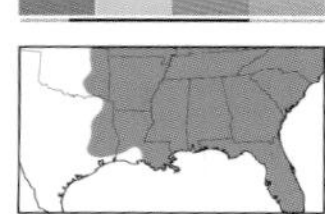

Eudryas unio 93-1964 (9299) **Common**

TL 21 mm. Resembles Wilson's Wood-Nymph, but ST area is crimson with scalloped ST line, and inner margin band does not extend into AM area. Spots and costal patch are wine red. **HOSTS:** Evening primrose, false loosestrife, grape, hibiscus, Virginia creeper, and others. **NOTE:** Two or more broods.

BEAUTIFUL WOOD-NYMPH

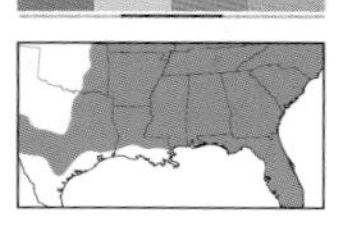

Eudryas grata 93-1966 (9301) **Common**

TL 24 mm. Resembles Pearly Wood-Nymph, but ST area is darker red with an evenly curved ST line. Inner portion of spots and costal patch is streaked olive. **HOSTS:** Peppervine, grape, and Virginia creeper. **NOTE:** Two or more broods.

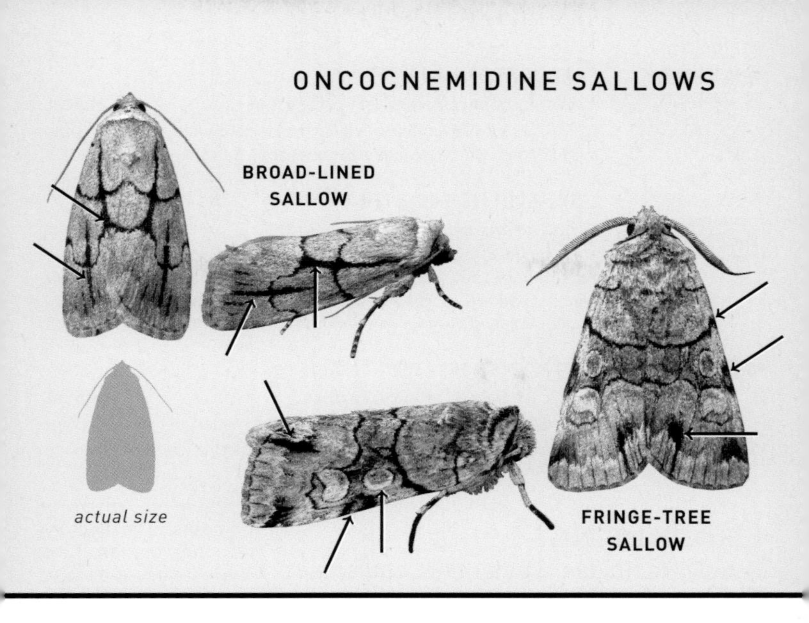
ONCOCNEMIDINE SALLOWS
BROAD-LINED SALLOW
actual size
FRINGE-TREE SALLOW

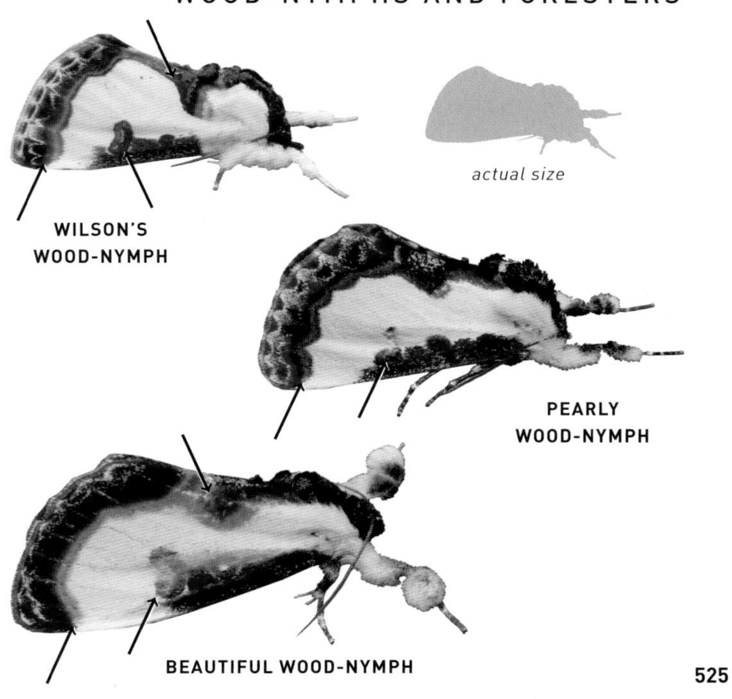
WOOD-NYMPHS AND FORESTERS
actual size
WILSON'S WOOD-NYMPH
PEARLY WOOD-NYMPH
BEAUTIFUL WOOD-NYMPH

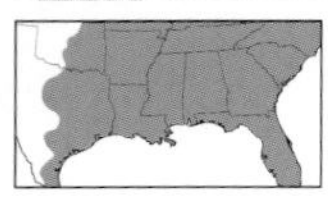

GRAPEVINE EPIMENIS

Psychomorpha epimenis 93-1975 (9309) Local

TL 12–14 mm. Velvety blackish FW has a bold white patch in outer median area. Black HW has a broad orange band. **HOSTS:** Grape.

EIGHT-SPOTTED FORESTER

Alypia octomaculata 93-1979 (9314) Common

TL 16–20 mm. Velvety black FW has two large cream-colored spots. HW is similar, but spots are white. Thorax is bordered by pale yellow bands. Legs are adorned with showy orange tufts. **HOSTS:** Peppervine, grape, and Virginia creeper. **NOTE:** One or more broods.

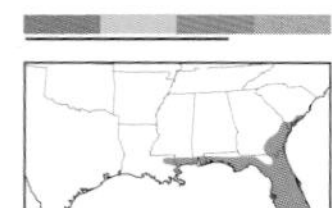

WITTFELD'S FORESTER

Alypia wittfeldii 93-1980 (9316) Local

TL 16–20 mm. Resembles Eight-spotted Forester, but FW spots are pale yellow and more bandlike in shape. **HOSTS:** Reported on Japanese persimmon.

GROUNDLINGS
Family Noctuidae, Subfamily Condicinae

A group of mostly small to medium-sized delta-shaped noctuids that rest with their wings flat or slightly tented. Found in woodlands and larger gardens, they are nocturnal and will come to lights in small numbers.

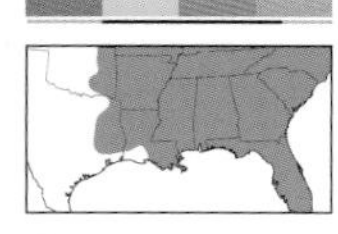

RED GROUNDLING

Perigea xanthioides 93-1986 (9689) Common

TL 13–18 mm. Pale orange FW is mottled rusty brown with inconspicuous scalloped lines. Median area is slightly darker. Orange reniform and orbicular spots are partly edged white along inner half. A gray streak passes through central median area. **HOSTS:** Ironweed and sweet joe-pye weed. **NOTE:** Two or more broods.

WHITE-DOTTED GROUNDLING

Condica videns 93-1989 (9690) Common

TL 13–18 mm. Tawny brown FW has a vertical blackish streak that passes through inner half of fragmented white reniform spot. Dark veins are peppered whitish. PM line is a series of white-edged black dots at veins. **HOSTS:** Composites, including aster and goldenrod. **NOTE:** Two or more broods.

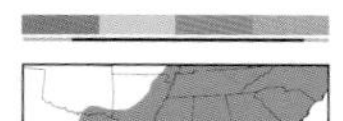

MOBILE GROUNDLING

Condica mobilis 93-1992 (9693) Common

TL 15–22 mm. Orange to reddish brown FW has a bold white dot in inner half of reniform spot. Dark veins are speckled whitish. All lines are indistinct and slightly scalloped; double AM and PM lines are filled with the lightest shade. **HOSTS:** Spanish needle. **NOTE:** Two or more broods.

WOOD-NYMPHS AND FORESTERS

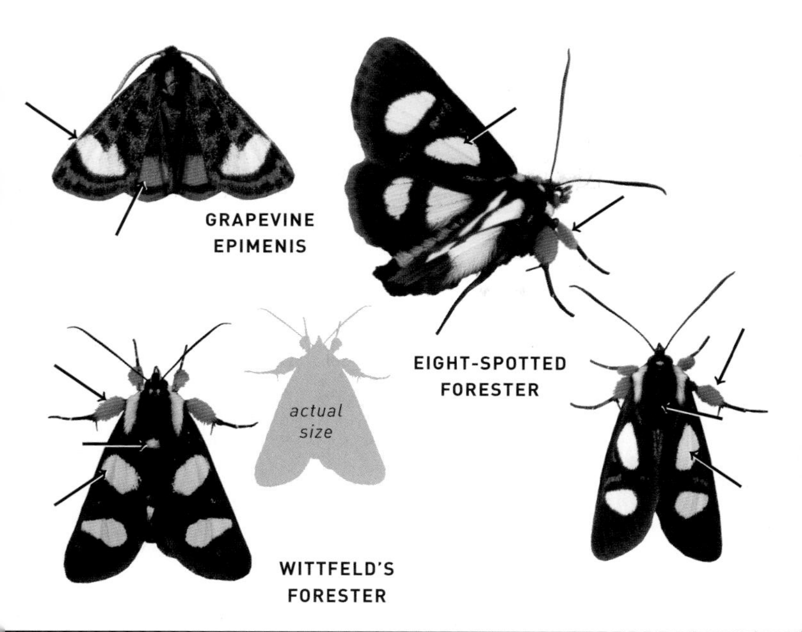

GROUNDLINGS

DUSKY GROUNDLING

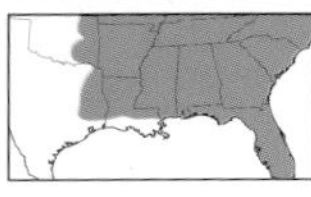

Condica vecors 93-1995 (9696) **Common**

TL 16–18 mm. Resembles Mobile Groundling, but FW is usually dark brown. Orbicular spot is usually partly outlined tan on upper edge. Black claviform spot is often noticeable. Fragmented pale ST line is accented with black wedges between veins. **HOSTS:** *Eupatorium* species. **NOTE:** Two or more broods.

CONCISE GROUNDLING

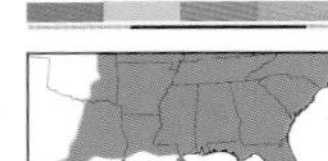

Condica concisa 93-1997 (9698) **Local**

TL 16–18 mm. Resembles Dusky Groundling, but FW is lighter brown and pale reniform spot lacks bold white dot in inner half. Black orbicular spot is usually reduced to a small loop. Double AM and PM lines are filled with light shading. **HOSTS:** Possibly beggar tick.

THE COBBLER

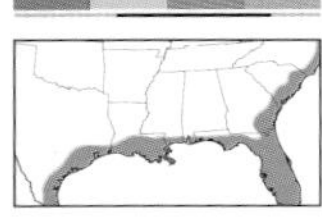

Condica sutor 93-1998 (9699) **Uncommon**

TL 16–18 mm. Resembles Mobile Groundling, but FW is less mottled and reniform spot lacks white dot in inner half. A diffuse dark streak connects orbicular and reniform spots. Veins sometimes lack black scaling. Terminal line is a series of white dots at veins. **HOSTS:** Celery, marigold, and various composites. **NOTE:** Multiple broods in south.

SPLOTCHED GROUNDLING

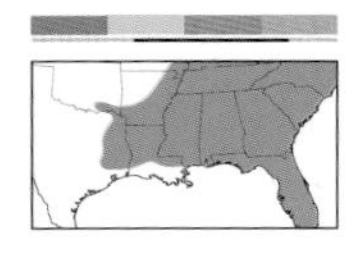

Condica cupentia 93-2014 (9713) **Uncommon**

TL 18–22 mm. Sexually dimorphic. Male has tan FW with sharply defined chocolate brown patches in basal and ST areas. Pale terminal line is interrupted by brown blotch at midpoint. Orbicular and reniform spots are outlined slightly paler tan. Female is brown with less contrast between ground color and dark patches. **HOSTS:** Unknown.

THE CONFEDERATE

Condica confederata 93-2015 (9714) **Uncommon**

TL 18–22 mm. Sexually dimorphic. Resembles Splotched Groundling but has a more scalloped appearance. Double AM and PM lines are filled creamy and have a creamy patch in basal area. Claviform spot is black. Female is darker with brown shading through pale median and ST areas. **HOSTS:** Climbing hempvine. **NOTE:** Multiple broods in south.

COMMON PINKBAND

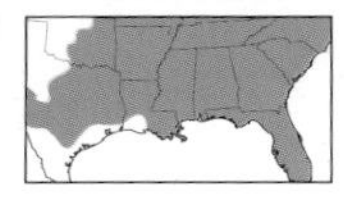

Ogdoconta cinereola 93-2018 (9720) **Common**

TL 12–13 mm. Speckled grayish to brown FW has paler, pinkish ST area; basal area is sometimes also pale. All lines and outlines to spots are thin and whitish. Reniform spot is pinched at middle, forming a figure eight. **HOSTS:** Burdock, wingstem, and various composites, such as ragweed and sunflower. **NOTE:** Three or more broods.

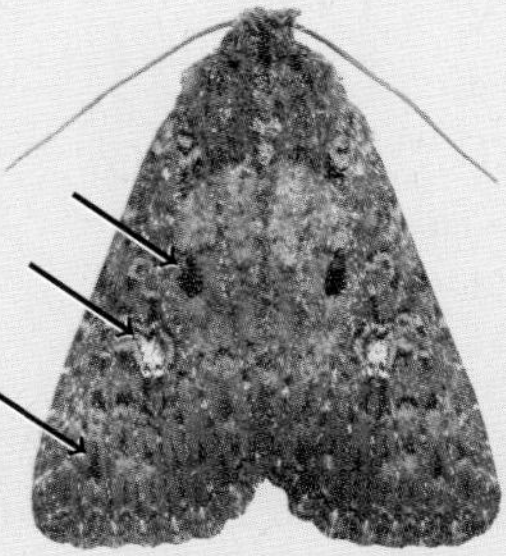

GROUNDLINGS

DUSKY GROUNDLING

CONCISE GROUNDLING

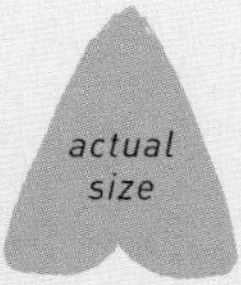

actual size

THE COBBLER

SPLOTCHED GROUNDLING

THE CONFEDERATE

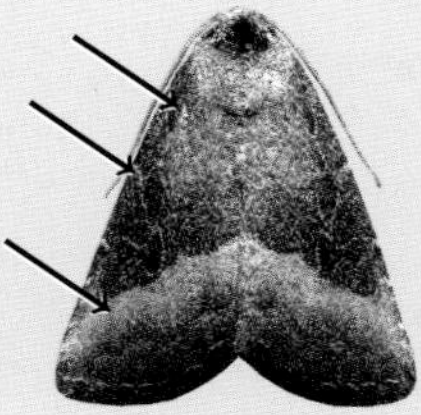

COMMON PINKBAND

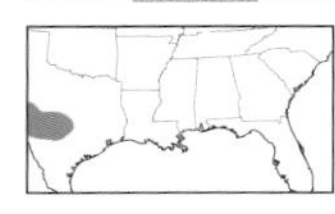

DARK NEARLY-EIGHT

Ogdoconta tacna 93-2022 (9724) **Local**

TL 13–15 mm. Resembles Common Pinkband but lacks pale ST and basal areas. Inner reniform spot has a tooth that almost touches orbicular spot. **HOSTS:** Unknown.

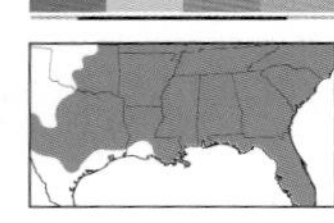

WATER-LILY MOTH

Homophoberia cristata 93-2024 (9056) **Uncommon**

TL 14–17 mm. Sexually dimorphic. Brown to slate gray FW has pale ST area beyond wavy black PM line. Orbicular and reniform spots are outlined pale tan and filled brownish. Male has brown shading bordering pale ST line and bipectinate antennae; female lacks shading and has filiform antennae. **HOSTS:** Yellow pond lily.

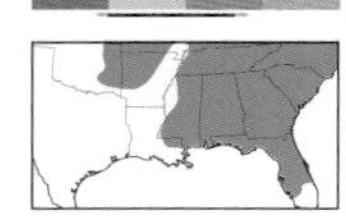

BLACK WEDGE-SPOT

Homophoberia apicosa 93-2025 (9057) **Common**

TL 11–13 mm. Resembles female of larger Water-lily Moth, but round reniform spot is larger and paler. Orbicular spot is darkly shaded. Claviform spot is a black outward-pointing wedge. Thorax has a white spot behind a short crest. Often rests with tip of abdomen raised above wings. **HOSTS:** Smartweed. **NOTE:** Two or more broods.

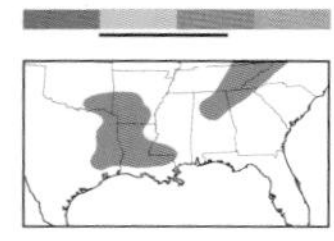

GREEN LEUCONYCTA

Leuconycta diphteroides 93-2026 (9065) **Common**

TL 15–16 mm. Pale green FW has square blackish patches in outer median and basal areas. White-filled double AM and PM lines are strongly scalloped into a chain of spots. Indistinct orbicular and reniform spots are outlined white. Worn individuals may appear whitish. **HOSTS:** Composites, including goldenrod and aster. **NOTE:** Two broods.

MARBLED-GREEN LEUCONYCTA

Leuconycta lepidula 93-2027 (9066) **Local**

TL 16 mm. Resembles Green Leuconycta but has more extensive black shading, including a blackish saddle in inner median area. Orbicular and reniform spots are usually strongly outlined white. **HOSTS:** Unknown.

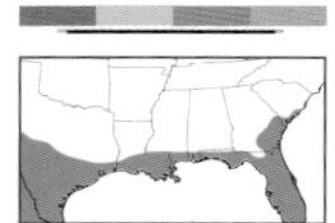

TIGER LANTANA MOTH

Diastema tigris 93-2028 (9067) **Local**

TL 13–14 mm. Straw-colored FW is strongly brindled with parallel pale orange, whitish, and rusty lines. Dark crimson basal, AM, and median bands are fragmented into chunky spots. **HOSTS:** Lantana. **NOTE:** Multiple broods.

DARK NEARLY-EIGHT

WATER-LILY MOTH

BLACK WEDGE-SPOT

GREEN LEUCONYCTA

actual size

MARBLED-GREEN LEUCONYCTA

TIGER LANTANA MOTH

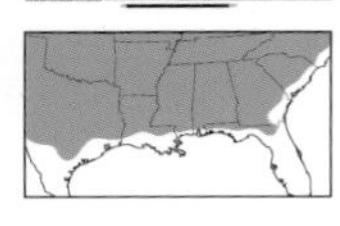

VERBENA MOTH

Crambodes talidiformis 93-2030 (9661) **Uncommon**

TL 16 mm. Straw-colored FW has streaky pattern of brown lines between veins. Costa and inner median area are dusky or brown. Bar-like reniform spot is pale, partly edged blackish. Thin, scalloped blackish PM line is sometimes indistinct. Fringe is checkered black and white. **HOSTS:** Vervain. **NOTE:** Two broods.

TRIPLEX CUTWORM

Micrathetis triplex 93-2031 (9644) **Common**

TL 9–13 mm. Speckled FW is variably straw-colored to light brown, with inconspicuous dotted lines. Blackish reniform spot is usually joined to costa by a dark smudge. ST area is shaded blackish. **HOSTS:** Unknown.

FLOWER MOTHS

Family Noctuidae, Subfamily Heliothinae

Small to medium-sized, often beautifully patterned noctuids. Many species are similar in appearance, and the HW color and pattern are often important for identification. Some are regularly encountered during the day taking nectar from flowers. A few, such as Corn Earworm, are prone to irruptive northerly movements in late summer and fall. Most species, even the day-fliers, are attracted to lights in small numbers.

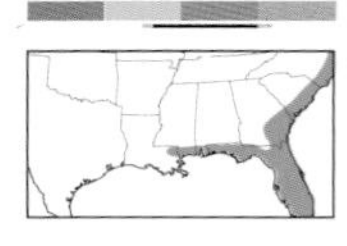

PINK STAR *Derrima stellata* 93-2039 (11055) **Local**

TL 10–12 mm. Golden yellow FW is bordered pink along costa and outer margin. Dark-edged claviform and reniform spots are filled white (sometimes gray). PM line is a series of white dots. **HOSTS:** Trees and plants, including alder, cabbage, rose, sumac, and walnut. **NOTE:** Two broods.

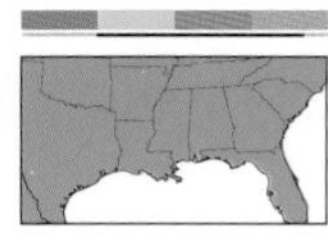

CORN EARWORM

Helicoverpa zea 93-2045 (11068) **Common**

TL 18–22 mm. Yellowish tan to dull orange FW has a darker band beyond scalloped PM line. Round orbicular spot has a dark dot in center. Reniform spot typically has a dusky dot in inner half. HW is whitish with blackish veins and dark terminal line. **HOSTS:** Low plants and crops, including corn, cotton, tomato, tobacco, and many others. **NOTE:** Multiple broods in south.

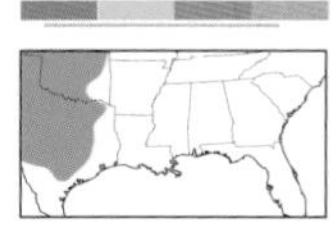

DARKER-SPOTTED STRAW MOTH

Heliothis phloxiphaga 93-2046 (11072) **Uncommon**

TL 17–20 mm. Pale tan FW has a dusky reniform spot at midpoint of strongly angled chestnut median band. Pale-edged ST line ends as a dark brown patch at costa. HW has a dark terminal line with a narrow, pale patch at midpoint. **HOSTS:** Flowers and seedheads of various herbaceous plants.

GROUNDLINGS

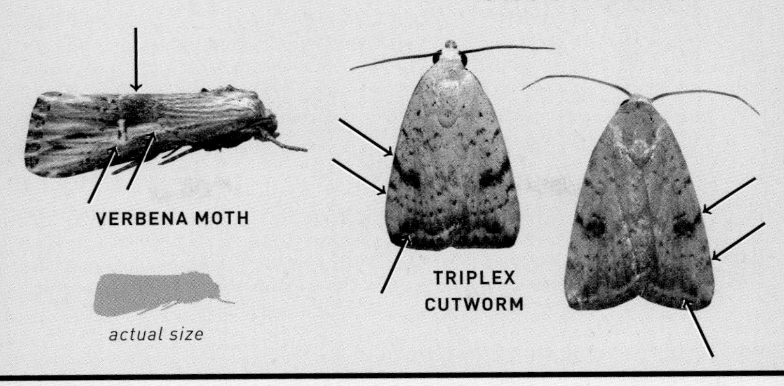

FLOWER MOTHS

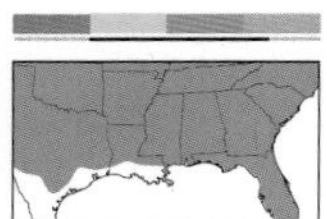

TOBACCO BUDWORM

Chloridea virescens 93-2054 (11071) **Common**

TL 15–19 mm. Greenish FW has parallel, dark, diffuse lines that are edged white above. PM line ends in subapical area. White HW has a dark ST band. **HOSTS:** *Abutilon*, cotton, geranium; also ground cherry, tobacco, and other members of nightshade family. **NOTE:** Multiple broods in south.

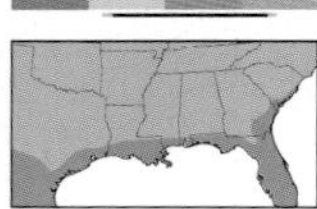

SUBFLEXUS STRAW MOTH

Chloridea subflexa 93-2055 (11070) **Uncommon**

TL 15–19 mm. Resembles Tobacco Budworm, but PM line curves to reach apex. Costa is usually paler. Whitish edging to lines is wider, and veins are often traced faintly whitish, making moth appear paler overall. **HOSTS:** Ground cherry and nightshade.

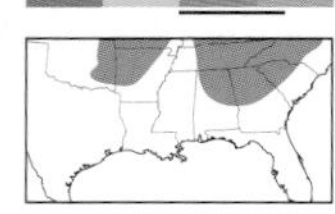

LUPATUS STRAW MOTH

Heliocheilus lupatus 93-2056 (11073.1) **Uncommon**

TL 15–19 mm. Yellowish FW has a large, blackish reniform spot at midpoint of dark median line. Distal half of FW is usually shaded orange, often with darker shading toward inner margin. Scalloped PM line often appears as a series of white-edged black dots. HW is yellow with a thin median line. **HOSTS:** Purpletop tridens and possibly other grasses.

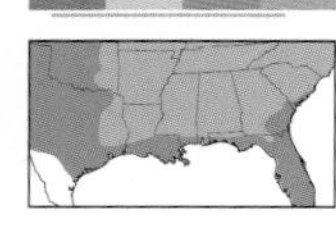

PARADOXICAL GRASS MOTH

Heliocheilus paradoxus 93-2058 (11074) **Local**

TL 13–15 mm. Sexually dimorphic. Straw-colored FW is often mostly unmarked aside from a faint dusky streak down center. PM line is a row of tiny white dots, sometimes bordered blackish. Male has a semi-translucent bulge in outer median area. Female is often darker, with a dark ST area and jagged PM and median lines. HW has a dark terminal line with a pale patch at midpoint. **HOSTS:** Grass.

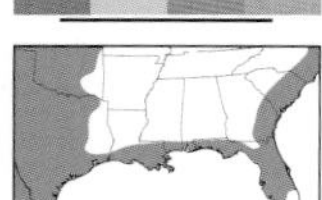

JAGUAR FLOWER MOTH

Schinia jaguarina 93-2073 (11132) **Common**

TL 14–16 mm. Light brown FW has warm brown basal and ST areas. Smooth AM line is strongly angled at midpoint. PM line meets costa perpendicularly. Diffuse, dusky reniform spot is often faint. Golden yellow HW has a dark terminal line with a pale patch at midpoint. **HOSTS:** *Psoralea* species.

CLOUDED CRIMSON

Schinia gaurae 93-2083 (11168) **Common**

TL 14–16 mm. Pale yellow FW has a slash of crimson shading angling inward from apex. Basal area and terminal line are suffused pinkish. Often has golden yellow wash in median area. White HW has a flush of pink in outer terminal line. **HOSTS:** Gaura. **NOTE:** One or two broods.

FLOWER MOTHS

SUBFLEXUS STRAW MOTH

TOBACCO BUDWORM

LUPATUS STRAW MOTH

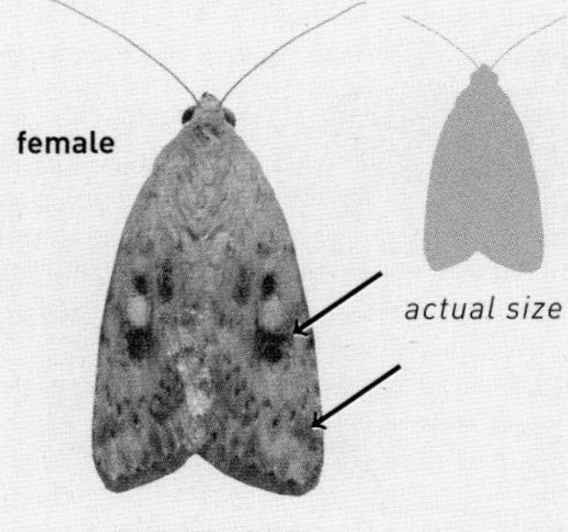

PARADOXICAL GRASS MOTH

JAGUAR FLOWER MOTH

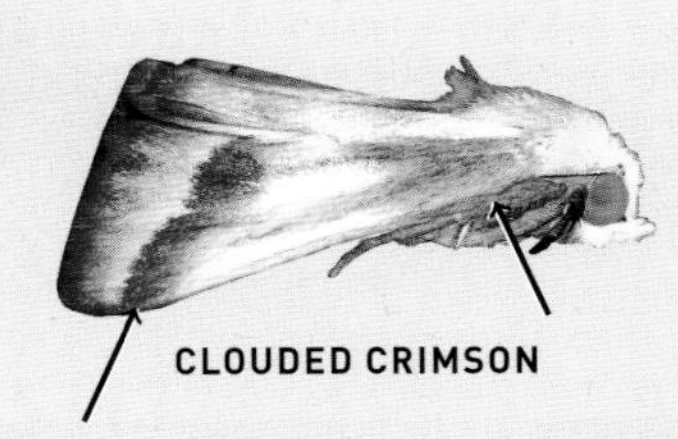

CLOUDED CRIMSON

PAINTBRUSH FLOWER MOTH

Schinia cupes 93-2084 (11134) **Local**

TL 14–16 mm. Brown FW has scalloped, double AM and PM lines filled tan. Veins are traced lightly whitish. Dusky reniform spot has a pale crescent in center. Straw-colored HW has brown veins and terminal line with a diffuse pale spot. **HOSTS:** Indian paintbrush.

RAGWEED FLOWER MOTH

Schinia rivulosa 93-2091 (11135) **Common**

TL 14–16 mm. Resembles Jaguar Flower Moth, but FW is usually grayish with dark brown to blackish basal and ST areas. White ST line is well defined. Dusky reniform spot is often noticeable. AM and PM lines sometimes meet at midpoint. **HOSTS:** Ragweed.

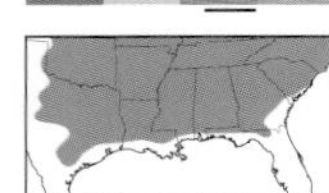

SLENDER FLOWER MOTH

Schinia gracilenta 93-2092 (11147) **Common**

TL 14–16 mm. Resembles Jaguar Flower Moth, but AM line is more rounded and PM line curves up to meet costa. Pale HW has a faint terminal line. **HOSTS:** Jesuit's bark and annual marsh elder.

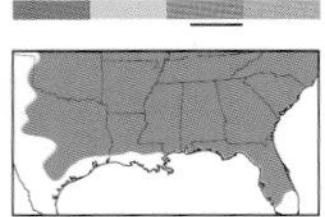

THOREAU'S FLOWER MOTH

Schinia thoreaui 93-2093 (11141) **Uncommon**

TL 16–19 mm. Resembles Jaguar Flower Moth, but AM line is sharply angled. AM and PM lines are narrow and wavy. Reniform spot is simply a blotch of diffuse shading. **HOSTS:** Great ragweed.

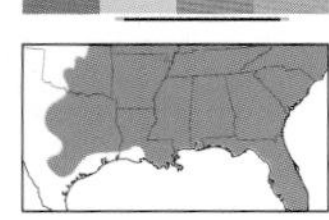

THREE-LINED FLOWER MOTH

Schinia trifascia 93-2096 (11149) **Common**

TL 12–17 mm. Sexually dimorphic. Resembles Tobacco Budworm, but AM line curves strongly upward before reaching costa, framing dark green basal patch. All lines are whitish and broad but diffuse, edged dark green. HW is mostly white in male, white with an olive ST band in female. **HOSTS:** False boneset and joe-pye weed.

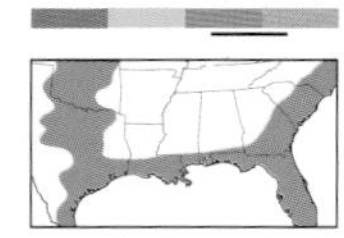

SORDID FLOWER MOTH

Schinia sordidus 93-2116 (11112) **Common**

TL 10–12 mm. Warm brown FW has darker basal and ST areas. Pale, lightly scalloped AM and PM lines are often indistinct; PM line curves strongly up and widens at costa. Diffuse reniform spot is blackish. Some individuals are darker brown with blackish edging to AM and PM lines and mottled dark shading in median area. Black HW has whitish spot and fringe. **HOSTS:** Goldenrod.

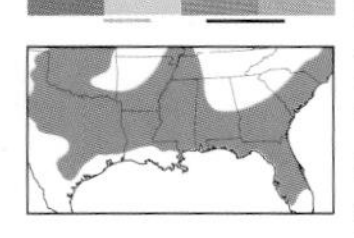

CAMPHORWEED FLOWER MOTH

Schinia nubila 93-2117 (11137) **Common**

TL 10–12 mm. Resembles Sordid Flower Moth but lacks dusky reniform spot. AM and PM lines are smooth; PM line curves only slightly upward at costa. Black HW has a pale median band and white fringe. **HOSTS:** Camphorweed.

FLOWER MOTHS

PAINTBRUSH FLOWER MOTH

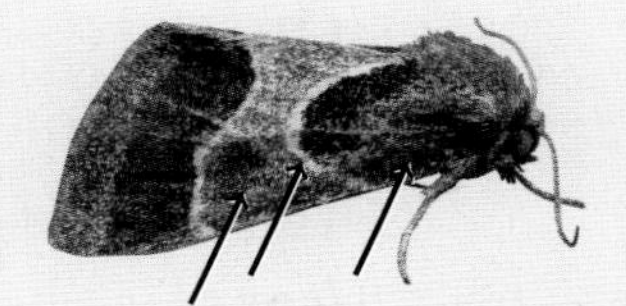

RAGWEED FLOWER MOTH

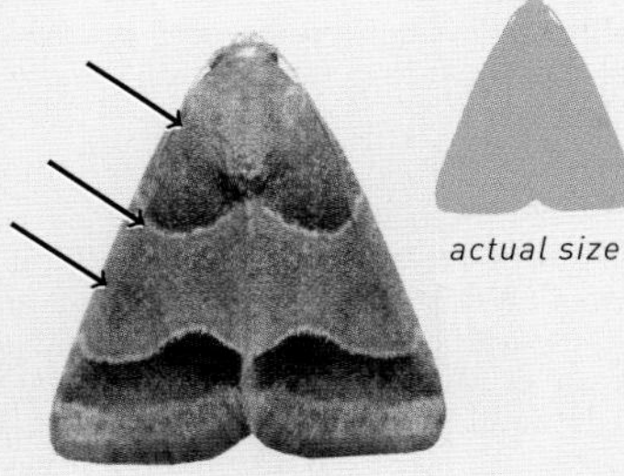

SLENDER FLOWER MOTH

actual size

THOREAU'S FLOWER MOTH

THREE-LINED FLOWER MOTH

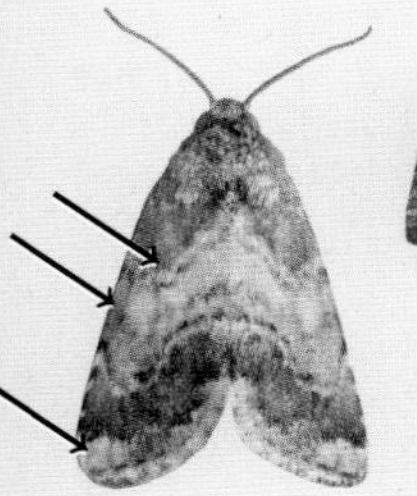

SORDID FLOWER MOTH

CAMPHORWEED FLOWER MOTH

ERIGERON FLOWER MOTH

Schinia obscurata 93-2119 (11118) **Common**

TL 9–11 mm. FW has broad, medium to dark brown basal and ST bands. Whitish AM and PM lines often blend into pale median area; AM line is strongly angled. Dark shading around dusky reniform spot reaches from ST area to costa. Blackish HW has a yellow basal patch and whitish fringe. **HOSTS:** Fleabane.

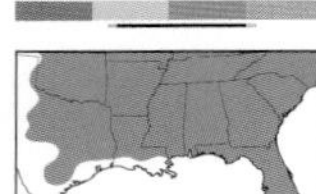

LYNX FLOWER MOTH

Schinia lynx 93-2120 (11117) **Common**

TL 10–12 mm. Resembles Erigeron Flower Moth, but AM and PM lines are distinct. Dark shading around reniform spot blends in to rest of lightly shaded median area. Yellow-orange HW has broad black terminal line and discal spot. **HOSTS:** Fleabane and camphorweed.

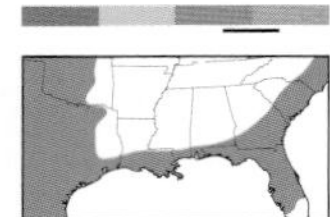

ALURING FLOWER MOTH

Schinia siren 93-2123 (11115) **Common**

TL 12–13 mm. Straw-colored FW has orange shading in basal and ST areas. Pale AM and PM lines are lightly toothed. Often has diffuse orange median line. Veins are usually faintly traced whitish. Thorax is orange. Rare individuals have variable blackish peppering in basal, median, and ST areas. All-black HW has a white fringe. **HOSTS:** Camphorweed.

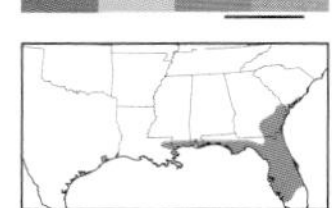

SPINOSE FLOWER MOTH

Schinia spinosae 93-2129 (11104) **Local**

TL 12–13 mm. Maroon FW has a speckled, pale gray median area with golden brown shading above a dusky, indistinct reniform spot. Wavy AM, PM, and ST lines are white. Whitish HW has a black terminal line and discal spot. **HOSTS:** Coastal joinweed.

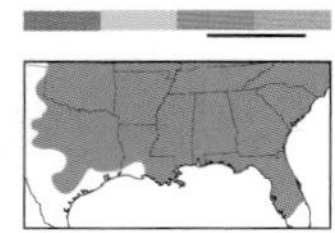

ARCIGERA FLOWER MOTH

Schinia arcigera 93-2134 (11128) **Common**

TL 12–13 mm. Sexually dimorphic. Maroon to warm brown FW has a paler median area and terminal line. White AM line is evenly curved. Nearly straight PM line is broadly bordered by blackish shading above. Male has yellow HW with a black terminal line and white fringe; female's HW is all black with white fringe. **HOSTS:** Aster, camphorweed, horseweed, and sea lavender.

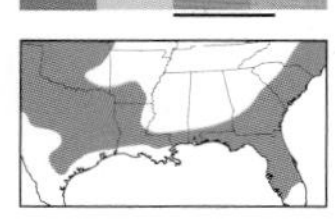

BROWN FLOWER MOTH

Schinia saturata 93-2135 (11140) **Common**

TL 12–13 mm. Peppery brown FW has slightly jagged white AM and PM lines sometimes edged with diffuse black. Median area is usually slightly paler. Reniform spot may be dusky or almost absent. HW is yellowish, occasionally with a diffuse brownish terminal line. **HOSTS:** Camphorweed and silkgrass.

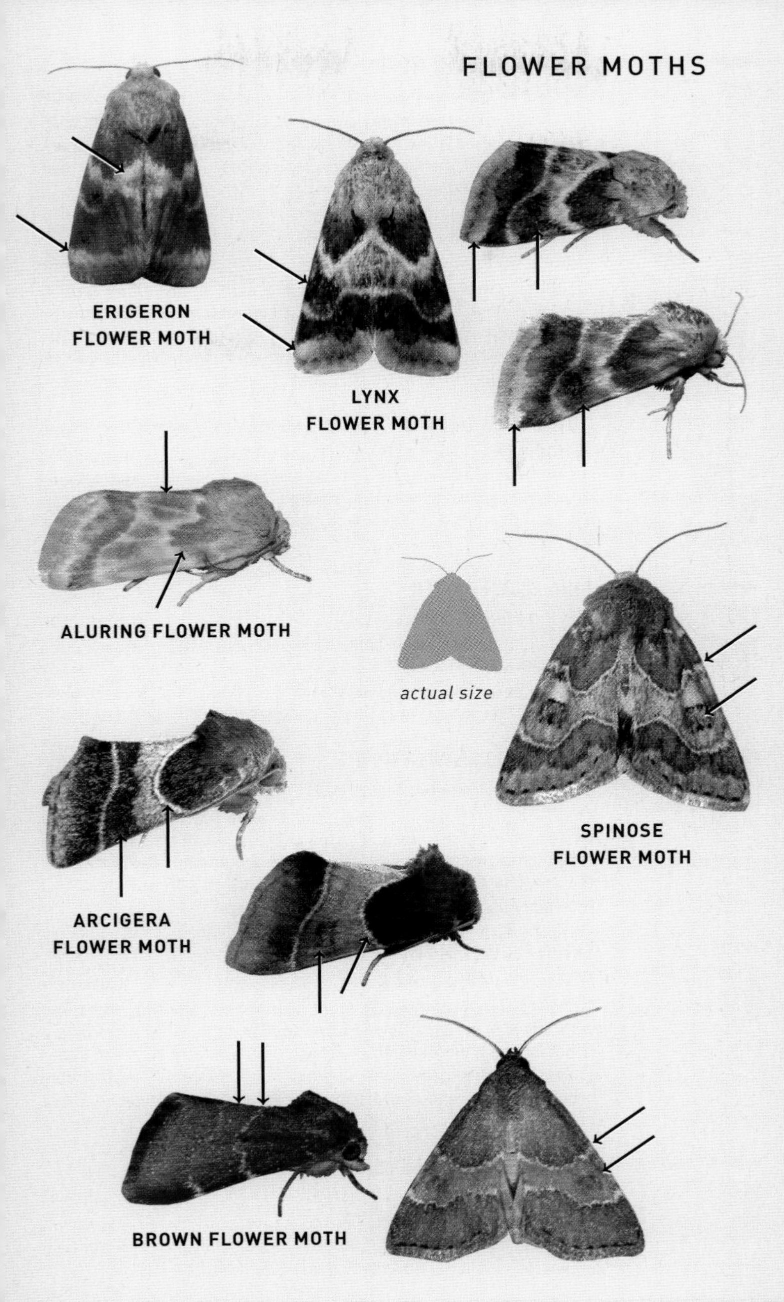

ERIGERON FLOWER MOTH
LYNX FLOWER MOTH
ALURING FLOWER MOTH
actual size
SPINOSE FLOWER MOTH
ARCIGERA FLOWER MOTH
BROWN FLOWER MOTH

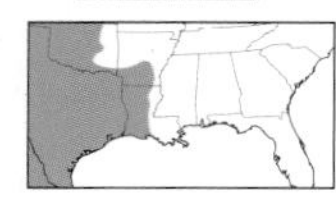

PAINTED FLOWER MOTH

Schinia volupia 93-2139 (11106) **Local**

TL 11–12 mm. Bright crimson pink FW is boldly patterned with fragmented, toothed white AM and PM lines. White ST line is diffuse and often indistinct. Thorax is yellow-orange. Rosy HW has a pale fringe. **HOSTS:** Indian blanket.

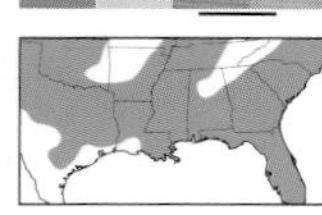

BLEEDING FLOWER MOTH

Schinia sanguinea 93-2145 (11173) **Common**

TL 15–18 mm. Raspberry pink FW has a paler median area variably shaded pinkish to tan. Veins are traced whitish, especially in terminal area and along AM and PM lines. Dark pink reniform spot typically is diffuse. HW is variably all brown to whitish with a broad brown ST band. **HOSTS:** Blazing star.

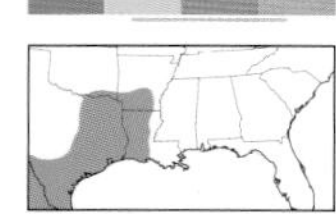

ROYAL FLOWER MOTH *Schinia regia* 93-2147 (11166) **Local**

TL 13–16 mm. Pale FW has pink basal and ST areas. Sharply angled white AM line and smooth PM line are broad. Median and terminal areas are tan, often suffused pinkish, especially in reniform area. Whitish HW has a diffuse pink terminal line and whitish fringe. **HOSTS:** Texas ironweed.

BINA FLOWER MOTH

Schinia bina 93-2150 (11105) **Uncommon**

TL 11–13 mm. Wine red FW has tawny ST area. Median area is often lighter. Whitish AM and PM lines are frequently indistinct. Black HW has two pale spots in median area and a white fringe. **HOSTS:** Various composites.

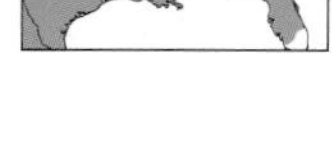

GOLDENROD FLOWER MOTH

Schinia nundina 93-2156 (11177) **Common**

TL 13–15 mm. Olive FW has irregular white AM and PM lines and white patches in median and terminal areas. Basal area is often suffused bronzy. Hourglass-shaped reniform spot and small orbicular dot are black. White HW has a diffuse brown patch at apex and brownish discal crescent. **HOSTS:** Goldenrod.

THIRD FLOWER MOTH

Schinia tertia 93-2158 (11179) **Common**

TL 12–14 mm. Pale FW has fine black AM and PM lines bordered with marbled bluish gray and brown in basal and ST areas. PM line is broadly edged brown above. Bluish gray reniform spot and indistinct orbicular spot are partly outlined black. White HW has a diffuse brown terminal line and discal crescent. **HOSTS:** Probably dotted blazing star.

CHRYSELLUS FLOWER MOTH

Schinia chrysellus 93-2171 (11199) **Local**

TL 11–14 mm. Satin white FW has sharply defined brassy brown median and terminal lines. Thorax is brown. White HW has a diffuse brown terminal line. **HOSTS:** Prairie broomweed.

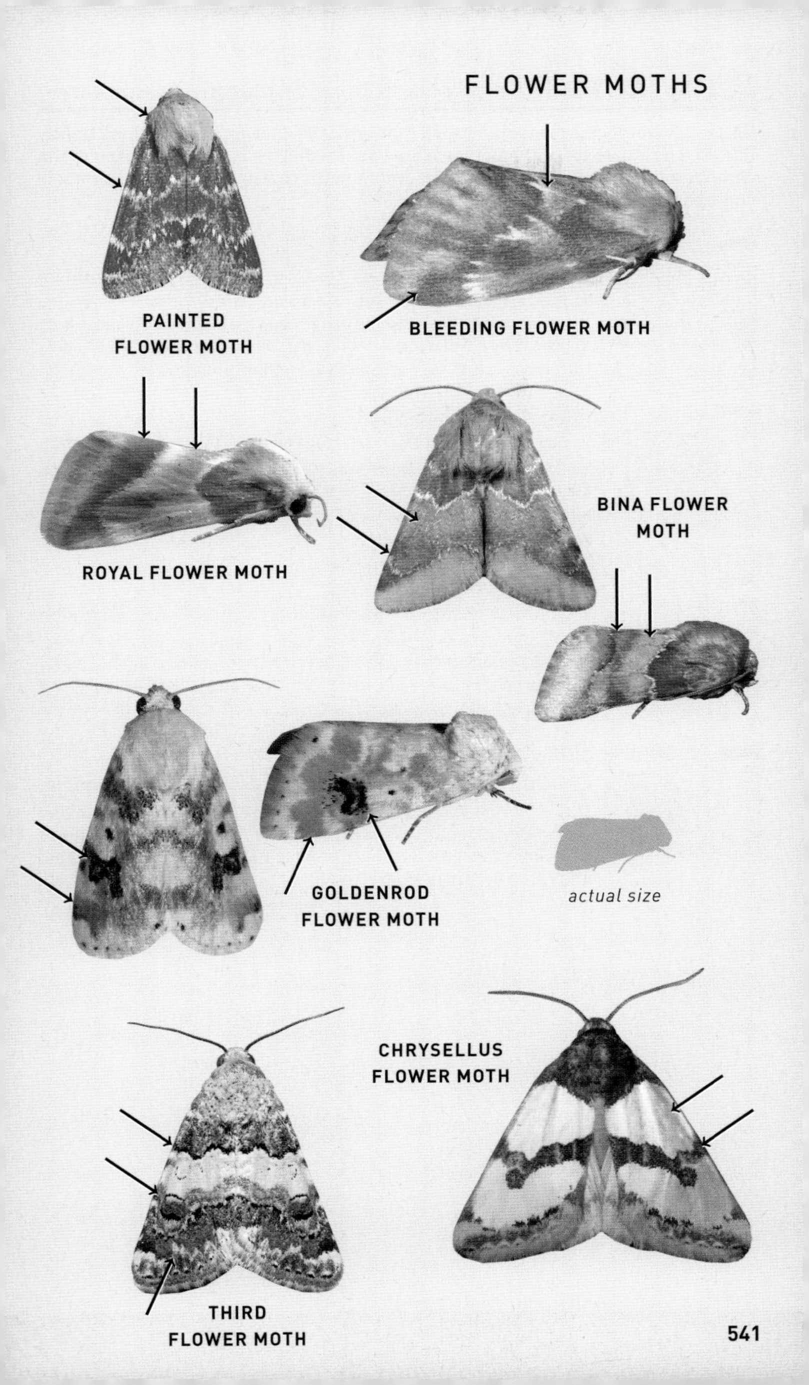
FLOWER MOTHS
PAINTED
FLOWER MOTH
BLEEDING FLOWER MOTH
ROYAL FLOWER MOTH
BINA FLOWER
MOTH
GOLDENROD
FLOWER MOTH
actual size
CHRYSELLUS
FLOWER MOTH
THIRD
FLOWER MOTH

FERN MOTHS Family Noctuidae, Subfamily Eriopinae

Small, strikingly patterned noctuids that typically rest with their wings tightly folded. They have tufts of hairlike scales on the thorax and inner margin of the wing that stick up at rest. These woodland moths are nocturnal and regularly come to light.

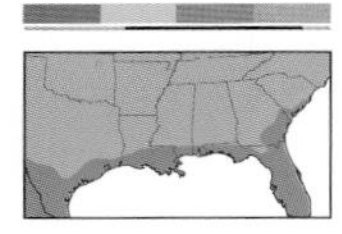

FLORIDA FERN MOTH

Callopistria floridensis 93-2190 (9630) **Common**

TL 15–16 mm. Brown FW has a white-edged, thin brown median line that widens dramatically into a bold brown triangle at costa. ST area is gray with a warm brown subapical patch. Veins are often lightly traced paler. **HOSTS:** Ferns. **NOTE:** Multiple broods.

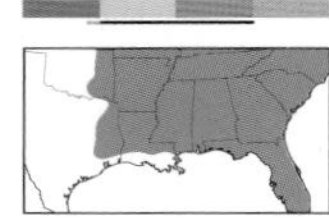

PINK-SHADED FERN MOTH

Callopistria mollissima 93-2192 (9631) **Common**

TL 15 mm. Mottled, warm brown FW has bright pink bands along slightly jagged black AM and PM lines. Orbicular and especially reniform spots are partly outlined white. **HOSTS:** Ferns, including New York fern. **NOTE:** Two broods.

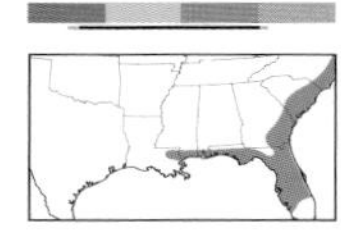

GRANITOSE FERN MOTH

Callopistria granitosa 93-2193 (9632) **Common**

TL 15 mm. Warm brown FW is boldly patterned with scalloped, white-filled double AM and PM lines. Large reniform spot is partly outlined with black-edged white. Orbicular and claviform spots, and ST area, are often pink. **HOSTS:** Ferns.

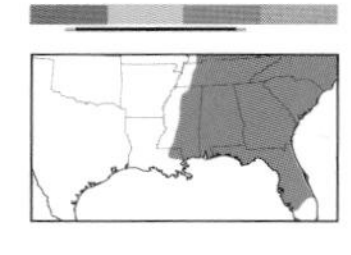

SILVER-SPOTTED FERN MOTH

Callopistria cordata 93-2194 (9633) **Common**

TL 15–16 mm. Rusty brown FW has large satin white claviform and subreniform spots. Orbicular and reniform spots are partly outlined with white blotches. Scalloped, double PM line is filled satin white. **HOSTS:** Ferns. **NOTE:** Two broods.

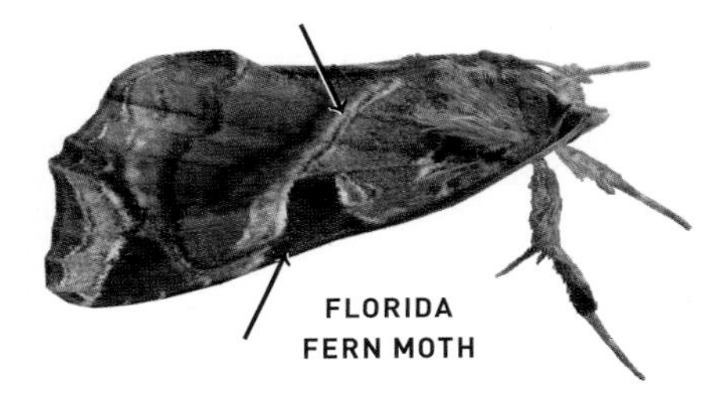

FLORIDA
FERN MOTH

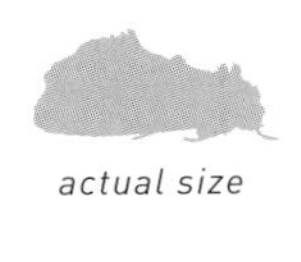

actual size

PINK-SHADED
FERN MOTH

GRANITOSE
FERN MOTH

SILVER-SPOTTED
FERN MOTH

Phosphilas and Armyworms

Family Noctuidae, Subfamily Noctuinae, Tribes Pseudeustrotiini, Phosphilini, and Prodeniini

Small to medium-sized noctuids commonly found in woodlands, gardens, and open areas. The phosphilas are delta-shaped and rest with their wings flat. The armyworms have long, narrow wings that they hold tight to the body. All are nocturnal and will come to light. The armyworms are also frequent visitors to sugar bait.

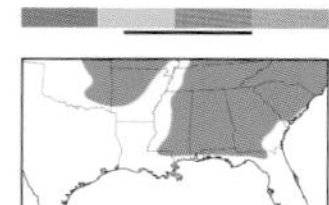

PINK-BARRED PSEUDEUSTROTIA

Pseudeustrotia carneola 93-2205 (9053) **Common**

TL 11–13 mm. Mottled brown FW is paler beyond inconspicuous PM line. An oblique pink bar slants from costa to inner edge of brown reniform spot. Irregular white ST line is broadly edged light brown. **HOSTS:** Dock and smartweed.

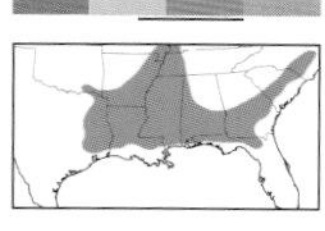

BROWN PSEUDEUSTROTIA

Pseudeustrotia indeterminata 93-2206 (9054) **Uncommon**

TL 11–13 mm. Resembles Pink-barred Pseudeustrotia in pattern, but pale areas are instead a slightly lighter shade of brown. Sometimes slanting orbicular bar is most obvious as a pale orbicular spot. **HOSTS:** Unknown.

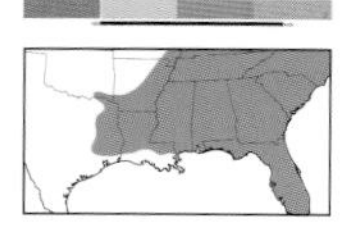

TURBULENT PHOSPHILA

Phosphila turbulenta 93-2208 (9618) **Uncommon**

TL 15–19 mm. Mottled brown FW is paler beyond toothed blackish PM line and along inner margin. Reniform spot is usually pale and indistinctly edged. Black dash at anal angle is conspicuous. **HOSTS:** Greenbriar. **NOTE:** Two or more broods.

SPOTTED PHOSPHILA

Phosphila miselioides 93-2209 (9619) **Uncommon**

TL 18 mm. Mottled green-and-black FW has a large reniform spot that is either white or warm brown. Orbicular spot is sometimes pale but more often indistinct. Often has a diffuse blackish streak along inner median area. Toothed PM line is marked with tiny white dots at veins. **HOSTS:** Greenbriar. **NOTE:** Multiple broods.

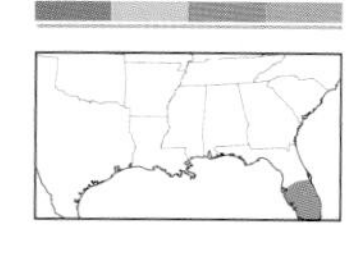

NORTHERN PHUPHENA

Phuphena tura 93-2212 (9634) **Local**

TL 13–16 mm. Light coppery brown FW has gently curved double AM and PM lines filled light brown. Large inconspicuous reniform spot is outlined light brown. ST and terminal lines are dotted. ST area is sometimes paler. **HOSTS:** Unknown.

PINK-BARRED PSEUDEUSTROTIA

BROWN PSEUDEUSTROTIA

TURBULENT PHOSPHILA

actual size

NORTHERN PHUPHENA

SPOTTED PHOSPHILA

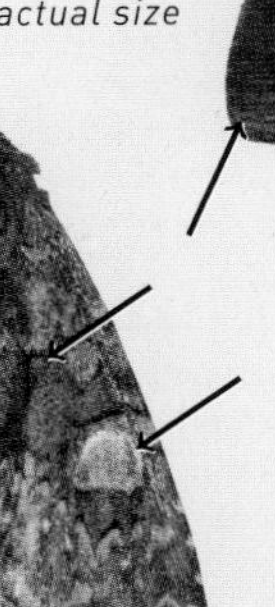

CHOCOLATE MOTH

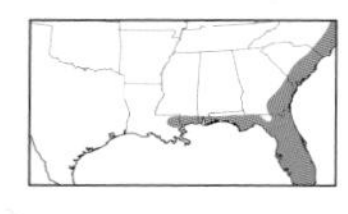

Acherdoa ferraria 93-2213 (9636) **Uncommon**

TL 11–14 mm. Brownish gray FW has rich brown median area bounded by wavy AM and PM lines edged white along inner half. Pale orange reniform spot is edged white. Costa is brownish gray. Raised abdomen has a brushy tip. Antennae are broadly pectinate in male. **HOSTS:** Unknown.

BEET ARMYWORM

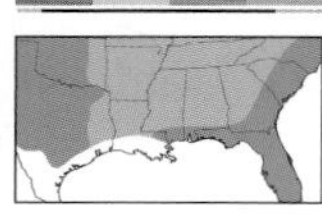

Spodoptera exigua 93-2215 (9665) **Common**

TL 15–16 mm. Grayish brown FW has a round, pale orange orbicular spot and gray reniform spot, both outlined whitish. Jagged AM and PM lines are double. Diffuse whitish ST line is edged brownish below. **HOSTS:** Grass and crops, such as bean, beet, corn, and potato. **NOTE:** Multiple broods. Also known as Small Mottled Willow Moth.

FALL ARMYWORM

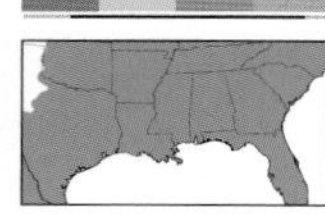

Spodoptera frugiperda 93-2216 (9666) **Common**

TL 16–19 mm. Sexually dimorphic. Male has rusty brown orbicular spot and dark reniform spot, both ringed pale and connected by a whitish teardrop; apex is whitish, and inner ST area typically has a blackish patch. FW of female is uniform grayish brown with indistinct markings; pale-ringed orbicular spot is slanting, and AM and PM lines are double. **HOSTS:** Grass, crops, and low plants. **NOTE:** Multiple broods.

YELLOW-STRIPED ARMYWORM

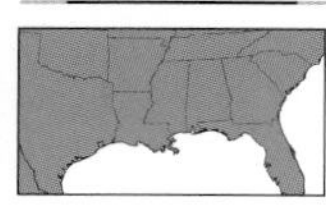

Spodoptera ornithogalli 93-2219 (9669) **Common**

TL 18–24 mm. Sexually dimorphic. Resembles Fall Armyworm, but orbicular spot blends in to slanting pale median streak, and ST area is more extensively grayish. Veins are traced whitish in median area. Male has a rusty tan saddle in inner median area. Female has a very shallow tooth at midpoint of white terminal line. **HOSTS:** Wide variety of grasses, crops, and low and woody plants. **NOTE:** Multiple broods.

VELVET ARMYWORM

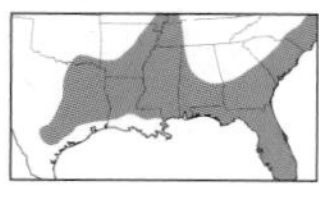

Spodoptera latifascia 93-2220 (9670) **Common**

TL 20–22 mm. Sexually dimorphic. Resembles Yellow-striped Armyworm, but male has uniformly gray ST band and a rusty brown reniform spot. Female has straight white terminal line, and basal portion of costa is often paler. **HOSTS:** Various grasses, crops, and low plants. **NOTE:** Multiple broods.

DOLICHOS ARMYWORM

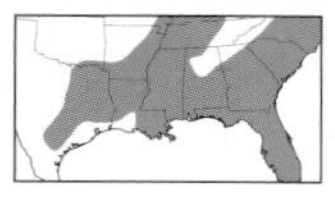

Spodoptera dolichos 93-2221 (9671) **Common**

TL 20–24 mm. Dark brown FW has a tan inner margin adjoining a tan central stripe on thorax. Long orbicular and reniform spots slant perpendicular to one another. Grayish apical patch often bleeds up to indistinct PM line. **HOSTS:** Various grasses, crops, and low plants. **NOTE:** Multiple broods.

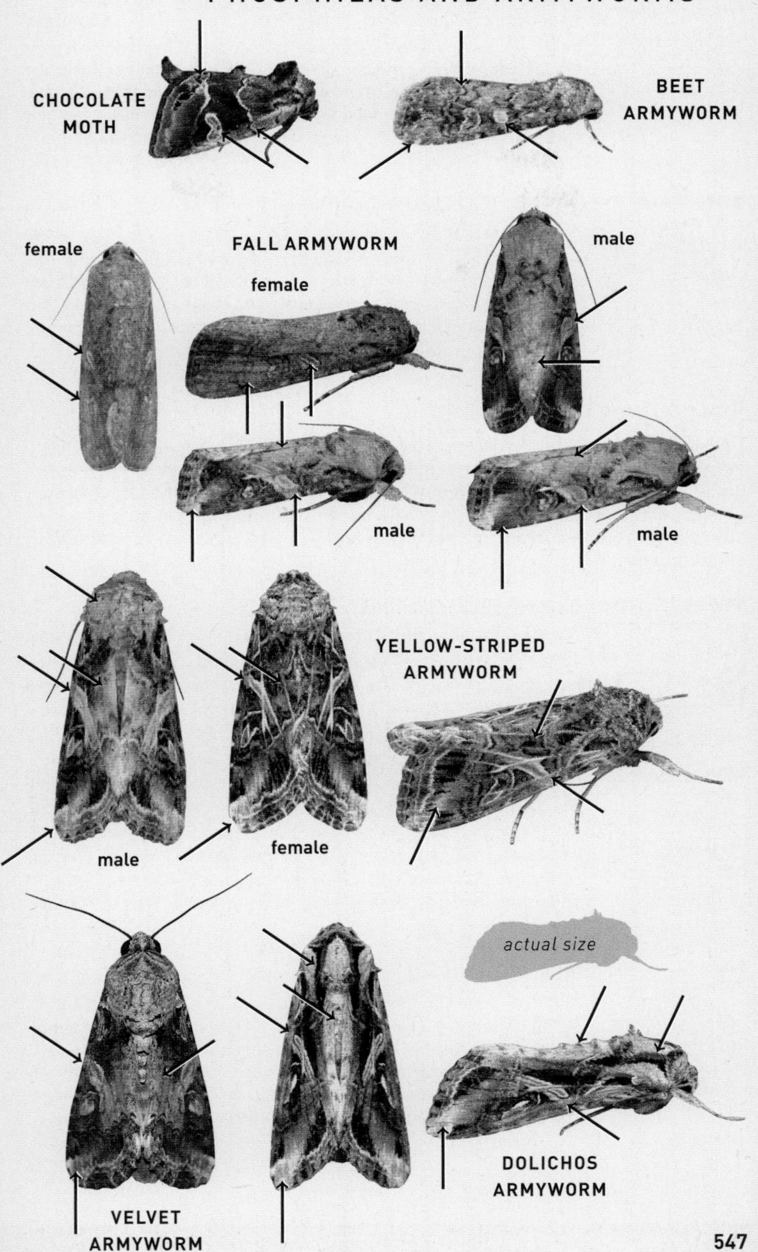
CHOCOLATE MOTH
BEET ARMYWORM
female
FALL ARMYWORM
female
male
male
male
YELLOW-STRIPED ARMYWORM
male
female
actual size
DOLICHOS ARMYWORM
VELVET ARMYWORM

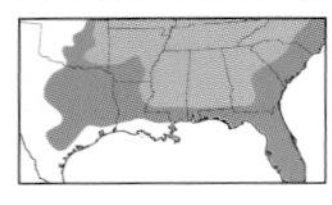

SOUTHERN ARMYWORM

Spodoptera eridania 93-2223 (9672) **Common**

TL 18–24 mm. Streaky pale tan FW has a grayish inner margin. Broad, diffuse brown median band is sometimes faint or nearly absent. Small black reniform spot is indistinct. Some individuals have a bold blackish bar extending from reniform spot to outer margin. **HOSTS:** Various grasses, crops, and low plants. **NOTE:** Multiple broods.

GRAY-STREAKED ARMYWORM

Spodoptera albula 93-2224 (9673) **Local**

TL 18–24 mm. Resembles Southern Armyworm but usually has grayish shading in AM area and not along inner margin. Narrow, dark gray orbicular spot is ringed pale and is almost reached by thin black basal dash. **HOSTS:** Various grasses, crops, and low plants. **NOTE:** Multiple broods.

MIDGETS

Family Noctuidae, Subfamily Noctuinae, Tribe Elaphriini

Small moths that have round wings and are mostly clad in shades of brown with contrasting patterns. They usually rest with their wings flat but occasionally fold them tented. Most are nocturnal and will visit lights and sugar bait in small numbers.

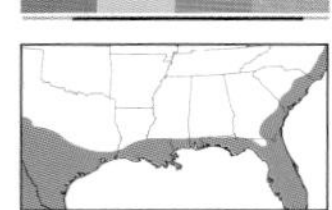

BROWN-SPOTTED MIDGET

Elaphria fuscimacula 93-2225 (9675) **Common**

TL 12–13 mm. Brown FW has slightly darker median area and suffusion along ST line. Curvy AM and PM lines are double and filled lighter. Round orbicular and hourglass-shaped reniform spots are black. **HOSTS:** Unknown. **NOTE:** Two or more broods.

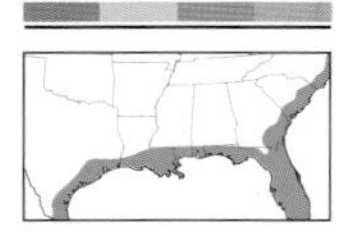

SUGARCANE MIDGET

Elaphria nucicolora 93-2226 (9676) **Local**

TL 12–13 mm. Light brown FW has darker veins speckled with white. A narrow dark patch connects wavy AM and PM lines, passing behind pale reniform spot. Base of thorax has a white tuft. **HOSTS:** Sugarcane, bahiagrass, rye, and other grasses. **NOTE:** Multiple broods.

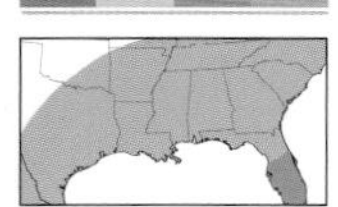

BICOLORED MIDGET

Elaphria agrotina 93-2227 (9677) **Local**

TL 10–12 mm. Brown FW has a broad, pale costal streak that wraps across inner basal area and thorax. Inner median area is tinged warm brown. Small claviform spot is whitish, and apex has whitish patch. Some individuals lack white claviform spot and show only slightly paler brown costa and thorax. **HOSTS:** Unknown.

PHOSPHILAS AND ARMYWORMS

actual size

SOUTHERN ARMYWORM

GRAY-STREAKED ARMYWORM

VARIEGATED MIDGET

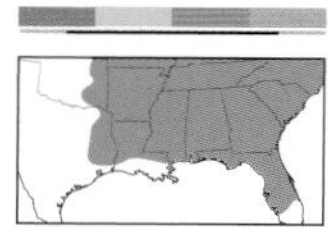

Elaphria versicolor 93-2228 (9678) **Common**

TL 14 mm. Mottled brownish FW has a kinked white PM line that widens toward costa. Indistinct orbicular and reniform spots are separated by a small black patch. Gray ST area has a pale orange apical patch. **HOSTS:** Unknown. **NOTE:** Two or three broods.

CHALCEDONY MIDGET

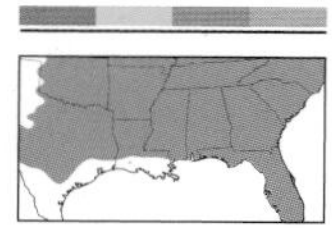

Elaphria chalcedonia 93-2230 (9679) **Common**

TL 12–14 mm. Resembles Variegated Midget but has a broad, pale yellowish costal streak that wraps across inner basal area and thorax. PM line is more diffuse at costa. **HOSTS:** Snapdragon. **NOTE:** Presumably two or more broods.

DELTOID MIDGET

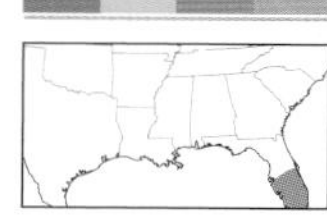

Elaphria deltoides 93-2231 (9679.1) **Local**

TL 10–12 mm. Brown FW has a large, dark brown patch crisply edged white in inner median area, forming a broad saddle. Reniform spot is a small black dot. **HOSTS:** Snapdragon.

GEORGE'S MIDGET

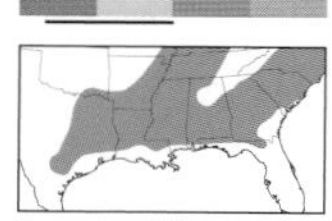

Elaphria georgei 93-2232 (9680) **Uncommon**

TL 12 mm. Brown FW has a wide, dark median band. AM line is edged white along inner half. PM line is broadly edged with diffuse whitish shading. Brown reniform spot is edged whitish, blending into PM line. **HOSTS:** Possibly green algae or fungi.

FESTIVE MIDGET

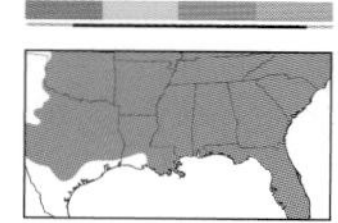

Elaphria festivoides 93-2233 (9681) **Common**

TL 14 mm. Straw-colored FW has contrasting warm brown saddle in inner median area. Curved pale reniform spot is open toward costa and backed by a blackish patch. Dusky brown ST area contrasts with pale apical patch. Thorax has a dark brown collar. **HOSTS:** Box elder and possibly green algae. **NOTE:** Two broods.

EXESA MIDGET

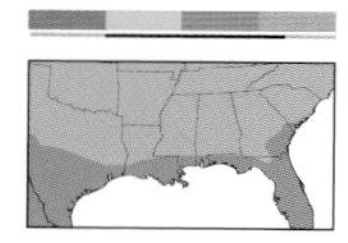

Elaphria exesa 93-2236 (9682) **Local**

TL 14 mm. Resembles Chalcedony Midget, but double AM line is whitish. Reniform spot is pale in lower half. Lacks a distinct costal stripe. Basal and ST areas have a mottled appearance. ST line is often visible as fragmented pale yellow spots. **HOSTS:** Unknown. **NOTE:** Two or more broods.

GRATEFUL MIDGET

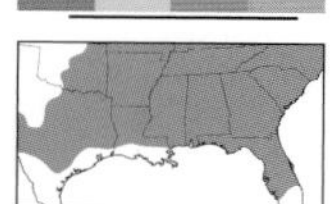

Elaphria grata 93-2238 (9684) **Common**

TL 13 mm. Reddish brown FW is peppered with white scales along dark veins. Slightly curved AM and PM lines are whitish. Orbicular spot is a small black dot. Hourglass-shaped reniform spot contains two black dots. **HOSTS:** Possibly green algae or fungi. **NOTE:** Multiple broods.

MIDGETS

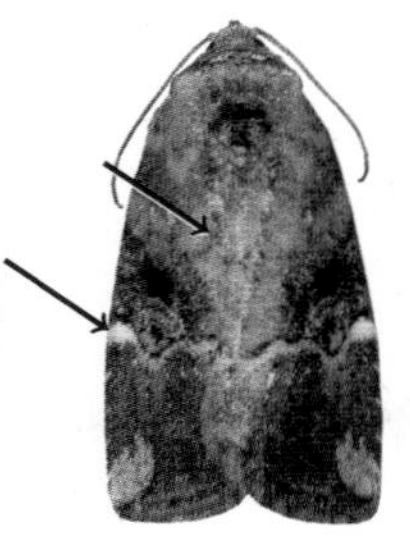

VARIEGATED MIDGET

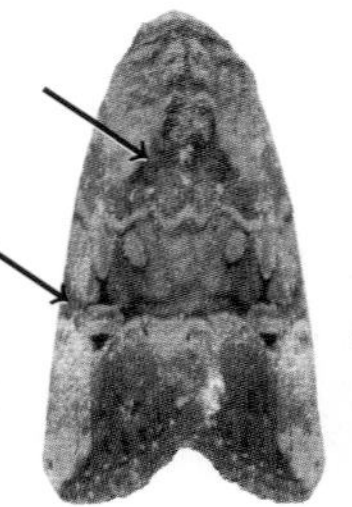

CHALCEDONY MIDGET

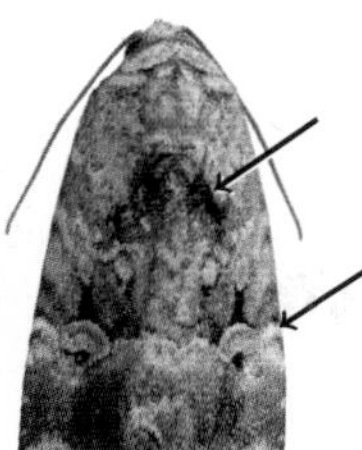

DELTOID MIDGET

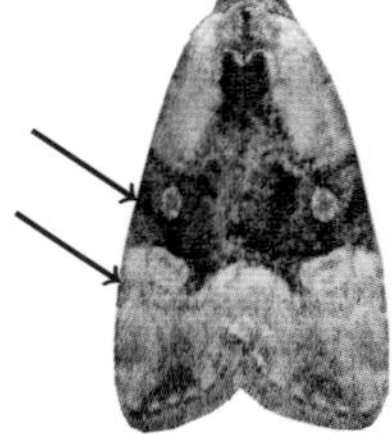

GEORGE'S MIDGET

FESTIVE MIDGET

EXESA MIDGET

actual size

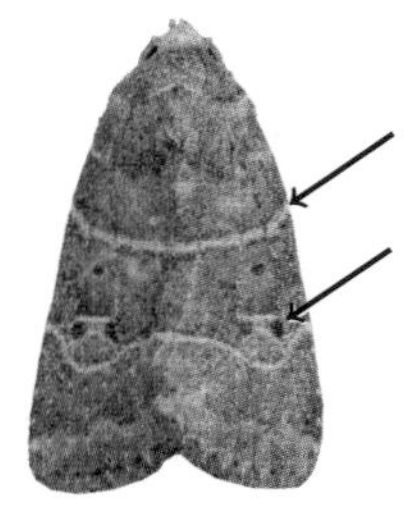

GRATEFUL MIDGET

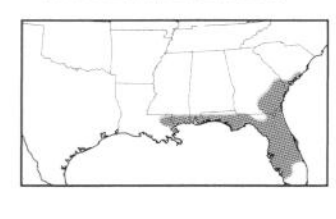

GREEN MIDGET

Elaphria cyanympha 93-2239 (9297.2) **Local**

TL 12–13 mm. Pale green to green-tinged white FW has a bold pattern of fragmented white-edged black AM, PM, and basal lines that widen at costa. A black bar marks central median area. Terminal line and fringe are partly checkered black. **HOSTS:** Unknown.

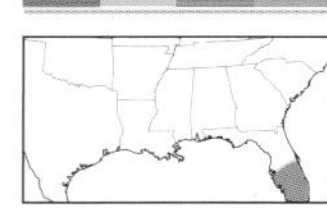

FLUID MIDGET

Gonodes liquida 93-2248 (9687) **Local**

TL 13–14 mm. Light gray FW is darker bronzy brown in median area, below slanting white median line. Narrow white PM line bulges around white-edged, hourglass-shaped reniform spot. Apex is pale gray. Base of thorax is bronzy brown. **HOSTS:** Unknown.

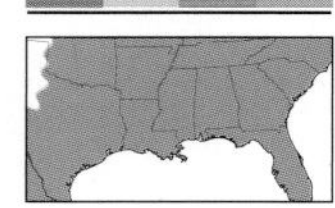

THE WEDGLING

Galgula partita 93-2249 (9688) **Common**

TL 11–13 mm. Sexually dimorphic. Variable. Male has yellowish tan to rusty brown FW with a large blackish blotch at midpoint of costa. Gently curved, whitish PM line is edged slightly darker above and often widens slightly at inner margin. Round orbicular and hourglass-shaped reniform spots are outlined whitish. Female is similar, but FW is darker in color, maroon to dark brown. **HOSTS:** Wood sorrel. **NOTE:** Three or more broods.

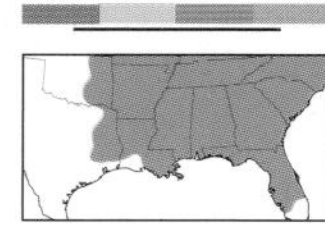

CLOAKED MARVEL

Chytonix palliatricula 93-2249.5 (9556) **Common**

TL 15–17 mm. Dimorphic. Grayish brown FW has a thick black bar in inner median area that ends as a short white dash. Scalloped, black AM and PM lines are edged whitish. PM line curves evenly, touching inner margin at a slight angle. Basal area is shaded darker. "Cloaked" individuals have white orbicular and reniform spots bleeding into white inner median area. **HOSTS:** Gilled fungi. **NOTE:** One or two broods.

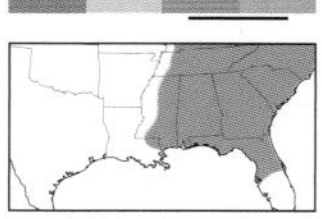

MASKED MARVEL

Chytonix sensilis 93-2249.5 (9557) **Uncommon**

TL 14–16 mm. Resembles typical form of Cloaked Marvel, but PM line bends sharply and touches inner margin at a right angle. Inner reniform spot has a dark spot. Does not have a white-cloaked form. **HOSTS:** Fungi.

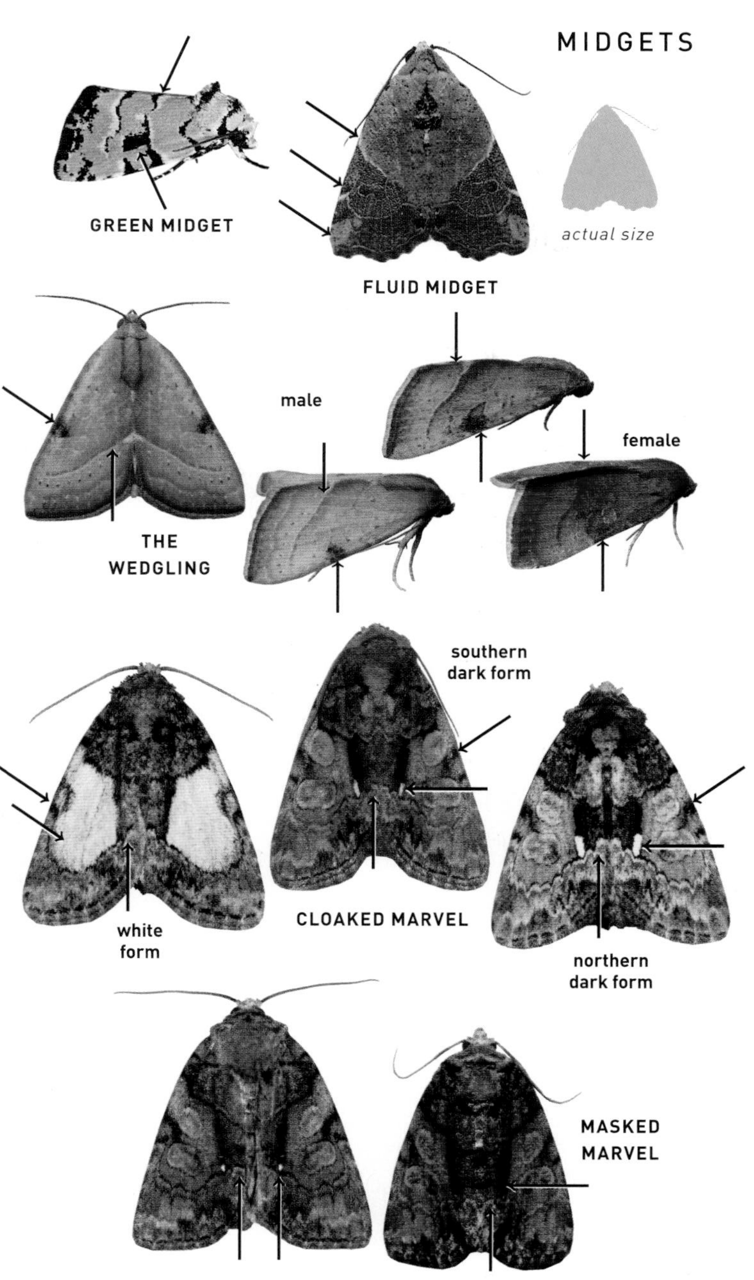
MIDGETS
GREEN MIDGET
FLUID MIDGET
actual size
male
female
THE WEDGLING
southern dark form
white form
CLOAKED MARVEL
northern dark form
MASKED MARVEL

Assorted Noctuids

Family Noctuidae, Subfamily Noctuinae, Tribes Caradrinini, Dypterygiini, and Actinotiini

A varied group of small to medium-sized noctuids, some well patterned. They are found mostly in woodlands, fields, and gardens, though White-eyed Borer occurs in a variety of wetland habitats. They will come to light in small numbers.

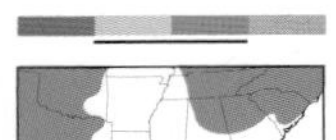

MIRANDA MOTH

Proxenus miranda 93-2266 (9647) **Local**

TL 13–15 mm. Shiny grayish brown FW has a small white reniform spot. Dark AM and PM lines and orbicular dot are often indistinct or absent. HW is whitish. **HOSTS:** Low plants, including alfalfa, dandelion, and strawberry. **NOTE:** Two broods.

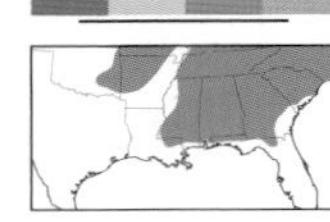

THE SLOWPOKE

Athetis tarda 93-2269 (9650) **Common**

TL 14–15 mm. Tan to violet brown FW has slightly fragmented, blackish AM and PM lines; AM line has a tooth near costa. Diffuse median band is sometimes present. Orbicular spot is a tiny black dot. Reniform spot has a dotted yellow outline. ST line is edged yellow. **HOSTS:** Dead oak leaves. **NOTE:** Two broods.

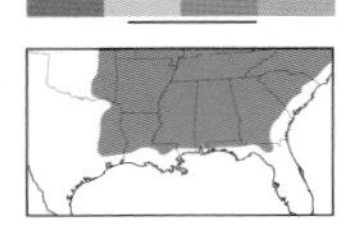

AMERICAN BIRD'S-WING

Dypterygia rozmani 93-2272 (9560) **Uncommon**

TL 20 mm. Smooth, slate gray FW has crisp, curvaceous black AM and PM lines, and outlined spots. A streaky patch of brownish tan and white shades anal angle and inner margin, merging with raised dorsal crest on thorax. **HOSTS:** Dock and climbing false buckwheat. **NOTE:** One or two broods.

BLACK BIRD'S-WING

Dypterygia patina 93-2274 (9561) **Uncommon**

TL 14–16 mm. Dark grayish brown FW has a diffuse paler brown streak along inner margin. Reniform spot is partly edged white along lower edge. Lines are indistinct except where PM line reaches inner margin. **HOSTS:** Unknown.

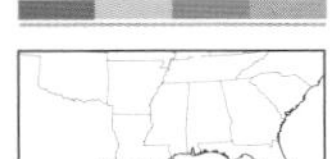

ORBED NARROW-WING

Magusa orbifera 93-2281 (9637) **Local**

TL 15–19 mm. Variable. Many individuals are virtually identical to Variable Narrow-Wing except in size. Bicolored form is typically rusty brown with long, bold white basal dash. A whitish patch is often present in PM area of paler inner half. Mottled form lacks pale bar extending from apical patch to PM line. **HOSTS:** Leadwood and Humboldt coyotillo. **NOTE:** Unlike Variable Narrow-Wing, does not wander.

MIRANDA MOTH

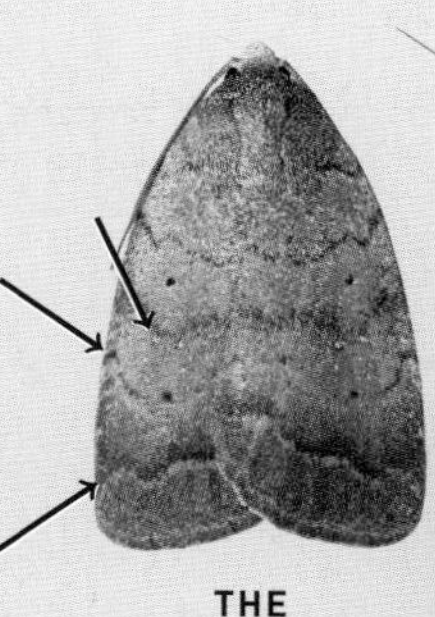

THE SLOWPOKE

AMERICAN BIRD'S-WING

BLACK BIRD'S-WING

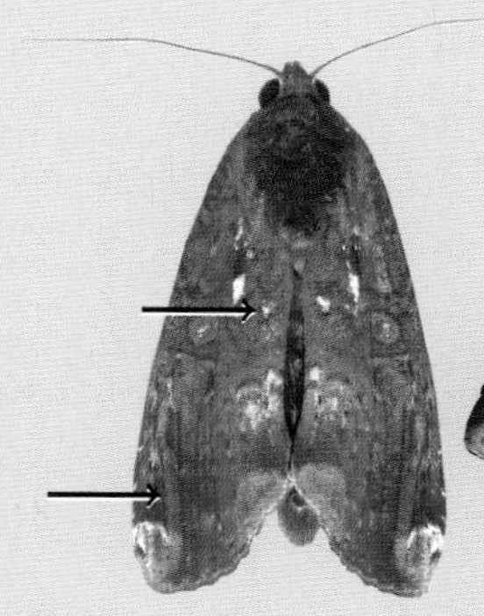

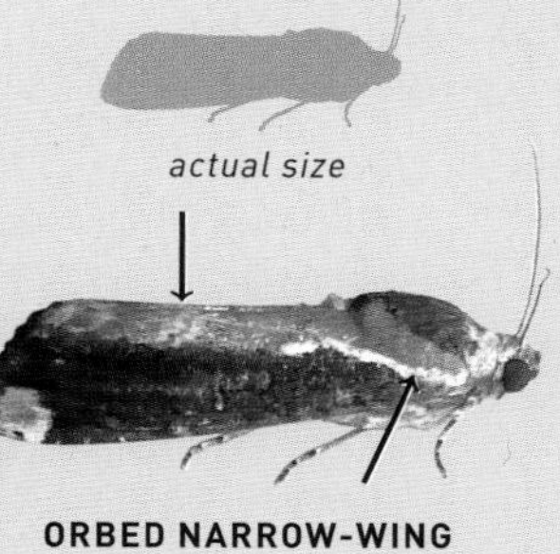

actual size

ORBED NARROW-WING

VARIABLE NARROW-WING

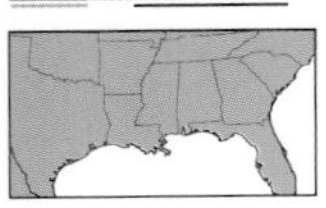

Magusa divaricata 93-2282 (9637.1) **Uncommon**

TL 18–22 mm. Variable. Long FW ranges from tan to chocolate brown and may be smooth or mottled. Round, pale apical patch is always present but sometimes faded. Commonly bicolored; darker half is usually chocolate brown. Halves are often separated by a thin white basal dash. A diffuse, pale bar typically extends from apical patch to PM line on mottled individuals. **HOSTS:** Buckthorn and coyotillo. **NOTE:** A vagrant from Mexico and Cen. America that wanders north each fall, sometimes as far as Canada.

GRAY HALF-SPOT

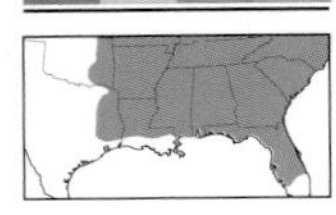

Nedra ramosula 93-2283 (9582) **Common**

TL 15–25 mm. Light grayish brown FW has a streaky woodgrain pattern. A band of blackish shading runs from base to apex, passing behind inner half of kidney-shaped brown-and-white reniform spot. **HOSTS:** Saint John's wort. **NOTE:** Two or more broods.

WHITE-EYED BORER

Iodopepla u-album 93-2287 (9522) **Uncommon**

TL 18 mm. Rosy brown to lilac gray FW has black and chestnut shading in central median area. Brown reniform spot is edged with white on inner half. Scalloped AM and PM lines are diffuse. Whitish ST line is edged darker below. **HOSTS:** Saint John's wort. **NOTE:** Two or more broods.

ANGLE SHADES

Family Noctuidae, Subfamily Noctuinae, Tribe Phlogophorini

Medium-sized noctuids that are commonly found in woodlands and large gardens. They have the distinctive habit of rolling the outer third of the FW while at rest. All are boldly patterned with dark chevrons in the median area. They regularly come to light in small numbers.

AMERICAN ANGLE SHADES

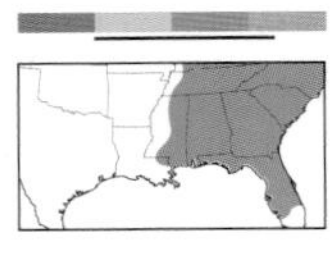

Euplexia benesimilis 93-2290 (9545) **Common**

TL 17 mm. Dark brown FW has a pale tan to pinkish band beyond PM line. Double AM line is filled brown. Large reniform spot is partly filled and outlined creamy whitish. **HOSTS:** Ferns, deciduous trees, and low plants, including aster, huckleberry, and willow. **NOTE:** Two or three broods.

OLIVE ANGLE SHADES

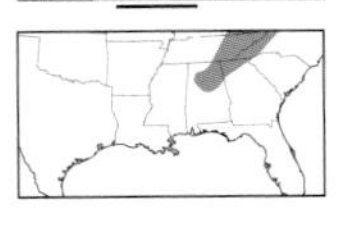

Phlogophora iris 93-2291 (9546) **Local**

TL 25 mm. Olive green FW has pale, angled AM and PM lines that are diffusely edged brownish, and touch at inner margin. Grayish brown orbicular and reniform spots fuse together to form a V-shape. Inner half of basal area has a large blackish patch. Pale ST line is edged with diffuse dark shading above. **HOSTS:** Trees and low plants, including alder, dandelion, dock, and thistle.

ASSORTED NOCTUIDS

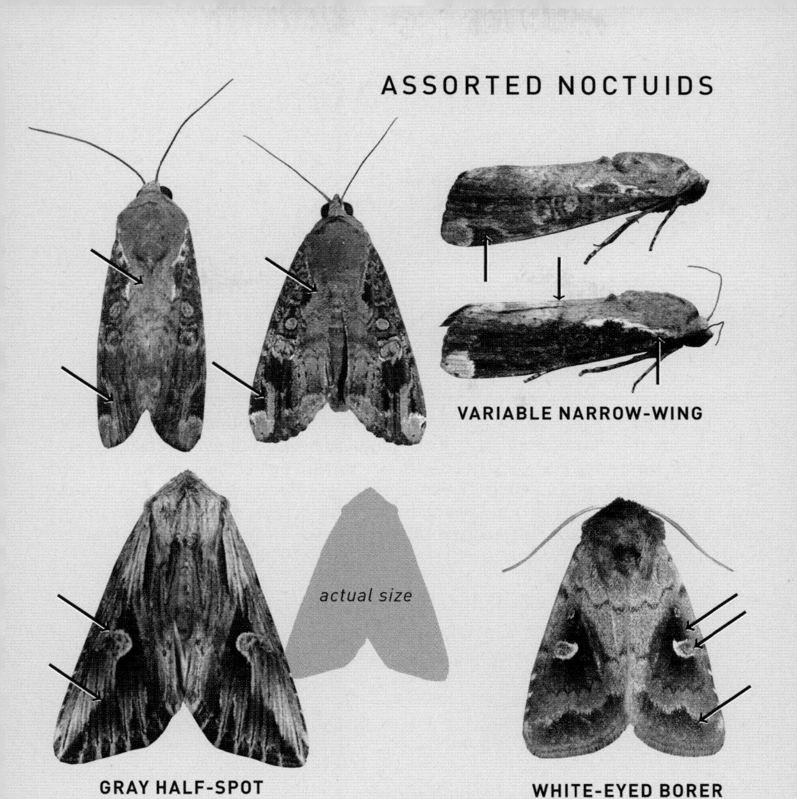

ANGLE SHADES

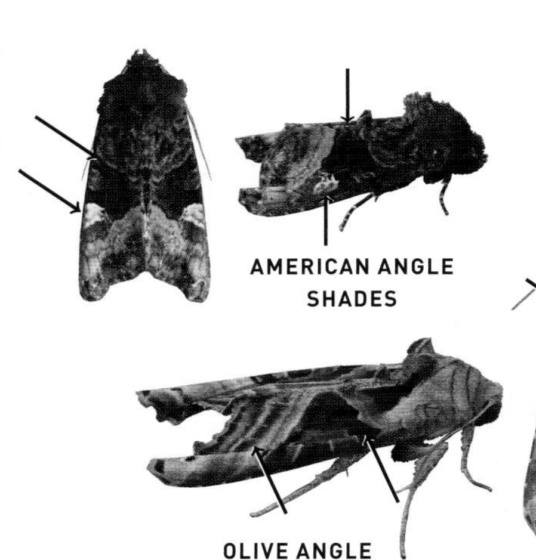

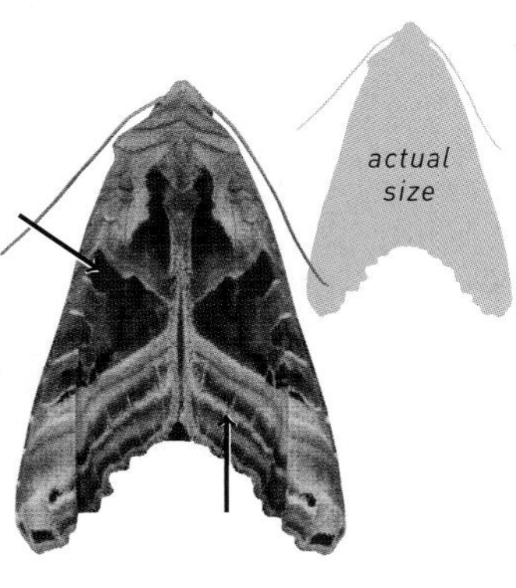

BROWN ANGLE SHADES

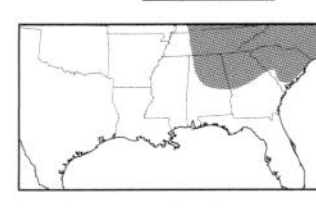

Phlogophora periculosa 93-2292 (9547) **Common**

TL 27 mm. Dimorphic. Resembles Olive Angle Shades, but FW is pinkish brown and median area is either medium or dark brown. AM line is typically more horizontal. Lacks dark basal patch, and shading along ST line is lighter. **HOSTS:** Trees and plants, including alder, balsam fir, cranberry, and plum. **NOTE:** The two forms were once thought to be male and female but are actually not sex-linked.

APAMEAS, BROCADES, AND BORERS

Family Noctuidae, Subfamily Noctuinae, Tribes Apameini and Arzamini

A large and varied group of noctuids mostly found in woodlands and old-fields. Many are highly host-specific. The apameas are medium-sized and rest with flat wings. Some are plainly attired with woodgrain patterns. The brocades are typically smaller, often with a darker median area and well-defined orbicular and reniform spots. The late-season borers are bright, medium-sized moths; most are yellow-orange or bronzy with fractured white spots and can be difficult to separate. All are nocturnal and will come to light in small numbers.

AIRY APAMEA *Apamea vultuosa* 93-2303 (9341) **Local**

TL 18–22 mm. Yellowish brown FW has darker brown shading along costa and behind pale-outlined spots. Orbicular spot slants toward reniform spot, which has a black dot in inner half. Veins in ST area are marked with tiny, paired black dots. Broad, dark wedges mark subapical and anal dashes. Blackish thorax has a slightly paler central stripe. **HOSTS:** Grass.

BROKEN-LINED BROCADE

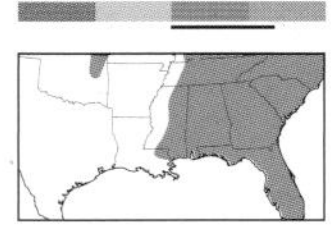

Mesapamea fractilinea 93-2370 (9406) **Common**

TL 12–16 mm. Variable. FW is pale yellowish to brown, sometimes peppered, with darker brown shading in outer median area. Veins are speckled black and white. Reniform spot is filled and partly edged whitish. **HOSTS:** Corn and grass.

YELLOW-SPOTTED BROCADE

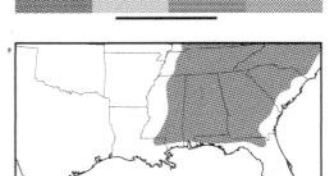

Oligia chlorostigma 93-2377 (9402) **Uncommon**

TL 14–16 mm. Peppery grayish brown FW has a whitish streak along basal section of inner margin, merging with white-edged inner AM line. Orbicular and reniform spots are shaded yellowish. Two tufts protrude above wings when folded. **HOSTS:** Grass.

BLACK-BANDED BROCADE

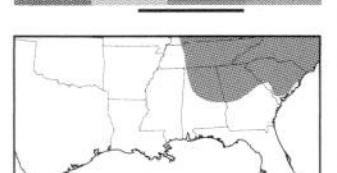

Oligia modica 93-2378 (9404) **Common**

TL 15–17 mm. Pale tan or gray FW has a dark brown outer median area behind large, dusky orbicular and reniform spots. Veins of ST area are marked with thin black dots. Thorax has projecting "epaulets." **HOSTS:** Unknown.

ANGLE SHADES

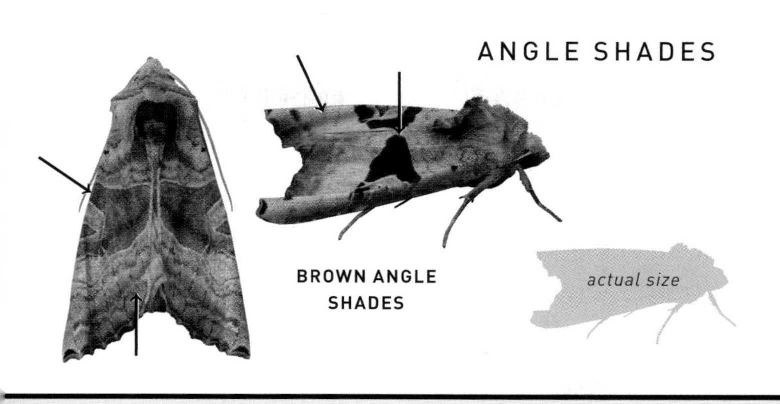

BROWN ANGLE SHADES

APAMEAS, BROCADES, AND BORERS

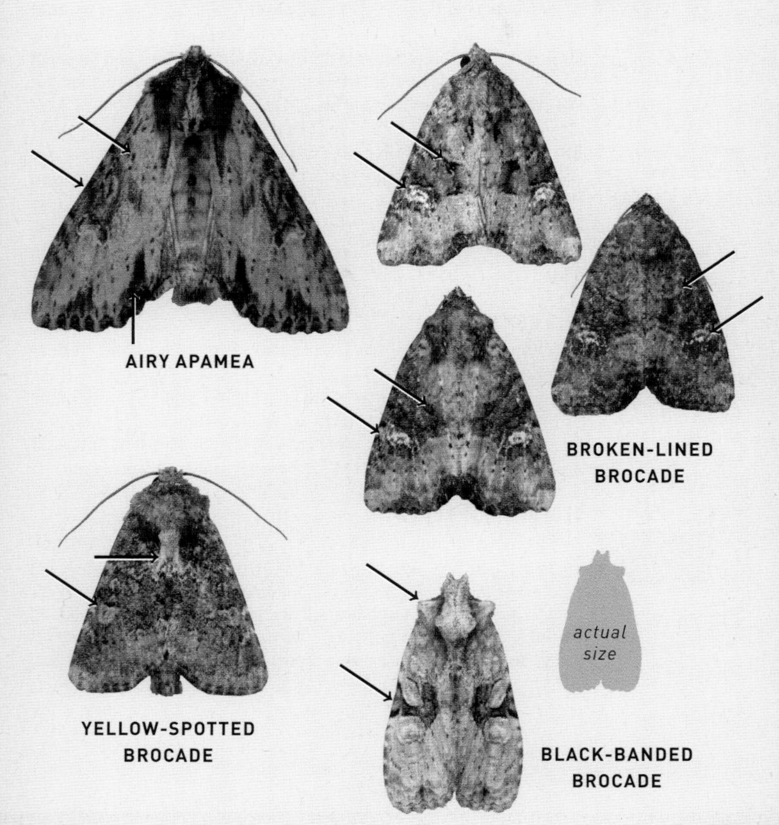

AIRY APAMEA

BROKEN-LINED BROCADE

YELLOW-SPOTTED BROCADE

BLACK-BANDED BROCADE

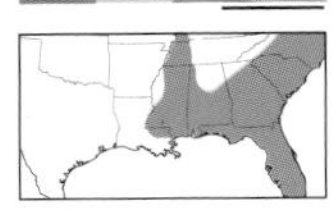

COMELY BROCADE

Meropleon cosmion 93-2409 (9425) **Uncommon**

TL 16–18 mm. Grayish brown FW is shaded rusty through median area, sometimes extensively. A long, thin white basal dash reaches to indistinct reniform spot, and is connected to grayish costa by whitish orbicular spot and pale basal area. ST line is a row of black dots. **HOSTS:** Stems of canary grass, sugarcane, and sugarcane plumegrass.

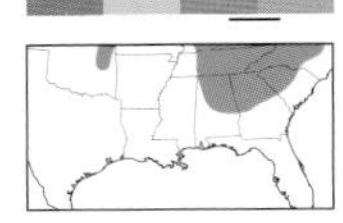

MULTICOLORED SEDGEMINER

Meropleon diversicolor 93-2413 (9427) **Local**

TL 16–18 mm. Bicolored FW is whitish below straight median line. Upper median area is dark brown. Double AM line forms a large bulge near inner margin, which is usually filled rusty brown. Slanting orbicular spot is edged white. A peppery whitish patch cuts through midpoint of PM line. **HOSTS:** Stems of woolgrass.

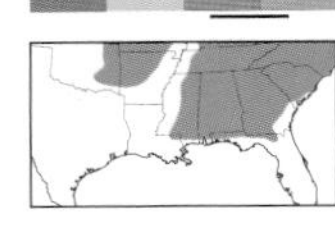

NEWMAN'S BROCADE

Meropleon ambifusca 93-2414 (9428) **Local**

TL 13 mm. Resembles Multicolored Sedgeminer, but ST area is strongly shaded brownish. AM line is wavy or scalloped and does not form large, rounded bulge. Often lacks peppery patch at midpoint of PM line. **HOSTS:** Stems of big bluestem.

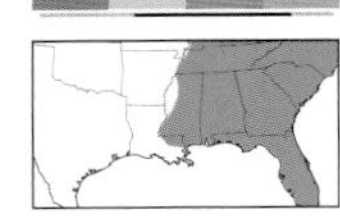

OBLONG SEDGE BORER

Capsula oblonga 93-2438 (9449) **Uncommon**

TL 25–27 mm. Peppery pale tan FW has faint light gray and darker tan streaks through median area. PM line and terminal line are rows of tiny black dots. Tufted abdomen projects from beneath folded wings. **HOSTS:** Cattail and bulrush.

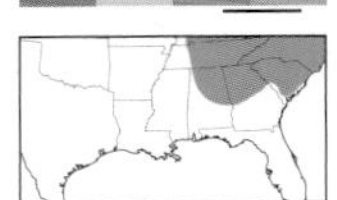

IRIS BORER

Macronoctua onusta 93-2442 (9452) **Local**

TL 25–27 mm. Pale tan FW has a broad, slate gray band along costa that bleeds over top of black-edged orbicular and reniform spots. Toothed double AM and PM lines are finely etched in black. Veins in terminal area are shaded slate gray; gray shading fills entire ST area in rare individuals. **HOSTS:** Larvae bore into roots of iris.

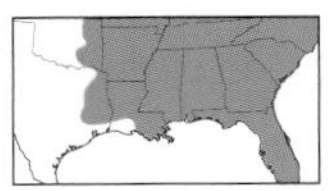

ELDER SHOOT BORER

Achatodes zeae 93-2444 (9520) **Common**

TL 13–17 mm. Variable. Gray FW has rusty red basal, median, and ST bands that range from faint to dominating. ST line is a row of orange dots that meet large orange apical patch. PM line consists of paired black dots along veins. Thorax has a fiery orange dorsal stripe. **HOSTS:** Elderberry; also reported on corn, dahlia, and grass.

APAMEAS, BROCADES, AND BORERS

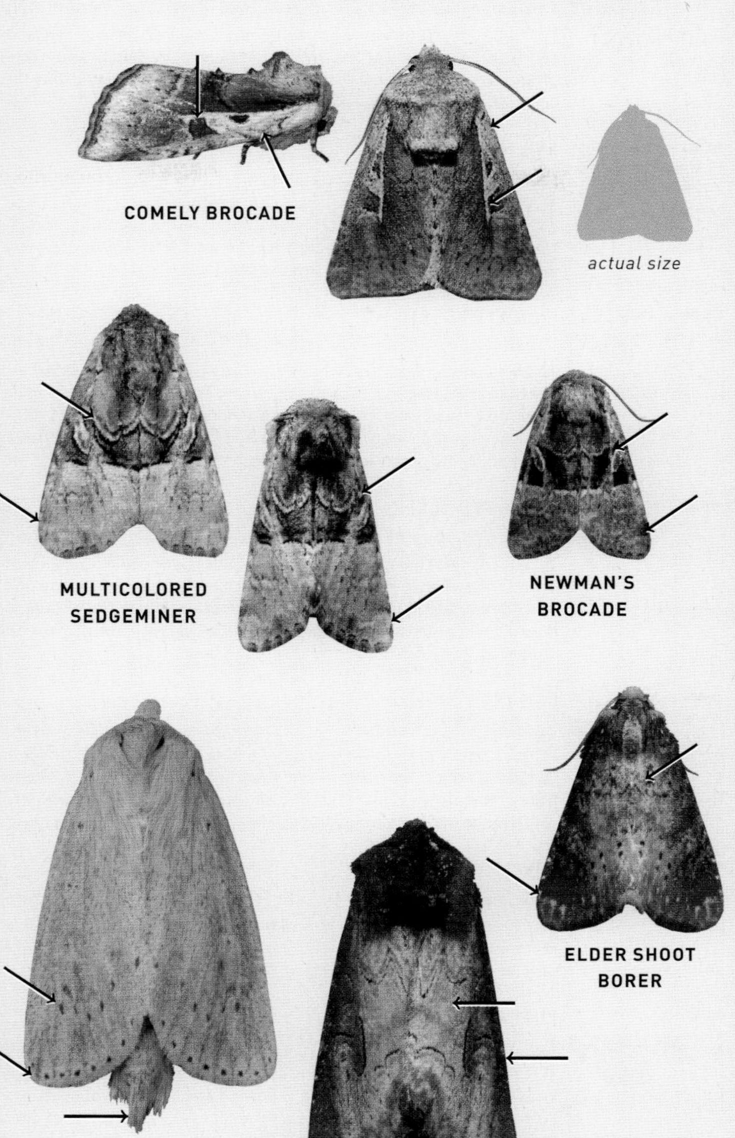

BUFFALO BORER

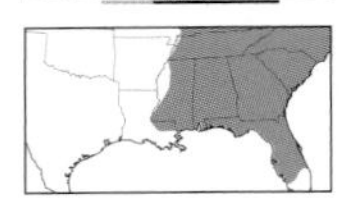

Parapamea buffaloensis 93-2464 (9463) **Uncommon**

TL 17–23 mm. Dark rusty brown FW is shaded violet gray in basal and ST areas. Wavy AM and straight PM lines are double and curve inward at costa. Orbicular and reniform spots are rusty brown or white, fractured into spots. **HOSTS:** Lizard's tail.

ASH TIP BORER

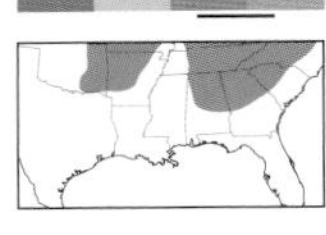

Papaipema furcata 93-2468 (9495) **Uncommon**

TL 20–22 mm. Brownish yellow FW has diffuse brown median and double PM lines. Fragmented white spots are large and rounded; reniform spot has curving, yellowish line at center. Basal area has one or two small white spots. **HOSTS:** Ash and box elder.

IRONWEED BORER

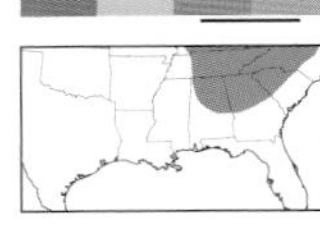

Papaipema cerussata 93-2470 (9505) **Local**

TL 20–22 mm. FW has brown median area that is bronzy toward inner margin. Basal and ST areas are violet gray. Apical patch is usually bronzy. Claviform, orbicular, and reniform spots appear as relatively narrow lines of white (rarely orange) spots. Base of FW has a cluster of white spots. **HOSTS:** New York ironweed.

BRICK-RED BORER

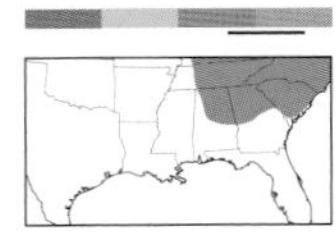

Papaipema marginidens 93-2472 (9492) **Uncommon**

TL 20–22 mm. Uniform reddish brown FW has indistinct scalloped brown lines. Fractured reniform spot is large with a curvy white line at center. Costa is marked with small white patches. Basal area is boldly scalloped white. Apical patch is pale yellowish. **HOSTS:** Unknown.

OSMUNDA BORER

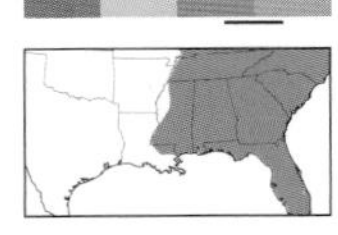

Papaipema speciosissima 93-2475 (9482) **Uncommon**

TL 23–27 mm. Orange FW is shaded brown to pale violet in ST area. Diffuse median and double PM lines are sharply angled. White (rarely blackish) claviform, orbicular, and reniform spots are narrow bars. **HOSTS:** Ferns, including cinnamon, interrupted, and royal.

SENSITIVE FERN BORER

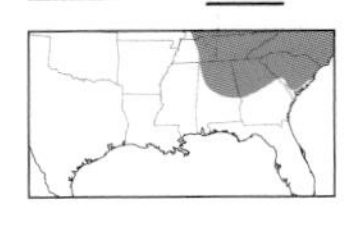

Papaipema inquaesita 93-2476 (9483) **Common**

TL 17–21 mm. Resembles Osmunda Borer, but spots are round and filled orange (rarely, claviform and orbicular are white). Veins are often dark. **HOSTS:** Sensitive fern.

BURDOCK BORER

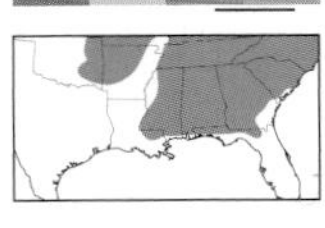

Papaipema cataphracta 93-2497 (9466) **Uncommon**

TL 17–22 mm. Peppery golden brown FW has violet shading in basal and ST areas, and a contrasting yellowish apical patch. All spots are large and round; fragmented reniform spot has curving line at center. Basal area has a cluster of yellowish spots. **HOSTS:** Burdock; also crownbeard, thistle, and others.

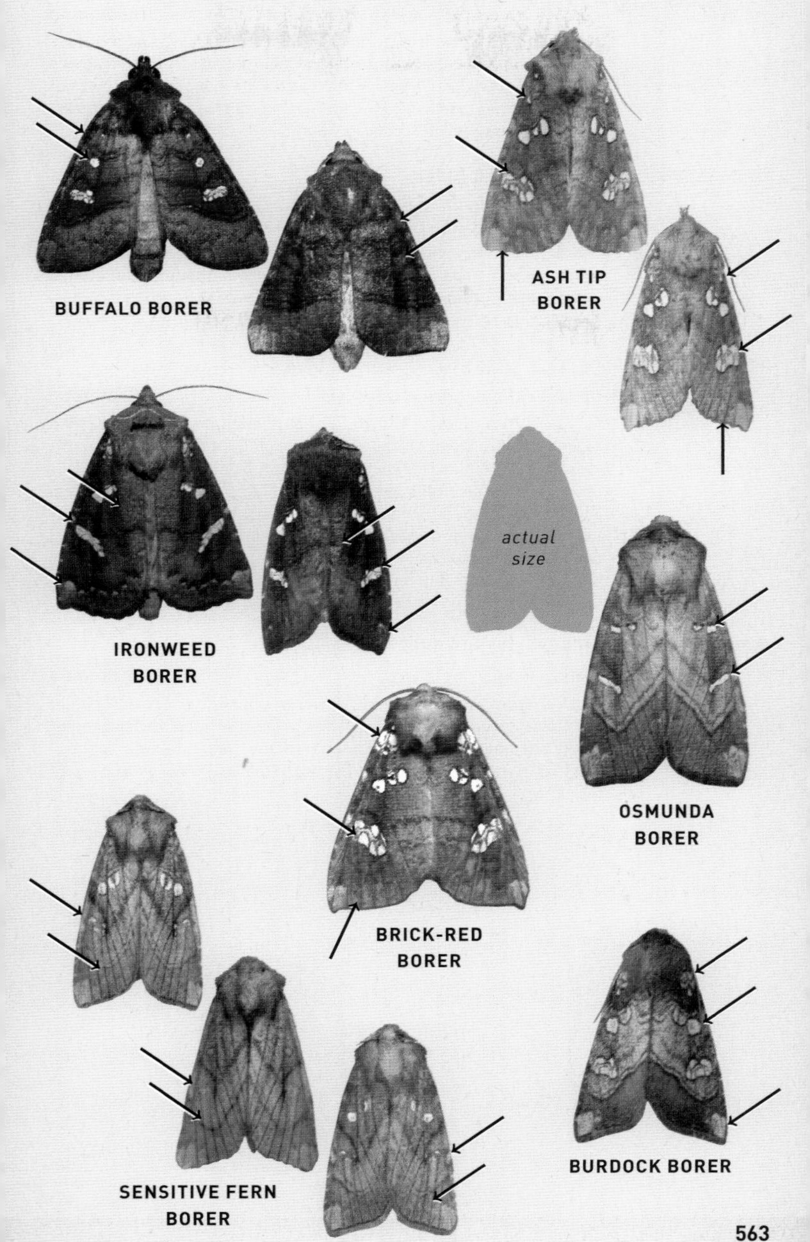
BUFFALO BORER
ASH TIP BORER
IRONWEED BORER
actual size
OSMUNDA BORER
BRICK-RED BORER
SENSITIVE FERN BORER
BURDOCK BORER

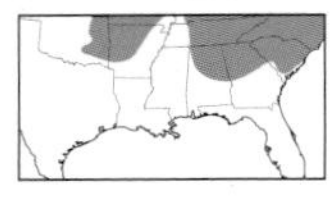

STALK BORER

Papaipema nebris 93-2501 (9496) **Uncommon**

TL 16–20 mm. Hoary brown FW is shaded lighter in ST and basal areas. Indistinct AM line is strongly curving. Almost straight PM line is edged white toward inner margin. Spots are usually indistinct but are sometimes partly filled white or (less often) blackish. **HOSTS:** Burdock, ragweed, and other low plants.

WHITE-TAILED DIVER

Bellura gortynoides 93-2513 (9523) **Uncommon**

TL 19–27 mm. Tan to pale orange FW has strongly angled, rusty AM and median lines and toothed PM line. Orange-filled reniform and orbicular spots are surrounded by dark brown shading, sometimes extensively. ST line is a row of black spots or wedges. **HOSTS:** Yellow pond lily. **NOTE:** Two broods.

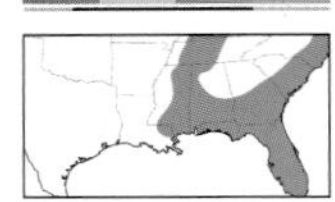

CATTAIL BORER

Bellura obliqua 93-2517 (9525) **Uncommon**

TL 22–28 mm. Grayish brown FW has a slanting pale basal patch extending onto sides of thorax and merging into pale costa. Slanting, narrow, pointed reniform spot is orange, edged whitish. Orbicular spot is teardrop-shaped. Apex is pointed. **HOSTS:** Cattail. **NOTE:** Two broods. Genetic studies suggest that what is currently recognized in e. N. America as Cattail Borer may actually be a complex of two or more species. Many individuals with shorter, rounder wings and more uniform reniform spots may prove to be a separate species once the genus has been examined and revised.

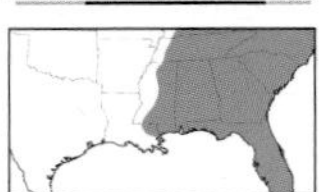

PICKERELWEED BORER

Bellura densa 93-2518 (9526) **Common**

TL 19–26 mm. Resembles Cattail Borer, but shorter FW has a blunt apex. Orange orbicular and reniform spots are round, and backed by dusky shading along veins. Basal area and costa often do not contrast as markedly with rest of FW. **HOSTS:** Pickerelweed and water hyacinth.

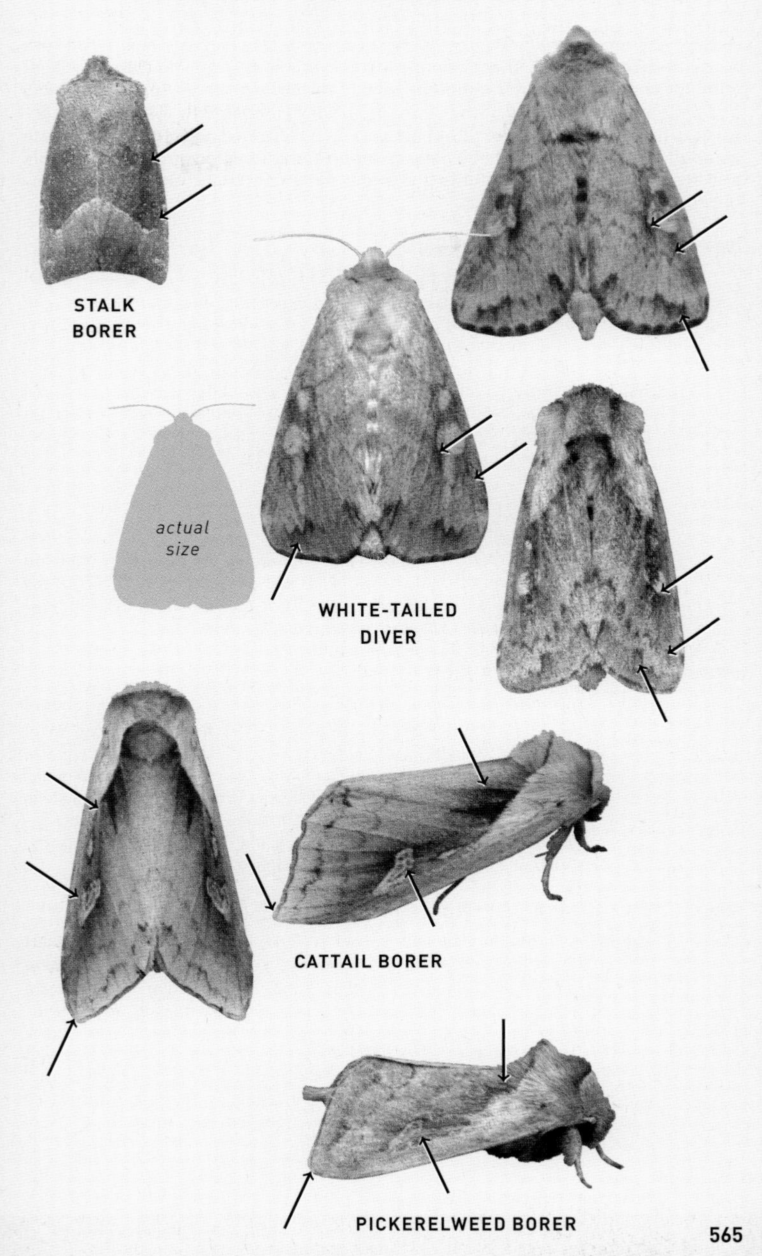

STALK BORER

WHITE-TAILED DIVER

CATTAIL BORER

PICKERELWEED BORER

Pinions and Xyleninine Sallows

Family Noctuidae, Subfamily Noctuinae, Tribe Xylenini

Medium-sized noctuids that are typically grayish or orange. Most species rest with flat wings, though a few pinions hold their narrow wings close to the body and appear tubular. Many of these are cold-hardy and among the earliest noctuids to fly in spring, and the latest in the fall, in colder regions; most hibernate as adults through the winter. They are inhabitants of woods and weedy fields and are nocturnal, visiting lights and sugar bait in small but sometimes regular numbers.

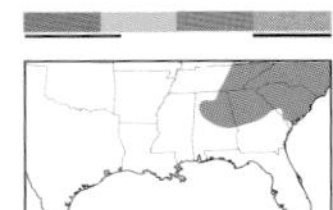

DIMORPHIC PINION

Lithophane patefacta 93-2532 (9886) **Local**

TL 18–20 mm. Dimorphic. Pale individuals have tan FW with light brown bands in median and ST areas. Rectangular black patch in inner median area is strongly defined. AM and PM lines are rows of tiny paired black dots along veins. Dark individuals have chocolate brown FW with diffusely defined spots, outlined rusty, that blend into pale yellowish costa. Black patch in median area is usually noticeable. Thorax has projecting "epaulets." **HOSTS:** Blueberry, cherry, hickory, maple, oak, and others.

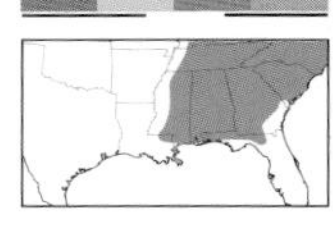

BETHUNE'S PINION

Lithophane bethunei 93-2533 (9887) **Local**

TL 18–20 mm. Resembles pale individuals of Dimorphic Pinion but has brown shading surrounding pale-edged reniform spot. Dusky patch in inner median area is faint. **HOSTS:** Deciduous trees, including ash, birch, chokecherry, elm, maple, and willow.

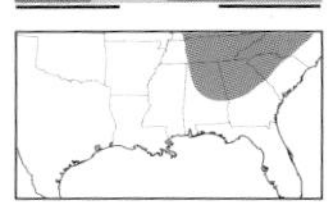

WANTON PINION

Lithophane petulca 93-2536 (9889) **Local**

TL 18–20 mm. Dimorphic. Resembles Dimorphic Pinion, but FW of pale individuals lacks strong median line, and orbicular and reniform spots are outlined in brown. Faint dusky patch is often present in inner median area. FW of dark individuals is dark brown, and costa is straw-colored and checkered dark brown. Pale spots are crisply defined and outlined brown. Dusky patch is usually indistinct. **HOSTS:** Alder and birch; also ash, cherry, maple, oak, and willow.

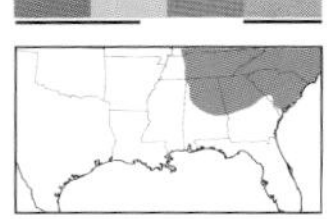

SIGNATE PINION

Lithophane signosa 93-2542 (9895) **Local**

TL 18–20 mm. Dimorphic. Resembles Dimorphic Pinion, but FW of pale individuals lacks median band, and orbicular and reniform spots are indistinct. Veins are speckled blackish along length, lending a streaky appearance. Dusky patch in inner median area is usually strong but narrow. FW of dark individuals is dark brown with blackish veins. Costa is straw-colored, checkered light brown. Orbicular and reniform spots are well defined, with some brown shading inside. Dusky patch is indistinct. **HOSTS:** Sycamore.

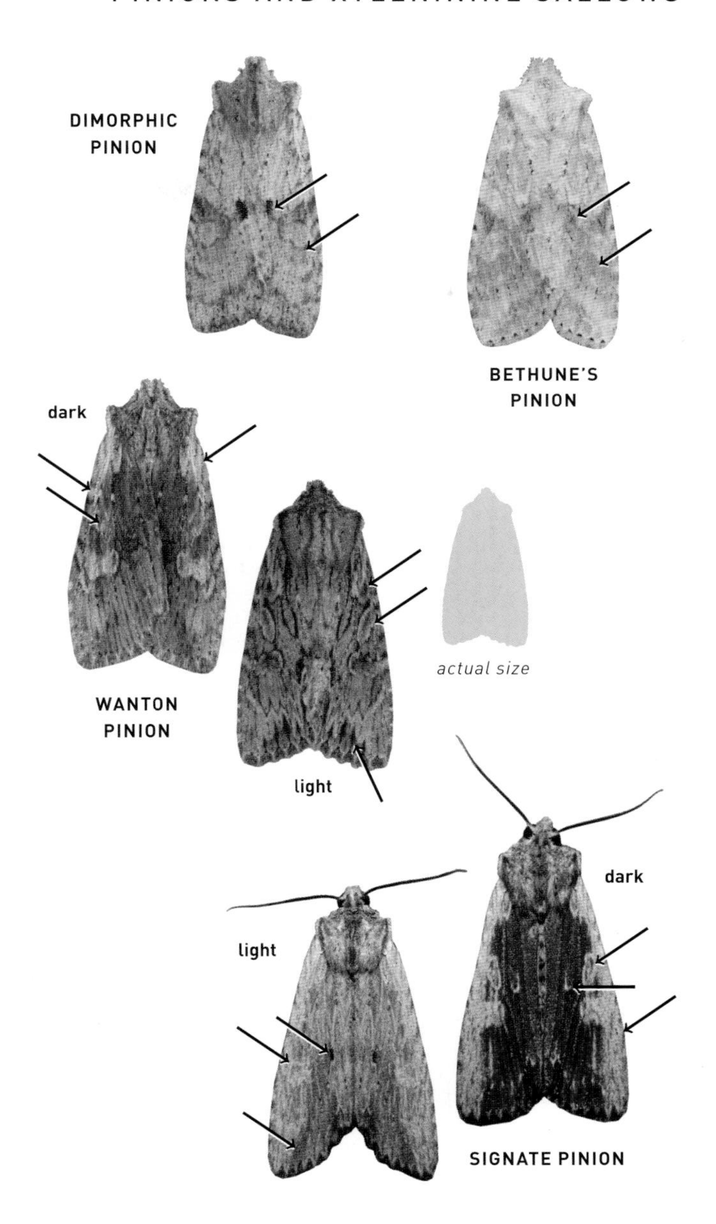
DIMORPHIC PINION
BETHUNE'S PINION
dark
WANTON PINION
light
actual size
light
dark
SIGNATE PINION

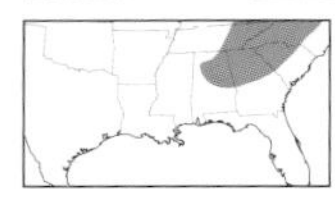

SHIVERING PINION

Lithophane querquera 93-2550 (9904) **Local**

TL 20–21 mm. Pale gray FW has a bold black bar at midpoint of costa. Black basal dash is bordered by white. Reniform spot has fragmented chestnut center. ST line has pairs of black dots at midpoint and anal angle. **HOSTS:** Deciduous trees, including apple, birch, cherry, and hickory.

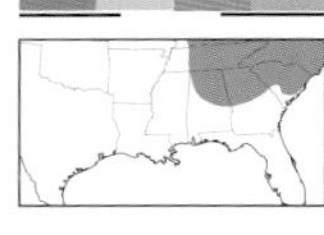

PALE GREEN PINION

Lithophane viridipallens 93-2552 (9905) **Local**

TL 18–20 mm. Light gray FW is tinged pale green when fresh. Scalloped, double AM and PM lines are often inconspicuous. A bold blackish patch sits between pale orbicular and reniform spots. **HOSTS:** Shrubs, including greenbrier and eastern hornbeam. **NOTE:** More commonly encountered in spring.

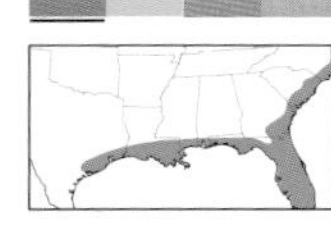

BALD CYPRESS PINION

Lithophane abita 93-2555 (9928.1) **Local**

TL 18–20 mm. Mottled, light gray FW has a diffuse rusty spot in reniform area. A thin black dash runs parallel to inner margin in median area. A narrow median line often passes along lower edge of large orbicular spot. **HOSTS:** Bald cypress.

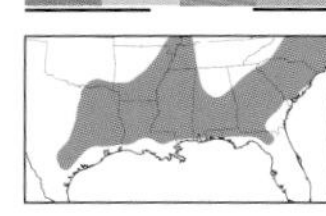

LEMMER'S PINION

Lithophane lemmeri 93-2566 (9899) **Local**

TL 18–20 mm. Light gray FW has fine blackish streaks along veins. Long, thin white dash in inner median area is edged black and bordered with a small rusty patch. White-edged orbicular and reniform spots are indistinct. ST area has dusky shading at midpoint. **HOSTS:** Atlantic white cedar and red cedar.

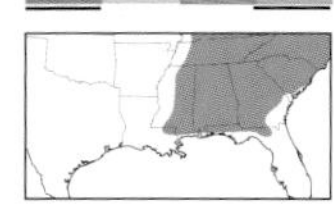

ASHEN PINION

Lithophane antennata 93-2577 (9910) **Common**

TL 23–25 mm. Light to medium gray FW has a black-edged reniform spot filled diffusely rusty brown. Hourglass-shaped orbicular spot is filled pale in outer half. Pale gray shoulder patch is bordered on inner edge by thin black basal dash. **HOSTS:** Deciduous trees, including apple, ash, elm, hickory, oak, and willow.

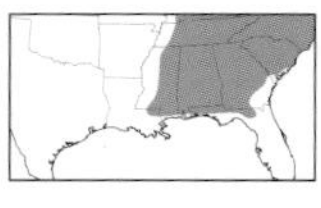

MUSTARD SALLOW

Pyreferra hesperidago 93-2583 (9929) **Local**

TL 16–17 mm. Peppery yellowish or orange FW has nearly parallel, bold, rusty AM, median, and PM lines. Veins in lower median and ST area are traced rusty. Hourglass-shaped reniform spot has a dusky spot in inner half. **HOSTS:** Witch hazel.

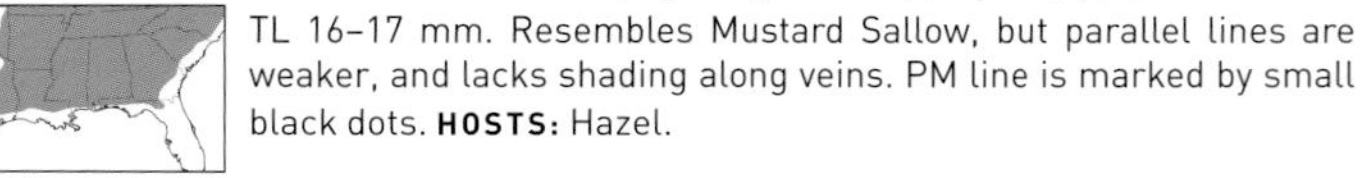

PETTIT'S SALLOW *Pyreferra pettiti* 93-2586 (9932) **Local**

TL 16–17 mm. Resembles Mustard Sallow, but parallel lines are weaker, and lacks shading along veins. PM line is marked by small black dots. **HOSTS:** Hazel.

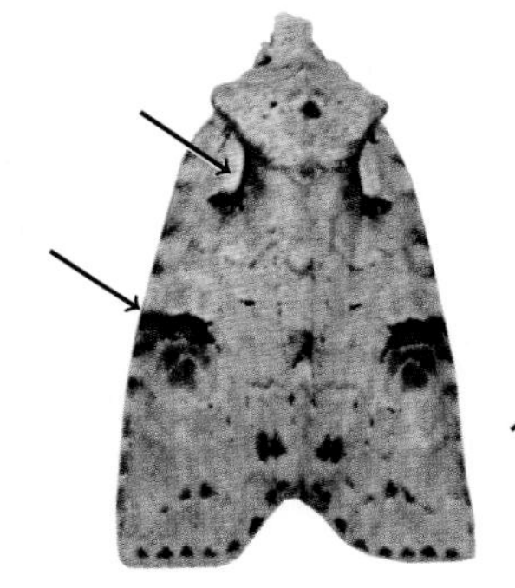
SHIVERING PINION

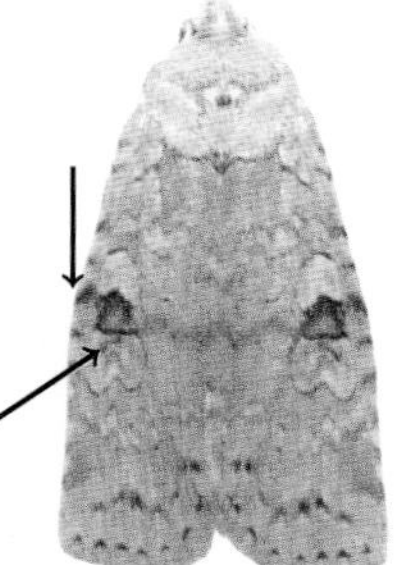
PALE GREEN PINION

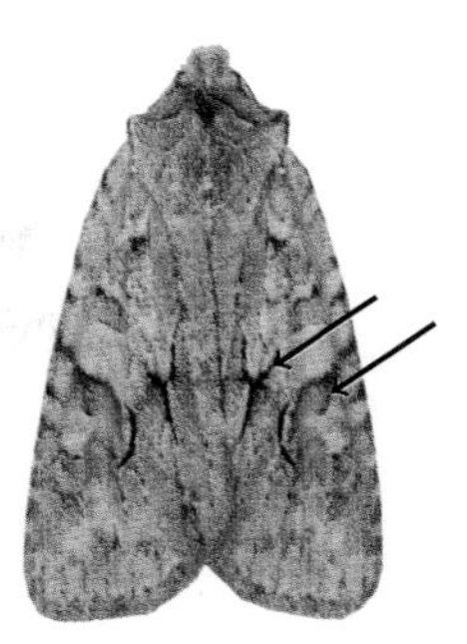
BALD CYPRESS PINION

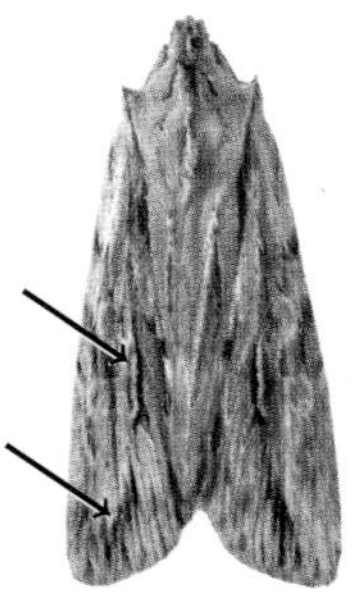
LEMMER'S PINION

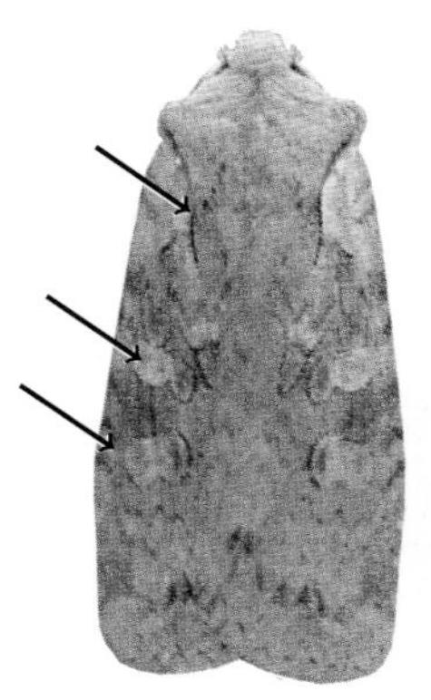

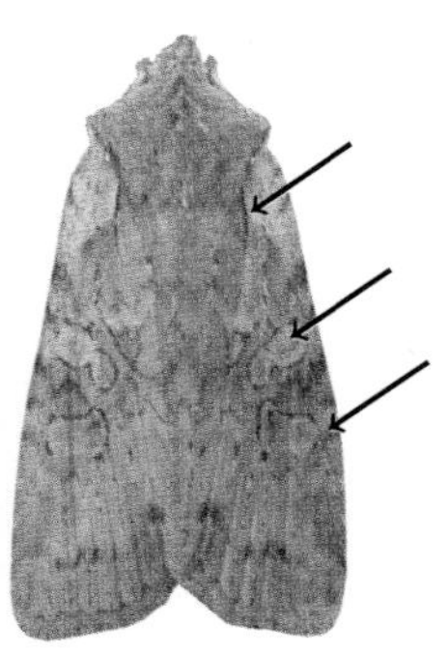
ASHEN PINION

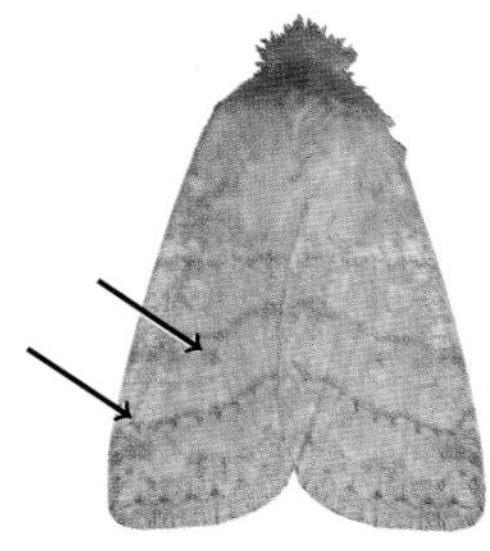
MUSTARD SALLOW

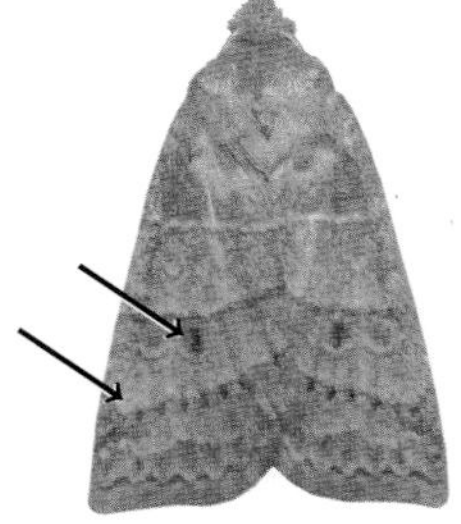
PETTIT'S SALLOW

actual size

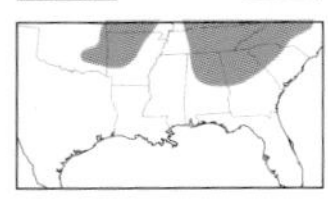

STRAIGHT-TOOTHED SALLOW

Eupsilia vinulenta 93-2587 (9933) Common

TL 18–20 mm. Orange to rusty brown or maroon FW has a large orange or white reniform spot abutted by two tiny white dots. Orbicular spot, when present, is orange. PM line is slightly toothed. **HOSTS:** Deciduous trees, including cherry, maple, and oak. **NOTE:** This species' common name refers to the toothed edge of FW scales when viewed under a microscope; all other *Eupsilia* species, including Franclemont's Sallow, have curled teeth.

FRANCLEMONT'S SALLOW

Eupsilia cirripalea 93-2589 (9934) Local

TL 17–19 mm. Virtually identical to Straight-toothed Sallow, and many individuals may not be identifiable in the field. FW generally has a smoother, more uniform look and tends toward reddish to reddish orange. Veins are reliably shaded blackish in terminal area and along PM line. Orbicular spot is always absent. **HOSTS:** Woody plants, including birch, blackberry, cherry, hickory, maple, oak, persimmon, and walnut.

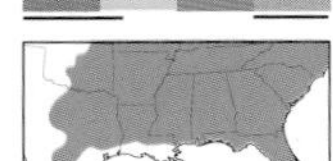

VARIABLE SALLOW

Sericaglaea signata 93-2595 (9941) Common

TL 18–23 mm. Pale orange to brown FW is often peppery and is overlaid with pale veins. Often has darker ST area contrasting with paler terminal line. Double AM and PM lines are filled pale and edged dark tan or brown. Diffuse median band is dark brown. Indistinct reniform spot has black dot in inner half. **HOSTS:** Deciduous trees, including ash, cherry, hickory, oak, and others.

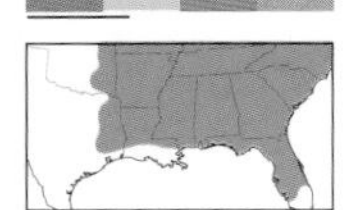

RED-WINGED SALLOW

Xystopeplus rufago 93-2596 (9942) Common

TL 17–19 mm. Reddish brown FW has large, yellow-edged orbicular and reniform spots that are often open toward costa. AM and PM lines are dotted blackish and often indistinct. Apex is pointed. **HOSTS:** Blueberry, plum, hazel, oak, and others.

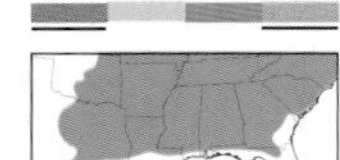

ROADSIDE SALLOW

Metaxaglaea viatica 93-2598 (9944) Common

TL 22–28 mm. Reddish brown FW has bands of purplish brown shading beyond scalloped AM and PM lines. Large, pale-edged orbicular and reniform spots converge and sometimes touch; reniform spot has faint dusky spot in inner half. Usually has a single pale vein parallel to inner margin. **HOSTS:** Apple, crab apple, mountain ash, and cherry.

FOOTPATH SALLOW

Metaxaglaea semitaria 93-2599 (9945) Uncommon

TL 24–26 mm. Resembles Roadside Sallow, but purplish brown shading is lighter, sometimes almost absent along AM line. Diffuse brown median band is usually dark and connects with dusky patch in inner reniform spot. Converging orbicular and reniform spots do not touch. **HOSTS:** Blueberry, chokecherry, and oak.

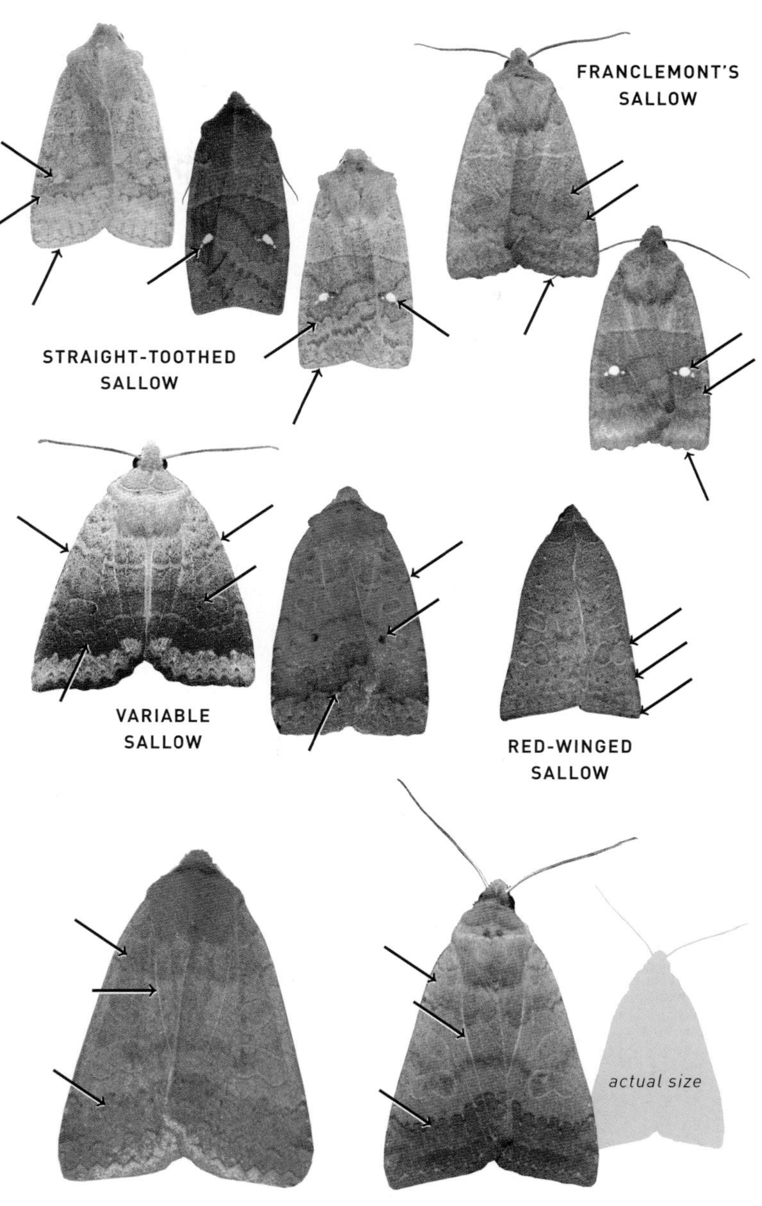
FRANCLEMONT'S SALLOW
STRAIGHT-TOOTHED SALLOW
VARIABLE SALLOW
RED-WINGED SALLOW
actual size
ROADSIDE SALLOW
FOOTPATH SALLOW

SOUTHERN SALLOW

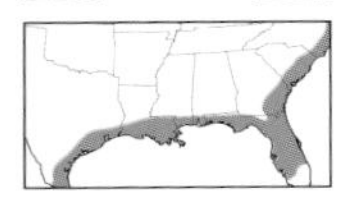

Metaxaglaea australis 93-2600 (9945.1) **Uncommon**

TL 19–23 mm. Resembles larger Footpath Sallow, but shading in ST area is darker and more uniform, and is absent along AM line. Median band is often faint. ST line is not well defined. **HOSTS:** Unknown, but has been reared on cherry and oak.

HOLLY SALLOW

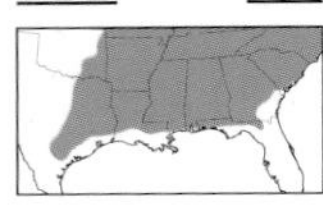

Metaxaglaea violacea 93-2601 (9945.2) **Uncommon**

TL 22–24 mm. Resembles larger Footpath Sallow, but shading in ST area is darker and more uniform. Entire basal area is usually lightly shaded. Orbicular and reniform spots are outlined rusty brown. Rusty ST line is well defined. **HOSTS:** American holly.

SLOPING SALLOW

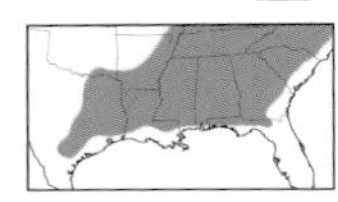

Epiglaea decliva 93-2602 (9946) **Uncommon**

TL 22–26 mm. Chestnut brown FW has slightly toothed PM line that has blackish dots at each vein. Large, pale-edged orbicular and reniform spots are only slightly sloping, and are often filled darker than ground color; reniform spot has blackish smudge in inner half. ST line is warm brown edged pale yellow, sometimes fragmented. **HOSTS:** Apple, cherry, maple, oak, and others.

TREMBLING SALLOW

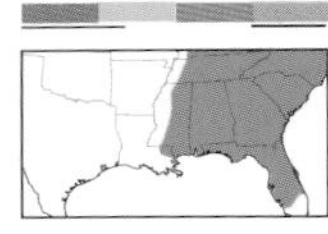

Chaetaglaea tremula 93-2606 (9949) **Common**

TL 20–22 mm. Speckled brown FW has gently curving dusky AM line. Median area is slightly darker, and orbicular and reniform spots appear paler; reniform spot has a tiny black dot in inner half. ST line has a blackish smudge at midpoint. Inner margin is thinly edged pale yellow. **HOSTS:** Lowbush blueberry, huckleberry, oak, sweet fern, and others.

SILKY SALLOW

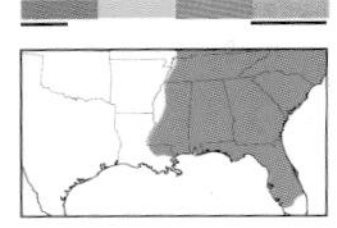

Chaetaglaea sericea 93-2607 (9950) **Common**

TL 20–23 mm. Peppery brown FW has thin, pale yellowish veins. Curving, dark brown AM and PM lines are edged pale. Orbicular and reniform spots are darker, edged pale. ST line is pale. Inner margin is distinctly edged pale yellowish. **HOSTS:** Blueberry and oak; also aspen, cherry, oak, and others.

SCALLOPED SALLOW

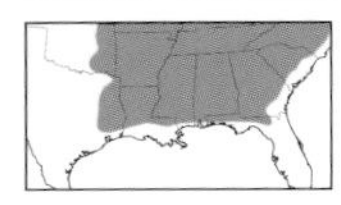

Eucirroedia pampina 93-2609 (9952) **Common**

TL 21–25 mm. Orange FW has yellow-edged rusty AM and PM lines that converge toward inner margin. Large orbicular and reniform spots are brownish with yellow outline. Outer margin is strongly scalloped. Typically rests in a shallow headstand position. **HOSTS:** Deciduous trees and plants, including cherry, oak, and poplar.

SOUTHERN SALLOW
HOLLY SALLOW
SLOPING SALLOW
TREMBLING SALLOW
actual size
SCALLOPED SALLOW
SILKY SALLOW

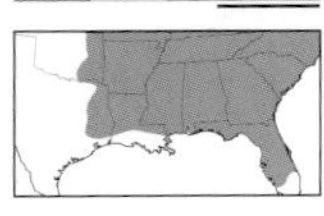

BICOLORED SALLOW

Sunira bicolorago 93-2616 (9957) **Common**

TL 18–20 mm. Peppery pale orange FW has bold, strongly wavy, brownish AM, median, and PM lines. Orange ST line is bordered with diffuse brownish shading and accented with faint black dashes along veins. Reniform spot has a blackish dot in inner half. Some "bicolored" individuals have dusky shading in lower half. **HOSTS:** Deciduous trees, including cherry, elm, maple, oak, and others.

AMERICAN DUN-BAR

Cosmia calami 93-2672 (9815) **Common**

TL 15–17 mm. Variable. Tan to brownish FW has whitish, strongly slanting AM and curved PM lines that are often edged brownish. Blurry brown median line may be absent; when present, it angles between white-outlined orbicular and reniform spots, which are sometimes dark. Rare individuals have dark brown shading in ST and basal areas, sometimes extensive. Terminal line is sometimes a row of dark dots. **HOSTS:** Oak; also caterpillars of other lepidopteran species.

TAPESTRY CALICO

Properigea tapeta 93-2723 (9592) **Common**

TL 13–15 mm. Mottled, rusty brown FW has bold white orbicular dot and curving reniform spot. Fragmented, wavy, black AM, PM, and ST lines are edged light brown. Thorax often has a tawny stripe down center. **HOSTS:** Larvae have been reared on resurrection fern.

MARSH FERN MOTH

Fagitana littera 93-2749 (9629) **Uncommon**

TL 15–16 mm. Rusty brown FW has bold white reniform spot shaped like a carpentry nail. Smoothly curving AM and PM lines are whitish, edged with subtle brown shading. Costa is edged thinly whitish. **HOSTS:** Ferns, including eastern marsh, Virginia chain, and royal.

PINIONS AND XYLENININE SALLOWS

Spring Quakers, Woodlings, and Woodgrains

Family Noctuidae, Subfamily Noctuinae, Tribe Orthosiini

Sturdy delta-shaped noctuids that are mostly clad in shades of brown. The *Orthosia* are sometimes confusing but can be reliably identified by the combination of orbicular spot and ST line. All of these moths are single-brooded and are on the wing in spring and early summer, though Fluid Arches is active though the summer. Usually found in wooded areas, though some are tolerant of highly urban environments. They are nocturnal and will visit lights; some can be very common.

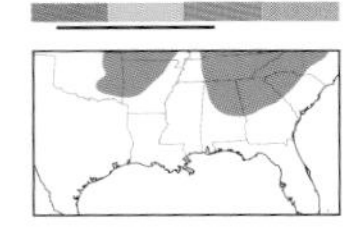

RUBY QUAKER *Orthosia rubescens* 93-2770 (10487) Local

TL 20 mm. Variable. Pale orange to brown FW is speckled reddish brown, sometimes heavily. Basal area is usually paler. Round orbicular spot is pale yellow and edged below with reddish shading. Reniform spot has a blackish dot in inner half. ST line is indistinct but edged with rusty blotches at midpoint and costa. Antennae of male are bipectinate. **HOSTS:** Deciduous and coniferous trees, including aspen, cherry, hemlock, maple, and oak. **NOTE:** Frequently confused with Speckled Green Fruitworm Moth; note ST line and orbicular spot.

GARMAN'S QUAKER

Orthosia garmani 93-2771 (10488) Local

TL 18–22 mm. Grayish brown to reddish brown FW has a pale terminal line. Inner section of irregular ST line is accented with blackish wedges. Diffusely edged orbicular and reniform spots are pale. Thorax is densely hairy. **HOSTS:** Deciduous trees and other woody plants, including blueberry, chokecherry, dogwood, and sugar maple.

GRAY QUAKER

Orthosia alurina 93-2774 (10491) Uncommon

TL 19–21 mm. Lightly speckled, pale gray FW has diffuse gray AM and PM lines. Brownish median band is often noticeable. Pale ST line is edged reddish and fragmented into uneven dashes. Large reniform spot has a small black dot in inner half. **HOSTS:** Various deciduous trees, including basswood, cherry, elm, maple, oak, and willow. **NOTE:** Sometimes confused with Speckled Green Fruitworm Moth; note strong AM and PM lines.

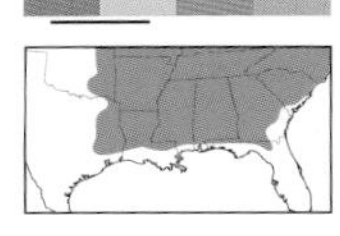

SPECKLED GREEN FRUITWORM MOTH

Orthosia hibisci 93-2778 (10495) Common

TL 20–23 mm. Extremely variable. Resembles Ruby Quaker, but ST line is pale yellow edged thinly reddish, and orbicular spot is FW ground color. Speckled FW may be reddish brown to tan; basal area is not noticeably paler. Orbicular and reniform spots often touch or even merge. **HOSTS:** Deciduous and coniferous trees, including apple, chokecherry, elm, hickory, poplar, spruce, tamarack, and willow.

SPRING QUAKERS, WOODLINGS, AND WOODGRAINS

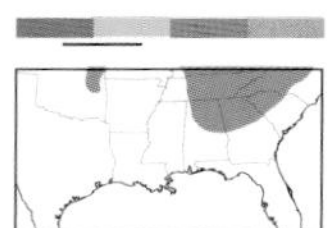

NORMAN'S QUAKER

Crocigrapha normani 93-2784 (10501) **Local**

TL 18–19 mm. Speckled, tan to reddish brown FW has an evenly curved, pale-edged AM line. Median area is usually slightly darker. Orbicular and reniform spots are thinly edged pale; reniform spot has black dot in inner half. ST line is faint. Diffuse apical patch is whitish. **HOSTS:** Deciduous and coniferous trees, including ash, birch, cherry, ironwood, maple, spruce, tamarack, and willow.

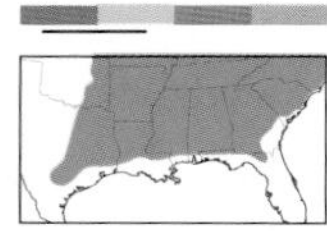

INTRACTABLE QUAKER

Himella fidelis 93-2785 (10502) **Common**

TL 13–18 mm. Brown FW has pale yellowish ST line and large spots that are well defined. AM line is blackish, edged faintly pale; AM and PM lines often have blackish dots near inner margin. Veins are sometimes thinly traced pale in ST area. **HOSTS:** Deciduous trees, including apple, elm, hickory, oak, and witch hazel.

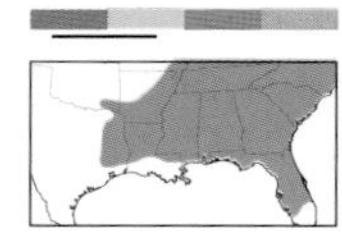

ALTERNATE WOODLING

Egira alternans 93-2799 (10517) **Common**

TL 18–21 mm. Gray FW has a thin black AM line bordered broadly with a pale gray band that includes orbicular spot. Reniform spot and inner basal area are shaded rusty. Often has a thick black bar in inner median area. Thorax has a pale collar. **HOSTS:** Woody plants, including blueberry, honeysuckle, and willow.

DISTINCT QUAKER

Achatia distincta 93-2800 (10518) **Common**

TL 20 mm. Pale gray FW is marked with thin black, double AM and PM lines that are connected in inner median area with a bold, kinked black bar. Large orbicular and reniform spots are usually partly shaded warm brown. **HOSTS:** Deciduous trees and shrubs, including ash, birch, butternut, grape, maple, and oak.

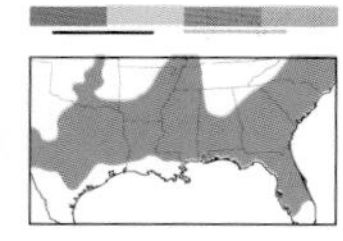

GRAY WOODGRAIN

Morrisonia mucens 93-2801 (10519) **Common**

TL 17–20 mm. Pale gray FW has a barklike pattern of warm brown streaks. Inner median area has a thick dusky bar. Often has dark smudges at anal and subapical dashes, and a thin black basal dash. Usually has dark shading between indistinct orbicular and reniform spots. **HOSTS:** Unknown. **NOTE:** Two broods in FL.

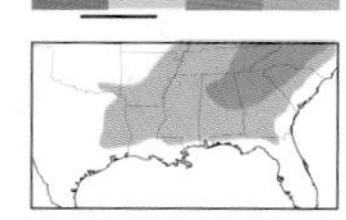

BICOLORED WOODGRAIN

Morrisonia evicta 93-2802 (10520) **Common**

TL 19–21 mm. Grayish brown FW has a broad blackish band along inner margin. Slanting pale orbicular spot is usually fused to brown-filled reniform spot. Anal dash is whitish. Some individuals lack blackish band. **HOSTS:** Deciduous trees and shrubs, including cherry, dogwood, hazel, and spirea.

SPRING QUAKERS, WOODLINGS, AND WOODGRAINS

NORMAN'S QUAKER

INTRACTABLE QUAKER

ALTERNATE WOODLING

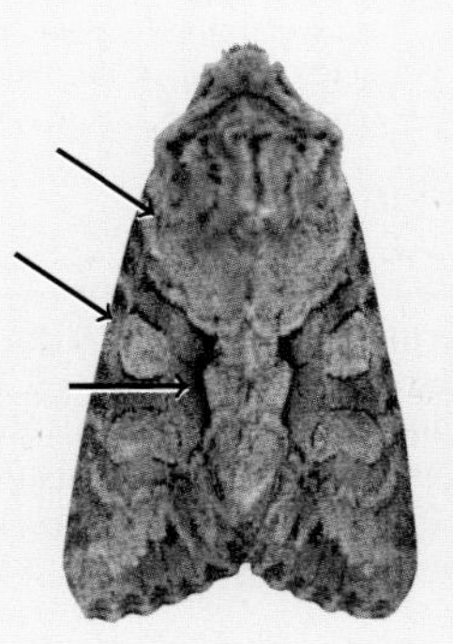

DISTINCT QUAKER

GRAY WOODGRAIN

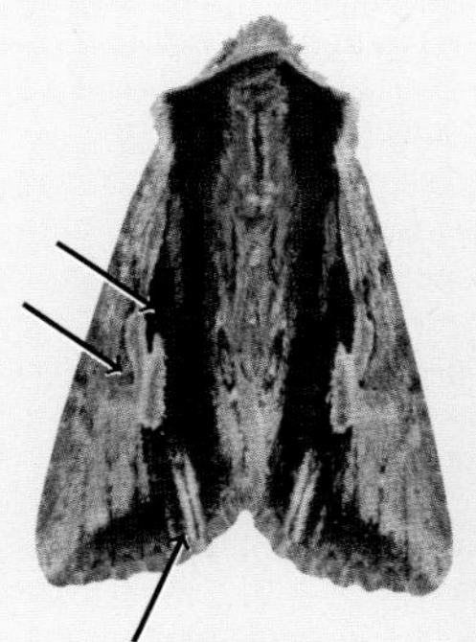

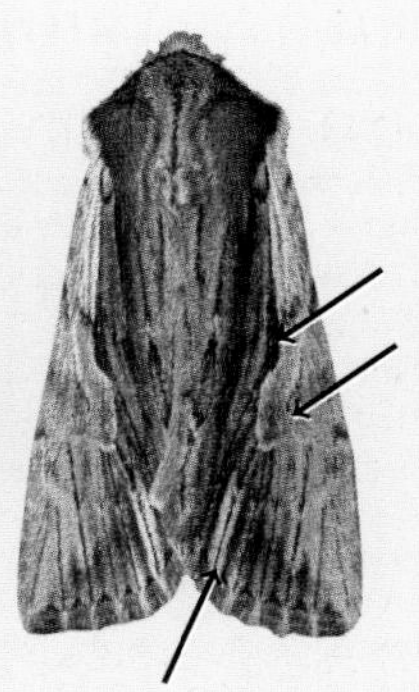

BICOLORED WOODGRAIN

actual size

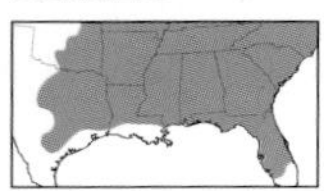

CONFUSED WOODGRAIN

Morrisonia confusa 93-2803 (10521) **Common**

TL 18–22 mm. Resembles Gray Woodgrain, but FW is tan and lacks dark shading between orbicular and reniform spots. Black dash in inner median area, and dark smudges in ST area, are reduced or absent. Veins are usually traced lightly blackish, creating a streakier appearance. **HOSTS:** Deciduous trees and shrubs, including beech, birch, elm, cherry, maple, and oak.

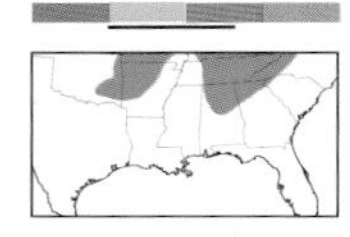

TRIANGULAR WOODGRAIN

Morrisonia triangula 93-2804 (10521.1) **Common**

TL 18–22 mm. Pale gray FW has a bold dark brown triangular patch at midpoint of costa. Narrow black dashes mark inner basal area and anal angle. Veins are lightly speckled black. **HOSTS:** Unknown.

FLUID ARCHES *Morrisonia latex* 93-2805 (10291) **Local**

TL 22–26 mm. Gray FW has slanting black median line that connects to a series of black dashes reaching down to midpoint of outer margin. Upper median area and orbicular spot are pale. Reniform spot is usually partly shaded brownish. Outer basal area is dark. **HOSTS:** Deciduous trees and shrubs, including birch, elm, maple, and oak.

LARGE ARCHES Family Noctuidae, Subfamily Noctuinae, Tribes Tholerini and Hadenini

Medium-sized midsummer noctuids that often display intricate FW patterns. They mostly occur in woodlands and old-fields but also in large gardens. Wheat Head Armyworm and The Pink-Streak resemble wainscots. All typically rest with their wings in a shallow tent. They are attracted to light in small to moderate numbers.

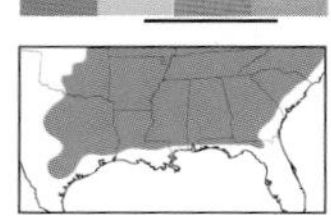

BRONZED CUTWORM

Nephelodes minians 93-2810 (10524) **Common**

TL 21–28 mm. Pinkish to grayish brown FW often has a violet sheen when fresh. Reddish brown median area is edged with curved double AM and PM lines. Orbicular and reniform spots are pale. ST area is often crossed with blackish veins. **HOSTS:** Grass and low plants.

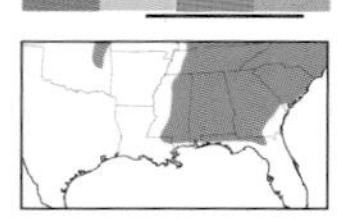

STRIPED GARDEN CATERPILLAR MOTH

Trichordestra legitima 93-2886 (10304) **Common**

TL 20 mm. Pale lilac gray FW has reddish brown shading in outer median and basal areas. Round orbicular spot is pale. Claviform spot is a black wedge. Reniform spot has gray dot in inner half. **HOSTS:** Wide variety of crops and low plants.

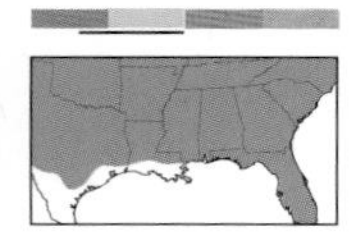

CAPSULE MOTH *Hadena capsularis* 93-2911 (10317) **Local**

TL 15–17 mm. Brownish gray FW has double black AM and PM lines connected by black wedge-shaped claviform spot. Large round orbicular spot is white with a brownish center. Whitish ST area has three black dashes near anal angle. **HOSTS:** Seed capsules of pinks.

SPRING QUAKERS, WOODLINGS, AND WOODGRAINS

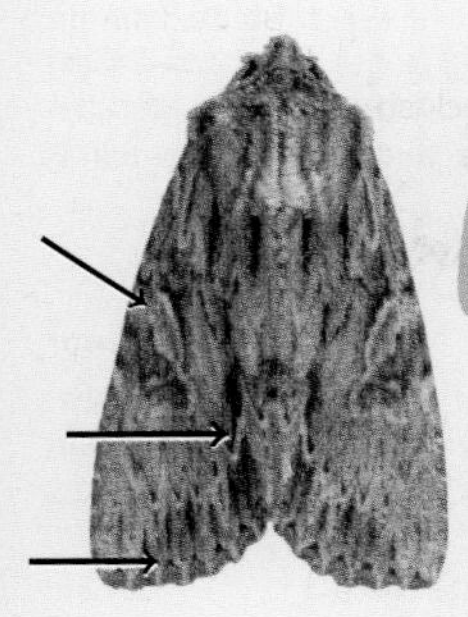

CONFUSED WOODGRAIN

actual size

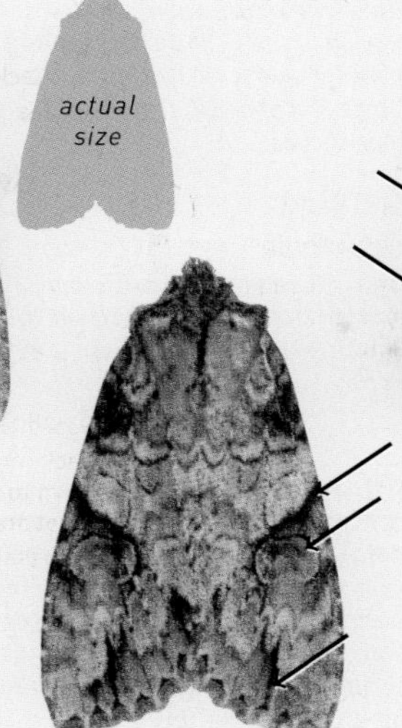

FLUID ARCHES

TRIANGULAR WOODGRAIN

LARGE ARCHES

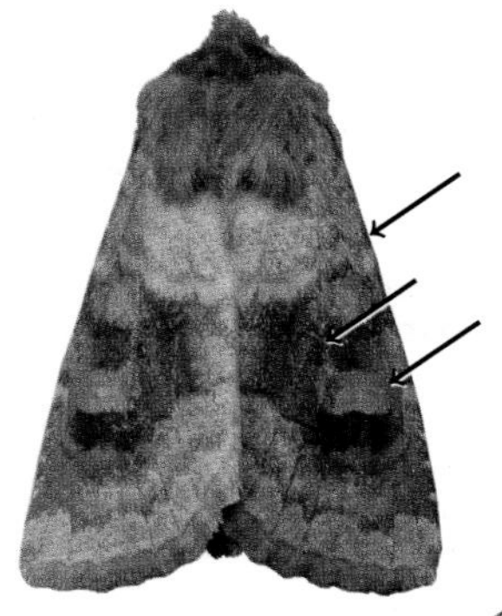

BRONZED CUTWORM

actual size

STRIPED GARDEN CATERPILLAR MOTH

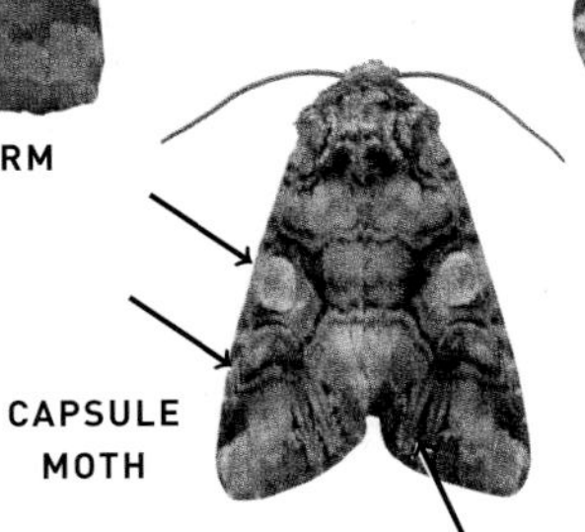

CAPSULE MOTH

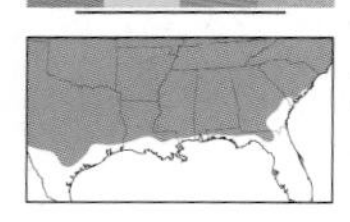

WHEAT HEAD ARMYWORM

Dargida diffusa 93-2928 (10431) **Common**

TL 15–19 mm. Straw-colored FW has a thick white central vein bordered by broad dark brown shading that reaches gray outer margin. Costa is grayish. Usually has a black reniform dot. Thorax has three whitish stripes. Legs have light tufts of hairlike scales on foretibia. **HOSTS:** Grass seedheads, including many cereal grains. **NOTE:** Three or more broods.

THE PINK-STREAK

Dargida rubripennis 93-2931 (10434) **Uncommon**

TL 18–21 mm. Resembles Wheat Head Armyworm, but dark areas are rosy pink. Thorax is cinnamon with whitish dorsal stripe. Legs have heavy tufts of hairlike scales on foretibia. **HOSTS:** Switchgrass.

WAINSCOTS

Family Noctuidae, Subfamily Noctuinae, Tribe Leucaniini

Sleek tan or brownish noctuids that are often nondescript and can be confusing and difficult to identify; HW color is often useful for separating species. Most common in midsummer months, when most species can be encountered throughout the season. Primarily found in old-fields, open areas, and wetlands, though some are tolerant of highly urban environments. They are mostly nocturnal and will visit lights in small numbers.

THE WHITE-SPECK

Mythimna unipuncta 93-2935 (10438) **Common**

TL 20–25 mm. Variable. Peppery pale tan or orange brown FW has a tiny white dot at center, usually edged with dark shading. Orbicular and reniform spots are pale orange and often indistinct. Slanting apical dash touches dotted ST line. HW is usually entirely sooty gray with dark veins. **HOSTS:** Generalist on grass; also woody plants and cereal grains. **NOTE:** Three or more broods.

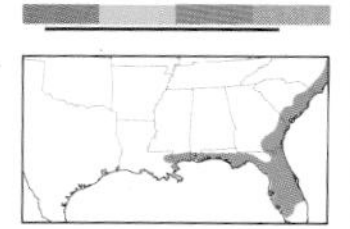

LINEN WAINSCOT *Leucania linita* 93-2938 (10440) **Local**

TL 17–18 mm. Pale tan FW is lightly striated brown along whitish veins. Central vein is edged with darker brown and ends at a white dot. PM line is a complete row of tiny dots; terminal line is absent. Apex is pointed. White HW has a very faint brownish ST band. **HOSTS:** Grass, including switchgrass, cord grass, and common reed. **NOTE:** Three or more broods.

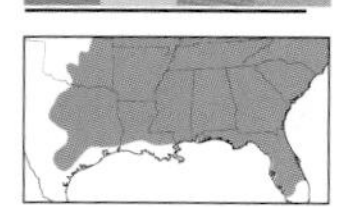

PHRAGMITES WAINSCOT

Leucania phragmitidicola 93-2943 (10444) **Uncommon**

TL 19–20 mm. Tan FW is traced with pale veins and usually thinly peppered with black scales. Thick, white central vein is strongly edged with blackish or brownish shading and is abutted by a black discal dot where vein splits. PM line is dotted; terminal line is tiny dots. Central median area is often shaded rusty. White HW has a faint brown terminal line. **HOSTS:** Grass and sedge. **NOTE:** Two broods.

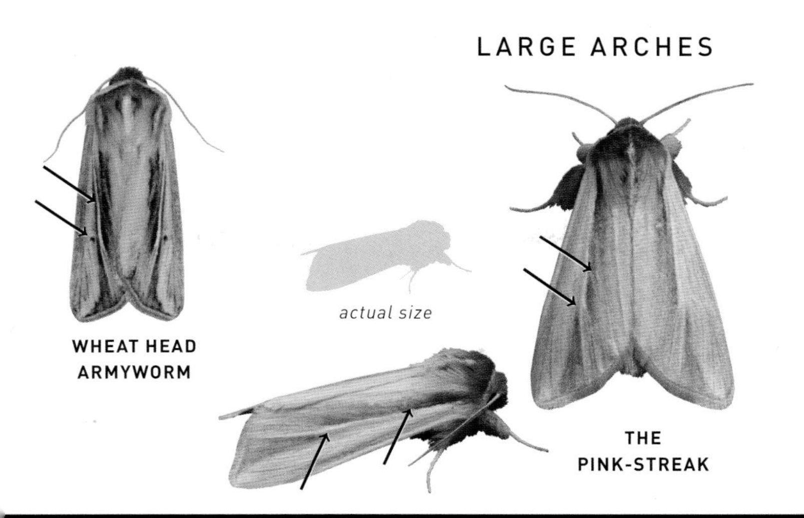
LARGE ARCHES
WHEAT HEAD
ARMYWORM
actual size
THE
PINK-STREAK

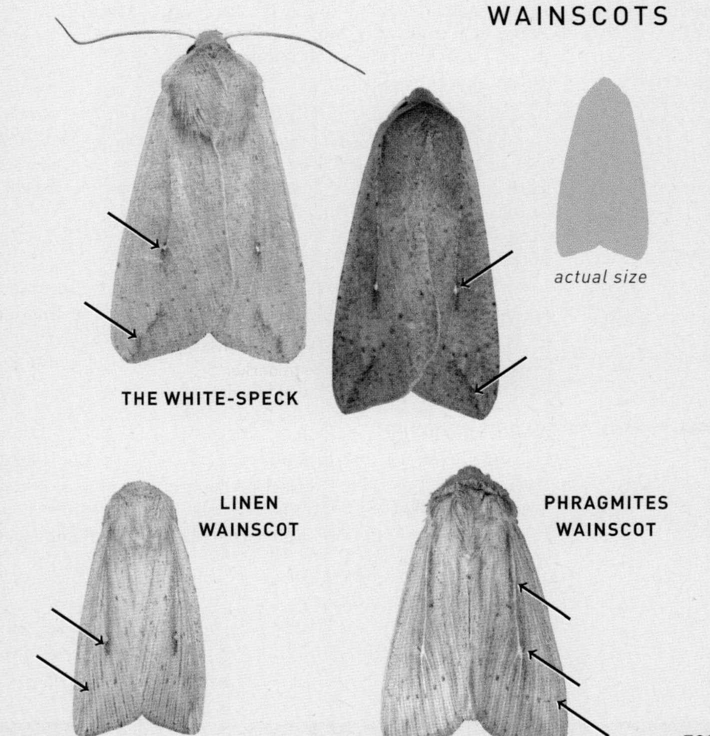
WAINSCOTS
actual size
THE WHITE-SPECK
LINEN
WAINSCOT
PHRAGMITES
WAINSCOT

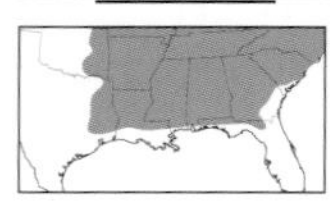

LINDA'S WAINSCOT

Leucania linda 93-2944 (10445) **Common**

TL 20–21 mm. Resembles Phragmites Wainscot, but FW has a small black dot in AM area abutting weak, darkly shaded, white central vein. Tan FW often has a distinct reddish tinge. Pearly gray HW is heavily suffused grayish brown with darker veins. **HOSTS:** Unknown. **NOTE:** Two broods.

MANY-LINED WAINSCOT

Leucania multilinea 93-2945 (10446) **Common**

TL 19–20 mm. Tan FW has white veins, marked in between by brown lines; central vein is heavily shaded dark brown. A triangle of three black dots marks lower FW: one at fork of central vein, and two marking otherwise-absent PM line. Usually lacks a dot in AM area. White HW has a narrow dusky ST band. Thorax has collar of three parallel dusky lines. **HOSTS:** Grass and sedge.

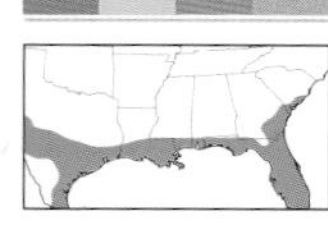

UNKNOWN WAINSCOT

Leucania incognita 93-2950 (10450) **Local**

TL 18–21 mm. Pale tan FW has a bold dusky median streak with a comma-shaped white discal spot. PM and terminal lines are an even row of blackish dots. White HW has a distinctive dotted terminal line. Thorax has brown lateral stripes. **HOSTS:** Unknown, but presumably grass and sedge.

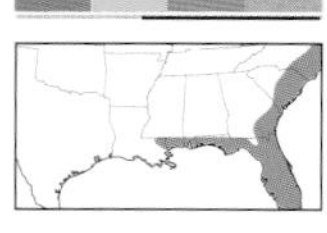

WHITE-DOTTED WAINSCOT

Leucania subpunctata 93-2955 (10453.1) **Local**

TL 18–20 mm. Dark brown FW has a dusky median streak with a bold white discal dot. Costa is pale. Slanting apical dash borders paler apex. White HW has dusky veins and terminal line. **HOSTS:** Grass and sedge. **NOTE:** Multiple broods.

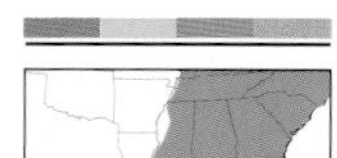

SCIRPUS WAINSCOT

Leucania scirpicola 93-2957 (10455) **Common**

TL 18–21 mm. Resembles Unknown Wainscot, but central streak of FW is lighter brown and often not as long, and terminal line is faint or nearly absent. Whitish HW has a darker terminal line. **HOSTS:** Grass and sedge. **NOTE:** Multiple broods.

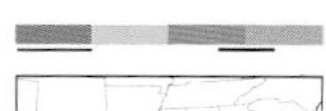

WANING WAINSCOT

Leucania senescens 93-2958 (10455.1) **Local**

TL 16–18 mm. Tan FW has pale veins thinly edged brown. Central median area is shaded brown from base to black-dotted PM line. White discal spot is marked with a black dot. Outer margin has a triangle of brown shading near apex. Satin white HW has a narrow dusky terminal line. **HOSTS:** Unknown, but presumably grass and sedge.

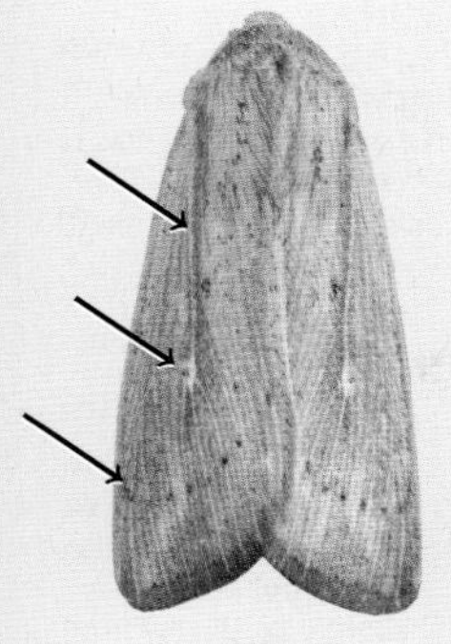

LINDA'S WAINSCOT

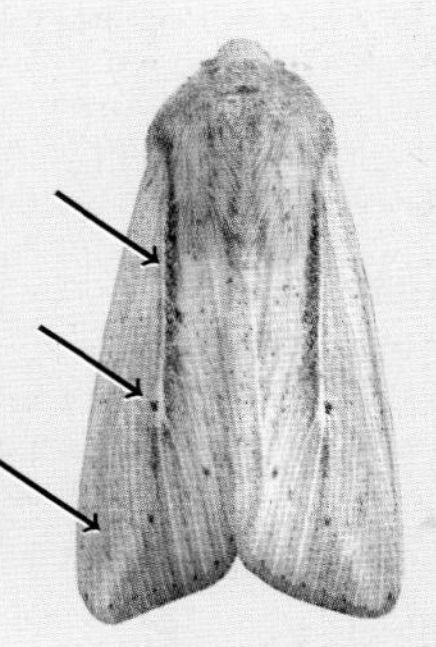

MANY-LINED WAINSCOT

UNKNOWN WAINSCOT

WHITE-DOTTED WAINSCOT

actual size

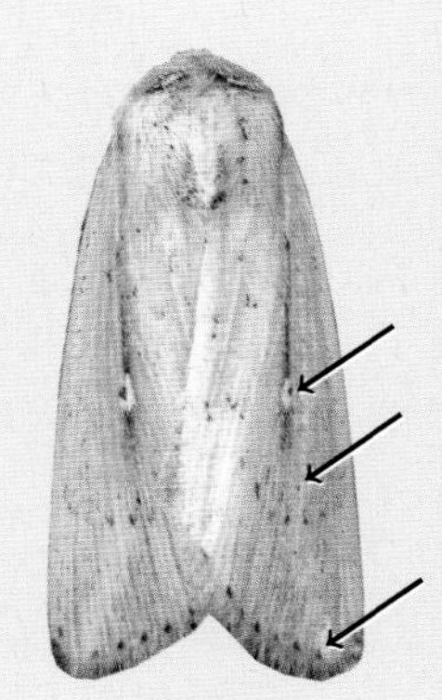

SCIRPUS WAINSCOT

WANING WAINSCOT

ADJUTANT WAINSCOT

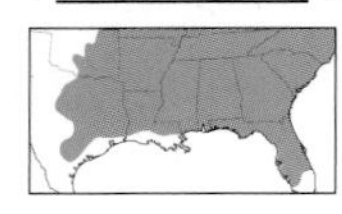

Leucania adjuta 93-2960 (10456) **Common**

TL 18–20 mm. Resembles Many-lined Wainscot, but central vein is edged warm brown. Usually has a well-defined small black dot in AM area abutting central vein. Pearly gray HW has heavy dusky suffusion, especially in female. Thorax has peppery gray bands behind head. **HOSTS:** Grass. **NOTE:** Multiple broods.

UNARMED WAINSCOT

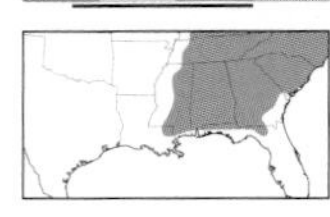

Leucania inermis 93-2963 (10459) **Uncommon**

TL 16–18 mm. Peppery tan FW has brown shading below diffusely edged, dot-centered reniform spots. Zigzag PM line appears as two rows of blackish dots. Foretibia is slightly hairy in both sexes. **HOSTS:** Grass. **NOTE:** Two broods.

URSULA WAINSCOT

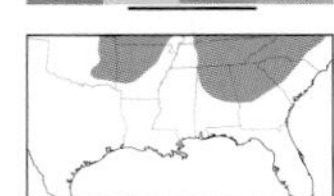

Leucania ursula 93-2965 (10461) **Uncommon**

TL 18–19 mm. Resembles Unarmed Wainscot, but foretibia of male has large tufts of hairlike scales. Brown shading extends behind both orbicular and reniform spots. Females lack foretibia tufts and many may not be separable in the field. **HOSTS:** Grass and sedge. **NOTE:** Three broods.

FALSE WAINSCOT

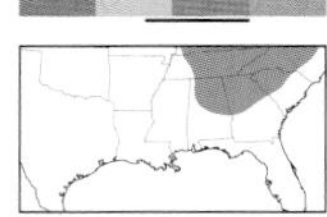

Leucania pseudargyria 93-2966 (10462) **Uncommon**

TL 24–26 mm. Resembles smaller Ursula Wainscot, but FW often has a strong reddish tint, especially in central median area. Shading behind orbicular and reniform spots is lighter. Foretibia of male has massive tufts of hairlike scales. **HOSTS:** Grass.

ZIPPERED WAINSCOT

Leucania pilipalpis 93-2967 (10463) **Local**

TL 23–26 mm. Peppery tan FW has a zigzag PM line. A bold blackish patch passes behind indistinct, pale, brownish-centered reniform spot. Apex is pointed. Pearly HW is heavily suffused brown. **HOSTS:** Crabgrass and honeysuckle. **NOTE:** Three or more broods.

WAINSCOTS

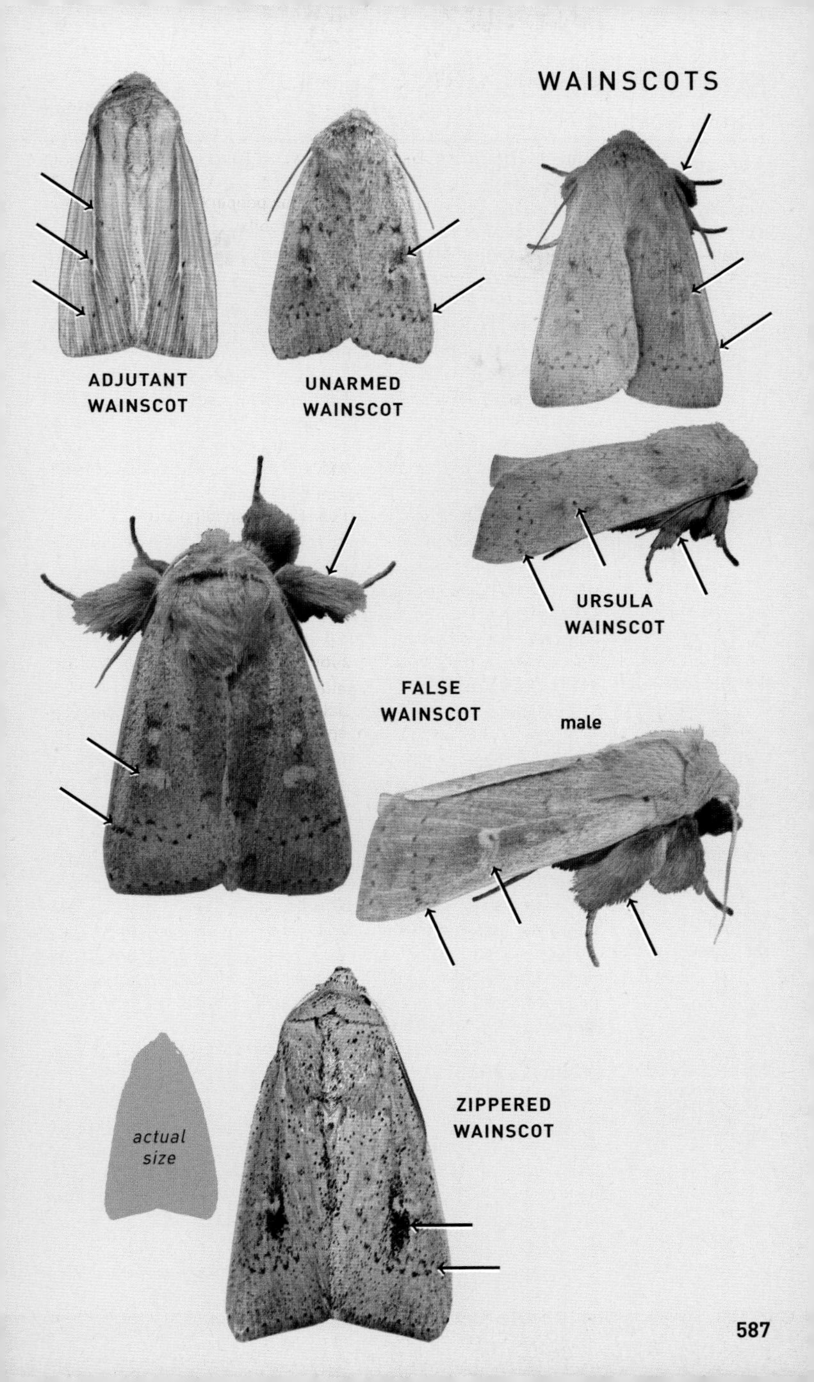
ADJUTANT WAINSCOT
UNARMED WAINSCOT
URSULA WAINSCOT
FALSE WAINSCOT
male
ZIPPERED WAINSCOT
actual size

Small Arches and Summer Quakers
Family Noctuidae, Subfamily Noctuinae, Tribe Eriopygini

A group of small noctuids that are on the wing in summer and early fall. Typically they are found in woodlands and old-fields, but sometimes in gardens. The arches are mostly intricately patterned and brightly colored with defined spots. The small summer quakers are mostly brown or chestnut and typically show a more rounded wing shape. All are nocturnal and regularly come to light in small or moderate numbers.

THE THINKER

Lacinipolia meditata 93-3016 (10368) **Uncommon**

TL 15–17 mm. Variable. Peppery brownish to purplish gray FW often has warm brown shading in inner median area and below pale-edged reniform spot. Scalloped, double AM and PM lines are usually filled grayish. Costa is checkered along lower half. **HOSTS:** Low plants, including clover, dandelion, and tobacco. **NOTE:** Two broods.

ARMED ARCHES

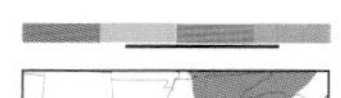

Lacinipolia teligera 93-3039 (10393) **Common**

TL 14–16 mm. Lilac gray FW has warm brown shading in inner median area and below black-edged reniform spot. Claviform spot is a narrow, bold black V connecting scalloped, double AM and PM lines. Veins are lightly traced black in ST area. **HOSTS:** Unknown.

BRISTLY CUTWORM

Lacinipolia renigera 93-3044 (10397) **Common**

TL 14–15 mm. Purplish gray FW has lime green accents in basal area, white-edged reniform spot, and inner PM line. Claviform spot is black. **HOSTS:** Wide variety of crops and low plants. **NOTE:** Multiple broods.

ERECT ARCHES

Lacinipolia erecta 93-3050 (10403) **Uncommon**

TL 15–17 mm. Peppery brown FW has whitish AM and PM lines and outlines to spots; reniform spot has a black dot in inner half. ST area is often shaded grayish along wavy, pale yellowish ST line. Inner margin is grayish, or pale brown on darker individuals. Diffuse, angled median line is indistinct on darker individuals. **HOSTS:** Common dandelion. **NOTE:** Two broods.

LAUDABLE ARCHES

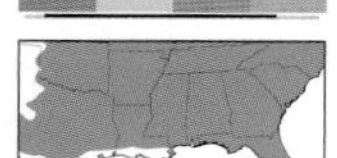

Lacinipolia laudabilis 93-3065 (10411) **Common**

TL 15–17 mm. Pale green FW has dusky shading in central median area, with a chestnut patch between inconspicuous orbicular and white-edged reniform spots. Scalloped, double AM and PM lines are filled white. White ST line is edged with black spots. **HOSTS:** Low plants, including common dandelion. **NOTE:** Two broods.

SMALL ARCHES AND SUMMER QUAKERS

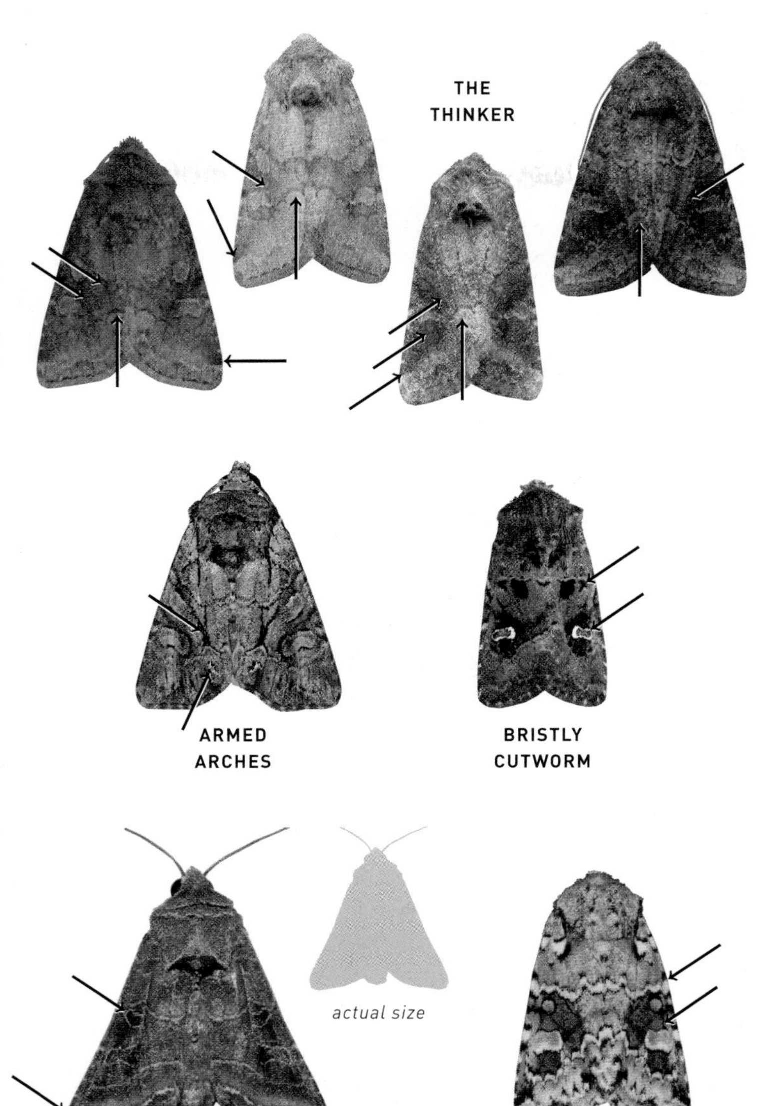

EXPLICIT ARCHES

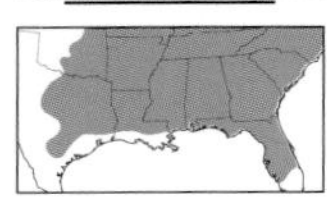

Lacinipolia explicata 93-3067 (10413) **Common**

TL 15–17 mm. Resembles Laudable Arches, but median area is entirely dusky, and chestnut shading bleeds into reniform spot and basal area. AM line is not scalloped. ST line is often indistinct. **HOSTS:** Low plants, including clover and lettuce. **NOTE:** Two broods.

IMPLICIT ARCHES

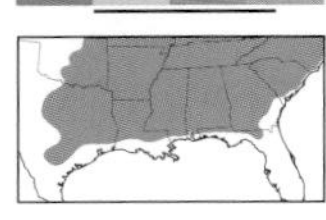

Lacinipolia implicata 93-3068 (10414) **Uncommon**

TL 15–17 mm. Resembles Laudable Arches, but entirely dusky median area is narrower at inner margin, and AM line is not scalloped. ST line is usually indistinct. **HOSTS:** Reported on common dandelion and other low plants. **NOTE:** Two broods.

NORTHERN SCURFY QUAKER

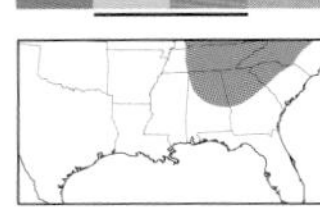

Homorthodes furfurata 93-3088 (10532) **Local**

TL 13–14 mm. Pale tan or reddish brown FW is weakly peppered grayish along veins. Scalloped, double AM line is inconspicuous; PM line has tiny black dots at veins. Pale-edged reniform spot has black dot in inner half. An angled dusky median band is sometimes present. **HOSTS:** Deciduous trees, including cherry and maple.

SOUTHERN SCURFY QUAKER

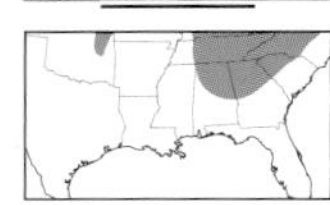

Homorthodes lindseyi 93-3089 (10532.1) **Local**

TL 13–14 mm. Resembles Northern Scurfy Quaker, but black dot of reniform spot is boldly edged white. PM line is marked by white-edged black dots at veins. **HOSTS:** Unknown.

RUDDY QUAKER

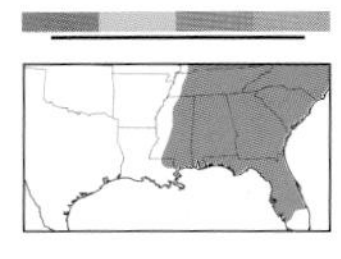

Protorthodes oviduca 93-3113 (10563) **Uncommon**

TL 14–16 mm. Grayish brown to chestnut FW has veins traced dark gray. Double AM and PM lines are relatively even and filled pale. Dark orbicular and reniform spots are crisply outlined pale; inner half of reniform spot is blackish. Wavy ST line is pale orange. **HOSTS:** Dandelion, grass, plantain, and other low plants. **NOTE:** Two broods.

TEXAS QUAKER

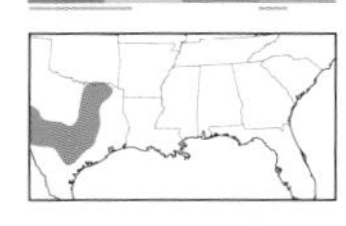

Protorthodes orobia 93-3114 (10565) **Local**

TL 12–15 mm. Resembles Ruddy Quaker, but FW is dark grayish brown and peppery. Slightly darker orbicular and reniform spots are narrowly outlined whitish. Terminal line has pale dots at veins. **HOSTS:** Unknown.

SHEATHED QUAKER

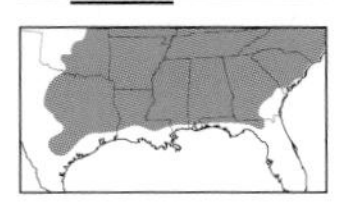

Ulolonche culea 93-3118 (10567) **Common**

TL 16–18 mm. Yellowish brown FW has even, yellowish AM and PM lines edged black. Reniform and orbicular spots are outlined pale; inner half of reniform spot is blackish. Zigzag, pale ST line is accented with tiny black dots. **HOSTS:** Oak.

SMALL ARCHES AND SUMMER QUAKERS

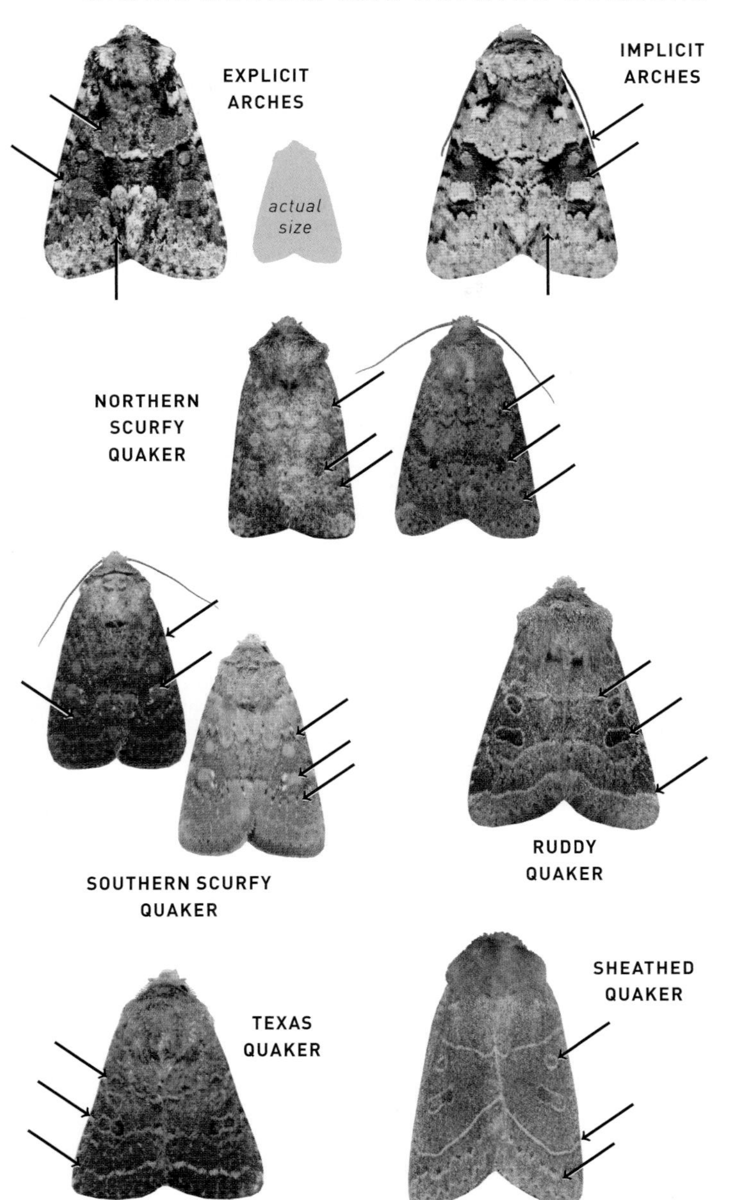

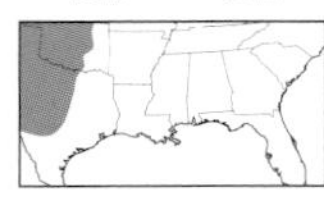

BICOLORED QUAKER

Ulolonche disticha 93-3124 (10573) **Local**

TL 13–15 mm. Bicolored FW has light gray basal area separated from darker grayish brown distal half by a slanting blackish median line. Reniform spot is sometimes shaded orange and has a blackish dot in inner half. Inconspicuous ST line ends as a black spot at costa. **HOSTS:** Unknown. **NOTE:** Two broods.

RUSTIC QUAKER

Orthodes majuscula 93-3136 (10585) **Common**

TL 18–20 mm. Peppery brown FW has crisp pale yellow lines and outlines to spots. AM and PM lines kink upward near costa. PM line is edged with paired black spots along veins. Large, pale-edged orbicular and reniform spots usually touch. **HOSTS:** Low plants, including aster, dandelion, goldenrod, plantain, and sensitive fern. **NOTE:** Two or more broods.

CYNICAL QUAKER

Orthodes cynica 93-3138 (10587) **Common**

TL 15–17 mm. Reddish brown FW has pale-edged, even black AM and PM lines that are bordered by black dots at veins. Large, white-edged orbicular and reniform spots do not touch. Sometimes has darker, diffuse median band. Wavy ST line is pale but sometimes indistinct. **HOSTS:** Dandelion and goldenrod. **NOTE:** Two broods.

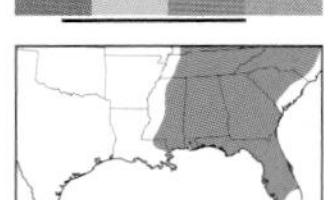

GOODELL'S ARCHES

Orthodes goodelli 93-3141 (10289) **Uncommon**

TL 17–18 mm. Brown FW has grayish veins peppered white. Scalloped, double AM and PM lines are filled light brown. Wedge-shaped claviform spot is sometimes visible only as a black spot. Reniform spot is partly outlined and filled white. Terminal line has pale dots at veins. **HOSTS:** Unknown. **NOTE:** Two broods.

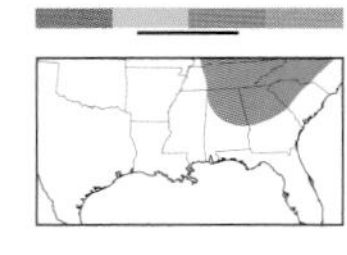

DISPARAGED ARCHES

Orthodes detracta 93-3146 (10288) **Local**

TL 17–18 mm. Mottled grayish brown FW has rusty shading in inner basal and median areas, and pale patch at base. Scalloped, double AM and PM lines are indistinct. Black claviform spot is wedge-shaped. Dusky orbicular and reniform spots are outlined whitish. **HOSTS:** Trees and low plants, including blueberry, clover, hickory, and oak.

SIGNATE QUAKER

Tricholita signata 93-3193 (10627) **Common**

TL 15–18 mm. Brown to reddish brown FW has indistinct darker lines and grayish veins. Reniform spot has a thick, fractured white outline and orange brown center. Diffuse, angled median band is often dark. **HOSTS:** Low plants, including dandelion and plantain.

SMALL ARCHES AND SUMMER QUAKERS

BICOLORED QUAKER

RUSTIC QUAKER

CYNICAL QUAKER

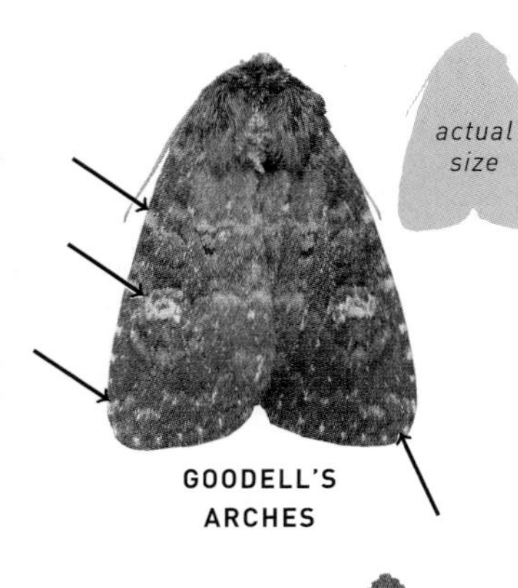

GOODELL'S ARCHES

DISPARAGED ARCHES

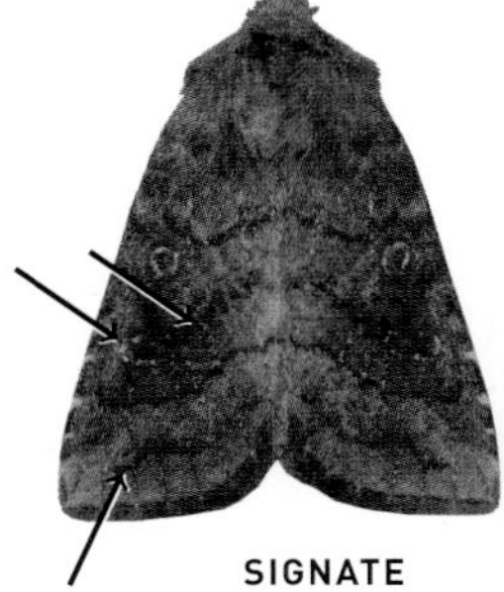

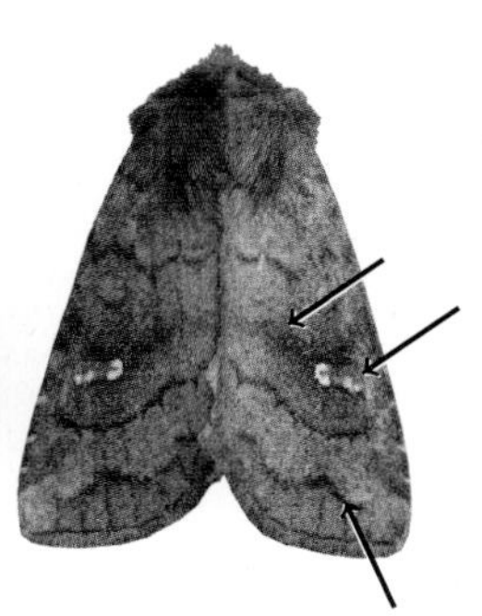

SIGNATE QUAKER

Spanish Moth

Family Noctuidae, Subfamily Noctuinae, Tribe Glottulini

A distinct species that cannot be confused with any others. The pink, orange, and black FW may bring to mind traditional Spanish patterning. The larvae are thought to be toxic. It is occasionally attracted to lights. What was previously considered a single species, *Xanthopastis timais*, that ranged throughout the Americas is now recognized as a complex of at least six species. Only *X. regnatrix* occurs north of Mexico.

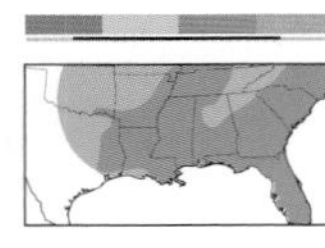

SPANISH MOTH

Xanthopastis regnatrix 93-3210 (10640) **Common**

TL 21–24 mm. Pink (sometimes whitish) FW has black shading along costa and outer median area. Basal dash and edging to orbicular and reniform spots are bright golden yellow. ST line is checkered black and gold. Spiky thorax is black. **HOSTS:** Amaryllis, spider lily, narcissus, and others.

Darts

Family Noctuidae, Subfamily Noctuinae, Tribe Noctuini

Medium-sized moths with a long, moderately narrow FW. Many are late-season fliers found in old-fields and gardens. They are mostly nocturnal and will come to light, but a few can be found taking nectar from flowers during daylight. Some, such as Ipsilon Dart and Pearly Underwing, can frequently be found at sugar bait.

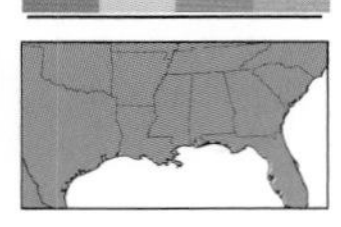

PEARLY UNDERWING

Peridroma saucia 93-3211 (10915) **Common**

TL 22–27 mm. Variable. Mottled FW is pale tan to chestnut brown. Jagged, double AM and PM lines are often indistinct except at costa. Large oval orbicular spot is sometimes boldly pale. Pearly gray HW has darker veins. Thorax has raised frosty dorsal stripe. **HOSTS:** A generalist on a wide variety of trees, crops, and low plants. **NOTE:** Multiple broods.

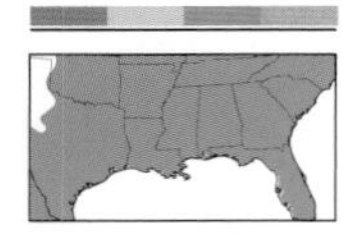

GREEN CUTWORM *Anicla infecta* 93-3212 (10911) **Common**

TL 17–20 mm. Peppery pale tan to lilac gray FW has rusty terminal line crossed by wavy whitish ST line. Fragmented black reniform spot is outlined pale yellowish and rusty. Thorax has thick black collar behind head. **HOSTS:** Grass; also low plants, including beet, clover, and tobacco. **NOTE:** Multiple broods.

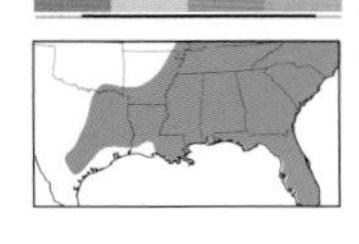

SLIPPERY DART *Anicla lubricans* 93-3214 (10901) **Common**

TL 17–20 mm. Pale gray FW has reddish brown shading below PM line. AM, median, and PM lines are faint, most obvious as black spots along costa. Fragmented black reniform spot lacks outline. Pearly gray HW has a darker terminal line and white fringe. Thorax has narrow black collar. **HOSTS:** Unknown. **NOTE:** Two broods.

SPANISH MOTH

DARTS

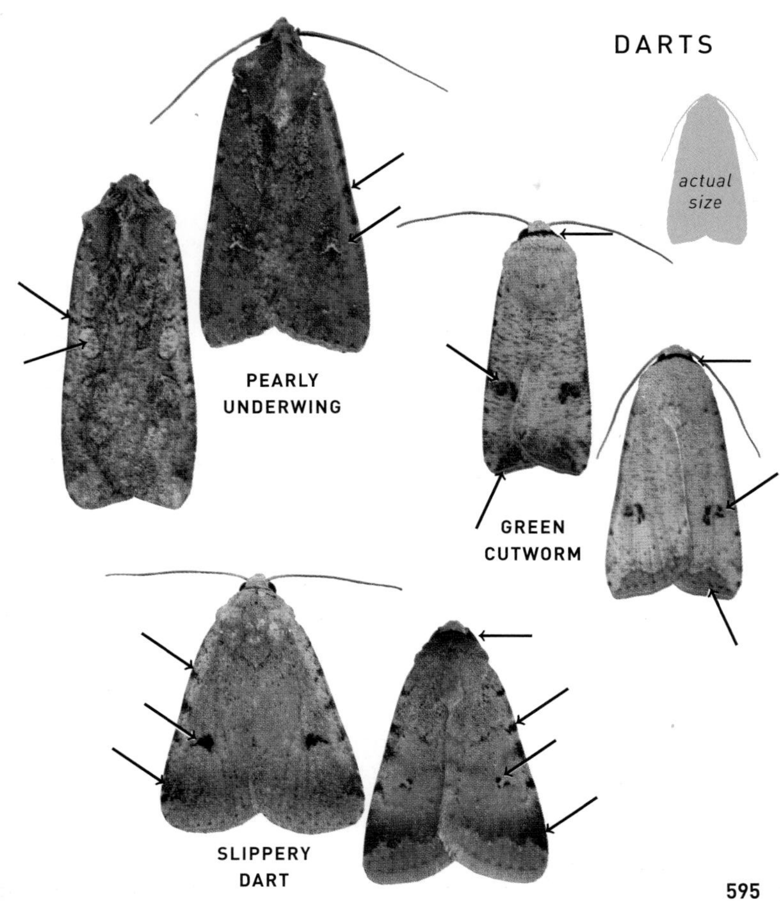

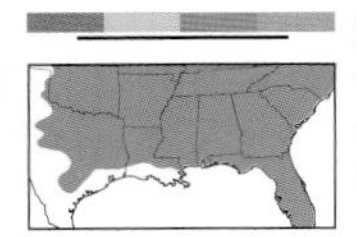

SNOWY DART *Anicla illapsa* 93-3216 (10903) Common

TL 17–19 mm. Similar to Slippery Dart, but pale FW is suffused reddish. Light AM and PM lines are brownish, becoming blackish spots on costa. Fragmented black reniform spot is small. Strikingly white HW lacks any darker shading. **HOSTS:** Grass and sedge. **NOTE:** Two or more broods.

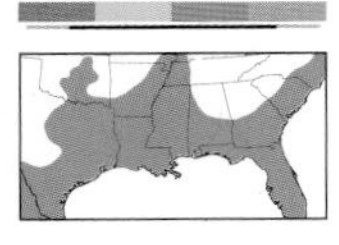

SIMPLE DART *Anicla simplicius* 93-3217 (10907) Common

TL 17–20 mm. Peppery light gray FW lacks any obvious lines apart from a dotted terminal line. ST and terminal areas are shaded brownish. Inner median area sometimes has a smudgy black dash extending up from fragmented reniform spot. White HW lacks any darker shading. **HOSTS:** Unknown.

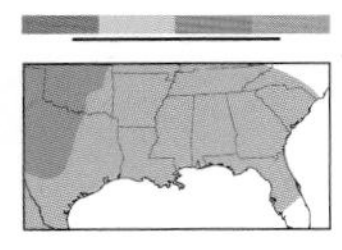

WORTHY DART *Anicla digna* 93-3219 (10908) Local

TL 14–19 mm. Lightly peppered whitish gray FW is slightly darker and more densely speckled in ST area. Fragmented black reniform spot is often small. Thorax has narrow black collar behind head. HW is totally white. **HOSTS:** Unknown.

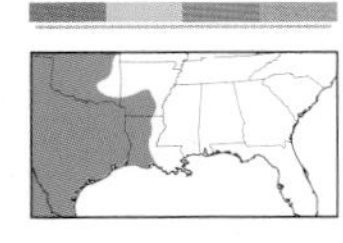

CREAKY DART *Hemieuxoa rudens* 93-3227 (10914) Common

TL 14–16 mm. Dimorphic. Typical form has brown FW with pale basal area edged by a thick black basal dash. Round, pale orbicular and gray-centered reniform spots are connected by a thick black bar. Speckled form lacks black bars; speckled FW has a faint AM line, blackish centers to both spots, and darker shading along costa. White HW has a narrow black terminal line. **HOSTS:** Snakeweed.

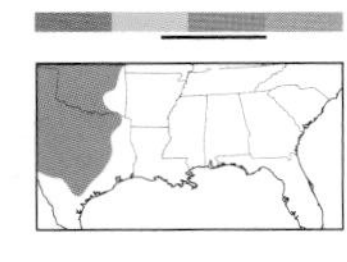

INCLINED DART *Dichagyris acclivis* 93-3232 (10870) Common

TL 17–19 mm. Slate gray FW has paler warm brown shading along basal half of costa. Pale reniform spot is backed by a broad black bar. Short claviform spot is edged black. ST line is a row of pale dashes. Some individuals have brown shading in ST area. **HOSTS:** Switchgrass. **NOTE:** General pattern looks like that of some *Agrotis* species.

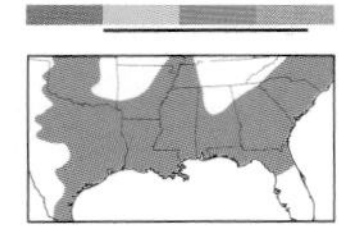

FRINGED DART
Eucoptocnemis fimbriaris 93-3257 (10694) Common

TL 14–18 mm. Light gray to reddish brown FW has tiny black and white dots on veins along PM line. Bar-shaped, pale reniform spot is often bordered below by dark brown shading. Orbicular spot is a black dot. **HOSTS:** Unknown, but probably various low plants.

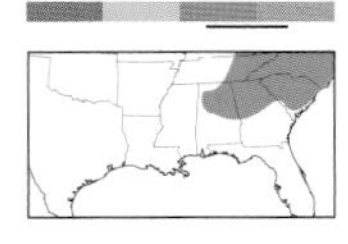

RUBBED DART *Euxoa detersa* 93-3461 (10838) Local

TL 17–18 mm. Variable. Grayish brown FW has a broad brown bar in central median area, behind pale, brown-centered orbicular and reniform spots. Costa is grayish to whitish. Pale yellowish ST line is usually edged with black wedges. Veins are sometimes traced grayish. ST area sometimes has a wide grayish band. Inner median area sometimes has a black-edged claviform spot and pale streak parallel to inner margin. **HOSTS:** Corn, cranberry, tobacco, and many other low plants and crops.

DARTS

SNOWY DART

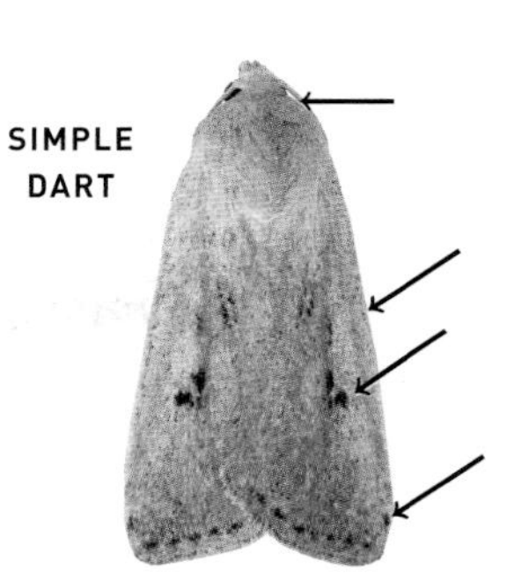

SIMPLE DART

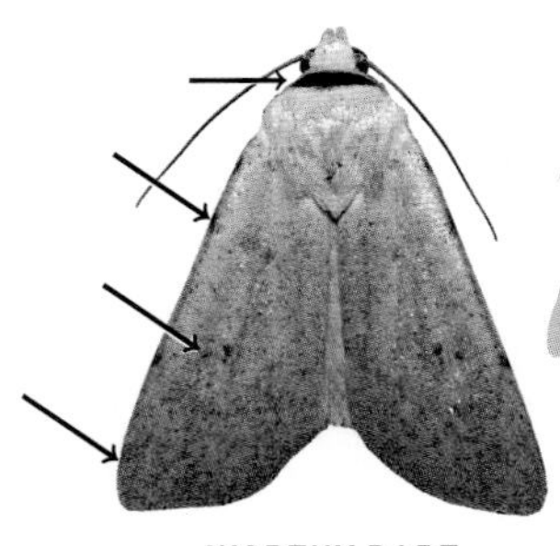

WORTHY DART

actual size

CREAKY DART

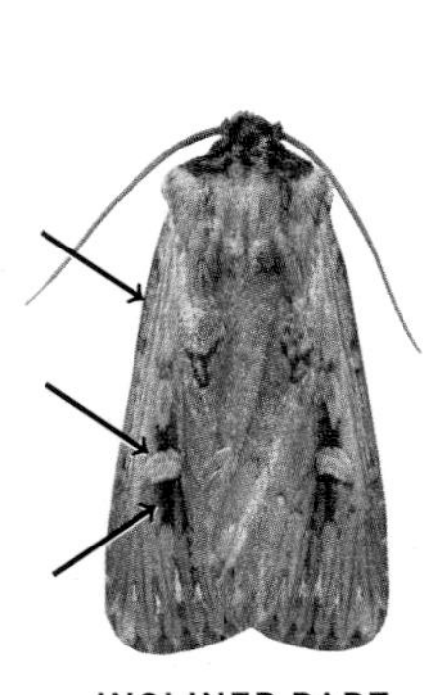

INCLINED DART

FRINGED DART

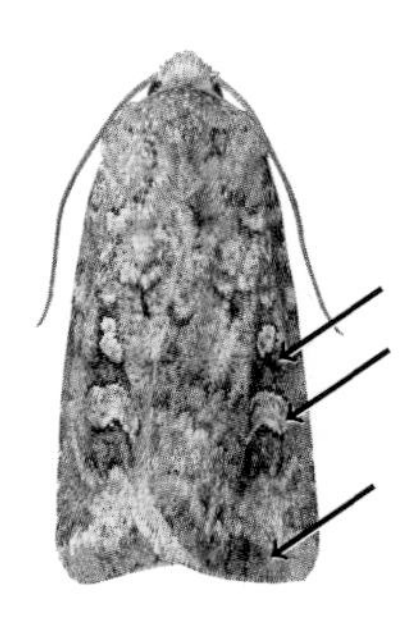

RUBBED DART

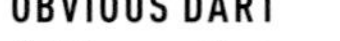

OBVIOUS DART

Feltia manifesta 93-3494 (10666) **Uncommon**

TL 18–20 mm. Dark grayish brown to reddish brown FW has narrow, slightly scalloped AM and PM lines; inner AM line often has a small dark wedge. Black orbicular and reniform spots are conspicuous. **HOSTS:** Low plants such as clover; also dead leaves.

KNEE-JOINT DART

Feltia geniculata 93-3495 (10680) **Common**

TL 18–19 mm. Grayish brown FW has thick double AM and PM lines filled light gray. Whitish orbicular and reniform spots are open-ended toward costa and separated by a bold black patch. Claviform spot is edged black. **HOSTS:** Unknown; probably a generalist on low plants.

DINGY CUTWORM

Feltia jaculifera 93-3498 (10670) **Common**

TL 21 mm. FW has thick grayish edging to dark veins and along inner margin, costa, and ST line. Reniform spot is usually pale orange. Tip of claviform spot reaches beyond gray orbicular triangle. **HOSTS:** Generalist on various shrubs, crops, and grasses.

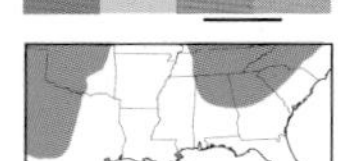

SUBGOTHIC DART

Feltia subgothica 93-3501 (10674) **Common**

TL 19–20 mm. Resembles Dingy Cutworm, but ST area is darker and claviform spot does not extend beyond orbicular triangle. Dusky shading in outer ST area does not usually extend beyond inner edge of reniform spot. Midpoint of ST line is often edged with small black wedges. **HOSTS:** Generalist on many low plants and crops.

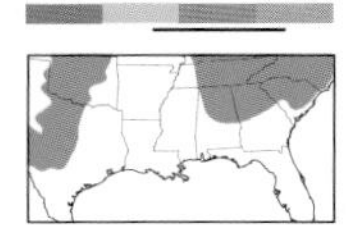

TRICOSE DART *Feltia tricosa* 93-3502 (10675) **Common**

TL 18–19 mm. Resembles Subgothic Dart, but dusky shading in outer ST area usually extends beyond inner edge of reniform spot, and ST line lacks black wedges. Some individuals may be difficult to separate in the field. **HOSTS:** Unknown; probably a generalist on crops and low plants.

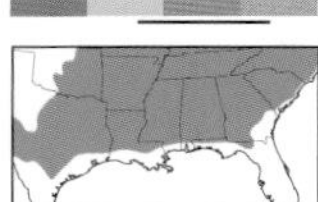
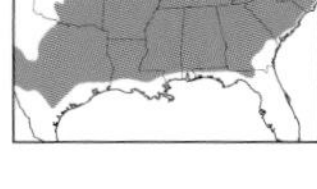

MASTER'S DART *Feltia herilis* 93-3503 (10676) **Common**

TL 22–24 mm. Resembles Subgothic Dart, but ST area and lower half of inner margin are entirely grayish. Black PM line is often visible at inner margin. **HOSTS:** Generalist on various low plants and crops.

SUBTERRANEAN DART

Feltia subterranea 93-3504 (10664) **Common**

TL 21–24 mm. Dimorphic. Dark grayish brown FW has tan terminal line and upper costa. Small, pale orbicular and reniform spots are backed by a thick black band. Double AM line swoops low to touch orbicular spot. Midpoint of terminal line has two black wedges. Light form is similar but tan instead of dark grayish brown. **HOSTS:** Generalist on grass, crops, and low plants. **NOTE:** Multiple broods.

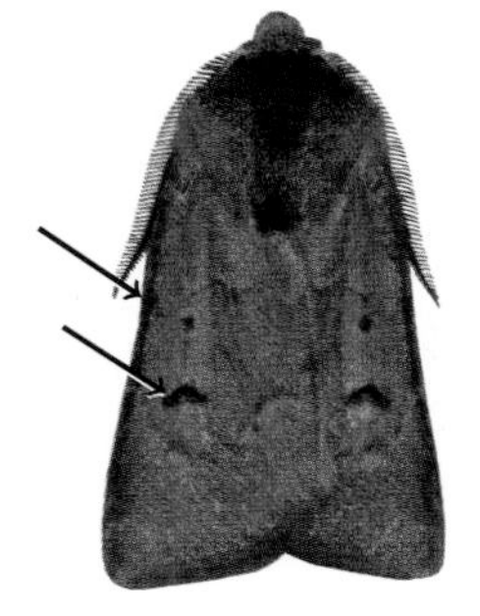

OBVIOUS DART

KNEE-JOINT DART

actual size

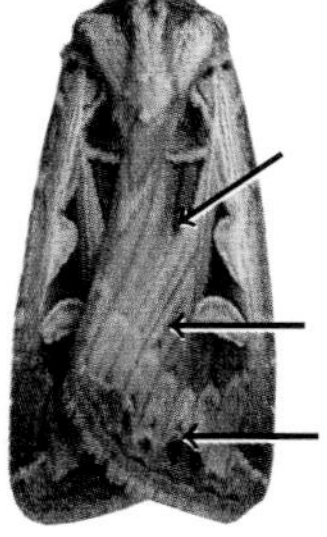

SUBGOTHIC DART

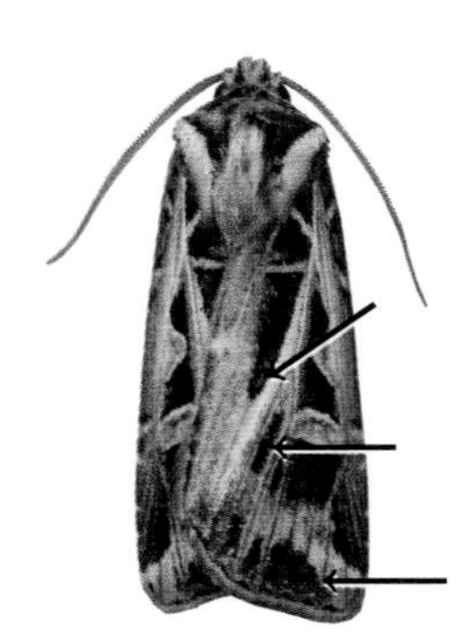

DINGY CUTWORM

TRICOSE DART

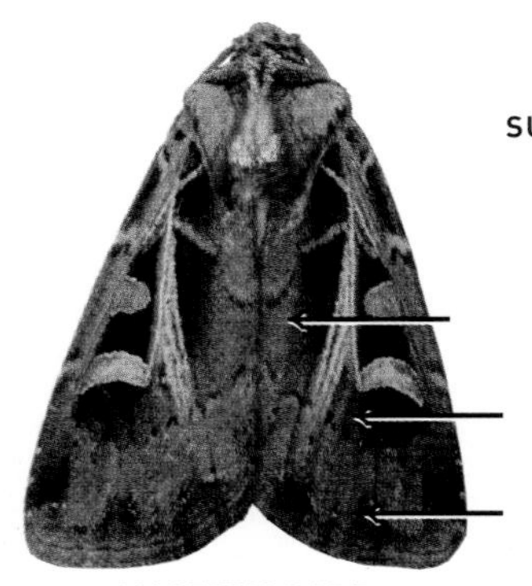

MASTER'S DART

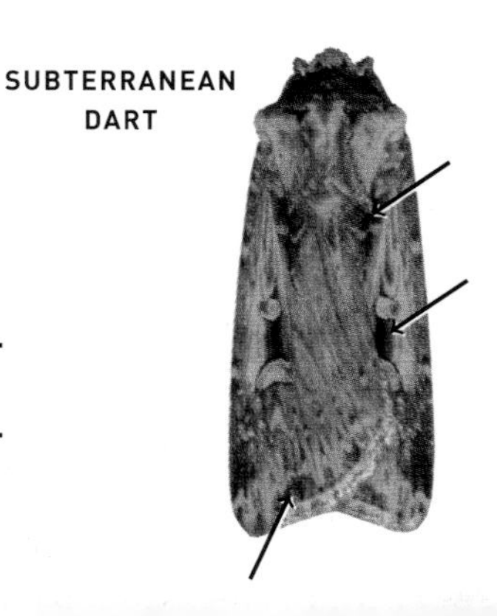

SUBTERRANEAN DART

OLD MAN DART

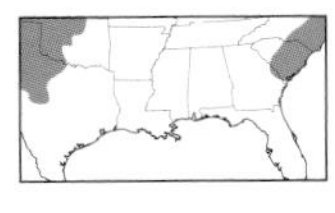

Agrotis vetusta 93-3506 (10641) **Common**

TL 18–24 mm. Light grayish brown FW has zigzag AM and PM lines usually visible only as white-edged black dots. Inconspicuous reniform spot has a large black spot in inner half. Orbicular spot is sometimes blackish. Terminal line is a row of black dots. **HOSTS:** Bean, corn, lettuce, tobacco, tomato, and other crops and low plants.

SWORDSMAN DART

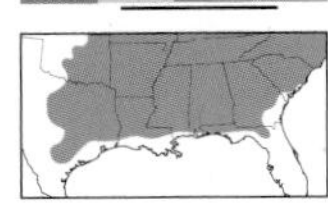

Agrotis gladiaria 93-3515 (10648) **Common**

TL 18–20 mm. FW has dark veins thickly edged pale, and lighter-shaded costa, inner margin, and sometimes ST area. Dark gray reniform and small orbicular spots are outlined pale. AM line is absent. Pale ST line is fragmented by veins. **HOSTS:** Bean, corn, tobacco, tomato, and other crops and low plants.

VENERABLE DART

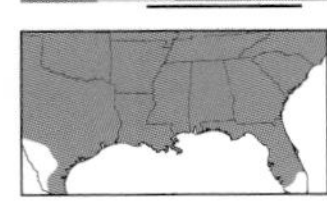

Agrotis venerabilis 93-3516 (10651) **Common**

TL 18–22 mm. Tan to brown FW has dark gray costal streak and blackish veins. Basal dash is a long, blackish loop. Gray reniform and orbicular spots are partly outlined pale and sometimes connected by a blackish patch. Midpoint of terminal line has a dark wedge. Thorax has blackish collar and two dark central stripes. **HOSTS:** Generalist on low plants and crops, including alfalfa, chickweed, corn, tobacco, and tomato.

RASCAL DART

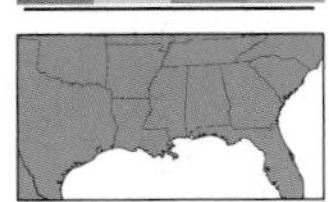

Agrotis malefida 93-3526 (10661) **Common**

TL 22–25 mm. Resembles Venerable Dart, but basal dash is absent; claviform spot is filled blackish. Orbicular spot is often thickly outlined pale. **HOSTS:** Generalist on low plants and crops, including clover, corn, and tomato.

IPSILON DART

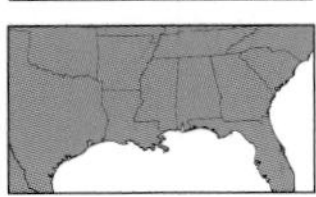

Agrotis ipsilon 93-3528 (10663) **Common**

TL 21–24 mm. Variable. Slate gray FW has tan ST and inner basal areas, or brownish tan FW has slate gray costa. Orbicular, claviform, and reniform spots are dark gray outlined black; reniform spot has a sharp dark wedge on lower edge. Thorax is gray; head is tan. **HOSTS:** Generalist on low plants and crops, including bean, corn, potato, and tobacco. **NOTE:** Multiple broods.

FLAME-SHOULDERED DART

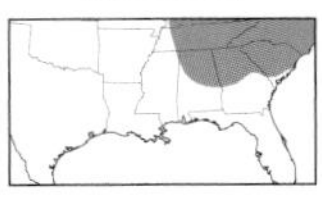

Ochropleura implecta 93-3529 (10891) **Common**

TL 14–15 mm. Chocolate brown FW has contrasting straw-colored costal streak. PM line is a row of black dots. Long black basal dash almost connects to thick black bar behind white-edged orbicular and reniform spots. ST line is a faint row of orange dashes. **HOSTS:** Aster, clover, dock, and other low plants. **NOTE:** Two broods.

DARTS

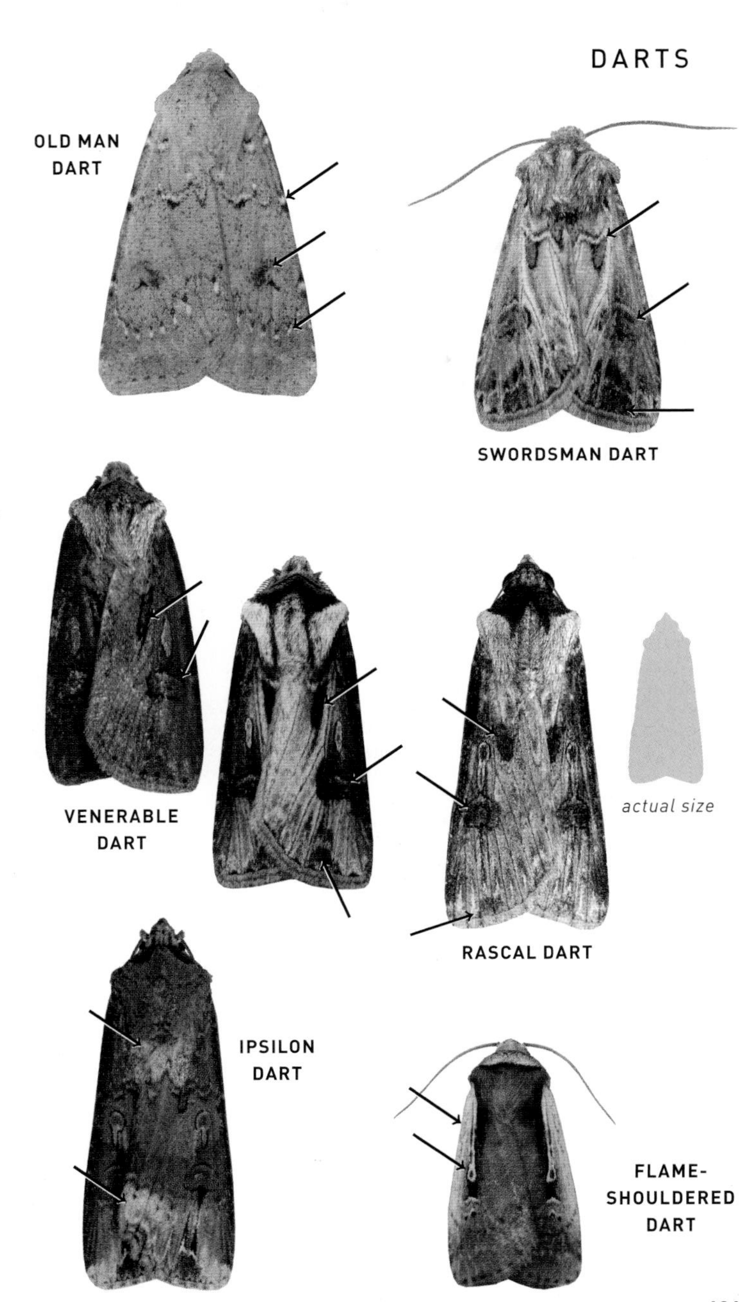
OLD MAN DART
SWORDSMAN DART
VENERABLE DART
RASCAL DART
actual size
IPSILON DART
FLAME-SHOULDERED DART

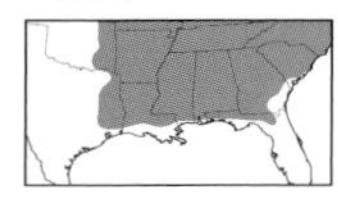

REDDISH SPECKLED DART

Cerastis tenebrifera 93-3536 (10994) **Common**

TL 18–19 mm. Brick red FW has pale yellowish to gray orbicular and reniform spots; reniform spot has black shading in inner half. Dark brick AM and PM lines are sometimes edged grayish. Costa often has grayish patches in median area. Wavy ST line is pale yellow. **HOSTS:** Cherry, dandelion, and lettuce.

BENT-LINE DART

Choephora fungorum 93-3543 (10998) **Uncommon**

TL 21–27 mm. Pale tan to orange brown FW has square dark brown patch at midpoint of costa. Crisp brown AM and PM lines curve upward at costa. **HOSTS:** Clover, dandelion, tobacco, and other low plants.

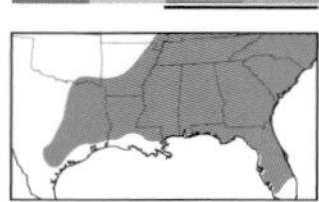

LARGE YELLOW UNDERWING

Noctua pronuba 93-3551 (11003.1) **Common**

TL 30–35 mm. Variable. FW may be pale tan, orange, warm brown, or dark chocolate brown, and smooth, mottled, or speckled. Round orbicular spot is pale. Reniform spot is dark; inner half is blackish. A narrow, pale-edged black bar marks costa at ST line. Double AM and PM lines are sometimes present. Golden yellow HW has a broad black ST band. **HOSTS:** Grass and low plants. **NOTE:** An introduced European species.

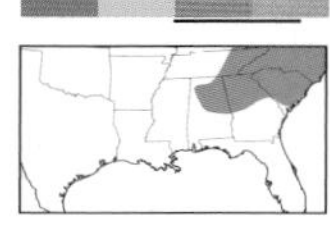

SOUTHERN VARIABLE DART

Xestia elimata 93-3583 (10967) **Common**

TL 18–21 mm. Lilac to ashy gray FW has large, lightly brown-shaded orbicular and reniform spots connected on inner edge by a thin blackish bar that is often bordered costally by warm brown shading. Toothed black AM and PM lines widen into black bars at costa. Claviform spot is sometimes edged black. **HOSTS:** Pines, including loblolly, longleaf, shortleaf, and Virginia.

NORTHERN VARIABLE DART

Xestia badicollis 93-3584 (10968) **Local**

TL 18–21 mm. Resembles Southern Variable Dart but has a thin black basal dash, and black line connecting spots extends above orbicular spot and AM line. AM and PM lines are usually strongly edged whitish. **HOSTS:** Pine.

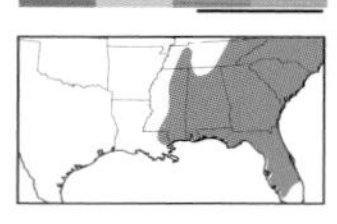

DULL REDDISH DART

Xestia dilucida 93-3586 (10969) **Common**

TL 18–21 mm. Reddish brown FW has bold pale yellowish reniform spot, and round brown orbicular spot that is sometimes edged pale. Costa is usually frosted light gray. Toothed AM and PM lines are indistinct except as bars at costa. **HOSTS:** Blueberry, huckleberry, and probably other heath species.

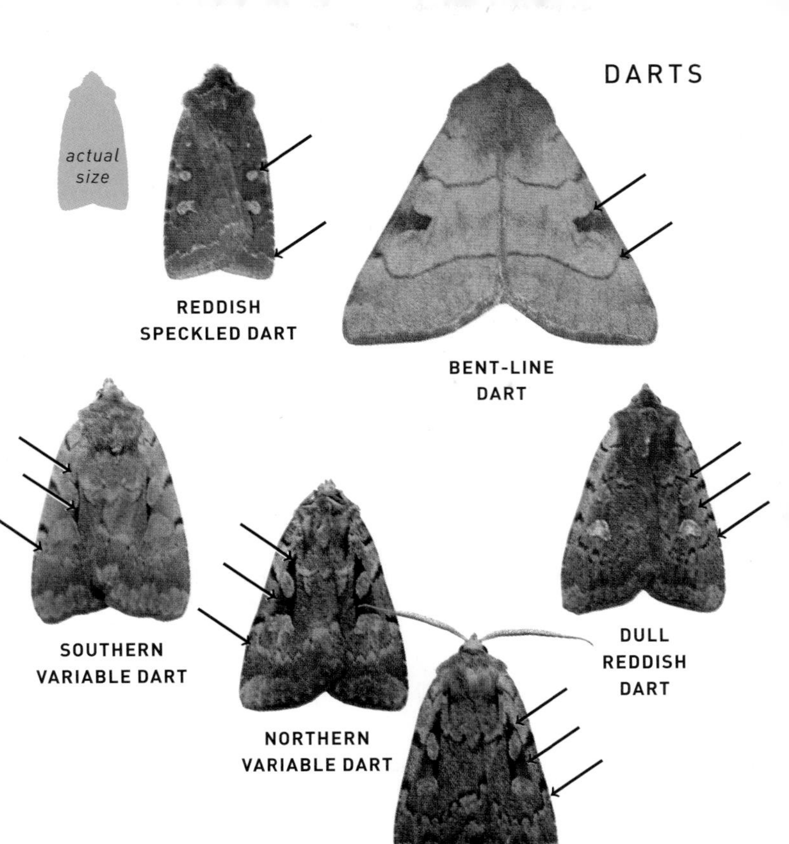
actual size
REDDISH SPECKLED DART
BENT-LINE DART
SOUTHERN VARIABLE DART
NORTHERN VARIABLE DART
DULL REDDISH DART

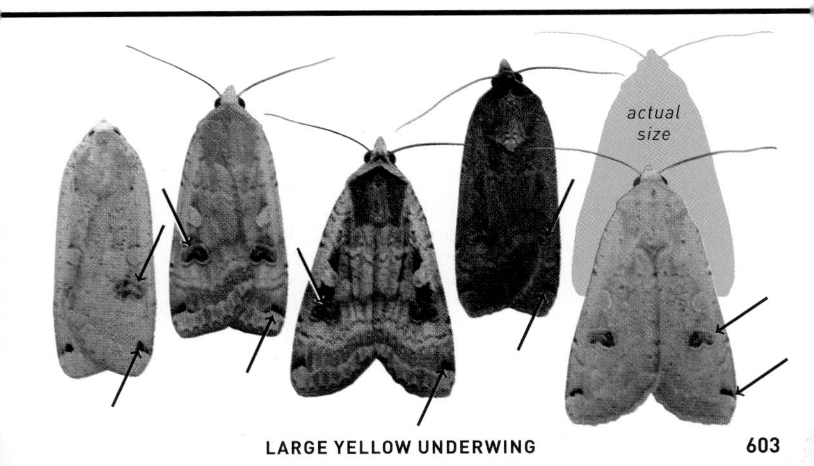
actual size
LARGE YELLOW UNDERWING

GREATER BLACK-LETTER DART

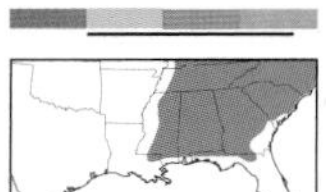

Xestia dolosa 93-3589 (10942.1) **Common**

TL 20–21 mm. Sexually dimorphic. Slate gray FW has a bold tan, bluntly triangular orbicular spot widening toward brown costa. Bicolored reniform spot is dark gray in inner half. A thick black bar connects the two spots. Inner median and ST areas are brown in female. **HOSTS:** A wide variety of trees, crops, and low plants, including barley, clover, and tobacco. **NOTE:** Two or three broods.

PALE-BANDED DART

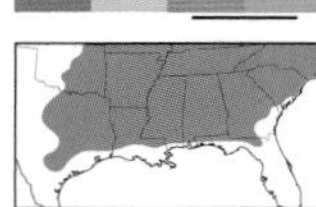

Agnorisma badinodis 93-3626 (10955) **Uncommon**

TL 20–22 mm. Brown FW has a darker brown patch between pale-outlined orbicular and reniform spots. Dark brown AM line has a small black dot at midpoint. Pale-edged PM line is evenly curving. ST area is shaded brown. Thorax has a thick black collar. **HOSTS:** Generalist on shrubs and low plants, including aster, chickweed, and dock.

SQUARE-SPOTTED DART

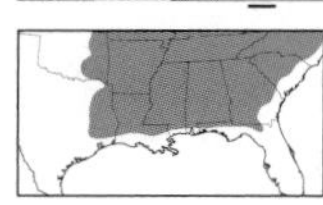

Agnorisma bollii 93-3628 (10956) **Uncommon**

TL 20–22 mm. Peppery gray to grayish brown FW has a bold, black square patch between inconspicuous orbicular and reniform spots, and a roundish black patch above orbicular spot. Whitish AM and PM lines are evenly curving. Thorax has a thick black collar. **HOSTS:** Unknown.

BROWN-COLLARED DART

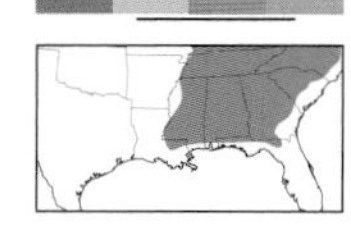

Protolampra brunneicollis 93-3649 (11006) **Common**

TL 20–22 mm. Brown to grayish brown FW has slightly paler spots outlined in brown. PM line is usually visible as a row of black dots. Double AM line is usually faint. Blackish subapical patch is often darkest marking. Thorax has a dark brown collar. **HOSTS:** Blueberry, clover, dandelion, sweet fern, tobacco, and other low plants. **NOTE:** Two broods.

GREATER RED DART

Abagrotis alternata 93-3680 (11029) **Common**

TL 20–23 mm. Peppery grayish brown to reddish brown FW has a dark ST band and pale terminal line. Dark orbicular and reniform spots are edged whitish; reniform spot is usually shaded blackish in inner half. Double AM and PM lines are often faint. **HOSTS:** Generalist on deciduous trees and low plants, including apple, blueberry, cabbage, hickory, oak, and walnut.

DARTS

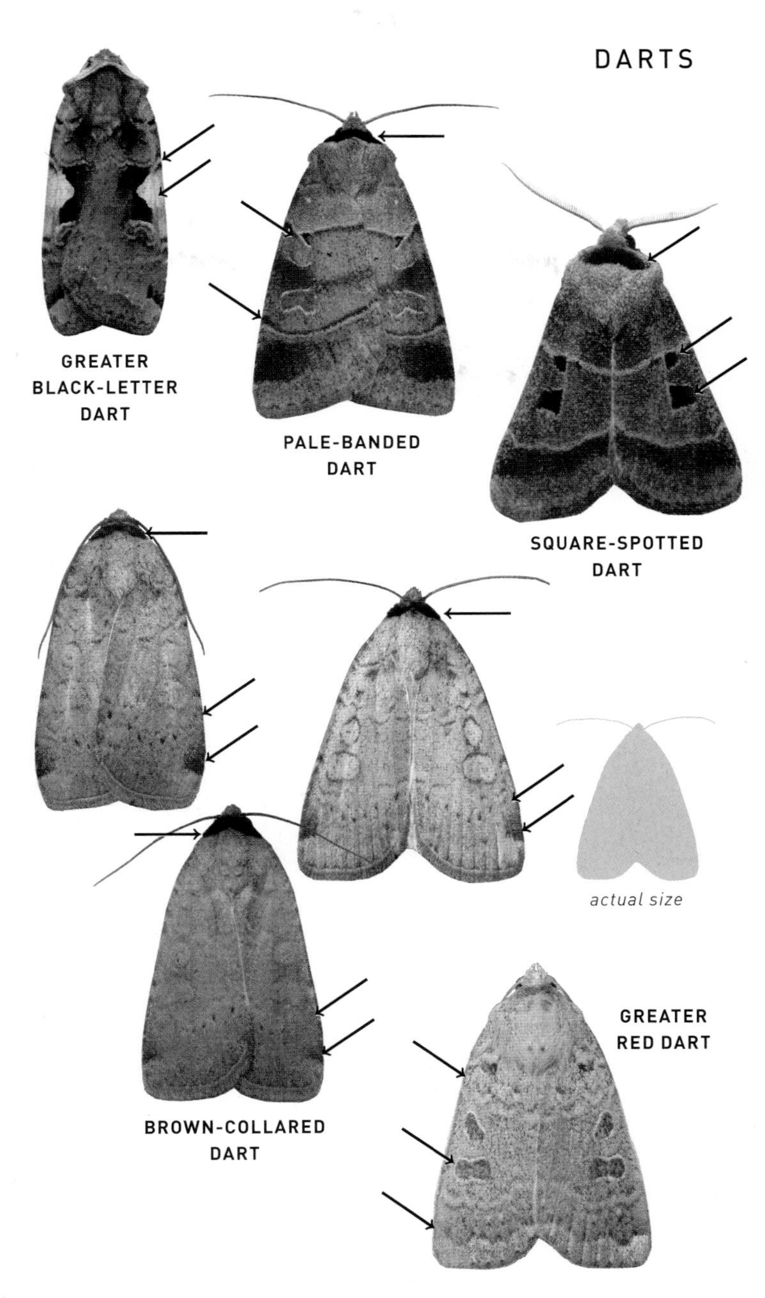

GREATER BLACK-LETTER DART
PALE-BANDED DART
SQUARE-SPOTTED DART
actual size
GREATER RED DART
BROWN-COLLARED DART

Orange-spotted Flower Moth

Brown-spotted Zale

Raspberry Wave

ACKNOWLEDGMENTS

Only two names are on the front of this guide, but it takes far more than that to bring a book to print. From concept to completion, this guide has taken nearly five years. So many people have been involved along the way, directly or indirectly, that I won't be able to thank everyone, but I'm incredibly grateful to all of them.

First and foremost, I must thank our editor, Lisa White. She was the one to approach us with the idea of doing another moth guide after the release of our first one. It turns out that birthing a book is not unlike birthing a child—you've forgotten the ordeal of the first by the time you decide to have another. Lisa, I hope I haven't given you cause to regret your offer too many times these few years. Thank you so much for your incredible, boundless patience and understanding—especially when I announced I was having a baby in the middle of the whole thing.

I also need to thank my husband, Dan. He was witness to the stress and late nights that came with the first book and would not have been wrong to call me crazy for considering doing another. But he offered me instead his support, and also his time. Dan, thank you for standing by me along this long and exhausting road, and for putting your own work on hold in order to look after the baby when I was on a deadline crunch. It's not an exaggeration to say that I couldn't have done this without you.

To my friend and coauthor, Dave: thanks for your willingness to jump on board this giant undertaking again, and all the work you once again put into the book. Feels like we're finally possibly starting to get the hang of things, doesn't it?

To the rest of the publication team at Houghton Mifflin Harcourt: thank you for everything—there is so much more to creating a book than just writing it. To Emily Snyder for tackling the flight bars for me and several other little tasks as needed. To Beth Burleigh Fuller for directing the organization of our materials and taking all our last-minute changes in stride. To Elizabeth Pierson for amazing attention to detail and catching so many errors during copyedits. To Eugenie S. Delaney for the super design and composition of the whole book. Thanks also to the proofreader, Ellen Fast, and indexer, Donna Riggs, and to Margaret Rosewitz, for technical assistance with the photos. To Taryn Roeder and Liz Anderson, thank you for helping make sure the guide gets into the hands that want it.

To our 79 photographers: thank you for happily donating your images to the book. There is no way Dave and I could have completed the guide without the contributions of other moth-ers, and we are indebted to all of you for your support. Special thanks go to Parker Backstrom, Ken Childs, Mark Dreiling, and Carol Wolf, for providing free access to virtually their entire photo collections, and for repeatedly coming to the rescue in filling last-minute gaps.

To Chris Schmidt and Charlie Covell: thanks for your willingness to lend your expertise to such a big undertaking, to help us try to put the best book out that we could.

To Greg Pohl: thank you for providing the latest version of your new checklist, and answering my taxonomy questions to make sure the book was as up-to-date as possible upon release.

To Mom: thank you for saying OK when I asked if you'd proof our species names and numbers—for the second book in a row. That's how I know how much you love me.

To Emma Richardson, Jeremy West, and Alyssa Hollingsworth: thank you for helping me out with image work at extremely short notice when I realized I wasn't going to make my deadline. You guys are lifesavers.

To our agent, Russell Galen: thanks for not only taking care of all the legal and financial stuff so we don't have to think about it, but also for being an ear and a sounding board when we've needed it.

To Mom, Dad, Gaelan, and Rheanna: thank you for your enthusiasm and support throughout, and for stepping in when I needed a hand. To Charlee Hoffman and Silvia Park, for the late-night productivity sprints that kept me on task. And to Lisa Hawkey, for the friendship and long conversations, and the childminding when I was trying to make my deadline.

To my two children: thank you for being such terrific sleepers. Also for brightening my days and making the world wonder-full again. I love you both.

And finally, to you, the moth enthusiast, whose interest in a weird but wonderful hobby is what made this guide possible in the first place. Here's to many late nights and amazing discoveries.

—SL

Who could resist the idea of working on such a guide? So when Seabrooke mentioned collaborating on a new guide to southeast moths, it wasn't such a difficult decision to make. I thank her for yet another moth-fueled ride!

This book would not have been possible without the help and encouragement of our editor, Lisa White, and the superb staff at HMH. Everyone involved did a truly remarkable job putting it all together—many thanks to you all. To our agent Russell Galen—thanks so much for believing we could do this all over again!

As is always the case there are a myriad of people that contributed both directly and indirectly to my work on this book. Many of the species covered in this new guide also occur in my home province of Ontario, so I would like to extend my gratitude to all the folk who have graciously allowed me to study moths on their properties. Peter Carson and Mary Gartshore, Dennis Barry and Margaret Carney, Lois Thomas, and the Freeman family have been particularly welcoming and supportive.

I would like to single out my good friend Michael King for his endless encouragement and great company during countless hours of fieldwork over the years.

Parker Backstrom and Merrill Lynch from North Carolina have both been crucial sounding boards for me throughout the course of this project. Their advice and encouragement have been tremendously helpful. I thank Merrill for arranging a short stay to catch moths at Hag's Head, North Carolina. I extend a huge thank you to Wilson Baker for arranging a superb few days of fieldwork in southern Georgia so Parker, Merrill, and myself could learn more about the moths of that region. Wilson's knowledge of all aspects of natural history from this region is truly remarkable. Phil Spivey, the plantation manager at our study site in Georgia, was wonderfully accommodating throughout our stay. I should add that Parker has graciously contributed many of his superb photographs for use in this guide.

I am fortunate indeed to have many good friends who share my passion for moths. Over the course of this project I have spent countless hours in the field with these like-minded souls. In particular I would like to thank Richard Aaron, Jon Curson, Phill Holder, Carolyn King, Steve LaForest, Phil Milton, Steve Pike, and Francis Solly.

I would like to thank Brad Hubley and Antonia Guidotti in the Natural History Department at the Royal Ontario Museum for allowing access the extensive Lepidoptera collections. This museum work was an essential element in the preparation of the species accounts.

Charles Covell, Donald Lafontaine, and Christian Schmidt gave freely of their time checking identifications and generally acted as quality control. Their collective expertise is immense and invaluable.

For me none of this would have happened had I not got interested in moths in the first place. Ian Hunter and the late Dennis Batchelor of Sandwich Bay Bird Observatory in Kent, UK, fueled the initial spark that has burned brightly for the past 35 years.

And, as always, I owe my greatest debt of gratitude to my infinitely patient wife, Katie, our son, James, and my family in Canada and UK for all the love and support a mothman could ever ask for.

—DB

Tersa Sphinx

Abdomen: The third, and terminal, body section of an insect; comprises 10 segments.

AM (antemedial) area: See *Basal area*.

AM line: The antemedial line that separates the basal and median areas of the forewing. This is often an important identification feature on noctuid moths.

Anal angle: The angle formed where the inner margin and outer margin of a wing converge.

Anal dash: A narrow line extending parallel to the inner margin near the anal angle of the forewing.

Antennae: Segmented sensory appendages positioned at the inside edge of each eye.

Apex: The pointed tip of the forewing.

Apical dash: A narrow line extending parallel to the costa near the apex of the forewing.

Basal area: The area at the base of the wings, often divided from the median area by a basal or antemedial (AM) line.

Basal dash: A narrow line running parallel to the costa in the basal part of the forewing.

Basal line: A dark line cutting across the basal section of the wing.

Bipectinate antennae: Featherlike antennae with two branches on most segments.

Brindled: Showing a lightly striped pattern.

Brood: A single generation.

Case: A silken, cocoonlike structure formed as a shelter by the larvae of some moth species.

Caterpillar: The larval form of all lepidopterans.

Claviform spot: A round or wedgelike spot positioned between the orbicular spot and the inner margin on the forewing. This is especially noticeable on some noctuid moths.

Cocoon: A protective covering made from silk or other materials by a caterpillar prior to pupation.

Collar: The dorsal part of the prothorax immediately behind the head of a moth.

Costa or costal margin: The leading edge of the wing; usually refers to the forewing.

Coxa: The basal segment of an insect leg.

Crepuscular: Active at dawn and/or dusk.

Cutworm: A popular name referring to the caterpillars of certain noctuid moths that are considered agricultural pests.

Dashes: Short black or dark lines in the forewing patterns of some moths, particularly dagger moths and underwings.

Dimorphic: Having two distinctly different forms within a species; the difference may be in size, shape, color, or pattern. Often used to refer to the difference between the male and female of a species (sexual dimorphism).

Discal spot or dot: A small dotlike marking in the center of a wing, especially noticeable on many geometric moths.

Diurnal: Active during daylight hours.

Dorsal: Referring to the upper surface of the body.

Exoskeleton: The hard outer body armor of an insect.

Falcate: Referring to the swept-back shape of the apex on the forewing of some moths, such as Arched Hooktip.

Fasciculate antennae: Threadlike antennae with clusters of short bristles along the length of one side.

Femur: The segment of an insect leg between the coxa and the tibia.

Filiform antennae: Threadlike antennae without projections. Also called simple antennae.

Flight period: The seasonal window of time that the adult moth of a species can be encountered.

Forelegs: The front legs of an insect.

Frass: The pelletlike excrement of caterpillars.

Fringe: The hairlike scales that form the outer margin of the wings.

FW: The forewing of a moth.

Habitat: The surroundings, specialized or general, in which a species lives.

Hindlegs: The rear legs of an insect.

Hooked: See *Falcate*.

Host plants (or hosts): The particular plants on which a caterpillar feeds.

HW: The hindwing of a moth.

Immigrant: A species that is prone to erratic or regular flights in response to drought or favorable weather conditions.

Inner margin: The inner edge of the wing closest to the abdomen.

Labial palps: A pair of sensory appendages extending forward, and often curving upward, from the lower part of the head. They are particularly noticeable on many delta-shaped noctuid moths such as the *Zanclognatha* species.

Larva: The caterpillar stage in the life cycle of lepidopterans.

Leaf mine: The excavations in plant tissue that result from tunneling moth larvae.

Macromoth: A broad term covering all of the Noctuoidea and many other larger groups of moths (such as sphinx moths), though some species are smaller than many micromoths.

Median area: The central portion of the forewing.

Median line: A line that passes through the median area of the forewing, usually between the orbicular and reniform spots.

Micromoth: A term that refers to the tiny or small moths that are not included in the Noctuoidea. However, note that some micromoths, such as the carpenterworm moths, are larger than most macromoths.

Moth trap: A means of catching moths, usually using lights.

Neotropics: The geographical region that includes South and Central America.

Nocturnal: Active at night.

Orbicular spot: A round spot or outline in the inner median part of the forewing. This is often an important identification clue for many species of noctuid moths.

Outer margin: The rear edge of the wings.

Palps: See *Labial palps*.

Pectinate antennae: Antennae that are lined with long projections along one side, resembling half a feather.

Peppered: Referring to a finely speckled pattern present on the wings of many moths, especially geometer species.

Pheromone: A chemical substance secreted by animals, often as sex attractants. Commercially available pheromones can sometimes be used to attract rarely seen moths, such as the clearwing borers in the family Sesiidae.

PM (postmedial) area: The area between the ST line and the PM line.

PM line: The postmedial line that separates the median area from the subterminal (ST) area on the forewing. This is often an important identification feature on noctuid moths.

Prothorax: The first of three thoracic segments on an insect; the one closest to the head.

Pupa: The final stage of metamorphosis whereby larval features are replaced with adult characteristics.

Reniform spot: A spot or outline, often kidney-shaped, in the outer median part of the forewing. This can be an important identification clue for many species of noctuid moths.

Scales: Modified setae or bristles that are flattened or hairlike. Scales normally cover the body and wings of most lepidopterans.

Serrate antennae: The saw-shaped antennae of certain moth species.

Setae: Bristles that emerge from the exoskeleton. They are modified to form scales in adult lepidopterans.

Sexually dimorphic: Showing a distinct difference in pattern or color between males and females of the same species.

Shading: Broad areas of color or tinting on the wings of a moth.

Simple antennae: Threadlike antennae without branches. Also called filiform antennae.

ST (subterminal) line: The subterminal line that is often present between the postmedial (PM) line and the terminal line. This is often an important identification feature on noctuid moths.

Stigma: A spot, usually white or silver, in the middle of the median area. Seen mostly in loopers.

Subreniform spot: A small spot or outline positioned between the reniform spot and the inner margin of the forewing. This is an important identification feature on some noctuid moths.

Tarsus: The small terminal segments on an insect's leg.

Terminal line: The outermost line on the wings of a moth before the fringe.

Thorax: The middle of the three main body segments on an insect.

Tibia: The long, narrow leg segment (sometimes spiny) between the femur and the tarsus.

TL: The total length of a moth as measured from the front of the head to the tip of the wings or abdomen.

Vagrant: A rare immigrant species that is not to be expected within a certain area.

Veins: Branching tubular structures that support the wing membranes.

Ventral: Referring to the lower surface of the body.

WS: The wingspan of a moth as measured from one wingtip to the other. This is an important measurement for most geometrid moths and some spread-winged noctuid moths.

Beautiful Grass-Veneer

If you've been bitten by the mothing bug, here are some additional resources that may help you expand your identification skills and familiarize yourself with what can be found in your area. Some of these publications may be out of print, but used copies should be readily available online with a little digging.

Another suggestion is to try searching for resources specific to your area. Many moth enthusiasts have started posting online photographs and lists of moths. A web search for "moths of <your state/city>" will often turn up local websites. Many of these moth-ers are happy to offer their help, and may even be willing to get together for a night of mothing. Some states have official annotated checklists published by the state government, and these too will come up in such a web search (also try "Lepidoptera of <your state>").

There are several regional Facebook groups for moth enthusiasts; it is worth searching to see if there is one for your area.

Also see if there are entomologist or naturalist associations for your state or city. These groups can be amazing fonts of knowledge, with friendly, enthusiastic people who are delighted to welcome new people to their passion. Many will hold public moth nights that you can attend.

Printed Guides and Checklists

Beadle, David, and Seabrooke Leckie. 2012. *Peterson Field Guide to Moths of Northeastern North America.* Boston: Houghton Mifflin Harcourt.

Covell, Charles V., Jr. 2005. *A Field Guide to Moths of Eastern North America.* Martinsville: Virginia Museum of Natural History.

Eaton, Eric R., and Kenn Kaufman. 2007. *Kaufman Field Guide to Insects of North America.* Boston: Houghton Mifflin. [This excellent guide has a small but informative section on moths; the entire book is a rich resource for the general naturalist.]

Himmelman, John. 2002. *Discovering Moths: Nighttime Jewels in Your Own Backyard*. Camden, ME: Down East Books.

Hodges, R. W., et al. 1983. *Check List of the Lepidoptera of America North of Mexico*. London: E. W. Classey.

Holland, William Jacob. 1968. *The Moth Book: A Guide to the Moths of North America*. New York: Dover Publications.

Leverton, Roy. 2001. *Enjoying Moths*. London: Academic Press. [Although this book uses British moths as examples, the information it contains is applicable anywhere.]

Pohl, Gregory R., Bob Patterson, and Jonathan P. Pelham. 2016. *Annotated taxonomic checklist of the Lepidoptera of North America, North of Mexico*. Working paper uploaded to ResearchGate.net. 766pp. DOI: 10.13140/RG.2.1.2186.3287.

Powell, Jerry A., and Paul A. Opler. 2009. *Moths of Western North America*. Berkeley: University of California Press.

Wagner, David L. 2005. *Caterpillars of Eastern North America: A Guide to Identification and Natural History*. Princeton, NJ: Princeton University Press.

Internet Resources

BugGuide.Net. Hosted by Iowa State University Department of Entomology. An incredible resource for all insects, not just moths. Includes user-submitted photographs as well as information on identification, range, food plants, etc., where it is available. An "ID Request" feature allows you to upload your own photographs for expert identification.
bugguide.net

Butterflies and Moths of North America. A searchable database of Lepidoptera records for the United States. Provides maps, written accounts, and photographs for many species. Butterflies are very well covered, but moths receive some treatment as well.
butterfliesandmoths.org

iNaturalist. This citizen science observation-reporting website covers all living organisms but is a great tool for organizing and sharing your moth records. The auto-identification feature is a great tool to help with IDs, and other users will help with confirmation. The platform also offers the ability to filter records (including your own) by species,

date, or location. A free app is available for submitting and managing records from mobile devices.
inaturalist.com

The Lepidopterists' Society. Some information is contained at the site itself, but it also provides an excellent list of additional resources, both in print and online. Although this is not strictly an identification or information website, the Lepidopterists' Society provides an excellent opportunity to learn more from like-minded people. This is the largest North American lepidopteran organization and is continental in scope. You can also look for more regional societies or associations in your area.
lepsoc.org

LepSnap. A citizen science website dedicated specifically to moths, which uses an auto-identification tool to help you label your images. Records are organized taxonomically, and the app allows you to filter and search within your sightings. A free app is available for submitting and managing records from mobile devices. The platform supports cross-sharing to iNaturalist.
lepsnap.org

Microleps.org. An outstanding web resource for micromoths. In addition to photographs of spread specimens, this site also provides information on host plants, feeding habits, flight period, etc.
microleps.org

Mothing and Moth-watching. One of the first Facebook groups dedicated to sharing the enjoyment of moths. Membership is open, and contributors are friendly and supportive, helping with identifications where they can. Although the group has a North American focus, images are often shared from elsewhere in the world, showcasing some outstanding diversity.
facebook.com/groups/137219092972521

Moths of Canada. A series of publications and online pages produced by the Canadian Biodiversity Information Facility of the Government of Canada. Excellent photographs of pinned specimens are accompanied by metric rulers for size reference.
cbif.gc.ca/spp_pages/misc_moths/phps/mothindex_e.php

Moths of the Eastern United States. Another public Facebook group with a large membership. This one focuses on just the eastern half of the continent.
facebook.com/groups/MothsoftheeasternUS

North American Moth Photographers Group (MPG). Hosted by the Mississippi Entomological Museum at Mississippi State University. The best moth identification website out there; contains a compilation

of images for the majority of North American species accompanied by range maps displaying recorded observations.
mothphotographersgroup.msstate.edu

Tortricid.net. Originally begun as an identification resource for the large micromoth family Tortricidae. That role is now more thoroughly filled by MPG, and this website instead serves as an excellent source of additional information about the Tortricidae.
tortricidae.com

Public Events

Mothapalooza. Begun in 2013, this now-annual weekend event is held in southern Ohio and attracts moth-ers from across the continent. The event is so popular that registration is currently capped at about 175 attendees. Participants get to moth alongside numerous experts at night and look for day-flying moths (and many other species) on afternoon field trips through the beautiful Appalachian foothills. Afternoon breakout sessions and evening keynote speeches are often by prominent figures in the field.
mothapalooza.org

National Moth Week. Despite its name, this is now a global event that invites moth-ers to record the moths seen on any given night over the course of a week, usually held in July. Data can be submitted through the website and are used to build a citizen science database accessible to the scientific community. Participants can count on their own in their backyard or can join a public moth night in their area; a list of public events is available on the website. Any and all skill levels are welcomed.
nationalmothweek.org

White-tipped Black

Many photographers made valuable contributions to this guide by donating excellent material for the plates. They are listed here in alphabetical order, with the species they provided images for noted by P3 numbers.

Lyn Atherton, 91-1456

Seth Ausubel, 93-1155

Parker Backstrom, 30-0083, 30-0100, 42-0027, 58-0031, 62-0200, 62-0596, 62-0695, 62-0795, 64-0022, 64-0047, 66-0001, 66-0026, 66-0033, 66-0053, 66-0063, 70-0003, 70-0004, 80-0061, 80-0092, 80-1023, 80-1044, 80-1055, 80-1164, 80-1292, 80-1454, 89-0009, 89-0010, 89-0017, 89-0023, 89-0086, 89-0090, 89-0092, 89-0096, 89-0105, 89-0110, 89-0135, 89-0177, 89-0179, 91-0006, 91-0545, 91-0672, 91-0803, 91-0867, 91-0898, 91-1007, 91-1076, 91-1084, 91-1146, 91-1148, 91-1158, 91-1160, 91-1170, 91-1276, 91-1331, 91-1376, 93-0034, 93-0077, 93-0154, 93-0178, 93-0460, 93-0559, 93-0629, 93-0700, 93-0729, 93-0730, 93-0731, 93-0762, 93-0764, 93-0769, 93-0770, 93-0783, 93-0789, 93-0796, 93-0826, 93-0833, 93-0837, 93-0857, 93-0858, 93-0864, 93-0954, 93-0955, 93-0959, 93-1075, 93-1095, 93-1101, 93-1166, 93-1169, 93-1252, 93-1292, 93-1343, 93-1429, 93-1430, 93-1444, 93-1681, 93-2054, 93-2056, 93-2091, 93-2116, 93-2123, 93-2409, 93-2517, 93-2595, 93-2598, 93-3067, 93-3118, 93-3257, 93-3504, 93-3543

C. D. Barrentine, 93-2073

Giff Beaton, 64-0093

Thomas Bentley, 21-0117, 89-0178

Betsy Betros, 64-0062

John M. Bills Jr., 89-0012

Michael Boone, 91-1014

Mark A. Brogie, 93-0830

Mark H. Brown, 91-0816

Valerie G. Bugh, 70-0005, 70-0010, 89-0025, 89-0078, 89-0153, 89-0154, 89-0186, 91-0157, 91-0506, 91-0548, 91-0620, 91-1125, 91-1159, 91-1309, 91-1376, 93-0118, 93-0177, 93-0183, 93-0632, 93-0720, 93-0943, 93-1050, 93-1244, 93-1248, 93-1250, 93-1254, 93-1260, 93-1275, 93-1308, 93-1312, 93-1353, 93-1363, 93-1375, 93-1385, 93-1386, 93-1408, 93-1431, 93-1440, 93-1519, 93-1575, 93-1606, 93-1640, 93-1682, 93-1689, 93-1770, 93-1963, 93-2022, 93-2028, 93-2084, 93-2139, 93-2150, 93-2171, 93-2224, 93-2281, 93-2566, 93-2596

Ken Childs, 16-0079, 30-0006, 30-0028, 30-0069, 30-0071, 30-0106, 30-0115, 30-0118, 30-0140, 30-0157, 30-0172, 30-0186, 30-0203, 30-0205, 30-0213, 33-0155, 33-0287, 33-0361, 33-0378, 42-0009, 42-0055, 42-0102, 42-0127, 42-0215, 42-0227, 42-0249, 42-0260, 42-0329, 42-0468, 42-0476, 42-0508, 42-0671, 42-1086, 42-1140, 42-1330, 42-1352, 42-1622, 42-1713, 46-0054, 58-0006, 62-0004, 62-0027, 62-0044, 62-0144, 62-0152, 62-0257, 62-0259, 62-0262, 62-0281, 62-0313, 62-0331, 62-0333, 62-0379, 62-0384, 62-0393, 62-0412, 62-0425, 62-0637, 62-0926, 62-1028, 62-1066, 62-1117, 62-1189, 62-1292, 62-1302, 62-1304, 62-1391, 64-0109, 64-0132, 66-0011, 66-0015, 66-0035, 66-0037, 66-0051, 66-0054, 66-0055, 66-0068, 66-0069, 80-0062, 80-0092, 80-0117, 80-0123, 80-0130, 80-0233, 80-0235, 80-0258, 80-0270, 80-0313, 80-0321, 80-0324, 80-0372, 80-0447, 80-0479, 80-0486, 80-0557, 80-0641, 80-0651, 80-0658, 80-0685, 80-0707, 80-0718, 80-0742, 80-0795, 80-0802, 80-0805, 80-0815, 80-0835, 80-0877, 80-0906, 80-0907, 80-1055, 80-1072, 80-1188, 80-1191, 80-1215, 80-1219, 80-1273, 80-1276, 80-1322, 80-1324, 80-1354, 80-1373, 80-1378, 80-1418, 80-1470, 80-1521, 80-1554, 87-0002, 89-0187, 91-0515, 91-0521, 91-0529, 91-0532, 91-0853, 91-0893, 91-1076, 91-1084, 91-1160, 91-1261, 93-0034, 93-0319, 93-0476, 93-0494, 93-0497, 93-0499, 93-0509, 93-0529, 93-0531, 93-0538, 93-0554, 93-0657, 93-0667, 93-0685, 93-0700, 93-0784, 93-0813, 93-0942, 93-1029, 93-1145, 93-1241, 93-1268, 93-1299, 93-1382, 93-1391, 93-1460, 93-1468, 93-1651, 93-2015, 93-2092, 93-2119, 93-2206, 93-2223, 93-2249.6, 93-2513, 93-2542, 93-2552, 93-2586, 93-2589, 93-2955, 93-3089, 93-3583, 93-3628

Richard Crook, 93-0349

Rob Curtis / theearlybirder.com, 91-1125

John Davis, 93-1438

Tony DeSantis, 42-0362, 93-0042, 93-1499, 93-2550, 93-2804

Jason J. Dombroskie, 11-0019, 93-1186

Mark Dreiling, 30-0141, 33-0113, 36-0187, 42-0007, 42-0020, 42-0030, 42-0245, 42-0387, 42-0410, 42-0413, 42-0496, 42-0534, 42-0576, 42-0577, 42-0594, 42-0635, 42-0650, 42-0681, 42-0698, 42-0762, 42-0889, 42-1823, 42-1841, 48-0006, 58-0002, 62-0413, 62-0477, 62-1110, 62-1157, 62-1267, 62-1324, 64-0059, 64-0117, 66-0003, 66-0006, 66-0043, 66-0050, 80-0002, 80-0004, 80-0052, 80-0165, 80-0225, 80-0260, 80-0371, 80-0482, 80-0523, 80-0626, 80-0838, 80-1047, 80-1131, 80-1167, 80-1195, 80-1225, 80-1421, 80-1444, 80-1469, 80-1533, 87-0001, 91-0077, 91-0500, 91-0511, 91-0657, 91-0683, 91-0697, 91-0755, 91-1101, 91-1138, 91-1156, 91-1379, 93-0036, 93-0077, 93-0125, 93-0126.1, 93-0189, 93-0218, 93-0242, 93-0309, 93-0398, 93-0402, 93-0470, 93-0668, 93-0683, 93-0725, 93-0754, 93-0785, 93-0848, 93-0878, 93-0885, 93-0900, 93-1151, 93-1264, 93-1277, 93-1302, 93-1313, 93-1329, 93-1396, 93-1460, 93-1463, 93-1478, 93-1480, 93-1688, 93-1797, 93-2046, 93-2093, 93-2135, 93-2156, 93-2158, 93-2215, 93-2249-6, 93-2771, 93-2785, 93-2801, 93-3216, 93-3217, 93-3502, 93-3526

Lynette Elliott, 93-0343

Bill Evans, 91-1088

Mary E. Gartshore, 80-1334

Gary J. Goss, 23-0001, 42-0890, 42-1151, 58-0034, 62-0114, 80-0050, 80-0698, 80-0747, 80-0779, 80-1007, 80-1009, 80-1025, 80-1216, 80-1217, 80-1370, 80-1384, 89-0108, 89-0161, 89-0188, 91-0541, 91-0545, 91-0642, 91-0646, 91-0660, 91-1034, 91-1232, 91-1237, 91-1348, 91-1415, 91-1435, 93-0241, 93-0448, 93-0459, 93-0493, 93-0963, 93-0964, 93-1255, 93-1335, 93-1980, 93-1997, 93-2039, 93-2083, 93-2135, 93-2464, 93-2555

Donald Gudehus, 89-0169, 91-0643

Jennifer W. Hanson, 80-0817

Randy Hardy, 62-1327, 91-0536, 91-0538, 91-0740, 91-1121, 93-0433, 93-1001, 93-1767, 93-2058

Ann Hendrickson, 62-0149, 64-0009, 64-0010, 70-0006, 70-0010, 80-1018, 80-1345, 80-1352, 80-1377, 87-0002, 87-0026, 89-0113, 89-0132, 91-0549, 91-0818, 91-0832, 91-1126, 93-0246, 93-0399, 93-0658, 93-0700, 93-0763, 93-0793, 93-0816, 93-1244, 93-1248, 93-1288, 93-1304, 93-1658, 93-1678, 93-1726, 93-2055, 93-2147, 93-2150, 93-2227, 93-3114

John Himmelman, 64-0065

Jeff Hollenbeck, 93-0658

Peter Homann, 33-0155, 80-0073, 91-1208, 91-1303, 93-0243, 93-0945, 93-1165, 93-1287, 93-1551

Scott Housten, 93-0765

Ali Iyoob, 89-0018

Téa Kesting-Handly, 91-1014

Michael King, 64-0081

Bernie Knaupp, 93-3494

Curtis Lehman, 89-0040, 89-0170

Timothy J. Lethbridge, 93-2145

Tony Leukering, 91-0566

Larry Line, 91-1437

Steven Long Photography, 64-0088

Merrill Lynch, 91-1267, 93-0066, 93-0341, 93-2470

Chrissy McClarren, 64-0114

Jim McCulloch, 64-0067

Cynthia Mead, 89-0194

Marcia Morris, 91-0771, 91-1258

Roy F. Morris II, 91-0609, 93-0364, 93-0945

Tom Murray, 42-0483, 93-0253, 93-1420, 93-1434, 93-1463, 93-2024, 93-2208, 93-2960

Steve Nanz, 42-0399

Randy Newman, 89-0184, 91-1235

Marcie O'Connor, 93-0830, 93-1551, 93-2672

J. Forman Orth, 93-0278

Nelson Poirier, 89-0082, 93-2502, 93-2966

Peter W. Post, 89-0040, 89-0194

Mike Quinn / texasento.net, 36-0242, 80-1131, 89-0006, 93-1108

Jon Rapp, 60-0001, 62-0745, 80-1277, 80-1366, 89-0014, 91-1157, 93-0942, 93-1464

Greg Raterman, 85-0022, 91-1436, 93-0307, 93-0660.97, 93-1156

Glenn Richardson, 89-0079

Jason D. Roberts, 66-0060, 89-0072, 89-0081, 91-1208, 93-0783, 93-1552, 93-2472, 93-2599, 93-3067

Fran C. Rutkovsky, 42-0028, 80-1008, 93-0602

John B. Schneider, 89-0150, 93-0070

Nolie Schneider, 89-0206, 91-1154, 93-1130, 93-1351

Dixie Searfoss, 64-0099

Robert D. Siegel, MD, PhD, Stanford University, 93-3210

Marvin Smith, 93-0242, 93-0871, 93-1107

R. G. Snider, 93-0759

Owen Strickland, 89-0209

Candice Talbot, 89-0106

Jeremy Tatum, 93-1142

Heidi Trudell, 93-0420

Renn Tumlison, 93-0853

Royal Tyler, 93-2601

Steve Walter, 93-2129

Wade Wander, 93-0245, 93-2771

Dave Wendelken, 91-0688

Carol Wolf, 07-0001, 21-0044, 30-0011, 30-0025, 30-0052, 30-0058, 30-0060, 33-0079, 33-0152, 33-0353, 42-0012, 42-0021, 42-0029, 42-0038, 42-0065, 42-0183, 42-0200, 42-0232, 42-0255, 42-0259, 42-0532, 42-0537, 42-0631, 42-0647, 42-0727, 42-1006, 42-1157, 42-1686, 42-1687, 42-1807, 46-0015, 46-0060, 46-0072, 46-0107, 46-0110, 46-0112, 46-0125, 46-0153, 46-0157, 62-0004, 62-0112, 62-0144, 62-0158, 62-0164, 62-0248, 62-0296, 62-0298, 62-0366, 62-0379, 62-0384, 62-0412, 62-0436, 62-0483, 62-0485, 62-0493, 62-0509, 62-0519, 62-0541, 62-0596, 62-0597, 62-0622, 62-0662, 62-0663, 62-0696, 62-0710, 62-0737, 62-0832, 62-0909, 62-1065, 62-1083, 62-1116, 62-1129, 62-1181, 62-1182, 62-1330, 62-1388, 64-0019, 64-0115, 66-0004, 66-0011, 66-0026, 66-0031, 66-0047, 66-0052, 66-0059, 70-0009, 80-0006, 80-0033, 80-0069, 80-0093, 80-0122, 80-0342, 80-0349, 80-0370, 80-0483, 80-0506, 80-0662, 80-0666, 80-0695, 80-0699, 80-0705, 80-0738, 80-0741, 80-0744, 80-0766, 80-0787, 80-0816, 80-0824, 80-0850, 80-0866, 80-1006, 80-1014, 80-1019, 80-1032, 80-1059, 80-1061, 80-1067, 80-1178, 80-1202, 80-1255, 80-1261, 80-1268, 80-1286, 80-1295, 80-1298, 80-1304, 80-1311, 80-1315, 80-1316, 80-1319, 80-1328, 80-1337, 80-1348, 80-1351, 80-1353, 80-1381, 80-1391, 80-1393, 80-1413, 80-1414, 80-1420, 80-1470, 80-1479, 80-1498, 80-1508, 80-1516, 80-1531, 80-1551, 80-1560, 87-0036, 89-0087, 89-0117, 89-0138, 89-0146, 89-0154, 89-0156, 89-0160, 89-0163, 89-0166, 89-0174, 91-0007, 91-0050, 91-0133, 91-0334, 91-0503, 91-0508, 91-0513, 91-0515, 91-0524, 91-0527, 91-0530, 91-0531, 91-0532, 91-0548, 91-0557, 91-0559, 91-0560, 91-0564, 91-0568, 91-0583, 91-0584, 91-0585, 91-0587, 91-0590, 91-0591, 91-0608, 91-0626, 91-0627, 91-0632, 91-0673, 91-0689, 91-0692, 91-0698, 91-0700, 91-0741, 91-0752, 91-0763, 91-0772, 91-0795, 91-0832, 91-0833, 91-0837, 91-0865, 91-0898, 91-0998, 91-1004, 91-1073, 91-1076, 91-1137, 91-1151, 91-1167, 91-1171, 91-1172, 91-1178, 91-1179, 91-1255,

91-1291, 91-1331, 91-1441, 91-1457, 93-0060, 93-0069, 93-0076, 93-0101, 93-0102, 93-0119, 93-0123, 93-0133, 93-0137, 93-0144, 93-0150, 93-0153, 93-0185, 93-0188, 93-0220, 93-0294, 93-0300, 93-0308, 93-0319, 93-0376, 93-0384, 93-0387, 93-0389, 93-0396, 93-0430, 93-0439, 93-0443, 93-0446, 93-0458, 93-0462, 93-0463, 93-0465, 93-0494, 93-0504, 93-0505, 93-0518, 93-0522, 93-0527, 93-0537, 93-0543, 93-0546, 93-0550, 93-0580, 93-0617, 93-0618, 93-0627, 93-0628, 93-0629, 93-0634, 93-0635, 93-0644, 93-0646, 93-0658, 93-0659, 93-0660, 93-0660.97, 93-0694, 93-0695, 93-0697, 93-0704, 93-0727, 93-0755, 93-0758, 93-0775, 93-0776, 93-0780, 93-0781, 93-0835, 93-0869, 93-0870, 93-0871, 93-0872, 93-0874, 93-0891, 93-0925, 93-0931, 93-0943, 93-0945, 93-0951, 93-0954, 93-0969, 93-0972, 93-0977, 93-0982, 93-0992, 93-1015, 93-1025, 93-1027, 93-1034, 93-1056, 93-1061, 93-1063, 93-1075, 93-1084, 93-1086, 93-1089, 93-1091, 93-1095, 93-1096, 93-1115, 93-1122, 93-1129, 93-1153, 93-1164, 93-1240, 93-1252, 93-1262, 93-1267, 93-1268, 93-1276, 93-1301, 93-1305, 93-1311, 93-1322, 93-1382, 93-1633, 93-1658, 93-1781, 93-1986, 93-2031, 93-2056, 93-2193, 93-2212, 93-2213, 93-2220, 93-2223, 93-2224, 93-2225, 93-2227, 93-2231, 93-2233, 93-2236, 93-2239, 93-2248, 93-2274, 93-2282, 93-2409, 93-2513, 93-2595, 93-2600, 93-2606, 93-2723, 93-2950, 93-2958, 93-2967, 93-3039, 93-3050, 93-3124, 93-3219, 93-3227

Brandon Woo, 64-0134

Karen Yukich, 64-0155

Robert Lord Zimlich, 30-0152, 42-0398, 42-1688, 62-1015, 66-0022, 66-0037, 80-0055, 80-0410, 80-1025, 80-1217, 80-1330, 80-1367, 80-1417, 89-0012, 91-0514, 91-1232, 93-0036, 93-0165, 93-0279, 93-0952, 93-1048, 93-1136, 93-1460, 93-2607

Ornate Moth

Page references in **bold** refer to illustrations.

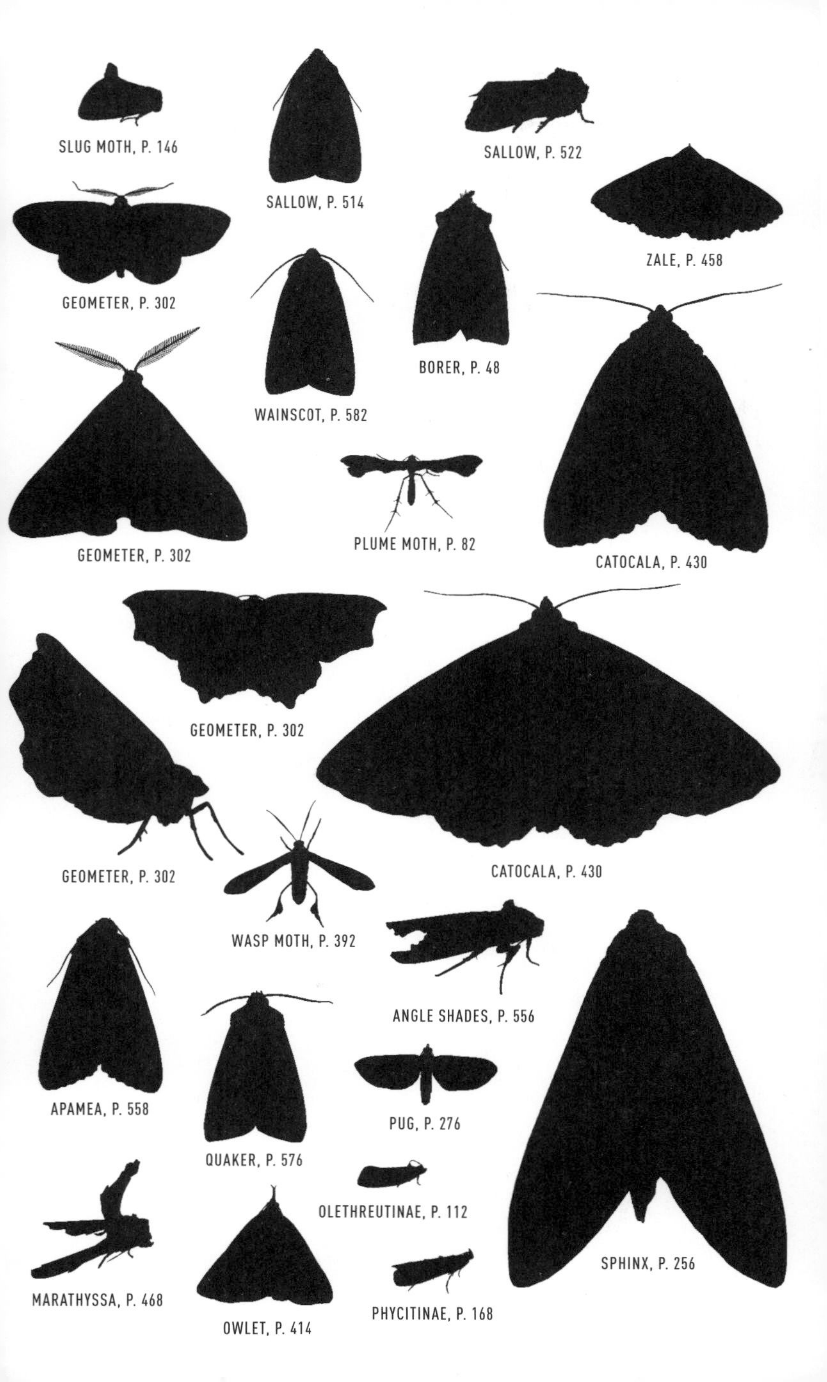
SLUG MOTH, P. 146
SALLOW, P. 514
SALLOW, P. 522
GEOMETER, P. 302
ZALE, P. 458
BORER, P. 48
WAINSCOT, P. 582
GEOMETER, P. 302
PLUME MOTH, P. 82
CATOCALA, P. 430
GEOMETER, P. 302
GEOMETER, P. 302
WASP MOTH, P. 392
CATOCALA, P. 430
ANGLE SHADES, P. 556
APAMEA, P. 558
QUAKER, P. 576
PUG, P. 276
OLETHREUTINAE, P. 112
SPHINX, P. 256
MARATHYSSA, P. 468
OWLET, P. 414
PHYCITINAE, P. 168

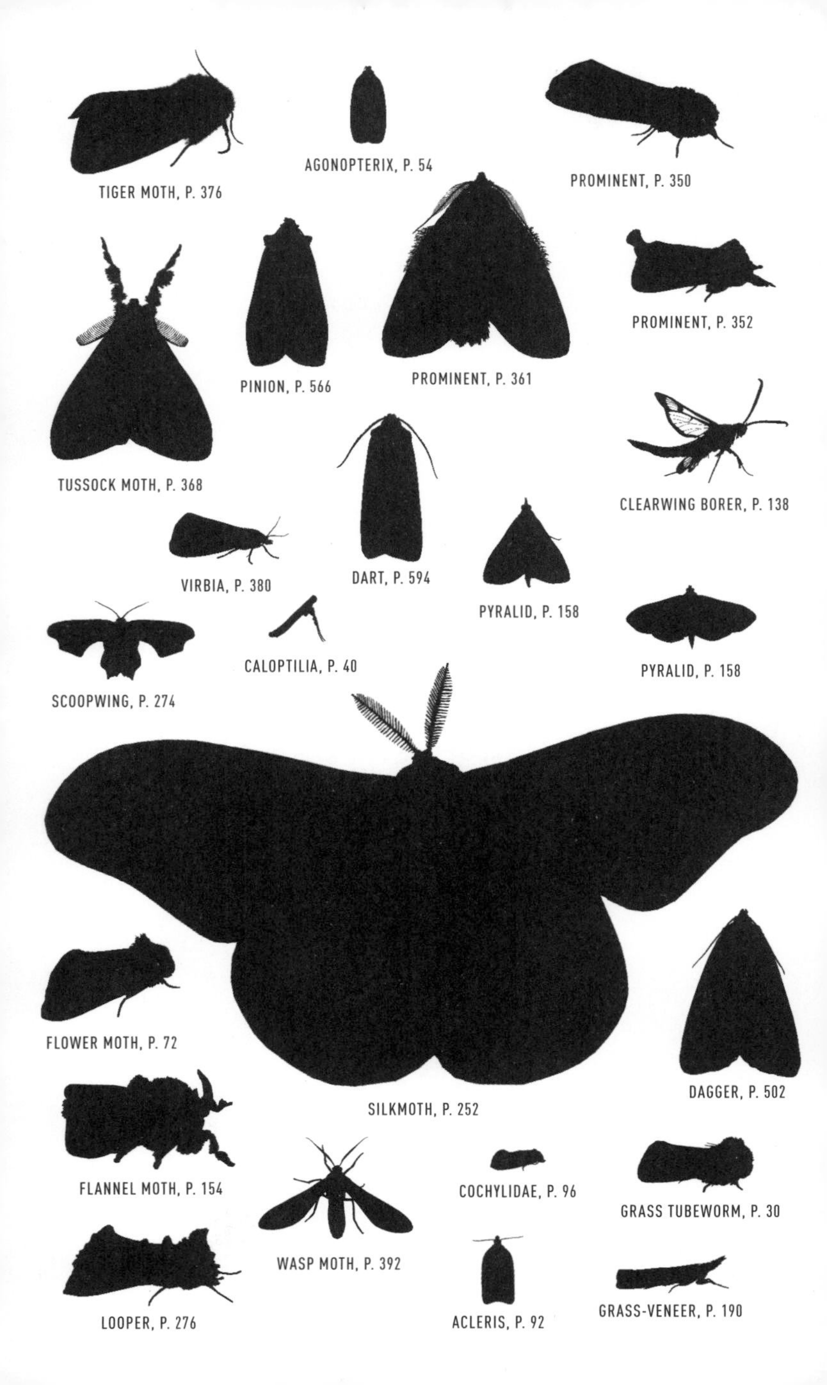
TIGER MOTH, P. 376
AGONOPTERIX, P. 54
PROMINENT, P. 350
PROMINENT, P. 352
TUSSOCK MOTH, P. 368
PINION, P. 566
PROMINENT, P. 361
CLEARWING BORER, P. 138
VIRBIA, P. 380
DART, P. 594
PYRALID, P. 158
SCOOPWING, P. 274
CALOPTILIA, P. 40
PYRALID, P. 158
FLOWER MOTH, P. 72
SILKMOTH, P. 252
DAGGER, P. 502
FLANNEL MOTH, P. 154
COCHYLIDAE, P. 96
GRASS TUBEWORM, P. 30
WASP MOTH, P. 392
LOOPER, P. 276
ACLERIS, P. 92
GRASS-VENEER, P. 190